David R. Salem

Structure Formation in Polymeric Fibers

David R. Salem (Editor)

Structure Formation in Polymeric Fibers

With Contributions from

N. Aminuddin, D.G. Baird, C.W.M. Bastiaansen, J. Blackwell, S.Z.D. Cheng, B. Clauss, J.A. Cuculo, H. Fong, M.W. Frey, R.V. Gregory, F.W. Harris, J.W.S. Hearle, J.F. Hotter, D.J. Johnson, R.K. Krishnaswamy, P.J. Lemstra, C.Y. Li, F. Li, K.W. McCreight, N.S. Murthy, S. Rastogi, D.H. Reneker, D.R. Salem, J.M. Schultz, J.E. Spruiell, N.Vasanthan, Y. Yoon, Q. Zhou

HANSER

Hanser Publishers, Munich

Hanser Gardner Publications, Inc., Cincinnati

The Editor:
David R. Salem, TRI/Princeton, 601 Prospect Avenue, P.O. Box 625, Princeton NJ 08542, USA

Distributed in the USA and in Canada by
Hanser Gardner Publications, Inc.
6915 Valley Avenue, Cincinnati, Ohio 45244-3029, USA
Fax: (513) 527-8950
Phone: (513) 527-8977 or 1-800-950-8977
Internet: http://www.hansergardner.com

Distributed in all other countries by
Carl Hanser Verlag
Postfach 86 04 20, 81631 München, Germany
Fax: +49 (89) 98 12 64

Library of Congress Cataloging-in-Publication Data
Salem, David R.
Structure formation in polymeric fibers/David R. Salem.
p. cm.
Includes bibliographical references and index.
ISBN 1-56990-306-9 (hardcover)
1. Textile fibers, Synthetic, 2. Polymers. I. Title.

TS1548.5 .S314 2000
677′.4--dc21

Die Deutsche Bibliothek – CIP-Einheitsaufnahme
Structure formation in polymeric fibers / David R. Salem. – Munich:
Hanser; Cincinnati : Hanser Gardner, 2000
ISBN 3-446-18203-9

Production Manager: Sandra Gottmann, Bonn
Typeset in England by Marksbury Multimedia Ltd., Bath
Printed and bound in Germany by Druckhaus Thomas Müntzer, Bad Langensalza

Contributors

Norman Aminuddin, KoSa, P.O. Box 4, Highway 70W, Salisbury, N.C. 28147, USA

Donald G. Baird, Department of Chemical Engineering, Polymer Materials and Interfaces Laboratory, Virginia Polytechnic Institute and State University, Blacksburg, VA 24061-0211, USA

Cees W.M. Bastiaansen, ETH Zürich, Institut für Polymere, ETH Zentrum, UNO C15 Universitätstrasse 41 CH-8092 Zurich, Switzerland

John Blackwell, Department of Macromolecular Science, Case Western Reserve University, Cleveland OH 44106-7202, USA

Stephen Z. D. Cheng, Maurice Morton Institute and Department of Polymer Science, The University of Akron, Akron, Ohio 44325-3909, USA

Bernd Clauss, Institut für Chemiefasern, Korschtalstr. 26, D-73770 Denkendorf, Germany

John A. Cuculo, North Carolina State University, Fiber and Polymer Science — Graduate Program, Raleigh, North Carolina 27695-8301, USA

Hao Fong, Department of Polymer Science, The University of Akron, Akron, OH 44325 USA

Margaret W. Frey, Champlain Cable Corporation, 12 Hercules Dr. Colchester, VT 05446, USA

Richard V. Gregory, School Fiber and Polymer Science, Clemson University, Clemson, SC, USA

Frank W. Harris, Maurice Morton Institute and Department of Polymer Science, The University of Akron, Akron, OH 44325-3909, USA

John W.S. Hearle, The Old Vicarage, Mellor, Stockport SK6 5LX, U.K.

Joseph F. Hotter, PPG Industries, 473 New Jersey Church Road, Lexington, NC 27293, USA

David J. Johnson, Fiber Science, University of Leeds, Leeds, LS2 9JT, U.K.

Rajendra K. Krishnaswamy, Department of Chemical Engineering, Polymer Materials and Interfaces Laboratory, Virginia Polytechnic Institute and State University, Blacksburg, VA 24061-0211, USA

Piet J. Lemstra, Dutch Polymer Institute/Eindhoven University of Technology, Faculty of Chemical Engineering and Chemistry, P.O. Box 513, 5600 MB Eindhoven, The Netherlands

Christopher Y. Li, Maurice Morton Institute and Department of Polymer Science, The University of Akron, Akron, OH 44325-3909, USA

Fuming Li, Maurice Morton Institute and Department of Polymer Science, The University of Akron, Akron, OH 44325-3909, USA

Kevin W. McCreight, Maurice Morton Institute and Department of Polymer Science, The University of Akron, Akron, OH 44325-3909, USA

N. Sanjeeva Murthy, Honeywell International Inc. Morristown, NJ 07962, USA

Sanjay Rastogi, Dutch Polymer Institute/Eindhoven University of Technology, Faculty of Chemical Engineering and Chemistry, P.O. Box 513, 5600 MB Eindhoven, The Netherlands

Darrell H. Reneker, Department of Polymer Science, The University of Akron, Akron, OH 44325, USA

David R. Salem, TRI/Princeton, 601 Prospect Avenue, P.O. Box 625, Princeton NJ 08542, USA

Jerold M. Schultz, Department of Chemical Engineering and Materials Science Program, University of Delaware, Newark, DE 19716, USA

Joseph E. Spruiell, Department of Materials Science and Engineering, University of Tennessee, Knoxville, TN 37996-2200, USA

Nadarajah Vasanthan, TRI/Princeton, 601 Prospect Avenue, P.O. Box 625, Princeton, NJ 08542, USA

Yeocheol Yoon, Maurice Morton Institute and Department of Polymer Science, The University of Akron, Akron, OH 44325-3909, USA

Qiang Zhou, North Carolina State University, Fiber and Polymer Science — Graduate Program, Raleigh, NC 27695-8301, USA

Contents

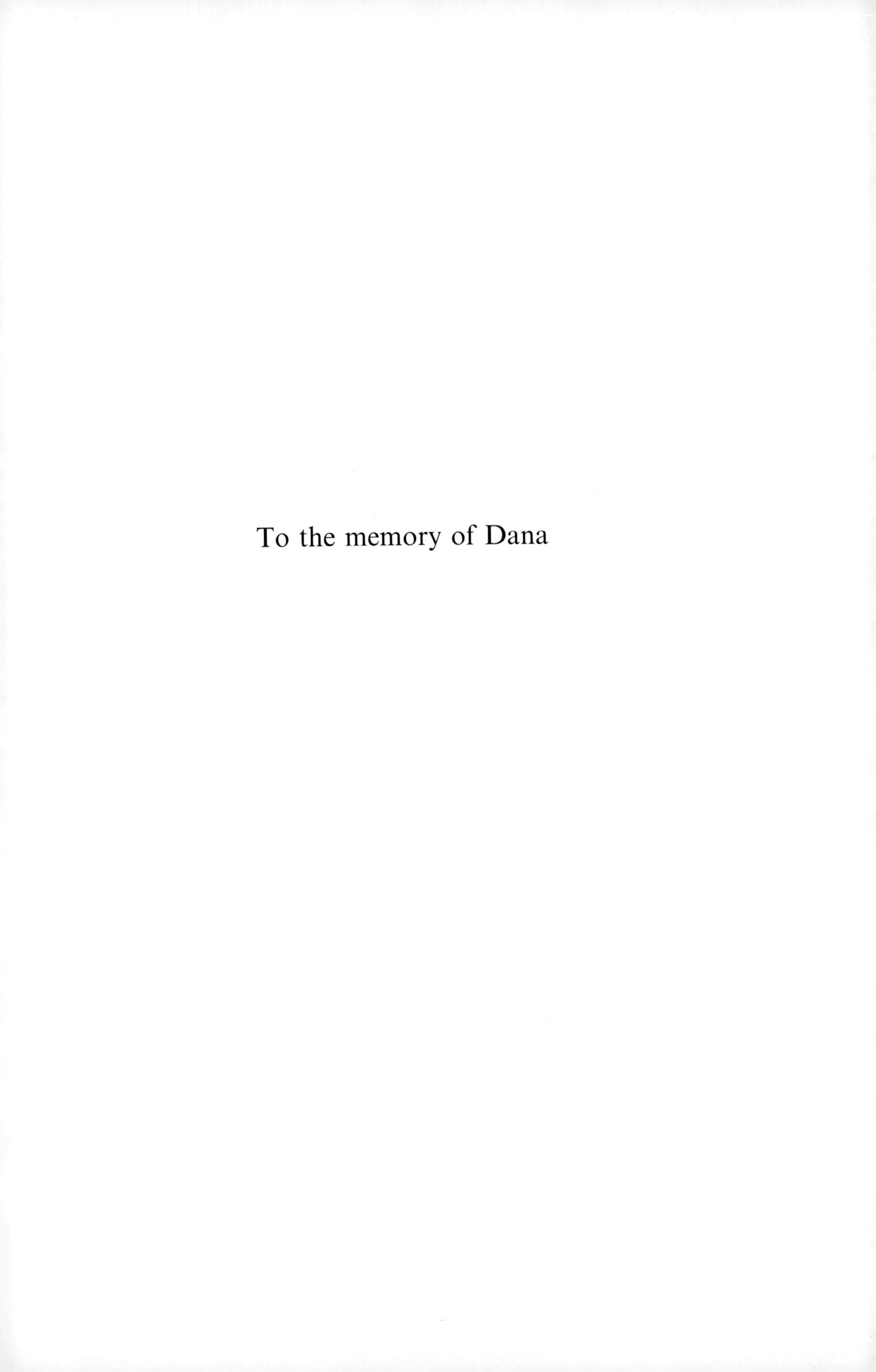

To the memory of Dana

Preface

This book is intended to provide a comprehensive and critical review of the science of fiber formation. It focuses on the evolution of microstructure (on the scale of nanometers to micrometers) in relation to forming conditions and molecular properties, and it surveys the connection between fiber structure and macroscopic properties. The book should also be of value to those interested in the general phenomena of polymer deformation, orientation, and structure formation.

Although the properties, processes and applications of polymeric fibers have become highly diverse, it was considered preferable to cover the subject in a single volume. The approach has been to review the key areas in depth, emphasizing fundamental principles and recent advances, while indicating sources for further reading on less central topics. Although each chapter is self-contained and can be read independently, they have been arranged in an order that would provide a sense of direction to the reader who is patient enough to start at the beginning and proceed to the end.

It will be clear from this book that there has been intense and highly productive research activity on the physics of fiber formation and polymer orientation over the last forty years, in both industrial and academic laboratories, and that there remains enormous potential for further advances. For those entering the field, this book should provide the foundation necessary to understand and contribute to the development of the subject in the 21st century. It will be of use to all scientists and engineers involved in the research and practice of fiber formation (as well as other polymer deformation processes) and to undergraduate and graduate students taking courses in polymer physics, polymer engineering, materials science and engineering, fiber science, and chemical engineering.

I wish to thank Professor Martin Sussman of Tufts University for suggesting this project to me. Due to personal circumstances Martin was unable to contribute to the book himself, but his belief in it was a constant source of encouragement. I would like to express my sincere appreciation to the contributors, for the quality of their work and for their enthusiasm and commitment. It is also important to acknowledge the thorough efforts of the reviewers of the draft manuscripts, including Peter Barham, Henry Chanzy, Dan Edie, John Hearle, Karl Jacob, Maurits Northolt, Ludwig Rebenfeld and Andrzej Ziabicki.

It would not be possible, or at least it would be very unwise, to take on a project of this type without the whole-hearted support of one's family, which I am fortunate to have had. And after the death of my wife, Dana, I could not have completed the book without the knowledge that she would strongly wish it to be published, or without the full backing of my daughter, Julia. I am deeply grateful to them both. I am also thankful to Betsy and Fred Levinton for their moral and practical support, in ways that freed-up my time for the book at some critical moments.

David Salem
Princeton, October 2000

1 Variations on a Theme of Uniaxial Orientation: Introductory Remarks on the Past, Present and Future of Fiber Formation

David R. Salem
TRI/Princeton, Princeton, New Jersey, USA

The formation of polymeric fibers through human invention (rather than natural processes) was achieved decades before the molecular structure of polymers was understood. Regenerated cellulose fibers were produced, albeit by a slow and polluting process, at the end of the 19th century. The strong resistance to Hermann Staudinger's idea, first put forward in the early 1920s, that small molecules could link together by covalent bonds into long polymer chains, and the eventual acceptance of this kind of molecular complexity about ten years later, is a well-known story. Wallace Carothers was an early proponent of Staudinger's view, and his experimental studies on polymer synthesis at DuPont furnished the proof that polymers are giant molecules and not, as previously supposed, colloidal aggregates of small molecules. It was this breakthrough in understanding that lead quite rapidly to the synthesis of polyamide (nylon) 66 and its formation into fibers of considerable commercial importance. Throughout the remainder of the 20th century numerous new polymers were synthesized that could be formed into fibers, covering an enormous range of properties.

It became increasingly evident in the 1950s and 60s that the properties of polymers are as strongly dependent on the physical organization of the macromolecules as on their chemical constitution. For example, a flexible-chain polymer of a given molecular weight and molecular weight distribution can exhibit a broad spectrum of mechanical properties (brittle, ductile, elastic, tough ...), mirroring the equally diverse arrangements that the molecules can assume. There emerged, therefore, a research imperative to understand how an assembly of macromolecules, each containing thousands of atoms, develops structure; how specific molecular arrangements can be induced; and how these structures are related to properties. This is the realm of polymer physics, where the fundamentals of fiber formation largely reside.

To form fibers from flexible-chain polymers, a randomly oriented (isotropic) melt or solution of the polymer must be converted into solid filaments having a high degree of preferred orientation along the fiber axis. This structural transformation, achieved through various spinning and drawing methods, entails a number of complex molecular processes. Some of the factors involved are time and temperature dependent molecular motions, crystallization and other phase transitions under high-stress, entanglement constraints, and various interchain interactions. Thus the final state of molecular order in a fiber from a given polymer is highly dependent on process variables such as stress, strain, temperature and time, and also on the length and length-distribution of the molecules.

Research on structure formation in polymeric fibers has both benefited from and contributed to the rapid advances in theoretical and experimental polymer physics over the last few decades – especially in the areas of rheology, network deformation, orientation phenomena, and crystallization. This has lead to a growing ability to control structure development in fiber and film forming processes, so that properties can be engineered to meet the specific demands of an ever-widening range of applications. Furthermore, the increasing precision with which the chemical constitution and conformational structure of macromolecules can be controlled has created an expanding array of molecular properties with which to explore the evolution of order during fiber formation, and to produce fibers that optimally reflect those properties.

The emergence of fibers with exceptional tensile modulus and strength, combined with their intrinsic low density, is one of the results of this research activity. For example: polymer scientists have designed and synthesized a variety of stiff, aromatic main-chain polymers exhibiting liquid-crystalline properties, and have learned how to orient them into highly ordered fiber structures with specific strength and modulus far exceeding that of steel: it has been discovered how to reorganize entangled, partially-crystalline assemblies of (flexible-chain) polyethylene molecules (previously used only for low-strength fibers, plastic bags, inexpensive toys, etc.) into extended-chain 'high performance' fibers with specific modulus and strength similar or higher than those of liquid-crystalline polymer fibers: and methods have been found to convert polyacrylonitrile and mesophase-pitch fibers into carbon or graphite fibers that retain their outstanding mechanical properties at extremely high temperatures. Frequently, high strength/high modulus fibers are used as reinforcing materials in polymer composites, with diverse applications that include bone replacements and other biomedical implants, tennis rackets, bridges, automobiles, racing bikes, airplanes and space vehicles. Some examples of non-composite applications are heavy-duty ropes, satellite tethers, high-performance sails, and bullet-proof vests.

Research has brought about and enhanced numerous other remarkable fiber properties: chemical, oxidative, and UV resistance, electrical conductivity, biodegradability, nano-diameter fibers, etc. And the quest for new high-technology fibers is continuing unabated due to growing demands from the biomedical, construction, engineering, telecommunications, and electronics sectors. There have also been considerable improvements in the properties and functionality of fibers for the consumer textile market, such as high elastic recovery, high moisture absorption, bulk and texture, flame resistance and microfibers.

A number of these novel properties, for both specialty and commodity applications, have been realized by forming fibers from polymer blends and copolymers, taking advantage of the fact that properties of different polymers can be combined in a single material, sometimes with important synergistic effects. This has given rise to the concept of molecular (or *in situ*) composites, an example of which is the blending of flexible chain and liquid crystalline polymers to produce oriented fibers that combine high modulus with toughness.

Research and development efforts have also helped to provide major productivity gains. High quality textile fibers, with uniform and reproducible properties, are produced at throughput speeds of 100 m/s, making fiber manufacturing one of the most efficient and productive industries. These high throughput speeds have, in turn, revealed much fundamental information about the kinetics of polymer orientation and crystallization processes. A current challenge is to further increase production rates without compromising fiber properties, or even while enhancing them.

Polymer physics and the study of fiber formation are no longer in their infancy, as they were forty years ago, but much remains to be learned. Some aspects of structure development have been clearly established through systematic experimental studies, but others remain stubbornly obscure due to the inability of characterization methods to provide the level of detail required. Theories of molecular motion, network deformation and crystallization have provided important conceptual insights and some useful predictive power, but they are unable to reliably quantify structure development during most fiber formation processes. It is arguable that this is because fiber formation usually occurs under conditions that are far from thermodynamic equilibrium.

Certainly, there are few substances more complex than polymers. In a polymer above its glass-transition temperature, different parts of a single molecule may be simultaneously in a mobile liquid-like state, a crystalline solid-state, and in various other states of order between these extremes. Moreover any event involving one part of the chain will influence the 'destiny' of neighboring parts, and will ultimately have repercussions at very long distances along the chain and along interacting chains. Of course, it is the complexity of polymers that provides their versatility, since the large range of (meta)stable states into which they can be driven provides a corresponding range of properties. But detailed characterization of these states and prediction of the conditions giving rise to them are areas of research that will occupy scientists for decades to come. It is likely that polymer physics, including the study of fiber formation, will benefit from and contribute to advances in theories of nonequilibrium thermodynamics and complex systems, and that these theories will become somewhat more tractable with the continuing gains in computational power. Continued developments in the sophistication and resolution of analytical instruments for characterizing polymer structure will also contribute to improved understanding of molecular organization in these materials.

Computational chemistry is now being used to design molecular structures suitable for fiber formation and it is possible to determine the intrinsic axial modulus and some other mechanical properties of real and 'virtual' molecules from quantum mechanical calculations. Molecular dynamics modeling of fiber formation processes, however, is very much in the preliminary stages. The intricacies involved in following the dynamics of about a million atoms over a period of half a second or more in the simulated forming process, requires enormous computing power (even with some significant approximations in the model). It is possible, however, that computer simulations will eventually be able to predict the evolution of fiber morphology under an almost limitless range of spinning and drawing conditions.

As mentioned earlier, one of the remarkable developments in fiber formation at the end of the previous century was the development of ultra-high modulus polyethethylene fiber via gel-spinning and ultra-drawing. Attempts to produce ultra-high modulus fibers from polymers with stronger (polar) chain interactions (e.g. polyesters and polyamides) have been unsuccessful, but the impetus to do so remains strong. A breakthrough may require approaches involving *in-situ* control of intermolecular interactions. Significant improvements in the mechanical properties of these fibers may also arise from further control of spinline dynamics, systematic studies of novel drawing sequences using the Incremental

Drawing Process (IDP), in which temperature and strain-rate can be varied over a wide range *during* the deformation, and advances in computer simulations of spinning and drawing processes.

A valuable area for further fundamental and applied research is the electrospinning process, in which nanofibers are produced by application of an electric charge to a polymer melt or solution. An important objective of electrospinning studies is to increase throughput rates, since the current technology is too slow for most commercial applications.

The complexity of macromolecules that can be synthesized will increase dramatically over the next decade, as a result of research on molecular self-assembly and the production of polymers by genetically engineered enzymes. A number of current projects are specifically focused on producing novel polymers for fiber formation via biological systems that can synthesize proteins and polyesters. Organizing these molecules into fibers with properties that take proper advantage of the molecular design will require particular ingenuity. It may require, for example, a better understanding of how organization and self-assembly of complex macromolecules is achieved in biological systems.

Research on molecular machines (or nano-machines), based on very recent advances in the synthesis of 'discotic' molecules, is likely to impact fiber formation processes in the not-too-distant future. Current work in this area is too preliminary to justify a section in the book, but Karl Jacob and Malcolm Polk at the Georgia Institute of Technology have taken significant steps towards the synthesis of discotic molecules with rotational characteristics suitable for use in molecular machines for fiber production. The basic idea is that discotic molecular columns contained in a membrane are made to rotate in a magnetic field in a synchronized fashion. The rotation will pull down polymer molecules supplied at the top of the membrane, and organize and orient them as they pass through to the bottom surface, where they are released. The positioning of the discotics and the sequential interactions between the discotic end groups and molecular segments of the polymer must be precisely designed to pull segments of polymer in the prescribed sequence. It is clear that this approach would provide the potential for precise control over molecular ordering processes.

Examples of polymer fiber properties currently sought-after are ultra-high modulus and high strength *combined* with high compressive and shear strength/modulus, high strength combined with elastic recovery from very high deformations, high strength combined with high electrical conductivity and, for optical fiber applications, low-loss photon transmission. Development of nano-engineered 'smart' fibers that would change properties in response to environmental, mechanical or electrical stimuli is being pursued, and so is research on the formation of defect-free carbon nanotube fibers that would exhibit exceptional strength, flexural properties, electrical conductivity and transport properties.

As in any field of worthwhile endeavor, some concepts that look promising today will lead nowhere and others will emerge unexpectedly and succeed. There is no question, however, that the special properties arising from the geometry (e.g. high specific surface), low density, and structural versatility of polymeric fibers will continue to spur important discoveries in polymer science, fiber formation, and related areas, and will result in further growth of the range and sophistication of fiber applications.

2 Structure Formation During Melt Spinning

Joseph E. Spruiell
University of Tennessee, Knoxville, Tennessee, USA

2.1 Concepts and Theories

2.1.1 Introduction

The origin of the melt spinning process apparently dates to the 1845 English patent of R. A. Brooman [1] who conceived the basic concept as a method to produce filaments from gutta percha. But it was Carothers and Hill [2] who first described the process in the modern era. The work of Carothers and his associates [3] led to successful commercial application of melt spinning in 1939 as a process to produce polyamide 66 ("nylon") filaments and yarns.

The melt spinning process involves melting and extrusion of the material to be processed through a multihole capillary die, called a spinneret, followed by cooling and solidification to form filaments that can be wound on a bobbin or otherwise processed. A tensile force is usually applied to the molten extrudate in order to draw the filaments down to a desired diameter. This draw-down causes the filament velocity to increase along the spinning path until it reaches a final velocity, called the take-up velocity or spinning speed. The process is most commonly applied to polymer resins or inorganic glasses. In the present chapter, we will discuss only organic polymers.

A basic form of the melt spinning process is illustrated in Figure 2.1. Polymer, usually in the form of dried granules or pellets, is fed into an extruder where it is melted and conveyed to a positive displacement, metering pump. The metering pump controls and ensures a steady flow of polymer to the "spin pack" where the polymer is filtered and forced through the capillaries of a multifilament spinneret. The extruded filaments

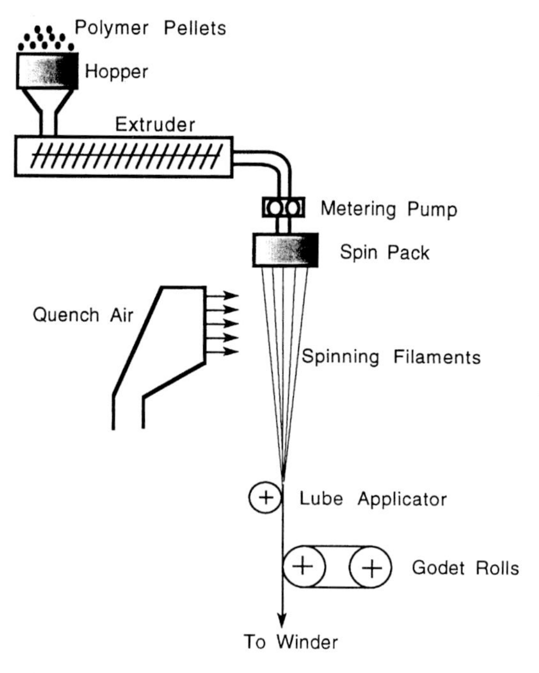

Figure 2.1 The melt spinning process.

are drawn down to smaller diameters, i.e., finer deniers, by the action of a godet roll, while they are simultaneously being cooled (quenched) by air blowing across the filament bundle. The resulting filaments are either wound onto a bobbin or they are passed directly to another processing step such as "drawing" (see Chapter 4) or texturing (see Chapter 12).

Many modifications of the basic process have been developed in order to achieve specific new products. An example is the "spunbonding" process for producing non-woven webs, illustrated in Figure 2.2. In the particular process illustrated, the extruded filaments are cooled and drawn down by air that enters the enclosed chamber near the spinneret and which is sucked out of the chamber below a conveyor belt on which the filaments are deposited. A venturi increases the air speed in order to provide more air drag on the filaments and greater filament draw-down. The deposited filaments are conveyed to a bonding system, commonly a thermal bonding calender, whose purpose is to bond the fibers together at specific points so that the fibrous web becomes a non-woven fabric.

The earliest published papers dealing with the dynamics and structure development of filaments as a result of changes in melt spinning variables were those of Ziabicki and Kedzierska [4–8] published in 1959–1962. These authors gave the first quantitative description of the process and established the influence of various process and materials variables. A major review of available literature on the process that was available at the time it was published (1976) together with a further analysis of the process was provided in a classic book by Ziabicki [9].

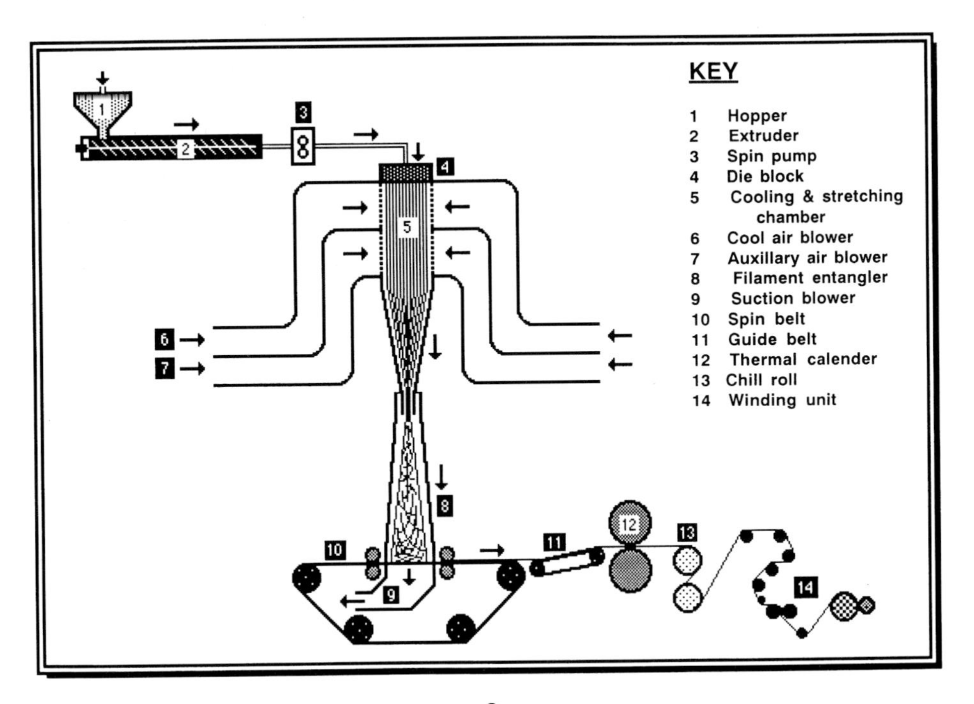

Schematic of the Reicofil® Spunbonding Process.

Figure 2.2 One type of spunbonding process.

The major process variables for melt spinning are:
(1) extrusion temperature,
(2) mass flow rate of polymer through each spinneret hole,
(3) take-up velocity of the wound-up or deposited filaments,
(4) the spinline cooling conditions,
(5) spinneret orifice shape, dimensions and spacing, and
(6) the length of the spinline.

These variables are not entirely independent of each other. For example, the length of the spinline will generally be controlled by the efficiency of the cooling conditions along the spinline. More efficient cooling allows shorter spinlines. Spinline cooling is largely controlled by the velocity, temperature and distribution of the cooling air, but it is also affected by factors such as spinneret configuration, mass throughput and the specific design of the cooling system.

One of the most important variables of the melt spinning process is take-up velocity. For a given final filament diameter or linear density and a fixed number of holes in the spinneret, the take-up velocity controls the productivity of the spinline. It also has a marked effect on the structure and properties of the melt spun filaments and how the spun filaments behave in subsequent processing steps such as drawing and texturing. For about three decades after melt spinning was introduced as a commercial process, spinning speeds increased gradually to about 1500 m/min. Then, beginning in the 1970s, maximum commercial spinning speeds increased rapidly due to the development of winders capable of winding at higher speeds. The primary driving force for the development of these winders was increased productivity. Other important factors associated with the development of higher winding and spinning speeds included (1) the use of partially oriented feed yarns, produced in the range of 2750 to about 4000 m/min, for the draw texturing process, and (2) the desire to eliminate processing steps, such as drawing, by forming fully oriented yarns in a one step process [9, 10]. The study of the effects of "high speed spinning" attracted much attention in the 1970s and 1980s, as we shall see. Recent efforts in the 1990s include process changes such as the addition of "hot tube" spinning and/or "hydraulic drag" spinning. In the former case, the spinline is reheated, after the initial quench and before it reaches the first godet. The latter case passes the filaments through a liquid bath prior to the first godet that cools the filaments and allows higher spinline stresses to be achieved due to the hydraulic drag in the spinline. Both of these types of modifications influence the dynamics and thermal history of the treadline and can lead to improvements in the mechanical or dyeing properties of the yarns [11–13].

In addition to the process variables described above, there are a large number of materials variables that affect the spinnability and the structure and properties of melt spun filaments. Generally, the materials variables can be divided into two major categories: 1) variables that affect the rheology of the polymer melt, and 2) variables that affect the solidification behavior of the polymer. Variables of the former type include molecular weight, molecular weight distribution (MWD), chain stiffness, branching, additives, fillers, etc. which affect or control the resin's viscoelastic properties. Most fiber forming polymers are semicrystalline and solidification for these materials generally refers to crystallization. In this case, factors that affect the crystallization (and melting) temperature, crystallinity or crystallization kinetics are important. Such factors include composition and stereoregularity of the molecule (e.g., tacticity, comonomer content, branching, etc.), molecular weight, and the influence of additives such as nucleating agents, antioxidants, pigments, etc. Of course, for fully amorphous polymers or polymers whose crystallization kinetics are too slow to allow crystallization during melt spinning, solidification refers to vitrification. In this case the glass transition temperature is an important parameter.

2.1.2 An Engineering Analysis of the Process

A detailed engineering analysis of the melt spinning process involves a treatment of the dynamics of melt spinning, selection of an appropriate rheological constitutive equation for the melt being spun, application of material and energy balances, and a treatment of molecular orientation development and crystallization in the presence of molecular orientation. It is clear that any such analysis will contain numerous assumptions and approximations, as our understanding of several aspects of the problem is relatively poor. Nevertheless, even an imperfect analysis can be extremely valuable in helping us understand the influence of the many variables and the interactions among variables that occur in the melt spinning process. In the analysis that follows, we will limit ourselves to the description of a steady state spinning process. We will largely ignore radial variations within the spinning filament on the basis that (1) the filaments are thin and undergo a nearly pure extensional flow field and (2) thermal gradients across the radius of the filament are small. Other major assumptions will be described along the way.

2.1.2.1 Force and Momentum Balance

A schematic illustrating the forces acting on a single filament in a spinline is shown in Figure 2.3. An overall force balance on a single filament of the spinline may be written as [9]

$$F_{rheo} = F_o + F_{inert} + F_{drag} - F_{grav} + F_{surf} \tag{2.1}$$

where F_{rheo} is the rheological force in the fiber at a distance z from the spinneret, F_o is the rheological force at the exit of the spinneret ($z = 0$), F_{inert} is the inertial force produced by the acceleration of the polymer mass along the spinline, F_{drag} is the the drag force caused by the fiber moving through the cooling medium (usually air), F_{grav} is the gravitional force acting on the spinline, and F_{surf} is the surface tension force at the fiber/air interface. Surface tension is usually small compared to other components of the force except for very low viscosity materials. Thus for most polymeric materials, we may neglect it. Assuming a circular filament cross section of diameter D, the other terms may be expressed by [9]

$$\begin{aligned} F_{inert} &= W(V - V_0) \\ F_{drag} &= \int_0^z \pi D \sigma_f \, dz \\ F_{grav} &= \int_0^z \rho g \left(\frac{\pi D^2}{4} \right) dz \end{aligned} \tag{2.2}$$

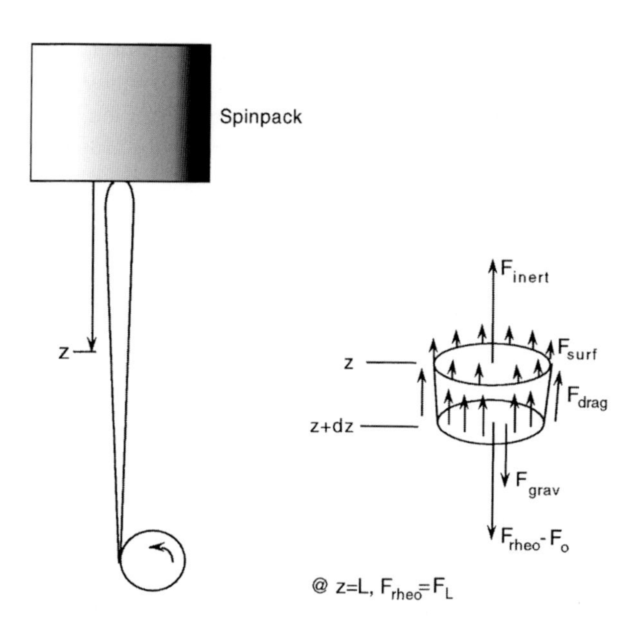

Figure 2.3 Schematic illustration of the forces acting on a spinning filament.

where D and V are the diameter and velocity at a given point in the spinline; subscript zero refers to the exit of the spinneret at z = 0. Here, σ_f is the shear stress at the fiber/air interface due to aerodynamic drag, ρ is filament density, g is acceleration due to gravity, and W is the mass throughput rate per spinneret hole. Continuity (material balance) requires that

$$W = \rho AV = \rho\pi(D/2)^2V \tag{2.3}$$

and we may write

$$\sigma_f = \frac{1}{2}\rho_a V_r^2 C_d \tag{2.4}$$

where ρ_a is the density of air, V_r is the relative axial velocity of the spinning filament and the cooling medium, and C_d is called the drag coefficient. If the cooling air has no component of velocity parallel to the spinning filament, then V_r is just the filament velocity V. Various methods of evaluating C_d have been described in the literature [14–21]; the interested reader is referred there for details. A common result correlates the air drag coefficient with the cooling air Reynolds number, Re:

$$C_d = K(Re)^{-n} \tag{2.5}$$

Values of K and n determined by various investigators have been tabulated and compared by Shimizu et al. [21, 22]. The exponent n typically lies in the range 0.6–0.8. With n fixed at 0.61, K ranges from 0.23 to >1, depending on the investigator and the method used to obtain the result.

With the above assumptions, the gradient of axial tension along the spinline can be written as,

$$\frac{dF_{rheo}}{dz} = W\frac{dV}{dz} + \frac{1}{2}\rho_a C_d V^2 \pi D - \frac{Wg}{V} \tag{2.6}$$

The rheological force is directly related to the axial spinline stress, σ_{zz}, by

$$\sigma_{zz} = F_{rheo}/\pi(D/2)^2 \tag{2.7}$$

The value of spinline stress at the point in the spinline at which solidification is occurring is a very important quantity that, for a given material, largely determines the final structure and properties of the spun filaments. We can obtain the spinline stress at any point in the spinline using a modified version of the force balance of Eq. 2.1 which expresses F_{rheo} from the perspective of the measureable quantity F_L, the tension in the spinline at distance L from the spinneret (near the take-up device):

$$F_{rheo}(z) = F_L - \int_z^L \rho_a C_d V^2 \pi D dz - W[V(L) - V(z)] + \int_z^L \rho g \pi (D/2)^2 dz \tag{2.8}$$

A number of investigators [e.g., 9, 22–27] have discussed the relative magnitudes of the various force components of Eqs 2.1 and 2.2 as a function of spinning speed. As already mentioned the surface tension force is negligible except when spinning low molecular weight materials. Further, the gravity force is small compared to other forces except when spinning thick filaments at low speeds. Thus, under the usual spinning conditions for a high polymer, the inertial and air drag forces are the major components of the rheological force. The way the forces develop along the spinline is illustrated in Figure 2.4, which shows calculated values of the component forces as a function of distance from the spinneret. For these calculations, a mass throughput of 2.5 g/min per hole and a take-up velocity of 6600 m/min were assumed. The polymer density was taken as that of

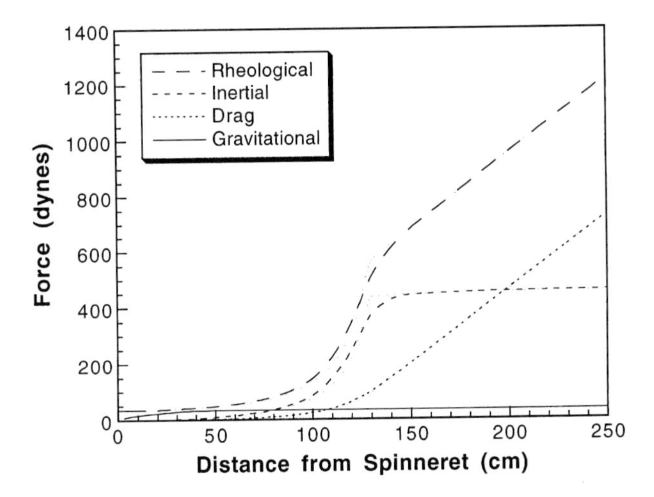

Figure 2.4 Calculated rheological, air drag, inertial, and gravitational force profiles for polyamide 66 at a spinning speed of 6600 m/min.

polyamide 66. This figure shows that at high take-up speeds, the inertial force increases in the upper part of the spinline and then levels out as the filament freezes and its diameter stops drawing down. The air drag component develops later, increasing with increase in fiber velocity, and continues to increase, after the velocity (and diameter) levels out, due to the increase in length of the spinline contributing to the drag. As first noted by Shimizu et al. [21], this suggests that the inertial force plays the dominant role in the determination of the structure of the filaments. It is the dominant force in the region of the spinline in which the structure is being developed, just prior to the leveling out of diameter, velocity and the inertial force.

2.1.2.2 Constitutive Equation

Molten polymers are viscoelastic fluids and exhibit both elastic and viscous response to applied forces. Several authors [e.g., 28–32,22] have examined the use of viscoelastic constitutive equations to describe the melt spinning process. The influence of elastic effects become more important as spinning speed increases due to the increase in the rate of elongation of the filament. Polymers with high molecular weight, broad molecular weight distribution and branched chains, e.g. certain polyolefins, are also more viscoelastic and require a viscoelastic constitutive equation. In spite of this, it is known that the omission of elastic effects still provides a reasonable approximation to the actual behavior in many spinning processes and materials such as polyesters, polyamides, and spinning grade polypropylene [25–27,33,34]. Possible reasons for this are (1) the polymers used for spinning are often chosen to have modest molecular weight and, hence, elasticity, and (2) the rapid decrease in temperature of the spinning filaments and consequent rapid increase in viscosity may overshadow the effect of elasticity. Thus, it is common to use a purely viscous constitutive equation to describe behavior during melt spinning. We define a uniaxial "apparent elongational viscosity," η, as

$$\eta(T,\dot{\varepsilon}) = \{\sigma\}_{zz} \Big/ \left(\frac{dV}{dz}\right) \tag{2.9}$$

Here, η may be a function of temperature, T, and strain rate, $\dot{\varepsilon}$. If η is assumed independent of strain rate, then Equation 2.9 describes the constitutive equation for a Newtonian fluid. On the other hand, a generalized power law equation of the Cross/Carreau [35, 36] type has been used [37] to describe the viscous behavior in elongation for a material whose viscosity decreases with strain rate. In this case the viscosity is described by an equation of the form

$$\eta(T,\dot{\varepsilon}) = \frac{\eta_o}{1 + (a\eta_o\dot{\varepsilon})^b} \tag{2.10}$$

where η_o is the zero strain rate viscosity and a and b are empirical constants. The viscosity is usually assumed to follow Arrhenius temperature dependence in the high temperature range well above the melting point:

$$\eta_o = A \exp(B/T) \tag{2.11}$$

As temperature approaches the glass transition temperature, the temperature dependence of amorphous polymers no longer follows Equation 2.11 well, and it is advisable to shift to the Williams-Landell-Ferry (WLF) equation [38] to describe the temperature dependence of viscosity.

A major problem in the treatment of melt spinning is the influence of crystallization on the rheology (i.e. the viscosity) of the spinliine. It was recognized early that the viscosity increases rapidly as crystallization sets in. To account for this effect Shimizu et al. [22] introduced an empirical expression:

$$\eta(T,\dot{\varepsilon},\theta) = \eta(T,\dot{\varepsilon})\exp(c\theta^{d}) \tag{2.12}$$

where c and d are appropriately chosen constants and θ is the relative crystallinity.

Ziabicki [39] later introduced the idea that, due to crystallization, the long chains in a polymer melt become gradually interconnected, small crystals acting as 'physical crosslinks.' He suggested that, when the number of these physical crosslinks reaches some critical value, the system loses fluidity and converts into an elastic solid. With further increase in crystallinity (i.e. crosslinking) the solid becomes rigid. This led him to propose the following equation to describe the temperature and crystallinity dependent viscosity:

$$\eta(T,\theta) = f(T) \Big/ \left[1 - \frac{\theta}{\theta_{cr}}\right]^{\alpha} \tag{2.13}$$

where f(T) is a function of temperature only and θ_{cr} is the critical crystallinity needed for sufficient 'crosslinking' of the melt that $\eta \rightarrow \infty$. Experimental evidence has recently been published [40] that suggests that the level of crystallinity corresponding to θ_{cr} may be quite low, of order 2–3%.

Clearly, the effect of even small amounts of crystallinity will drastically change the dynamics of melt spinning. Therefore, it is very important that further studies be performed to develop a better understanding of the interaction of crystallization and rheology.

2.1.2.3 Energy Balance

The variation of fiber temperature as a function of distance from the spinneret is determined from an energy balance on the spinline. Heat is transferred from the filaments by convection, radiation, and conduction along the fiber to cooler sections of fiber and to objects in contact with the fiber such as the take-up godet. Conduction is easily shown to be negligible in comparison to other mechanisms. Radiation is very dependent on the temperature of the filament; it can be very important for spinning of inorganic glasses or metals where spinline temperature can be quite high. But, because spinline temperatures for organic polymers rarely exceed 300 °C, radiation is often neglected or is incorporated into the convective contribution through the value of the heat transfer coefficient chosen [9,22,23,26,27,34,41–43].

Early investigators [23,26,41–43] also ignored the heat of crystallization which is converted to sensible heat when the polymer crystallizes. One reason for this is that these early treatments were aimed largely at analyzing the behavior of PET, which did not crystallize in the spinline under spinning conditions available at the time the treatments were carried out. However, for spinning of polyolefins or for high-speed spinning of PET, other polyesters and polyamides, it is necessary to include this term in the analysis. Neglecting radial temperature variations within the filament and including the heat of crystallization, the differential energy balance is [22,27,34,44]

$$\frac{dT}{dz} = -\frac{\pi Dh(T - T_a)}{WC_p} + \frac{\Delta H_f}{C_p}\frac{dX}{dz} \tag{2.14}$$

where T is fiber temperature, T_a is the temperature of the cooling medium, ΔH_f is the heat of fusion, X is crystalline fraction, h is the heat transfer coefficient, and C_p is the polymer heat capacity.

The key parameter in Equation 2.14, that is required to compute temperature profiles along the spinline, is the heat transfer coefficient, h. Using both theory and empiricism, many authors, e.g. [23,45–47], have developed relationships describing h as a function of spinning conditions through use of dimensionless groups such as Nusselt (Nu), Reynolds (Re), Prandtl (Pr), and Grashof (Gr) numbers. Many of these expressions relating heat transfer coefficient to process variables were summarized by Ziabicki [9]. The correlation developed by Kase and Matsuo [23,45] is one of the most commonly used:

$$\frac{hD}{k_a} = 0.42\left(\frac{\rho_a V_r D}{\mu_a}\right)^{1/3}\left[1 + \left(8\frac{u_c}{V_r}\right)^2\right]^{1/6} \tag{2.15}$$

where k_a is thermal conductivity of air, μ_a is kinematic viscosity of air, u_c is the component of air velocity perpendicular to the filament (called "cross blow"), and all other quantities were previously defined.

It is, perhaps, worthy of note that some melt spinning is done into water baths or other cooling media. Water baths often are used in order to rapidly cool large diameter filaments which otherwise may require very long spinlines to achieve enough cooling for take-up.

2.1.2.4 Multifilament Effects

The description of the dynamics and heat transfer given above applies to each filament of a multifilament melt spinning process, but it is important to recognize that the air flow velocities, air temperature and the boundary conditions, may vary from one filament to another in the multifilament bundle. In particular, the hot filaments on the cooling air inlet side of the bundle will heat the inlet cooling air as it passes through the bundle. The air velocities can also vary from one side of the bundle to the other. The axial velocity of the running filaments will impart an axial component to the incoming air stream and reduce the velocity component in the cross-blow direction. Due to changes in both the cooling air temperature and velocity, the quench conditions on the cooling air exit side of the bundle may be substantially different than on the inlet side. When this is the case, it leads to substantial differences in the dynamics as well as the heat transfer as a function of position within the multifilament bundle. Barovskii et al. [48] studied the temperature and velocity distributions within a 140 filament bundle of nylon 6 filaments with cross-flow cooling. They observed that, near the spinneret, the air velocity was reduced by 60% while the air temperature increased by 200 °C on passing through the bundle. The reduction in air velocity and increase in temperature was less, further from the spinneret. Such differences in the dynamics and heat transfer can lead to significant differences in structure and properties of filaments from different parts of the multifilament bundle.

Several authors have discussed and/or attempted to model multifilament effects. Yasuda et al. [49] subdivided the space containing the fiber bundle into individual cells to which they applied mass and energy balances. Ishihara et al. [50], Dutta [51] and Schoene and Bruenig [52] gave later treatments, using a somewhat similar approach. The interested reader is referred to these original works for more detail. While multifilament effects raise an important industrial problem of how to reduce such effects, so that the filament structures and properties are uniform throughout the bundle, our interest will be focused on the structure developed with a given set of boundary conditions. Hence, we will consider a monofilament approach in subsequent analysis, but it is important for the reader to realize that, in an actual multifilament spinning processes, the boundary conditions may differ within the filament bundle.

2.1.2.5 Development of Orientation and Birefringence

Molecular orientation is generated as a result of polymer deformation, whether the deformation is carried out in the melt or the solid state. The main requirement is that the molecular relaxation time of the deformed molecules is long compared to the experimental time allowed for relaxation. This means that the temperature range in which the orientation is developed, the cooling rate and the deformation rate are critical parameters.

Maxwell first noted development of birefringence due to molecular orientation in flowing polymer melts in 1873 [53]. By definition, birefringence in a fiber or filament, Δn, is equal to the difference between the index of refraction parallel and perpendicular to the fiber axis: $\Delta n = n_z - n_r$.

Several investigators [54–56] have noted that well above the glass transition temperature and at relatively low stress levels, the birefringence of a filament being spun from an amorphous polymer is proportional to the applied tensile stress:

$$\Delta n = C_{op}\sigma_{zz} \tag{2.16}$$

This is simply one form of the stress-optical law, and C_{op} is called the stress optical coefficient. According to the theory of rubber elasticity [57,58]

$$C_{op} = \frac{2\pi}{45kT}\frac{(n^2+2)^2}{n}(\alpha_1 - \alpha_2) \tag{2.17}$$

Here, n is the average index of refraction of the polymer and $(\alpha_1 - \alpha_2)$ is the difference in polarizability parallel and perpendicular to the chain segment. It is noteworthy that

Equation 2.17 indicates that C_{op} is inversely proportional to the absolute temperature. However, it is frequently found that C_{op} can be treated as a constant over the range of temperature where the birefringence of the spinning polymer is developing. Note also that C_{op} can be positive or negative depending on the relative values of the two components of polarizability.

Specification of uniaxial orientation is normally done in terms of orientation factors defined by Hermans and his coworkers [59,60]. They defined an orientation factor, f (also denoted $\langle P_2(\cos\theta) \rangle$), for the molecular chains relative to a reference axis (the fiber axis) in terms of the anisotropy of the polarizability tensor. Assuming cylindrical symmetry and a one phase system:

$$f = \frac{\overline{\alpha_{zz} - \alpha_{rr}}}{\alpha_1 - \alpha_2} = \frac{3\overline{\cos^2\phi} - 1}{2} \quad (2.18)$$

where $\overline{\alpha_{zz} - \alpha_{rr}}$ is the mean value of the difference between the components of the polarizability in the axial and radial directions in the fiber and ϕ is the angle between the chain axis (the 1 direction) and the fiber axis. According to this definition, $f=1$ when all molecules are aligned parallel to the fiber axis, $f=0$ when they are randomly dispersed (isotropic system), and $f=-0.5$ if all molecules are perpendicular to the fiber axis. Based on this definition it is readily shown [61] that

$$f = \frac{\Delta n}{\Delta^o} \quad (2.19)$$

where Δ^o is the *intrinsic birefringence* of the material. The intrinsic birefringence is the maximum possible birefringence corresponding to all molecules aligned parallel to the fiber axis (see also Section 13.3.2).

In semicrystalline polymers we must specify the orientation of both crystalline and amorphous phases. In this case Stein and Norris [62] have shown that

$$\Delta n = (1 - X_c) f_a \Delta_a^o + X_c f_c \Delta_c^o + \Delta n_{form} \quad (2.20)$$

where X_c is the crystalline fraction and the subscripts c and a refer to the crystalline and amorphous phases. Δn_{form} is a form birefringence due to interaction of the two phases with light. In semicrystalline, homopolymer fibers it is small and is usually neglected [62,63]. Generally, the intrinsic birefringences have been assumed to be constants independent of the level of orientation and crystallinity in the sample, though some recent work has suggested that this may not be entirely true.

For crystalline materials, the concept of the Hermans' orientation factor can also be extended to describe the orientation of each of the crystallographic axes, a, b, and c, with respect to the fiber axis as described by Stein and Norris [62]. Since the chain axis is usually parallel to the c-crystallographic axis by convention, the c-axis orientation factor is equivalent to the crystalline orientation factor f_c described above. The use of a- and b-axis orientation factors is useful for obtaining a better understanding of the detailed morphology present in the fiber. The orientation factors for the three crystallographic axes are referred to as Hermans-Stein orientation factors.

2.1.2.6 Crystallization in the Spinline

In general, crystallization is a nucleation and growth process. At low supercooling, i.e., near the melting point, the driving force for nucleation is low and the material crystallizes slowly. On the other hand, at temperatures near the glass transition temperature, the growth of crystals is slow due to lack of molecular mobility. These effects produce a maximum in the overall crystallization rate at an intermediate temperature. But even at the temperature of maximum crystallization rate, a finite amount of time is required for crystallization to be completed. During melt spinning the filament cools continuously through the temperature range where crystallization is possible, spending a limited amount of time at any given temperature. Increasing the cooling rate decreases the amount of time available for crystallization at any given temperature. Thus, increasing cooling rates tend to suppress the amount of crystallization that can occur. Ultimately, if we cool sufficiently fast, crystallization does not have time to occur, and the material simply vitrifies into a noncrystalline glass when cooled below its glass transition temperature. Thus, the nature of crystallization in the spinline depends on a balance between the various factors that affect the polymer's crystallization kinetics and those that control the cooling rate.

The rate of crystallization and its temperature dependence varies greatly from one polymer to another. It is also well established that the rate of crystallization of polymers is greatly enhanced by the presence of molecular orientation. The reasons for this will be described in more detail below. Assuming that this is true, we can obtain a good qualitative understanding of the important relationships between cooling conditions and crystallization kinetics using the concept of a "continuous cooling transformation diagram" [64]. The concept is illustrated schematically in Figure 2.5. Here, cooling curves, numbered 1–5, are plotted for the material on temperature versus log time axes. The "c-

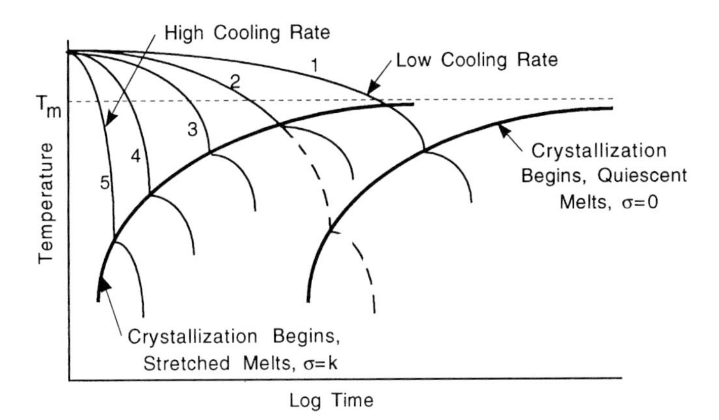

Figure 2.5 Schematic continuous cooling transformation diagram illustrating the influence of cooling rate and stress on crystallization behavior.

curve" on the right shows the start of crystallization under quiescent conditions as determined, for example, by cooling at different rates in a differential scanning calorimeter. It simply illustrates that faster cooling produces greater supercooling of the melt before crystallization occurs. The location and shape of this curve relative to the time and temperature axes is determined by the crystallization kinetics. A faster crystallizing material would have its "c-curve" displaced to shorter times and the c-curve for a slower crystallizing one would be shifted to longer times. Cooling at a rate that misses the "nose" of the curve will result in quenching the material to an amorphous, glassy state. Also, if stress in the spinline increases the crystallization rate, then there should be another curve, illustrated by the left-hand "c-curve" in Figure 2.5, which represents the start of crystallization under a given stress. Since the crystallization rate is expected to be a function of the level of molecular orientation in the melt, and orientation is a function of stress in the spinline, there is, in principle, a different "c-curve" for each stress level. According to this analysis, stress would also be expected to increase the temperature at which crystallization takes place for a given cooling rate as is illustrated by cooling curve 2. Figure 2.6 shows an actual continuous cooling transformation diagram for HDPE constructed by Spruiell and White [64] from the data of Dees and Spruiell [65] and other data from the literature [66–68]. It is obvious from this figure that crystallization on the spinline occurs orders of magnitude faster than under quiescent conditions, especially in the higher temperature range where the crystallization is under

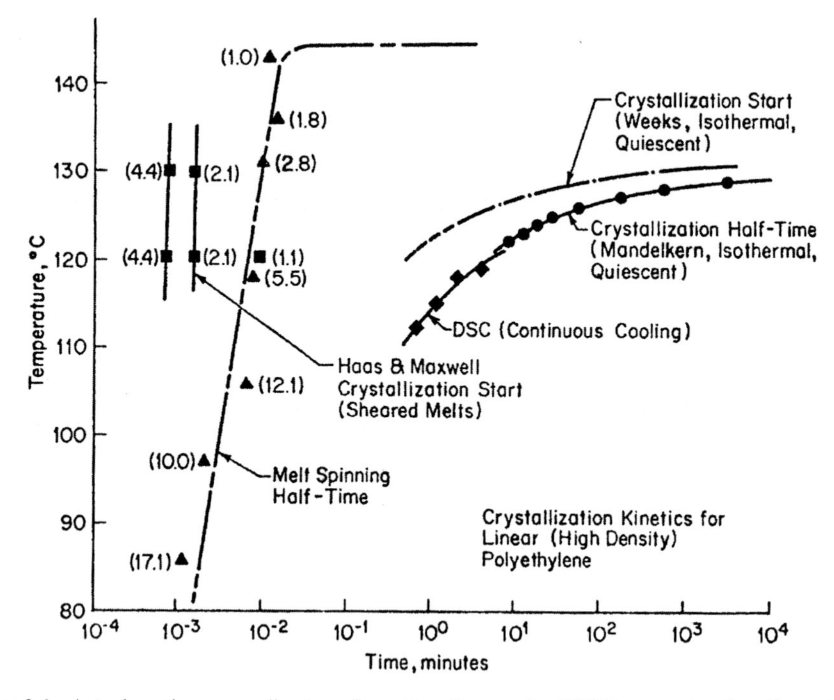

Figure 2.6 Actual continuous cooling transformation diagram for HDPE comparing the effects of stress to quiescent crystallization behavior. Numbers in parentheses are the stress in dynes/cm^2 $\times$ 10^{-6}.

nucleation rate control. The melt spinning data are comparable to the data obtained by Haas and Maxwell [68] for melts crystallized under shear between two glass plates.

The quantitative treatment of crystallization in the spinline is fraught with difficulties due to the non-isothermal conditions and the influence of molecular orientation, i.e., stress-induced crystallization. The analysis of non-isothermal crystallization even in the absence of molecular orientation is not well established, and the situation is even worse for crystallization in the presence of molecular orientation. Other than direct measurements on the spinline, no experimental techniques are available for measurement of crystallization kinetics in the presence of molecular orientation and no reliable theory is available for guidance.

Most investigators [9,22,44,69–77] have started from the classical treatment of isothermal crystallization, and they have modified it in some way in an attempt to deal with these factors. Most authors have used an equation similar in form to the well-known Avrami equation to describe the progress of crystallization. For isothermal crystallization this equation is [78]

$$\theta = 1 - \exp(-kt^n) \tag{2.21}$$

where k is the crystallization rate constant and n is the Avrami index. Using the Avrami theory as a basis, Nakamura et al. [71] incorporated an "isokinetic approximation" and derived the following equation to describe the transformation process under non-isothermal conditions:

$$\theta = 1 - \exp\left[-\left(\int_0^t K(T)dt'\right)^n\right] \tag{2.22}$$

Here, n is the Avrami index derived from isothermal experiments, and K(T) is related to the isothermal crystallization rate through the relation

$$K(T) = [k(T)]^{1/n} \tag{2.23}$$

For process modeling, a differential form of Equation 2.22 is often most useful [77]. It reads

$$\frac{d\theta}{dt} = nK(T)(1-\theta)\left[\ln\left(\frac{1}{1-\theta}\right)\right]^{\frac{n-1}{n}} \tag{2.24}$$

Several other mathematical descriptions of non-isothermal crystallization have been presented [69,73–75]. Patel and Spruiell [77] examined the various possibilities with the aim of choosing among them for process modeling. They concluded that the Nakamura et al. model described above was probably the best available at the time of their paper. In the case of melt spinning, we must interpret the value of K(T) to include the effects of molecular orientation. Thus it is no longer given by Equation 2.23, but must be written K(T,f) where f is a measure of molecular orientation in the melt such as the Hermans' orientation function [59,60].

Ziabicki [9] was the first to attempt a phenomenological analysis of the influence of molecular orientation. He proposed that K(T,f) could be written as a series expansion

$$K(T,f) = K(T,0) + a_1f + a_2f^2 + a_3f^3 + \ldots \tag{2.25}$$

The linear term in the above equation drops out for symmetry reasons. For not too high orientations, we can obtain

$$\ln[K(T,f)/K(T,0)] = A(T)f^2 \tag{2.26}$$

or

$$K(T,f) = K(T,0)\exp(Af^2) \tag{2.27}$$

Since A(T) is assumed always positive, Equation 2.27 predicts a monotonic increase in K with increase in orientation. We can consider A(T) to be an empirical parameter to be determined from experimental data. Wasiak et al. [79] found the value of A(T) for PET to be 210 at 95°C and 940 at 115°C. Thus, crystallization of PET is extremely sensitive to even very small amounts of molecular orientation present in the melt. Ziabicki also suggested that the quiescent crystallization rate could be approximated by

$$K(T) = K_{max}\exp\left[(-4)\ln(2)\frac{(T - T_{max})}{D^2}\right] \tag{2.28}$$

where K_{max} is the value of the rate constant at the temperature of the maximum crystallization rate, T_{max}, and D is the half-width of the curve of K versus T. Substituting Equation 2.28 in Equation 2.27, we arrive at

$$K(T,f) = K_{max}\exp\left[(-4)\ln(2)\frac{(T - T_{max})}{D^2} + Af^2\right] \tag{2.29}$$

A somewhat different approach was taken by Katayama and Yoon [44]. They and others have attempted to include the effect of orientation into the kinetic equation based on a combination of rubber elasticity theory and the quiescent crystallization theory of Hoffman et al. [80,81]. Following Kobayashi and Nagasawa [70] they argued that the change in crystallization rates was a result of the thermodynamic effects of deformation of the molecular entanglement network. If ΔF_{iso} is the difference in free energy between the crystal and the amorphous melt for the isotropic (quiescent) state, and ΔF_{or} is the free energy change on crystallization from the oriented melt, then the difference in the free energy change due to deformation and orientation of the melt, ΔF_{def}, can be written

$$\Delta F_{def} = \Delta F_{or} - \Delta F_{iso} \tag{2.30}$$

Expressing the values of ΔF_{iso} and ΔF_{or} in terms of enthalpy and entropy differences and assuming that $\Delta H_{iso} \cong \Delta H_{or}$,

$$\Delta F_{def} = -T(\Delta S_{or} - \Delta S_{iso}) = -T\Delta S_{def} \tag{2.31}$$

Here, ΔS_{def} is interpreted as the entropy difference between the isotropic and oriented amorphous melt. According to rubber elasticity theory, ΔS_{def} may be related to the molecular extension ratio, λ:

$$\Delta S_{def} = k_1\left(\lambda^2 + \frac{2}{\lambda} - 3\right) \tag{2.32}$$

and the birefringence, Δn, is given by

$$\Delta n = k_2\left(\lambda^2 - \frac{1}{\lambda}\right) \tag{2.33}$$

Thus for small extensions with $\lambda \cong 1$,

$$\Delta S_{def} = k_3(\Delta n)^2 \tag{2.34}$$

Since $f = \Delta n/\Delta^{o}$ we obtain

$$\Delta F_{or} = \Delta F_{iso} + CTf^2 \tag{2.35}$$

Substituting ΔF_{or} for ΔF_{iso} in the classical nucleation and growth theories [80–82] and simplifying, we arrive at

$$\dot{N}(T,f) = \dot{N}_o \exp\left[-\frac{U^*}{R(T - T_\infty)}\right] \exp\left[-\frac{C_1}{T\Delta T + CT^2f^2}\right] \tag{2.36}$$

and

$$G(T,f) = G_o \exp\left[-\frac{U^*}{R(T - T_\infty)}\right] \exp\left[-\frac{C_2}{T\Delta T + CT^2f^2}\right] \tag{2.37}$$

where N(T,f) is the heterogeneous nucleation rate in the presence of orientation, G(T,f) is the linear growth rate in the presence of orientation. In the above equations, U^* is the activation energy for segmental jumping, R is the gas constant, $T_\infty = T_g - 30(K)$, $\Delta T = T_m^0 - T$ and C_1 and C_2 are grouped constants involving ΔH_{iso}, R, surface energies and interchain distance. The values of C_1 and C_2 can be determined, in principle, from quiescent nucleation and growth rate data. Encouraged by the similarity of the above equations, Katayama and Yoon [44], and Patel and Spruiell [77] assumed that the overall crystallization rate constants have the same form. Thus

$$K(T,f) = K_o \exp\left[-\frac{U^*}{R(T - T_\infty)}\right] \exp\left[-\frac{C_3}{T\Delta T + CT^2f^2}\right] \tag{2.38}$$

Again, K_o and C_3 can be obtained from quiescent crystallization kinetics data. The value of C can only be determined by back calculation from data taken from the melt spinline. Thus, we may consider it an adjustable parameter required to bring the values of crystallization rate constant in the presence of molecular orientation into agreement with experiment.

As already noted, crystallization during melt spinning involves very high cooling rates (up to 10^3–10^4 deg/s). The isokinetic approximation of Nakamura et al. assumes that crystallization kinetics is a function of temperature alone and does not depend on cooling rate. There is now experimental evidence that at high cooling rates, crystallization kinetics become a function of cooling rate as well as temperature [83,84]. The reasons for this dependence are explained in a new model of crystallization kinetics in variable external conditions proposed by Ziabicki [85]. This model specifically deals with situations when temperature, pressure or stress change with time, and it includes transient and athermal effects associated with the change in external conditions. The athermal effects increase the overall crystallization kinetics with increase in cooling rate. This is due to the fact that embryos (clusters) of molecules which are too small to be stable at high temperature may become stable crystal nuclei at lower temperature and greater supercooling, if the cooling rate is fast enough to prevent their relaxation before reaching the crystallization temperature. This new approach to crystallization kinetics under variable external conditions would appear to be ideally suited for application to the modeling of the melt spinning process. However, to date, it has only been applied to simple cases such as non-isothermal crystallization in the absence of stress and orientation [84,86].

2.1.2.7 Simulation of Dynamics and Structure Development

Using analyses similar to those discussed in preceding sections, a number of investigators [22,23,26–30,32,34,41–44] have mathematically simulated certain aspects of the melt spinning process. The reason for the great interest in simulating melt spinning is the large number of variables that affect the resulting filament properties, as discussed previously. Especially noteworthy among these efforts was the pioneering work of Kase and Matsuo [23,25,26] who first simulated the dynamics of both steady state spinning and time dependent phenomena.

Most of the early attempts to simulate spinning [23,25,26,28–30,41–44] omitted crystallization phenomena because of the difficulties of treating non-isothermal crystallization and the effects of molecular orientation. The omission of crystallization phenomena made it difficult to use the models to develop an understanding of the final structure and properties of the spun filaments. The models of Shimizu et al. [22] and Katayama and Yoon [44] were the first to simulate crystallization in the spinline, though Ziabicki [9] had previously laid the fundamental basis for such treatments. Later simulations emphasizing the treatment of crystallization and structure development include those of Spruiell and coworkers [27,34,37]. To date, simulations have been carried out for a variety of polymers, including PET [22,41–44], polyamide 66 [27], polyamide 6 [34] and polypropylene [37].

A typical example of simulated results is shown in Figure 2.7. These results were simulated using the working equations described in Table 2.1. The physical properties and crystallization kinetics used in the simulation were consistent with those of a polyamide 6 with a viscosity average molecular weight of 25,000. It was further assumed that the polymer was spun into ambient air without cross-blow. The mass throughput was taken to be 2.5 g/min per hole for these results. Because of its slow crystallization kinetics, the simulation predicts that polyamide 6 will not crystallize in the spinline at spinning speeds less than about 5000 m/min. This result is reasonably consistent with experimental data as described in the second section of this chapter. Similar results would be obtained for PET, which also exhibits slow quiescent crystallization kinetics.

Simulations carried out for polypropylene, a polymer that crystallizes faster than polyamide 6 or PET, are illustrated in Figures 2.8 through 2.11. These calculations were carried out using a somewhat more sophisticated model developed by Ding [37]. In this model Equation 2.10 is used to describe the effect of strain rate on viscosity and the differences resulting from differing molecular weight and MWD. A somewhat different method of dealing with crystallization kinetics was also used.

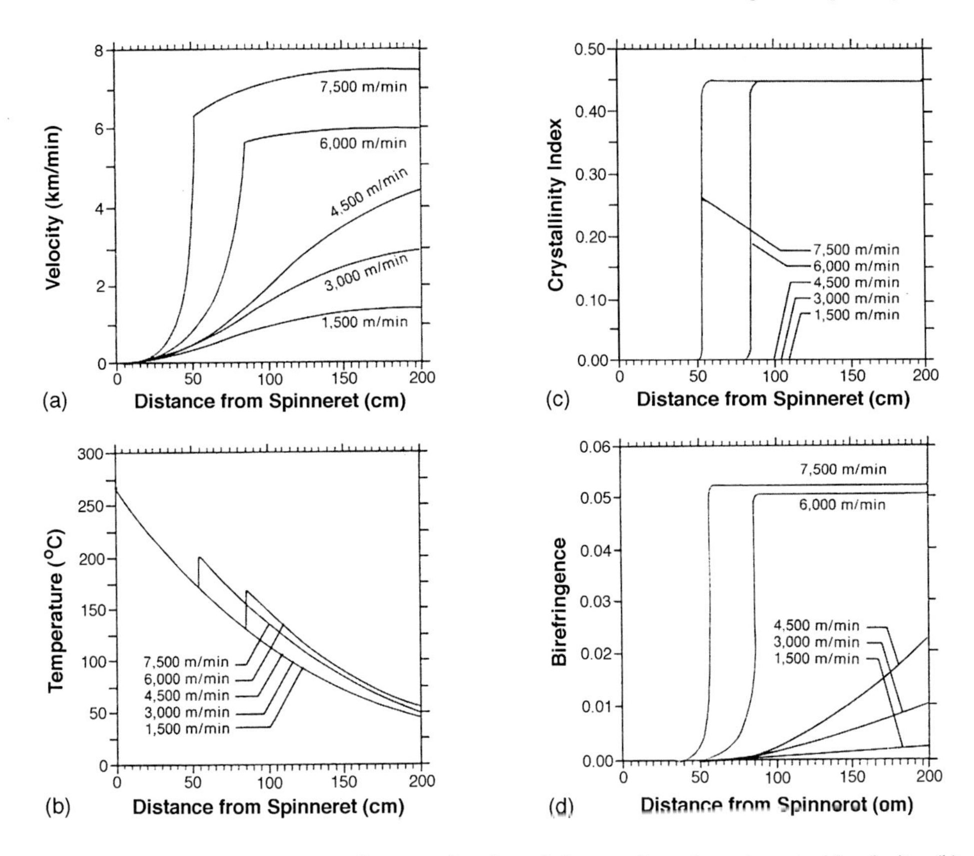

Figure 2.7 Simulated spinline profiles as a function of distance from the spinneret. (a) velocity; (b) temperature; (c) crystallinity; (d) birefringence. A mass flow rate of 2.5 g/min per hole and physical properties similar to those of polyamide 6 with a viscosity average molecular weight of 25,000 were assumed.

In Figure 2.8 is shown simulated diameter, temperature, birefringence and crystallinity profiles for a polypropylene homopolymer with $M_w = 142{,}000$ and $M_w/M_n = 2.8$. The extrusion temperature is taken as 210 °C, mass throughput is 1.68 g/min per hole and the ambient air temperature is 30 °C. In Figure 2.8 the changes in the profiles are shown with change in take-up velocity. Figure 2.9 shows the effect of changing only the mass throughput of polymer at a constant spinning speed of 3500 m/min. Figure 2.10 shows the effect of changing the weight average molecular weight of the polymer at constant polydispersity, and Figure 2.11 shows the effect of changing the polydispersity at constant weight-average molecular weight. Note that all the calculations shown in Figures 2.8 through 2.11 relate to a fixed modest cooling condition corresponding to extrusion into stationary air (no cross-blow) at 30 °C, and that all results would change if we change the cooling condition. In particular, introducing cross blow will cause the crystallization and structure development to occur much closer to the spinneret.

The mathematical simulation makes it quite clear that crystallization in the spinline is controlled by a balance between factors that increase crystallization kinetics and the tendency for cooling rate to suppress crystallization, as previously discussed. The crystallization kinetics is determined, primarily, by the nature of the polymer and the level

Table 2.1 Working Equations for a Simple Mathematical Model of the Melt Spinning Process

1. Continuity

$$W = (\pi D^2/4)\ \rho V \qquad (1)$$

$$V(0) = V_0 \text{ with } D = D_0$$

2. Momentum balance

$$dF_{rheo} = dF_{inert} + \delta F_{drag} - \delta F_{grav} \qquad (2)$$

$$F_{rheo}(0) = ?\ (\text{Guess}) \qquad (3)$$

$$dF_{inert} = WdV \qquad (4)$$

$$\delta F_{drag} = \pi \rho_a C_d V^2 D dz \qquad (5)$$

$$\delta F_{grav} = (Wg/V)dz \qquad (6)$$

3. Energy balance

$$\frac{dT}{dZ} = \frac{-h\pi D(T - T_a)}{WC_p} + \frac{\Delta H d\theta}{C_p dz} \qquad (7)$$

$$T(0) = T_0 \text{ (Extrusion Temperature)}$$

4. Rheological equation

$$\frac{dV}{dz} = \frac{\sigma}{\eta} \qquad (8)$$

$$\sigma = F_{rheo}\ (\pi D^2/4) \qquad (9)$$

For $T > T_m$

$$\eta = A(M_w)^{3.55} \exp\left[\frac{B}{T + 273}\right] \qquad (10)$$

For $T < T_m$

$$\eta = A(M_w)^{3.55} \exp\left[\frac{B}{T + 273}\right] \left\{a\left(\frac{\theta}{\theta\infty}\right)^b\right\} \qquad (11)$$

5. Birefringence and orientation

$$\Delta n_{am} = C_{0p}\sigma \qquad (12)$$

$$\Delta n = (1 - \theta)\Delta n_{am} + \theta f_c \Delta_c^0 \qquad (13)$$

$$f_a = \Delta n_a / \Delta^o{}_a \qquad (14)$$

6. Crystallization kinetics

$$\frac{d\theta}{dz} = \frac{n\theta\infty K}{V} \left[\int (K/V)dz'\right]^{n-1} \exp\left[-\left(\int (K/V)dz\right)^n\right] \qquad (15)$$

$$\theta(0) = 0$$

$$k(T, f_a) = C_1 \exp\left[\frac{-C_2}{T + 71.6}\right] \exp\left[\frac{-C_3}{T\Delta T + CT^2 f_a^2}\right] \qquad (16)$$

of molecular orientation developed. The major variables affecting the molecular orientation are those that have the greatest effect on spinline stress, namely polymer viscosity (i.e., molecular weight), spinning speed and mass throughput. Molecular

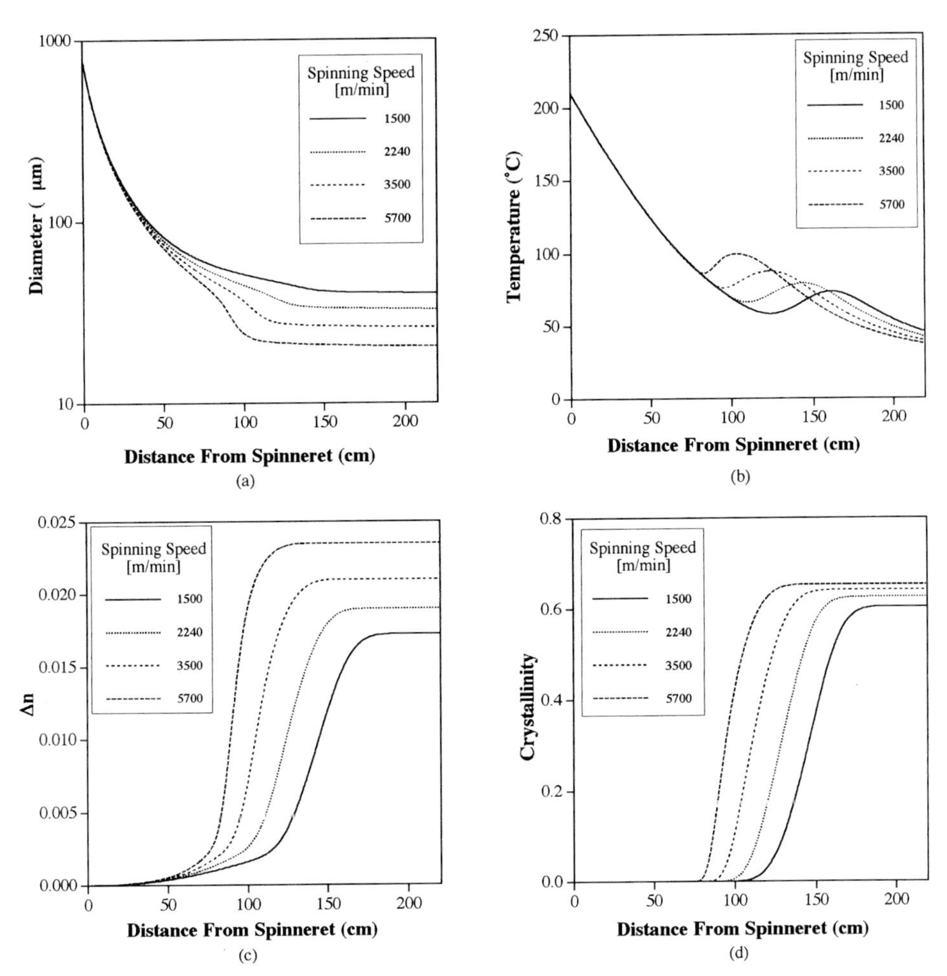

Figure 2.8 Simulated spinline profiles as a function of distance from the spinneret for a polypropylene homopolymer having an $M_w = 142{,}000$, $M_w/M_n = 2.8$, melt temperature = 210 °C, mass troughput = 1.68 g/min per hole, ambient air temperature = 30°C. (a) diameter, (b) temperature, (c) birefringence and (d) crystallinity.

orientation is affected to a lesser extent by other spinning variables such as extrusion temperature and cooling air cross-blow velocity. The cooling rate is largely controlled by polymer mass flow rate, spinning speed and cooling air temperature and velocity. The balance of the competition between crystallization kinetics and cooling rate determines if crystallization occurs in the spinline and the temperature at which it occurs. Under appropriate conditions an increase in spinning speed can lead to either an increase or a decrease in the temperature at which crystallization occurs. Whenever the increase in spinning speed produces a major increase in the crystallization kinetics, we would expect the crystallization temperature to increase. But if the crystallization kinetics is saturated, an increase in take-up velocity will still increase the cooling rate and that will lower the crystallization temperature. Increasing the molecular weight increases the polymer viscosity

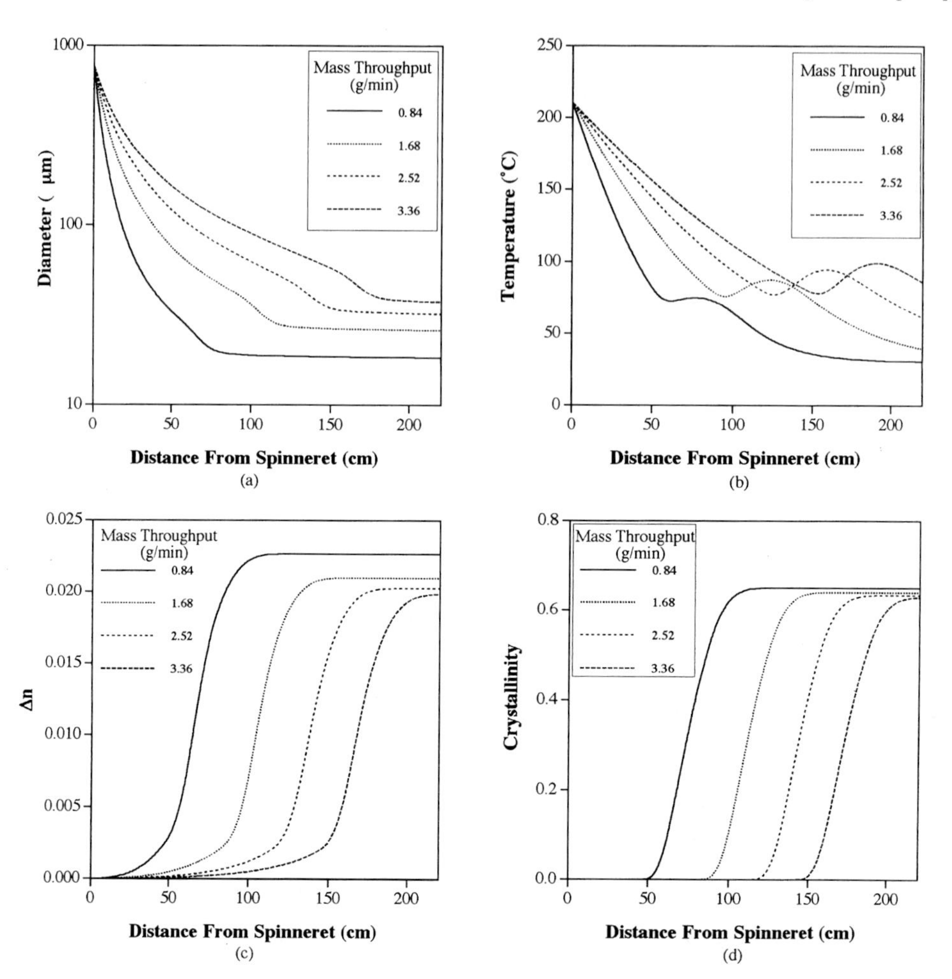

Figure 2.9 Simulated spinline profiles for same polypropylene as in Fig. 2.8 showing the effect of changing only the mass throughput. (a) temperature, (b) diameter, (c) birefringence, and (d) crystallinity.

and leads to a greater stress and molecular orientation in the spinline; this usually leads to higher crystallization temperature. Furthermore, the final orientation developed in as-spun fibers is strongly influenced by the orientation developed in the melt just prior to the onset of crystallization due to oriented nucleation and growth of crystals from the oriented melt.

We can rationalize many other experimental observations based on the results of such simulations. However, not all observations, especially the morphology and properties of the filaments, are predictable from such simulations, and it is necessary to examine carefully the experiments carried out for each polymer.

2.1.2.8 Effects of Radial Temperature Distribution

In the analysis given above we have used the so-called thin filament approximation throughout, neglecting radial distributions in temperature, stress, viscosity, etc. While

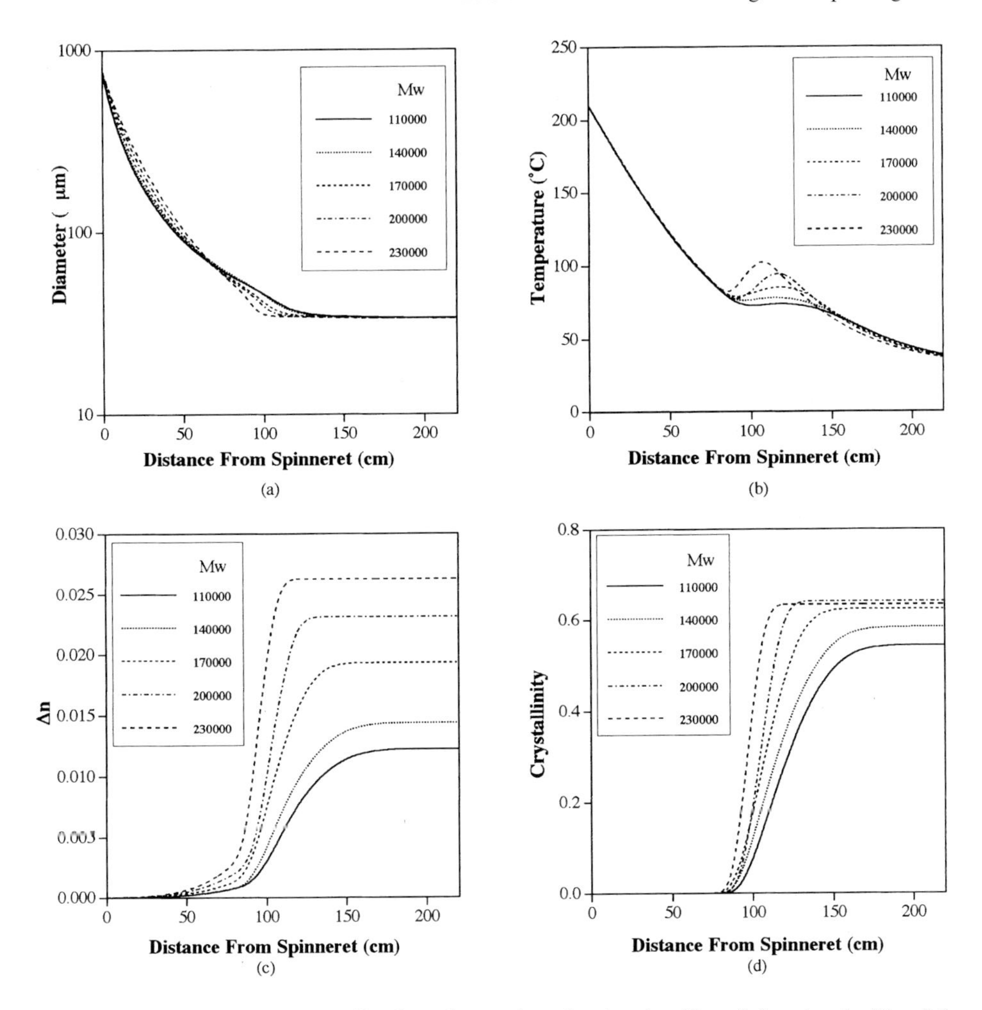

Figure 2.10 Simulated spinline profiles for polypropylene showing the effect of changing the M_w of the resin at fixed $M_w/M_n = 3.8$. Take-up speed = 2000 m/min, melt temperature = 210 °C, mass throughput = 1.60 g/min per hole, ambient air temperature = 30 °C. (a) diameter, (b) temperature, (c) birefringence, and (d) crystallinity.

this is a reasonable approximation for most aspects of melt spinning, it must be noted that certain experimental observations, to be discussed later, indicate that radial distribution of temperature can produce significant radial variation of the structure of filaments under certain conditions. Analyses of radial temperature and velocity distributions were carried out by Matsuo and Kase [87], Shimizu et al. [22] and by Katayama and Yoon [44].

It was concluded that, under most conceivable spinning conditions, the velocity field within the filaments is essentially flat, and, for all practical purposes, purely extensional. However, the computed temperature profiles showed that there can be a significant temperature differential across the spinning filament. This temperature differential

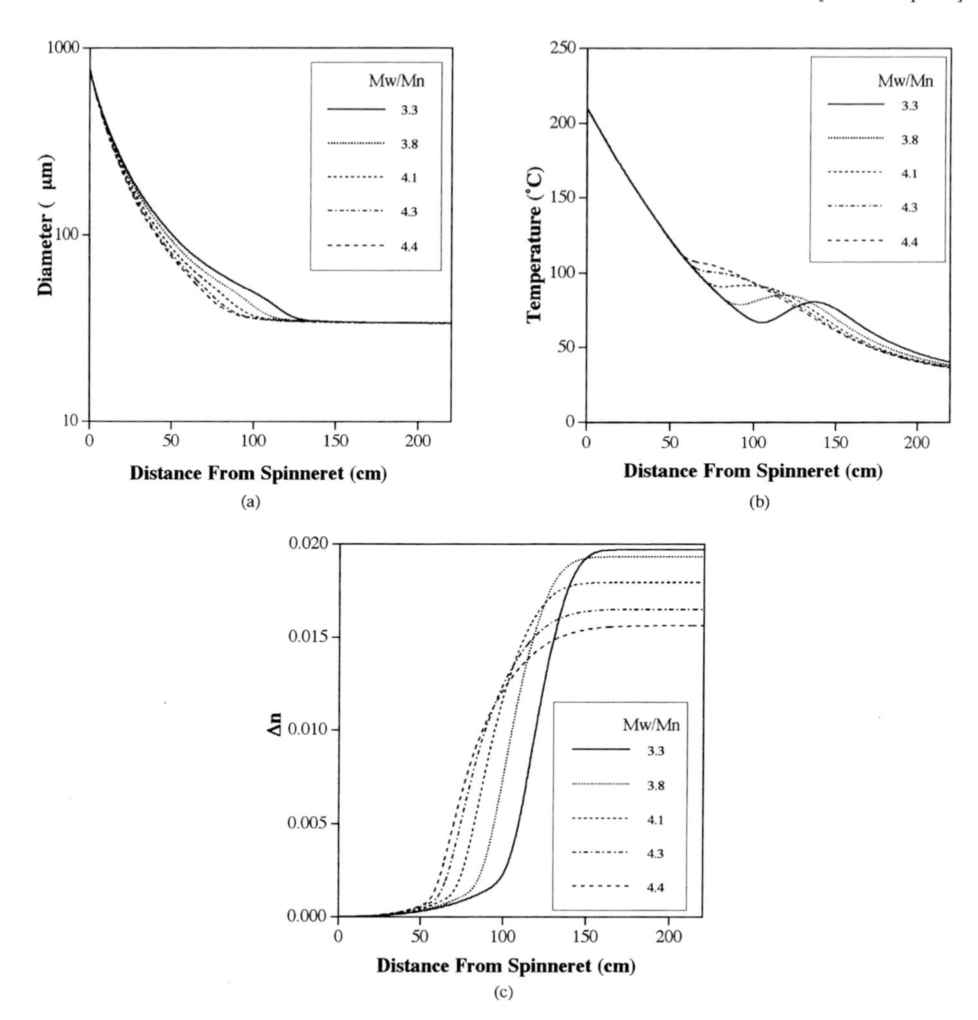

Figure 2.11 Simulated spinline profiles for polypropylene showing the effect of changing M_w/M_n at constant $M_w = 170{,}000$. Other conditions are the same as for Fig. 2.10. (a) diameter, (b) temperature, and (c) birefringence.

reaches a maximum at a short distance below the spinneret and then decreases thereafter as the average filament temperature further decreases. The temperature distribution is approximately parabolic in shape and the differential may amount to as much as 10% of the difference between the temperatures of the fiber and the cooling air (e.g., about 10–15 degrees at the point of maximum differential for PET). This radial variation in temperature within the filament also implies that there is a radial variation in viscosity and stress across the filament. According to reference 44 the onset of crystallization perturbs the radial temperature distribution due to the heat of crystallization and vice versa, crystallization occurring first at the surface of the filament and gradually moving inward toward the center.

2.2 Experimental Observations and Discussion

2.2.1 Polyolefins

2.2.1.1 Early Observations of Structure Development in Polyolefins

As synthetic fiber forming materials, polyolefins developed much later than polyamides and polyesters. Materials that were sufficiently stereoregular to develop the properties demanded of fibers date to the Nobel Prize winning research and the development of coordination catalysts by Ziegler [88] and Natta [89] in the early to mid 1950s. The bulk of the work on polyolefins relates to polyethylene and polypropylene, with only a small amount of work being done on other polymers. Of the polyolefins, polypropylene has proven to be quite versatile as a melt spun fiber, finding uses in a variety of yarns, ropes, and in woven and non-woven fabrics covering such applications as garments, disposable diapers, hygiene products, and many others. For this reason there has been significantly more melt spinning research done on polypropylene than on any other polyolefin.

Although polyethylene is of limited commercial importance as a melt spun fiber, the studies of melt spinning of polyethylene are significant because of the role that this polymer has played in developing our understanding of the morphological structure of polymers. This is based on the pioneering work of Keller [90] and Till [91] on polymer single crystals in 1957, and in the later work of Keller and coworkers [92,93] and Pennings et al. [94–96] on crystallization in the presence of molecular orientation. This understanding of polymer morphology developed during and just before the period that much of the early work on polyolefin fiber formation from the melt was being done. As a consequence of this and the fact that polyethylene is one of the simplest polymers from the point of view of crystal and morphological structure, the early work on polyethylene and polypropylene incorporated these concepts of morphology into the description of the fiber forming process. This has contributed substantially to our current understanding of the structure of melt spun fibers. For this reason we begin our discussion of experimental work here. As an aside, we note that polyethylene has proved to be an important, high strength, high modulus fiber when prepared by the gel spinning technique. However, we will not deal with this process or fibers made from it in this chapter, but see Chapter 5.

The first major study dealing with structure development during melt spinning of polyolefins was carried out by Katayama, Amano and Nakamura [97]. They developed techniques for on-line measurement of filament surface temperature, diameter, birefringence, and both wide angle and small angle X-ray diffraction patterns as a function of distance from the spinneret on a running spinline. They applied these techniques to study the melt spinning of high-density polyethylene, polypropylene, and polybutene-1. An example of their results is shown in Figure 2.12 that shows a plot of filament diameter, birefringence and surface temperature as a function of distance from the spinneret for polypropylene. The experimental results are quite consistent with the overall appearance of the simulated profiles shown in Figure 2.8, bearing in mind that the process conditions are very different. The diameter draws down rapidly in the upper part of the spinline and levels out near its final diameter at a spinline position at which the

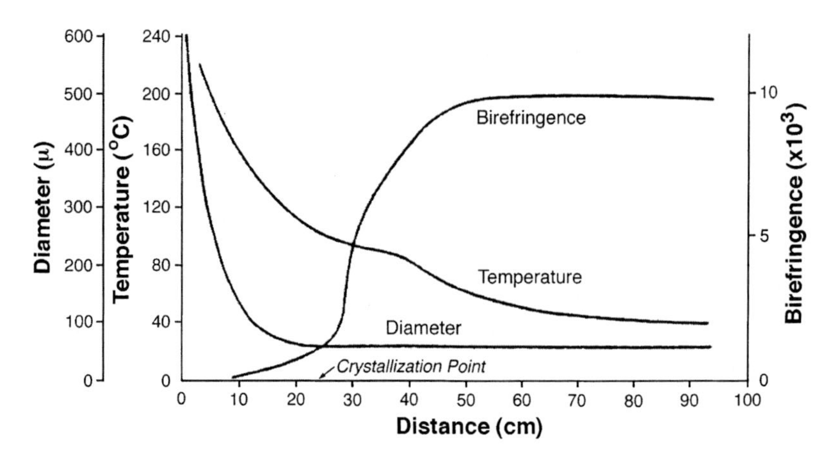

Figure 2.12 Experimentally measured changes in diameter, birefringence and temperature along the spinning way for iPP. (Data of reference [97]).

temperature profile exhibits a plateau and the birefringence rises rapidly. The on-line wide-angle X-ray patterns showed that crystallization is occurring in this part of the spinline, with crystalline reflections being first detected at the point marked "Crystallization point" in Figure 2.12. The plateau in the temperature profile is created by the heat of crystallization that is evolved during solidification of the polymer. The birefringence rises slightly before the start of crystallization, increases more rapidly during crystallization, and finally levels out near its final value as crystallization is completed. This sequence of events clearly shows that crystallization occurs in the presence of molecular orientation, but also suggests that oriented nucleation and growth of crystals contributes strongly to the development of the final orientation developed in the filament. The latter point was fully established by Oda, White and Clark [54].

Katayama et al. [97] also observed a two-point small angle X-ray pattern for polyethylene, after crystallization was largely completed, in agreement with off-line measurements made on as-spun filaments. Interestingly, they also observed a periodicity developing parallel to the fiber axis at a position in the spinline prior to that at which crystalline reflections could be detected in wide-angle diffraction patterns. The nature and significance of this observation is not yet fully resolved, but additional insight may be obtained from recent work carried out on polyvinylidene fluoride described in Section 2.2.4.1.

2.2.1.2 Further Study of Polyethylene

Additional studies of structure development of melt spun high density polyethylene filaments were published by Fung and Carr [98], Abbot and White [99], White, Dharod and Clark [100], Nadkarni and Schultz [101] and Dees and Spruiell [65]. The latter authors studied a HDPE with a melt index of 5.0 and examined the influence of take-up velocity, mass throughput, and extrusion temperature. They also carried out on-line measurements similar to those of Katayama et al. They monitored the tension in the spinline and were able to compute the stress at any position in the filament. Figure 2.13

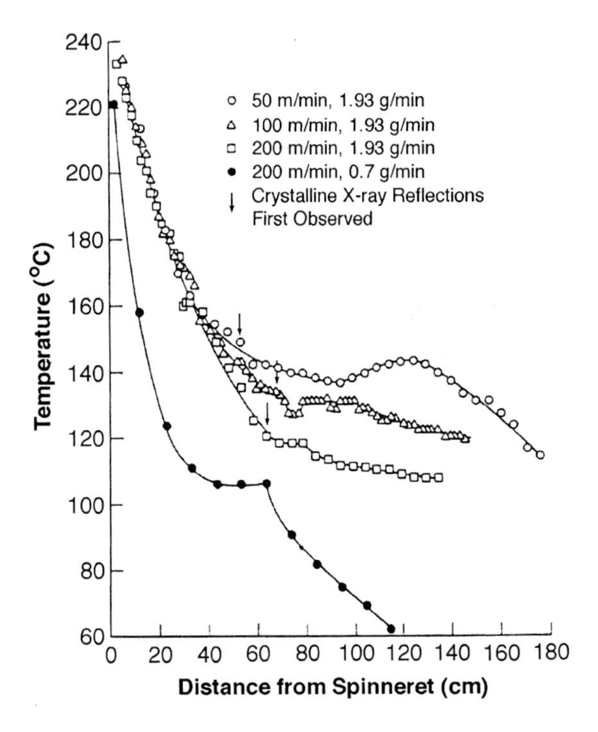

Figure 2.13 Measured fiber surface temperature profiles for HDPE spinning filaments. (After reference [65]).

shows the fiber surface temperature as a function of distance from the spinneret. From this figure it is clear that the crystallization temperature is a strong function of both take-up velocity and mass throughput. Under constant cooling air temperature and flow rate, these two factors have a major influence on the cooling rate of the polymer, as already discussed. Increasing the mass throughput at constant take-up velocity increases the filament diameter and the amount of material that must be cooled, resulting in a slower cooling rate. Increasing the take-up velocity at constant mass throughput reduces the filament diameter and increases the velocity of the filament running through the cooling air, resulting in an increase of cooling rate.

Dees and Spruiell computed crystallinity as a function of distance from the spinneret from on-line X-ray patterns and converted the distance scale to a time scale knowing the velocity profile. They found that the crystallization rate increased somewhat with take-up velocity, a fact that could be explained by the greater supercooling and lower crystallization temperature. However, they were able to show that the crystallization rate on the spinline was orders of magnitude faster than under quiescent conditions. This is best illustrated, as discussed previously, by the use of the "continuous cooling transformation diagram." This concept was illustrated schematically in Figure 2.5, and Figure 2.6 showed an actual continuous cooling transformation diagram for HDPE constructed by Spruiell and White from the data of Dees and Spruiell and other data from the literature [65–68].

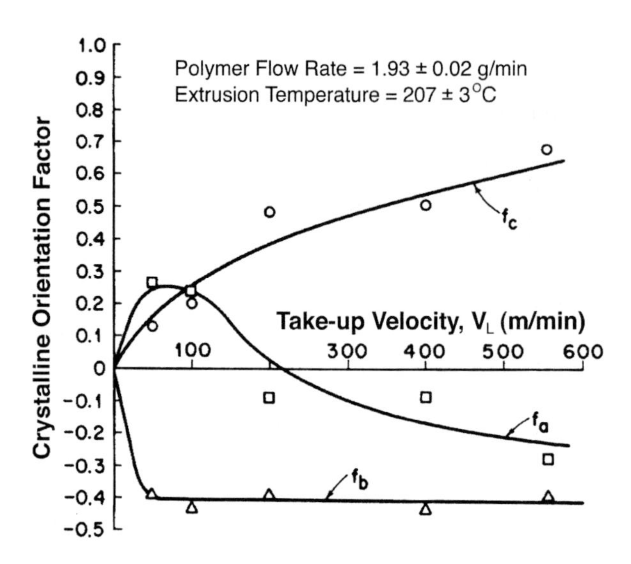

Figure 2.14 Hermans-Stein crystalline orientation functions as a function of take-up velocity for as-spun HDPE filaments. (After reference [65]).

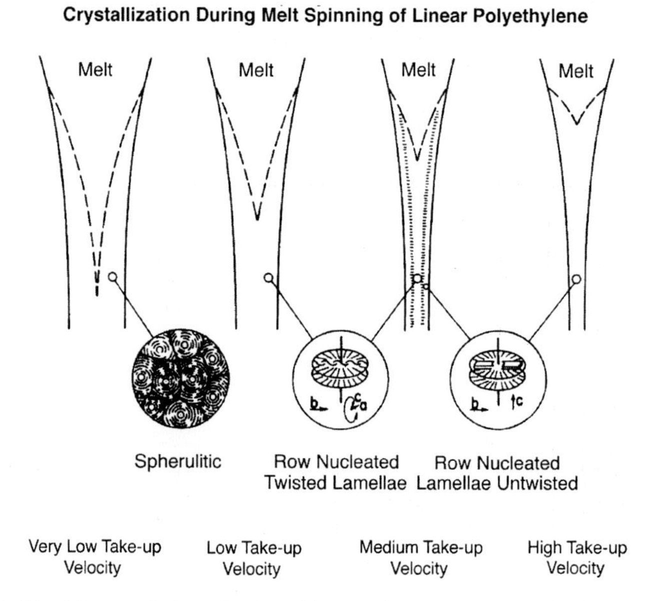

Figure 2.15 Model of the morphology developed in as-melt spun HDPE filaments. (After reference [65]).

Several investigators have studied the molecular orientation developed in melt spun polyethylene fibers [65,97,99–101]. Figure 2.14 shows Hermans-Stein orientation functions for as-spun polyethylene filaments as a function of take-up velocity as reported by Dees and Spruiell [65]. The polymer chains are parallel to the c-axis in the polyethylene unit cell; consequently, the gradual increase of the c-axis orientation function with increase of take-up velocity suggests that the chains are becoming more aligned with the fiber axis. The b-axis rapidly becomes nearly perpendicular to the fiber axis, while the a-axis orientation function first increases and then decreases.

Dees and Spruiell proposed the morphological model shown in Figure 2.15 to describe the structure of melt spun HDPE filaments. The model is based on a detailed interpretation of the orientation data of Figure 2.14, the observation of two-point small angle X-ray patterns (slightly modified at the lowest spinning speeds), and SEM photomicrographs showing lamellar texture perpendicular to the fiber axis. This model is based on the concept of row structure developed earlier by Keller and Machin [92] to describe the results from laboratory samples crystallized while stretching the melt. Ribbon-like, lamellar polyethylene crystals, similar to those that grow in the radial direction in spherulites, are nucleated by and grow epitaxially on, fibril nuclei generated by the elongational straining and molecular orientation of the melt. The fibril nuclei are not necessarily extended-chain crystals, but they may contain an appreciable number of chain-folds. Since the lamellar crystals can form all along the fibril nuclei, and the heat transfer from the filament is primarily in the radial direction, there is a tendency for them to grow perpendicular to the fibril and fiber axis. This results in the morphology referred to by Keller as a "row structure," but has also been called "cylindritic." The growth direction of the lamellar crystals is the b-axis of the polyethylene unit cell, which explains the rapid alignment of the b-axis in the direction perpendicular to the fiber axis in the melt spun fibers. If the stress is low the lamellae may exhibit twist, as they grow outward from the fibril nucleus, just as they do in 'banded' polyethylene spherulites grown from quiescent melts. If the lamellae twist about the b-axis, the a- and c-axes of the

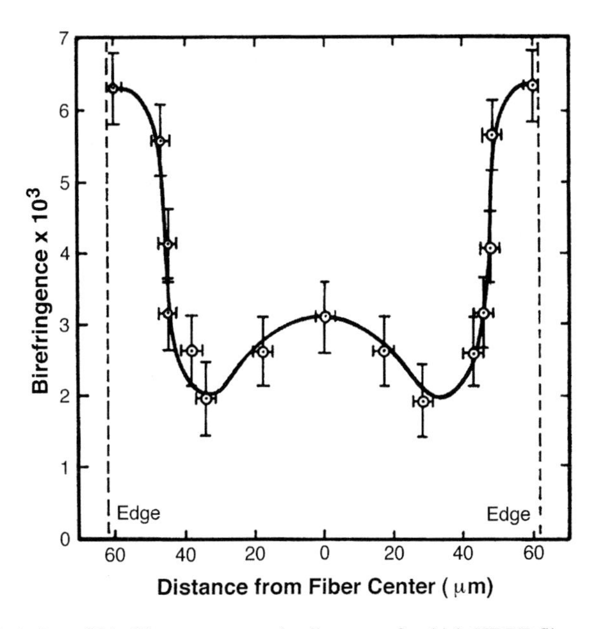

Figure 2.16 Variation of birefringence across the diameter of a thick HDPE filament. (After reference [98]).

polyethylene unit cell will rotate about the b-axis, and the orientation of the a- and c-axes with respect to the fiber axis will be equal, as is approximately true at a spinning speed of about 50 m/min in Figure 2.14. In this case, $f_b \cong -0.5$, $f_a = f_c \cong 0.25$*. When the stress is higher there is less twisting of the lamellar crystals with the result that the c-axis becomes more aligned with the fiber axis while the a-axis tends to become more nearly perpendicular to it. In this case f_c increases while $f_a \to -0.5$.

A skin-core structure was found to exist under certain conditions of spinning, with higher orientation in the skin than in the core, as is illustrated in Figure 2.15. The experimental basis for this suggestion was a decrease of orientation with increased distance from the spinneret in the initial stages of crystallization, a behavior that was also observed by Katayama et al. [97]. Radial variations of birefringence in thick melt spun polyethylene filaments were observed by Fung and Carr [98], Figure 2.16. They suggested that the variations were the result of radial temperature gradients and crystallization under higher stress levels in the outer layer of the filament.

2.2.1.3 Further Investigations of Melt Spinning of Polypropylene

The early published studies of the melt spinning of polypropylene were those of Capuccio et al. [102], Compostella et al. [103], and Sheehan and Cole [104]. These authors established the basic characteristics of both as-melt spun filaments and filaments that had been drawn after spinning. In particular, Sheehan and Cole studied polypropylenes having weight average molecular weights that ranged from 245,000 to 470,000 (as estimated from correlation with intrinsic viscosity determined in decalin), and melt flow index (MFR) ranging from 9.37 to 0.61. The extrusion temperatures ranged from 235 °C to 280 °C. The filaments were spun into air at 25 °C or into water ranging in temperature between 10–90 °C. Their take-up speeds were very low, in the range 5–100 m/min, with consequently very low spin draw-down ratios (typically 3:1). They found that the high extrusion temperatures resulted in marked thermal-oxidative degradation of the polypropylenes, especially the higher molecular weight resins. The degradation was much worse in the presence of air than in its absence. Quenching into water, especially at the higher spinning temperature and lower water temperature, resulted in the formation of the paracrystalline form of polypropylene referred to as the "smectic" form by previous researchers [105,106]. Extrusion into air, under the conditions studied, resulted in the monoclinic α-crystalline form. Higher molecular weight and lower extrusion temperature resulted in greater orientation in the spun filaments. The primary purpose of Sheehan and Cole's work was to develop filaments with tenacity's greater than 12 g/den. They were successful using high molecular weight resin that was spun to give the smectic phase and drawn to very high draw ratios.

Spruiell and White [64] showed that the development of orientation and morphology in polypropylene was fairly similar to that in polyethylene when spinning is carried out in air and the monoclinic α-phase is formed. Figure 2.17 shows the crystalline orientation factors, for a polypropylene with a MFR of 6.6, for three different extrusion temperatures

*Here, the symbols f_a and f_c do not refer to amorphous and crystalline orientation, but to the orientation of the crystallographic a and c axes with respect to the fiber axis.

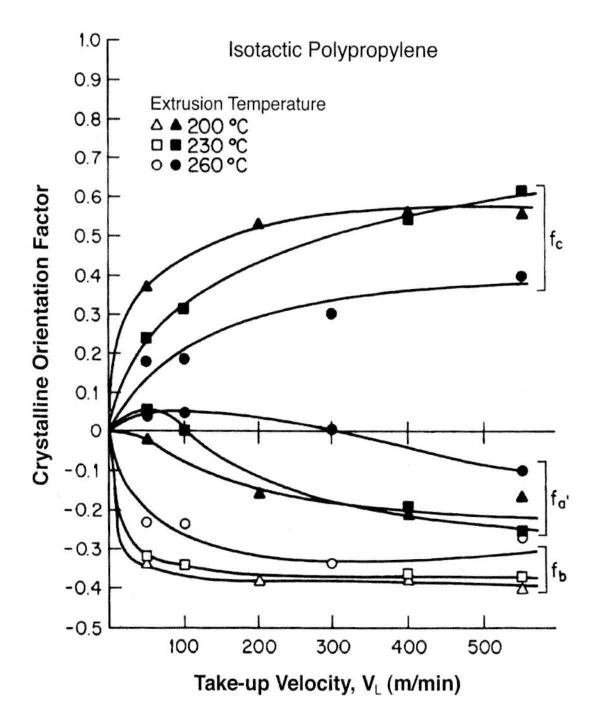

Figure 2.17 Crystalline orientation functions versus take-up velocity for as-spun iPP with MFR = 6.6. (After reference [64]).

plotted versus take-up velocity. Note that lower extrusion temperature produces higher molecular orientation at a given spinning speed. The resemblance of Figure 2.17 to Figure 2.14 is quite striking; the major difference being that the a'-axis orientation factor does not reach as high a positive value at low spinning speeds as in the case of polyethylene. This was interpreted to mean that there is a lower tendency for lamellae to twist and rotate about their growth direction in polypropylene than in polyethylene.

Spruiell and White and Nadella et al. [107] showed that both the crystalline orientation and the total orientation as measured by birefringence could be correlated with the spinline stress. This is illustrated in Figure 2.18 for crystalline orientation. Here the orientation factors of Figure 2.17 are plotted versus spinline stress rather than take-up velocity. Nadella et al. showed that this sort of correlation also held when comparing resins with different MFR.

Using on-line measurements, Nadella et al. showed that crystallization of polypropylene in the spinline occurred at lower temperatures than for polyethylene, in spite of the higher melting point of polypropylene. They also showed that lower spinline stresses and higher cooling rates tend to enhance the formation of smectic phase. These features can be interpreted in terms of the schematic continuous cooling transformation diagram of Figure 2.19. Cooling rate 1 produces the expected result that polypropylene crystallizes at a higher temperature than polyethylene. For cooling rate 2 the faster crystallization kinetics of polyethylene results in a higher crystallization temperature for polyethylene than for polypropylene. Rapid cooling, as for cooling rate 3, completely misses the nose

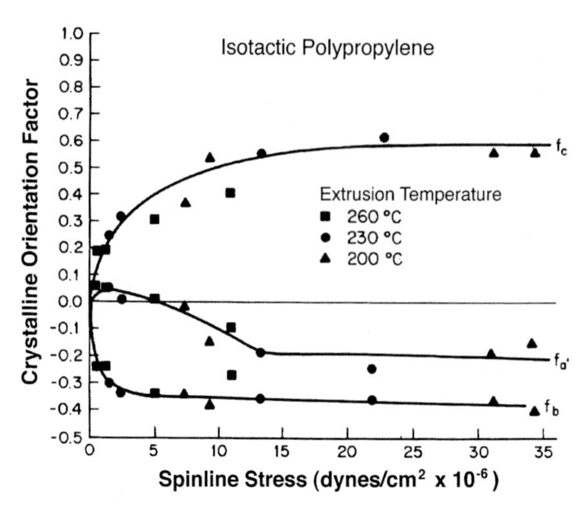

Figure 2.18 Crystalline orientation functions versus spinline stress for as-spun iPP with MFR = 6.6. (After reference [64]).

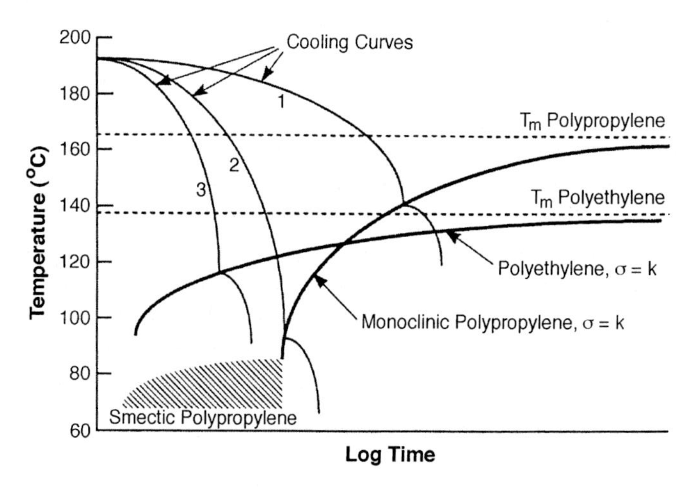

Figure 2.19 Schematic CCT diagram illustrating the relative behavior in melt spinning of HDPE and iPP.

of the CCT curve for the formation of α-monoclinic polypropylene, thus producing the smectic phase.

A number of investigators [97,103,107–110] observed that highly oriented polypropylene crystallized under stress frequently exhibits a "bimodal" crystal texture. This texture is characterized by one crystalline component, called the primary component, oriented with the c-axes of the crystals parallel to the fiber axis. Another component, called the secondary component, has the c-axes of the crystals nearly perpendicular to the fiber axis and the a- or a′-axis parallel to the fiber axis. (The a′-axis is a hypothetical one which is defined to be perpendicular to both the c- and b-axes, and is therefore at an angle of about 9.3° to the a-axis in the monoclinic unit cell). Estimates of the relative amount of the secondary component indicate that it typically composes 10–40% of the sample.

According to Andersen and Carr [109], the secondary component exhibits broader X-ray diffraction peaks and cannot be easily imaged in dark field electron microscopy. This suggests that the crystals of the secondary component are smaller than the crystals making up the primary component. On-line X-ray patterns [97] showed that the reflections for the primary component appear first and the secondary component appears afterward at a somewhat lower position on the spinline. The evidence seems to indicate that crystallization occurs in three stages. In the first stage some form of fibril nuclei form. In the second stage crystal lamellae nucleate on the fibril nuclei and grow radially outward with their c-axes parallel to the fiber axis, creating a row or cylindritic type structure. Finally, in the third stage, the secondary component is nucleated by and grows epitaxially on the row nucleated lamellae to create the bimodal orientation. Based on these ideas, Clark and Spruiell [110] proposed the rough morphological model shown in Figure 2.20 for the structure of flow crystallized polypropylene. A similar model was suggested independently by Andersen and Carr. More recently, Lotz and Wittman [111] suggested that the source of the bimodal orientation in fibers is common to that which produces polypropylene lamellar branching and "quadrites" grown from solution [112] and is related to the extensive lamellar branching and unusual optical properties of melt grown spherulites [113,114]. They conclude that this effect is a natural consequence of a well-defined homoepitaxy on the structurally favorable lateral (010) faces of lamellar crystals of α-phase.

Researchers at DuPont [115,116] and Celanese [117–119], motivated by the discovery that polypropylene filaments may be melt spun in such a way as to produce remarkably elastic filaments, reached rather similar conclusions about the structure of melt spun polypropylene filaments to those described above. The properties of these so-called 'hard elastic' filaments superficially resembled those of rubbery materials such as Spandex-type elastomers. Their studies also gave further insight into the nature of interlamellar tie molecules.

Typical curves for room temperature engineering stress versus strain are shown in Figure 2.21 for as-meltspun polypropylene filaments. The two examples illustrate the difference in behavior of a filament spun at low speed, with rather low molecular orientation ($f_c = 0.18$), and one spun at a higher speed, with considerably higher

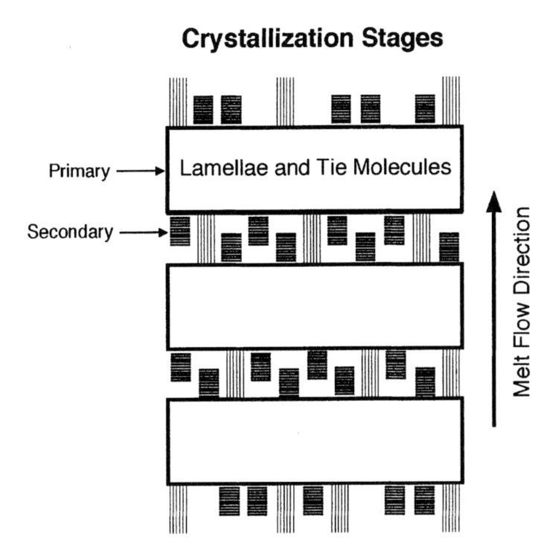

Figure 2.20 Model of the bimodal texture that occurs in flow crystallized iPP. (After reference [110]).

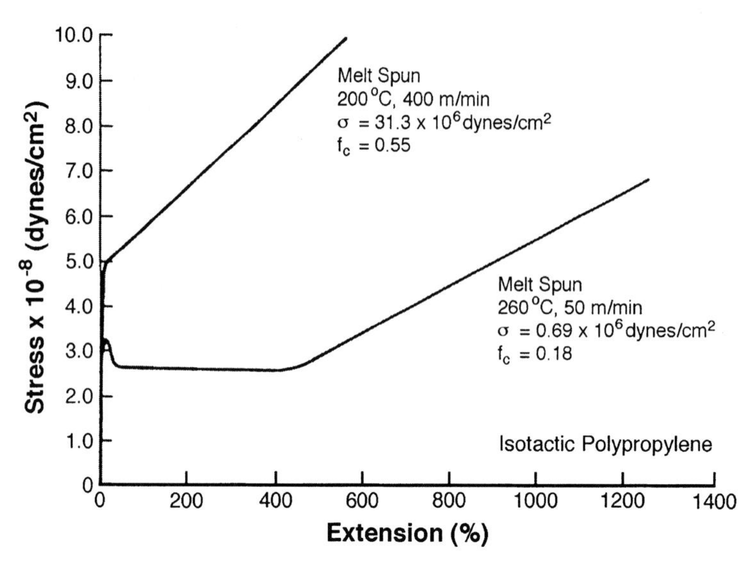

Figure 2.21 Typical stress versus elongation curves for melt spun isotactic polypropylene filament.

orientation ($f_c = 0.55$). The filament with low orientation exhibits a "yield point," filament necking and extension at essentially constant load to about 450% elongation (the so-called 'natural draw'), followed by a period of work hardening and high elongation to break. Because of its high ductility, this filament can be drawn to high draw ratios. The filament with higher orientation does not exhibit a marked yield point or neck down. It does have a higher yield strength, tenacity and lower elongation to break.

The elastic recovery after 100% extension of two filaments having similar properties to those in Figure 2.21 is illustrated in Figure 2.22. The more oriented sample with the well-developed row structure exhibits high elastic recovery, while the elastic recovery of the filament with low orientation is much smaller. The elastic recovery of the row structured filaments can be further improved by an annealing treatment, which increases the long period spacing as measured by SAXS and further perfects the row structure. The elastic filaments also exhibit a reversible decrease in density when stretched which is caused by the formation of numerous voids and surface connected pores. They also show a negative temperature coefficient of retractive force on stretching and high deformability with good elastic recovery at liquid nitrogen temperature. The latter behavior is not characteristic of true elastomers, as they generally become brittle below their glass transition temperature. These features suggest that the elastic recovery of these filaments is "energy driven" rather than "entropy driven," as in the case of true elastomers. Clark [116] and Sprague [119] suggested that this behavior could be explained by a structural model of the type illustrated schematically in Figure 2.23. The basic idea of the model is that a row structure exists in which lamellae are only connected to each other at certain tie points. When stretched, the lamellae bend elastically and voids are opened between lamellae. When the stress is released the lamellae regain their original shape, producing the elastic recovery and reduction in void volume. Electron microscopy observations of stretched, row-structured, elastic polypropylene films were shown to be generally consistent with this model [119].

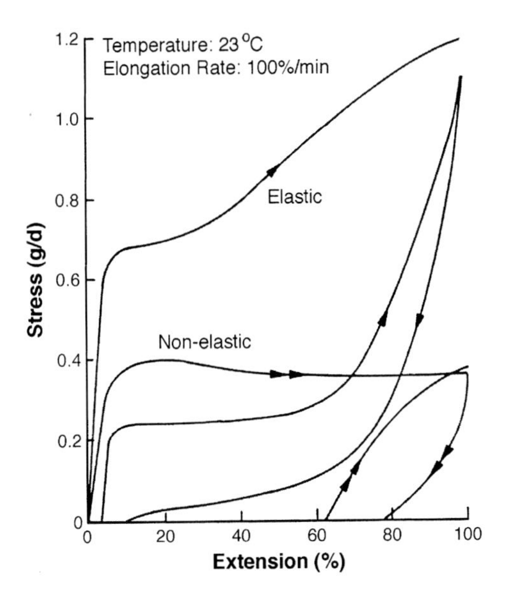

Figure 2.22 Elastic recovery after 100% elongation of low and high orientation iPP filaments. (After reference [119]).

Studies described up to this point have dealt with spinning of polyolefins at relatively low spinning speeds, i.e., take-up velocities less than about 1000 m/min. The early studies were limited to these low take-up velocities for two reasons. First, winding devices were not commonly available for winding speeds much above 1000 m/min until the late 1960s and early 1970s. In fact, it was the late 1970s before Barmer Maschinenfabrik A. G. (BARMAG) began to market the first commercial high speed winder capable of reaching speeds approaching 6000 m/min [120]. Second, the early commercial polyolefin resins manufactured using Ziegler-Natta catalysts were highly viscoelastic due to high molecular weights and broad MWDs. These materials failed by cohesive fracture in the spinline when attempting to spin them at high speeds. In the 1970s patents began to appear, e.g. [121], for preparation of polypropylenes with "controlled rheology." These materials were and continue to be prepared primarily by "visbreaking" of reactor resins. The process of

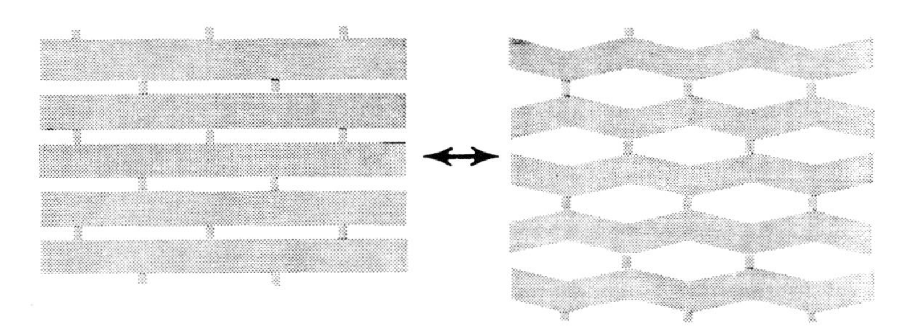

Figure 2.23 Model illustrating reversible deformation of row structure present in highly oriented melt spun iPP. (After reference [99]).

visbreaking involves reducing the molecular weight and narrowing the MWD by chain scission due to addition of a peroxide initiator and extensive shearing and mixing in an extruder at elevated temperature.

The first studies of high speed spinning of polypropylene were carried out by Shimizu et al. [122–124]. They spun polypropylene at speeds ranging from 500 m/min to 6,000 m/min, but little information was given about the nature of the resin used in the study. It was found that both density and birefringence increased rapidly with increase in take-up velocity as shown in Figure 2.24. Higher spinning temperatures required higher take-up velocities to reach the same birefringence or density, but this dependence on extrusion temperature

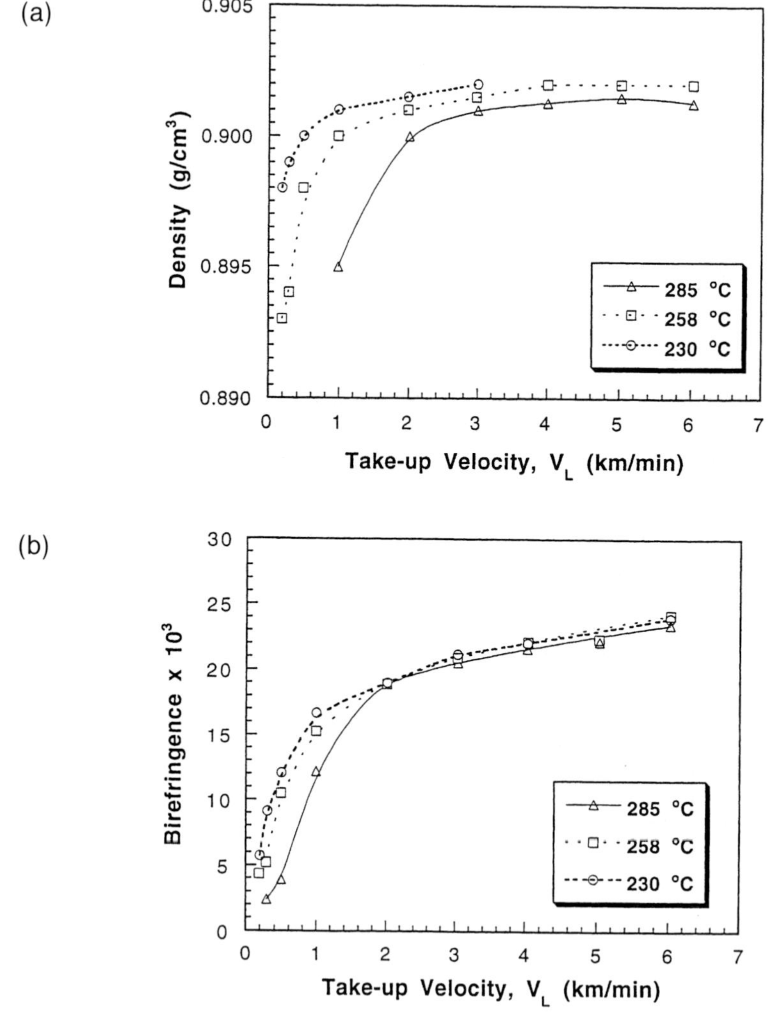

Figure 2.24 Development of (a) density and (b) birefringence of iPP as a function of take-up velocity for three different extrusion temperatures. (Data of reference [123]).

largely disappeared above about 2000 m/min. At higher spinning speeds the birefringence and density continued to increase but at a much slower rate. From the results of density, birefringence and X-ray diffraction data, they computed the c-axis (chain axis) crystalline orientation factor f_c and the amorphous orientation factor, f_a. These are shown in Figure 2.25 as a function of take-up velocity.

Above 2000 m/min the amorphous orientation increases faster than the crystalline orientation. The relatively high amorphous orientation at high spinning speeds correlated with a rapid increase of Young's modulus for the filaments as a function of spinning

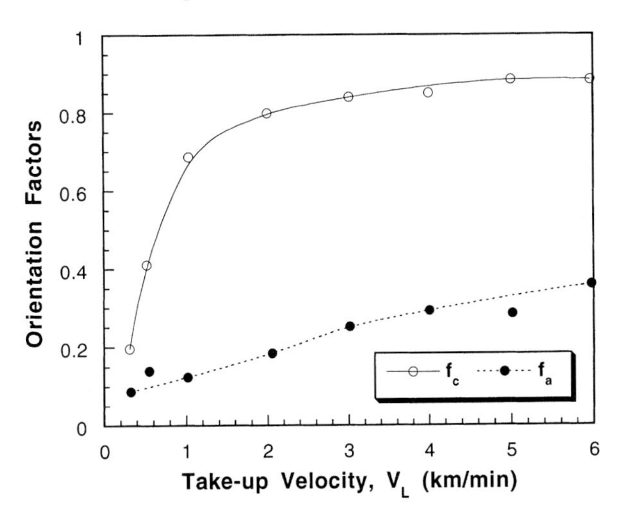

Figure 2.25 Crystalline and amorphous orientation factors of iPP as a function of take-up velocity. (Data of reference [124]).

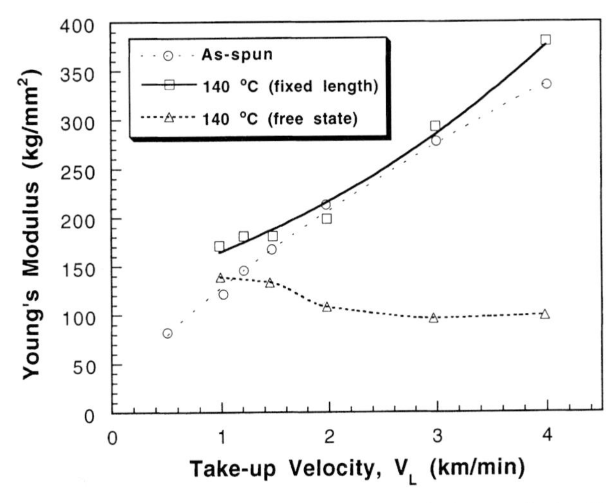

Figure 2.26 Young's modulus of iPP filaments as a function of take-up velocity. Data are shown for as-spun filaments and for filaments that were annealed at 140°C at fixed length or in a free state. (Data of reference [124]).

speed, Figure 2.26. Annealing at 140 °C in a free state produced substantial filament shrinkage (up to about 10%) that increased rapidly above 1000–2000 m/min, presumably due to the relaxation of the amorphous orientation. Such annealing also caused a major reduction in Young's modulus as shown in Figure 2.26.

Kloos [125], Fan et al. [126] and Spruiell and coworkers [127–131] have studied the influence of polypropylene resin characteristics on the high-speed melt spinning behavior and filament properties. Kloos and Fan et al. prepared resins with a wide range of polydispersity by peroxide degradation of broad distribution precursors. They also examined the influence of high molecular weight tails in the distribution by blending higher molecular weight material into lower molecular weight, narrower distribution resins. Spruiell and coworkers studied materials that had been specially prepared by a commercial polypropylene supplier to examine specific resin characteristics. These resins were also prepared by visbreaking techniques in most instances. All of these investigators found that narrowing the MWD improved spinnability compared to a broader MWD material of similar weight average molecular weight or MFR. This allowed spinning to much higher spinning speeds without incurring significant breaks in the spinline.

Changing the molecular weight and MWD also produced changes in the structure and properties of the filaments when prepared under similar spinning conditions. This is illustrated by the data shown in Figures 2.27, 2.28 and the X-ray patterns in Figures 2.29–2.31. These data are for three different resins having MFR values of 12, 35, and 300, corresponding to M_w values of 238,000, 170,500, and 124,400 g/mole, and polydispersities of 4.76, 2.16, and 2.85, respectively. Figure 2.26 shows that the density of the filaments decreases with increase of MFR or with decrease of molecular weight. This also corresponds to the fact that filaments produced from the 12 MFR, broader MWD resin

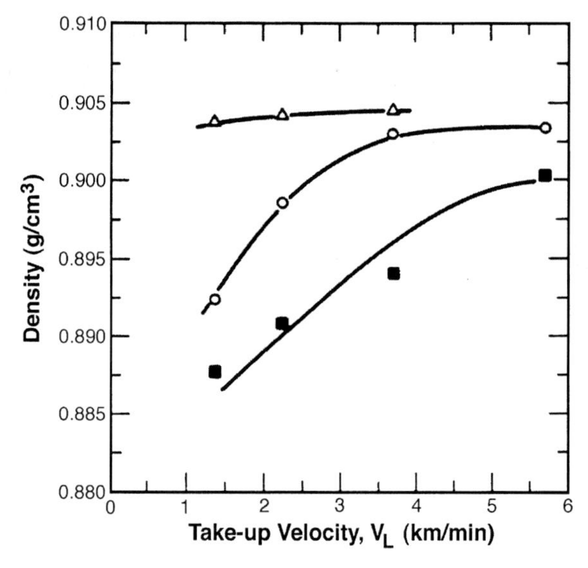

Figure 2.27 Density versus take-up velocity for three iPP resins of different MFR. △ – 12MFR; ○ – 35MFR; ■ – 300MFR. (Data of reference [127]).

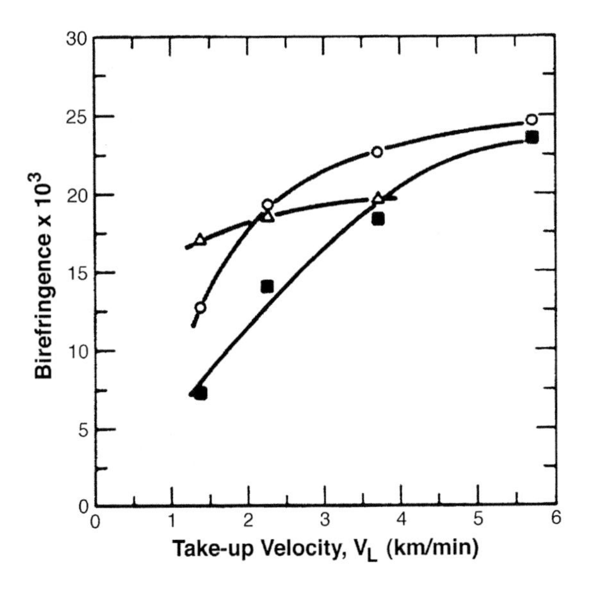

Figure 2.28 Birefringence versus take-up velocity for three iPP resins of different MFR. (Data of reference [127]).

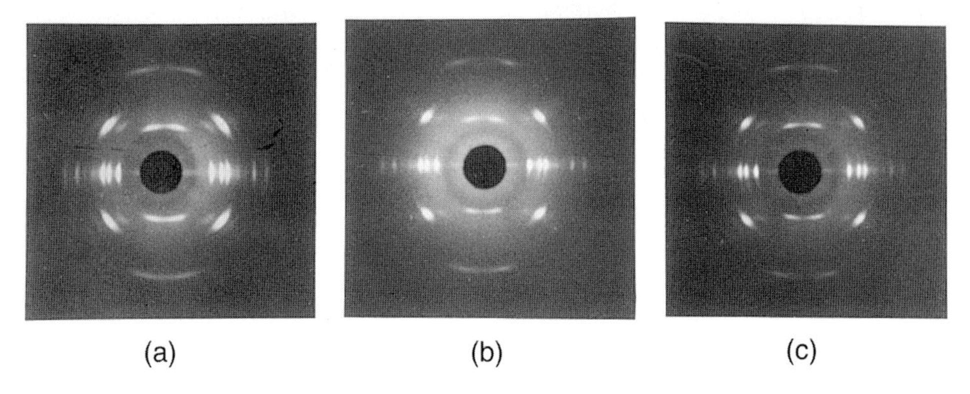

Figure 2.29 X-ray patterns for 12 MFR iPP resin at different spinning speeds. (a) 1360 m/min; (b) 2500 m/min; (c) 3500 m/min. (Data of reference [127]).

exhibited highly oriented, α-monoclinic structure at all spinning speeds above 1000 m/min, Figure 2.29. The higher MFR, narrower MWD samples were smectic at lower spinning speeds, and underwent a transition to α-monoclinic structure as take-up velocity increased (see Figure 2.30 and 2.31). This transition occurred at higher spinning speeds with increase in MFR or with increase of spinning temperature.

On-line measurement [128] showed that this transition is related to the temperature at which crystallization occurs in the spinline. Figure 2.32 shows the temperature at which crystallization begins as a function of resin and spinning conditions as estimated from the plateau in the on-line temperature profiles. Crystallization occurs at higher spinning temperatures at higher spinning speeds, due to stress-induced crystallization, in spite of the

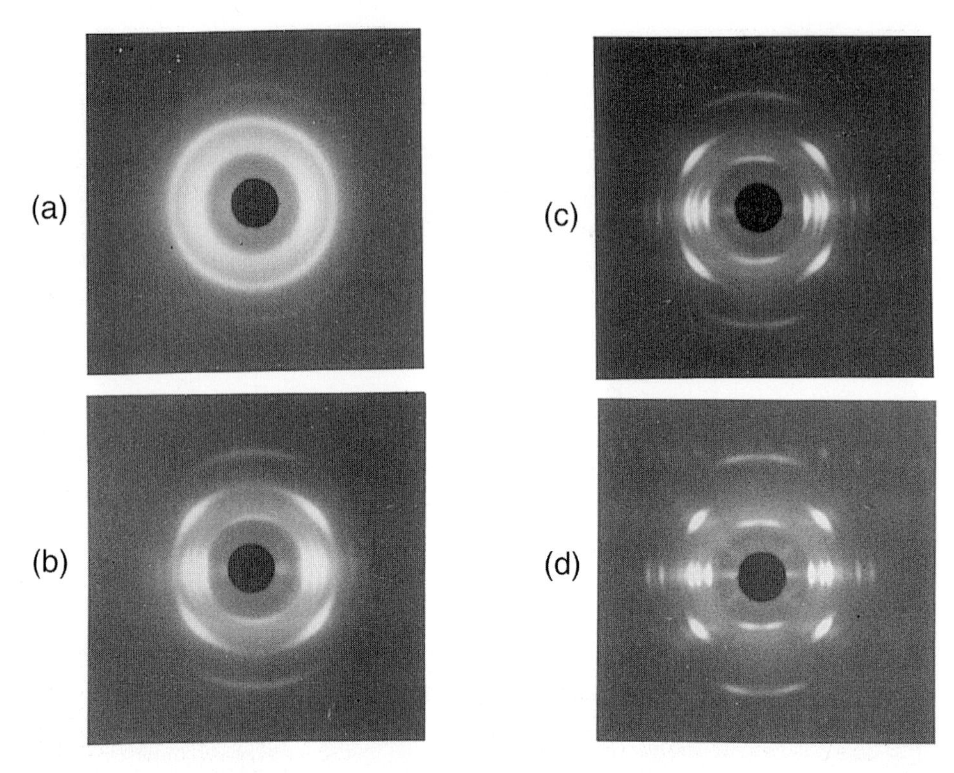

Figure 2.30 X-ray patterns for 35 MFR iPP resin at different spinning speeds. (a) 1430 m/min; (b) 2240 m/min; (c) 3300 m/min; (d) 5700 m/min. (Data of reference [127]).

fact that higher spinning speeds correspond to higher cooling rates. When crystallization occurs below about 70–80 °C, the structure is smectic. This is in agreement with the concept discussed above in terms of CCT curves, Figure 2.19. It appears that higher molecular weight produces greater spinline stress that enhances the crystallization kinetics and causes crystallization to occur at higher temperatures in the spinline, producing α-phase. Based on data for blends of high molecular weight fractions with lower molecular weight, narrow distribution resins, it seems that the high molecular weight tails in the distribution are especially significant in enhancing the crystallization rate [125,126]. This appears to be due to the fact that the long chain molecules in the high molecular weight tail serve to develop the fibril nuclei that begin the stress-induced crystallization process [130].

Kloos [125] suggested, based on broadening of small angle X-ray reflections, that the morphology changes from the well developed row structure (described above) for the broad distribution polymer to a more fibrillar structure, reminiscent of that of a drawn filament, for narrow MWD filaments that are spun at high speeds. This effect may also be associated with the formation of the latter structures at lower temperatures than the former ones.

In order to separate the effect of average molecular weight from that of MWD, Misra et al. [130] studied a matrix of resins. At each of three different MFR levels, different resins were spun ranging from narrow to relatively broad MWD. Broadening the distribution at a given MFR level led to higher density but lower birefringence. The effect of both MFR

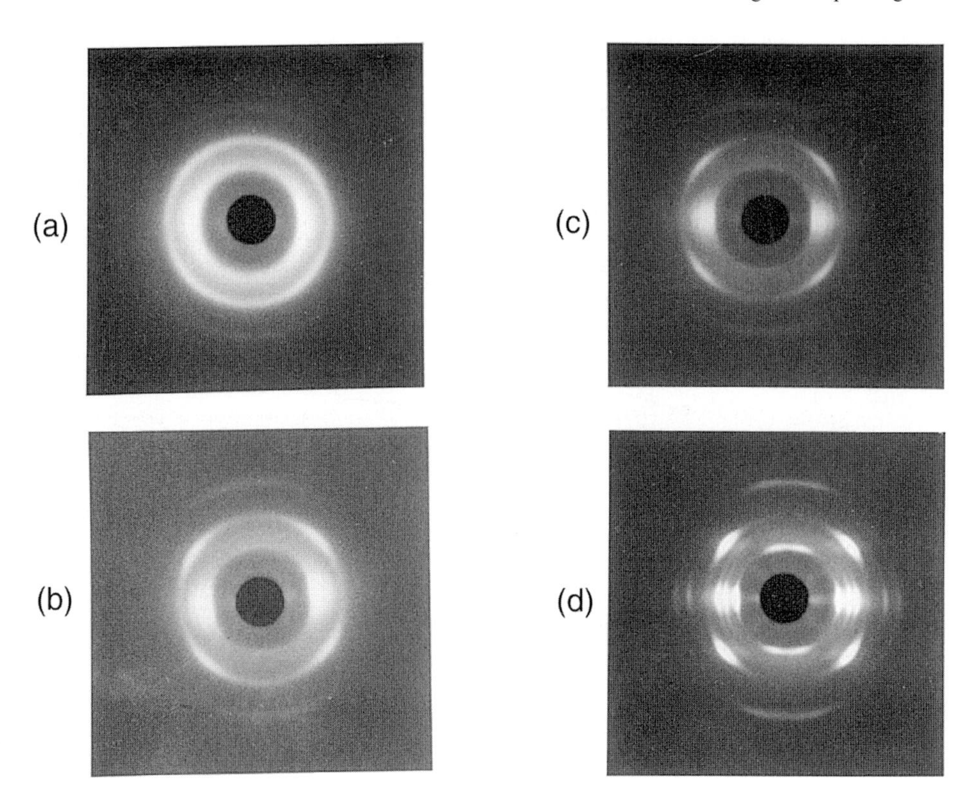

Figure 2.31 X-ray patterns for 300 MFR iPP resin at different spinning speeds. (a) 1430 m/min; (b) 2140 m/min; (c) 3300 m/min; (d) 5600 m/min. (Data of reference [127]).

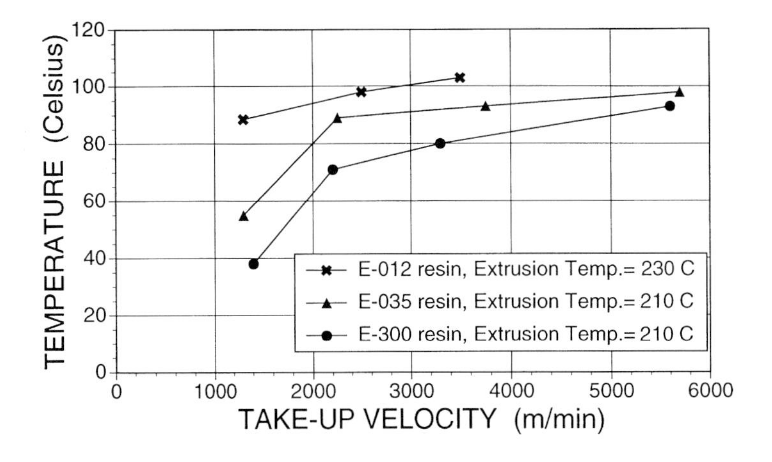

Figure 2.32 Temperature of crystallization as a function of spinning speed for three iPP resins of different MFR. (Data of reference [128]).

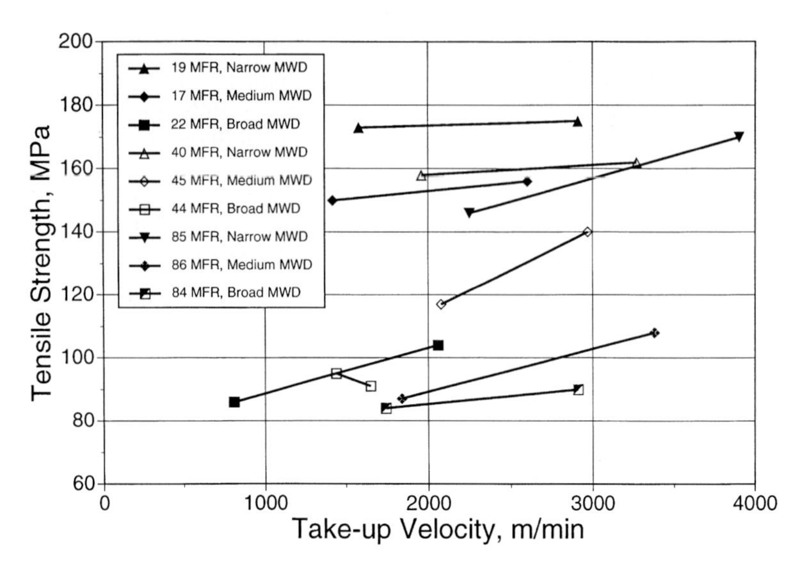

Figure 2.33 Tensile strength of iPP resins of differing MFR and MWD as a function of take-up velocity. All resins were spun at 210 °C. (Data of reference [130]).

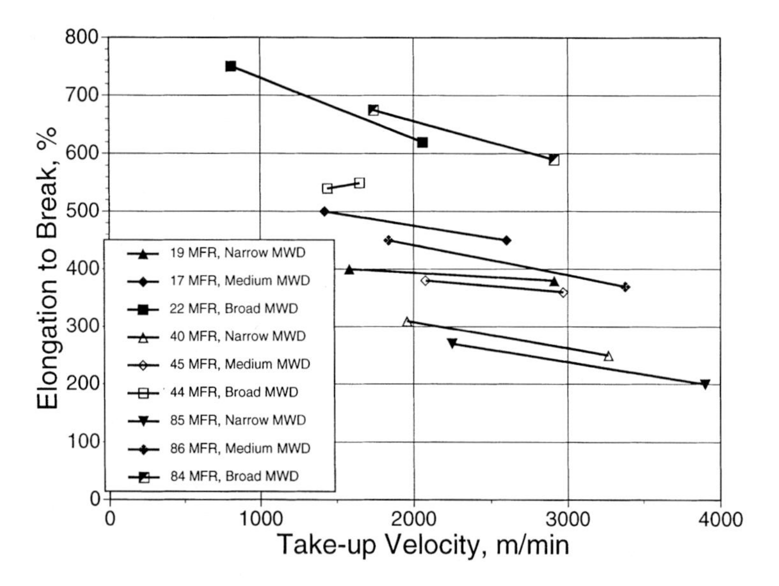

Figure 2.34 Elongation to break of iPP resins of differing MFR and MWD as a function of take-up velocity. All resins were spun at 210 °C. (Data of reference [130]).

(corresponding roughly to weight average molecular weight) and MWD on the mechanical properties of as-spun filaments can be ascertained from Figures 2.33–2.35. From these figures it is clear that increasing molecular weight (decreasing the MFR level) tends to increase the tensile strength and modulus as might be anticipated. But the MWD

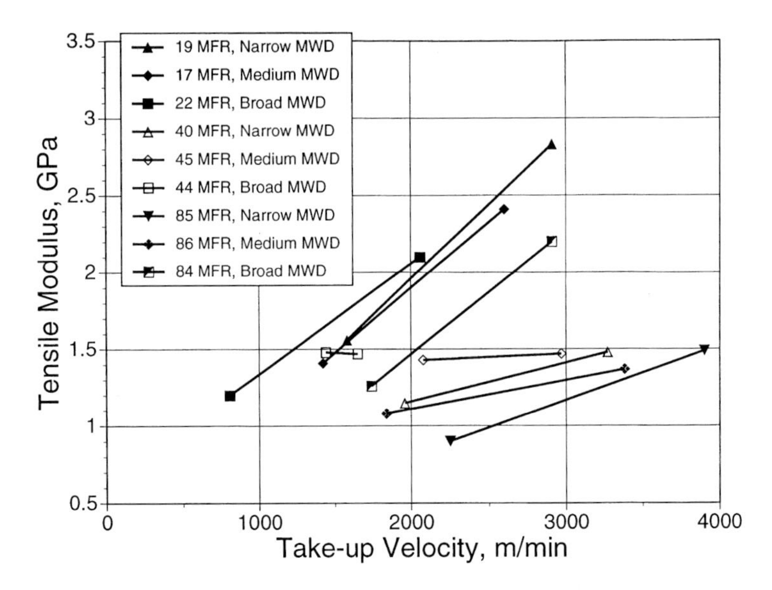

Figure 2.35 Young's modulus of iPP resins of differing MFR and MWD as a function of take-up velocity. All resins were spun at 210 °C. (Data of reference [130]).

is of equal or greater significance; the narrow MWD resins develop higher tensile strength and lower elongation-to-break than their broader distribution counterparts, at similar MFR level. The density and birefringence results helped to explain the mechanical property data on the basis that tensile strength is little affected by crystallinity, but increases with increased molecular orientation as measured by birefringence. The

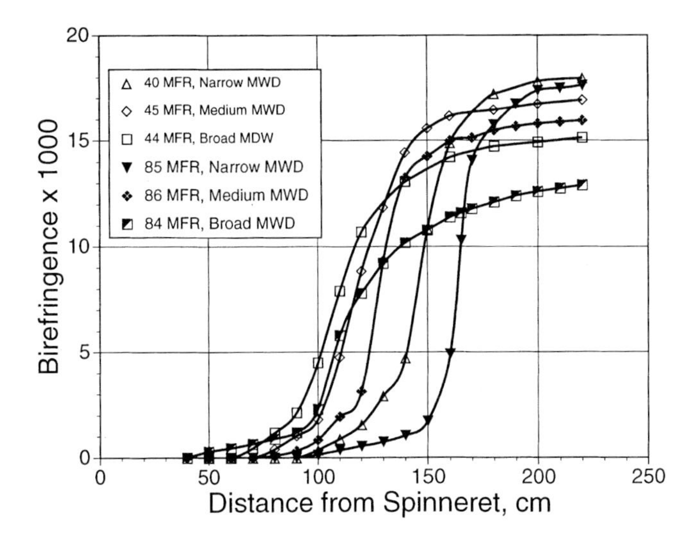

Figure 2.36 On-line birefringence profiles for iPP resins of differing MFR and MWD spun under similar conditions. (Data of reference [130]).

modulus, on the other hand, is a function of both molecular orientation and crystallinity (density). An increase in either produces an increase in modulus. This explains the observation that, at similar spinning conditions, the modulus tends to be higher for the broad distribution samples, since they exhibit higher densities.

On-line measurements showed that the broad distribution samples began crystallizing closer to the spinneret, and thus at higher temperatures, than the equivalent narrow distribution samples. This is readily seen from Figure 2.36 which shows birefringence profiles along the spinline for two sets of resins with similar MFR but differing MWD. As previously mentioned, the rapid rise in birefringence is caused by oriented nucleation and growth of crystals. The effect of broadening the MWD can be explained by the greater susceptibility to stress-induced crystallization of the broad distribution resins as discussed above.

Figure 2.36 also shows that the birefringence of the broad distribution samples is initially well above that of the narrow distribution resins, but it eventually crosses over and ends up at a lower level in the final as-spun filaments. X-ray patterns showed that the broad distribution samples contained a substantially higher fraction of the secondary component of the bimodal orientation than the narrow distribution resins, thus explaining the lower final birefringence. This effect is also responsible for the unexpected behavior of the birefringence data for the broader distribution sample shown in Figure 2.28.

Aside from molecular weight and MWD, polypropylene resins differ in several other respects. Among these are level of isotacticity, additive packages such as antioxidants, nucleating agents, pigments, etc., and comonomer identity and content. The influence of

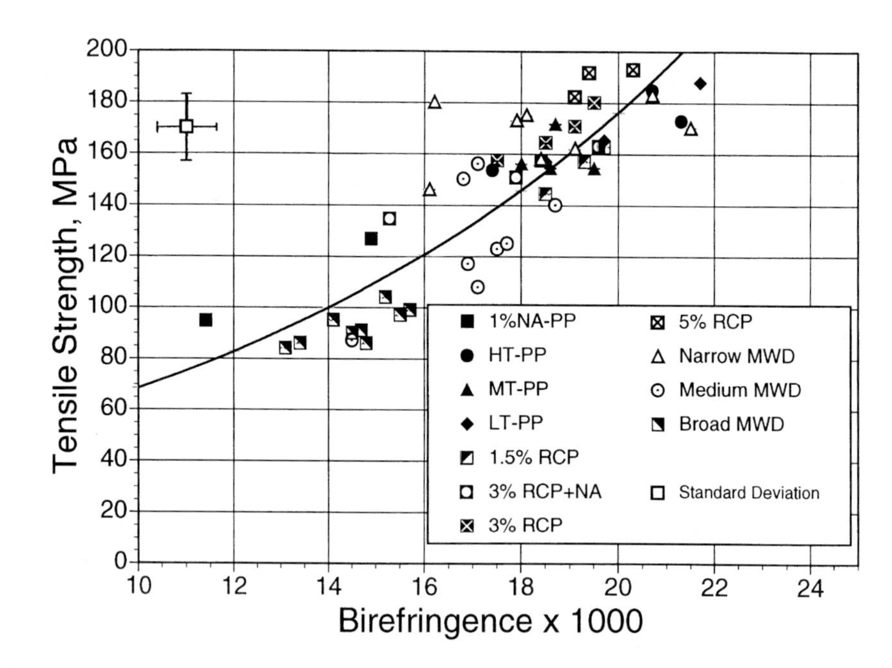

Figure 2.37 Correlation of tensile strength with birefringence of as-spun filaments prepared from a wide range of iPP resins. The resins include variations in molecular weight, polydispersity, isotacticity, ethylene copolymer content, and nucleating agent additive. (Data of reference [131]).

isotacticity level, ethylene comonomer content, and nucleating agent additions on the structure and properties of melt spun filaments was recently studied by Spruiell et al. [131]. In general, increasing the degree of isotacticity increases the crystallinity and tensile modulus of the spun filaments, while increasing the ethylene content has the opposite effect. Nucleating agent additions also lead to greater crystallinity, but, under certain conditions, can lead to lower tensile modulus in spite of the greater crystallinity. For given spinning conditions, tensile strength increased slightly with increased ethylene content but was little affected by tacticity in the range studied. Nucleating agents lowered the tensile strength of the spun filaments. The effects of nucleating agent on the tensile strength and modulus were traced to their ability to raise the crystallization temperature and reduce the level of molecular orientation generated in the filaments. Overall, it was found that tensile strength exhibited a strong correlation to birefringence for polypropylenes having a wide range of molecular weight, MWD, isotacticity, comonomer content, and even nucleating agent additions, as shown in Figure 2.37. A similar, though inverse, correlation also exists between elongation-to-break and birefringence. However, no such correlation is possible for tensile modulus, since it is strongly affected by crystallinity as well as molecular orientation.

The effects of pigment additions have been studied by Wu [132] and by Lin et al. [133]. It was found that some pigments could nucleate the crystallization of polypropylene and have nearly the same effect as the addition of a nucleating agent. However, the nucleating activity varies greatly depending on the particular pigment. The nucleating ability of several different pigments, as judged by the influence of the pigment on the DSC crystallization temperature of a polypropylene homopolymer, is listed in Table 2.2 [134]. Note that organic pigments generally have greater ability to nucleate polypropylene than do inorganic pigments, presumably as a result of better epitaxial matching with the polypropylene structure. The influence of selected pigments with differing nucleating ability on the tenacity of as-spun polypropylene filaments is presented in Figure 2.38 and compared to the tenacity of unpigmented filament. As the amount of pigment and its ability to nucleate crystallization increases, the tenacity of the as-spun filaments decreases and the elongation-to-break (not shown) increases.

Polypropylene melt spun filaments are also known to exhibit a variation of structure across their diameters in certain instances. The most common observation is for the orientation, as measured by birefringence, to be higher at the surface than at the center of the filament in a manner similar to that shown for polyethylene in Figure 2.16. However, the reverse effect, namely higher orientation in the center and lower on the surface of the filament has been claimed by Kozulla [135]. Such a filament is said to exhibit greatly improved thermal bonding to other filaments when used for thermally bonded non-wovens such as diaper coverstock. Such filaments can be made, according to the patent, by taking advantage of the tendency for polypropylene to undergo thermal oxidative degradation when spun in the presence of oxygen. By careful control of stabilizers and processing procedures they have produced a "skin-core" structure in which the outer layer is thermally degraded, low molecular weight, low birefringence polypropylene while the inner layer retains a much higher molecular weight and higher birefringence. The outer layer or "skin" softens and forms a bond to other fibers and filaments at temperatures below that at which the inner layer melts and/or loses its strength. Consequently a strong bond is formed without greatly weakening the strength of the filament in the bonded region, resulting in much improved strength of the non-woven web.

Table 2.2 Crystallization Temperature for a 45 MFR iPP Containing 0.5 wt.% of Various Pigments

Pigment Designation	Peak* Temperature (°C)	Onset* Temperature (°C)
Molydate Orange	108.1	114.2
D1937 PE Black	108.2	113.7
D1159 Chrome Yellow	107.8	113.1
D1059 Rutile TiO_2	111.7	116.0
SEC6758 Rutile TiO_2	113.6	117.7
SEC6759 Cabot 660 Furn. Black	116.8	120.7
D1137 Chromophthal Yellow	116.9	120.7
D1346 Fe_2O_3 *Red 101*	116.9	121.3
D1367 Perylene Scarlet Red	119.0	123.5
D1620 Carbazol Violet	119.5	124.1
D1822 Phthalocyanine Green	121.5	125.9
D1229 Chromophthal Orange 4B	121.9	125.8
D1724 Phthalocyanine Blue 8	121.9	126.6
D1345 Chromophthal Red-Brn	122.8	126.5
D1340 Berylene Red	122.8	127.1
D1726 Phthalocyanine Blue B	124.5	128.6
SEC6760 Ciba Geigy Red-Brn	124.8	127.4
D1621 Quinacridone Red 192	125.0	129.8
SEC6761 Phthalyl Blue	126.9	129.6
D1624 Quinacridone Violet 19	127.4	131.5

*Measured by DSC technique.

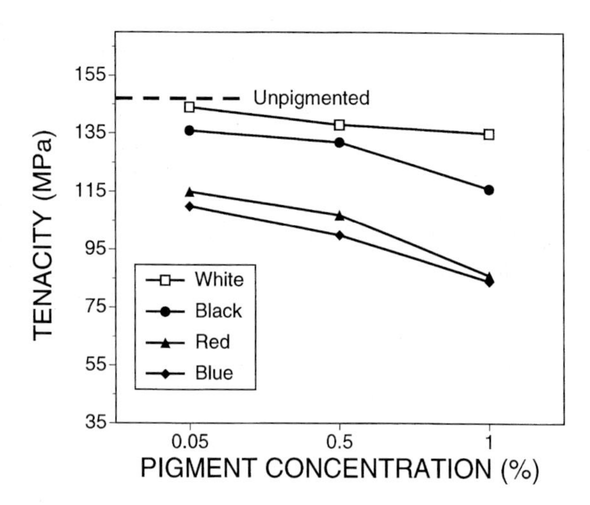

Figure 2.38 Tenacity of iPP filaments spun under fixed spinning conditions (take-up velocity = 1400 m/min) as a function of certain pigment additions and concentration.

2.2.2 Polyesters

2.2.2.1 Poly(ethylene terephthalate)

Interest in polyesters for fiber formation goes back to the pioneering work of Corothers who examined polyesters along with polyamides as potential fiber forming materials [2,3]. Poly(ethylene terephthalate) or PET is certainly the most investigated polyester and may be the most investigated of all polymers with respect to fiber forming properties. Although it was not commercialized as early as the polyamides, PET became a commercial fiber in 1953. Early studies, confined to relatively low spinning speeds, showed that as-spun PET filament was quenched to an amorphous state due to PET's slow crystallization kinetics; crystallization was brought about during the post-spinning drawing step [9]. The drawing process involves heating the amorphous, as-spun filament to a temperature slightly above its glass transition temperature and stretching it up to several times its original length in order to obtain high levels of molecular orientation and extension, and greatly improved filament tenacity and modulus (see Chapter 4). Much of this early work was reviewed by Ludewig [136].

Spinning speeds up to about 4000 m/min were investigated by Liska [137]. He found an indication that crystallization could occur during spinning for his highest spinning speed for PET with a relatively high intrinsic viscosity, i.e., high molecular weight. Early systematic studies of structure development in as-spun PET filaments that included very

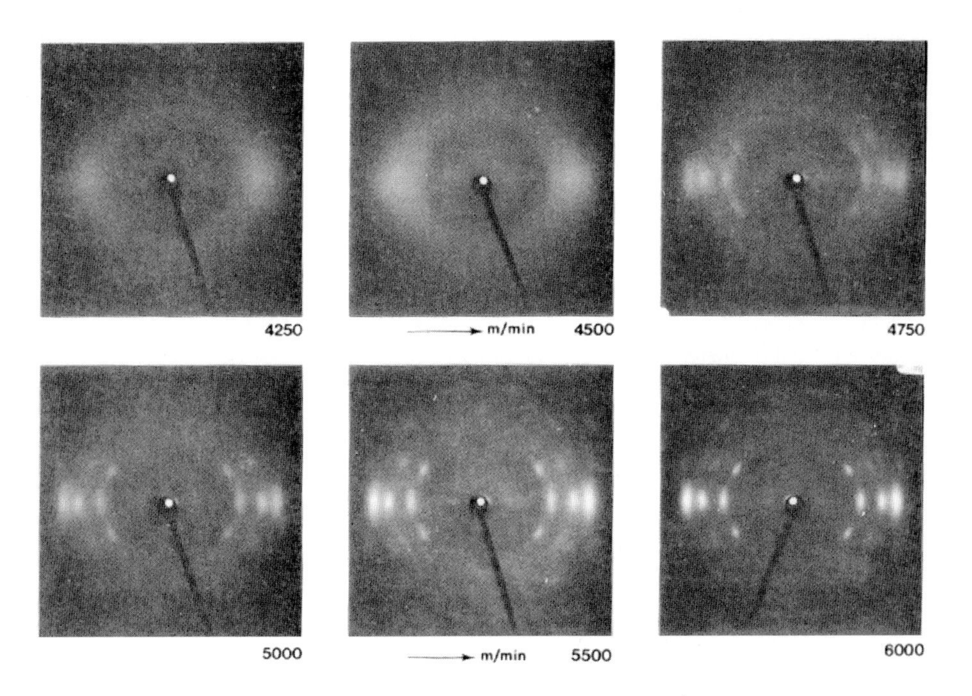

Figure 2.39 X-ray patterns of melt spun PET as a function of take-up velocity. (Data of reference [138]).

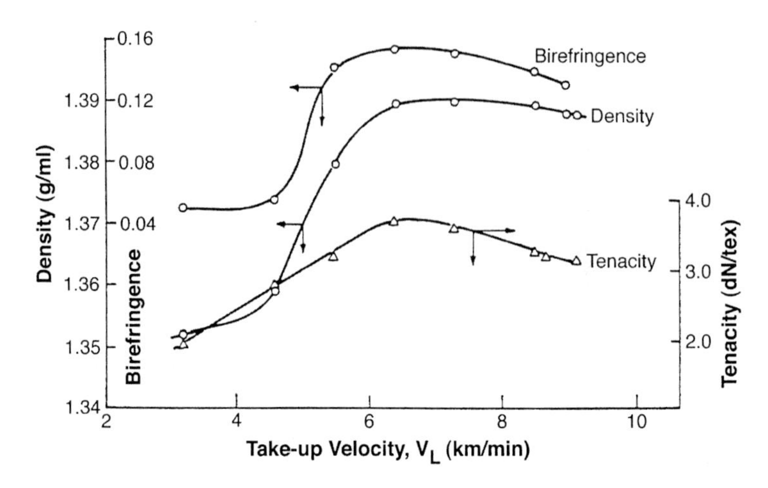

Figure 2.40 Properties of as-spun PET yarns as a function of spinning speed. (Data of reference [10]).

high spinning speeds were published by Shimizu et al. [138] in 1977 and by Heuvel and Huisman [139] in 1978. The latter results, which included spinning speeds up to 6000 m/min, clearly showed crystalline reflections in X-ray patterns from filaments spun at take-up velocities above about 4500 m/min as shown in Figure 2.39. They also found that the density of the as-spun filaments remained almost equal to that of fully amorphous PET until a spinning speed of 3500 m/min was reached. Above this spinning speed the density increased rapidly, suggesting a change from amorphous to partly crystalline material. A similar conclusion was reached based on differential thermal analysis of filaments spun at different spinning speeds.

Numerous investigators, including Shimizu et al [124,140–142], Heuvel and Huisman [143], Vassilatos et al. [10], Perez [144], Yasuda [145], George [146], Matsui [147] and others, have since published extensive studies. For example, Figure 2.40 shows birefringence, density and tenacity versus take-up velocity published by Vassilatos et al. [10] for a 0.65 dl/g I.V. PET. Note that they spun filaments at speeds as high as 9000 m/min and held the filament denier divided by natural draw ratio constant (so as to produce a fiber with constant final diameter) rather than holding the mass throughput per filament constant as most other investigators have. A rapid rise in both crystallinity and birefringence begins at about 4000 m/min, saturates at about 7000 m/min and then decreases slightly at higher take-up velocities. The filament tenacity also exhibits a maximum in the neighborhood of 7000 m/min. Figure 2.41 shows that boiling water shrinkage of these yarns first increases, at low spinning speeds, due to gradual development of molecular orientation and subsequent relaxation of the amorphous chains during boiling water treatment. Shrinkage reaches a maximum at about 2500 m/min, just before crystallinity develops in the filaments and then drops rapidly as crystallinity increases. Presumably, crystallinity has a stabilizing effect on molecular orientation and shrinkage.

The spinning speed range in which crystallinity develops is a function of resin molecular weight [141]. Higher molecular weight allows the crystallinity to develop at lower take-up velocities as a result of the higher spinline stresses and higher melt

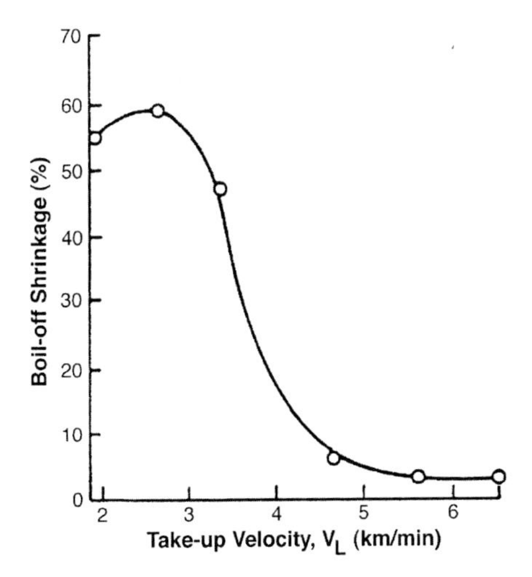

Figure 2.41 Boiling water shrinkage of as-spun PET yarns as a function of spinning speed. (Data of reference [10]).

orientation developed. The mass flow rate per hole also affects the spinning speed at which crystallinity develops [141]. Although the differences are small, the data consistently show that higher spinning speeds are required to achieve a given level of crystallinity when the mass flow rate is increased.

The above results together with other investigations [9,70,79,147-149] show that the rate of crystallization of PET at high spinning speeds is orders of magnitude faster than the maximum quiescent crystallization rate exhibited by PET. The behavior of PET during spinning can be qualitatively understood from the schematic CCT diagram of Figure 2.42. If cooling curve 1 represents the cooling rate occurring in the spinline at low spinning speed, cooling at this rate would produce an amorphous filament if crystallization occurred at or near the same rate as for quiescent crystallization. In this case the cooling curve does not intersect the "c-curve" representing the beginning of crystallization for quiescent crystallization. Increasing the spinning speed increases cooling rate slightly, but also increases molecular orientation and crystallization kinetics. However, it is not until the spinning speed exceeds about 3500 m/min, depending on molecular weight, that the crystallization kinetics becomes high enough so that crystallization can occur in the spinline. Still higher spinning speeds lead to faster crystallization kinetics and higher crystallization temperatures.

Several investigators [139,124,143,144] obtained both crystalline and amorphous orientation factors by combining birefringence or sonic modulus, x-ray diffraction and density crystallinity data. For PET, the value of f_a can also be determined directly from measurements of polarized "natural" fluorescence [150]. Results of such measurements are shown in Figure 2.43 where the $\Delta n/\Delta^\circ_c$ (with $\Delta^\circ_c = 0.220$) is taken as a measure of overall average orientation in the filaments, f_a is the amorphous orientation measured from polarized fluorescence, and f_a $(\Delta^\circ_a/\Delta^\circ_c)$ is a measure of amorphous orientation obtained from birefringence measurements. Plotting the latter quantity avoids the use of the poorly

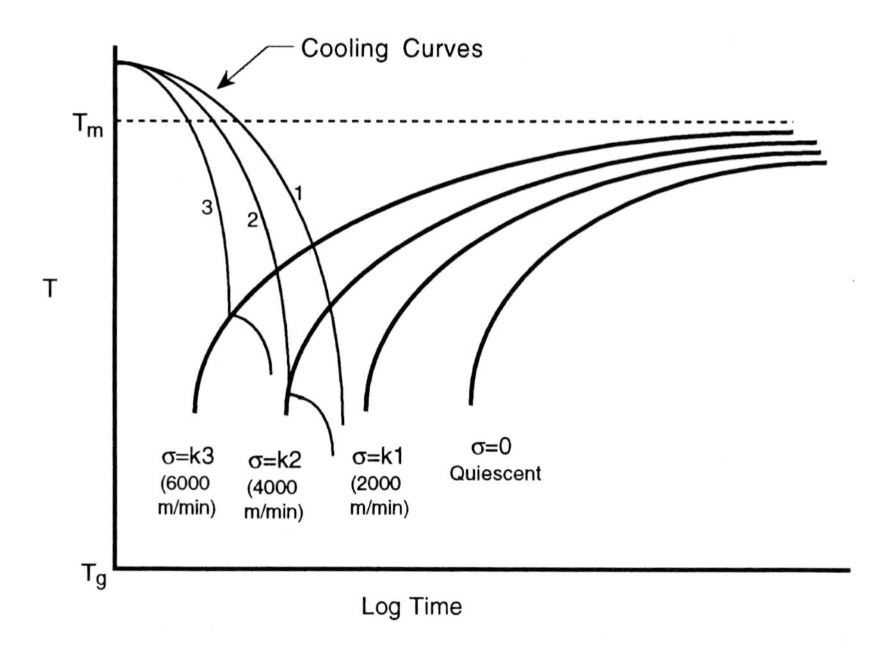

Figure 2.42 CCT diagram representing the crystallization behavior of PET during melt spinning.

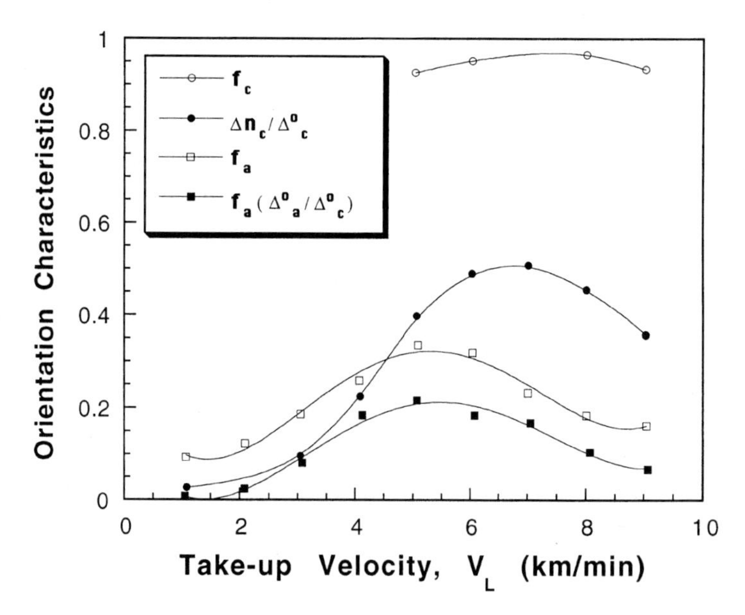

Figure 2.43 Orientation developed in as-spun PET filaments as a function of spinning speed. (After reference [150]).

known value of Δ°_a in determining f_a. The results of these measurements indicate that the crystalline orientation in high speed spun PET filaments is quite high at all speeds where crystallization occurs, with f_c approaching unity as spinning speed increases. The value of f_a first increases, reaches a maximum, and then decreases as spinning speed increases. This behavior correlates well with the maximum in tenacity, Figure 2.40.

The maximum in the amorphous orientation as a function of spinning speed was recently confirmed by Clauss and Salem [151] using the natural or "chain-intrinsic" fluorescence technique. These authors also compared the amorphous orientation in high speed spun fibers to that which is present in fibers that were spun at low speeds and then drawn to total orientations similar to those of the high speed as-spun fibers. This comparison clearly showed that the amorphous orientation in the high speed spun fibers is lower than in the spun and drawn fibers.

Wide-angle and small-angle X-ray scattering patterns of PET fibers are quite interesting and revealing. At low spinning speeds the wide-angle patterns, Figure 2.39, show only a uniform diffuse ring of scattering associated with the unoriented amorphous PET phase. As spinning speed increases, but prior to the onset of crystallization, the intensity of this broad maximum becomes nonuniform in the azimuthal direction, i.e., the intensity increases on the equator and decreases on the meridian. Shimizu et al. [124,152] argued that this represented the development of an oriented PET "mesophase" prior to crystallization to the usual PET triclinic crystal form of Daubeny et al. [153]. At spinning speeds high enough to allow crystallization to the triclinic phase, the breadth of the wide-angle reflections decrease with increase in spinning speed, indicating that the crystallite size increases. Crystallite sizes computed from line breadths are shown in Table 2.3. This behavior suggests that the crystallization temperature increased with increasing spinning speed. This would be consistent with the interpretation that higher spinning speeds lead to higher spinning stresses and higher crystallization kinetics that result in higher crystallization temperatures (compare Figure 2.42).

Figure 2.44 shows a schematic representation published by Shimizu et al. [124] of the small angle patterns obtained at different spinning speeds. Of course, at low spinning speeds the amorphous filaments give no appreciable scattering pattern. As crystallinity

Table 2.3 Crystal Dimensions Λ_{hkl} for As-Spun PET Filaments[1])

	Λ_{hkl} (Å)		
Preparation	(010)	($\bar{1}$ 10)	(100)
5000 m/min	49.4	29.7	29.5
6000 m/min	59.6	49.5	46.0
7000 m/min	67.2	53.9	53.3
8000 m/min	72.1	60.9	63.9
9000 m/min	81.1	69.3	65.2
Drawn and annealed[2]	64.1	47.8	41.0

[1]) Adapted from Shimizu et al. [124], values obtained using Scherrer formula
[2]) Spun 1000 m/min, drawn 3.75X and annealed 1 hr @ 175 °C at constant length

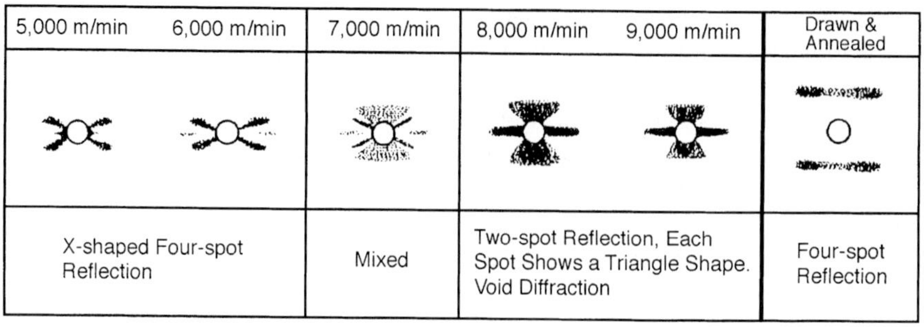

5,000 m/min
6,000 m/min
7,000 m/min
8,000 m/min
9,000 m/min
Drawn & Annealed
X-shaped Four-spot Reflection
Mixed
Two-spot Reflection, Each Spot Shows a Triangle Shape. Void Diffraction
Four-spot Reflection

Figure 2.44 Schematic illustration of SAXS patterns for as-spun PET filaments. (After reference [124]).

develops the pattern has a distinctive X-shaped four-point pattern. At higher speeds there is a transition to what appears to be a distorted two-point meridional pattern with an equatorial streak caused by fibrillation and/or void formation. Such voids would be elongated parallel to the fiber axis. Scanning electron micrographs of "peeled" PET filaments also indicate an increasingly fibrillar structure with increasing take-up velocity [144]. The small-angle patterns, together with the data from wide-angle patterns, have been interpreted in terms of a structural model like that shown in Figure 2.45 [141]. In this model the final high-speed spun fiber consists of microfibrils. Each microfibril contains alternating crystalline and amorphous regions parallel with the fiber axis. The dimensions

β = 0.1
γ = 1.2
170Å
70Å
φ
210Å
35Å
tanφ = 1.4
5,000 m/min
β = 0.2
γ = 2.0
30Å
90Å
φ
140Å
70Å
tanφ = 0.5
9,000 m/min

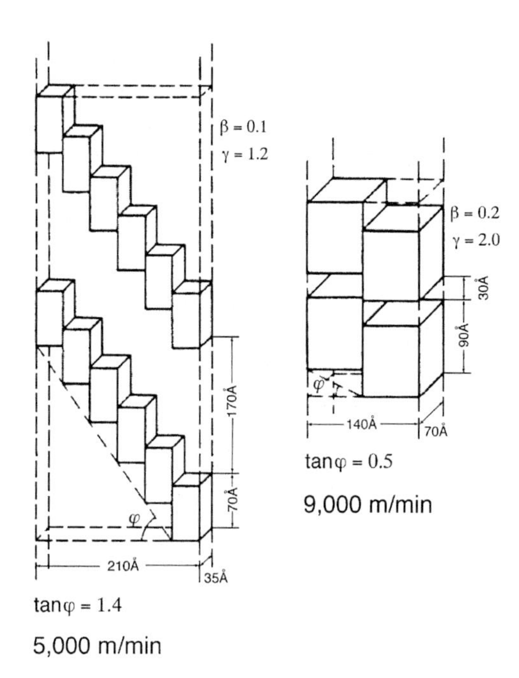

Figure 2.45 Structural models of PET filaments spun at 5000 and 9000 m/min. (After reference [124]).

of the crystal blocks depend on the spinning speed. Adjacent microfibrils are shifted parallel to each other forming a larger region in which the crystals form an angle φ with the normal to the fiber axis.

As early as 1979 Perez and Lecluse [154] reported that PET filaments spun at speeds in the neighborhood of 5,000 m/min exhibit a rapid draw-down or "necking" in the diameter profile along the spinline. The change in diameter often occurs over a relatively small distance of no more than a few filament diameters. This abrupt necking is somewhat analogous to the necking that occurs during drawing or tensile testing of filaments that have been spun at low take-up velocities. It can readily be observed in the on-line diameter profiles shown in Figure 2.46 [155]. The neck becomes more abrupt and more evident with increase in spinning speed above 4500 m/min. Figure 2.46 also shows on-line birefringence measurements in comparison to the diameter measurements. It is seen that the birefringence rises abruptly at the point in the spinline at which the neck occurs. The necking can also be seen in rapidly quenched "grab samples" taken from the spinning filament. These samples are obtained using a special device to cut a short segment from the running spinline. Figure 2.47 shows both diameter and birefringence variation of a sample cut from the region containing the neck. The zero on the distance scale in this figure corresponds to the point on the spinline at which the neck is just beginning and the neck develops over a distance of less than a centimeter. Both on-line and "grab sample" data show that the birefringence approaches the value for the as-spun filament as the diameter draws down and approaches its final value. X-ray diffraction patterns on grab samples show only an amorphous halo above the neck, but patterns made just below the

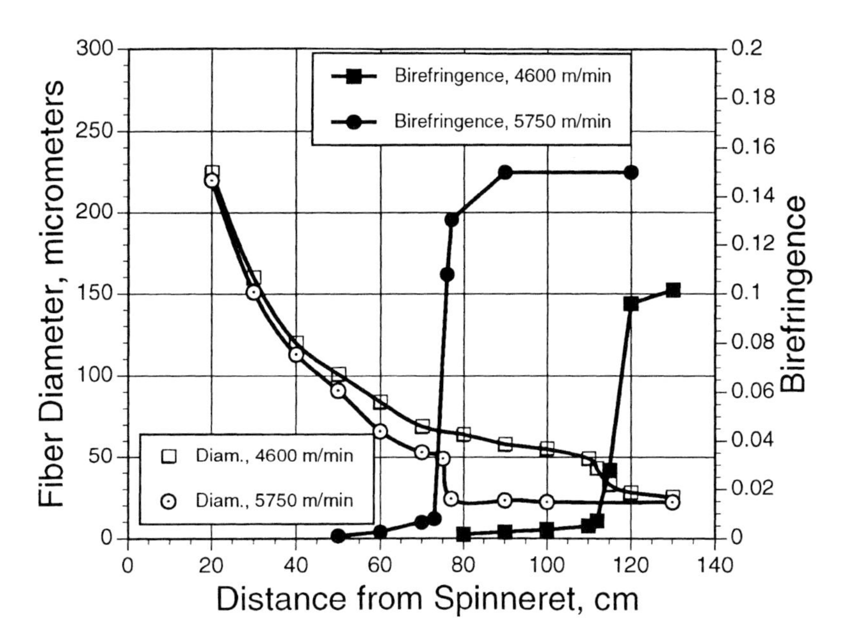

Figure 2.46 On-line diameter and birefringence profiles for PET. (Data of reference [155]).

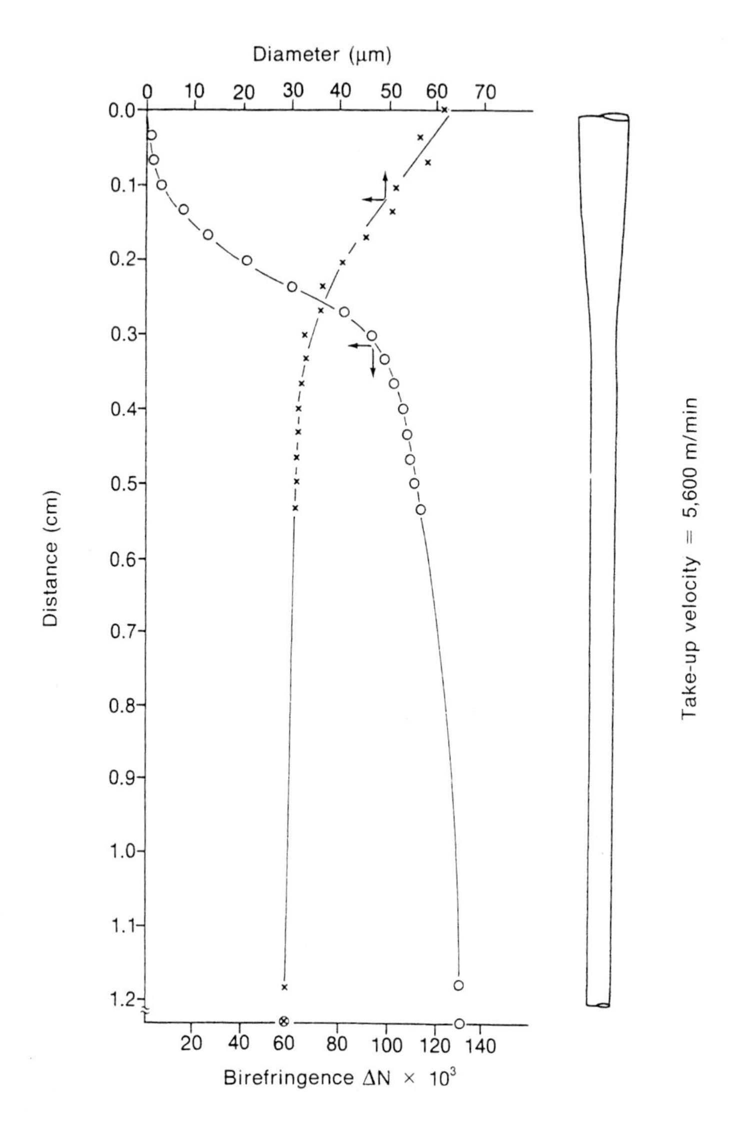

Figure 2.47 Neck observed in "grab" sample of PET. (After reference [155]).

neck exhibit crystalline reflections indicating a high level of orientation and crystallinity approaching that of the as-spun filament [144,155].

The phenomenon of neck formation has attracted considerable attention, but it is not readily explained. This unusual behavior has also been seen in other polymers, including poly(butylene terephthalate) or PBT [156], polyamide 6 [34,157], polyamide 66 [157] and, under certain conditions, in polypropylenes [158]. When necking is observed it is invariably associated with crystallization in the spinline, frequently occurring at a point on the spinline just before crystallization is detected. However, crystallization in the spinline is itself not a sufficient condition to cause necking, as numerous examples exist of crystallization occurring in the spinline without noticeable neck formation.

Kase [26] concluded that the existence of the necking effect likely requires a constitutive equation in which a maximum in the tension versus strain curve is followed by a minimum and subsequent increase of the tension beyond that of the first maximum. This would amount to a point along the spinline at which the viscosity decreases momentarily rather than continuing to increase with increased distance from the spinneret. This conclusion was based on detailed considerations of the dynamics of the spinline and on the early work of Vincent [159] on necking during cold drawing. It is, in fact, possible to simulate necking in the spinline under conditions where such a variation of viscosity with distance is allowed. However, it is not clear why such a variation should occur. Some authors [10,22,156,157] have suggested that the heat of fusion supplies enough energy to cause a local increase in temperature of the spinline resulting in local viscosity decrease. Others, e.g. [124], have suggested that the viscosity may decrease due to formation of an oriented mesomorphic phase.

Ziabicki [160] has carried out an extensive analytical investigation of the various factors that affect "necking" in the spinline. He concludes that the main factor responsible for "concentrated deformation" is a positive velocity gradient along the spinline. Polymer elasticity and elongation rate dependent viscosity do not contribute to neck formation, but his analysis strongly supports the hypothesis that the high viscosity gradient resulting from rapid stress-induced crystallization is largely responsible for the "neck-like" deformation. In spite of all the effort expended to investigate this phenomenon, there does not seem to be documented proof of the detailed physical cause of spinline necking at the present time.

Shimizu et al [124] have suggested the schematic morphological model shown in Figure 2.48 for fiber formation during high speed melt spinning of PET. This model is roughly

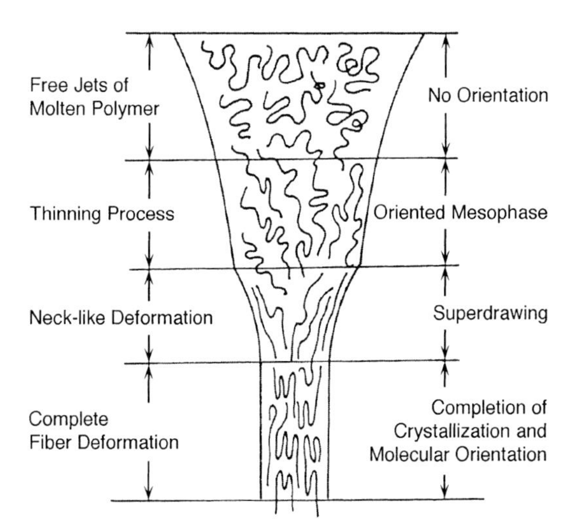

Figure 2.48 Schematic model of the development of structure during high speed melt spinning of PET. (After reference [124]).

consistent with the data described above, including the development of an oriented mesophase, neck-like deformation, and crystallization into a microfibrillated structure. Presumably, the details of the final fiber structure are consistent with the schematic model of Figure 2.45.

Wunderlich and his coworkers [161–163] have recently applied full-pattern X-ray diffraction refinement and thermal analysis techniques in an attempt to develop an improved model of PET fiber structure. The full-pattern X-ray diffraction technique involves collecting quantitative X-ray intensity data over the entire fiber pattern. The crystalline diffraction is removed from this pattern by appropriate calculation, leaving the noncrystalline contribution. In the course of doing this, considerable detail about the crystal structure and morphology are determined, including lattice parameters, atomic coordinates, crystallite sizes, estimated crystalline fraction, average tilt of the crystals with respect to the fiber axis, and a measure of the orientation distribution. Although they have applied the techniques to only a few samples, and even fewer as-spun samples, they conclude that a three-phase model is required to describe the structure and properties of PET fibers. The three phases consist of the triclinic crystalline phase, an isotropic amorphous phase, and an anisotropic noncrystalline or intermediate phase. An example of the separated contributions to the diffraction pattern are shown in Figure 2.49 for a fiber which was spun at 4000 m/min and then annealed in nitrogen at 200 °C for 84 hours at fixed length. The integrated intensity fractions of the differing components provide the following estimates of the phase fractions for the as-spun fiber and for the annealed fiber shown in Figure 2.49:

	PET 4 (as-spun)	PET 4A (annealed)
crystalline (%)	12.4	19.0
anisotropic (%)	22.5	42.5
isotropic (%)	65.1	38.5

Thermal analysis data were also interpreted in terms of three phases: crystalline, mobile amorphous, and rigid amorphous. The latter quantity presumably corresponds roughly to the anisotropic noncrystalline fraction defined by x-ray diffraction analysis. A two-phase model would divide the anisotropic fraction between the crystalline and isotropic fractions; this would tend to over-estimate both fractions. The noncrystalline anisotropic phase would seem to be equivalent to the mesophase of Shimizu et al., though Wunderlich et al. argue that it should not be called a mesophase as it is not independently stable without the surrounding crystalline and amorphous phases.

An aspect of PET filaments spun at high speeds not illustrated in the schematics of Figures 2.45 and 2.48 is the presence of radial variations in structure. Figure 2.50 shows the variation in birefringence of filaments spun at different speeds, indicative of a radial variation in molecular orientation. The birefringence and molecular orientation tend to be higher near the filament surface than at its center, especially at the highest spinning speeds. Radial variations in crystallinity have also been inferred from variation in the mean index of refraction and selected area electron diffraction patterns [144]. Crystallinity also tends to be higher at the surface of the fiber than at the center. As with polyolefins, these variations have often been referred to as a "skin-core structure" of the filament. The presence of the radial distribution is accentuated by factors that produce or enhance a

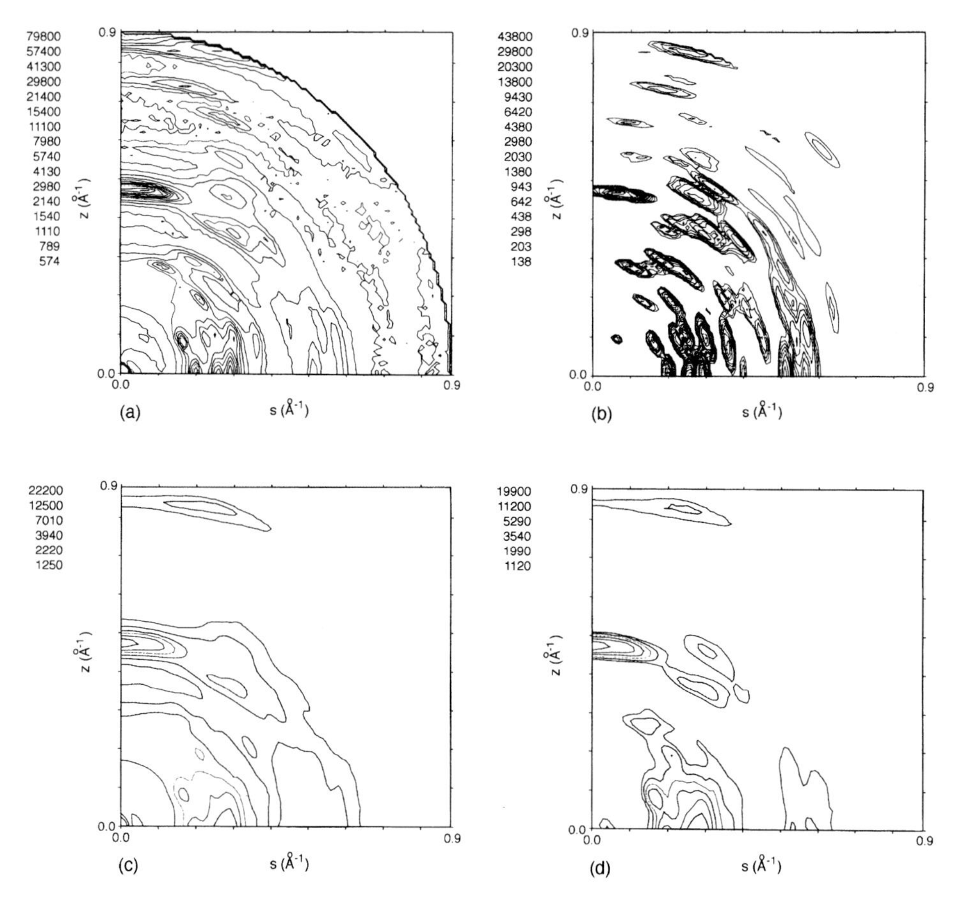

Figure 2.49 Contour plots of the total and separated intensities for a PET fiber sample spun at 4000 m/min and then annealed at 200 °C for 84 hours at fixed length. Values listed at the upper left of each figure are for the intensity contours. (a) Total x-ray diffraction intensity corrected for absorption, polarization and Compton scattering; (b) crystalline phase diffraction pattern; (c) noncrystalline scattering; (d) the anisotropic component of the noncrystalline scattering.

radial temperature distribution in the filaments, such as increasing quench air velocities or lower quench air temperatures [164]. Based on such observations, Shimizu et al. [124,164] suggested the mechanism illustrated in Figure 2.51 for the creation of the radial distribution of structure. In essence, the radial temperature distribution produces a radial variation of viscosity and spinning stress across the filament. Since molecular orientation in the melt is proportional to the stress, this leads to higher molecular orientation near the surface of the filament than at its center. The presence of the higher molecular orientation and lower temperature causes oriented crystallization to occur at the surface before it can occur in the center of the filament. The heat of crystallization given up when the surface crystallizes may contribute to some molecular relaxation in the interior of the filament which, combined with the very high cooling rates for such filaments, allows the core to quench without reaching the high levels of crystallinity formed near the surface.

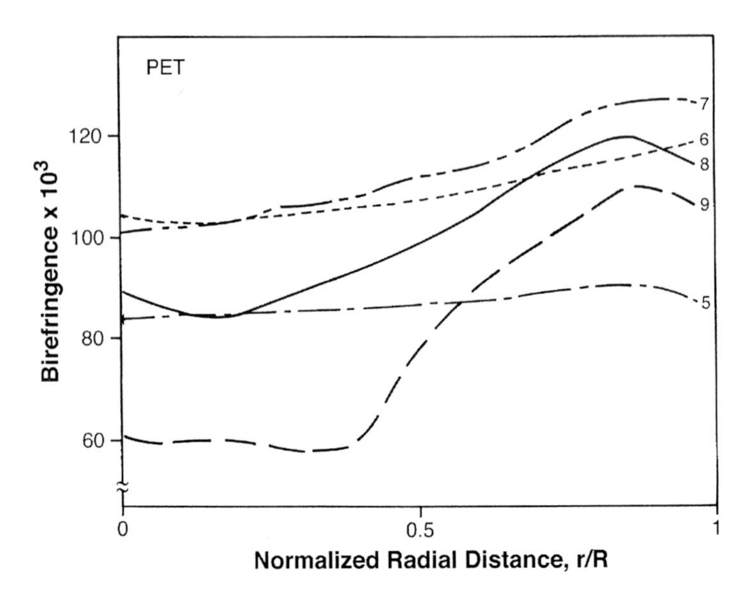

Figure 2.50 Radial variation of birefringence in high speed spun filaments of PET. Take-up velocities are indicated in km/min. (Data of reference [124]).

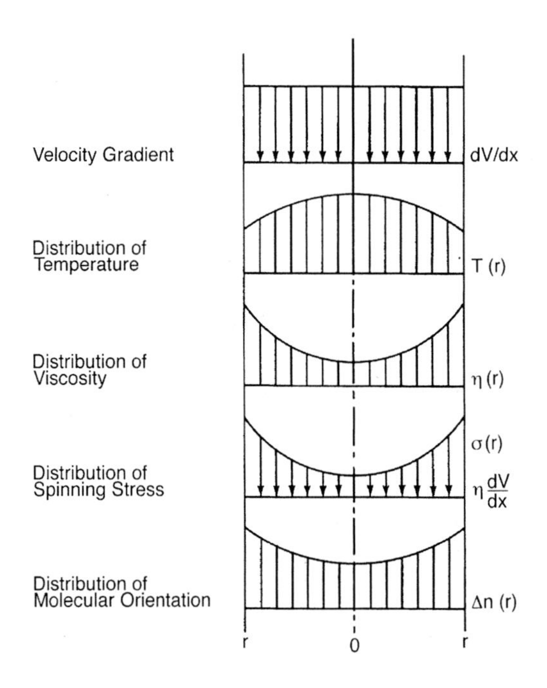

Figure 2.51 Mechanism of formation of radial variations in high speed spun filaments. (After reference [124]).

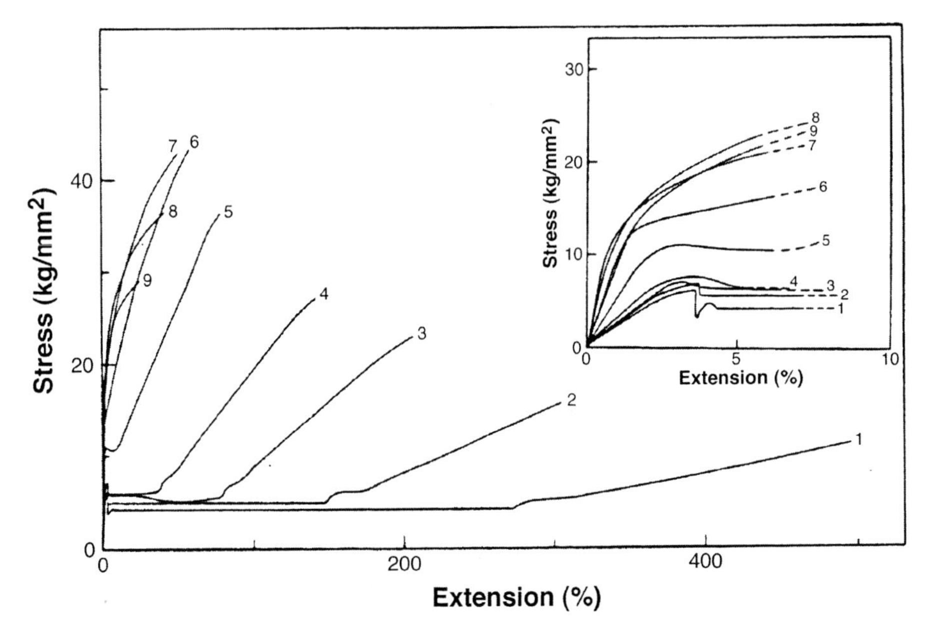

Figure 2.52 Stress vs strain curves for as-spun PET filaments. Take-up velocity is indicated in km/min. (After reference [124]).

The major changes in structure with spinning speed of melt spun PET filaments causes the mechanical and physical properties to vary markedly. We have already pointed out the tremendous influence of spinning speed and degree of crystallinity on the filament shrinkage in boiling water in Figure 2.41. Figure 2.52 shows stress versus extension curves for filaments spun at speeds ranging from 1000 m/min to 9000 m/min [124] with a constant mass throughput. Note that these data show the extension decreasing with increasing spinning speed and tenacity going through a maximum at about 6000–7000 m/min in agreement with the tenacity data shown in Figure 2.40. The "natural draw" decreases with spinning speed and disappears above 5000 m/min. The decrease in tenacity at high spinning speeds seems to be connected to the formation of the skin-core structure described above. The lower orientation in the core of the filament, as shown in Figure 2.50, would seem to explain this decrease, at least in part.

Vassilatos et al. [10] have discussed the differences between the properties of high speed spun PET filaments and filaments that are made by the traditional process of low speed spinning followed by a post spinning drawing process. Although it is possible to obtain tenacity values for high-speed as-spun filaments in the same range as for some drawn filaments, the filaments do exhibit substantial differences in other properties. In particular, the as-spun filaments exhibit enhanced dyeability. This appears to be related to lower amorphous orientation in the spun yarn than in the drawn yarn, at the same tenacity [10,151].

2.2.2.2 Other Polyesters

In addition to PET, a number of other polyesters have been melt spun. Poly(butylene terephthalate) or PBT is of particular interest because of its similarities and differences from PET. PBT has been used extensively as an engineering thermoplastic molding resin because it crystallizes faster than PET while exhibiting similar physical and mechanical properties. PBT has also attracted some attention as a fiber forming material because of its ability to be processed into filaments with high resilience and dyeability. Early researchers [165–177] discovered that PBT exists in two crystalline forms, and they studied the relationship of these two forms to the mechanical properties of PBT fibers. The two forms, called α and β-forms, both have triclinic unit cells. They differ primarily in the fact that the chain-axis repeat distance (c-axis) is longer in the β-form than in the α-form (about 1.26 nm compared to 1.18 nm). This difference occurs because the conformation of the four methylene group sequence is more extended in the β-form than in the α-form.

Drawing of filaments that consist mainly of α-form produces an increase in the amount of β-form at the expense of the α-form [176].

The structure and properties of as-melt spun PBT filaments were studied by Chen et al. [156,178]. At very low spinline stresses, corresponding to high mass-throughputs and low take-up velocities, the filaments could be quenched to the amorphous phase. Increasing the take-up velocity above about 1000 m/min produced the onset of ordering of the structure in the spinline. This ordering is probably a form of mesophase, paracrystallinity, or noncrystalline anisotropic phase analogous to that discussed above for PET. The amount and perfection of crystallinity increased with further increase in spinning speed, as did the crystalline orientation and crystallite size. A decrease in mass throughput, and other process factors that increase spinline stress also increased these latter quantities. Changes in molecular weight also influenced the crystallinity of the polymer. The observed effects of increasing the molecular weight seemed to be a balance between the crystallization enhancing effects of higher spinline stress and the crystallinity retarding effect of increased entanglements. X-ray diffraction showed that the as-spun filaments contained primarily α-form crystals, but also contained a small amount of β-form crystals, whose amount increased with spinning speed. The PBT as-spun filaments exhibited unexpected boiling water shrinkage behavior. At spinning speeds of 1000–1500 m/min the filaments actually grew in length by 2–3% rather than shrinking. At higher take-up velocities the filaments began to shrink, with the amount of shrinkage leveling off at between 2 and 3% at spinning speeds in the neighborhood of 3000 to 5000 m/min, depending on the molecular weight and other process variables. The modest shrinkage at high spinning speeds appears to be due to the stabilizing effect of crystallinity, but the explanation for filament growth at lower spinning speeds was not fully explained. It was suggested that it might be related to further crystallization of the filaments and transformation of the β-phase present to α-phase on exposure to boiling water.

Shimizu et al. [124] have described the spinning behavior of the polyester-ether polyethylene-1,2-diphenoxyethane-p,p′-dicarboxylate (PEET). This polymer behaves in many respects similar to PBT as it crystallizes more readily than PET and, hence, develops crystallinity at lower spinning speeds. Based on density data some form of ordering begins about 1000 m/min. Again, X-ray data indicate that the structure should be thought of as pseudocrystalline until spinning speeds in excess of 3000 m/min are reached. Annealing of the amorphous or pseudocrystalline filaments at 200 °C results in crystallization to a relatively stable crystal form referred to as the α-form. Above

3000 m/min the as-spun filaments contain a different crystal phase, referred to as the β-form. This form anneals to the α-form at spinning speeds of about 4000 m/min, but is stable to annealing when formed by spinning at speeds above 6000 m/min. The shrinkage behavior of this PEET is also similar to that of PBT. The amorphous filaments formed at very low spinning speeds shrink, but, as the pseudocrystallinity develops, the filaments stop shrinking and begin to grow in length. At still higher speeds, where there is a relatively high level of crystallinity, the filaments again shrink, but only a small amount.

Polyarylate, a wholly aromatic polyester based on polymerizing bisphenol-A with terephthalic and m-phthalic acids, was studied by Song et al. [179]. The melt spun filaments were found to be totally non-crystalline, but they exhibited significant molecular orientation as measured by birefringence, which increased with draw-down ratio. A linear correlation was found between the birefringence of the filament and the spinline stress in agreement with expectation based on the rheo-optical law. The slope of this curve was interpreted as the stress-optical coefficient for the melt. The glass transition temperature of this material is 175 °C; this represents a practical upper temperature limit to use of filaments prepared from it. See also Section 2.2.4.2 for additional comments.

Recently, interest in poly(ethylene-2,6-naphthalenedicarboxylate) (PEN) has developed due to potentially lower cost for the dimethyl-2-6-naphthalenedicarboxylate needed for its preparation. The polymer has both a higher glass transition temperature ($\approx$ 120 °C) and higher melting temperature ($\approx$ 265 °C) than PET. These properties give PEN considerable potential in the packaging field for application as hot-filled containers.

The fiber forming characteristics of PEN have recently been studied by Jager et al. [180] and by Cakmak and Kim [181]. This material behaves in many respects similar to PET. Low speed spinning results in amorphous filaments, while spinning at speeds in the neighborhood of 2000 m/min produces an oriented mesophase. Spinning at speeds above 2500 m/min results in stress-induced crystallization. Thus a slightly lower spinning speed is required for the development of crystallinity in PEN than in PET, at least in the resins that have been studied. Up to a spinning speed of about 3500 m/min the crystalline domains in PEN are primarily α-phase. Fibers processed at 3500 m/min up to 4000 m/min contained a second crystal phase, β, along with the α-phase. Highly crystalline fibers could be obtained by heat-set annealing. Tenacities of the PEN yarns are comparable to those of PET, but the modulus is substantially higher, and it is retained over a wider temperature range. Filament shrinkage is comparable to PET, though it passes through its maximum at somewhat lower spinning speed ($\approx$ 2000 m/min compared to 3300 m/min for PET).

Other aliphatic polyesters such as polyglycolide, polylactide and their copolymers have also been melt spun, and the fibers applied to make bioabsorbable sutures, for example. Except for a few papers [182–184] that described the preparation of high strength poly (L-lactic acid) fibers by a very slow melt spinning process followed by hot drawing, little detail about the processing and structure development during spinning was available in the open literature until the recent paper of Mezghani and Spruiell [185]. They found that poly (L-lactic acid) could be melt spun at speeds up to at least 5000 m/min, and the behavior was remarkably similar to that of PET. The major difference was that the maximum in the crystallinity, birefringence, elastic modulus, and tenacity occurred at lower spinning speeds than for PET, at about 2000–2500 m/min for the molecular weight and spinning conditions studied.

Studies of copolyesters, including certain wholly aromatic thermotropic liquid crystal co-polyesters also have been reported. Some discussion of copolyesters and of blends containing polyesters is given in later sections.

2.2.3. Polyamides

Here we restrict our attention to the aliphatic polyamides, as the aromatic polyamides are usually not melt spun, and they lie outside the scope of this chapter. The most common melt spun polyamides are polyamide 6 (nylon 6) and polyamide 66 (nylon 66). Although polyamide 66 was the first commercial melt spun synthetic fiber, it has not received as much attention in the open literature as polyamide 6. Consequently, we begin this discussion with polyamide 6.

2.2.3.1 Polyamide 6

Due in part to its slow crystallization kinetics, structure formation in polyamide 6 filaments exhibits many similarities to PET. However, polyamide 6 is more complex than PET due to the existence of an array of crystalline and pseudocrystalline forms and due to the influence of water absorption in the filaments on the glass transition temperature.

Before proceeding to discuss the structure of melt spun polyamide 6 filaments, it is necessary to consider the current understanding of the various polymorphic forms that have been described for polyamide 6. A review of the literature shows that a plethora of crystalline, paracrystalline and/or mesomorphic forms have been reported, with different authors sometimes using different notation for the same or very similar forms. The situation has been discussed by Gianchandani et al. [186]. Briefly, they suggest that there exist an amorphous phase and two basic types of crystalline or paracrystalline forms, as illustrated in Table 2.4, referred to as α-type or γ-type. The ideal prototype for the α-type structures is the

Table 2.4 Suggested Notation for Polymorphic Forms of Polyamide 6

I. Amorphous (vitrified liquid)
II. α-phase type **A.** α monoclinic (stable equilibrium phase of Holmes et al. [187]) **B.** α′ monoclinic (paracrystalline α-form of Roldan and Kaufman [190], metastable α form of Parker and Lindenmeyer [191]) **C.** α* pseudohexagonal (same as Illers et al. [192] γ* form, low orientation β form of Ziabicki [193], pleated α form of Stepaniak et al. [194,195], probably same as γ "nematic form" of Roldan and Kaufman)
III. γ-phase type **A.** γ monoclinic (prototype, stable or metastable phase of Arimoto [188,189] or Kinoshita [196], same as Bradbury [197,198] orthorhombic) **B.** γ* pseudohexagonal (β form of Roldan and Kauffman, γ of Vogelsong [199] and Ota [200], high orientation β form of Ziabicki)

generally accepted, equilibrium structure of Holmes, et al. [187]. This structure is monoclinic with a = 0.956 nm, b = 1.724 nm (chain axis), c = 0.801 nm, and $\beta = 67.5°$. The chains are fully extended and occur in antiparallel hydrogen bonded sheets. The prototype for the γ-type structures is the monoclinic γ-form of Arimoto [188,189] with a = 0.933 nm, b = 1.688 nm (chain axis), c = 0.933 nm, and $\beta = 59°$. The shortened chain-axis repeat distance is interpreted to correspond to a "kinked" or pleated hydrogen bonded sheet structure in which the hydrogen bonding occurs between parallel rather than antiparallel chains.

The α' paracrystalline monoclinic and the α^* pseudohexagonal readily anneal to the stable α monoclinic phase due to similar hydrogen bonding, whereas the γ type phases do not.

Numerous investigators have studied fiber formation in polyamide 6. Ziabicki and Kedzierska [4,7] measured birefringence and X-ray diffraction patterns of as-spun polyamide 6 filaments spun at spinning speeds up to about 3500-4000 m/min and equilibrated with the ambient conditions. The birefringence increased with take-up stress, and the X-ray data exhibited a broad equatorial reflection that was interpreted to indicate the presence of a poorly developed pseudohexagonal form. Hamana et al. [201,202], Ishibashi et al. [203,204], Hirami and Tanimura [205], and Bankar et al. [206] made on-line birefringence or wide-angle X-ray diffraction measurements. On-line X-ray diffraction measurements showed that, in the low take-up velocity regime investigated, as-spun polyamide 6 filaments are initially amorphous, but crystallization to a pseudohexagonal form takes place on the bobbin as the filaments equilibrate with the environment.

Later investigators [207–209,124,143,34,157] found that, at sufficiently high spinning speeds, polyamide 6 will crystallize in the spinline. This can be observed in on-line diameter, temperature and birefringence measurements as illustrated in Figure 2.53 for a

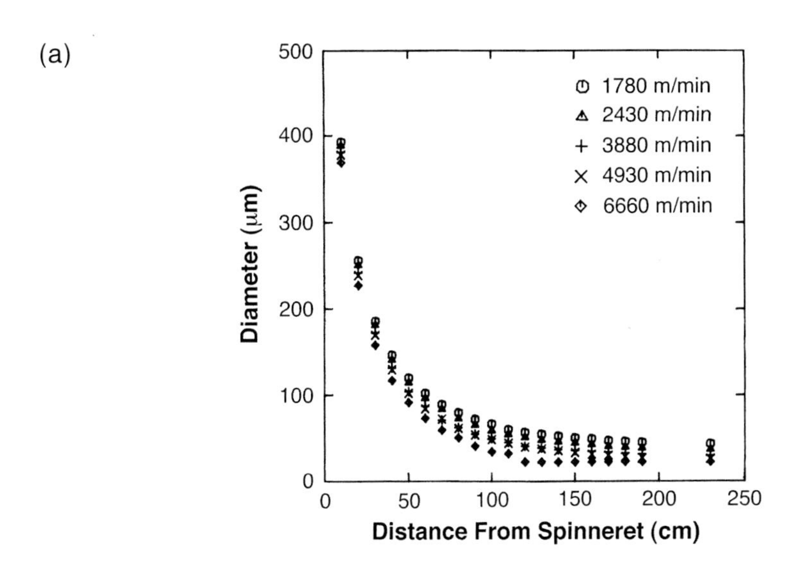

Figure 2.53 Experimentally measured spinline profiles for polyamide 6. (a) diameter; (b) temperature; (c) birefringence. Data are for a sample with viscosity average molecular weight of 53,000, mass throughput of 3.0 g/min per hole. (Data of reference [33]).

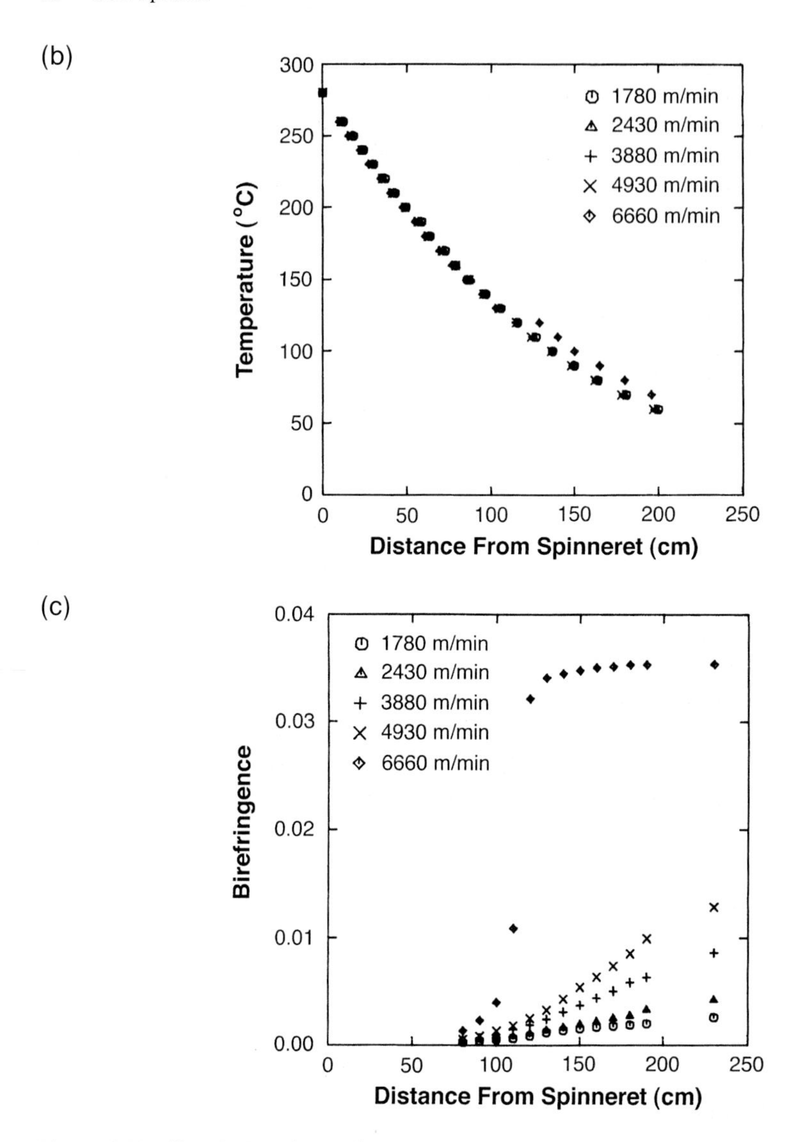

Figure 2.53 (Cont.) Experimentally measured spinline profiles for polyamide 6. (a) diameter; (b) temperature; (c) birefringence. Data are for a sample with viscosity average molecular weight of 53,000, mass throughput of 3.0 g/min per hole. (Data of reference [33]).

polyamide 6 having a viscosity average molecular weight of 53,000 g/mol. At the highest spinning speed of 6660 m/min, a slight necking is observed in the diameter profile. At the position on the spinline where the neck is observed there is a rapid increase of birefringence which presumably occurs as a result of the onset of oriented crystallization. On-line temperature measurements [34,157] also indicate a plateau or recalescence in the measured temperature versus distance from the spinneret at sufficiently high spinning speeds, see Figure 2.53(b).

The change of birefringence with time, immediately following spinning, is illustrated in Figure 2.54. For the filament spun at the highest speed, where crystallization appears to have occurred on-line, there is little change in birefringence with conditioning time. However, for all other spinning speeds, the birefringence of the filament changes greatly with conditioning time, indicating that oriented crystallization to a highly imperfect, pseudohexagonal phase is occurring. Presumably, absorption of moisture by the filament during conditioning lowers the effective T_g of the filament and provides the mobility needed to allow this to occur. This process also leads to growth in the length of the polyamide 6 filaments during conditioning, especially when spun at speeds below that at which crystallization occurs in the spinline.

Haberkorn, et al. [157] carried out simultaneous on-line diameter and temperature profiles for polyamide 6 in the high spinning speed range in order to correlate the two measurements.

Their results indicated a somewhat better developed necking in the spinline under their spinning conditions than is shown in Figure 2.53. They also did on-line X-ray diffraction patterns and correlated these with the diameter and temperature measurements. They showed that the bulk of the crystallization appears to occur just beyond the point where necking occurs. The X-ray patterns clearly showed that, under these high-speed spinning conditions, crystallization in the spinline produces highly oriented γ-form crystals.

Conditioned polyamide 6 filaments tend to be mixtures of poorly developed α and γ type phases. Several investigators [143,186,194,195,124,210] have attempted to determine the relative amounts of α and γ-type phases present in the polyamide 6 filaments as a function of spinning and conditioning treatments. There is reasonable consensus that, with increase in spinning speed and molecular orientation, the relative amount of γ-type phase increases and the amount of α-type phase decreases. The perfection of the γ-type

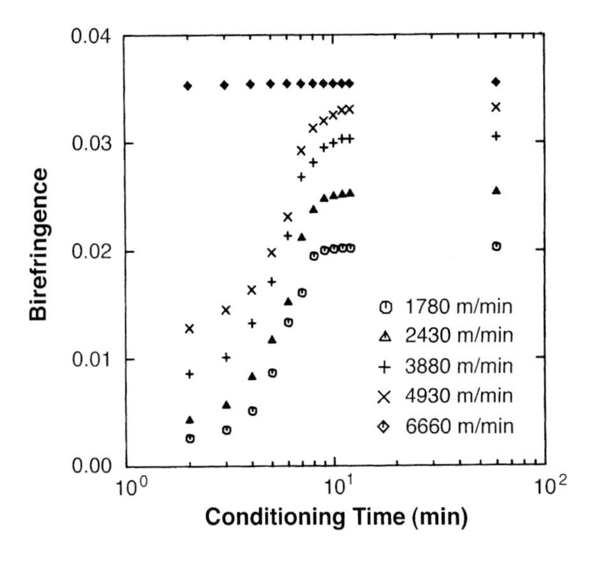

Figure 2.54 Birefringence as a function of conditioning time in humid air for polyamide 6 filaments whose spinline profiles were shown Fig. 2.52. (Data of reference [33]).

phase gradually increases with spinning speed, but does not achieve the prototype γ-monoclinic form. Presumably, crystallization or development of high orientation in the spinline in the dry state leads to γ-type structure. However, drawing, after conditioning, of filaments initially containing substantial amounts of γ-type phase produces a decrease in the γ-type content and an increase in α-type content. Sufficiently high draw ratios and draw temperatures can produce the α'-monoclinic or even the α-monoclinic form. The influence of annealing treatment is very much affected by the severity of the annealing treatment as well as the initial structure of the filament. Filaments spun at low speeds readily transform to α'-monoclinic phase when annealed in boiling water. However, filaments spun at high speeds, where substantial γ-type form is produced, require much more severe annealing treatment, e.g. boiling 20% aqueous formic acid solution, or high temperature steam autoclaving are required to remove the γ-type phase.

Structure development for polyamide 6 filaments is also quite sensitive to molecular weight [34,157,209], as is illustrated by the data in Figure 2.55, which shows birefringence versus conditioning time for filaments of differing molecular weight spun at 1000 m/min and 3.55 g/min per hole mass throughput. It appears that the high molecular weight resin (HMW) has crystallized on the spinline at a take-up velocity of only 1000 m/min, while the other resins with lower molecular weight have not.

Mechanical property data [207] for high speed spun polyamide 6 filaments having a moderate viscosity average molecular weight are shown in Figure 2.56. The general trends

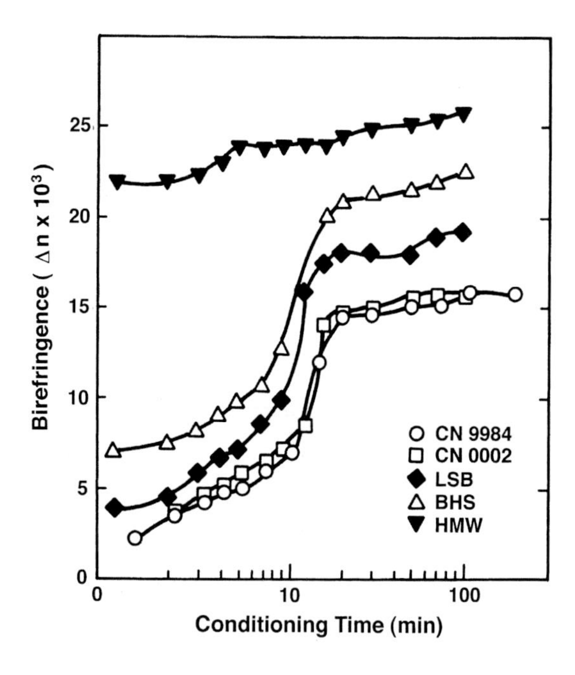

Figure 2.55 Birefringence as a function of conditioning time in humid air for polyamide 6 filaments of varying molecular weight. Viscosity average molecular weights of the different samples are: CN 9984 = 25,250; CN 0002 = 29,400; LBS = 36,400; BHS = 53,200; HMW = 72,900. (Data of reference [209]).

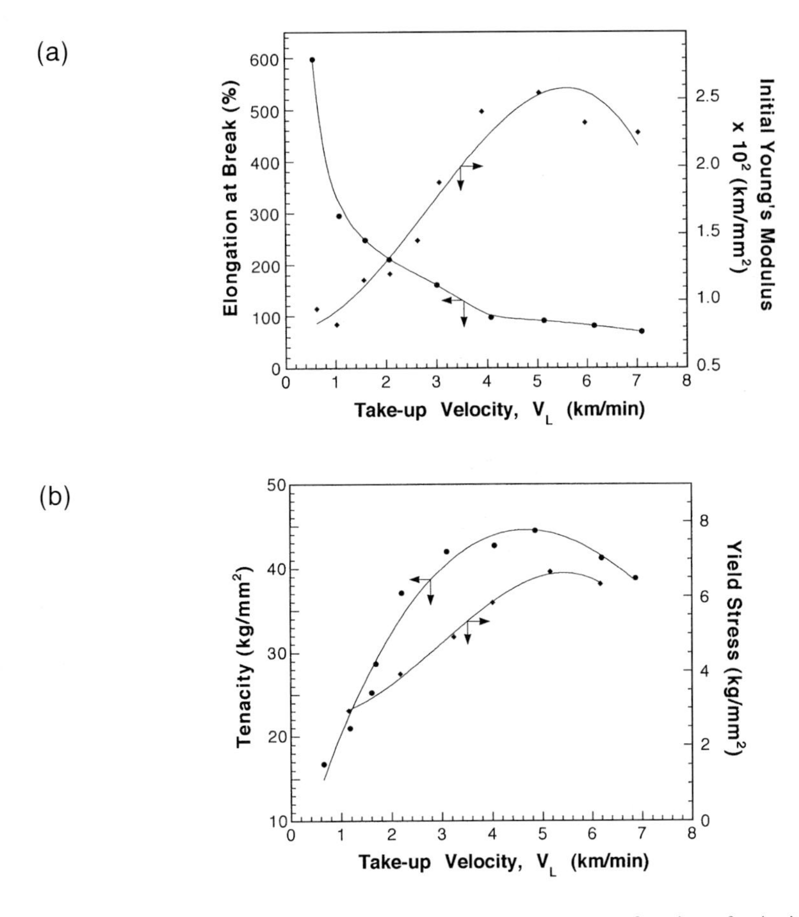

Figure 2.56 Mechanical properties of polyamide filaments as a function of spinning speed. (a) initial modulus and elongation at break; (b) tenacity and yield stress (Data of reference [207]).

are similar to those of PET. The decreases in initial modulus, yield strength and tenacity beginning at about 5000 m/min seem to also correlate with the development of radial variations of orientation in the filaments. Shimizu et al. [124,207] also observed a tendency for γ-type structure to decrease again in this range of spinning conditions.

2.2.3.2 Polyamide 66

The development of structure in polyamide 66 exhibits both similarities and differences from that of polyamide 6. It is important to note that the rate of quiescent crystallization of polyamide 66 is about an order of magnitude greater than that of polyamide 6 [211]. Under similar undercooling, the quiescent crystallization rate of polyamide 66 typically lies between that of polyethylene and that of polypropylene. This higher rate of crystallization might be expected to produce crystallization in the running spinline in a manner similar to the spinning of the polyolefins. However, it must be recalled that rheological factors also play an important role since enhancement of crystallization rates by molecular orientation will likely be necessary for crystallization under most common spinning conditions. Due to the narrow MWD and lower molecular weights of typical polyamide 66 resins used for spinning, in comparison to polyolefin resins, there might be a lower tendency to form row nuclei to help develop the crystallinity. Consequently, it is necessary to examine this question experimentally.

Chappel et al. [212], Danford et al. [213] and Li [214] carried out on-line measurements of birefringence of polyamide 66 filaments. Chappel et al., Lecluse [215], and Haberkorn et al. [157] made on-line X-ray patterns, while Li and Haberkorn measured on-line temperature profiles.

The studies of Chappel et al. and Danford et al. involved spinning speeds of only a few hundred meters per minute and modest cooling rates, but they did show that, under these conditions, crystallization occurred in the spinline. X-ray patterns of as-spun filaments, after equilibration with the ambient conditions, showed several reflections that could be identified with the α-triclinic structure of Bunn and Garner [216]. At these low spinning speeds the orientation in the spinline was relatively low and the crystallization seems to be largely thermally driven.

Lecluse made on-line X-ray patterns for filaments spun at much higher spinning speeds of 3000 to 5000 m/min. These patterns exhibited a single equatorial diffraction maximum at a d-spacing of about 0.417 nm, indicating that under these conditions a paracrystalline structure was formed during spinning similar to the high temperature γ-form [217,218]. Subsequent aging in humid conditions produced a tendency for the formation of α-form crystals and an increase in crystallinity. Figure 2.57 shows on-line birefringence profiles measured by Li for a polyamide 66 sample having a number average molecular weight of 15,600 g/mol at three different spinning speeds. The gradual development of final birefringence with increasing distance from the spinneret is

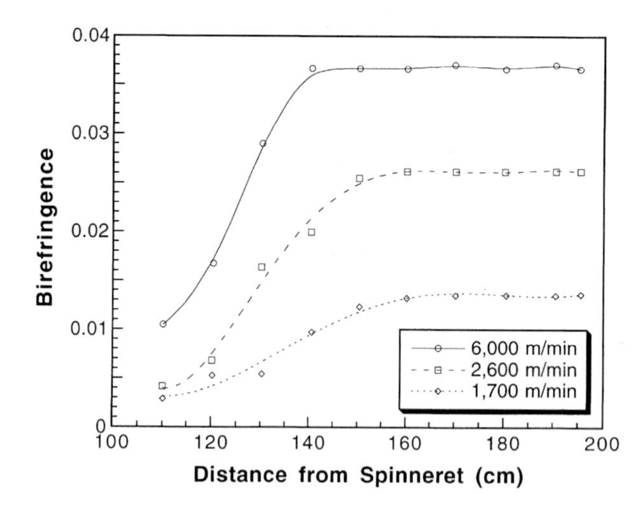

Figure 2.57 Experimental birefringence profiles for polyamide 66. (Data of reference [214]).

reminiscent of the behavior of polyolefins. If we assume that crystallization begins at the point in the spinline where the birefringence shoots up, on-line temperature profiles (not shown), measured by Li, indicate that this corresponds to a crystallization temperature of about 165 °C for filament spun at 6000 m/min, and, perhaps, a temperature as low as 135 °C for filament spun at 1700 m/min. The on-line diameter profiles showed no evidence of necking at the lower spinning speeds, but at about 6000 m/min the profile showed signs of neck development. The later work of Haberkorn et al. [157] carefully correlated diameter and temperature profiles measured on-line with on-line X-ray patterns taken at various distances from the spinneret. They confirmed the observation of a single equatorial X-ray reflection and the formation of a γ-like crystal form developing at a point in the spinline where the diameter was leveling out and the temperature profiles exhibited recalescence. They observed necking in the diameter profile at spinning speeds above about 4500 m/min.

Shimizu et al. [124] point out that the structure of as-spun polyamide 66 filaments can be interpreted in terms of the crystalline perfection index of Dismore and Statton [218]. This index is defined as:

$$\mathrm{CPI} = \frac{d_{100}/d_{010} - 1}{0.189} \times 100 \qquad (2.39)$$

where d_{hkl} are the corresponding interplanar spacings and the factor 0.189 is the value of the numerator for the well crystallized α-triclinic structure as reported by Bunn and Garner. According to Shimizu et al. the value of CPI has a value of about 55% at 1000 m/min and increases gradually with take-up velocity, reaching values in the neighborhood of 80% at take-up velocities of 7000–8000 m/min. They also note that the long period measured by SAXS increases with take-up velocity, a fact that suggests that the temperature of crystallization is increasing. These results are quite consistent with the changes in crystallization temperature observed by Li [214] and Haberkorn et al. [157].

The density of the as-spun and conditioned filaments increases with spinning speed, but is always substantially above the density of the amorphous phase, even at low spinning speeds. However, polyamide 66 filaments spun at spinning speeds below about 3500 m/min actually grow in length by several percent with time of conditioning in the ambient environment. Above this spinning speed the filaments undergo a small and almost immediate contraction on removal from the wind-up bobbin, after which they remain stable [124]. This suggests that, in the lower range of spinning speeds, the structure of polyamide 66 is undergoing a reorganization due to absorption of moisture from the environment in a manner that is analogous to that of polyamide 6 filaments.

The mechanical properties of polyamide 66 as-spun and conditioned filaments vary with spinning speed in a manner that is rather similar to that of polyamide 6 except that they do not exhibit a maximum and subsequent drop in the tenacity and modulus at spinning speeds above 5000 m/min. Instead, the tenacity and modulus continue to increase, at least to spinning speeds of 8000 m/min [124]. This suggests that radial variations of structure are not as significant in polyamide 66 filaments as in polyamide 6 filaments due to the higher crystallization kinetics of polyamide 66.

2.2.3.3 Other Polyamides

Melt spun filaments have been formed from several other polyamides, notably polyamides 4, 7, 11, 12, 46, and 610 [219–221]. Filaments have also been made from polyamide 3, but these were made by solution spinning, since it degrades before it melts [222]. Polyamide 11 has been commercially melt spun into filaments, while development of polyamide 4 and 610 has been attempted. Polyamide 46 fiber is currently undergoing development.

The melting points of the polyamide-m series decreases with increase in the value of m. Also, when m is odd, the polymer melts at higher temperatures than for m even. Thus, polyamides 4 and 7 have higher melting points (about 260 °C and 233 °C, respectively) than polyamide 6 (which melts about 223 °C) and polyamides 11 and 12 have lower melting points (about 192 °C and 179 °C). For the polyamide-mn series, polyamide 46 melts about 307 °C, polyamide 66 about 267 °C and polyamide 610 about 215 °C.

The crystal structures of all these polymers have been studied [199,216,223–226]. In general the structures of polyamides 4, 7, 11 and 12 are rather similar to that of polyamide 6, though polyamides 7 and 11 have triclinic unit cells. An α-form exists consisting of hydrogen bonded sheets containing extended, antiparallel chains. Other polymorphs also occur. In polyamide 12 a γ-form, similar to that occurring in polyamide 6, occurs and seems to be more stable than the α-form. The structure of polyamide 46 is apparently similar to that of polyamide 66 except for the difference in chain axis repeat.

There is relatively little information concerning structure development during melt spinning in any of these polyamides. In the melt spun and drawn state, polyamide 46 filaments are reported to exhibit structures similar to those of polyamide 66 filaments with improved properties at elevated temperature compared to polyamide 66 filaments [227]. They also exhibit reduced shrinkage compared to polyamide 66 [221].

2.2.4 Other Homopolymers

2.2.4.1 Polyvinylidene Fluoride

Recent studies by Cakmak et al. [228,229] described the melt spinning of polyvinylidene fluoride (PVDF). The primary aim of these studies was to obtain greater understanding of oriented crystallization in polymers through the use of a synchrotron radiation source to obtain on-line small-angle and wide-angle X-ray patterns on the spinning filaments. Apparently, PVDF was chosen for this study because (1) it has high crystallization rates and (2) it has high X-ray scattering power due to the presence of the fluorine atoms. These authors also chose to spin tapes rather than circular cross section filaments in order to get maximum sample volume in the X-ray beam. However, because of the rheological properties of the sample that was spun, spinning speeds were limited to 30 m/min or less.

They found that discrete small-angle X-ray patterns (long-period reflections) appeared earlier than did the crystalline wide-angle reflections. This behavior is similar to that observed for HDPE by Katayama et al. [97]. Cakmak et al. noted that the on-line long-periods increase with increase in take-up velocity and decrease with distance from the spinneret, the average off-line long-periods being substantially shorter than the on-line values. As the take-up velocity increased, the discrete scattering patterns converted from roughly circular to a teardrop shape; this suggested an increased spread in the distribution of long-period spacing as take-up velocity increased. This latter feature is somewhat masked in the final off-line scattering patterns. Lamellar orientation was also calculated using the azimuthal spread in the long-period reflection. These data first increased with distance from the spinneret and then decreased. The authors suggested that their results could be explained by the presence of two crystallization stages. In the first stage the periodic density fluctuations are established in regions that will eventually become the fibril nuclei or "shish" structures. In the second stage, a volume filling crystallization takes place between the existing crystallites along the "shish" structure together with a simultaneous radial over-growth of crystallites to form the cylindritic or "kebab" structure. The latter stage causes the observed reduction in long period with distance from the spinneret. Idealized models of the suggested structure development process are illustrated in Figure 2.58.

2.2.4.2 Engineering Thermoplastics

In recent years, studies of melt spinning of several "engineering thermoplastics" have appeared. Among these are polycarbonate (PC) [230], polyetherimide (PEI) [179], polyarylate (PAR) [179], poly(aryl ether sulfone)(PAES) [231], poly(p-phenylene sulfide) (PPS) [232], and poly(aryl ether ketone)s (PEEK and PAEK) [233–237]. Of this group, PC, PEI, PAR, and PAES belong to a group of polymers that exhibit little or no ability to crystallize, but they exhibit fairly high glass transition temperatures and substantial toughness below this temperature due to the presence of low temperature relaxation

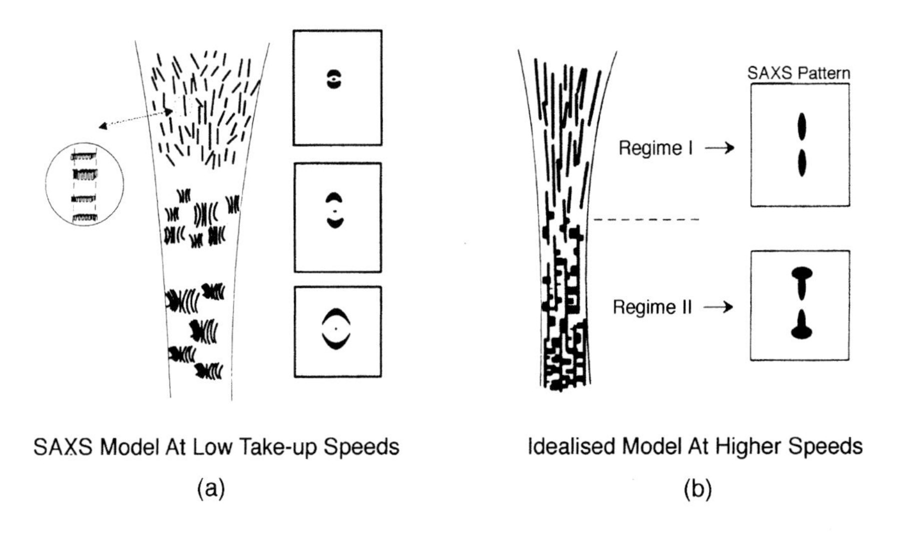

Figure 2.58 Models for the development of structure in PVDF melt spun filaments suggested by Cakmak et al. [229].

mechanisms. Melt spinning of these materials generally resulted in non-crystalline filaments. Their behaviors were all similar to that of polyarylate as described in Section 2.2.2.2. Though they did not crystallize in the spinline, they exhibited significant molecular orientation, as measured by birefringence, which increased with draw-down ratio. A linear correlation was found between the birefringence of the filament and the spinline stress in agreement with expectation based on the rheo-optical law. The slope of this curve was interpreted as the stress-optical coefficient for the melt. The tensile strength and Young's modulus, measured at room temperature, increased with drawdown ratio while elongation-to-break decreased. It is to be noted that elongation-to-break was generally higher than for common unoriented glassy polymers such as polystyrene and poly(methyl methacrylate), but not nearly so high as for as-spun polypropylene filaments. Modulus and tensile strength were substantially higher than for as-spun polypropylene.

PPS was found to produce essentially non-crystalline melt spun filaments at low draw-down ratios (low spinning speeds), but, at moderate draw-down ratios, the filaments possessed a small amount of crystallinity which increased slightly with increase in draw-down ratio. The crystallinity and orientation of the filaments could be increased substantially by drawing at temperatures in the neighborhood of 150 °C. The mechanical properties followed similar trends to those of the other engineering thermoplastics described above.

The poly(aryl ether ketone)s are aromatic ketone based high performance polymers that can exhibit substantial crystallinity under appropriate conditions. The best known member of this group is poly(ether ether ketone) (PEEK). PEEK has two ether groups to each ketone group. Both PEEK and another poly(aryl ether ketone) with a higher ketone/ether ratio have been melt spun. The melt spinning behavior of the two materials was quite similar. Both were non-crystalline at low drawdown ratios, but developed low levels of crystallinity when spun with moderately high drawdown ratios. Drawing or annealing at fixed length at about 160 °C and above produced highly crystalline filaments that were stable to temperatures near their melting point of 335 to 340 °C.

Hsiung and Cakmak [237] made SAXS patterns of the PAEK filaments. The melt spun and annealed filaments exhibited two point patterns, suggesting that the morphology consists of stacked lamellae, similar to many other melt spun filaments. Drawing produced a four-point pattern similar to that observed in drawn and annealed PET filaments. Thus, fiber formation in the poly(aryl ether ketone)s is similar, in many respects, to that of PET.

2.2.5 Copolymers Developed for Specialty Applications

2.2.5.1 Copolyester-Ether Elastic Filaments

The formation of elastic filaments from polyester-ether segmented copolymers (thermoplastic elastomers) were studied by Abhiraman et al. [238] and by Richeson and Spruiell [239]. These materials are segmented block copolymers whose molecules have so-called "hard" segments consisting of a crystallizable polyester and "soft" segments consisting of flexible, noncrystallizing segments containing the ether linkage. The soft segment provides a low glass transition and elastic behavior at ambient temperatures; the hard segment crystals act as physical cross-links that tie the ends of the soft segments together.

Abhiraman et al. point out that the ideal morphology to obtain optimum performance for such elastic filaments consists of:

(a) Complete phase separation between hard and soft segments to ensure good anchoring of soft segments and to eliminate the reduction of molecular mobility in the noncrystalline domains as a result of hard and soft segment mixing in these domains. This also implies complete crystallization of the hard segments.
(b) Perfect orientation of chains in the hard-segment crystals parallel to the fiber axis in order to provide maximum stability of these crystals to applied forces.
(c) Highly coiled soft segments–which should minimize the elastic modulus, maximize the elongation-to-break, and reduce internal stresses in the absence of external forces.
(d) A fibrillar morphology of alternating hard and soft-segment domains which should minimize irreversible shear of the hard segment domains.
(e) A high degree of continuity or "bridging" of hard segment domains by coiled soft segments with the elimination or reduction of chain folding in order to avoid loops that will not contribute to elasticity.

The polymers studied by Abhiraman et al. had poly(butylene terephthalate) hard segments (also referred to as poly(tetramethylene terephthalate) (PTMT)) and 5,5-dimethyl-1,3-poly(oxyethylene) hydantoin terephthalate soft segments (HPOE). The choice of hard segment was based on the rapid crystallization kinetics of PBT. Two fairly similar ratios of hard to soft segment contents were examined, 50/50 and 45/55 by weight. Richeson and Spruiell studied poly(butylene terephthalate)/ poly(tetramethylene ether glycol)-terephthalate copolymers (PBT/PTMEG-T). They studied five materials with hard segment contents ranging from 100 wt.% down to 44 wt.%. The latter authors also studied the quiescent crystallization kinetics of their materials and found that the rate of crystallization decreased rapidly with increased soft segment content. By a concentration of soft segment equal to 25 wt.%, the crystallization rate of the copolymer was reduced to that of PET, while higher soft segment contents crystallized substantially slower than PET.

Both Abhiraman et al. and Richeson and Spruiell found that it was possible to spin these materials using a carefully designed melt spinning process. Both sets of investigators spun into a water bath at low to moderate take-up velocities of 500–1000 m/min for Abhiraman et al. and 275–450 m/min for Richeson and Spruiell. The latter authors explained that the water bath was necessary in order to eliminate sticking together of filaments on the bobbin, especially for the materials with the lowest hard segment content and slowest crystallization kinetics. This occurs because the glass transition temperature of the melt is much below room temperature and in the absence of crystallinity, the filaments will stick to each other. The water bath reduces this tendency, presumably due to the drag and thermal conditions in the bath producing a small amount of crystallinity in the filaments. The low glass transition temperature and high mobility at ambient temperatures of these materials aid this. Richeson and Spruiell measured birefringence as a function of elapsed time after initial contact of the filament with the water bath and showed that the birefringence continues to increase noticeably for up to about 10 minutes

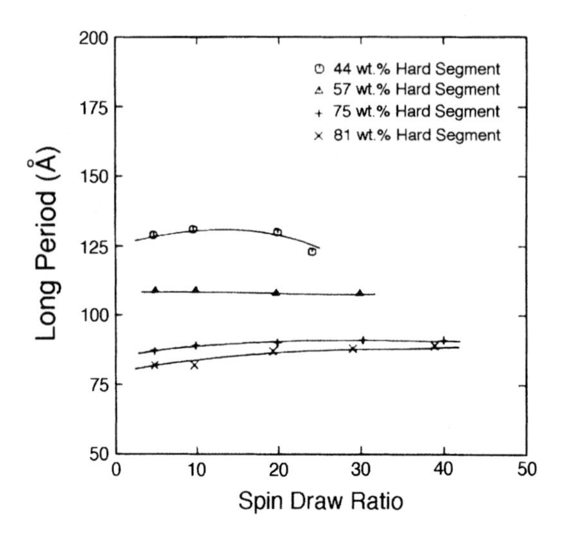

Figure 2.59 Long period of as-spun PBT/PTMEG-T filaments as a function of draw-down ratio and hard segment content. (Data of reference [239]).

after contact with water as a result of continued oriented crystallization. Also, it was necessary to exercise special care in order to take-up the filaments without stretching them between the first godet and the winder.

Both sets of investigators found that it was possible to obtain approximately the desired morphology through melt spinning. Based on WAXS and SAXS experiments, both types

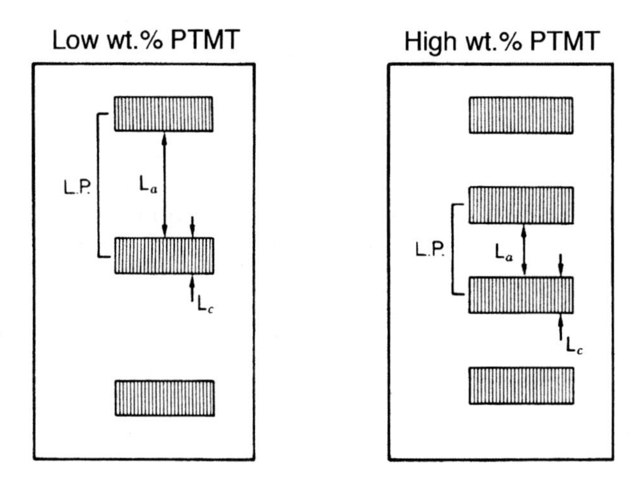

Figure 2.60 Model of the domain structure of PBT/PTMEG-T filaments explaining the effect of hard segment content on the long period. L_c is the thickness of the crystallites while L_a is distance between adjacent crystallites. (After reference [239]).

of filaments consisted of PBT hard segment lamellar domains stacked approximately along the fiber axis and separated by noncrystalline domains. The molecular orientation of the hard segment domains increased with spinning speed and/or spin draw-down ratio and reached a fairly high degree of orientation ($f_c \approx 0.7$) at the highest spinning speeds and draw-down ratios. The long-period measured by SAXS for the PBT/PTMEG-T copolymers was found to be nearly independent of draw-down ratio, but it increased with decrease in hard segment content as shown in Figure 2.59. Since crystallinity increased with hard segment content, this suggested that the increase of long period was due to increased separation of hard segment domains or an increased size of the soft-segment domains as illustrated in Figure 2.60.

2.2.5.2 High Modulus Fibers Prepared from Thermotropic Liquid Crystalline Polymers

Certain polymers can exist in a liquid crystalline state that is intermediate in order between a random amorphous melt and a three dimensionally ordered crystalline phase (see also Chapter 7). In one common type of liquid crystal phase, referred to as nematic, relatively stiff molecules pack together, to form domains in which the molecules are aligned side-by-side, but they are not otherwise ordered. Under quiescent conditions the different domains have different orientation with respect to sample reference axes, so that the overall sample properties are isotropic. However, under flow conditions the domains can rotate and the molecules in different domains tend to become aligned, creating a global preferred orientation of the molecules. High levels of orientation are more readily achieved in such cases than in the case of amorphous, entangled melts. Further, it has been found that orientation decay exhibits long relaxation times [240]. Thus, orientation, once developed, can be trapped in the material fairly readily.

There are two types of liquid crystal behavior, lyotropic and thermotropic. Lyotropic liquid crystal polymers exhibit liquid crystal behavior when dissolved at certain concentrations in a solvent. The thermotropic liquid crystal polymers exhibit liquid crystal behavior in a certain temperature range without the need for any solvent. The lyotropic materials (e.g., poly(p-phenylene terephthalamide)) are noted for their use in preparing ultrahigh modulus fibers through solution spinning from liquid crystalline solutions [241]. These materials are generally very stiff chain polymers that degrade before they melt (or flow) in the absence of a solvent; for this reason they cannot be melt spun. However, solution spinning is generally more complicated than melt spinning, due to solvent handling problems, and several investigators have examined the possibility of obtaining ultrahigh modulus fibers through the use of polymers that exhibit thermotropic liquid crystal phases at high temperature. Early in the development of these materials, Ide an Ophir [242] established that a high level of molecular orientation could be readily developed in elongation but not in shear flow.

The molecules developed to have thermotropic liquid crystal character are also rather stiff, but they are generally somewhat less stiff than those used to obtain lyotropic behavior. This characteristic allows them to soften and flow in a thermotropic liquid crystal state at temperatures below those at which they decompose [242]. Many of these materials are copolymers and their compositions are chosen to control the tendency to crystallize so that a liquid crystalline rather than the fully ordered and rigid crystalline phase is obtained in the appropriate temperature range.

One of the first thermotropic liquid crystal polymers disclosed, and one of the most studied, is a copolyester of PET with 60 mol% p-hydroxybenzoic acid, referred to as PET/60PHB. The early study of Sugiyama, et al. [243] showed that extrudates and fibers spun from this material were highly fibrillated. Their data suggested that moderately high levels of orientation could be obtained by extrusion through a capillary die with high L/D ratio; high spindraw ratios had marginal ability to increase the orientation further. However, these authors did not measure mechanical properties of the fibers, and later investigators [244–249] established the need for melt drawdown in order to optimize mechanical properties. The required spindraw ratio decreased with increasing L/D of the die. The orientation and modulus were found to be independent of the extrusion rate and spinline cooling conditions [249]; this was presumably as a result of the long relaxation times of the liquid crystal polymer.

Some investigators found the modulus to be a function of extrusion temperature, though there was controversy over whether the modulus increased, decreased or was unaffected by extrusion temperature. Acierno et al. [244] found that modulus decreased with increase of extrusion temperature, while Cuculo and Chen [247] and Muramatsu and Krigbaum [245, 246] found that modulus increased with increase of extrusion temperature. In a later paper the latter authors [250] and Lee et al. [249] found that the orientation and modulus were essentially independent of extrusion temperature. These discrepancies were largely explained [248,250] in terms of the amount of PHB crystallinity retained in the polymer at the extrusion temperature due to the relative value of the extrusion temperature compared to the melting point. The results became essentially independent of extrusion temperature when all crystallinity was destroyed and extrusion was carried out from the nematic melt. The relative independence of orientation and modulus on factors other than drawdown ratio is illustrated in Figures 2.61 and 2.62, respectively.

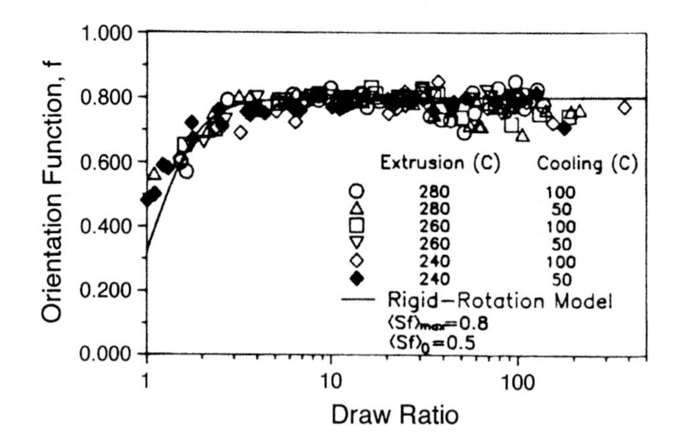

Figure 2.61 Molecular orientation parameter for PET/PHB60 filaments as a function of draw-down ratio and various other processing conditions. (Data of reference [249]).

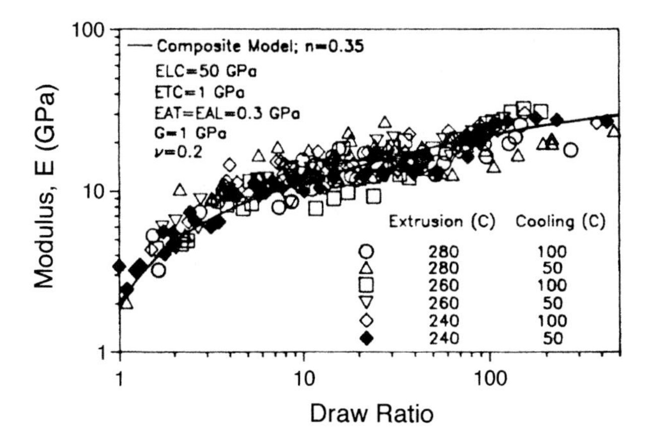

Figure 2.62 Young's modulus of PET/PHB60 filaments as a function of draw-down ratio and various other processing conditions. (Data of reference [249]).

The data in Figures 2.61 and 2.62 also show that the orientation plateaus at the relatively low draw-down ratio of about 3, while the modulus continues to increase to draw-down ratios of at least several hundred. The rapid development of the orientation was explained to be a result of the liquid crystalline nature and the rapid rotation and alignment of the domains with the fiber axis [245–250]. According to Lee et al. [249], the more gradual increase in the modulus with drawdown ratio can be attributed to the deformation of both the nematic domains and the "grain boundary" phase between the domains. The modulus increase at low drawdown ratios is presumably a result of the orientation of the nematic domains. Further increase in drawdown ratio does not have appreciable effect on the measured orientation, but it increases the domain aspect ratio in the fiber direction. According to composite theory this should result in an increase of fiber modulus, assuming that the domains themselves have very high longitudinal moduli. The polymer chains in the "grain boundary" phase also undergo extension at high drawdown ratios, which further contributes to the increase in fiber modulus.

The fiber-forming behavior of several other thermotropic liquid crystal polymers has also been studied. Among these are the random copolyester of p-hydroxybenzoic acid and 2-hydroxy-6-napthoic acid (HNA) containing 42 mol% of the HNA units [251], copolyesters prepared by acidolysis reactions from p-acetoxybenzoic acid, p,p-biphenol bis-acetate, terephthalic acid and isothalic acid [252], and a polymer consisting of five aromatic components: HBA, HQ (hydroxyquinone), TA (terephthalic acid), HBP (4-4′ dihydroxybiphenyl), and IA (isothalic acid) [249]. Generally, the processing behavior for all these materials is rather similar to that described above for PET/60PHB, but the ultimate modulus values obtained seem to be controlled by the chain stiffness. Values of tensile modulus as high as 100 GPa were obtained for wholly aromatic copolyesters at drawdown ratios above 500 [252].

Kenig [253] attempted to model the orientation behavior of fibers produced from thermotropic liquid crystalline polymers based on the assumption that the liquid crystal domains rotate in the flow field in a manner analogous to short fiber suspensions in a molten polymer matrix. Lee et al. [249] expanded this idea and combined it with the Halpin-Tsai equations in order to describe the development of the modulus. The solid lines in Figs. 2.61 and 2.62 show the relatively good fit of this approach to the experimental data for PET/60 PHB. It must be noted, however, that there are a number of parameters involved in such a fit, including the mechanical properties of the individual domains and those of the "grain boundary" phase.

2.2.6 Polymer Blends and Bicomponent Fibers

2.2.6.1 Basic Concepts

Polymer blends are mixtures of two or more polymers. Since the number of possible combinations is quite large, no attempt will be made here to discuss all the blends that have been studied. Rather, a few selected cases will be described to illustrate the important features of melt spun fibers produced from blends. A good review of polyblend fibers was published earlier by Hersh [254]. "Bicomponent fibers" are a special class of polyblend fibers that are prepared by special spinning techniques as discussed below. An early review of the basics of this technology was published by Jeffries [255].

The purpose of blending polymers is to achieve improved processing or properties for some specific application. In general, properties can be achieved by blending that cannot be obtained from homopolymers.

Polymer blends can be either "compatible" or "incompatible." In the case of compatible blends the result is a relatively homogeneous, single-phase "solution" of the polymer molecules in each other, at least in some temperature range. Incompatible blends are heterogeneous mixtures of two or more phases; i.e., the polymers generally remain as separate phases in the blend, though some partial solution may occur. Most blends of two or more polymers are incompatible. Compatible blends are relatively rare, though a number of examples have been cited [256]; these generally occur as a result of some specific interaction, such as hydrogen bonding, between the polymer molecules.

Fibers can be melt spun from either compatible or incompatible blends. The spinning of compatible blends is generally similar to that of homopolymers, though the processing and properties of the blend will normally be different from that of either homopolymer in the blend. The spinning of incompatible blends provides both additional opportunities and challenges compared to the spinning of homopolymers. In particular, control over the morphology of the phases present in the blend is of utmost importance, as it can lead to many new and exciting properties of the blend fiber. Examples of the potential morphologies are illustrated in Figure 2.63. In the case of melt spun fibers, control of the phase morphology is most often achieved by control of the way in which the two polymers are brought together. In one case the polymers are combined to form an intimate mixture of the phases in the melt. One of the blend's phases will be continuous while the other phase(s) will be distributed as tiny particles in the continuous phase. This mixture is then extruded through the spinneret in a manner similar to that of spinning a

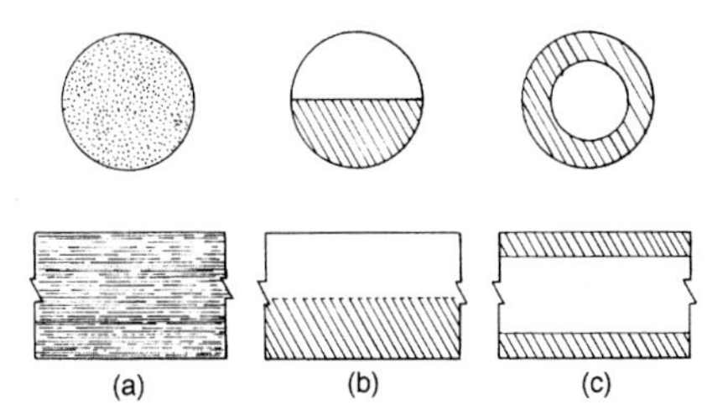

Figure 2.63 Schematic illustration of longitudinal and cross-sectional views of (a) biconstituent matrix/fibril polyblend fibers, (b) side-by-side bicomponent fibers, and (c) sheath-core bicomponent fibers.

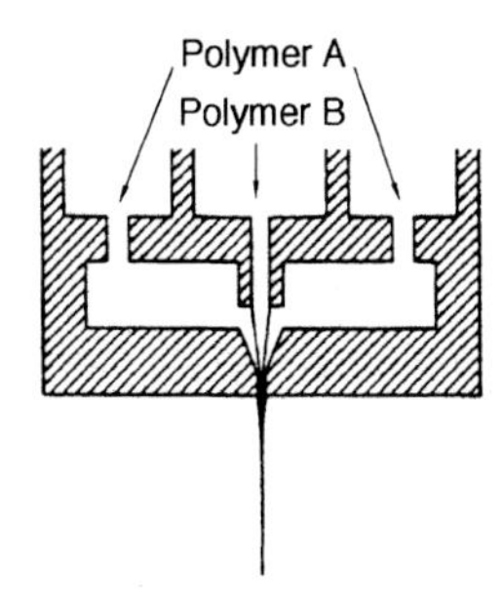

Figure 2.64 Illustration of a typical spinneret used for preparing sheath-core bicomponent filaments.

homopolymer. Due to the elongation of the filament during the spinning process, this produces a final filament consisting of many microfilaments in a matrix of the continuous phase, referred to as matrix-fibril (M/F) morphology, Figure 2.63 (a). Fibers produced from such intimately mixed phases have also been referred to as "biconstituent fibers" in the literature [239,240] to distinguish them from "bicomponent fibers." The latter term is usually used to describe fibers that are composed of two polymers but have either a side by side (S/S) or a sheath-core (S/C) structure as illustrated in Figure 2.63 (b) and (c). Numerous variants of the basic morphologies illustrated in Figure 2.63 exist. These include cases where (1) the phases of bicomponent fibers are very asymmetric in quantity or location, (2) there are three or more sectors of the bicomponent filaments, and (3) the interface between the sectors is intermingled or diffused in an effort to improve the bonding between the phases [257].

The preparation of S/S or S/C bicomponent fibers is based on careful control of the flow of the individual component streams from separate extruders into specially designed spinnerets which combine the two streams to form the bicomponent extrudate. An example of such a spinneret system for producing S/C filaments is illustrated in Figure 2.64.

2.2.6.2 Control of Phase Morphology

The high viscosity of polymers helps substantially to reduce the tendency for polymers to intermix inadvertently. However, this same effect requires substantial input of work to produce a fine distribution of one phase in another. Further, since the amount of interface increases as the particle size decreases and reduction of interfacial energy serves as a driving force to cause coalescence and growth of the particles, it is best to avoid excessive heating after the fine distribution of particles has been formed. In addition, the morphology of biconstituent fibers depends on the relative amounts of each phase and the relative viscosity of the two polymers at the extrusion temperature. This is illustrated by the data of Tsuji [258] for blends of nylon 6 with polystyrene. It was found that for nylon 6/polystyrene pairs having the same viscosity ($\eta_{pst}/\eta_N = 1$) the continuous phase tended to be the polymer which was present in greatest proportion. But when the viscosity of one polymer was greater than the other, the continuous phase tended to be the polymer with the lowest viscosity even when the high viscosity polymer was present in substantially greater amounts–up to about 80% of the total. Likewise, when two polymers are flowing together in a capillary or pipe, the lower viscosity polymer tends to encapsulate the higher viscosity one due to the fact that the low viscosity polymer is more easily deformed in the high shear regions near the wall [259–261]. This means that, when making S/C bicomponent fibers, the sheath will normally be the lower viscosity polymer and the core will be the higher viscosity.

Some control over the final diameter of the fibrils in a M/F type biconstituent fiber can be achieved through control of the particle size of the discontinuous phase entering the spinneret, the amount of drawdown during the spinning process, and through post-drawing the as-spun filaments. Papero et al. [262] studied the preparation and properties of polyamide 6/PET biconstituent fibers. Fibrils of PET formed in the polyamide 6 matrix at concentrations up to and including 50% PET. At 75% PET the morphology consisted of polyamide fibrils in a PET matrix. Fibril cross section dimensions ranged from 0.06 to ~ 2 μm, with the larger dimensions occurring near the 50/50 composition where the fibril cross sections were rather irregular. Away from this composition the cross sections tended to be circular in shape with diameters averaging a few tenths of a μm. The length of the fibrils was of order 100–200 μm.

In addition to the distribution and shape of the components (phases) in polyblend fibers, the properties of the fibers depend on the structure of the individual components (phases). Kikutani et al. [263] have recently shown that the orientation and crystallinity of each component of bicomponent melt spun filaments can be very different than that of homopolymer filaments spun under similar conditions. The PET core of PP/PET sheath/core bicomponent filaments prepared by high speed spinning was found to have enhanced orientation and crystallinity development as compared to that when the same PET was spun alone under the same conditions. Unlike fibers spun from 100% PP, the PP sheath of the bicomponent filaments developed much lower orientation and remained in a pseudo-hexagonal (smectic) form even at high take-up speeds. They used a numerical simulation, similar to that discussed in Section 2.1.2.6, to help analyze their experimental results. The simulation suggested that the solidification stress in the PET increased while that in the PP component decreased as a result of differences in the activation energy of elongational viscosity between the two polymers. They also noted that if PET crystallizes first, then the stress in the still molten PP could relax since the spinline is no longer deforming. The entire stress can be supported by the PET phase in the part of the spinline where the structure of the PP phase is being developed. These results clearly show that the presence and properties of the other component that is present will influence the structure of both components in bicomponent fibers.

A similar argument could be made that the molecular orientation and crystallinity of the individual components in biconstituent fibers would depend on the relative viscoelastic properties of the two components in the melt and on their relative crystallization (solidification) temperatures. For example, it has been shown [264] that for blends of liquid crystal polymers (LCPs) with the flexible chain PBT, the molecular orientation in the PBT phase increases relative to pure PBT as the content of the LCP phase increases. Further, the molecular orientation of the LCP phase decreases as the content of PBT increases. However, a detailed analysis of these effects was not attempted.

Several authors [e.g., 264–268] have studied M/F fibers formed from blends of liquid crystal polymers with flexible chain polymers in the hope of getting high modulus due to reinforcement of the flexible chain polymer by highly oriented, high modulus, fibrillar domains of LCP. Recent advances in this area are reported by Krishnaswamy and Baird in Section 11.5.

2.2.6.3 Applications of Polyblend Fibers

Hersh [254] has given the following list of actual and potential uses of polyblend fibers:

1. Increased dyeability
2. Decreased static charge
3. Decreased flatspotting of nylon tire yarn
4. Decreased flammability
5. Improved hand and luster (more silklike fibers)
6. Increased water absorption
7. Preparation of artificial leather and suede
8. Increased bulk and covering power
9. Increased light and heat stability
10. Increased whiteness
11. Reduced soiling
12. Reduced pilling
13. Preparation of ultrafine fibers for papermaking
14. Modifying shrinkage characteristics
15. Preparing woollike fibers
16. Preparing fibers for ion-exchange and separation applications
17. Improved transparency
18. Improved strength and elastic recovery
19. Preparing absorptive fibers for protective clothing
20. Improved crazing and oil resistance of fibers

To this list we would add:

21. Preparation of thermal bonding fibers for non-wovens
22. Improved modulus

For a detailed discussion of many of the above applications, the reader is referred to Hersh [254] and other available references, including a substantial patent literature. We will simply note a few examples.

One of the earliest applications of bicomponent fibers was to obtain natural crimp in the fiber or yarn similar to that of wool. This increases the bulk, cover and/or stretch of the yarn [255,269,270]. In order to produce crimp, the two components must be present in the S/S or in an eccentric S/C configuration. The development of crimp depends on differential shrinkage of the two components when the fiber is relaxed from an oriented condition [255,269,270]. Essentially any variable that will lead to such differential shrinkage can be used as a basis to develop crimped fibers. Thus, not only differences in polymer composition between the two components, but other differences such as differences in molecular weight, structural differences such as crystallinity, etc. can produce the desired crimping [270–272]. In order to obtain crimp, the two components must adhere to one another. Generally, adherence increases as the chemical difference between the components decreases, but this normally leads to a decrease of the differential shrinkage required to crimp the fiber. Thus, a compromise must be found which will provide both adherence and differential shrinkage.

Fiber for thermal bonded non-wovens has been produced as either S/C or S/S bicomponent. The two components have different melting points. The higher melting component serves to hold the fiber shape and properties while the lower melting component melts as the non-woven web passes through the thermal bonding calender, thus bonding the web into a non-woven fabric [273,274].

The M/F type polyblend or biconstituent fiber has also found several applications. Fine fibers for preparing artificial leather and papers can be produced by dissolving the matrix of M/F type fibers leaving tiny fibers whose dimensions are similar to those required for these applications [262]. Early attempts to make artificial leather or suede used blends of either PET or polyamide 6 with polystyrene [275–277]. The continuous polystyrene phase was dissolved away leaving PET or polyamide 6 fibers with diameters of order 0.3–1 μm. Permanently antistatic carpet fibers have been produced by incorporating a conductive polymer as fibrils into a polyamide matrix which provides the other properties required for carpets [278,279]. This morphology has also been used to produce tire cord yarns that substantially reduce the tire flatspotting that occurs for polyamide tire cord [262]. This was accomplished by blending about 30% PET with the polyamide 6. This increased the modulus and lowered the moisture regain of the yarns, which led to a decrease of tire flatspotting. Other approaches have also been used to produce nonflatspotting tire yarns [254].

From the examples cited above, it is clear that the number of applications of polyblend fibers is quite large and is continuing to grow.

2.3 Concluding Remarks

Melt spinning is by far the most common method of producing synthetic fibers. It is also a very versatile method that can prepare fibers in many forms for innumerable applications.

It is clear that the properties of melt spun fibers are controlled by the conditions of spinning as well as the composition, rheological properties and crystallization kinetics of the resin being spun. Over a period of approximately 50 years, the art and science of the melt spinning process has been advanced to a high level. We now have a reasonable understanding of the influence of the many process and materials variables on the structure and properties of melt spun filaments. However, because of the great versatility of the process and continued development of new and better resins and new applications, there is likely to be continued work in this field for many years to come.

Acknowledgments

The author wishes to thank the many individuals who contributed to this chapter by supplying information, figures, and suggestions. He is especially indebted to his graduate students who, over the years, have carried out much of the research which stimulated the author to write this chapter. Finally, he wishes to thank Pitt Supaphol for preparing many of the figures.

2.4 References

1. Brooman, R.A. English Patent, 10,582 (1845)
2. Carothers, W.H. and Hill, J.W. *J. Amer. Chem. Soc.* 1932, 54, p. 1579
3. *Collected Papers of W. H. Carothers on High Polymeric Substances*, Mark, H. and Whitby, G.S., eds., Interscience, New York, 1940
4. Ziabicki, A. and Kedzierska, K., *J. Appl. Polym. Sci.*, 1959, 2, p. 14
5. Ziabicki, A. and Kedzierska, K., *Kolloid-Z.*, 1960, 171, p. 51
6. Ziabicki, A., *Kolloid-Z.*, 1961, 175, p. 134
7. Ziabicki, A. and Kedzierska, K., *J. Appl. Polym. Sci.*, 1962, 6, p. 111
8. Ziabicki, A. and Kedzierska, K., *J. Appl. Polym. Sci.*, 1962, 6, p. 361
9. Ziabicki, A., *Fundamentals of Fiber Formation*, Wiley, London, 1976
10. Vassilatos, G., Knox, B.H. and Frankfort, H.R.E., in *High Speed Fiber Spinning–Science and Engineering Aspects*, Ziabicki, A. and Kawai, H., eds., Wiley: New York, 1985, p. 383
11. Kurita, K. and Teramoto, Y., US Patent No. 4,923,662 assigned to Toyobo Co.
12. Cuculo J., US Patent No. 5,171,504 assigned to North Carolina State University, Rayleigh, NC
13. Wang, H-H., Hu, Y-J., *Textile Res. J.*, 1997, 67(6), p. 428
14. Thompson, A. B., in *Fibre Structure*, Hearle, J. W. S. and Peters, R. H., eds., Butterworth, London, 1953, p. 480
15. Higuchi, K. and Katsu, T. *Sen-i Gakkaishi*, 1960, 13, p. 1
16. Sakiadis, B. C., *AIChE J.*, 1961, 7, p. 467
17. Fukuda, K. *Sen-i Gakkaishi*, 1966, 22, p. S-3
18. Sano, Y. and Orii, K., *Sen-i Gakkaishi*, 1968, 24, p. 212
19. Matsui, M., *Trans. Soc. Rheol.*, 1976, 20, 465

20. Gould, J. and Smith, F. S., *J. Text. Inst.*, 1980, 71, p. 38
21. Shimizu, J., Okui, N. and Tamai, K., *Sen-i Gakkaishi*, 1983, 39, p. T-398
22. Shimizu, J., Okui, N. and Kikutani, T., in *High Speed Fiber Spinning–Science and Engineering Aspects*, Ziabicki, A. and Kawai, H., eds., Wiley: New York, 1985, p. 173
23. Kase, S. and Matsuo, T., *J. Appl. Polym. Sci.*, 1967, 11, p. 251
24. Ziabicki, A., in *High Speed Fiber Spinning–Science and Engineering Aspects*, Ziabicki, A. and Kawai, H., eds., Wiley: New York, 1985, p. 23
25. Kase, S. and Matsuo, T., *J. Polym. Sci.*, A, 1965, 3, p. 2541
26. Kase, S., in *High Speed Fiber Spinning–Science and Engineering Aspects*, Ziabicki, A. and Kawai, H., eds., Wiley: New York, 1985, p. 67
27. Zeminski, K. F. and Spruiell, J. E., *J. Appl. Polym. Sci.*, 1988, 35, p. 2223
28. Denn, M. M., Petrie, C. J. S. and Avenas, P., *AIChE J.*, 1975, 21, p. 791
29. Fisher, R. J. and Denn, M. M., *AIChE J.*, 1977, 23, p. 23
30. Gagon, D. K. and Denn, M. M., *Polym. Engr. Sci.*, 1981, 21, p. 844
31. Matsui, M. and Bogue, D. C., *Polym. Engr. Sci.*, 1976, 16(11), p. 735
32. Papanastasiou, T. S., Alaie, S. M. and Chen, Z., *Intern. Polymer Processing*, 1994, 9, p. 148
33. Bheda, J. H. and Spruiell, J. E., *J. Appl. Polym. Sci.*, 1990, 39, p. 447
34. Patel, R. M., Bheda, J. H. and Spruiell, J. E., *J. Appl. Polym. Sci.*, 1991, 42, p. 1671
35. Cross, M. M., in *Polymer Systems: Deformation and Flow*, Wetton and Whorlow, eds., McMillan, London, 1968
36. Carreau, P. J., Ph. D. Dissertation, Univ. of Wisconsin, 1979
37. Ding, Z., Ph. D. Dissertation, Univ. of Tennessee, 1996
38. Williams, M. L., Landel, R. F. and Ferry, J. D., *J. Am. Chem. Soc.*, 1955, 77, p. 3701
39. Ziabicki, A., *J. Non-Newtonian Fluid Mech.,* 1988, 30, 157
40. Pogodina, N. V. and Winter, H. H., *Macromolecules*, 1998, 31, p. 8164
41. Yasuda, H., Sugiyama, H. and Hayashi, S., *Sen-i Gakkaishi*, 1979, 35, p. 370
42. George, H. H., *Polym. Engr. Sci.*, 1982, 22, 292
43. Dutta, A. and Nadkarni, V. M., *Text. Res. J.*, 1984, 54, 35. See also *Polym. Engr. Sci.*, 1987, 27, p. 1050
44. Katayama, K. and Yoon, M. in *High Speed Fiber Spinning–Science and Engineering Aspects*, Ziabicki, A. and Kawai, H., eds., Wiley: New York, 1985, p. 207
45. Kase, S. and Matsuo, T., *J. Polym. Sci.*, 1965, A3, p. 2541
46. Moutsoglou, A., *Trans. ASME, J. Heat Transfer*, 1983, 105, p. 830
47. Bos, C., *Int. J. Heat Mass Transfer*, 1988, 31, p. 167
48. Barovskii, V. R., Klochko, N. F., Pankeev, A. M. and Polovets, L. M., *Khim. Volokna*, 1974, 3, p. 23
49. Yasuda, H., Sugiyama, H. and Hayashi, S., Sen-i-Gakkaishi, 1984, 40, p. T-227
50. Ishihara, H., Hayashi, S. and Ikeuchi, H., Intern. Polymer Processing, 1989, 4, p. 91
51. Dutta, A., *Textile Res. J.*, 1987, 57, p. 13
52. Schoene, A. and Bruenig, H., *Arch. Mech.*, 1990, 42, p. 571
53. Maxwell, J. C., *Proc. Roy. Soc.*, 1873, 22, p. 46
54. Oda, K., White, J. L. and Clark, E. S., *Polym. Engr. Sci.*, 1978, 18, p. 53
55. Janeschitz-Kriegl, H., *Polymer Melt Rheology and Flow Birefringence*, Springer-Verlag, New York, 1983
56. Van Krevelen, D. W., *Properties of Polymers*, Elsevier, New York, 1976
57. Kuhn, W. and Grun, F., *Kolloid Z.*, 1943, 101, p. 248
58. Treloar, L. R. G., *Physics of Rubber Elasticity*, 2nd Ed., Dover, New York, 1958
59. Hermans, P. H. and Platzek, P., *Kolloid Z.*, 1939, 88, p. 68.
60. Hermans, P. H., Hermans, J. J., Vermaas, D. and Weidinger, A., *J. Polym. Sci.*, 1947, 3, p. 1
61. White, J. L. and Spruiell, J. E., *Polym. Engr. Sci.*, 1981, 21, p. 859
62. Stein, R. S. and Norris, F. H., *J. Polym. Sci.*, 1956, 21, p. 381
63. Stein, R. S., *J. Polm. Sci.*, 1958, 31, 327
64. Spruiell, J. E. and White, J. L., *Polym. Engr. Sci.*, 1975, 15, p. 660
65. Dees, J. R. and Spruiell, J. E., *J. Appl. Polym. Sci.*, 1974, 18, p. 1053
66. Weeks, J. J., *J. Res. Nat. Bur. Stand. A. Phys. Chem.*, 1963, 67A,p. 441
67. Mandelkern, L. in *Growth and Perfection of Crystals*, Doremus, R. H.; Roberts, B. W. and Turnbull, D., eds, Wiley: New York, 1958, p. 478

68. Haas, T. W. and Maxwell, B., *Polym. Eng. Sci.*, 1969, 9, p. 225
69. Ozawa, T., *Polymer*, 1971, 12, p. 150
70. Kobayashi, K. and Nagasawa, T., *J. Macromol. Sci.*, (B), 1970, 4, p. 331
71. Nakamura, K., Watanabe,T., Katayama, K., and Amano, T.,. *J Appl. Polym. Sci.*, 1972, 16, p. 1077
72. Nakamura, K., Katayama, K. and Amano, T., *J. Appl. Polym. Sci.*, 1973, 17, p. 1031
73. Dietz, W., *Colloid and Polym. Sci.*, 1981, 259, p. 413
74. Kamal, M. R. and Chu, E., *Polym. Engr. Sci.*, 1983, 23, p. 27
75. Velisaris, C., and Seferis, J., *Polym. Engr. Sci.*, 1986, 26, p. 1574
76. Cebe, P., *Polym. Engr. Sci.*, 1988, 28, p. 1192
77. Patel, R. M. and Spruiell, J. E., *Polym. Engr. Sci.*, 1991, 31, p. 730
78. Avrami, M., *J. Chem. Phys.*, 1939, 7, p. 1103
79. Wasiak, A. and Ziabicki, A., *Appl. Polym. Symp.*, 1975, 27, p. 111
80. Hoffman, J. D. and Lauritzen, J. I., *J. Res. Natl. Bur. Std.*, A, 1961, 65, p. 297
81. Hoffman, J. D., Davis, G. T., Lauritzen, J. I., in *Treatise on Solid State Chemistry: Crystalline and Non-crystalline Solids*, Hannay, J. B., ed., Plenum, New York, 1976, vol. 3, Chap. 7
82. Fischer, J. and Turnbull, D., *J. Chem. Phys.*, 1949, 17, p. 71
83. Hieber, C. A., *Polymer*, 1995, 36, p. 1455
84. Sajkiewicz, P., *Polymer*, 1999, 40, p. 1433
85. Ziabicki, A., *Colloid and Polymer Sci.*, 1996, 274, p. 209
86. Ziabicki, A., *Colloid and Polymer Sci.*, 1996, 274, p. 705
87. Matsuo, T. and Kase, S., *Sen-i Gakkaishi*, 1968, 24. 512. Also Kase, S., *J. Appl. Polym. Sci.*, 1974, 18, p. 3267
88. Ziegler, K.; Holzkamp, E.; Breil, H. and Martin, H. *Angew. Chem.* 1955, 67, p. 426
89. Natta, G.; Pino, P.; Corradini, P.; Danusso, F.; Mantica, E.; and Mazzanti, G. and Moranglio, G. *J. Am. Chem. Soc.* 1955, 77, p. 708-1710
90. Keller, A. *Phil. Mag.* 1957, 2(8), p. 1171-1175
91. Till, Jr., P. H. J. *Polym. Sci.* 1957, 24, p. 301-306
92. Keller, A. and Machin, M. J. *J. Macromol Sci.-Phys.*, 1967 B1 (1), p. 41
93. Hill, M. J. and Keller, A. *J Macromol Sci.-Phys.* 1969, 133, p. 153
94. Pennings, A. J., van der Mark,; J. M. Boij, and H. C. *Kolloid-Z.*, 1970, 236, p. 99
95. Pennings, A. J. and Pijpers, M. F. J., *Macromolecules*, 1970, 3, p. 261
96. Pennings, A. J., van der Mark, J.M.A.A. and Keil, A. M. *Kolloid-Z. Z. Polym.*, 1970, 237, p. 336
97. Katayama, K., Amano, T. and Nakamua, *K. Kolloid-Z. Z. Polym.*, 1968, 226, p. 125
98. Fung, P. Y. F. and Carr, S. H., *J. Macromol. Sci.-Phys.*, 1972, B6, p. 621
99. Abbott, L. E. and White, J. L., *Appl. Polym. Symp.*, 1973, 20, p. 247
100. White, J. L., Dharod, K. C., and Clark, E. S., *J. Appl. Polym. Sci.*, 1974, 18, p. 2539
101. Nadkarni, V. M., and Schultz, J. M., *J. Polym. Sci.-Phys.*, 1977, 15, p. 2151
102. Capuccio, V., Loen, A., Bertinotti, F. and Lonti, W., *Chim. End. (Milan)*, 1962, 44, p. 463
103. Compostella, V., Coen, A. and Bertinotti, F., *Angew, Chem.*, 1962, 74, p. 618
104. Sheehan, W. C. and Cole, T. B., *J. Appl. Polym. Sci.*, 1964, 8, p. 2359
105. Natta, G., Peraldo, M. and Corradini, P., *Atti. Accad. Nazl. Lincei., Rend.*, 1959, 26, p. 14
106. Wyckoff, H. W., *J. Polym. Sci.*, 1962, 62, p. 83
107. Nadella, H. P., Henson, H. M., Spruiell, J. E. and White, J. L., *J. Appl. Polym. Sci.*, 1977, 21, p. 3003
108. Awaya, H. and Zasshi, N. K., *Nippon Kagaku Zasshi*, 1961, 82, p. 1575
109. Andersen, P. G. and Carr, S. H., *J. Mater. Sci.*, 1975, 10, p. 870
110. Clark, E. S. and Spruiell, J. E., *Polym. Eng. Sci.*, 1976, 16, p. 176
111. Lotz, B. and Wittmann, J. L., *J. Polym. Sci. B. Phys.*, 1986, 24, p. 1541
112. Khoury, F., *J. Res. Natl. Bur. Stand.*, 1966, 70A, p. 29
113. Padden, H. D. and Keith, H. D., *J. Appl. Phys.*, 1959, 30, p. 1479
114. Binsbergen, F. L. and de Lange, B. G. M., *Polymer*, 1968, 9, p. 23
115. Herman, A. J., U. S. Patent No. 3,256,258, assigned to E. I. duPont de Nemours and Company, 1966
116. Clark, E. S., in *Structure and Properties of Polymer Films*, Lenz, R. W. and Stein, R. S., eds., Plenum.: New York, 1973, p. 267
117. Quyn, R. G. and Brody, H. *J. Macromol Sci. -Phys.*, 1971, B5(4), p. 721

118. Noether, H. D. and Whitney, W., *Kolloid Z. Z. Polym.*, 1973, 251, p. 991
119. Sprague, B. S., *J. Macromol. Sci.-Phys.*, 1973, B8, p. 157
120. *Fiber Producer*, 1982, 10(5), p. 19
121. Steinkamp, R. A., Grail, T. J., U. S. Patent 3,862,265, assigned to Exxon Research and Engineering,1975
122. Shimizu, J., Toriumi, K. and Tamai, K., *Sen-i Gakkaishi*, 1977, 33, p. T-208
123. Shimizu, J., Okui, N. and Imai, Y., *Sen-i Gakkaishi*, 1979, 35, p. T-405
124. Shimizu, J., Okui, N and Kikutani, T., in *High Speed Fiber Spinning*, Ziabicki, A. and Kawai, H., eds. Wiley, New York, 1985, p. 429
125. Kloos, F. in *Proceedings of the 4th International Conference, Polypropylene Fibers and Textiles*, Nottingham, U.K., 1987, p. 6/1
126. Fan, Q., Xu, D., Zhao, D. and Qian, R., *J. Polym. Engr.*, 1985, 5(2), p. 95
127. Lu, F. M. and Spruiell, J. E., *J. Appl. Polym. Sci.*, 1987, 34, p. 1521
128. Lu, F. M. and Spruiell, J. E., *J. Appl. Polym. Sci.*, 1987, 34, p. 1541
129. Lu, F. M. and Spruiell, J. E., *J. Appl. Polym. Sci.*, 1993, 49, p. 623
130. Misra, S., Lu, F.M., Spruiell, J. E. and Richeson, G. C., *J. Appl. Polym. Sci.*, 1995, 56, p. 1761
131. Spruiell, J. E., Lu, F. M., Ding, Z. and Richeson, G. C., *J. Appl. Polym. Sci.*, 1996, 62, p. 1965
132. Wu, Z., *International Fiber Journal*, 1988, p. 85
133. Lin, Y., Zhou, J., Spruiell, J. E. and Stahl, G. A., *Proceedings, Soc. Plastics Engineers Annual Technical Conference*, 1991, 37, p. 1950
134. Lin, Y., M. S. Thesis, The University of Tennessee, 1991
135. Kozulla, R.E., U.S. Patent No. 5,318,735 assigned to Hercules, Inc., 1994
136. Ludewig, H., *Polyester Fibers -Chemistry and Technology*, Wiley-Interscience, New York, 1971
137. Liska, E., *Kolloid-Z.*, 1973, 251, p. 1028
138. Shimizu, J. Toriumi, K. and Tamai, K., *Sen-i Gakkaishi*, 1977, 33, p. T-208
139. Heuvel, H. M. and Huisman, R., *J. Appl. Polym. Sci.*, 1978, 22, p. 2229
140. Shimizu, J., Okui, N., Keneko, A. and Toriumi, K., *Sen-i Gakkaishi*, 1978, 34, p. T-64
141. Shimizu, J., Okui, N., Kikutani, T. and Toriumi, K., *Sen-i Gakkaishi*, 1978, 34, p. T-93
142. Shimizu, J., Okui, N., Kikutani, T. and Takaku, A., *Sen-i Gakkaishi*, 1981, 37, p. 29
143. Heuvel, H. M. and Huisman, R., in *High Speed Fiber Spinning–Science and Engineering Aspects*, Ziabicki, A. and Kawai, H., eds., Wiley: New York, 1985, p. 295
144. Perez, G., in *High Speed Fiber Spinning–Science and Engineering Aspects*, Ziabicki, A. and Kawai, H., eds., Wiley: New York, 1985, p.333
145. Yasuda, H., in *High Speed Fiber Spinning–Science and Engineering Aspects*, Ziabicki, A. and Kawai, H., eds., Wiley: New York, 1985, p. 363
146. George, H., in *High Speed Fiber Spinning–Science and Engineering Aspects*, Ziabicki, A. and Kawai, H., eds., Wiley: New York, 1985, p. 271
147. Matsui, M., in *High Speed Fiber Spinning–Science and Engineering Aspects*, Ziabicki, A. and Kawai, H., eds., Wiley: New York, 1985, p. 137
148. Smith, F. S., and Steward, R. D., *Polymer*, 1974, 15, p. 283
149. Alfonso, G. C., Verdona, M. P. and Wasiak, A., *Polymer*, 1978, 19, p. 283
150. Ziabicki, A. and Jarecki, L., in *High Speed Fiber Spinning–Science and Engineering Aspects*, Ziabicki, A. and Kawai, H., eds., Wiley: New York, 1985, p.225
151. Clauss, B. and Salem, D. R., *Polymer*, 1992, 33, p. 3193
152. Shimizu, J., Kikutani, T., Takaku, A. and Okui, N., *Sen-i Gakkaishi*, 1984, 40, p. T-177
153. Daubeny, R. P., Bunn, C. W., and Brown, C. J., *Proc. R. Soc. London*, 1954, 226A, p. 531
154. Perez, G. and Lecluse, C., 18th International Man-Made Fibre Conference, Dornbirn, Austria, June 20-22, 1979
155. Bai, C. C., Ph. D. Dissertation, University of Tennessee, Knoxville, 1984
156. Chen, S., Yu, W. and Spruiell, J. E., *J. Appl. Polym. Sci.*, 1987, 34, p. 1477
157. Haberdorn, H., Hahn, K., Breuer, H., Dorrer, H.-D. and Matthies, P., *J. Appl. Polym. Sci.*, 1993, 47, p. 1551
158. Zhou, J. and Spruiell, J. E., in *Nonwovens-An Advanced Tutorial*, TAPPI Press, Atlanta, GA, 1989
159. Vincent, P. I., *Polymer*, 1960, 1, p. 7

160. Ziabicki, A., *J. Non-Newtonian Fluid Mech.*, 1988, 30, 141; Ibid. p. 157
161. Fu, Y., Busing, W. R., Jin, Y., Affholter, K. A. and Wunderlich, B., *Macromolecules*, 1993, 26, p. 2187
162. Fu, Y., Busing, W. R., Jin, Y., Affholter, K. A. and Wunderlich, B., *Macromol. Chem. Phys.*,1994,195, p. 803
163. Fu, Y., Annis, B., Boller, A., Jin, Y. and Wunderlich, B., *J. Polym. Sci., Polym. Phys.*, 1994, B32, p. 2289
164. Shimizu, J., Okui, N. and Kikutani, T., *Sen-i Gakkaishi*, 1981, 37, p. T-135
165. Boye, Jr., C. A. and Overton, J. R., *Bull. Am. Phys. Soc.*, 1974, 19, p. 352
166. Jakeways, R., Ward, I. M., Wilding, M. A., Hall, J. H., Desborough, I. J. and Pass, M. G., J. *Polym. Sci., Polym. Phys.*, 1975, B13, p. 799
167. Mencik, Z., *J. Polym. Sci., Polym. Phys.*, 1975, 13, p. 2173
168. Jakeways, R., Smith, T., Ward, I. M. and Wilding, M. A., *J. Polym. Sci., Polym. Letters*, 1976, 14, p. 41
169. Yokouchi, M., Sakakibara, Y., Chatani, Y., Tadokoro, H., Tanaka, T. and Yoda, K., *Macromolecules*, 1976, 9, p. 266
170. Hall, I. H. and Pass, M. G., *Polymer*, 1976, 17, p. 807
171. Brereton, M. G., Davies, G. R., Jakeways, R., Smith, T. and Ward, I. M., *Polymer*, 1978, 19, p. 17
172. Ward, I. M., Wilding, M. A. and Brody, H., *J. Polym. Sci., Polym. Phys.*, 1976,14, 263
173. Stambaugh, B., Koenig, J. L. and Lando, J. B., *J. Polym. Sci., Polym. Phys.* 1979, 17, p. 1053
174. Tashiro, K., Nakai, Y., Kobayashi, M. and Tadokoro, H., *Macromolecules*, 1980, 13, p. 137
175. Stein, R. S. and Misra, A., *J. Polym. Sci., Polym. Phys.*, 1980, 18, p. 327
176. Rong, S. and Williams, H. L., *J. Appl. Polym. Sci.*, 1985, 30, p. 2575
177. Lu, F. and Spruiell, J. E., *J. Appl. Polym. Sci.*, 1986, 31, p. 1595
178. Chen, S. and Spruiell, J. E., *J. Appl. Polym. Sci.*, 1987, 33, p. 1427
179. Song, S. S., Cakmak, M. and White, J. L., *Intern. Polym. Proc.*, 1991, 6, p. 332
180. Jager, J., Juijn, J. A., Van Den Heuvel, C. J. M. and Huijts, R. A., *J. Appl. Polym. Sci.*, 1995, 57, p. 1429
181. Cakmak, M. and Kim, J. C., *J. Appl. Polym. Sci.*, in press
182. B. Eling, S. Gogolewski and A. J. Pennings, *Polymer*, 1982, 23, p. 1587
183. J. P. Penning, H. Dijkstra and A. J. Pennings, *Polymer*,1993, 34, p. 942
184. L. Fambri, A. Pegoretti, R. Fenner, S. D. Incardona and C. Migliaresi, *Polymer*, 1997, 38, p. 79
185. Mezghani, K. and Spruiell, J. E., *J. Polym. Sci. B. Polym. Phys.*, 1998, 36, p. 1005
186. Gianchandani, Spruiell, J. E. and Clark, E. S., *J. Appl. Polym. Sci.*, 1982, 27, p. 3527
187. Holmes, D. R., Bunn, C. W. and Smith, D. J., *J. Polym. Sci.*, 1955, 17, p. 159
188. Arimoto, H., *J. Polym. Sci.*, A, 1964, 2, p. 2283
189. Arimoto, H., Ishibashi, M., Hirai, M. and Chatani, Y., *J. Polym. Sci.*,A, 1965, 3, p. 317
190. Roldan, L. G. and Kaufman, H. S., *J. Polym. Sci., Polym. Letters*, 1963, B1, p. 603
191. Parker, J. P. and Lindenmyer, P. H., *J. Appl. Polym. Sci.*, 1977, 21, p. 821
192. Illers, H. K., Haberkorn, H. and Sinak, P., *Makromol. Chem.*, 1972, 158, p. 285
193. Ziabicki, A., *Kolloid-Z.*, 1959, 167, p. 132
194. Stepaniak, R. F., Garton, A., Carlsson, D. J. and Wiles, D. M., *J. Polym. Sci. Polym. Phys.*, 1979, 17, p. 987
195. Stepaniak, R. F., Garton, A., Carlsson, D. J. and Wiles, D. M., *J. Appl. Polym. Sci.*, 1979, 23, p. 1747
196. Kinoshita, Y., *Makromol. Chem.*, 1959, 33, p. 1
197. Bradbury, E. M. and Elliott, A., *Polymer*, 1963, 4, p. 47
198. Bradbury, E. M., Brower, I. Elliott, A. and Parry, D. A. D., *Polymer*, 1965, 6, p. 465
199. Vogelsong, D. C., *J. Polym. Sci.*, A, 1963, 1, p. 1055
200. Ota, T., Yoshizaki, O. and Nagai, E., *J. Polym. Sci.*, A, 1964, 2, p. 4865
201. Hamana, I., Matsui, M. and Kato, S., *Meilland Textilber.*, 1969, 4, p. 382
202. Hamana, I., Matsui, M. and Kato, S., *Meilland Textilber.*, 1969, 5, p. 499
203. Ishibashi, T., Aoki, K. and Ishii, T., *J. Appl. Polym. Sci.*, 1970, 14, p. 1597
204. Ishibashi, T. and Ishii, T., *J. Appl. Polym. Sci.*, 1976, 20, p. 335
205. Hirami, M. and Tanimura A., *J. Macromol. Sci.-Phys.*, 1981, B19, p. 205

206. Bankar, V. G., Spruiell, J. E. and White, J. L., *J. Appl. Polym. Sci.*, 1977, 21, p. 2341
207. Shimizu, J., Okui, N., Kikutani, T., Ono, A. and Takaku, A., *Sen-i Gakkaishi*, 1981, 37, p. T-143
208. Heuvel, H. M. and Huisman, R., *J. Appl. Polym. Sci.*, 1978, 26, p. 713
209. Koyama, K., Suryadevara, J. and Spruiell, J. E., *J. Appl. Polym. Sci.*, 1986, 31, p. 2203
210. Salem, D. R., Moore, R. A. F. and Weigmann, H.-D., *J. Polym. Sci., B: Polym. Phys.*, 1987, 25, p. 567
211. *Polymer Handbook*, Brandup, J. and Immergut, E. H., eds., Wiley, 1989, p. VI p. 311
212. Chappel, F. P., Culpin, M. F., Gosden, R. C. and Tranter, T. C., *J. Appl. Chem.*, 1964, 14, p. 12
213. Danford, M. D., Spruiell, J. E. and White, J. L., *J. Appl. Polym. Sci.*, 1978, 22, p. 3351
214. Li, X., M.S. Thesis, University of Tennessee, 1987
215. Lecluse, C., *High Speed Spinning of Polyamide 66* Topical Report, Rhone-Poulenc Fibres, Venissieux (France), 1982
216. Bunn, C. W. and Garner, E. V., *Proc. Roy. Soc. London Ser. A*, 1947, 189, 39
217. Slichter, W. P., *J. Appl. Phys.*,1955, 26, p. 1099
218. Dismore, P. F. and Statton, W. O., in *Small Angle Scattering from Fibrous and Partially Oriented Systems*, Marchessault, R. H., ed., Interscience, New York, 1966, p. 133
219. Kitao, T., Kobayashi, H., Ikegami, S. and Ohya, S., *J. Polym. Sci. Polym Chem. Ed.*, 1973, 11, p. 2633
220. W. H. Carothers, U. S. Patent No. 2,130,948 (1938)
221. Kudo, K., Mochizuki, M., Shunichi, K. Watanabe, M. and Hirami, M., *J. Appl. Polym. Sci.*, 1994, 52, p. 861
222. Masamoto, J., Sasaguri, K., Ohizami, C., Yamaguchi, K. and Kobayashi, H., *J. Appl. Polym. Sci.*, 1970, 14, p. 667
223. Kinoshita, Y., *Makromol. Chem.*, 1959, 33, p. 1
224. Slichter, W. P., *J. Polym. Sci.*, 1959, 36, p. 259
225. Inoue, K. and Hoshino, S., *J. Polym. Sci. Polym. Phys. Ed.*, 1973, 11, p. 1077
226. Slichter, W. P., *J. Polym. Sci.*, 1958, 35, p. 77
227. Mochizuki, M. and Kudo, K., *Sen-i Gakkaishi*, 1991, 47, p. 336
228. Yang, Y., Cakmak, M. and White, J. L., *J. Appl. Polym. Sci.*, 1985, 30, p. 2615
229. Cakmak, M., Teitge, A., Zachmann, H. G. and White, J. L., *J. Polym. Sci.:B. Phys.*, 1993, 31, p. 371
230. Kang, H. J. and White, J. L., *Intern. Polym. Proc.*, 1986, 1, p. 12
231. Yilmaz, F. and Cakmak, M., *Intern. Polym. Proc.*, 1994, 9, p. 141
232. Song, S., White, J. L., Cakmak, M., *Intern. Polym. Proc.*,1989, 4, p. 96
233. Zhen, X. J., Kitao, T., Kimura, Y., Taniguchi, I., *Sen-i Gakkaishi*, 1985, 41, p. T-1
234. Shimizu, J., Kikutani, T., Ookoshi, Y., Takaku, A., *Sen-i Gakkaishi*, 1985, 41, p. 461
235. Shimizu, J., Kikutani, T., Ookoshi, Y., Takaku, A., *Sen-i Gakkaishi*, 1987, 43, p. 7
236. Song, S., White, J. L., Cakmak, M., *Sen i Gakkaishi*, 1989, 45, p. 242
237. Hsiung, C. M. and Cakmak, M., *Polym. Engr. Sci.*, 1991, 31, p. 172
238. Abhiraman, A. S., Kim, Y. W. and Wagener, K. B., *J. Polym. Sci. B: Polym. Phys.*, 1987, 25, p. 205
239. Richeson, G. C. and Spruiell, J. E., *J. Appl. Polym. Sci.*, 1990, 41, p. 845
240. Suto, S., Shimamura, K., White, J. L. and Fellers, J. F., *Proceedings of Society of Plastics Engineers Annual Technical Conference*, 1982, 28, p. 42
241. Kwolek, S. L., U.S. Patent No. 3,671, 542, 1972; also Morgan, P. W., *Macromolecules*, 1977, 10, p. 1381
242. Ide, Y. and Ophir, Z., *Polym. Engr. Sci.*, 1983, 23, p. 261
243. Sugiyama, H., Lewis, D. N., White, J. L. and Fellers, J. F., *J. Appl. Polym. Sci.*, 1985, 30, p. 2329
244. Acierno, D., La Mantia, F. P., Polizzoti, G., Ciferri, A. and Valenti, B., *Macromolecules*, 1982, 15, p. 1455
245. Muramatsu, H. and Krigbaum, W. R., *J. Polym. Sci. B: Polym. Phys.*, 1986, 24, p. 1695
246. Muramatsu, H. and Krigbaum, W. R., *J. Polym. Sci. B: Polym. Phys.*,1987, 25, p. 803
247. Cuculo, J. A. and Chen, G.-Y., *J. Polym. Sci. B: Polym. Phys.*, 1988, 26, p. 179
248. Chen, G.-Y., Cuculo, J. A. and Tucker, P. A., *J. Polym. Sci. B: Polym. Phys.*, 1988, 26, p. 1677
249. Lee, W.-C., Dibenedetto, A. T., Gromek, J. M., Nobile, M. R. and Acierno, D., *Polym. Engr. Sci.*, 1993, 33, p. 156

250. Muramatsu, H. and Krigbaum, W. R., *J. Polym. Sci. B: Polym. Phys.*,1987, 25, p. 2303
251. Muramatsu, H. and Krigbaum, W. R., *Macromolecules*, 1986, 19, 2850
252. Itoyama, K., *J. Polym. Sci. B: Polym. Phys.*, 1988, 26, p. 1845
253. Kenig, S., *Polym. Engr. Sci.*, 1987, 27, p. 887
254. Hersh, S. P., in *Handbook of Fiber Science and Technology: High Technology Fibers*, vol. III, part A, ed. Lewin, M. and Preston, J., Marcel Dekker, Inc., New York, 1985, p. 1-44
255. Jeffries, R., *Bicomponent Fibres*, Merrow, Watford, England, 1971
256. Olibisi, O., Robeson, L. M., and Shaw, M. T., *Polymer-Polymer Misibility*, Academic Press, New York, 1979
257. Southern, J. H., Martin, D. H., and Baird, D. G., *Text. Res.* J., 1980, 30, p. 411
258. Tsuji, W., in *Cellulose and Fiber Science Developments: A World View*, ed. Arthur, Jr., J. C., ACS Symposium Series 50, American Chemical Society, Washington D. C., 1977, p. 252
259. Lee, B. L. and White, J. L., *Trans. Soc. Rheol.*, 1974, 18, p. 467
260. Han, C. D., *Rheology in Polymer Processing*, Academic Press, New York, 1976
261. Southern, J. H. and Ballman, R. L., *J. Polym. Sci. Polym. Phys. Ed.*, 1975, 13, p. 863
262. Papero, P. V., Kubu, E. and Roldan, L., *Text. Res. J.*, 1967, 37, p. 823
263. Kikutani, T., Radhakrishnan, J., Arikawa, S., Takaku, A., Okui, N., Jin, X., Niwa, F. and Kudo, Y., J. *Appl. Polym. Sci.*, 1996, 62, p. 1913
264. Moon, H.-S., Park, J.-K. and Liu, J.-H., *J. Appl. Polym. Sci.*, 1996, 59, p. 489
265. Qin, Y., Brydon, D. L., Mather, R. R. and Wardman, R. H., *Polymer*, 1993, 34(17), p. 3597
266. Qin, Y., *J. Appl. Polym. Sci.*, 1994, 54(6), p. 735
267. Joslin, S. L., Giesa, R. and Farris, R. J., *Polymer*, 1994, 35(20), p. 4303
268. Joslin, S. L., Jackson, W. and Farris, R. J., *J. Appl. Polym. Sci.*, 1994, 54(3), p. 289
269. Tippets, E. A., *Text. Res. J.*, 1967, 37, p. 524
270. Oppenlander, G. C., U.S. Patent 3,505,164 (1970)
271. Oppenlander, G. C., U.S. Patent 3,509,013 (1970)
272. Jurkiewitsch, G., U.S. Patent 3,533,904 (1970)
273. Tomioka, S. and Taniguchi, M., U.S. Patent 4,500,384 (1985)
274 Marcher, B., *Tappi J.* 1991, 74(12), p. 103
275. Nathan, C. N., *Text. Inst. Ind.*, 1976, 14, p. 299
276. von Falkai, B., *Chemiefasern/Texilind.*, 1977, 27, p. 37
277. Berger, W. and Daniel, E., *Textiltech.*, 1976, 26, p. 222
278. Magat, E. E. and Tanner, D., U.S. Patent 3,475,898 (1967)
279. Barnes, G. T., *Mod. Text.*, 1970, 51(7), p. 45

3 Advances in the Control of Spinline Dynamics for Enhanced Fiber Properties

John A. Cuculo, Joseph F. Hotter and Qiang Zhou
North Carolina State University, Raleigh, North Carolina, USA

3.1 Introduction

An unwillingness to accept the limited amount of structural development that takes place during fiber formation in a standard melt spinning process has inspired a tremendous amount of past and present research activity. The fact that fiber structure, and hence properties, can be intentionally manipulated through manufacturing process variables has firmly established the concept that synthetic fibers be considered engineered materials. While specific processes typically have a great many variable parameters, the impact of each of these parameters will generally fall within the realm of either a time, temperature or tension effect. The prospect for exploiting each of these principal effects will be briefly reviewed in Section 3.2. An additional factor that has continued to drive research activities, centers around the perceived potential for vast property improvements (actual vs. theoretical maxima), as well as the economic advantages associated with a more direct route to improved fiber properties. These basic concepts and the relevant patented advances in melt spinning will be discussed in more detail in Section 3.3. The presentation of this background information will have set the stage for the description of a uniquely modified melt spinning process that has only recently emerged. The means by which this modification achieves such judicious control of the time, temperature and tension, as well as the nature of the fiber structure and properties that result, will be expounded upon in Section 3.4. Finally, in Section 3.5, there will be presented a number of closing comments, including notions of how the evolution of this unique process can potentially bring the theoretically predicted ultimate fiber properties more realistically within our grasp through a simple, safe and direct melt spinning approach.

3.2 Moving Beyond Imposed Limits as Take-Up Speeds Increase

Factors that control structure development in a melt spinning threadline may be broadly categorized into tension, temperature and time effects. As take-up speeds increase the manner in which each of these effects is impacted has been thoroughly characterized through constantly improving on-line measurement capabilities and simulation logic. Threadline tension profiles derived through a simplified steady-state equation of the momentum balance, have allowed the individual force contributions to be quantified [1,2]. As engineering advancements allowed take-up speeds to increase, two of these force contributions emerged as dominant, and will be discussed later. As for the temperature and time effects, these are intimately and critically related to the overall crystallization process. In this case, the advent of higher take-up speeds changed the balance between time and temperature during the cooling of the threadline. Unfortunately, a complete understanding of the sole effects of temperature and time on the crystallization that occurs in an actual melt spinning threadline is limited by our inability to quantifiably account for the impact of increasing molecular orientation as the threadline develops. Later in this section, a semi-qualitative discussion of these effects on the crystallization process will be presented. The concept of transient effects has also been a topic of great importance in understanding the limitations of ultra-high take-up speeds. As a developing threadline is strained energy is imparted to it. As the strain rate increases, the component of that imparted energy that is stored, rather than dissipated, will eventually exceed the energetic capacity of the material resulting in rupture of the threadline [3,4]. Some additional comments regarding the way in which this transient effect is characterized will also be made later in this section.

3.2.1 The Dominant Forces

As discussed in Section 2.1.2, the individual forces which contribute to the development of threadline tension may be described in the following differential format

$$\frac{dF}{dz} = WdV - 2\pi R\sigma_s dR + 2\pi\sigma_d R - \rho g\pi R^2 \tag{3.1}$$

where W is the mass throughput, V is the axial velocity, R is the filament radius, (σ_s is the surface tension, σ_d is the shear stress resulting from (aerodynamic) drag, ρ is the polymer density, and g is the acceleration of gravity. When Eq. 3.1 is integrated from any point in the threadline, z, to the point of take-up, the following familiar expression for the overall force balance is generated

$$F_{ext} + F_{grav}(z) = F_{rheo}(z) + F_{inert}(z) + F_{surf}(z) + F_{aero}(z) \tag{3.2}$$

where F_{ext} represents the constant external force imposed by the take-up device and $F_{rheo}(z)$ represents the rheological resistance to yielding exhibited by the developing threadline. Figure 3.1 shows graphically the experimental distribution of each of these tensile force contributions for a PET single filament spinning process operated at 5000 m/min. It is clear that at this level of take-up speed, $F_{rheo}(z)$ increases significantly down the length of the threadline and is controlled predominantly by the inertial and aerodynamic force components. Note, the surface tension force component has been assumed negligible and hence is not shown in Fig. 3.1. It should be noted that this method of characterizing the

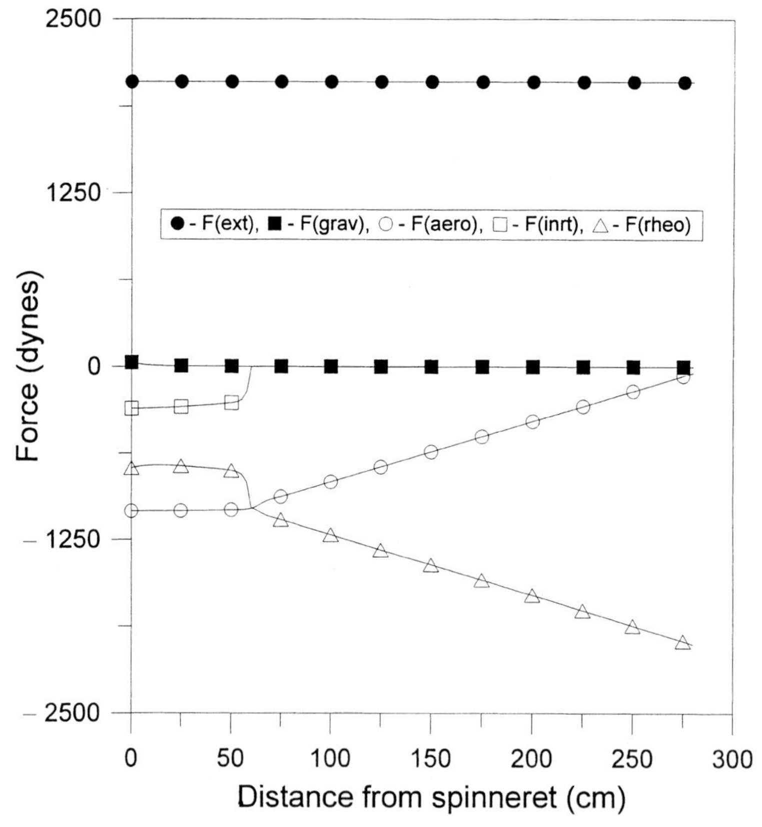

Figure 3.1 Rheological, air drag, inertial, and gravitational force contributions versus distance from the spinneret for PET filament (5 denier,chip IV = 0.97dl/g) spun at 5000 m/min.

development of threadline tension also provides a means of estimating $F_{rheo}(0)$, the rheological force experienced at the spinneret face, which has at times been erroneously assumed to be zero. Ziabicki has provided an excellent review and com-parison of some of the most notable attempts to accurately characterize this quantity [1]. The $F_{rheo}(z)$ profile can subsequently be used to generate a threadline tension profile as

$$\sigma_{th}(z) = F_{rheo}(z)/A(z) \tag{3.3}$$

where A(z) is the corresponding cross-sectional area profile of the developing threadline.

Figure 3.2 shows calculated threadline tension profiles for take-up speeds ranging from 2000 to 5000 m/min. While the use of higher take-up speeds clearly yields a higher overall tension, the level achieved is still considered quite low, approximately an order of magnitude less than the yield stress reached in the cold drawing of PET [3]. In addition, the well documented necking phenomenon reduces the inertial and aerodynamic contributions in the region from approximately 0 to 60 cm, which results in lower tensions in this critical structure development zone. However, the large and rapid strain imposed over the very short necking region has a strong orienting effect with

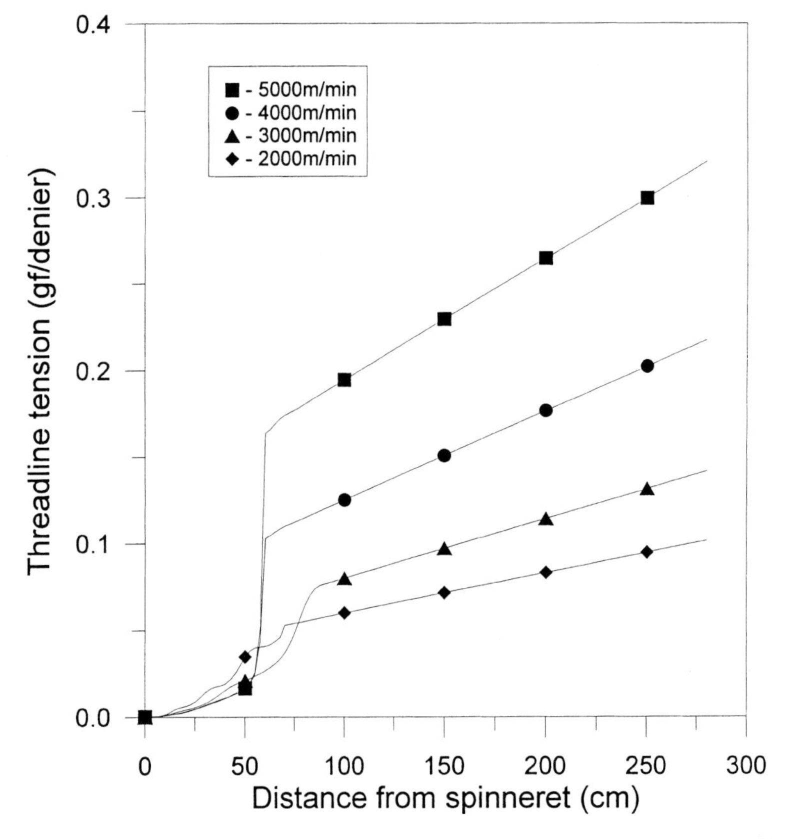

Figure 3.2 Calculated threadline tension profiles for PET filament (5 denier,chip IV = 0.97dl/g) spun at take-up speeds ranging from 2000 to 5000 m/min. (For PET, 1 gf/d is approximately equal to 120 MPa)

birefringence values increasing dramatically over this range of take-up speeds. As will be discussed later, the dominant nature of drag at the interface between the filament surface and the surrounding medium has, until recently, been a relatively untapped potential through which nearly an order of magnitude of greater threadline tension control can be achieved. Further, and of major importance, the location of the process perturbation described later which allows this drag component to be manipulated may be varied to coincide with the presumed position of structure development. The tension enhancement level can then be controlled by variation of the exposure distance in the medium.

3.2.2 Cooling and Crystallization

Most synthetic fibers are semi-crystalline materials. Crystallinity plays an important role in determining the strength, stiffness and dimensional stability of synthetic fibers, hence a desire to understand and control the manner in which it occurs has always been of paramount importance in the fiber spinning process. With the advent of higher spinning speeds, the significance of stress induced crystallization in the spinline became more evident, especially for the case of PET which is considered to be a relatively slow crystallizing material. A brief review of the governing aspects of crystallization [5–15] should provide the background information necessary to grasp the concept of how a particular type of threadline modification might be used to control both the extent and type of crystallization that occurs.

The most general feature that applies to all crystallizable polymeric materials is that crystallization occurs only within a specific temperature range. This range is bounded by the melting point and the glass transition temperature ($T_m > T > T_g$), which of course varies for different types of materials. As the threadline cools during its transition from the molten to the solid state, it passes through this temperature range. It should be noted that materials having a T_g below ambient temperature, such as polypropylene, will continue to crystallize even after the spinning process is complete. Strictly speaking, the term "crystallization" is a broad reference to both nucleation and crystal growth, however, these two facets of the overall crystallization process are often incorporated to arrive at a composite rate constant, as described by Evans [5]. Then, based on the classical theory of phase transformation [6–10], the extent to which the as-spun material crystallizes would simply be dictated by the composite rate constant and the time available for crystallization. A discussion of the theoretical treatments of crystallization during the spinning process is given in Section 2.1.2.6, and the subject will not be revisited here. We will only state the following general expression, which we consider to be a useful representation of both the orientation (f_a) and temperature (T) dependence of the crystallization rate [1,15]:

$$K(T, f_a) = \ln(2)[t_{1/2}(T_{max})]^{-1} \exp\left(\frac{-4\ln(2)(T - T_{max})^2}{D^2}\right) \exp(A(T) \cdot f_a^2) \qquad (3.4)$$

These temperature and orientation dependencies are shown schematically in Figure 3.3.

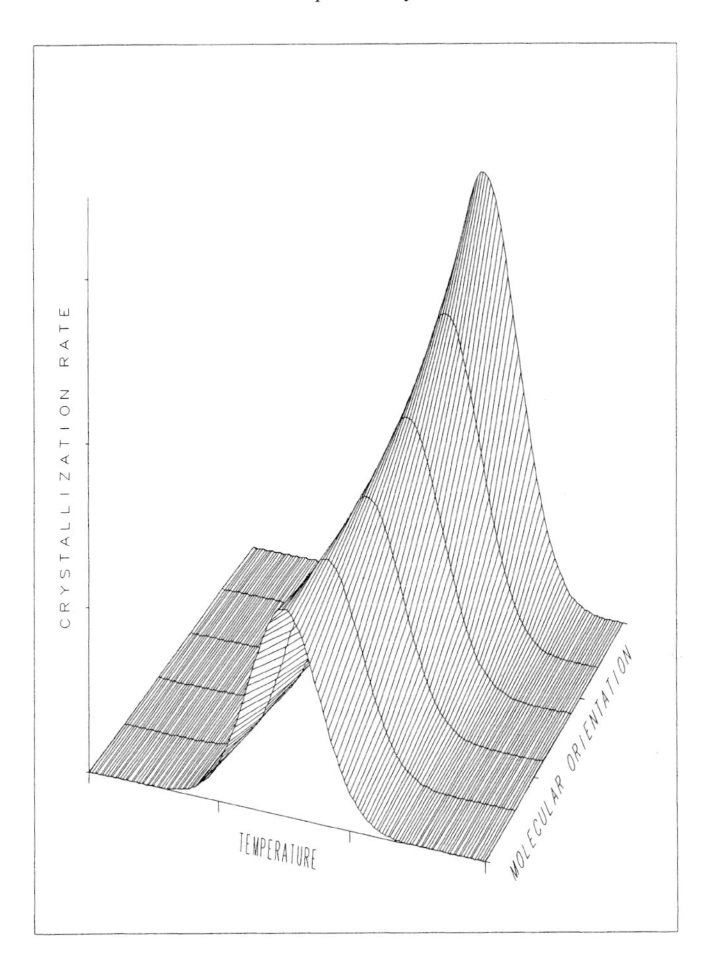

Figure 3.3 Schematic representation of the temperature and molecular orientation dependence of the crystallization rate.

Note that while an increase in molecular orientation produces the anticipated exponential rise in the crystallization rate, it also lowers the enthalpy of the melt, and even more so the entropy of the melt, to the extent that an increase in the apparent melting point occurs [16]. This increase in melting point allows the already accelerated crystallization process to begin occurring earlier within the cooling process and thereby further promotes the overall extent of crystallization. The connection between stress (tension), orientation and crystallization rate has led to the widely accepted concept of "stress induced crystallization". Among the most prominent examples of how significant this connection can be is in the case of PET, which in an isotropic state is a relatively slow crystallizing material. When melt spun at low-speeds, such that little stress and orientation develop during the spinning process the resulting fiber is essentially completely amorphous, however, under high-speed high stress spinning conditions PET can achieve crystallinities of nearly 50%.

While the focus thus far has been on the impact of orientation on the rate constant, the prospect of controlling the time available for crystallization to occur must not be neglected. As take-up speed is increased, while denier is held constant, a melt spinning threadline is drawn down to its final diameter more quickly. This causes the threadline to cool more rapidly and hence reduces the time available for crystallization to occur. The increase in crystallinity with increasing take-up speed cited above for PET occurs because the effect of an increasing crystallization rate far outweighs the effect of less time being available for crystallization to occur. However, as alluded to at the end of the previous section, recent efforts to more fully explore different spinning media, or environments other than air, that could be applied to melt spinning have identified another potential benefit that may offer additional latitude with regard to the overall cooling process. Consider the impact of a process perturbation, applicable virtually anywhere along the length of a developing threadline, that had heat transfer characteristics at least an order of magnitude superior to air, which is typically used as the standard quench medium. If such a perturbation could be employed, it might be possible to actually maintain a region of the developing threadline at, or very near, the temperature corresponding to the maximum crystallization rate for some extended period of time.

The effects of stress or flow induced orientation on the overall crystallization process also requires some comment with regard to the types of crystalline structures formed from a more oriented precursor melt. A stationary or isotropic melt will tend to form a spherulitic or folded chain type of crystalline structure, while a more oriented melt, typically achieved via high flow rates or large strain rates produces a more fibrillar extended chain type crystalline structure. The significance of a fibrillar extended-chain structure and its ability to enhance the load bearing performance of the semi-crystalline macroscopic fiber will be discussed in Section 3.4.3.

3.2.3 Transient Effects

Transient, or deformation-rate related, effects are important components in understanding the rheological behavior of polymeric materials in melt spinning. A constitutive equation describes the relationship between an imposed rheological stress and a deformational response. For materials which exhibit Newtonian (purely viscous) behavior, this relationship may be satisfactorily modeled by the Newtonian constitutive equation, and may be conveniently expressed for this particular application as

$$\sigma = \eta_e \cdot V' \tag{3.5}$$

where σ is the threadline stress, V' is the axial strain rate (or velocity gradient, (dV/dz)), and η_e is the elongational viscosity. While shown as a constant in Eq. 3.5, the elongational viscosity (η_e) is known to be a function of both temperature (T) and strain rate (V'), both of which vary along the length (z) of a developing threadline.

$$\eta_e(z) = \eta_e[T(z), V'(z)] \tag{3.6}$$

At low strain rates (below 600 s^{-1}), the elongational viscosity is predominantly a function of temperature only and is reasonably close to 3μ (where μ is the shear viscosity) predicted by the Newtonian model. However, when subjected to high strain rates (above 600 s^{-1}), such as those experienced under high-speed spinning conditions, the effect of strain rate on elongational viscosity can become quite significant with a strong elastic component now being present and the Newtonian model no longer being applicable. The simplest model capable of approximating "viscoelastic" rheological behavior is the Maxwell fluid model, which is composed of a spring (elastic) and dashpot (viscous) coupled in series. Forgoing a review of this well published model and again focusing on its most pertinent form for this particular application, leads to the following expression(s)

$$\sigma = \frac{\eta_e}{1 + \tau_f V'} \cdot V' \text{ or } V' = \frac{\sigma}{\eta_e - \tau_f \sigma} \tag{3.7}$$

where τ_f is the relaxation time, and all other variables are as defined previously for the Newtonian model. It should however be noted that Eq. 3.7 represents the most simplified case, with all contributions to the development of threadline tension (σ), aside from the imposed take-up force (F_{ext}), having been omitted (F_{inert}, F_{aero}, etc.). It should also be noted that generally speaking fiber forming polymers possess a distribution of molecular weights, and hence a distribution of relaxation times. While a weighted average for the relaxation time might seem appropriate, Ziabicki [3] has pointed out the existence of indirect evidence to suggest that the presence of small amounts of very high molecular weight species can contribute strongly to viscoelastic behavior, without affecting shear viscosity, but modifying elongational flow, die swell effects, melt fracture, and so on. Bearing in mind the positional dependence of the elongational viscosity, Eq. 3.7 indicates that as relaxation time and/or threadline stress increase the predicted deformation rate (spinning dynamics) has the potential to increase rapidly, which in turn can lead to instability, as well as the well documented "necking" phenomenon.

In order to incorporate both the viscous and elastic components into a single variable quantity, the apparent elongational viscosity (η_{app}) has been defined as the ratio of stress to the elongation rate:

$$\eta_{app} = \frac{\sigma}{V'} = \frac{\eta_e}{1 + \tau_f V'} = \frac{\eta_e}{1 + D_e} \tag{3.8}$$

Furthermore, the Deborah number (D_e), has been defined as the product of the relaxation time (τ_f), and the characteristic time of deformation or local strain rate (V'). As described by Ziabicki [3], in simple viscoelastic models, the Deborah number can be used as the criterion of the appearance of strong transient effects, where values approaching or greater than unity tend to correspond with the onset of significant viscoelastic behavior. In a comparison of the general classes of fiber forming polymers, the melts of polyolefins such as polyethylene can have relaxation times within the approximate range of 0.2–10 seconds (depending on molecular weight and molecular weight distribution), while those of polyesters such as polyethylene terephthalate are generally on the order of 2 to 3×10^{-3} seconds [17]. Based on this 2 to 3 order of magnitude difference in the characteristic relaxation times of these materials, the polyolefins would be expected to exhibit viscoelastic behavior at much lower strain rates (or take-up speeds) than in the case of the polyesters. As alluded to at the start of this section, this would force the polyolefins to exceed some critical level of stored energy associated with that necessary to rupture the threadline at a strain rate much lower than that required for the polyesters, with the net result being much lower attainable maximum take-up speeds for polyolefins compared to polyesters.

The impact and limitations associated with the dominant forces, cooling and crystallization, and transient effects have all been briefly reviewed. However, to this point the discussion has centered around how each of these critical aspects of the spinning process responds within a standard unperturbed spinning environment. The standard spinning environment being referred to is, of course, ambient air with no significant fluctuations in temperature or viscosity. Under such circumstances an abundance of experimentation has been performed and the manner in which threadline tension develops, crystallization occurs and transient effects begin to limit take-up speeds have all been well characterized for a variety of fiber forming polymers. Having crudely established the most prominent features that govern each of these phenomena and their interdependencies, this logically leads to a desire to achieve more discriminate and positive control over them. Consider the following questions:

1. How might the respective magnitude of inertial and drag forces be increased beyond their current limits, and what would be the resultant impact on as-spun orientation?
2. What level of crystallinity would result if more judicious control could be achieved of how and when molecular orientation develops and at what level and duration temperature is maintained along the entire length of the developing threadline?
3. What would be the impact on structure and properties if the strain rate and temperature profiles could be independently controlled such that transient effects no longer limited the use of ultra-high take-up speeds?

Clearly, the ongoing quest to answer questions such as these will require us to expand the latitude of innovation that we exercise in present and future melt spinning research.

3.3 Recent Development Efforts and Achievements

Over the past thirty years, very few fiber forming polymers have achieved more than a small fraction of their potential theoretical tensile strength. From one perspective, this limitation is considered inevitable and is accepted as a consequence of the highly isotropic and entangled nature of the precursor polymer melt. And of course, this ignores the minor and again inevitable imperfections introduced in a standard fiber extrusion process. From another perspective, it may be argued that if judicious control, well beyond that practiced in conventional spinning processes, can be achieved then it may be possible to push some of today's common commodity polymers well beyond their present limits. Regardless of the proper perspective, this section will begin with a brief review of predicted theoretical tensile strengths and hence provide a means of gauging our progress thus far towards these ultimate goals. Attention will then be focused on some of the additional incentives with regard to economics and operability that could potentially be achieved by producing high tenacity fibers in a one-step process through the use of controlled spinline dynamics, compared to the two-step processes currently used to obtain such fibers. Lastly, studies in the general areas of the melt spinning process which have resulted in some of the more prominent patented advances in the field of fiber melt spinning will be briefly highlighted.

3.3.1 Theoretical Tensile Strengths and the Potential for Vast Improvements

The prediction and knowledge of theoretical ultimate tensile strengths provide a means of gauging the relative extent of irregularities and/or imperfections when exploring new processing techniques. Table 3.1 lists the predicted theoretical ultimate tenacities for various polymers via classical, kinetic and thermodynamic calculation methods. While the basis on which each of these calculations has been performed will not be discussed here, some limited background information for each of these methods is given in Table 3.1. In addition to the predicted ultimate tenacities, Table 3.1 also shows a brief list of measured strength values for actual fiber samples produced from various types of spinning processes. The first two entries of Table 3.1 are PET and PE, which together represent a large proportion of the total volume of polymeric materials produced for use in the synthetic fiber industry. In both cases the predicted ultimate tensile strength declines as the assumed failure mechanism changes from primary to secondary bond rupture. These two entries illustrate well the significant strength enhancements that can be realized through the use of alternative spinning processes. While solution/gel spinning processes (followed by drawing) have clearly made greater strides than melt spinning processes (and have enormously enhanced tensile modulus, as discussed in Chapter 5), there is significant potential for vast tensile strength improvements in every case, regardless of which predicted ultimate tenacity is most accurate. Even though solution and gel spinning provide significant advantages, such as the reduction of entanglements, superior retention of molecular weight and the ability to process ultra-high molecular weight polymers, these same processes are also severely hampered by the required use of toxic solvents, very low spinning speeds and the need for post-treatment.

3.3.2 The Ideal One-Step Spinning Process

An ideal one-step melt spinning process would allow the structure (orientation/crystallinity) and properties (strength/modulus/shrinkage) required of the final yarn product to be achieved during the initial stage of fiber formation, and hence eliminate the need for subsequent drawing and/or heat-setting. This ideal one-step process would

Table 3.1 Predicted Ultimate and Presently Achieved Tensile Strengths for Various Fiber Forming Polymers.

Calculation Method		**Classical** [18]	**Kinetic** [19,20]	**Thermodynamic** [21]	**Presently Achieved** [22–24]		
Failure Mechanism		Primary (C-C) Bond Rupture	Primary/Secondary Bond Rupture	Secondary Bond Rupture	(Note, the properties of fibers of the same chemical type can vary widely. Hence, values shown here should be used for general comparison only.)		
Basis of Calculation		Chain Deformation	Rate Dependent Process	Melting Point Depression			
Polymer Type	T_m (°C)[a]	**Predicted Ultimate Tensile Strength** (gf/d)			**Measured Tensile Strength** (gf/d)		
					meltspun	gel spun	soln.spun
PET	273	232	99.7	37.9	9.5	—	15.5
PE	150	372	172.3	81.5	6.2	49.3	—
PP	185	218	80.5	33.6	11.6	12.8	—
Nylon 6	220	316	129.5	66.6	9.8	—	—
POM	180	264	76.4	—	13.7	—	—
Rayon	180d	—	98.4	—	—	—	3.2
PPT (Kevlar)	425d	235	118.1	—	—	—	22.0

[a] d = decomposes

POM = Polyoxymethylene

Note: 1 GPa ≈ 1 gf/d $\times 0.09\rho$. where ρ is polymer density in g/cm^3 (usually in the range 0.9–1.5 g/cm^3)

also be reliable, trouble free and amenable to uninterrupted operation at very high take-up speeds.

In the early stages of the investigation of high-speed spinning technology it was hoped that fully oriented structures, with properties approaching and/or exceeding those of fibers produced through conventional low-speed spinning and drawing might be achieved through the use of ultra-high take-up speeds. However, as mechanical engineering advancements made the utilization of these ultra-high take-up speeds possible the inferiority of the resulting mechanical properties with increasing take-up speeds, especially above ca. 6000 m/min, became evident. The most accepted reason for these inferior mechanical properties, as well as threadline instability and breakage, is arguably the presence of a radially non-uniform (sheath-core) structure that develops as a result of the large strain rates and rapid increase in cooling rate which occurs as take-up speeds increase. The fact that high-speed spinning under standard spinning conditions, without modification, had not and would not lead to the so-called "ideal one-step spinning process" resulted in diminished efforts to further investigate this technology.

Even though the prospect appeared unlikely, let us briefly consider a few of the potential advantages in the development of such an ideal process, if it were somehow possible through additional modification. From an economic perspective, the elimination of post-spinning drawing and heat-setting equipment would result in tremendous savings with regard to capital costs, maintenance costs, and energy consumption, while the use of very high take-up speeds (large throughputs) would lead to increased production rates and hence reduced manufacturing costs. From an operational perspective, a shorter route to the final product would decrease the number of process variables and likely reduce overall product variability. Lastly, and most crucially, from a structure/property perspective, an ideal one-step spinning process would increase the probability of forming the extended-chain type of crystalline structure required to surpass current tensile strength limits related to the folded-chain conformation. Note that it is believed near perfect orientation must be generated before the material begins to crystallize during solidification, if the undesirable folded-chain crystalline morphology is to be avoided.

3.3.3 Patented Advances in Fiber Melt Spinning

A brief review of some of the patents granted over the past fifty years will help identify recurring features and variables that have been of primary interest in the field of synthetic fiber development via melt spinning. Table 3.2 provides a partial listing and very brief description of those patents related to advances in the control of spinline dynamics. While the majority of these patents make claims regarding the invention of a process that would be applicable to all of the major classes of fiber forming polymers, such as polyesters, polyamides and polyolefins, some more specifically limit their claims to only a single type of polymer substance. Generally speaking, the patents listed in Table 3.2 focus on how a single change, or combination of changes, in the general areas of extrusion, quench and take-up speed have led to the creation of either an improved route to an existing product, or to a new and unique product altogether.

Table 3.2 Partial Listing and Brief Description of Patents Related to Advances in the Control of Spinline Dynamics

Issue Date	Patent #	Inventor	Assignee	Description
7/6/43	2,323,383	Dreyfus	Celanese	Production of synthetic filament as applied to a broad range of fiber forming polymers.
7/29/52	2,604,667	Hebeler	Du Pont	Useful as-spun properties via high-speed spinning with no threadline modification.
10/3/61	3,002,804	Kilian	Du Pont	Spinning coupled directly to stretching through a liquid drag bath produces uniformly oriented textile filaments.
9/11/62	3,053,611	Griehl et al.	Inventa	Useful as-spun properties via high-speed spinning where filaments are extruded into a heated shaft.
1/2/68	3,361,859	Cenzato	Du Pont	Delayed cooling yields an as-spun yarn of lower orientation that can be drawn to higher tenacity.
7/12/78	4,113,821	Russell et al.	Allied Chemical	Improved drawn yarn mechanical quality via controlled pretension applied prior to drawing.
1/16/79	4,134,882	Frankfort et al.	Du Pont	High shear spinneret leads to high-speed as-spun fiber with low differential Δn and improved dyeability.
3/25/80	4,195,161	Davis et al.	Celanese	Development of a unique as-spun internal structure that exhibits low shrinkage under a high degree of force.
11/8/83	4,414,169	McClary	Fiber Industries	High stress spinning of an as-spun filament yarn leads to a more dimensionally stable drawn yarn product.
1/10/84	4,425,293	Vassilatos	Du Pont	Liquid bath quench retards crystallization and produces an oriented amorphous yarn spun at high-speed.
3/20/90	4,909,976	Cuculo et al.	NCSU	Use of zone cooling and heating maintains an optimum temperature profile and promotes spinline crystallization.
5/8/90	4,923,662	Kurita et al.	Kaisha	Liquid bath quench retards crystallization and produces a high-speed as-spun yarn of improved drawability.
7/23/91	5,034,182	Sze et al.	Du Pont	Controlling draw down behaviour via unique aspiration leads to improved spinning continuity at high-speeds.
11/26/91	5,067,538	Nelson et al.	Allied-Signal	Very high stress spinning of an as-spun filament yarn leads to a more dimensionally stable drawn yarn product.
2/4/92	5,085,818	Hamlyn et al.	Allied-Signal	Higher as-spun yarn Δn at lower speeds and lower IV's via proper combination of quench, melt temperature, etc.

(continued)

Table 3.2 (continued)

Issue Date	Patent #	Inventor	Assignee	Description
9/22/92	5,149,480	Cuculo et al.	NCSU	Liquid bath controls tension and temperature to yield a highly oriented, crystalline and radially uniform as-spun fiber.
12/15/92	5,171,504,	Cuculo et al.	NCSU	Liquid bath controls tension and temperature such that post solidification in-line drawing yields a radially uniform fiber.
8/24/93	5,238,740	Simons et al.	Hoechst Celanese	Heated column to control threadline quench profile leads to unique properties in both the as-spun and drawn yarns.
11/8/94	5,362,430	Herold, II et al.	Du Pont	Aqueous quench produces an as-spun polyamide yarn capable of direct use in textile and carpet applications.

The area of extrusion is broadly considered here to include such factors as melt viscosity, flow induced orientation, and polymer composition. Melt viscosity is, for a given type of material, considered during processing to be a function of shear rate and temperature, both of which can be varied to achieve an optimum melt viscosity profile. As for the concept of flow induced orientation, attempts have been made through proper design of capillary dimensions and geometry to maximize the orientation resulting from both shear and elongational flow within the spinneret. While not specifically a subject of the patents shown in Table 3.2, polymer composition may also be used to alter flow behavior and nucleation rates, both of which are directly related to spinline dynamics.

Here, quench refers to the cooling profile over the entire length of the spinline. As described earlier, the rate of cooling plays a significant role in governing the resulting level of as-spun crystallinity. As also described earlier, the temperature profile of a developing threadline in part controls the elongational viscosity, which in turn dictates the response of threadline stress to the imposed strain rate. It is interesting to note that even in the earliest patent listed here, special attention was already being focused on the nature of the medium into which the freshly extruded filaments would first pass. While air related quenches are still predominant, liquid type quenches have gained considerable attention during the past few years. As will be described later, the use of a liquid type quench medium has led to the recognition of additional benefits with regard to improved heat transfer efficiency, and the ability to significantly augment tension control through the interaction between viscosity and drag.

As for take-up speed, this particular area of melt spinning has probably been the most publicized. While makeshift take-up devices helped lead to the relatively early identification of the potential benefits associated with high-speed spinning, these devices were not yet capable of continuous operation with high denier yarns, and hence allowed only for small experimental quantities of fiber to be generated. The substantial increase of as-spun property levels that accompanies increasing take-up speeds is directly attributable to the increase in spinline tension, which as described earlier, impacts both the resulting orientation and the crystallization rate. It should also be noted that many of the patented advances resulting from increased take-up speeds may have very likely been inadvertently realized through a simple quest to increase throughput rates and reduce production costs.

3.4 The Preferred Route to Judicious Control of Spinline Dynamics

Studies dealing with the correlation between structure and properties of fibrous materials have shown that a high degree of molecular orientation and a significant amount of crystallinity are prerequisites for those materials to achieve high axial strength and tensile modulus. In a high-speed melt spinning process, the nature of the neck-like deformation which occurs just before filament solidification, results in a sudden increase in the spinning stress to a near maximum value. Undesirably, the occurrence of this abrupt attenuation results in a loss of tension up-stream of the "neck" point (refer to Figure 3.2). The chain folding dominated structure thus obtained makes it very difficult, if not impossible, to further extend the molecular chains to the extent required for high performance reinforcement applications. Modifications and eventual radical changes to the spinline dynamics profiles will need to be made in order to achieve a quantum leap beyond the conventional limits. This will be the discussion focus in this section.

3.4.1 Judicious Control via Radical Change

One of the most intellectually stimulating innovations in melt-spinning is the attempt to gain total control of the threadline dynamics through introduction of on-line perturbations to the high-speed melt spinline, at the time the fiber structure, and hence the final properties, is incubating or being developed. As a moderate intrusion to the conventional melt spinning threadline, a temperature profile modifier has been utilized in the region adjacent to the quench stack. A typical example is the combination of on-line zone cooling and heating (OLZCH) imposed as a quench supplement to promote development of desirable fiber structure such as crystallization and crystalline orientation [25]. In this invention, the fresh filaments emerging from the spinneret are initially cooled to an optimum temperature, which is normally above the glass transition temperature of the polymer, then maintained near that temperature for a desirable period of time for crystallization and orientation to develop, and finally cooled below the solidification point for take-up. Proper configurations for the location and temperature of the cooling and heating and the duration of the filaments in the isothermal surroundings strongly depend on the spinning speed and the properties of the polymer under consideration. Change of the temperature profile could cause changes of the threadline tension due to the effectual interdependencies among the dynamic variables. Therefore, both molecular orientation and crystallization are expected to be affected as a result of the temperature optimization. This type of temperature modification as a means of threadline dynamics control has been practically applied in commercial production.

Individual responses of the key spinline dynamic parameters, such as time, tension and temperature are completely entangled with each other. Unintentional or passive responses of one or more dynamic parameters induced by changing others may divert the general objective held for the initial changes. There has been a strong incentive for the judicious control of the spinline dynamics, in which the responses from the key dynamic parameters could be more discriminatingly directed and controlled in a manner favorable to structure and property enhancement. The integration of a liquid isothermal bath (LIB) into the high-speed melt spinning process created a great potential to accomplish this goal [26,27]. Unlike the temperature profile modification, the on-line perturbation with an LIB could alter temperature, tension and time effects simultaneously and essentially independently. The primary benefit from utilization of an LIB is that filaments thus produced possess two typical characteristics: high birefringence indicative of a high level of molecular orientation, and a uniformly radial fine structure. At least four extra effective spinning variables associated with the LIB are now available for more specific and precise control of the melt spinning process: liquid medium, liquid temperature, liquid depth and LIB location. Versatility, flexibility as well as simplicity in the application of the LIB enable the melt spinning process to efficiently produce yarns with a variety of superior performance characteristics, for example, high tensile strength and low thermal shrinkage.

3.4.2 The Response of Spinline Dynamics

The merit of the advanced melt spinning process with totally controlled spinline dynamics is the ability to substantially increase the threadline tension, independent of temperature and time, as much as an order of magnitude higher than that ordinarily obtained in a conventional high-speed process. At the same time, the temperature and duration of exposure crucial to fiber structure development are being optimized to attain the desired responses. The success in uncoupling the major spinline dynamic parameters by proper manipulation of the LIB variables and spinning speed has laid the ground work for a great improvement of all prospective fiber properties. Research results demonstrated that for the LIB assisted melt spinning process, the spinline dynamics changed from the conventional mode governed by inertial and air drag forces to a process controlled by the manner in which the on-line perturbation is imposed [28]. The sensitivity of the spinline dynamics with respect to the control of spinning variables could be illustrated, as an example, by varying the LIB location along the spinline.

In an attempt to examine the effects of the LIB location on the fiber structure and properties, the melt spinline, starting at the exit of the spinneret and terminating at the take-up, has been divided into four segments referred to as LIB-I through LIB-IV [29]. LIB-I and LIB-II encompass a region where the filaments are believed to be molten when entering the liquid bath, with the LIB-I located closer to the spinneret than is the LIB-II.

LIB-III is thus chosen so that the filaments start solidifying inside the liquid bath, while in the case of LIB-IV, the filaments have already been solidified before entering the LIB. At different LIB locations not only do the spinline dynamics respond diversely, but also the resulting as-spun fibers perform differently. The diversity in the stress-strain behavior of as-spun fibers, as depicted in Figure 3.4, symbolizes the predominant role played by the location of the LIB along the melt spinning threadline.

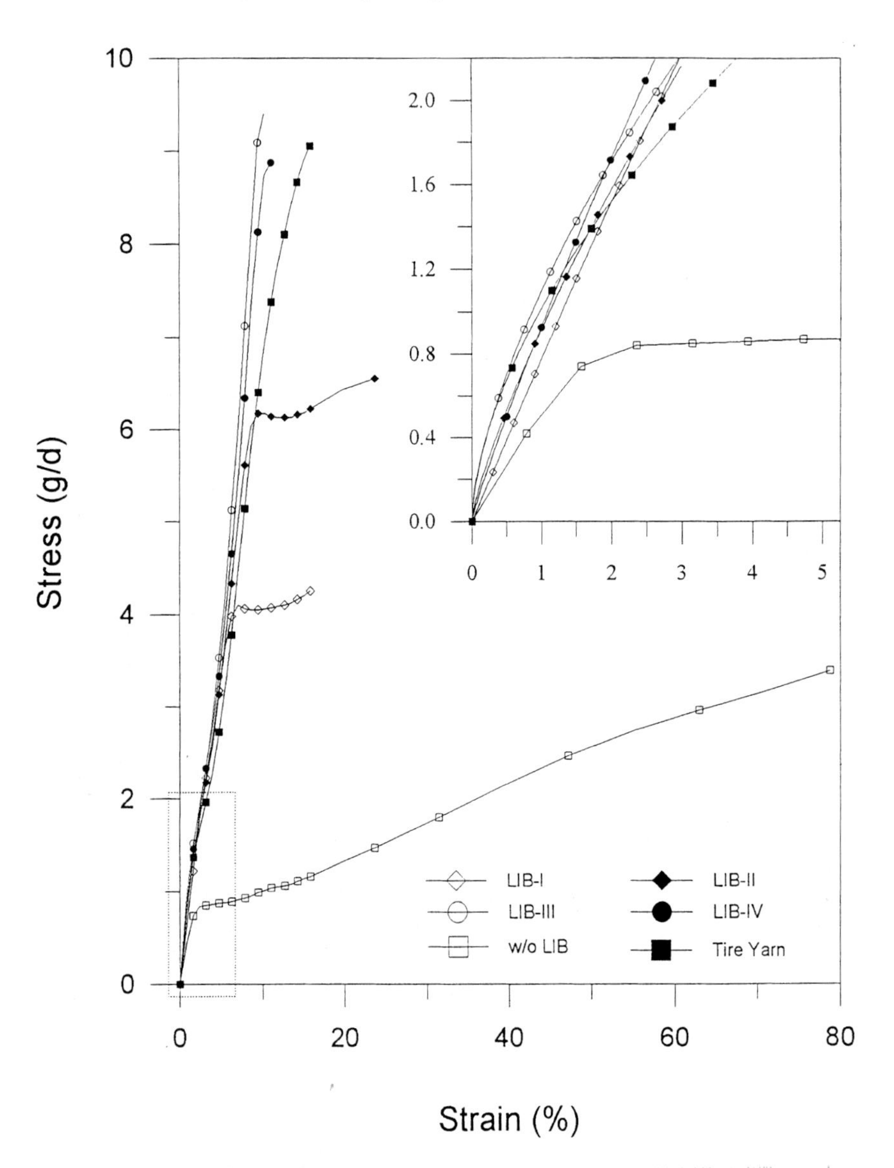

Figure 3.4 Stress-strain behavior of PET fibers produced without LIB, or with LIB at different threadline locations. A commercial tire cord yarn is also included for comparison. Ref. [29]. Reprinted by permission of John Wiley and Sons, Inc.

The availability of several pieces of state-of-the-art equipment and techniques for the on-line study of the dynamic parameters has provided the opportunity to look closely at the delicate relationships among the process variables, molecular structure and final fiber properties. A laser doppler anemometer, infra-red thermal image analyzer, optical diameter sensor and ultra-sensitive tensiometer have all been utilized to achieve real time observations of the spinline dynamics with minimum intrusion and/or turbulence to the process. With the help of this sophisticated instrumentation and technology, a critical point has been identified for the LIB location with respect to the spinline dynamics. When the LIB is located above the critical point close to the spinneret face, the tension of the entire spinline, especially in the region prior to the liquid, will be increased. Otherwise, the threadline tension in the region ahead of the liquid bath will be decreased, although the tension of the threadline in the region beyond the LIB can be dramatically increased. The critical point is thus defined as the lowest point for the top of the LIB in the spinline, in which the tension of the perturbed system is always higher than that of the unperturbed system at any comparable position in the spinline.

3.4.3 Unprecedented As-Spun Fiber Properties

The immediate gain from the control of spinline dynamics is the great opportunity to improve the final properties of the as-spun fibers using this advanced high-speed melt spinning process. Table 3.3 summarizes some of the most critical properties as well as some of the corresponding structural characteristics. Two characteristics are utilized as measures of the dimensional stability of the fibers: free shrinkage at elevated temperature of 177 °C and LASE-5, load at the specified elongation of 5 %. The average crystallite volume, and hence the number of crystallites per unit volume, was determined from wide-angle X-ray scattering (WAXS) patterns. Though the crystallite shape is unknown, it may be assumed that it reflects that of its repeating unit cell, and a simple equation to calculate the average crystallite volume for PET may be used [30]:

$$V_c = 19.16 D_{\bar{1}05} N^2 \tag{3.9}$$

where 19.16 is a geometric parameter calculated from the unit cell parameters, $D_{\bar{1}05}$ is the average crystallite dimension in a direction perpendicular to the $\bar{1}05$ plane (i.e., roughly along the chain axis direction), N^2 is the average product of crystallographic repeats in the lateral directions which may be calculated by

$$N^2 = \frac{1}{3}\left(N_{100}N_{010} + N_{100}N_{\bar{1}10} + N_{010}N_{\bar{1}10}\right) \tag{3.10}$$

and

$$N_{100} = D_{100}/d_{100},\; N_{010} = D_{010}/d_{010},\; N_{\bar{1}10} = D_{\bar{1}10}/d_{\bar{1}10} \tag{3.11}$$

Table 3.3 Properties and Corresponding Structural Characteristics of Various As-Spun PET Fibers.

Process	Spinning Speed (m/min)	Density (g/cm^3)	Apparent Crystal Size (nm)			No. of Crystals $\times 10^{18}/cm^3$
			010	100	$\bar{1}05$	
Normal	4000	1.3781	4.84	5.28	6.11	2.302
LIB-I	3000	1.3647	3.12	3.61	5.20	4.226
LIB-II	3000	1.3680	2.64	3.21	4.11	7.895
LIB-III	4000	1.3670	—	—	—	—
LIB-IV	4000	1.3600	2.30	2.51	3.40	10.61

Process	Tenacity (gf/d)[f]	Modulus (gf/d)	LASE-5[a] (gf/d)	Shrinkage[b] (%)	Δn[c]	f_a[d]	f_c[e]
Normal	4.35	51.1	0.92	—	0.115	0.22	0.90
LIB-I	5.11	83.4	2.81	5.9	0.152	0.47	0.93
LIB-II	5.33	96.1	3.35	7.3	0.170	0.58	0.91
LIB-III	6.82	101.1	3.61	10.5	0.205	0.76	0.88
LIB-IV	11.75	125.8	—	—	0.220	0.82	0.83

[a] LASE-5: Load at the specified elongation of 5 percent;
[b] Shrinkage: Free shrinkage at 177 °C hot air;
[c] Δn: Optical birefringence measured using polarized light microscope;
[d] f_a: Hermans' orientation function for non-crystalline phase;
[e] f_c: Hermans' orientation function for crystalline phase;
[f] For PET, 1 gf/d is approximately equal to 0.12 GPa.

where D_{hkl} is the average crystallite dimension perpendicular to the *hkl* plane and d_{hkl} is the Bragg *d*-spacing for this plane. Knowing the apparent crystallite volume and the volume fraction crystallinity derived from density, another morphologic characteristic, the average number of crystallites per unit volume, n, may be calculated from:

$$n = \frac{X_v}{V_c} V \tag{3.12}$$

where V is the chosen unit volume, e. g. Å^3 or cm^3. The average number of crystals per unit volume of material might be analogous to the "crosslink density" of a rubber like material in the analysis of extended chain structures and enhanced molecular connectivity (see also Section 4.6.1)

The most noticeable effect of the threadline modification is the diminished role of spinning speed on fiber structure and properties. In the spinning speed range of 3000 to 4000 m/min, the majority of the fiber properties are drastically influenced by variations of the LIB location along the threadline (Table 3.3). The average birefringence of PET as-spun fibers, Δn, for example, may be increased from 0.12, a partially oriented structure, to 0.22, a highly oriented structure. (See Section 13.3.2 for a discussion of orientation from birefringence measurements).

3.4.4 The Concept of Extended Chains and Enhanced Molecular Connectivity

The term extended chain crystals, described by Wunderlich et al. [31–36], may be applied to any crystal of a linear high polymer of sufficiently high molecular weight that shows a length in the molecular direction of at least 2000 Å. This definition is chosen because such a length corresponds to a molecular weight of at least 20,000, a value that is in the order of magnitude of the accepted lower limit of a high polymer molecule. It also represents a length above which effects due to the crystal surface at right angles to the chain axis or of end groups become negligible. In their work, chain extension in crystals was described as a two-step process: nucleation of each crystallizing molecule to a folded chain conformation, followed by an increase in fold length in a solid state reorganization step. This reorganization step was enhanced in the case of crystallization at high temperature under elevated pressure. It is thus believed that mechanical deformation during crystallization is also able to produce extended chain crystals. Of course, folding must be avoided completely to attain "fully or near-fully" extended chains.

Theoretical modeling of the formation kinetics of folded and extended-chain crystals (in the absence of mechanical stress) has been an active research area. Hikosaka has proposed a theory to explain the nucleation and growth of folded-chain crystals (FCC) and extended-chain crystals (ECC) of linear chain polymers, in which the generalized nucleation rate j for a linear sequential process is given by [37,38]:

$$j = \eta_o (kT/h) \exp[-\{G_m/kT + \Delta E_m/k(T - T_g)\}_{max}] P_s \tag{3.13}$$

Here, η_o is a constant, kT is the thermal energy at temperature T, h is Planck's constant, G_m is the free energy for forming a nucleus of the *m*th stage, $\Delta E_m = \Delta E$ (activation energy necessary for diffusion) from the *m*th to the $(m+1)$th stage, T_g is the glass transition temperature, and P_s is a survival probability of a nucleus defined as:

$$P_s = \left\{ \sum_{m=0}^{\infty} \exp[\{G_m/kT + \Delta E_m/k(T - T_g)\} - \{G_m/kT + \Delta E_m/k(T - T_g)\}_{max}] \right\}^{-1} \tag{3.14}$$

This illustrates the simple physical interpretation that *j* is given by a product of two probabilities, the probability of passing through the maximum activation barrier of formation of a nucleus, $\exp[-\{G_m/kT + \Delta E_m/k(T - T_g)\}_{max}]$, and the survival probability of the nucleus, P_s. The origin of ECC and FCC was shown to be related to the ease or difficulty of chain sliding diffusion within a crystal (nucleus or lamella), and it was concluded that there should not be a difference in the fundamental formation mechanism of these two morphologies. When chain sliding diffusion within a crystal is easy, both nucleus and lamella will easily thicken and grow two- and three-dimensionally, respectively, which will result in formation of ECC. When chain sliding diffusion is difficult, nucleus and lamella cannot easily thicken and thus they grow only one- and two-dimensionally, which will result in formation of FCC. This argument remains to be proved.

It is well known that the mechanical properties of polymers are influenced strongly by the structure of the non-crystalline regions. The degree of order, or the state of anisotropy in the non-crystalline region has been widely characterized by the degree of molecular orientation. For a fibrous product, optical birefringence and Hermans' orientation function of the noncrystalline regions (usually denoted f_a or $<P_2(\cos\theta)>_a$) are two of the most important structural characteristics of a fiber. However, the molecular orientation cannot be directly related to fiber tenacity, as is illustrated by the pronounced divergence of tenacity values corresponding to a specific birefringence value for PET fibers in the very high birefringence region of Figure 3.5. This indicates the inadequacy of using a mean measure of molecular orientation as the sole representative for the level of chain extension. Compared with the degree of molecular chain extension, the segmental orientation does not take into account effects of chain folding, chain ends and distribution of taut-tie chain lengths. The basis of the present discussion rests on the assumption that the molecular chains in a polymeric fiber may adopt any of the following conformations: regular folds, random coils or uniaxial extension. Normally, the

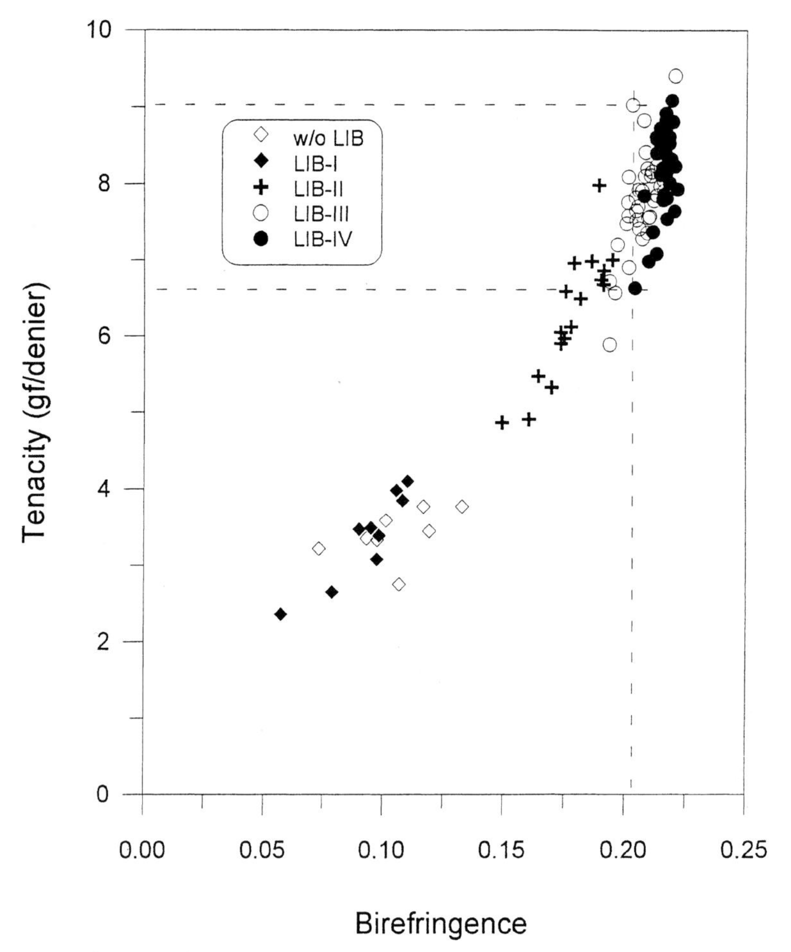

Figure 3.5 Tenacity vs. optical birefringence for PET fibers produced with or without LIB. Ref. [29]. Reprinted by permission of John Wiley and Sons, Inc.

crystallization process of polymers with flexible chains, requiring large conformational rearrangements and a significant heat of fusion, is accompanied by extensive back folding. Consequently, the mechanical properties are limited by the absence of molecular continuity. Furthermore, a high degree of non-crystalline orientation without an extensive connection with crystal blocks will not only bear external load less efficiently, but will also cause a reduction of dimensional stability.

Many structural models have been proposed to explain the remarkable enhancement of the mechanical properties of semi-crystalline polymers on drawing [39]. These models assume either crystal continuity (continuous crystals) or taut tie molecules, connecting crystallites longitudinally as a structural element, providing an efficient axial force transmission. Experimental results have also indicated that the thermal dimensional stability of a polymeric fiber may be significantly enhanced by the presence of a specific crystalline structure, rather than a large amount of crystallinity [40, 41]. "Molecular connectivity" appears to be a key determinant of important fiber properties, and it is an idea which can also be found in models devised to calculate some physicochemical properties of low molecular weight components [42,43]. To best describe an ideal molecular structure of a high performance fiber from semi-crystalline polymers, the fully stretched non-crystalline molecules should be confined tightly by the extended-chain crystal blocks, while at the same time these crystal blocks should be connected extensively by the non-crystalline molecules. The above statement describes the main concept of enhanced molecular connectivity.

3.5 Potential Applications and Future Development

The limiting spinning speed, imposed by the structural uniformity problems and the appearance of the neck-like deformation, seem to have been overcome by the inclusion of proper on-line perturbations. The great potential of this advanced melt spinning technology is not beyond attainment once a more complete understanding of the relationship between the fine molecular structure and fiber properties is achieved, coupled with judicious control of the spinline dynamics. The success of the technological advancement can also be attributed to the availability of modern machinery with substantially increased operating speeds and throughput capability, sophisticated instrumentation for precise on-line measurement and control and, especially, dramatic improvements in polymer quality.

There is strong evidence that the extended-chain macromolecular structure that we believe to exist in somc of these fibers could be revealed by several characterization methods. Spectroscopy and scattering techniques are particularly applicable to this purpose. It needs to be pointed out that currently there is no specific individual technique to prove unequivocally the existence of an extended-chain structure in polymeric fibers. However, through comparison, cross-examination and combination of data obtained from various characterization experiments, decisive reliable information can be deduced concerning molecular fine structure and structure–property correlations. In the case of the advanced melt spinning process employing the LIB, some powerful indications have been identified supporting the structural uniqueness of the fibers produced. There are, for example, high crystalline and noncrystalline orientation, laterally confined crystal growth, a large number of crystallites per unit volume of material and extensive extended chain segments. The results from thermodynamic calculations support the premise of the phase transition for a material like PET from polymer melt to extended-chain solid state. Research has been and is still being conducted with the ultimate goal of approaching this extended-chain structure during fiber formation through the route of perturbed high-speed melt spinning.

3.6 References

1. Ziabicki, A., *Fundamentals of Fiber Formation* (1976) J. Wiley & Sons, New York
2. Lin, C.Y., Ph.D. Dissertation, NCSU, Raleigh, NC (1990)
3. Ziabicki, A. in *High-Speed Fiber Spinning* (1985) Ziabicki, A. and Kawai, H., Eds., Wiley-Interscience, New York, Chap. 2
4. Reiner, M., *Deformation, Strain and Flow* (1960) H.K. Lewis, London
5. Evans, U.R., *Trans. Faraday Soc.* (1945) 41, p. 365
6. Avrami, M., *J. Chem. Phys.* (1939) 7, p. 1103
7. Avrami, M., *J. Chem. Phys.* (1940) 8, p. 212
8. Avrami, M., *J. Chem. Phys.* (1941) 9, p. 177
9. Kolmogoroff, A.N., *Izvestiya Akad. Nauk USSR, Ser. Math.* (1937) 1, p. 335
10. Mehl, R.F. and W.A. Johnson, *Trans. Am. Inst. Mining Met. Engrs.* (1939) 135, p. 416
11. Mandelkern, L., *J. Appl. Phys.* (1955) 26, p. 443
12. Nakamura, K., Watanabe, T. Katayama, K. and Amano, T. *J. Appl. Polym. Sci.* (1972) 16, p. 1077
13. Nakamura, K. Katayama, K. and Amano, T. *J. Appl. Polym. Sci.* (1973) 17, p. 1031
14. Nakamura, K. Watanabe, T. Amano, T. and Katayama, K. *J. Appl. Polym. Sci.* (1974) 18, p. 615
15. Ziabicki, A., *Colloid Polym. Sci.* (1974) 252, p. 207
16. Raghavan, J.S., Ph.D. Dissertation, NCSU, Raleigh, NC (1996)
17. Ishizuka, O. and Koyama, K. in *High-Speed Fiber Spinning* (1985) Ziabicki, A. and Kawai, H. Eds., Wiley-Interscience, New York, Chap. 6
18. Ohta, T., *Polym. Eng. and Sci.* (1983) 23, p. 13
19. Perepelkin, K.E., *Fiziko- Khimicheskaya Mekhanika Materialov* (1970) 2, p. 2
20. Termonia, Y. and Smith, P. *Polymer* (1986) 27, p.1845
21. Smith, K.J., *Polym. Eng. and Sci.* (1990) 30, p. 8
22. Bigg, D.M., *Polym. Eng. and Sci.* (1988) 28, p. 13
23. Ito, M. and Takahashi, K. and Kanamoto, T. *J. Appl. Polym. Sci.* (1990) 40, p. 1257
24. Billmeyer, F.W., *Textbook of Polymer Science* (1984) J. Wiley & Sons, New York
25. Cuculo, J. A., et al., U. S. Patent, 4,909,976 (1990)
26. Cuculo, J. A., et al., U. S. Patent, 5,149,180 (1992)
27. Cuculo, J. A., et al., U. S. Patent, 5,171,504 (1992)
28. Lin, C.-Y., Tucker, P. A. and Cuculo, J. A., *J. Appl. Polym. Sci.* (1992) 46, p. 531
29. Cuculo, J. A., et al., *J. Appl. Polym. Sci.* (1995) 55, p. 1275
30. Miller, R.W., Southern, J.H. and Ballman, R.L. *Text. Res. J.* (1983), 53, p. 670
31. Wunderlich, B. and Davidson, T., *J. Polym. Sci., Part A-2* (1969) 7, p. 2043
32. Davidson, T. and Wunderlich, B., *J. Polym. Sci., Part A-2* (1969) 7, p. 2051

33. Prime, R. and Wunderlich, B., *J. Polym. Sci., Part A-2* (1969) 7, p. 2061
34. Prime, R. and Wunderlich, B., *J. Polym. Sci., Part A-2* (1969) 7, p. 2073
35. Prime, R., Wunderlich, B. and Melillo, L., *J. Polym. Sci., Part A-2* (1969) 7, p. 2091
36. Gruner, C., Wunderlich, B. and Bopp, R., *J. Polym. Sci., Part A-2* (1969) 7, p. 2099
37. Hikosaka, M., *Polymer* (1987) 28, p. 1257
38. Hikosaka, M., *Polymer* (1990) 31, p. 458
39. Porter, R. S., et al., *J, Polym. Sci., Polym. Phys. Ed.* (1985) 23, p. 429
40. Rim, P. B. and Nelson, C. J., *J. Appl. Polym. Sci.* (1991) 42, p. 1807
41. Cuculo, J. A., et al., *J. Polym. Sci., Polym. Phys. Ed,* (1995) 33, p. 909
42. Pogliani, L., *J. Phys. Chem.* (1993) 97, p. 6731
43. Pogliani, L., *J. Phys. Chem.* (1994) 98, p. 1494

4 Draw-Induced Structure Development in Flexible-Chain Polymers

David R. Salem
TRI/Princeton, Princeton, New Jersey, USA

4.1 Introduction

The production of fibers from flexible-chain, organic polymers involves forming the polymer into filaments and extending them uniaxially in order to orient the molecules in the direction of the applied strain. To impart molecular orientation, the rate of molecular relaxation must be long compared to the time available for relaxation during filament extension. As discussed in Chapter 2, melt spinning at high speeds (typically in excess of 3000 m/min) produces sufficient tensile stresses in the rapidly cooling polymer to induce significant molecular orientation and the formation of oriented crystallites. At lower spinning speeds, however, little or no orientation is induced and a sequential drawing step is required to provide adequate alignment of the molecules.

Disadvantages of high-speed melt spinning are that orientation of molecules in the noncrystalline regions of the fiber is usually low compared with that attainable by drawing the filament after spinning at low (or high) speeds [1–3], and significant crystal chain folding is present. Although the resulting tensile properties are adequate for a number of textile applications, the tensile modulus and tenacity of fibers required for 'high performance' and advanced engineering applications (e.g.: fiber-reinforced composites for aerospace, automobiles and construction; tire cord; ropes; parachutes; satellite tethers; biomedical implants) can generally be obtained only by extending and orienting the molecules in a drawing operation following spinning.

Nevertheless, conventional drawing processes do not provide full alignment of molecular chains, and fiber modulus and tensile strength fall far short of the theoretical values calculated for fully extended chains, as discussed in Chapter 5. The (calculated) ultimate axial modulus of common, fiber-forming polymers is in the range 100–350 GPa, whereas the moduli achieved in conventional drawing processes are about 10% of these values. The ultimate tensile properties of these flexible-chain polymers are even more remarkable when considered in relation to their densities. The ultimate tensile modulus of polyamide 6, for example, would be similar to that of steel, but its specific modulus would be an order of magnitude higher.

The principle factors limiting chain extension appear to arise from constraints imposed by molecular entanglements, chemical interactions between chains, and crystallites. To obtain fully extended chains, a randomly coiled polymer with a molar mass of 10^6 would need to be strained to a draw ratio of about 80, whereas (in the absence of chain slippage) the maximum attainable draw ratio from a conventional drawing operation is about 5.

For apolar polymers such as polyethylene and polypropylene, where intermolecular forces are minimal, methods for overcoming this limitation were discovered in the 1970s [4–9] and 1980s [10–15]. In the method leading to the highest modulus values, ultra-high molecular weight polyethylene fibers are spun from semi-dilute solutions (gels), resulting in filaments that can be subsequently drawn to draw ratios in excess of 50, even after complete removal of the solvent prior to drawing [10–15]. The gel spinning and ultra-drawing route has provided commercial polyethylene fibers with tensile modulus in the region of 170 GPa and tensile strength of 3–4 GPa. The high chain extensions appear to result from solvent-induced dissolution of chain entanglements trapped between crystals, and from the ductility of PE crystals at drawing temperatures above the crystalline α relaxation temperature.

For polymers with significant polar interactions, however, the gel-spinning method does not provide suitable precursor material for ultra-drawing because inter-chain interactions remain active in the solution-crystallized polymer, and the crystals provide rigid crosslinks impeding chain extension. Since polar polymers have higher melting temperatures and higher creep resistance than polyethylene, there remains considerable interest in devising a drawing method that would result in fibers with tensile properties closer to the theoretical values for these polymers. Incremental progress has been made in this direction, but a breakthrough will probably require approaches involving greater control over the strength and distribution of inter-molecular interactions, and based on a fuller understanding of the interplay between drawing variables (e.g. temperature and strain rate) and microstructure evolution (e.g. the kinetics of crystallization and molecular relaxation). There would also be benefit from improved understanding of the influence of the evolved microstructure attained under one set of drawing conditions on the development of microstructure in a subsequent draw step with different conditions.

Although the achievement of ultimate tensile properties in fibers from flexible-chain polymers is an important goal, it is not the only (practical) reason to seek better understanding of structure formation during drawing. For many applications, a significant degree of molecular disorder is necessary. The diffusion of dyes, ultra-violet stabilizers and other auxiliaries into a fiber requires substantial polymer chain mobility at temperatures above T_g, which can only occur in noncrystalline regions having chains that are not fully extended [16]. The ability of a fiber to recover from reasonably large strains requires disordered regions distributed in a suitable crystallite network [17]. Heat-setting of fibers, essential in numerous commercial applications, involves relaxation of stresses in the disordered regions, and optimizing heat-setting efficiency requires control of the crystalline structure which tends to constrain this relaxation [18]. It has recently been found that the susceptibility of PET fibers to photodegradation is influenced by various details of the polymer microstructure [19]. For example, intermediate levels of molecular orientation provide better light stability than highly oriented chains because the latter result in catastrophic delamination in the chain axis direction.

Moreover, the aim is often to engineer a precise balance of various fiber properties rather than to achieve a single, ultimate property. Therefore, scientific understanding of process/structure/property relations in fiber drawing is not only of fundamental importance in the quest for enhanced tensile properties, but also in the control of structure and properties for specific applications.

The mechanisms of microstructure evolution during drawing of a given polymer are not only influenced by strain rate and the draw temperature, but are also dependent on the state of order in the precursor filament. For slow crystallizing polymers, the precursor usually consists of a randomly oriented (isotropic) distribution of uncrystallized molecules. For rapidly crystallizing polymers that cannot be quenched to the amorphous state, the precursor filament generally consists of crystalline spherulites or row-nucleated crystal structures. High-speed-spinning can provide partially oriented and partially crystalline precursors, and in multi-step drawing the fiber enters each step in a different microstructural state, which can range from unoriented and amorphous to highly oriented and crystalline.

This chapter reviews the physics of structure development during drawing of various fiber-forming, flexible-chain polymers from different states of precursor order. The aim is to convey principles and basic understanding, while indicating practical implications along the way. All fibers are not given equal treatment, and some are omitted completely. This reflects the fiber's commercial importance, the amount and quality of research reported on it, and the belief that principles are better conveyed from detailed information on fibers typifying a chemical type than from pieces of information on all the available fibers. Cellulosic fibers are not discussed, because they are covered in Chapter 8, and some aspects of polyethylene fiber drawing are treated cursorily, because a detailed review is provided in Chapter 5.

4.2 Overview of Stress-Strain-Structure Relationships

4.2.1 Modes of Deformation

Fiber drawing involves the imposition of large strains to spun polymer filaments, at strain rates and temperatures that generate molecular orientation along the chain axis and enhanced mechanical properties. Although drawing can involve various temperature and strain rate sequences, there are essentially three kinetic modes of deformation: (1) constant extension rate (CER); (2) constant strain rate (CSR); and (3) constant force (CF).

Fiber drawing in laboratory research has frequently been done at a constant rate of extension, typically using tensile testing apparatus with a fixed upper jaw and a moveable lower jaw, and a furnace surrounding the fiber. During this type of deformation, the lower jaw descends at a constant velocity, and the true strain-rate of the fiber continuously decreases from its nominal value such that:

$$\dot{\varepsilon} = (t + l_o/v)^{-1} \tag{4.1}$$

where t is the draw time, l_o is the original specimen length and v is the extension rate.

Constant strain rate drawing has been used in some studies of polymer drawing, in order to simplify analysis of time-dependent effects [20–22]. As with CER drawing, CSR drawing is usually performed on tensile testing equipment, but the extension rate (or lower jaw speed) is continuously adjusted to maintain a constant rate of strain and

$$\lambda = \exp(\dot{\varepsilon}t) \tag{4.2}$$

Whereas CER drawing is sometimes used in the industrial manufacture of polymer film (tenter process), neither CER nor CSR drawing are used in commercial fiber formation because they cannot provide a continuous drawing process for fibers. In most commercial processes, undrawn filaments are supplied at constant velocity via a rotating feed roll, and taken up on a roll (or rolls) at higher speed, thereby causing extension of the filaments in proportion to the ratio of the take-up and feed velocities. The filaments in the drawing zone are often heated by contact with a hot plate, or by passage through a hot liquid or gaseous medium. In this continuous process, the applied force is constant, and the deformation kinetics of the fiber can be rather complex, as will be discussed later. A reasonable laboratory simulation of drawing between rolls is deformation under a (high) constant load, i.e. a creep experiment at high strain rate.

A limitation of CER experiments is that the maximum strain rate attainable in conventional tensile testing instruments is an order of magnitude lower than that commonly reached in the continuous (CF) drawing process. Since the differences in the magnitude and the evolution of the deformation kinetics in CER and CF drawing have been found to give rise to significant differences in the evolution of microstructure, information obtained from CER drawing cannot be directly applied to CF drawing. Nevertheless, results from CER experiments have provided important scientific understanding of the factors controlling strain-induced structure formation, and they have greatly assisted the analysis and interpretation of microstructure data obtained under the more complex deformation kinetics of CF drawing (see Sections 4.2.3 and 4.3.2).

Since filaments entering the drawing process are usually cylindrical and have a high length to diameter ratio, the decrease in diameter is proportional to the increase in length and the geometrical mode of deformation is simple (uniaxial) extension. In film drawing, the deformation can generally be considered uniaxial if the length/width aspect ratio is greater than 5, and results obtained from uniaxial drawing of film are directly applicable to fiber drawing. Data obtained from polymer rods and 'dogbones' a few millimeters thick can also be relevant to fiber drawing, but adiabatic effects are more likely to occur due to slow heat transfer, and the low cooling rate at the center of the specimen may cause structural changes after the completion of drawing. Studies on film drawn at constant width (where the increase in length is at the expense of film thickness, with no change in width) can be relevant to fiber drawing, because results from PET film show the development of orientation and crystallization to be similar in constant width and uniaxial modes of deformation [3,23]. The only structural difference is that, in addition to chain axis orientation, one of the lateral crystal planes tends to align preferentially with the film surface during constant-width deformation, whereas crystallites formed during uniaxial drawing of PET have no preferred orientation about the chain axis (i.e. they possess cylindrical symmetry) [23,24]. The uncrystallized chains, however, seem to possess cylindrical symmetry in both deformation geometries [25], and the planar orientation of the crystallites in constant-width drawing of PET does not seem to significantly influence the crystallization kinetics or the crystallite dimensions obtained [3,23,26]. Therefore, this review does not confine itself to the literature on drawing of fibers, which would not provide adequate insights. Where applicable, it includes information from investigations of film drawing in uniaxial and constant-width modes, and from drawing studies on (thick) molded specimens.

4.2.2 Constant Extension Rate Deformation

Flexible-chain polymers used for fiber formation are generally viscoelastic materials, and their stress-strain response is highly dependent on their molar mass, their structural state prior to the deformation, the evolution of their structure during the deformation, and the draw-temperature and strain rate applied.

We will start by considering amorphous polymers that crystallize during drawing. Of the crystallizable polymers that are usually drawn in the amorphous state, poly(ethylene terephthalate) is commercially the most important, and has been studied the most thoroughly, but others include poly(butylene terephthalate) PBT, poly(trimethylene terephthalate) PTT, poly(ethylene naphthalate) PEN, poly(lactic acid) PLA, and various polyketones. This will be followed by descriptions of the drawing behavior of crystalline polymers (e.g. polyethylene PE, polypropylene PP and polyamides PA6 and PA66) and of amorphous polymers that do not crystallize (e.g. poly(methyl methacrylate) PMMA). We will then discuss the phenomena of 'necking' and self-heating, which are applicable to all these polymers.

4.2.2.1 Amorphous-Crystallizable Polymers

Figure 4.1 shows typical stress-strain curves for amorphous, isotropic poly(ethylene terephthalate) film, CER-drawn at 90 °C (~15 °C above T_g) at various strain rates. The first rise in stress reflects the deformation of a network of entangled chains and induces significant molecular orientation (Figure 4.2a) [3,23,27]. Crystallization onset E_1 always coincides with the inflection point in the true-stress vs. strain curve, and is likely to cause it [3,23,26,27].

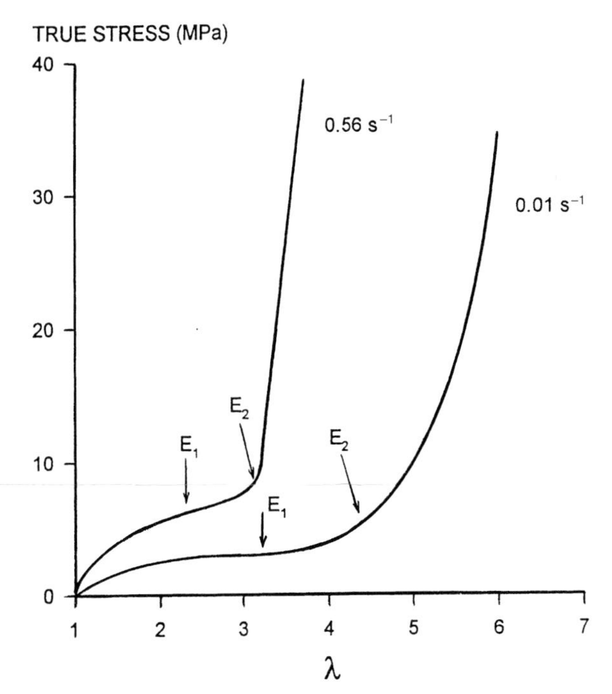

Figure 4.1 True stress versus draw ratio of PET film at two strain rates and a draw temperature of 90 °C. The onset of crystallization E_1 and the onset of regime 2 crystallization E_2 are indicated. (Adapted from Ref. [27]).

As shown in Figure 4.2, crystallinity and noncrystalline orientation develop rapidly between E_1 and E_2 (crystallization regime 1), while stress develops slowly (Figure 4.1). The sharp upturn in stress coincides with the transition to the second crystallization regime E_2. Here, a characteristic level of crystallinity (dependent on draw temperature but independent of strain rate) is reached and the rate of development of both crystallinity

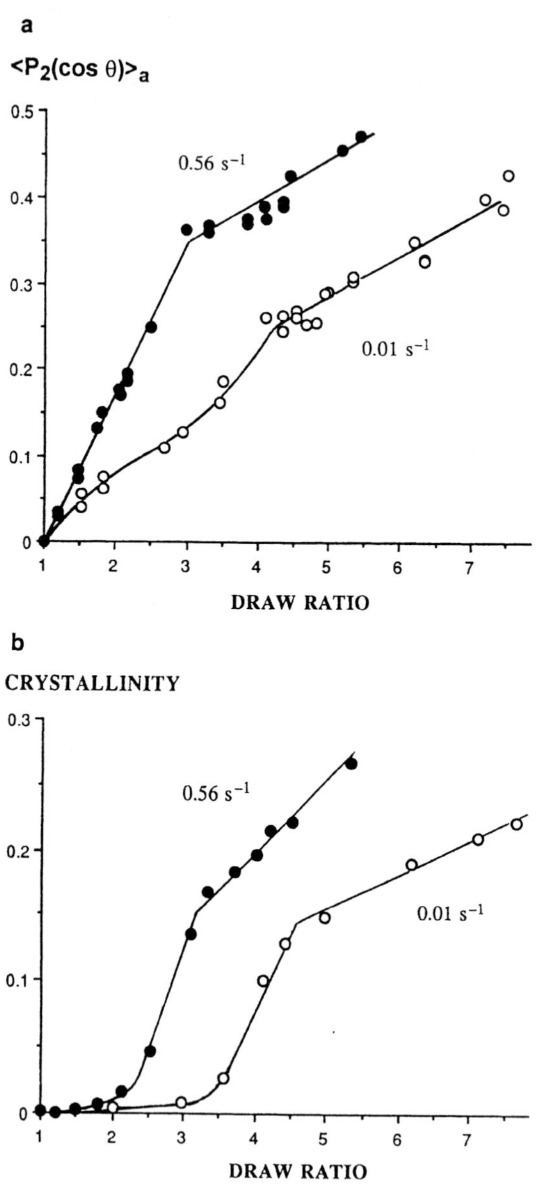

Figure 4.2 Development of (a) noncrystalline orientation (from intrinsic fluorescence) and (b) volume fraction crystallinity (from density), during drawing of PET film at 90 °C, showing the influence of strain rate. The critical orientation for onset of crystallization is evidently in the range 0.15–0.18. (Adapted from Ref. [3]).

and noncrystalline orientation abruptly decreases (Figure 4.2). There is now various evidence [23,27,28–30], including a comparison of the stress-strain-orientation behavior of PET with that of uncrystallizable poly(ethylene methyl terephthalate) PEMT, that a crystallite network providing rigid junction points has formed by E_2, and that the increase in stress largely results from an increase in polymer viscosity arising from the interconnection of crystallites [27]. Entanglements act as impermanent crosslinks that can slip and relieve stress in a time-dependent manner. At temperatures above T_g, there is sufficient chain mobility for slippage of entanglements to occur, especially at moderate to high draw ratios, and in the absence of the constraints to slippage imposed by crystallization, the upturn in stress would be much smaller. The slowing of orientation in the high-stress region may result from the formation of taut intercrystalline tie chains, which would halt the uncoiling of neighboring tie-chains that are not fully extended [3]. Deformation would then proceed via translational slippage between groups of crystallites held together by extended tie-chains (Figure 4.3). The fact that the upturn in stress occurs at a characteristic crystallinity, independent of strain rate, is indicative of a percolation transition and quantitative support for this notion has recently been obtained [26].

At temperatures above T_g, increasing strain rate increases the initial stress required to deform the entanglement network (Figure 4.1) [24] and accelerates the development of molecular orientation with draw ratio (Figure 4.2a) [3]. These effects arise from the reduction in time available for relaxation of stress (and orientation) with increasing rate of deformation. Faster development of orientation at higher strain rates leads to earlier onset of crystallization (E_1) (Figure 4.2b) [3,24,31] and a shift in the upturn of stress to lower draw ratios (Figure 4.1). Conversely, decreasing strain rate below a critical level dependent on draw temperature, will result in 'flow drawing' [24,32]. In this regime, the time available for relaxation is large enough that little or no orientation takes place, orientation-induced crystallization does not occur, and the upturn in stress is absent.

Increasing draw temperature has a similar, but not necessarily equivalent, effect to decreasing strain rate. With decreasing temperature, the initial rise in stress increases faster and reaches higher values and, in general, the final upturn in stress is shifted to lower draw ratios (Figure 4.4). As the temperature decreases below T_g, a downturn in

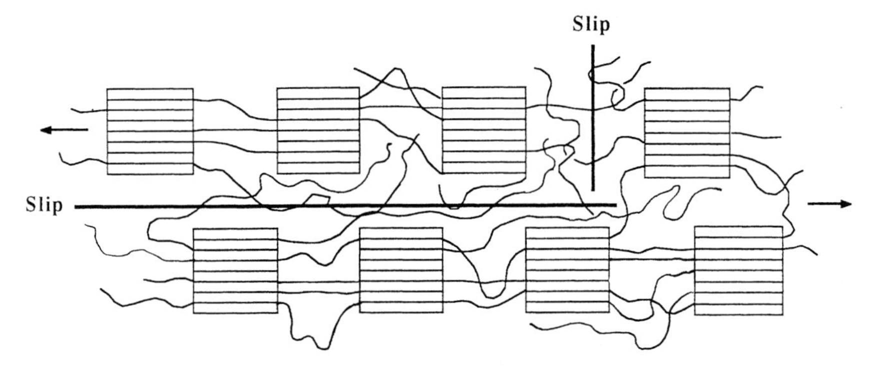

Figure 4.3 Proposed mechanism for extension of PET in the high stress regime, involving slippage of 'tie-segment blocks', in which crystals are linked by at least one extended tie chain [3]. Further alignment of the less oriented chains in the intecrystal regions is impeded. (This schematic illustrates a deformation mechanism and is not intended to provide a realistic representation of polymer morphology or topology.)

stress appears after the initial stress rise, which becomes more pronounced as temperature is decreased further. Well below T_g (usually around $T_g - 60\,^{\circ}C$), the polymer behaves like a brittle solid. These temperature effects reflect changes in molecular mobility due to thermal energy, i.e. higher temperature results in higher rates of molecular relaxation, decreasing the resistance of the polymer network to deformation. Temperature dependence of the relaxation rate causes (noncrystalline) molecular orientation to increase more slowly with draw ratio at higher temperatures [33,34] and the magnitude of the strain-rate dependent shift in the onset of crystallization increases with temperature [31]. Another effect of enhanced molecular mobility is an increase in the crystallization rate at a given level of noncrystalline orientation, and the interplay between the rate of molecular relaxation and the rate of crystallization has important consequences for structure development, as will be discussed in Section 4.3.2. It should be noted that at drawing temperatures below T_g, there is insufficient molecular mobility for entanglement slippage or stress-induced crystallization to occur, and the final upturn in stress does not result from the crosslinking and reinforcing effect of crystals, but from increasing molecular orientation in the plastically deformed material.

The influence on the stress-strain curve of increasing molecular weight is very similar to that of decreasing temperature (Figure 4.4). The higher the molecular weight the greater the entanglement density and the slower the rate of molecular relaxation. This results in a polymer that is more resistant to deformation, and in which molecular orientation develops more rapidly with draw ratio [34,35]. Consequently, crystallization is induced earlier, and the inflection point (E_1) and the final upturn in stress are shifted to lower draw ratios [34,35].

It is well known that certain chain conformations in the amorphous phase of a polymer may not be present in its crystalline phase and vice versa. The glycol unit in the amorphous phase of PET contains both *gauche* and *trans* conformations, whereas glycol units in a crystalline environment are always in the (extended) *trans* conformation [36]. Infrared studies have shown that drawing of amorphous PET imparts orientation mostly

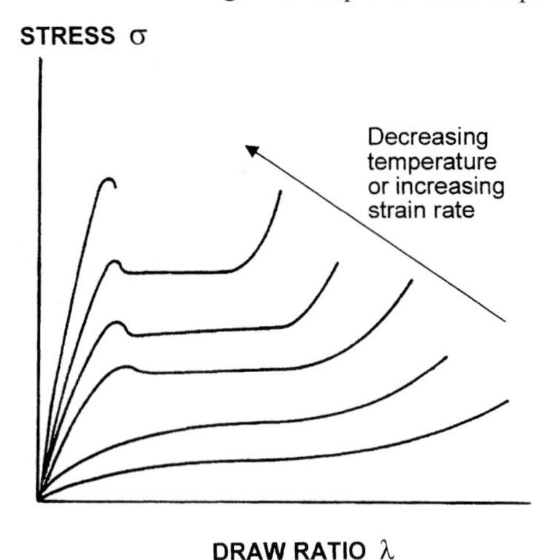

Figure 4.4 Typical stress-strain curves for an unoriented polymer, showing the influence of temperature and strain rate

to *trans* glycol units, as well as to the benzene rings, and that the *trans* fraction increases at the expense of the *gauche* content [28]. Orientation-induced crystal nuclei appear to be formed from neighboring chains in which the *trans* glycol units and the benzene rings have become highly oriented in the chain-axis direction.

Drawing of partially oriented PET has been investigated by Brody [37] and by Long and Ward [38]. Fibers with various levels of pre-orientation, obtained by spinning at a range of speeds, were drawn at room temperature and above T_g. Horizontal shifting of the stress-strain curves generated from cold drawing showed them to be essentially superimposable. This implies that the total network extensibility consisting of the strains imposed in the spinline and in subsequent cold-drawing (below T_g) is constant, and that the nature of the network is independent of spinning speed. The results of hot-drawing, however, indicate that there are in fact progressive changes in the network with spinning speed, which are only revealed through deformation mechanisms active above T_g. In particular, it is envisaged that the contour-length distribution of tie-molecules becomes less uniform as spinning speed increases. From our earlier discussions, it is also apparent that hot-drawing will modify the original network structure through entanglement slippage and orientation-induced crystallization.

4.2.2.2 Crystalline Polymers

So far we have considered the stress-strain response of amorphous polymers that crystallize in the course of deformation. In these polymers, stress-induced molecular orientation reduces the energy barrier between the amorphous material and the crystalline state, permitting crystallization at temperatures where none would occur in the unoriented polymer. Induction of high crystallinity levels in the undrawn polymer, by application of high temperatures or a suitable solvent [39], can result in a brittle material that is difficult or impossible to draw. In general, increasing crystallinity in any unoriented polymer increases the stress required to deform it, and shifts the stress-strain response towards brittle deformation.

Some polymers with high levels of unoriented crystallinity can, however, be readily drawn. Polyethylene and isotactic polypropylene, for example, cannot be quenched to the amorphous state due to their high crystallization rates, but the relatively high ductility of PE and PP crystals permits highly effective drawing of these polymers.

An unoriented, crystalline polymer generally consists of spherulitic structures, formed by radial growth of stacks of parallel crystal lamellae from a central nucleus (Figure 4.5a) [40-43]. Chains fold back at the surface of each lamella, and their extension in the chain-axis direction is in the region of 10 nm (Figure 4.5b). The noncrystalline component consists of crystal defects, free chain ends, chain folds and interlamallar tie-chains. Spherulites that have developed into complete spheres and are undeformed have no overall orientation, but each lamalla stack is highly anisotropic. Spherulite size varies from less than one micrometer to hundreds of micrometers, depending on the crystallization conditions, and spherulitic growth is often truncated by impingement with neighboring spherulites. The final shape of truncated spherulites is usually polygonal [44] and, when the nucleation density is very high, their development may not progress beyond the formation of randomly oriented stacks of parallel lamallae. If the stresses in

(a)

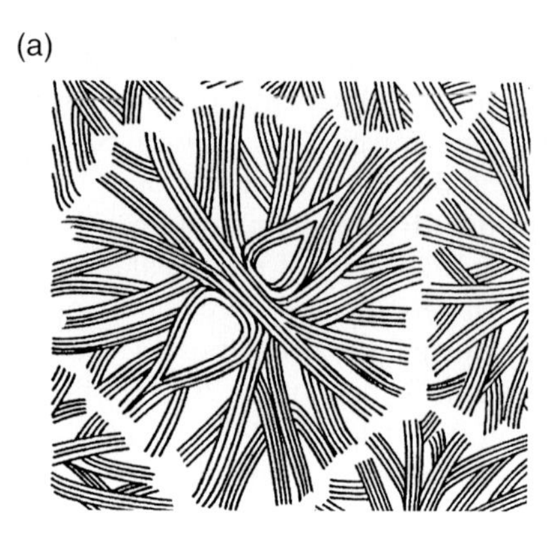

(b)

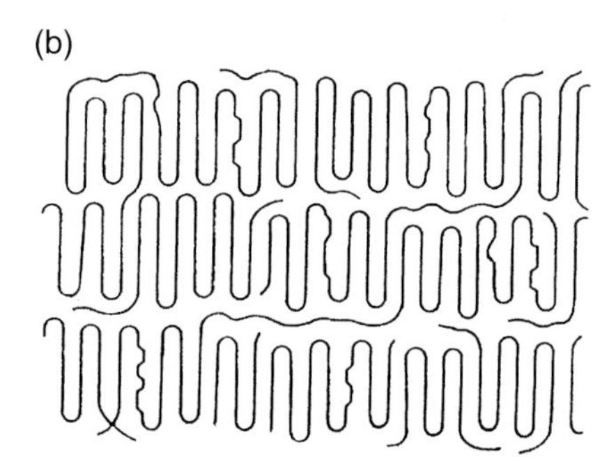

Figure 4.5 (a) Model of spherulite structure in an unoriented crystalline polymer, showing the stacks of parallel lamellae; (b) model of a stack of parallel lamellae of the spherulite structure. (Adapted from Ref. [42]).

fiber spinning are high enough, but not too high, row nucleated crystal structure are formed [45,46]. These consist of fibril nuclei oriented in the fiber axis direction, onto which chain-folded crystals grow epitaxially, in a direction perpendicular to the fibril axis (as described more fully in Section 2.2.1.2 and Figures 2.14 and 2.15). In some instances row-nucleated material and spherulites coexist.

The stress-strain response of crystalline polymers depends on temperature and strain rate in a similar way to amorphous polymers (Figure 4.4), although the molecular rearrangements taking place are different (Figure 4.6). In the early stages of drawing an unoriented crystalline polymer, spherulites become elongated in the draw direction [47–50], and in the region of the yield point, chain tilting and slipping occur within the lamellae of the chain-folded crystals. Then, if the drawing temperature is high enough, chains partially unfold and the lamellae break up into small crystallites connected to each other by uncrystallized tie molecules, forming 'microfibrillar' structures [41,42,51]. At high draw ratios, deformation involves the sliding motion of microfibrils past each other. Relatively recent studies, using neutron scattering experiments, indicate that the thinnest lamellae escape the breaking-up stage and are simply rotated and aligned, and that high

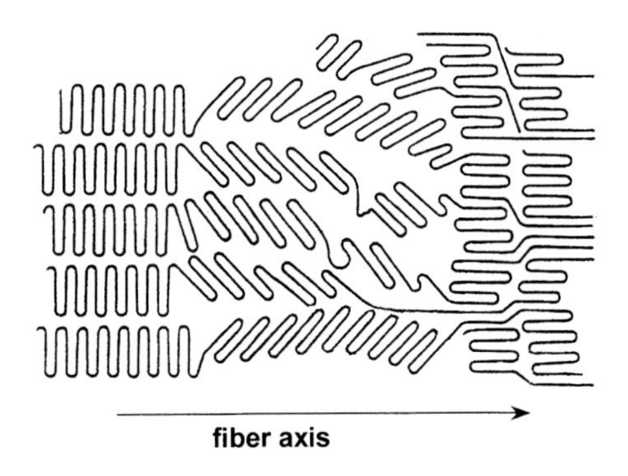

Figure 4.6 Molecular mechanism of plastic deformation of parallel lamellae in a polymer crystal [41].

draw temperatures cause some chain refolding and crystal thickening during the yield (and necking) process [52]. The strain-hardening effect at high draw ratios is a consequence of the large surface to volume ratio of the microfibrils, which provides strong resistance to their shearing displacement [43]. Within a certain temperature range, increasing the thermal energy during drawing will reduce the mechanical energy required to break-up and reorient the crystals, but at temperatures above this range, increased chain slippage may reduce the efficiency of the orientation process. In both cases, increasing temperature will result in a less steep rise in stress at the start of the deformation and to a lower yield stress. Chain slippage would extend the region of plastic deformation prior to the final upturn. In addition to these temperature effects, the molecular rearrangements giving rise to the stress-strain response of crystalline polymers are time dependent. Increasing the strain rate produces a faster rise in initial stress, a higher yield stress, and a shorter flow region.

The stress-strain behavior of crystalline polymers is also dependent on the strength of the intermolecular bonds and the crystal morphology. When drawn in a suitable temperature range, the crystallites in apolar polymers will provide little resistance to deformation and will unfold under low drawing stresses, whereas crystallites in polymers of high polarity will provide greater resistance to deformation at any temperature below the melting point. Similarly, some crystal modifications will offer more resistance to deformation than others. For example, more force is required to draw polypropylene when it has crystallized in the thermodynamically stable monoclinic form than when it mostly contains the less stable hexagonal crystal structure [53–57]. In certain crystalline polymers, drawing can cause transformation of one crystal morphology to another. This occurs in polyamide 6, where drawing stresses convert the twisted molecular configuration of the γ form to the extended α form (Figure 4.7) [58–62].

Polyamide 6 and 66 are usually partially crystalline prior to drawing, but in the as-spun state the level of crystallinity does not exceed about 35%. By contrast, the degree of crystallinity in as-spun PE and PP is generally between 50 and 80%. Polyamide fibers can be drawn effectively under a wide range of drawing temperatures. Below T_g, the efficiency of chain orientation during drawing of PA66 is high, but crystallinity stays the same [26] or may decrease due to disruption of the crystallites [63]. Chain orientation is less pronounced during drawing in the region of T_g and, while drawing above T_g does not produce higher chain orientation, significant crystallization is induced in this temperature

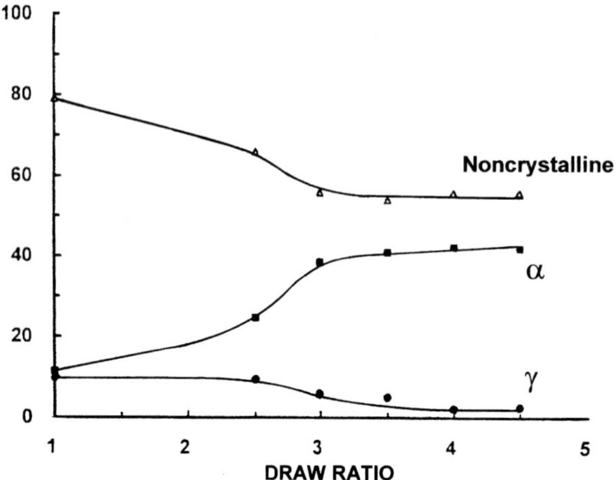

Figure 4.7 Change of phase content during drawing of polyamide 6 (adapted from Ref. [62]).

region. Figure 4.8 shows the development of crystalline and noncrystalline orientation during drawing of PA66 from the unoriented state at room temperature. The rapid development of crystalline orientation $<P_2(\cos\theta)>_c$ to almost full fiber-axis orientation compared with the slower develoment of orientation in the noncrystalline regions $<P_2(\cos\theta)>_a$ is common behavior for flexible-chain polymers. Similar behavior is shown for PA6 in Figure 4.9, where drawing at room temperature was performed on partially oriented fibers. Conventional drawing processes rarely cause $<P_2(\cos\theta)>_a$ to exceed 0.6, but this is substantially higher than the noncrystalline orientation values obtained in high-speed fiber spinning, where $<P_2(\cos\theta)>_a$ is usually less than 0.3.

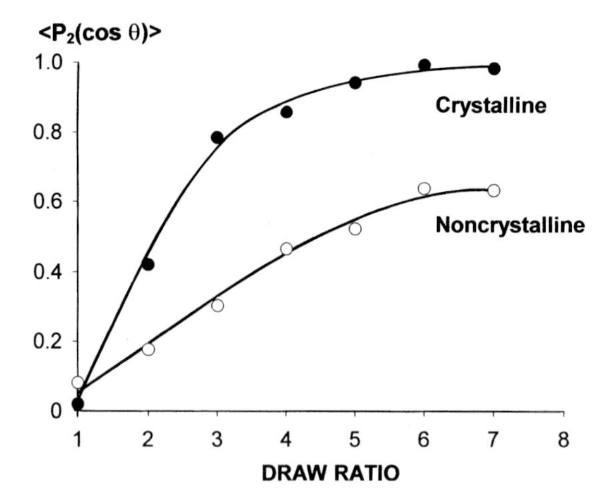

Figure 4.8 Development of molecular orientation in the crystalline and noncrystalline phases of polyamide 66 during drawing from the unoriented state at room temperature. (FTIR dichroic measurements)

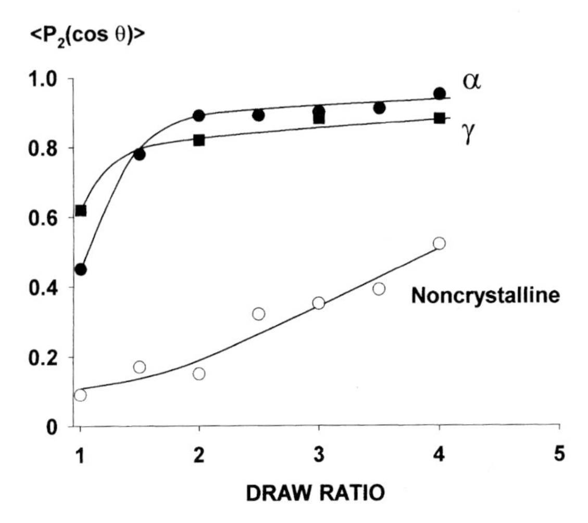

Figure 4.9 Development of molecular orientation in the α, γ and noncrystalline phases of polyamide 6 during drawing from a partially oriented state at room temperature. (WAXS and FTIR dichroic measurements)

Row nucleated structures with orientation of lamallae perpendicular to the fiber axis have been found to occur in as-spun PA6 fibers under suitable spinning conditions [59] (as well as in as-spun PP and PE fibers). Recent characterization of PA6 fibers spun at 800 m/min suggests that a fraction of the amorphous phase may also orient perpendicular to the fiber axis [64]. Subsequent drawing at 90 °C seems to cause the fraction of amorphous material having perpendicular orientation to increase at first and then to decrease rapidly once a draw ratio of 2 is reached. On the other hand, the fraction of amorphous material oriented along the chain axis increases continuously with draw ratio.

4.2.2.3 Uncrystallizable Amorphous Polymers

Some polymers do not crystallize at all, and the absence of strain-induced crystallization reduces the strength of the final upswing in stress and may eliminate it. For example, recent studies on poly(methyl methacrylate) PMMA, a noncrystallizing polymer, demonstrate that above T_g the strength of the stress upturn diminishes substantially with increasing temperature [65], reflecting increasing slippage of entanglements. By contrast, studies of (crystallizable) PET show that the rate of strain-stiffening is independent of draw temperature in the range 80–100 °C [26,66]. As mentioned earlier, this is because the formation of crystallites provides permanent network junction points, which increase the viscosity associated with entanglement slippage.

4.2.2.4 'Necking' Phenomenon

For both crystalline and amorphous polymers, a reduction in *nominal* stress following the initial stress rise (Figure 4.4) has often been associated with inhomogeneous strain, or 'necking'. Deformation involving neck propagation is often referred to as 'cold drawing', although at high enough strain rates and in some crystalline polymers, necking can occur at temperatures above T_g. Necking is characterized by localized extension and a corresponding decrease in cross-sectional area at some axial position along the specimen. In most polymers, this initial instability is followed by stable propagation of the neck up and down the specimen, which halts the decline in stress. A common misconception, however, is that a drop in nominal stress must necessarily be due to necking. The reduced local cross-section in the necked region inevitably leads to a decrease in nominal stress (until strain-hardening stabilizes the neck), but a similar decrease in stress can result from uniform strain-softening throughout the specimen. Homogeneous deformation accompanied by decreasing nominal stress has been observed during drawing of PET film at temperatures in the range 80–96 °C [3,26,66]. Recently Buckley and coworkers have monitored this behavior closely, using video recordings of the deformation [65]. At temperatures between 75 and 80 °C, some necking of the PET film occurred, which healed as drawing proceeded due to strain-hardening. At higher temperatures, and sufficiently high strain rates, the strain was essentially uniform even though nominal stress was initially falling after the yield point. A similar pattern was observed during drawing of PMMA above T_g.

Necking was first thought to be caused by the transformation of crystalline regions under the applied stress, until it was observed to also occur in amorphous polymers [67]. Another early interpretation was that thermal softening due to localized heating initiated neck formation, but it was later demonstrated that necking can occur under isothermal conditions [67,68]. Both these effects can be involved in the necking process, and will influence its progression, but they are not necessarily responsible for its initiation. Any material that strain softens may neck under certain conditions of deformation. Strain softening arises from yielding of the initial structure of the undrawn material, and the yielding mechanism depends on the nature of that structure, which might be spherulitic crystals, a network of entanglements, or some other molecular arrangement that resists deformation. However, most materials are not perfectly homogeneous and necking is the manifestation of an instability in the yielding process due to ‘defects’, on a molecular level or on a macroscopic level, which are more compliant than the surrounding material and act as areas of stress concentration.

One way to conceptualize the conditions that give rise to necking on the one hand and uniform deformation on the other is to consider the ‘specimen’ to consist of a distribution of defects (high-energy or ‘soft’ sites [69]). Under an applied stress these defects will expand and some will merge together to form an unstable region where the rate of strain is higher than in the neighboring material. If the temperature of deformation were increased, however, enhanced molecular mobility would cause structural relaxation, which would tend to annihilate the defects and lead to uniform strain. In amorphous polymers, the defects may be regions of higher free volume and the transition from necking to uniform strain is likely to occur in the region of the glass transition. For predominantly crystalline polymers, this transition is more likely to occur at temperatures close to the softening point of the crystals.

It may also be argued that in very rate dependent materials (or at temperatures where the material becomes very rate dependent) there will be rapid self-stabilization at the defect sites before they have a chance to expand, merge and form a neck [70]. Thus, as soon as a ‘soft-site’ starts to strain at a faster rate than the surrounding material, it will rapidly strain harden, due to accelerated structural evolution, until its modulus matches that of the neighboring material and the local strain rate decreases again. In these circumstances strain would essentially proceed uniformly, even under conditions where nominal stress is falling due to global yielding of the polymer.

4.2.2.5 Internal Heating Effects

When the deformation rate is high and/or the fiber is relatively thick, a significant fraction of dissipated energy can be absorbed by the material and cause self-heating, and this may be a strong effect in below-T_g drawing, where stress levels are relatively large. However, recent work using an infrared pyrometer to measure sample temperature indicated that for PET films with an initial diameter of ~30 μm, the increase in temperature during drawing above T_g in a constant force experiment (see below) does not exceed 2 or 3 °C at average strain rates as high as $15s^{-1}$ [26]. Remarkably low temperature rises have also been recorded recently in glass epoxies strained at 10^4 s^{-1} [65]. There are four factors which influence the degree of self-heating during deformation:

a. The rate of heat transfer to the surrounding medium.
b. The time-scale of deformation in relation to the time-scale for heat transfer. This determines whether full adiabatic heating can occur. Obviously, deformation will be fully adiabatic at lower drawing speeds in thicker samples.
c. The level of full adiabatic heating that can be generated by the material. This will increase with strain rate. The level of adiabatic heating is certainly less in polymers than would be expected if all the energy were dissipated, and it has been suggested that some of the energy is stored as structural changes.
d. The latent heat of crystallization (in the case of crystallizing polymers).

4.2.3 Constant Force Deformation

In the continuous drawing of fibers between two rolls, the force applied to the fiber is constant (CF drawing), whereas the resisting force exerted by the fiber (internal stress) as a function of deformation is similar to the stress-strain response in constant extension rate (CER) drawing. The crucial difference between CER and CF drawing is that the structural evolution of the fiber during CF drawing is reflected in the evolution of strain rate. The higher the viscosity of the polymer, the greater is its resistance to deformation and the lower is the strain rate under a given force. Since polymer viscosity tends to change in the course of deformation due to structural rearrangements (such as entanglement slippage, molecular orientation and crystallization), the strain rate evolves accordingly.

The residence time of the filament segment in the 'draw gap' is determined by the speed of the rollers and the distance between them. For a given draw ratio, increasing throughput speed (decreasing residence time) increases the drawing force and the *average* strain rate of the deformation. It is clear that neither the force nor the deformation kinetics of a filament segment traveling through the draw zone can be predetermined, and the strain-rate profile along the drawing line will evolve in such a way as to minimize the force required to reach the applied draw ratio. Due to the non-uniform strain-rate profile and the high maximum strain rates involved, necking occurs in CF drawing at temperatures where it would not take place in CER drawing. In a recent series of papers, rheological models have been developed to describe the mechanics of drawing between rolls, including new constitutive equations that incorporate large plastic deformations in hot and cold drawing [71–73].

Laboratory experiments in which a constant axial load is applied to a fiber (or film) specimen have provided information on the temperature and stress dependence of the deformation kinetics in polyamide 66 [74–76], polyamide 6 [77,78], polypropylene [79] and PET [79–83]. Figures 4.10 and 4.11 summarize the influence of increasing drawing force on the deformation kinetics of amorphous, unoriented, PET film at a temperature of 80 °C [82]. The overall deformation rate increases with increasing load. At a given load, the rate of deformation at the beginning of the deformation is low due to the resistance of a coherent entanglement network. As deformation proceeds, strain-softening, probably involving entanglement slippage, reduces the resistance to deformation and promotes a sharp increase in strain rate. At some point, orientation-induced crystallization leads to strain-hardening and to a rapid reduction in strain rate. Finally, a limiting draw ratio (plateau deformation) is reached where the resistance of the polymer to deformation is equal to the applied load. Increasing the load generally increases the plateau deformation, because the evolving crystallite network deforms more readily under a higher stress. However, there is evidence that at very high drawing temperatures, around 180 °C, application of a high load may cause crystallization to start so early that the plateau deformation is lower than that attained under smaller loads [83]. If the applied force and the associated deformation kinetics are below a critical level, the molecules do not orient significantly and ‘flow drawing’ occurs (Figure 4.10, curve a). This is because there is ample time for chain slippage, and the polymer will extend to failure [83] unless there is sufficient time for crystallization to be induced without the assistance of decreased entropy from molecular orientation (as eventually occurred in curve a of Figure 4.10).

The effect of increasing the draw temperature in CF drawing is to accelerate the kinetics of deformation. For example, the time taken to reach the plateau deformation of PET at a draw temperature of 97 °C is about 100 times shorter than at 80 °C [82].

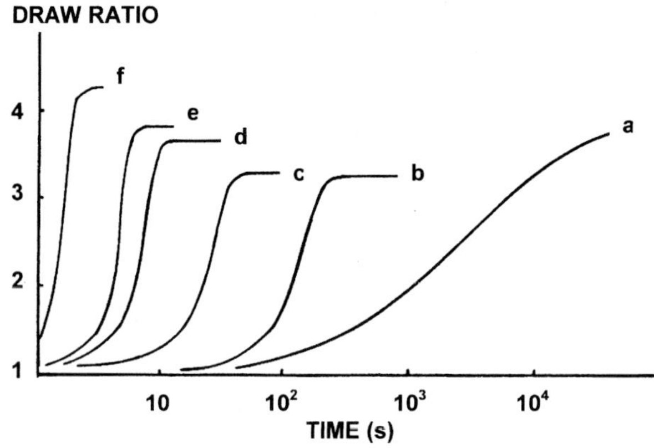

Figure 4.10 Kinetics of deformation of PET film drawn at 80 °C under various applied loads: (a) 1.5, (b) 3.8, (c) 5.0, (d) 6.3, (e) 7.5, (f) 10.5 MPa. (Adapted from Ref. [82]).

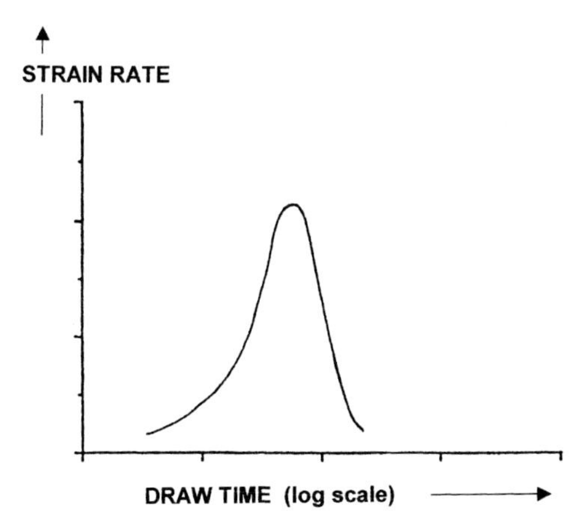

Figure 4.11 Schematic example of the evolution of strain rate during constant force drawing of uncrosslinked polymers that strain-harden, e.g. PET above T_g.

It follows from the above discussion that the rapid evolution of strain rate in CF drawing can cause excursions into deformation regimes that may not normally be entered in CER drawing. Deformation in these regimes influences the development of structure, which in turn influences the strain rate. It will be shown in Section 4.3.2 that the high maximum strain rates associated with CF drawing can have a significant impact on orientation and crystallization processes.

By installing a load-cell at the fixed upper-grip of a (constant force) laboratory drawing instrument, or from plots of draw ratio vs. time at various fixed loads, it is of course possible to obtain stress-strain curves at a given *average* strain-rate. It should be remembered, however, that these cannot be considered equivalent to stress-strain curves obtained under CER deformation at a similar *nominal* strain-rate, due to the large differences in the development of *true* strain-rate in CF and CER drawing. Draw ratio vs. time plots from CF drawing experiments can also be used to track the evolution of draw ratio with the applied load at different drawing times. This can provide useful information on the forces involved in drawing between rolls at given throughput speeds. Knowing the residence time of the deforming polymer in the draw gap between the rolls, it is possible to deduce the drawing force associated with a given draw ratio.

4.3 Orientation-Induced Crystallization

4.3.1 General Concepts

It is well known that the crystallization rate of a polymer in the oriented state is much higher than in the isotropic state, sometimes by several orders of magnitude [84–103]. The reason, in the simplest terms, is that the decrease in configurational entropy of oriented uncrystallized chains diminishes the entropy reduction required to reach the crystalline state. According to the classical theory of crystallization, the basic driving force for crystal nucleation is provided by the Gibbs free energy of the transformation of a given number of kinetic elements into a crystal cluster [104–106]. In polymer crystallization, a decrease in entropy due to orientation of molecular segments causes an increase in the free energy of crystallization and enhances the nucleation rate. Other factors arising from molecular orientation also contribute to increasing free energy of crystallization but to a much smaller degree, namely the enthalpy of polymer chains in the amorphous phase and the strain energy of the crystals [93,94]. Various theoretical interpretations of orientation-induced crystallization have been attempted [88–94], and the review presented in Section 2.1.2.6 provides a suitable theoretical background for the present discussions. Some additional remarks on developing predictive models of draw-induced (or stress induced) crystallization will be made later (Section 4.4.2).

Early interest in stress induced crystallization (SIC) was in relation to the deformation of crosslinked rubbers [84–87,107,108]. This was followed by intensive experimental studies of chain extension and crystallization in stressed polyethylene solutions [109–122]. Long fibrillar PE crystals with fully extended chains were found to form together with overgrowths of chain-folded crystals, resulting in the so-called 'shish-kebab' structure. (See also Section 5.3.1 and Figure 5.7). Other research showed that extended chain crystals without lamellar overgrowth can be obtained by drawing unoriented melt-spun fibers of polyethylene in the solid state and that, under suitable drawing conditions, fibers of very high modulus can be produced by this route [4–9]. (See also Section 4.7.1). However, the orientation mechanism involves unfolding the lamellar crystal structure that exists in the melt-spun filaments and, generally speaking, does not involve orientation-induced crystallization.

Studies on orientation-induced crystallization during drawing have more recently focused on uncrosslinked polymers that are amorphous, or of low crystallinity, after quenching from the melt, and crystallize during a solid-state drawing process. Poly(ethylene terephthalate), in particular, has been the subject of considerable research activity in this area [3,22,23,27,30–32,35,83,123–134]. Although most of the studies on draw-induced crystallization of PET have been concerned with film formation, they have direct relevance to fiber drawing. As mentioned earlier, commercial fiber drawing is always a constant force deformation, whereas PET film is sometimes drawn at a constant rate of extension. We will nevertheless review studies on crystallization during CER drawing of PET, because they provide a fundamental understanding of the interplay between the kinetics of deformation, orientational relaxation and crystallization, and they shed considerable light on results obtained from CF drawing, which will also be discussed.

4.3.2 The Case of Poly(ethylene terephthalate)

4.3.2.1 Conditions for Orientation-Induced Crystallization during Drawing

The drawing behavior of amorphous PET shows a dramatic change as the draw temperature is increased through the glass transition range. At temperatures below about

65 °C, the deformation involves necking, and the development of molecular orientation with strain is essentially independent of draw temperature and strain rate [135,136], as discussed in Section 4.4.1.4. In this temperature region, crystallization cannot occur, because there is insufficient molecular mobility, but an oriented mesophase can sometimes be formed [137,138], which disappears when higher temperatures are applied. On drawing at temperatures above T_g (about 75 °C), the strain in CER drawing (but not in continuous drawing) is essentially homogeneous throughout the specimen and, as a result of thermally activated relaxation processes, orientation increases more gradually than in 'cold drawing'. The strain-rate dependence of orientation development increases with temperature due to acceleration of the relaxation processes, and increased thermal

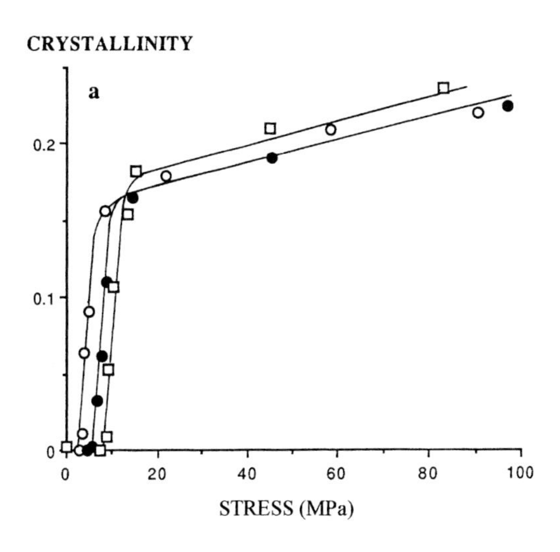

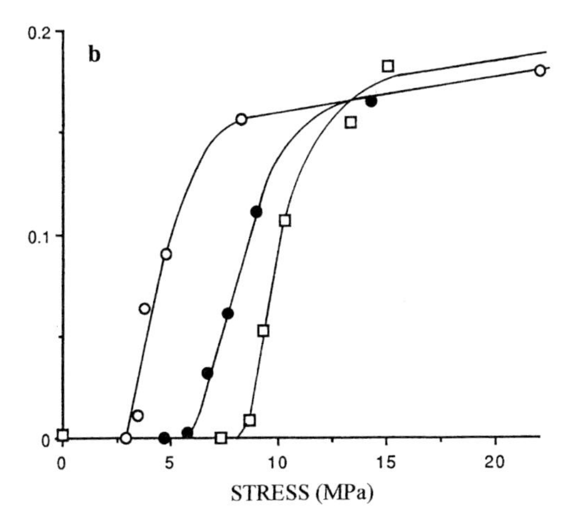

Figure 4.12 (a) Volume fraction crystallinity versus true stress in PET film at strain rates of 0.01; (○) 0.42 (•) and 2.1 s^{-1} (□); (b) enlarges the low stress region. (Adapted from Ref. [23]).

mobility of chain segments permits crystallization to take place on the timescale of the drawing operation once a sufficient level of orientation has been induced. This also applies to CF drawing, except that the high strain rates typically involved in this drawing mode can limit the amount of crystallinity that develops in the course of the deformation, as will be discussed later.

4.3.2.2 Effects of Stress in CER Drawing

Studies of constant extension rate (CER) drawing of PET film at strain rates ($\leqslant 2s^{-1}$ and temperatures between 80 and 100 °C have shown that, under this range of conditions, crystallization proceeds in two regimes [3,23,27,31]. As mentioned in Section 4.2.2.1, there is a low stress regime (crystallization regime 1) in which stress increases slowly with draw ratio and crystallinity increases rapidly; and a high stress regime (crystallization regime 2) in which stress increases rapidly and crystallinity increases slowly (Figures 4.1, 4.2 and 4.12). Figure 4.12b is an enlargement of the low stress region of Figure 4.12a and shows that crystallinity is not a unique function of stress. The critical stress for induction of crystallization σ_i increases with increasing strain rate and it has been found that the relationship between σ_i and (nominal) strain rate at various draw temperatures T_d is given by:

$$\sigma_i = a \log\dot{\varepsilon} + b(T_d) \tag{4.3}$$

where $a = 1.67$ and $b = (116 - T_d)/4.25$ [26]. The dominant cause of the strain-rate dependence of σ_i appears to be that the relationship between amorphous orientation and true stress is not unique, as would be predicted by conventional theories of rubber elasticity, but is strain rate dependent, probably as a result of a significant enthalpic contribution to the stress [139].

4.3.2.3 Effects of Strain Rate, Temperature, and Molecular Weight in CER Drawing

CER and CSR studies have revealed that the time available for molecular relaxation has a strong influence on crystallinity development [3,22,23,27,31,32,120,123]. At a given temperature, the time available for relaxation controls the level of molecular orientation attained at a given draw ratio which, in turn, determines the crystallization kinetics and the pseudo-equilibrium crystallinity level that can be reached. At strain-rates typical of CER drawing ($< 2s^{-1}$), crystallization is fast relative to the timescale of the drawing operation, and there will generally be time to reach, or approach, the pseudo-equilibrium crystallinity level corresponding to a particular combination of orientation and temperature.

Figure 4.13 shows schematics of the effects of strain rate, temperature and molecular weight on the development of noncrystalline orientation with draw ratio λ [3,22,33,34]. Noncrystalline orientation develops faster with increasing strain rate, with increasing molecular weight and with decreasing temperature. This is because the time available for molecular relaxation decreases with increasing strain rate, and the relaxation rate decreases with increasing molecular weight and decreasing temperature.

The time-dependence of orientation development is responsible for the strain-rate dependence of crystallization onset. At a given draw temperature, a critical molecular orientation must be reached to induce crystallization. For now, this critical orientation can be considered independent of strain rate, as indicated in the work of Le Bourvellec et al [22,123], although recent data have revealed evidence of some strain-rate dependence [127] and this will be discussed later. Since orientation develops more slowly at lower strain rates, the critical orientation, and therefore the onset of crystallization, is shifted to higher draw ratios as strain rate decreases (Figures 4.2b and 4.14) [22,23,27]. The effect of increasing temperature is to increase the magnitude of the strain-rate dependent shift in crystallization onset, because the rate dependence of noncrystalline orientation development increases with temperature (Figure 4.12) [22,31,33]. Decreasing molecular weight has a similar effect to increasing temperature [35].

For the moment we will continue to assume that, in the strain rate region studied, the degree of crystallinity at a particular draw ratio is largely dependent on the level of noncrystalline orientation obtained and, to a much smaller degree, on the time available

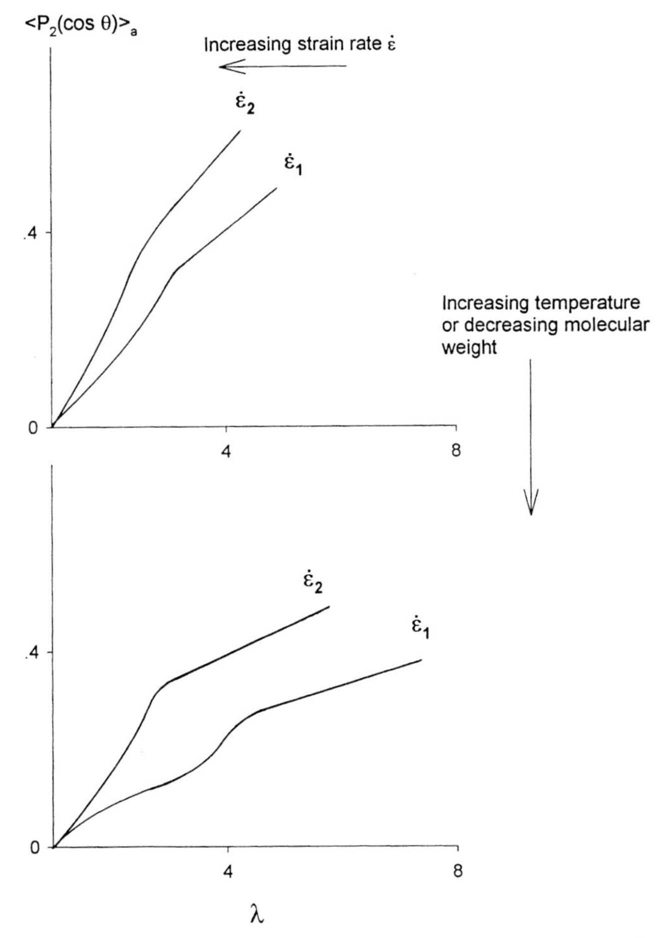

Figure 4.13 Schematic showing the influence of strain rate, temperature and molecular weight on the development of noncrystalline orientation during CER drawing of PET in the approximate nominal strain rate range 0.01 to 2 s^{-1} [3,22,26,27,33,128].

for crystallization. It can then be argued that, at a given level of noncrystalline orientation, there is a pseudo-equilibrium crystallization region where the crystallization kinetics are slower than in the initial crystallization stage, and that the level of crystallinity at the onset of the pseudo-equilibrium region increases with the degree of noncrystalline orientation attained [127]. Within the strain rate range typical of CER drawing, there is enough time available to approach the region of slower kinetics, so that the degree of crystallinity induced at each draw ratio essentially represents a pseudo-equilibrium value. The notion of a pseudo-equilibrium crystallinity level that increases with the degree of

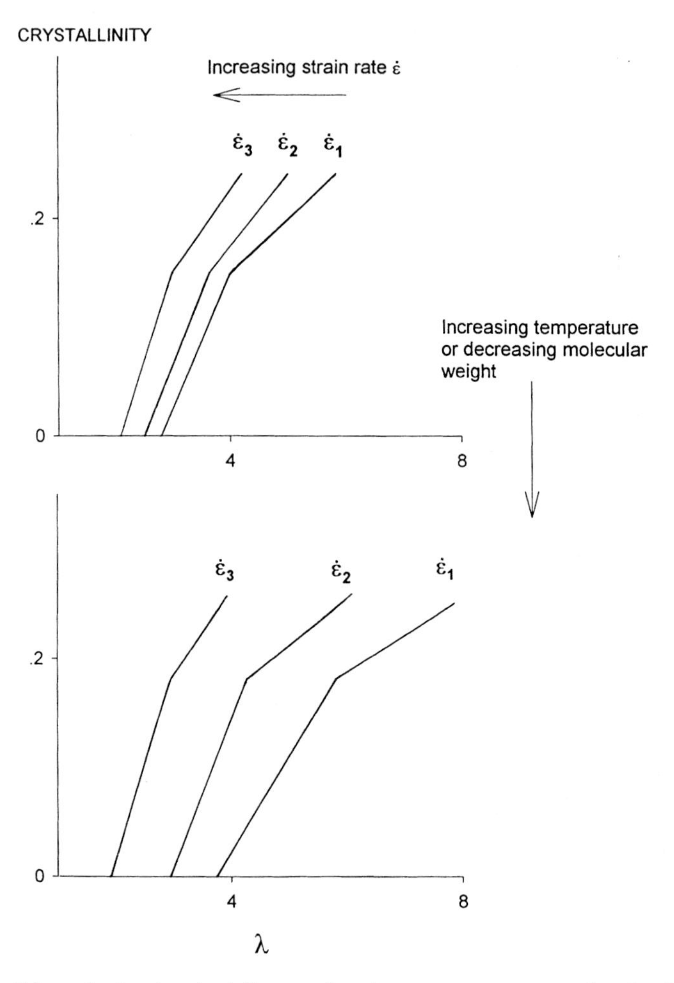

Figure 4.14 Schematic showing the influence of strain rate, temperature and molecular weight on the development of crystallinity during CER drawing of PET in the approximate nominal strain rate range 0.01 to 2 s^{-1} [22,23,26,31,35,125]. Note however that this is a slightly over-simplified representation because, although the characteristic crystallinity for onset of 'crystallization regime 2' (χ_r) increases with temperature (as shown), it is in fact independent of molecular weight [26,35].

noncrystalline orientation is consistent with the 80 °C (CSR) drawing data of Le Bourvellec et al. [22] and with much earlier work on vulcanized rubber (Figure 4.15) [84]. It also makes sense conceptually. The regime of fast crystallization involves the most highly oriented *crystallizable* material (i.e. highly oriented chain segments adjacent to other highly oriented chain segments) and crystallization would then proceed more slowly once this rapidly crystallizable material is depleted. The higher the overall level of noncrystalline orientation, the greater is the amount of highly oriented material that can participate in rapid crystallization, and the higher is the level of crystallinity associated with the pseudo-equilibrium regime. Since the stress and the noncrystalline orientation are continuously increasing during drawing, the pseudo-equilibrium crystallinity level continuously evolves to higher levels until a saturation regime is reached.

Recent data from CER drawing indicates, however, that the degree of crystallinity obtained at a given level of amorphous orientation increases quite significantly with decreasing strain rate [127]. This does not agree with earlier data obtained in CSR drawing studies [22], and additional studies are required to resolve this point. If the level of crystallinity obtained at a given draw ratio were not only dependent on the level of orientation attained, but also on the time available for crystallization in the strain rate region studied, it would not contradict the general concepts discussed above. It would however indicate that pseudo-equilibrium crystallinity values are not necessarily reached at all strain rates within the range 0.01 to 2 s^{-1}. In this case we may deduce that we are in a strain-rate/temperature regime where the time dependence of crystallization and the time dependence of orientational relaxation compensate each other, resulting in the observation that similar levels of crystallinity are attained irrespective of strain rate [31].

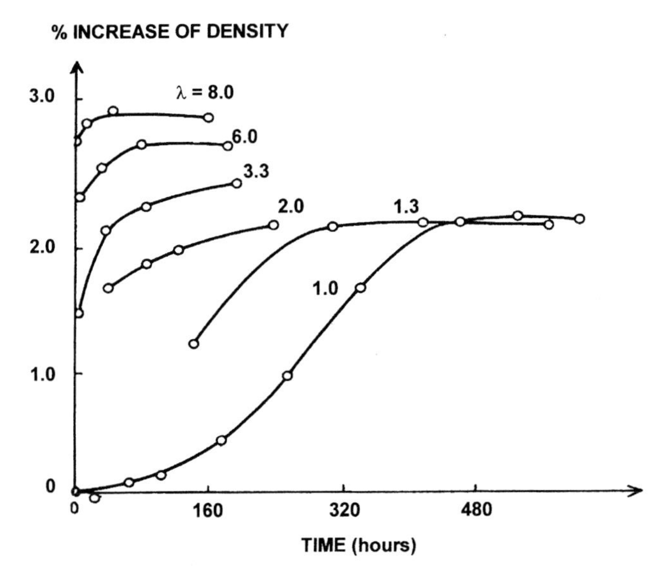

Figure 4.15 Effect of molecular orientation (which increases with draw ratio) on the post-draw crystallization kinetics of vulcanized rubber (adapted from Ref. [84]).

Salem has found that changing the strain rate of deformation at a given draw temperature simply shifts the crystallinity-time curves along the log-time axis without changing their shape, as shown for the draw temperature of 90 °C in Figure 4.16, and that the shift factor $A_{\dot{\varepsilon}}$ and the strain rate $\dot{\varepsilon}$ are related by a power law (Figure 4.17) [31,125]:

$$A_{\dot{\varepsilon}} = C\dot{\varepsilon}^{n} \tag{4.4}$$

The value of n increases with draw temperature and decreases with increasing molecular weight (Figure 4.18a) because n reflects the strength of the strain-rate dependent shift in the onset and kinetics of strain-induced crystallization, which is related to the rate of molecular relaxation [31,35,128]. If the crystallinity versus draw ratio relationship were

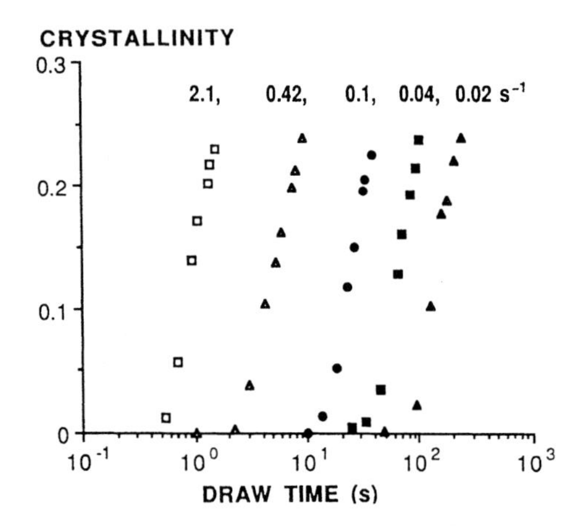

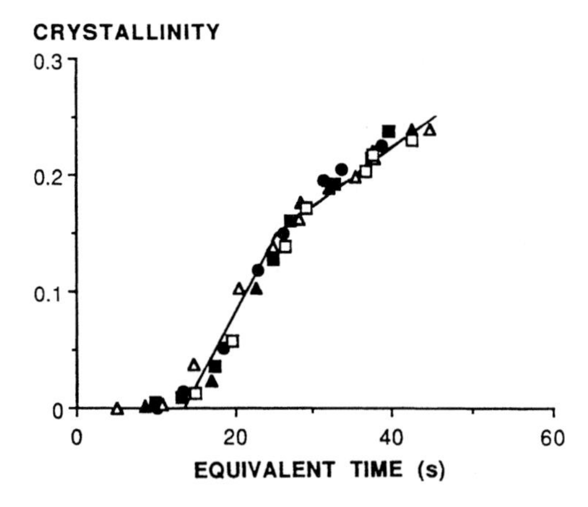

Figure 4.16 (a) Volume fraction crystallinity versus draw time for PET film at various strain rates for the draw temperature of 90 °C; (b) volume fraction crystallinity versus equivalent time, in which the data at 2.1, 0.42, 0.04 and 0.02 s^{-1} have been shifted along the time axis to superpose the data at the arbitrary reference strain rate of 0.1 s^{-1} [31]. Strain-rate/draw-time superposition is applicable to CER drawing of PET at draw temperatures in the approximate range 80–100 °C.

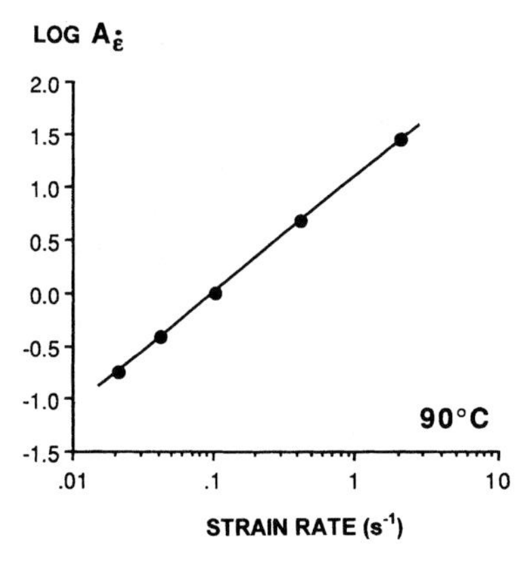

Figure 4.17 Relationship between shift factor $A_{\dot{\varepsilon}}$ and strain rate for PET film at a draw temperature of 90 °C, where n = 1.09(5) [31]. The power law relationship applies to all draw temperatures in the approximate range 80–100 °C.

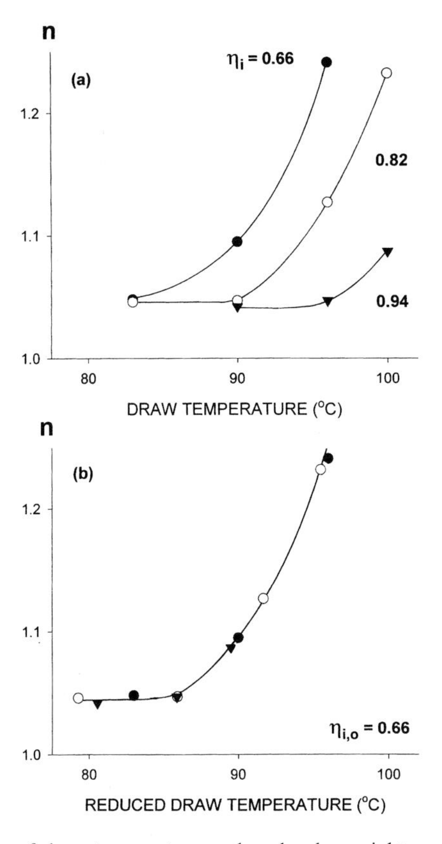

Figure 4.18 (a) Influence of draw temperature and molecular weight on n for PET film; (b) n versus reduced draw temperature, in which the data for intrinsic viscosity η_i of 0.82 ($M_n = 27000$) and 0.94 ($M_n = 33000$) have been shifted along the temperature axis to superpose the data for η_i of 0.66 ($M_n = 21000$). M_w/M_n was about 2.4 in each case.

independent of strain rate, n would equal unity. The faster the relaxation, the greater is the delay in crystallization onset and the more n exceeds unity. There is evidence that the effect of molecular relaxation is somewhat offset by the time-dependent nature of crystallization [125]. Thus, in the absence of molecular relaxation – in a highly crosslinked network for example – one could envisage n becoming less than unity.

The curves of n versus temperature at different molecular weights, in Figure 4.18a, shift to superpose each other (Figure 4.18b). Thus by fitting suitable functions to the shift factor vs. molecular weight relationship and to the reduced curve in Figure 4.18b, n can be obtained at any temperature or molecular weight. This means that, in CER drawing, the degree of crystallinity in regimes 1 and 2 can be predicted at any strain rate, draw temperature or molecular weight from the following relationships:

$$\chi_1 = \left(\frac{d\chi_1}{dt_{eq}}\right)_{T_d} (tA_{\dot{\varepsilon}} - t_1), \quad t_1 \leqslant A_{\dot{\varepsilon}} \leqslant t_2 \tag{4.5}$$

$$\chi_2 = \left(\frac{d\chi_2}{dt_{eq}}\right)_{T_d} (tA_{\dot{\varepsilon}} - t_2) + \chi_r(T_d), \quad tA_{\dot{\varepsilon}} \geqslant t_2 \tag{4.6}$$

where t is the real time, t_{eq} is the equivalent time, t_1 is the equivalent time for onset of crystallization and t_2 is the equivalent time for onset of regime 2 crystallization. χ_r, the characteristic level of crystallinity at the onset of regime 2, is independent of strain rate and molecular weight, and has a linear dependence on draw temperature [31].

From equation 4.4, it follows that the crystallization rate in regime 1 or 2 can be given by:

$$\frac{d\chi_{1,2}}{dt} = K_{1,2}\dot{\varepsilon}^n \tag{4.7}$$

as demonstrated by the experimental data in Figure 4.19. Two competing effects determine the temperature dependence of the crystallization kinetics, observed in the region of higher strain rates. Increasing draw temperature enhances (1) the rate of orientational relaxation of uncrystallized chains, and (2) the rate of crystallization at a given level of noncrystalline orientation. When strain rate is high, the latter effect predominates, because there is little time available for relaxation, but as strain rate decreases, this effect is increasingly offset by the relaxation process. At a given strain rate, the crystallization kinetics can be considered to be mainly controlled by the product of two parameters [123]:

$$\frac{d\chi}{dt} \propto \left(\frac{\delta\chi}{\delta \langle P_2\cos\theta\rangle_a}\right)_T \left(\frac{\delta \langle P_2\cos\theta\rangle_a}{\delta t}\right)_T \tag{4.8}$$

Increasing temperature causes $\delta\chi/\delta\langle P_2(\cos\theta)\rangle_a$ to increase and $\delta\langle P_2(\cos\theta)\rangle_a/\delta t$ to decrease. At low strain rates, these temperature dependent changes compensate each other, but at higher strain rates the second parameter becomes less temperature dependent and the overall crystallization kinetics increase with temperature due to the increase in $\delta\chi/\delta\langle P_2(\cos\theta)\rangle_a$ [126].

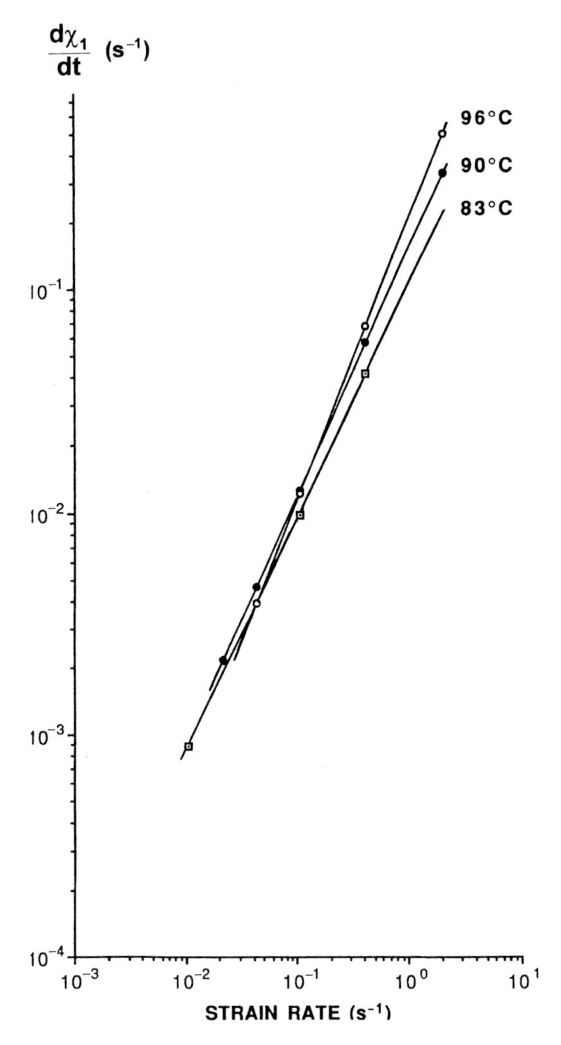

Figure 4.19 Influence of strain rate and draw temperature on crystallization kinetics of PET film during CER drawing (adapted from Ref. [31]).

These competing temperature effects also influence the onset of crystallization during drawing. At low strain rates, increasing draw temperature causes a pronounced increase in the draw ratio for onset of crystallization (λ_c), because higher temperatures enhance the rate of orientational relaxation [31–33]. However, the temperature-dependent shift in λ_c diminishes as strain rate increases and at sufficiently high strain rates, increasing temperature shifts λ_c to lower draw ratios (Figure 4.20) [31]. Since higher temperatures not only increase the rate of orientational relaxation, but also increase the rate of crystallization at a given level of noncrystalline orientation, the critical orientation for onset of crystallization is reduced [22,26,127]. Thus, when the time available for

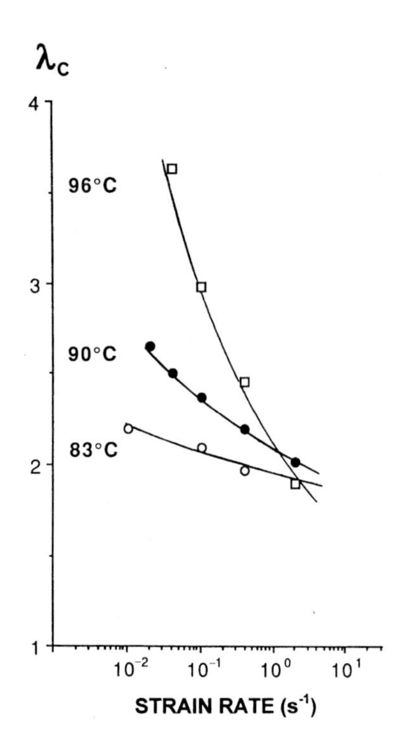

Figure 4.20 Influence of draw temperature and strain rate on the draw ratio for onset of crystallization λ_c in CER deformation [31].

relaxation becomes very short, at high strain rates, the effect of enhanced crystallization rate dominates [31,126].

4.3.2.4 Constant Force Drawing

Investigations of the fundamental phenomena involved in crystallization during CER drawing have been used to help interpret crystallization behavior during constant force (CF) drawing [127]. It should be remembered, as indicated in Sections 4.2.1 and 4.2.3, that constant force drawing at some average strain rate involves an excursion to strain rates that can be an order of magnitude higher than the average value (Figure 4.11). In the range of (average) strain rates from 5 to 17 s^{-1}, studies have shown that there is no influence of strain rate on orientation development or crystallinity development in CF drawing (Figure 4.21). Increasing temperature between 83 and 96 °C enhances rate of molecular relaxation and therefore decreases the rate at which the chains orient (Figure 4.21), but it has little influence on crystallinity development (Figure 4.22). This is because enhanced molecular mobility at higher temperatures results in more rapid crystallization at a given level of noncrystalline orientation.

The high strain rates in CF drawing reduce the time available for relaxation of orientation, so that the axial orientation of the noncrystalline chains evolves more rapidly in CF drawing at an average strain rate of 5 s^{-1} than in CER drawing at a nominal strain

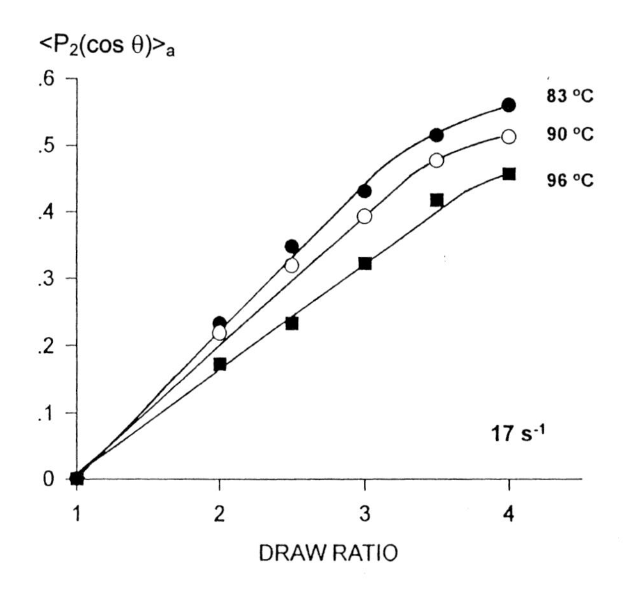

Figure 4.21 Noncrystalline orientation (from intrinsic fluorescence) versus draw ratio for continuous (constant force) drawing of PET film at three draw temperatures and an *average* strain rate of 17 s^{-1} [127].

rate of 0.56 s^{-1} (Figure 4.23) [127]. This would be expected to cause crystallinity to develop at lower draw ratios in CF drawing, but Figure 4.24 shows that the opposite is true. It can therefore be deduced that, at a given level of noncrystalline orientation, the degree of crystallinity is lower in CF drawing than in CER drawing (Figure 4.25). In high strain-rate CF drawing, there is insufficient time for enhanced crystallization rates associated with high noncrystalline orientation to compensate for the shorter time available for crystallization, which begins in the region of the strain rate maximum. On the other hand, as discussed earlier, there will often be sufficient time in CER drawing for crystallinity to reach pseudo-equilibrium levels associated with a given level of noncrystalline orientation. In high strain-rate CF drawing, this will only be possible when the deformation reaches the tail-end of the strain-rate spectrum, unless the sample is held at the draw temperature prior to quenching. It is clear that the high strain rates reached in typical CF drawing operations cause the deformation to enter a regime in which crystallinity development becomes strongly limited by the time available, and this has important implications for the control of fiber structure during drawing.

Real-time wide angle X-ray scattering (WAXS) studies at high CER strain rates of about 10s^{-1} suggest that very low levels of crystallinity are developed during drawing under these conditions [131], consistent with the notion that sufficient time is required to develop significant crystallinity. On the other hand, the crystallinity values reported are likely to be an underestimate because, when crystal size is small, WAXS cannot detect the presence of crystallinity at volume fractions less than about 0.1 by the method used.

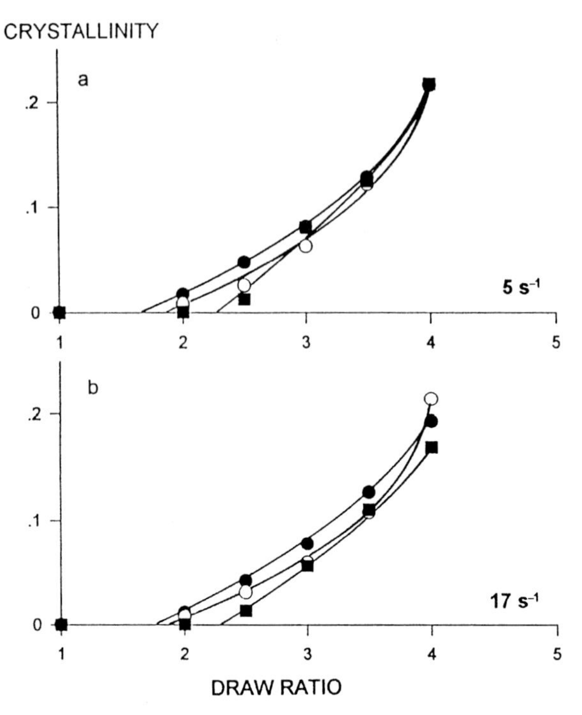

Figure 4.22 Volume fraction crystallinity (from density) as a function of draw ratio for continuous (constant force) drawing of PET film at draw temperatures of 83 °C (●), 90 °C (○) and 96 °C (■) and an *average* strain rate if 17 s^{-1} [127].

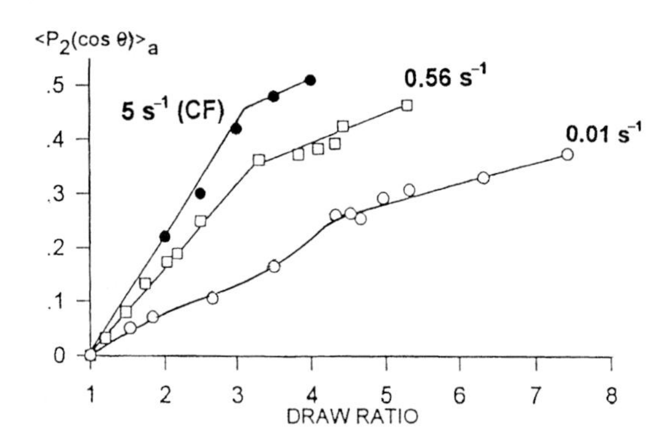

Figure 4.23 Noncrystalline orientation versus draw ratio for continuous (constant force) drawing of PET film, at 90 °C at an (average) strain rate of 5 s^{-1}, compared with film CER-drawn at strain rates of 0.56 s^{-1} and 0.01 s^{-1} (from [127]).

4.3.2.5 Constant Force Drawing above 100 °C

Drawing of PET is usually performed at temperatures below 100 °C, but drawing at higher temperatures also occurs in commercial practice, especially in multistage drawing

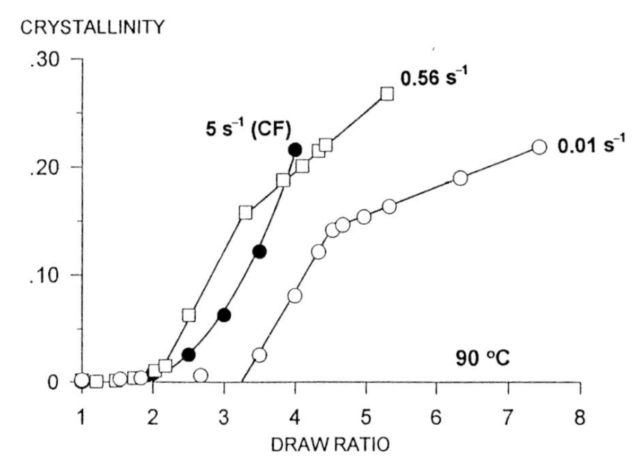

Figure 4.24 Volume fraction crystallinity versus draw ratio for continuous (constant force) drawing of PET film at 90 °C at an (average) strain rate of 5 s^{-1}, compared with film CER-drawn at strain rates of 0.56 s^{-1} and 0.01 s^{-1}.

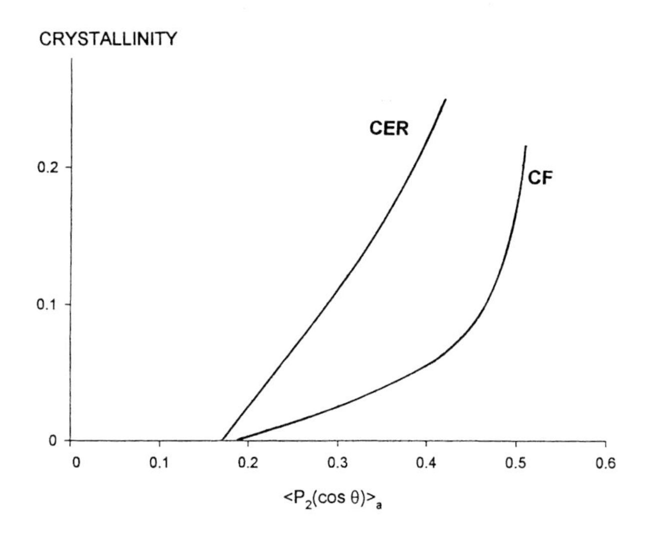

Figure 4.25 Volume fraction crystallinity versus noncrystalline orientation for constant extension rate (CER) drawing and constant force (CF) drawing at 90 °C. The nominal strain rate for CER drawing was 0.01 s^{-1} and the CF curve is from data obtained at (average) strain rates of 5 and 17 s^{-1}. (Adapted from Ref. [127]).

of partially-oriented fibers for tire cord (see Section 4.7.2). Some insight into the molecular rearrangements that take place in high-temperature drawing at high strain rates has been provided in a recent study on CF drawing of *unoriented* PET film in the temperature range 120–135 °C [128]. The initial stage of drawing at high temperatures is characterized by polymer flow where, as a result of high rates of molecular relaxation, neither molecular orientation nor crystallization occurs. Strain-rate increases sharply in the course of the deformation, which is attributable to a decrease in the entanglement network density. This reduces the time available for relaxation, and the chains start to orient at a draw ratio in the range of 2 to 3.5, depending on temperature. Orientation rapidly reaches a saturation level, which is much lower than at draw temperatures below 100 °C. Crystallization onset seems to lag only slightly behind orientation onset because the critical orientation for inducing crystallization is very low at these temperatures, and the final level of crystallinity is somewhat higher than that attained at conventional draw temperatures. By redrawing high-temperature drawn film along the same axis at 100 °C, total draw ratios as high as 9 could be achieved despite the high strain rates involved, and the level of noncrystalline orientation increased to values similar to those expected in single-stage drawing at 100 °C.

4.3.2.6 Crystal Size and Orientation

So far we have discussed changes in the degree of crystallinity during drawing of PET, but it is also of interest to know the size and orientation of the crystallites. Crytallite size can only be measured by X-ray diffraction techniques (see Chapter 13.1), and this is also the most suitable method for measuring crystallite orientation. As mentioned earlier, a limitation of X-ray diffraction is that the presence of crystallinity cannot generally be detected when volume fraction crystallinity is less than about 10%. At these crystallinity levels, the crystalline reflections are so weak that they are submerged in scattering from the amorphous fraction and even the most sophisticated curve-fitting procedures cannot unveil them (see also Section 13.3.1). By standard WAXS techniques, it is therefore not possible to monitor crystal growth (if any) from the onset of crystallization. This can only be done once the degree of crystallinity has reached about 15%, and studies of CER drawing at 90 °C indicate crystal widths in the region of 2.8 nm at this crystallinity level, increasing to 3.6 nm at the maximum draw ratio, where the volume fraction crystallinity is about 0.23 [23]. Samples drawn under constant force conditions appear to have similar lateral crystal dimensions at these levels of crystallinity [127]. It is less easy to measure crystal size accurately in the chain axis direction of PET, but crystal lengths in the region of 4.5 nm have been determined [26]. It is well known that post-draw heat treatments cause appreciable crystal growth, especially at temperatures exceeding 200 °C [18,140–142].

Brody has shown that a defining difference between high speed spun PET fibers and drawn PET fibers from low speed spun filament is the type of crystallinity induced [143]. In fiber drawing at temperatures around 90 °C, crystallization occurs as a consequence of orientation and the molecules in the crystal are load-bearing [2,143]. In this case, there is a direct relationship between the degree of crystallinity and the fiber modulus. Subsequent heat treatment of the drawn fibers produces thermally-induced epitaxial crystallization, which provides no additional load-bearing capability and no significant increase in modulus. In high speed spinning, orientation-induced crystallization appears to predominate between spinning speeds of 3000 and 4000 m/min and at higher speeds a combination of orientation-induced and thermally-induced crystallization takes place. These findings are supported by unpublished observations in our laboratory that the crystal chain-fold band is absent in the infrared spectra of drawn fibers, but is present in the spectra from fibers spun at speeds above 4000 m/min and in drawn fibers that have been subsequently annealed. Thus, chain folding appears to be a significant mode of crystallization in high speed spinning, but does not feature in draw-induced crystallization. Crystal width increases from about 3 nm at a spinning speed of 4000 m/min to about 6nm at 7000 m/min, and the crystal dimension in the chain axis direction appears to increase from about 6 nm to 10 nm in this winding speed range [26,144]. In general, the level of orientation in the noncrystalline regions is much higher in drawn fibers than in high speed spun fibers [2,3].

Stress-induced crystallization of PET and other polymers always results in highly oriented crystals even at very low levels of crystallinity, because the most highly oriented chains crystallize preferentially [89]. For drawn fibers, values of the crystal orientation function are typically in the region of 0.9, and since thermal treatments generally result in the growth of existing crystals, crystal orientation remains high after heat-setting.

4.3.3 Other Polymers

Recent studies on crystallization during CER drawing of poly(ethylene naphthalate) PEN provide interesting reinforcement of the PET results. Since the T_g of PEN is about 125 °C (50 °C higher than PET), draw temperatures of 140 °C and 150 °C were used to provide suitable comparison with the PET studies [34]. It was found that CER drawing of amorphous PEN at these temperatures produces crystallinity vs. draw ratio curves that are similar in shape to those of PET, i.e. a fast crystallization regime (during which little stress is developed) is followed by a slow crystallization regime (during which high stresses are developed). The influence of strain rate and draw temperature on crystallinity development in PEN is also consistent with previous observations on PET: decreasing strain rate shifts the curves to higher draw ratio, and increasing draw temperature magnifies this strain-rate dependence.

Of particular importance is the fact that, like PET, the plots of PEN crystallinity versus draw-time at various strain rates can be shifted along the log-time axis to superpose each other, and that the shift factor is given by Eq. 4.4. Changes in the exponent n as a function of draw temperature are given for both PET and PEN (of similar molecular weights) in Figure 4.26, and it can be seen that the curves are separated by about 50 °C, equal to the difference in T_g between the two polymers. These results hint at universal applicability of strain-rate/draw-time superposition to crystallizable polymers, or at least to flexible-chain polyesters. In the slow to moderate strain-rate range typical of CER drawing, the strain rate, draw-temperature and molecular weight dependence of strain-induced crystallization would then be entirely predictable from knowledge of the polymer's glass transition temperature. There is therefore some imperative to investigate, in a similar manner to the work on PET and PEN, the draw-induced crystallization behavior of other flexible chain polyesters, such as poly(butyl terephthalate), poly(trimethyl terephthalate) and polylactic acid.

There is a lack of information on draw-induced crystallization in polyamides. Whereas it is known that crystallinity in PA66 increases when drawn above T_g [63], effects of strain rate, draw temperature and molecular weight have not been studied systematically. However recent online X-ray diffraction studies by Hsiao et al. [145] have provided insights into the influence of draw temperature on unit cell dimensions and crystal size. A

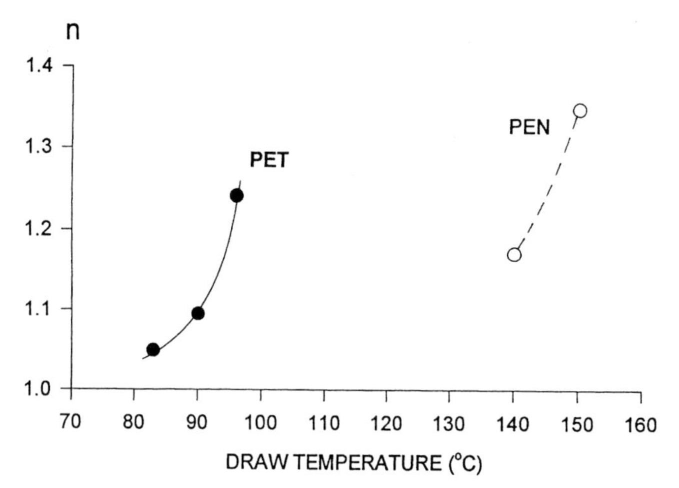

Figure 4.26 n versus draw temperature for PET and PEN film of similar molecular weights.

significant observation was that small crystals are annihilated during drawing of PA66, but the effect of strain is much less than the effect of temperature.

It has been reported that crystallinity increases during drawing of polyvinyl chloride [146] and polyvinyl alcohol [147] but, again, information on the influence of drawing conditions is virtually absent.

4.4 Theory and Modeling

4.4.1 Stress-Strain-Orientation Behavior

4.4.1.1 Deformation of Crosslinked Networks above T_g

Attempts to model polymer deformation at large strains began with the development of the kinetic theory of rubber elasticity [148–151]. There are numerous reviews of this well-established theory (e.g. Refs. [17,150–153]) and we will only recap its main predictions. The theory applies to linear, flexible chain polymers that are lightly crosslinked to form a network. Since these molecular chains will always tend to assume configurations corresponding to a state of maximum entropy, an external force is required to change the configurational arrangements of the chains and produce a state of strain. The stress-strain relationship of the crosslinked polymer is obtained by calculating the entropy of a single chain and then determining the change in entropy of the network of chains as a function of strain. In its simplest form the theory assumes that the displacements of the crosslinks are affine, i.e. linearly related to the macroscopic strain, and that a Gaussian distribution function definies the end-to-end separations of the network chains between crosslink points. For deformation in uniaxial extension, the Gaussian network model leads to the following relationship between true stress and draw ratio:

$$\sigma = nkT(\lambda^2 - \lambda^{-1}) \tag{4.9}$$

where n is the number of crosslinks per unit volume, k is Boltzmann's constant, and T is absolute temperature. This model can also predict the orientation distribution of the network chains as a function of strain, the second moment of the orientation function being given by:

$$<P_2(\cos\theta)> = \frac{1}{5N}(\lambda^2 - \lambda^{-1}) \tag{4.10}$$

where N is the number of rotatable segments between crosslink points.

Departures from behavior predicted by Eq. 4.10 have lead to various modifications of the theory. Rather than assuming affine displacements of the network junctions, the 'phantom' network model [154,155] permits large junction fluctuations and chain crossability, which reduce the modulus. Both the phantom and affine schemes predict the modulus in the region of moderate deformations to be independent of λ, whereas experimental studies frequently show modulus (or 'reduced stress' [σ*]) decreasing with increasing λ, as expressed by the empirical Mooney-Rivlin equation [151]:

$$[\sigma^*] = 2C_1 + (2C_2/\lambda) \tag{4.11}$$

The slope $2C_2$, which reflects the degree of departure from predicted behavior, decreases almost to zero when the network is sufficiently swollen by a solvent. This points to an influence of topological constraints in the form of molecular entanglements, because the entanglement density decreases with increasing dilution. It has been proposed that the extent to which the deformation departs from affine behavior is dependent on the degree of chain entangling around the crosslinks, since this will determine the firmness with which the crosslinks are embedded in the network structure. It can then be argued that stretching the network chains decreases the degree of entangling, reducing constraints on the junction fluctuations, and causing the modulus to decrease towards the value predicted for a phantom network [156–158]. In other words there is a gradual transition from nearly affine behavior at small strains to phantom behavior at larger strains.

Ball et al. [159], unhappy with a model that permits chains to pass through each other, proposed that entanglements between crosslinks can be modeled as 'slip links'. Here, two adjacent chains are considered to be constrained in a small ring, through which each chain is free to slide as far as the fixed crosslinks at either end. The free energy of deformation derived from this approach is expressed in terms of the degree of slippage η, the number of crosslinks N_c and the number of slip links N_s, and for uniaxial extension the stretching force is given by [160]:

$$\sigma = N_c kTD + N_s kT\left[(1+\eta)\left(\frac{\lambda}{(1+\eta\lambda^2)^2} - \frac{1}{(\lambda+\eta)^2}\right) + \eta\frac{\lambda D}{(1+\eta\lambda^2)(\eta+\lambda)}\right] \tag{4.12}$$

where $D = \lambda - \lambda^{-2}$. In the absence of slippage ($\eta = 0$) Eq. 4.12 predicts the free energy of a phantom network. Allowing slippage of entanglements between crosslinks, on the other hand, predicts a decreasing modulus with increasing extension, consistent with many experimental observations [161].

Further deviations from predictions of the Gaussian network theory occur at high elongations, where crosslinked polymers generally show an abrupt increase in stress. This phenomenon has been attributed to the limited extensibility of the network chains, because once the conformation of a chain becomes fully extended, Gaussian statistics no longer apply. Similarly, the Gaussian distribution function is only valid when the distances between crosslink points are much less than the extended chain length. It has been shown that these restrictions can be removed by using the inverse Langevin probability distribution (Figure 4.27), leading to an expression for the nominal stress in terms of the following series expansion [151]:

$$\sigma = nkT\left(\lambda - \frac{1}{\lambda}\right)\left[1 + \frac{3}{25N}\left(3\lambda^2 + \frac{4}{\lambda}\right) + \frac{297}{612N^2}\left(5\lambda^4 + 8\lambda + \frac{8}{\lambda^2}\right) + \ldots..\right] \tag{4.13}$$

And the second moment of the orientation function is given by:

$$<P_2(\cos\theta)> = \frac{1}{5N}\left(\lambda^2 - \frac{1}{\lambda}\right) + \frac{1}{25N^2}\left(\lambda^4 + \frac{\lambda}{3} - \frac{4}{3\lambda^2}\right) + \frac{1}{35N^3}\left(\lambda^6 + \frac{3\lambda^3}{5} - \frac{8}{5\lambda^3}\right) + \ldots \tag{4.14}$$

Edwards and Vilgis have shown that the slip-link model can also incorporate network inextensibility [160]. A fully extended chain between two fixed crosslinks will follow the shortest path between each entanglement (slip link), whereas the relaxed chain has the freedom to take a longer route. A parameter quantifying the inextensibility of the network

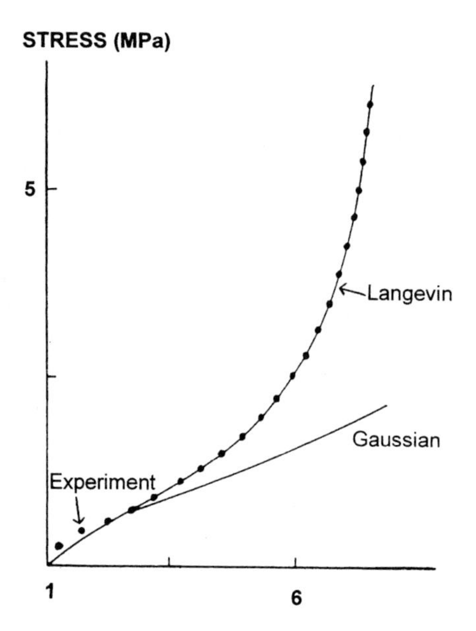

Figure 4.27 Experimental stress versus draw ratio data for a typical rubber compared with predictions from the Gaussian network theory and the inverse Langevin treatment with N = 75 and nkT = 0.27 N/mm^2 (adapted from Ref. [151]).

α is therefore defined as the ratio of the distance between entanglements and the length of a freely orienting segment. The upturn in stress at high extensions is modeled by a probability function with similar proprties to that of the inverse Langevin, and the force in uniaxial extension is then given by:

$$\sigma = N_c kTD\left[\frac{1-\alpha^2}{(1-\alpha^2\phi)^2} - \frac{\alpha^2}{1-\alpha^2\phi}\right] + N_s kT\left[\begin{array}{l} \dfrac{(1+\eta)(1-\alpha^2)\alpha^2 D}{(1-\alpha^2\phi)^2}\left(\dfrac{\lambda^2}{1+\eta\lambda^2}+\dfrac{2}{\lambda+\eta}\right) \\ + \dfrac{(1+\eta)(1-\alpha^2)}{1-\alpha^2\phi}\left(\dfrac{\lambda}{(1+\eta\lambda^2)^2}-\dfrac{1}{(\lambda+\eta)^2}\right) \\ + \dfrac{\eta D\lambda}{(1-\eta\lambda^2)(\lambda+\eta)} - \dfrac{D\alpha^2}{1-\alpha^2\phi} \end{array}\right] \tag{4.15}$$

where $\phi = \lambda^2 + 2\lambda^{-2}$ and $D = d\phi/2d\lambda (= \lambda - \lambda^{-2})$.

Since the abrupt increase in modulus at high elongations is generally observed only in networks that can undergo strain-induced crystallization, the attribution of this phenomenon to the limiting network extensibility has been challenged by some researchers [150,162]. It was argued that crystallization itself could account for the modulus increase, because the formation of crystallites would provide reinforcement and additional crosslinks in the network. Since available noncrystallizing networks were incapable of large deformations, distinguishing between the two interpretations required the synthesis of 'model' noncrystallizable networks with high extensibilities [162,163]. The stress upturn was found to be present in the noncrystallizable networks, but was much less pronounced than in the crystallizable networks used for comparison. Clearly, limited chain extensibility and crystallization may both contribute to the upturn, and the latter effect may dominate if the level of crystallinity induced is high enough.

4.4.1.2 Application of Network Models to Drawing of Uncrosslinked Polymers Above T_g

The models discussed above were devised for polymer networks containing permanent junction points, such as strong interchain chemical bonds, which prevent chains from sliding past each other. Most polymers used in fiber formation do not contain strong interchain bonds, and the applicability of rubber-elastic theories to the deformation of these materials cannot be assumed. However, Ward and coworkers observed that uncrosslinked amorphous PET behaves as a molecular network [164]. At temperatures in the range 80–85 °C (somewhat above T_g) its stress-strain-orientation behavior up to moderate draw ratios can approximately be described by Kuhn and Grün's theory of rubber elasticity [135,165,166] and, somewhat better, by the sliplink network model of Ball et al. [167]. It is due to the presence of molecular entanglements, which act as network junctions, that an uncrosslinked polymer may behave under certain temperature/time conditions as if it were crosslinked. However, rubber-elastic theory predicts a linear relationship between $<P_2(\cos\theta)>$ and stress (the linear stress-optical law) and, even in the region of low strains, significant deviations from linearity have been observed when amorphous polymers are stretched at temperatures above T_g [27,168]. Recent modifications of Kuhn and Grün's theory have taken into account anisotropic mean field effects and the polydispersity of the network [169]. This approach, involving application of rotational isomeric state formalism, appears to improve agreement with experimental data at low to moderate draw ratios.

Rubber-elastic models are not capable of describing the deformation of uncrosslinked polymers at moderate to high draw ratios [27,166]. We believe that the predominant cause of non-rubberlike behavior at moderate draw ratios is an increase in the number of monomers between network junctions during the deformation, arising from the onset of substantial entanglement slippage, especially at low strain rates and high temperatures. In the case of crystallizable polymers, entanglement slippage is eventually slowed down or arrested by the development of crystallinity. These views are supported by various phenomena occurring at moderate to high draw ratios in (crystallizable) PET above T_g: the shift in the stress upturn to higher draw ratios with increasing temperature, decreasing strain rate and decreasing molecular weight [23,26,29]; the much stronger upturn in stress compared with uncrystallizable polymers, such as PEMT [27,170,171] and PMMA [65]; the coincidence of the inflection point in the stress-strain curve with the onset of crystallization [3,23,27]; the draw-temperature independence of the slope of the stress-strain curve at high draw ratios, despite a strong dependence of (entanglement) crosslink density on temperature [26,29]; the rapid increase in strain rate during constant force drawing followed by an abrupt decrease after the onset of crystallization [127,129].

It would appear that the drawing of an amorphous crystallizable polymer above T_g involves a journey through various deformation regimes, from rubber-like, or even rubber-glass, behavior at low strains to rubber-flow behavior at higher strains, and finally to the deformation of an orientation-induced network of crystallites which constrains the flow process. The strain levels at which these transitions in deformation behavior occur obviously depend on temperature and strain rate, and when the strain rate is changing radically in the course of the deformation, as in constant force drawing, a further layer of complexity is added.

4.4.1.3 Polymer Melt Deformation: Application to Above-T_g Drawing

Since entanglement slippage seems to occur during drawing at temperatures quite close to T_g, it might be reasonable to consider the deformation in this region, as well as at higher temperatures, in terms of the extension of a polymer melt, where chain entanglements are represented as nonpermanent crosslinks at fixed points along the chain contour. The well known treatments of melt relaxation by Doi and Edwards [172–174] and Curtiss and Bird [175] are based on the notion that each chain in the melt can be considered to be confined in a tube created by the surrounding molecules [176]. The theory describes the relaxation of a melt after it has been subjected to an instantaneous (or very rapid) step strain. During deformation, the entanglement junction points move affinely to new positions, and the subsequent relaxation of the system back to the isotropic conformation occurs by three successive processes: a fast stress decay, a viscoelastic pseudoplateau, and a terminal flow zone. At short times the entanglements are considered to behave as fixed crosslinks, and the relaxation, which only involves parts of the chain whose motion is not restricted by the surrounding molecules, can be described by the Rouse model [177]. In the second time interval, entanglement constraints are released at a relatively slow rate and the deformed contour length of the chain retracts back towards its equilibrium contour length within the deformed tube. This retraction process results in a greater number of monomers between entanglements, i.e. a less dense network, and the relaxation time scales as the square of the molecular weight. The final relaxation process involves slow diffusion of the chain to its isotropic conformation by 'reptation' out of its oriented tube, and scales as the third power of the molecular weight. This process completes the disengagement of the chains from their primitive entanglements, begun in the pseudoplateau region.

In drawing operations, stresses are relaxing in the course of the deformation, whereas the mathematical treatment of Doi and Edwards is based on relaxation following a step strain. In order to understand which relaxation processes are activated in the course of drawing, it is useful to consider a simple drawing experiment at a constant strain rate $\dot{\varepsilon}$. According to the Deborah number criterion, a relaxation process with a relaxation time τ will be effective if $\dot{\varepsilon}\tau$ is about 1, it will be frozen out if $\dot{\varepsilon}\tau \gg 1$, and will have sufficient time to be completed if $\dot{\varepsilon}\tau \ll 1$. A consequence of applying large deformations is to decrease τ, so that relaxation processes will be activated at shorter times. The efficiency with which chains disentangle, for example, would be enhanced by increasing the strain. Thirion and Tassin [21] have applied the theories of Doi and Edwards, with some modification, to the calculation of stress-strain curves for progressive uniaxial extension at a constant strain rate. Their treatment was found to provide satisfactory agreement with stress-strain curves generated by constant strain-rate elongation of polystyrenes with various (narrow distribution) molecular weights at temperatures 30–50 °C above T_g [178]. It is noteworthy, however, that in addition to the entropic stress given by the Doi-Edwards model, Tassin, Thirion and Monnerie [178] deduced a significant contribution to stress from enthalpic processes, which can dominate stress-strain behavior at low draw ratios and which increases with increasing strain rate and increasing molecular weight [21,178]. Of course the calculated stress-strain curves of Tassin et al. predict no upturn in stress at high deformations, which is consistent with the experimental observations of amorphous polymers deformed at temperatures significantly above T_g. To model the above-T_g stress-strain behavior of crystallizable polymers after the onset of crystallization, it would be necessary to predict the development of crystallinity during drawing and its contribution to stress. In view of recent advances in our understanding of orientation induced crystallization, discussed in Section 4.3.2, this may be achievable.

4.4.1.4 Drawing of Amorphous Polymers Below T_g

The identification of an entanglement network in amorphous, uncrosslinked polymers provided an essential key to understanding their deformation behavior in the glassy state as well as above T_g. It was recognized that the deformation of polymer glasses appears to involve a flow process constrained by the elastic deformation of an entanglement network. Details of the molecular displacements involved in plastic flow below T_g are not known, but it is clear that they do not involve long-range molecular processes (i.e. chain disentanglement) because yield and necking associated with this type of flow are observed during drawing of highly crosslinked rubbers below their glass transition temperature. Apparently the flow process is associated with molecular rearrangements between points of entanglements and/or crosslinks, and can be considered a thermally activated rate process [179,180] in which the strain rate is given by:

$$\dot{\varepsilon} = \dot{\varepsilon}_o \exp - \left(\frac{\Delta H - \sigma_y \upsilon}{kT} \right) \tag{4.16}$$

where ΔH is the activation energy, υ is the activation volume, $\dot{\varepsilon}_o$ is a pre-exponential constant and σ_y is the flow stress. Harward and Thackay [181] proposed a mathematical model that describes the total flow stress generated during uniaxial deformation of a glassy polymer as the sum of an Eyring-type flow stress (Equation 4.16) and a rubber-elastic (inverse Langevin) stress (Equation 4.13). Renewed interest in this approach has lead to the development of three-dimensional constitutive models to describe deformation of polymer glasses [182–184], rubbery crosslinked polymers [185], and amorphous polymers in the glass-rubber region [29,186]. Edwards and Vilgis [187] have discussed how the slip-link concept might be extended to the deformation of polymers in the glassy state, but a quantitative analysis has not yet been developed.

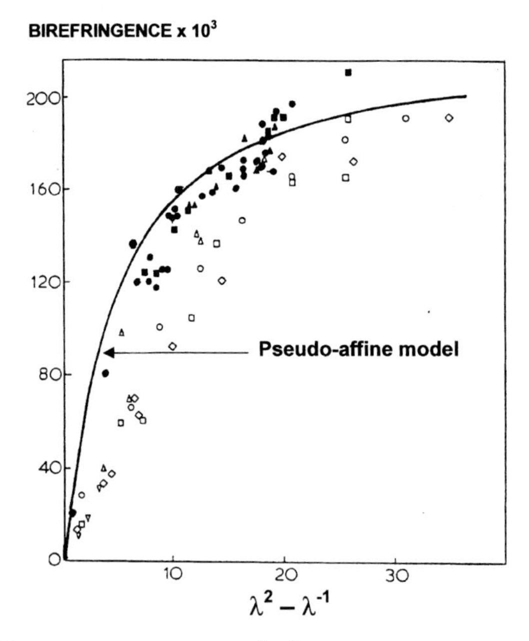

Figure 4.28 Birefringence of PET film versus λ^2-λ^{-1} at draw temperatures below T_g (closed symbols) and at 80 °C (open symbols) in comparison with predictions from the pseudo-affine model (continuous line). (Adapted from Ref. [135]).

In some amorphous polymers, such as PET and low density polyethylene (LDPE), orientation development during drawing at temperatures below T_g has been found to conform to the pseudo-affine scheme (Figure 4.28) [135], otherwise known as Kratky's floating-rod model. Here, the polymer is considered to consist of rod-like units which rotate towards the draw direction without changing the distance between crosslink points and

$$<P_2(\theta)> = \frac{1}{2}\left[\frac{2+k^2}{1-k^2} - \frac{3k\cos^{-1}k}{(1-k^2)^{3/2}}\right] \quad (4.17)$$

where $k=\lambda^{-3/2}$ [136,188,189]. In contrast to the affine scheme, the orientaton function depends only on strain and not, for example, on crosslink density, and the values of $<P_2(\cos\theta)>$ at a given strain are much greater than those predicted by the affine model. It has been noted that the pseudo-affine scheme predicts values of $<P_4(\cos\theta)>$ that are quite close to $<P_2(\cos\theta)>$, whereas in the affine deformation scheme $<P_4(\cos\theta)>$ is always very small [189]. In theory, therefore, measurements of $<P_4(\cos\theta)>$ using suitable techniques (e.g. polarized fluorescence, X-ray diffraction, NMR) can help distinguish deformation schemes, but in practice it is very difficult to obtain sufficiently accurate values of $<P_4(\cos\theta)>$.

The pseudo-affine model does not appear to be universally applicable to sub-T_g drawing of amorphous polymers because the orientation behavior of PMMA shows considerable temperature dependence (Figure 4.29) [190], unlike PET where a unique curve is obtained at all temperatures below T_g. Only at draw temperatures more than ~ 50 °C below T_g does the deformation of PMMA approach the pseudo-affine scheme and cease to become temperature dependent. This temperature dependent behavior below T_g may be associated with the nature and temperature position of the β relaxation in

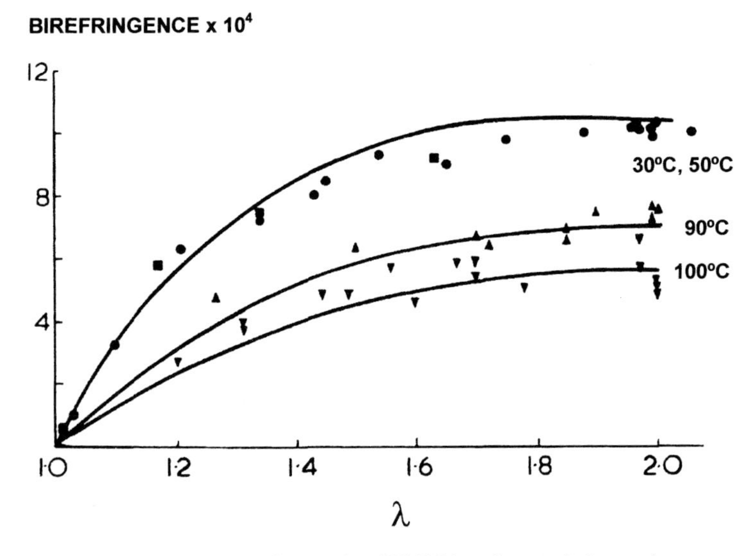

Figure 4.29 Birefringence versus draw ratio of PMMA rods extruded at various temperatures below T_g (adapted from Ref. [190]).

PMMA, and it has been suggested that there is progressive quenching out of the CH_3 rotations on the pendant side groups as temperature decreases [191]. It should be noted, however, that the data in Figure 4.29 was obtained from polymer rods with an initial diameter of 0.7 mm, and adiabatic heating may have had some influence on the deformation behavior.

4.4.1.5 Drawing of Crystalline Polymers

It has been found that in crystalline polymers, the relationship between the crystal orientation and draw ratio often conform quite well to the predictions of the pseudo-affine scheme [192–196], although deviations due to relaxation effects tend to occur in some temperature and strain-rate regions. In a drawing study of polypropylene above T_g, it was found that the noncrystalline phase also orients pseudo-affinely, as if the noncrystalline chains were embedded in the crystalline phase (see Section 13.3.3.4). Henneke et al. reported that for PET film crystallized in the undeformed, isotropic state, drawing at 78 °C (the T_g region) caused about 40% of the noncrystalline phase to orient cooperatively with the crystals in a pseudo-affine manner, while the rest oriented independently in a more or less affine manner (see Section 13.3.3.5). It may be noted that the volume fraction crystallinity of the undeformed PET sample was 0.27, compared to 0.45 for the PP sample.

There have been various attempts to model the load-elongation behavior of crystalline polymers [197–203]. Arridge and Barham modeled the postneck drawing stage by considering the polymer to consist of an assembly of stiff, needlelike crystalline fibrils, oriented in the draw direction and embedded in a softer amorphous, or partially crystalline, matrix [202,203]. The model is based on Cox's shear lag theory [204] with some modification to include a distribution of fibril lengths. By assuming that the ratio of fibril yield stress to matrix yield stress is temperature dependent, three drawing modes are defined: (1) fibrils plastically yield in an elastically deforming matrix; (2) fibrils and matrix plastically yield simultaneously; and (3) fibrils elastically deform in a plastically yielding matrix. The observed temperature dependence of the postneck drawing of polyethylene and polypropylene can then be explained by supposing that the matrix stress decreases with increasing temperature while the fibril flow stress is essentially independent of temperature. The model attributes the increase in tensile modulus of the fiber to an increase in aspect ratio of the homogeneously deforming fibrils, which increases their efficiency as reinforcing elements. This approach does not, however, involve a molecular interpretation of the deformation, and orientation effects and strain rate effects are not taken into account.

The approach of Ward and coworkers to modeling the deformation behavior of linear polyethylene is to represent the flow stress behavior by two thermally activated processes acting in parallel [198,199]. One process, which is associated with localized crystal slip (e.g. propagation of a Reneker defect through the crystal lattice [205]), has a relatively small activation volume and the other process, associated with the molecular network, has a relatively large activation volume. Application of this model to the creep behavior of ultra-high modulus polyethylene and polypropylene showed good agreement with experimental data. Amoedo and Lee [197] prefer to describe the inelastic deformation of crystalline polymers in terms of the rate equation of Lee and Zaverl, originally used to describe the flow behavior of metals by the thermally activated motion of dislocations. After conducting stress relaxation experiments on polypropylene samples to obtain the necessary material constants describing the rate and temperature dependent flow stress behavior of this material, simulated stress-strain curves were generated which provided satisfactory fits to the experimental results.

4.4.2 Crystallization

Even in the absence of molecular orientation, the nucleation stage of polymer crystallization is not well understood and is hampered by experimental difficulties. The classical theories of nucleation in polymers are reviewed in Wunderlich's book [206], and additional contributions to the theoretical development have been made since then, notably by Saddler [207] and Hikosaka [208]. However, some recent observations suggest that crystal nucleation from the amorphous state may primarily involve a process similar to phase separation by spinodal decomposition [209,210]. It was found that during thermally-induced crystallization of unoriented PET, long-range density fluctuations develop during the induction period (revealed by SAXS) while the overall density of the system remains unchanged from that of the amorphous state. The density fluctuation is highly regular with a characteristic length scale of 10 nm, and the dense regions, which do not involve local ordering of the chain segments, are considered to be the precursors of crystal nuclei. Crystal nuclei are presumed to form once the density fluctuation grows to a certain level, and the time for this to take place is the induction period. The SAXS features in the early stages of the induction period seem to develop in accordance with Cahn's linearized theory of spinodal decomposition [211], and the later stages can be described by Furukawa's scaling theory for the cluster growth regime [212]. SAXS/WAXS studies of the induction stage during spinning of polyvinylidene fluoride (PVDF) fibers [213] and melt extrusion of PET and PP films [214–216] indicate that this nucleation mechanism may also be involved in the stress-induced crystallization of polymers. However, these studies relied only on the use of WAXS to detect order at the polymer segment level, and it is well known that low levels of crystallinity can rarely be detected by this method when crystal size is small and/or contains significant packing imperfections. (The work of Wang et al. [217] does not help to resolve this problem, since their ability to detect 1% crystallinity in solutions of $C_{33}H_{68}/C_{12}H_{26}$ is entirely due to the very large size of the $C_{33}H_{68}$ crystals, which resulted in extremely narrow WAXS reflections). Further work will be required to substantiate the applicability of spinodal decomposition to crystal nucleation, with and without molecular orientation.

The kinetics of crystal growth in unoriented polymers under isothermal conditions can be predicted relatively well [218,219], and equations have been proposed for estimating the influence of molecular orientation on crystallization kinetics (see Section 2.1.2.6). However, these treatments do not succeed in predicting crystallinity development during drawing for a variety of reasons, most of which stem from the fact that orientation and crystallinity are developing simultaneously and are interdependent. Even in the absence of crystallization, models predicting strain-induced orientation in uncrosslinked polymers above T_g are unsatisfactory (see Section 4.4.1.2) and, once crystallization begins, additional complications arise. It is well known that crystallization influences the orientation distribution in the amorphous phase by preferential crystallization of the most highly oriented chain segments, and theoretical treatments of this phenomenon have been developed [220]. However, the various effects of the evolving crystallite network are much harder to predict. Crystals, acting as permanent junction points in an oriented network may promote or impede the development of noncrystalline orientation, depending on the distribution of chain contour lengths between crystals, the presence of taut tie molecules in the intercrystalline regions, the population density of crystals, and other factors. Also, the formation of a crystallite network can reduce the crystallization rate by restricting the molecular mobility and translational freedom of uncrystallized chains.

Under some drawing conditions, crystallization may be taking place non-isothermally due to adiabatic effects and/or the latent heat of crystallization (see Section 4.2.2.5), although recent data indicates that this may be a much smaller effect than was previously supposed [26]. Approximate models, based on modification of the Avrami theory, have been devised to predict crystallization under non-isothermal conditions (see Section 2.1.2.6) but, since the level of self-heating during drawing cannot be predicted reliably at present, they cannot be usefully applied to draw-induced crystallization.

Despite these problems, the experimental observations and the empirical models discussed in Section 4.3.2 have provided conceptual insights on deformation-induced crystallization that are proving helpful in the quest for a fundamental model applicable to crystallizable polymers. Another promising way forward is to use the growing computational power available for molecular dynamics modeling.

4.4.3 Molecular Dynamics Simulations

Rheological properties and draw induced structure development in polymeric fibers are entirely governed by the interchain and intrachain atomic interactions of the molecules. In principle, therefore, the most fundamental and comprehensive approach for predicting detailed structure evolution during drawing is the use of molecular modeling techniques. Molecular dynamics (MD) simulations, together with molecular mechanics (MM) simulations, can potentially predict the evolution of fiber morphology under an almost limitless range of simulated drawing conditions. In order to obtain accurate results, it is necessary to use about 500,000 atoms in the simulation, which requires parallel computing techniques at super-computer facilities. Jacob et al. [221] are currently pursuing this approach in conjunction with experiments using the Incremental Drawing Process (see Section 4.7.2), in order to assess the limiting orientation obtainable from multistage drawing sequences of various fiber-forming polymers. The basic procedure involves computation of the forces on every atom. The potential energy from bond stretching, angle bending, Lennard-Jones and Coulombic interactions are summed over all the atoms, MM simulations are used to minimize the potential energy (at 0 K) in the absence of kinetic energy, and MD simulations are then applied to follow the acceleration, velocity and position of each atom as a function of time. The present approach requires that the simulation be carried out every 10^{-15} s, corresponding to the C-H oscillation frequency. Clearly, this results in very long computing times, even on the most powerful computers, which is currently a major limitation to molecular simulations modeling.

Very preliminary results, using 40,000 atoms, have been obtained for PE at draw ratios up to about 4 and at draw-temperatures between the glass transition temperature and the melting point. Figures 4.30a and 4.30b shows the orientation development of a single PE

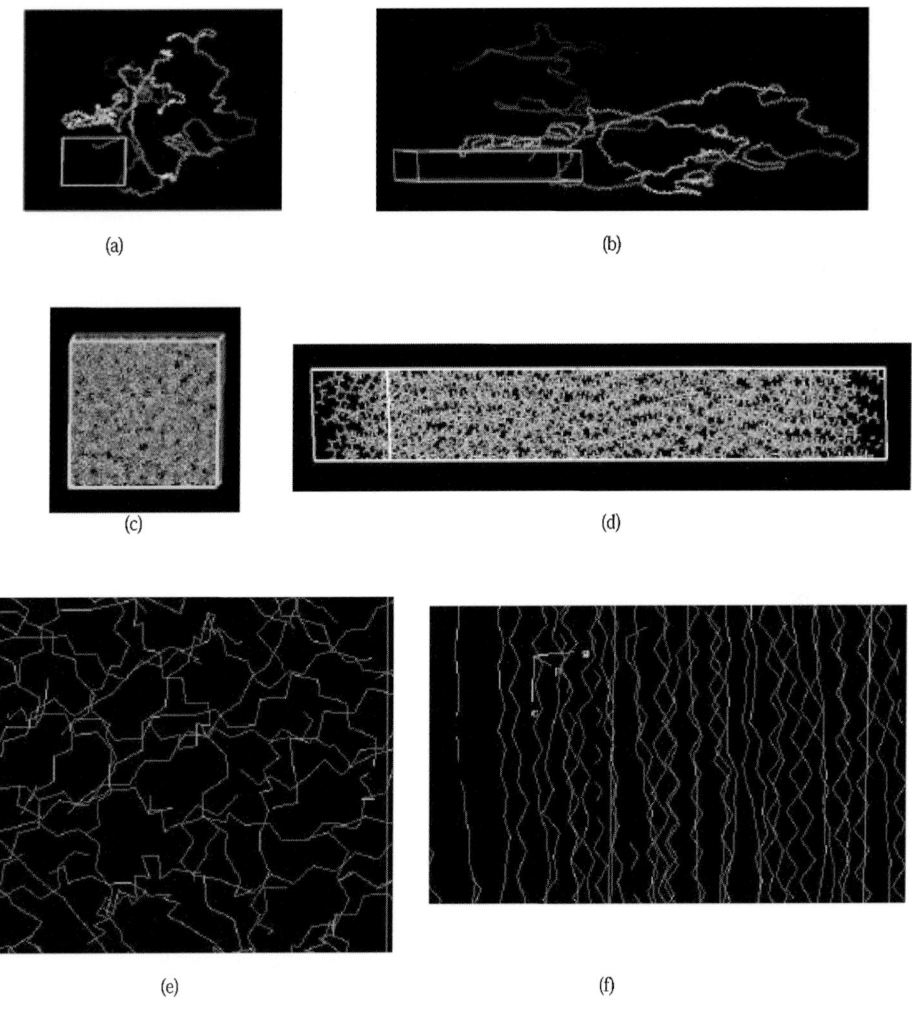

Figure 4.30 Computer simulations of a single PE molecule before drawing (a) and after drawing (b), and of a collection of PE molecules drawn from the (noncrystalline) isotropic state (c) to the oriented state (d) [221]. (e) and (f) show (magnified) details of the molecular arrangements in (c) and (d), but note that (f) is perpendicular to (d) – i.e. the draw direction is now vertical. The draw temperature was 27 °C and the applied stress was 100 MPa.

molecule before and after drawing at 27 °C, and Figure 4.30c and 4.30d show the orientation development of a collection of PE molecules from a noncrystalline, isotropic state under an applied stress of 100 MPa. Greater details of the isotropic and oriented state are shown in the 'close-up' pictures of Figure 4.30e and 4.30f. It should be pointed out that in order to reduce the computing time, the strain rates used in the simulation were more than an order of magnitude higher than are used in commercial processes. It is interesting, however, that the simulation indicates a molecular orientation of 0.65 at a draw ratio of about 4, and an increase in *trans* content from about 76% (in the isotropic state) to 88% in the oriented state. Despite the high levels of orientation there is no detectable crystalinity, as defined by the reported unit cell parameters. The absence of crystallinity is very likely a result of the high strain rates used. Even for a fast crystallizing polymer like PE, there would be insufficient time for the development of crystallinity during the deformation (see Section 4.3.2.4).

4.5 Morphology

4.5.1 Range of Order

We have referred to the development of noncrystalline (or amorphous) orientation, crystallinity and crystallite size, and have indicated that crystallites can crosslink the amorphous fraction. We have paid little attention, however, to the various arrangements in which the ordered and disordered regions may coexist in a drawn fiber (i.e. the polymer morphology); an area of some controversy and uncertainty. We do not provide a comprehensive review of this subject here, since it is covered rather fully by Hearle in Chapter 14, but we will highlight a few aspects that bear on current understanding of structure-property relationships.

Drawn fibers from flexible chain polymers generally contain crystallized and uncrytallized material. These are not separate phases in the thermodynamic sense, because molecules cross the 'phase' boundaries and they are therefore not independent of each other. This interdependence means that there are numerous ways in which the crystalline and noncrystalline material can be mutually arranged, and that these arrangements will depend on the properties of the molecule, and on the conditions of spinning, drawing and annealing. For example, there is evidence that highly oriented and highly crystalline polyethylene fibers possess crystalline continuity, such that 'intercrystalline bridges' traverse the disordered regions and connect the crystallites [222–224]. On the other hand, PET and polyamide fibers drawn at temperatures in the range T_g to T_g + 25 °C are usually only 20–30% crystalline, and they appear to consist of small crystallites irregularly dispersed in a partially oriented noncrystalline matrix. Whereas there are few crystal chain folds in drawn PET or polyamide fibers, annealing these polymers at high temperature causes substantial crystal growth and chain folding (see Figure 12.18), reducing the number of intercrystalline tie molecules. Moreover, if the fiber is permitted to shrink during the thermal treatment, the noncrystalline chains become disoriented, tending towards the properties of a true amorphous phase, and the morphology starts to resemble that of regularly stacked crystal lamellae connected by unoriented chains [225]. This does not mean that polyesters and polyamides are intrinsically incapable of forming highly oriented, predominantly crystalline fibers with crystal continuity, but no processing route has yet been devised to impart a structure of that kind (see Section 4.7.2).

It has long been known that both the crystalline and noncrystalline regions can have a range of order. Crystallites can have defects and lattice distortions, and the noncrystalline regions in drawn fibers invariably have a distribution of molecular orientation with preferred alignment along the draw direction.

Crystal lattice spacings, as well as the severity of crystal defects and lattice distortions, can vary with spinning, drawing and annealing conditions [60,61,144,226]. For example, molecules tend to pack closer together in the crystal when crystallization occurs during annealing at high temperature than when it is induced during drawing at temperatures close to T_g. Post-draw annealing at high temperature will generally reduce crystal imperfections and will sometimes increase the packing density in the crystal lattice.

4.5.2 Three Phase Models

The evolution of noncrystalline orientation with increasing strain and its dependence on draw temperature and strain rate have been discussed in earlier sections of this chapter. However, we have not yet mentioned that some researchers believe the noncrystalline fraction to be divided in two 'phases', a highly oriented phase and a poorly oriented (or unoriented) phase. This is the 'three phase' concept of fiber morpholgy, which has gained some popularity recently and requires brief discussion here.

While acknowledging an absence of observational confirmation, Prevorsek et al. considered their results to indicate that interfibrillar noncrystalline domains in PET and polyamide fibers contain highly extended molecules, and that chains in noncrystalline domains linking adjacent crystallites are less oriented [227]. These studies were on drawn fibers which had been annealed at high temperature and therefore contained a relatively high level of crystallinity ($\sim 50\%$).

The anisotropic nature of the WAXS 'amorphous halo' in drawn fibers has long been recognized, and it is possible to divide the amorphous scattering into isotropic and anisotropic components [137,138,225,228,229]. This does not in itself indicate that there are separate noncrystalline domains containing oriented and unoriented chains, it only indicates that some chains did not became oriented during the drawing process, while other attained various degrees of orientation. The distribution of orientation may be a continuum with oriented chains distributed more or less randomly amongst the unoriented material, or there may indeed be some kind of segregation of oriented and unoriented material. Similarly, DSC [230], NMR [231] and dielectric relaxation spectroscopy [232] results indicating mobile and rigid 'amorphous' fractions do not necessarily mean that these fractions occupy separate domains. The more highly oriented chains would have equally low mobility whether they were dispersed throughout the noncrystalline regions or whether they were in a neighborhood of similarly oriented chains.

However, Fu et al. offer noteworthy evidence for the existence of oriented and unoriented domains in the noncrystalline regions, which they term the 'intermediate phase' and the amorphous phase, respectively [233,234]. By calculating the crystalline phase diffraction from a structure model determined by a full-pattern X-ray diffraction analysis, the crystalline contribution was removed from the PET-fiber diffraction pattern, leaving the noncrystalline contribution, which was then divided into an amorphous phase and an oriented intermediate phase. Their procedure for subtraction of the crystalline contribution is more rigorous than procedures used in earlier studies, and their observation of rather sharp features in the noncrystalline scattering differs from the smooth, monotonic azimuthal intensity distribution obtained in previous analyses. The relatively complicated noncrystalline scattering pattern implies correlation between noncrystalline chains in some sort of partial, crystal-like order, suggestive of the mesomorphic phase observed in cold-drawn fibers (see Section 4.3.2.1). However, the mesomorphic phase is metastable and disappears after crystallization, whereas the 'intermediate phase' in the high speed spun fibers studied by Fu et al. seems to coexist with the crystalline phase. Fu et al. believe that the intermediate phase resides alongside the crystallites and that the noncrystalline regions linking adjacent crystallites (intercrystallite regions) are essentially isotropic. They use two main arguments to support this notion, based on the SAXS pattern: (1) the intercrystallite regions are of low electron density and must therefore be unoriented and (2) the length of the intercrystallite regions in their samples is short and would therefore crystallize easily if oriented.

However, this evidence is not conclusive. Firstly, it can be argued that short intercrystallite distances would impose a lack of translational and rotational freedom of noncrystalline chains tethered between adjacent crystallites, so that even partially-correlated chains would have difficulty falling in to crystalline register.

Secondly, chains in the intercrystallite regions need not be isotropic to provide sufficient electron density contrast to give rise to a distinct SAXS long period. Modeling [165] and experimental evidence [26] indicates that noncrystalline orientation can reach values at least as high as 0.45 before there is an important effect on polymer density (see also Section 13.3.1). In fact, in the samples studied by Fu et al., the *average* orientation of the intermediate phase does not much exceed 0.2 [234]. At this level of noncrystalline orientation, experimental data shows unambiguously that there is no effect on density [3,27,127]. The notion that the intercrystallite regions are essentially isotropic is also at odds with Prevorsek's observation of a significant decrease in the SAXS long period together with a decrease in 'amorphous orientation' after the initial stages of fiber shrinkage, indicating substantial contraction of chains in the intercrystallite regions [227]. Since the fibers in Fu et al.'s study were high speed spun fibers while Prevorsek et al. studied drawn fibers, orientation in the intercrystallite regions could well be lower in the former, but it is unlikely to be absent.

Another problem with locating the partially correlated chains in the interfibrillar regions is their apparent reluctance to crystallize. The formation of a highly oriented mesomorphic phase during cold drawing of PET is understandable, because the molecules are forced into alignment but do not have the molecular mobility to fall into crystalline register. It is difficult to understand, however, why the fraction of the 'intermediate phase' would almost double after constant length annealing of the as-spun fibers for 84 hours at 200 °C [234] (see also p. 60). Even unoriented chains crystallize rapidly at this temperature, and there is no obvious reason why partially correlated chains residing alongside crystals would not be incorporated rapidly into the existing crystals or form new ones.

Other researchers believe that semicrystalline fibers contain a crystal-amorphous interphase, and this could also explain the microstructure data of Fu et al. It is generally recognized that transition from a highly ordered crystalline phase to a random amorphous phase cannot occur abruptly [235]. Treatment of the crystal-amorphous transition region on the assumption that random configurations are adopted immediately upon emergence of the chains from the crystal leads to an excessive density close to the surface [236,237]. The crystal-amorphous interphase, which may consist of oriented, partially correlated chains and/or chain folds of unequal length, is 10–30 Å thick even in an unoriented system [239], and may well persist further in an oriented polymer. If the intermediate phase of Fu et al. actually corresponds to an amorphous-crystalline interphase, its fraction of the noncrystalline phase would clearly increase with increasing crystallinity, in accordance with their observations.

It may well be that the noncrystalline regions of certain polymers under specific spinning/drawing conditions arrange themselves into oriented interfibrillar regions and unoriented, intercryallite regions. Based on the same structural data, however, it is possible to envisage the existence of crystalline-amorphous interphase regions, with the remainder of the noncrystalline volume containing a distribution of order – from unoriented chains to taut tie molecules – that depends on the conditions of fiber formation. And other structure models may also be reasonable (see Chapter 14). The message is that despite enormous advances in the range and precision of instruments and analytical techniques for polymer structure characterization, it is still not possible to resolve some important details of fiber morphology, and many observations are open to an array of interpretations. Moreover, flexible chain polymers can adopt a spectrum of arrangements that are more limited by the engineering constraints of spinning/drawing processes than by the laws of physics. While this may make efforts to understand polymer structure formation seem daunting, simple and explicable patterns have sometimes emerged from systematic experimental studies. Major improvements in understanding are also likely to arise from molecular dynamics modeling of polymer deformation, since this potentially permits systematic exploration of simulated drawing sequences outside the limits imposed by current laboratory and commercial drawing methods.

4.6 Structure-Property Relationships

Control of fiber microstructure during spinning and drawing has certainly lead to major advances in fiber properties over the last four decades (see for example Figure 5.1). Room-temperature fiber properties that have received the most attention are initial modulus and tensile strength. However, extensibility (elongation at break), yield, work of rupture, dimensional stability (shrinkage and creep), heat-setting properties (see Chapter 12) and penetrant diffusion (dyeability) have also been the focus of many structure-property studies. Since the emphasis of the present review is on structure formation, the following discussion of structure-property relations provides a simplified outline of the subject, serving primarily as an introduction to further reading.

4.6.1 Tensile Modulus, Strength, Yield, and Elongation at Break

As discussed by Lemstra et al. in Section 5.2.1, the ultimate (axial) tensile modulus of a fiber, in which all the chains are perfectly oriented along the fiber axis, can be estimated from theory and from X-ray diffraction measurements of crystal deformation under load. In practice, the modulus of fibers from flexible chain polymers is usually much lower than the ultimate value because chains in the noncrystalline regions are far from fully aligned. The modulus of a polymer in the amorphous state E_a, with randomly oriented chains, is generally one to two orders of magnitude lower than its (chain axis) modulus in the crystalline state E_c [240,241]. If a simple series arrangement of amorphous and crystalline regions is assumed, with crystals fully oriented along the fiber axis, then the overall tensile modulus of the fiber is given by [242,243]:

$$\frac{1}{E} = \frac{(1-\chi)}{E_a} + \frac{\chi}{E_c} \tag{4.18}$$

and it is clear that the unoriented amorphous regions would have the dominant influence on fiber modulus, even at very high levels of crystallinity. If the crystalline regions are surrounded by the amorphous phase in a parallel-series arrangement, and $E_c >> E_a$, the influence of E_a is again predominant. Only when there is crystal continuity (amorphous and crystalline phases in parallel, or amorphous regions surrounded by crystalline regions) can E approach E_c. If however, E_a (parallel to the fiber axis) increases as a result of increased orientation in the noncrystalline regions and the formation of taut tie molecules, the fiber modulus can increase dramatically [244,245]. It may be noted that if the taut-tie molecules are assumed to be distributed uniformly over the cross-section of the intercrystalline regions and to act in parallel with unoriented amorphous material, the fraction of taut-tie molecules can be estimated from knowledge of E, E_c and E_a [246,247].

For ultra-high modulus fibers, the (room temperature) stress-strain curves are almost linear and breakage occurs abruptly, whereas lower modulus fibers yield before breakage and may display two yield points, as illustrated in Figure 4.31. Heuvel et al. attribute the first yield to the breaking of chains at entanglement points after straining of the entanglement network [248]. Subsequently, the crystals transfer most of the load and the tie molecules between successive crystals are considered to strain, causing an increase in modulus prior to their final breakdown. The values of both maxima in the modulus-strain curve are primarily dependent on the level of noncrystalline orientation, and the stain level at which the second maximum occurs decreases with increasing noncrystalline orientation.

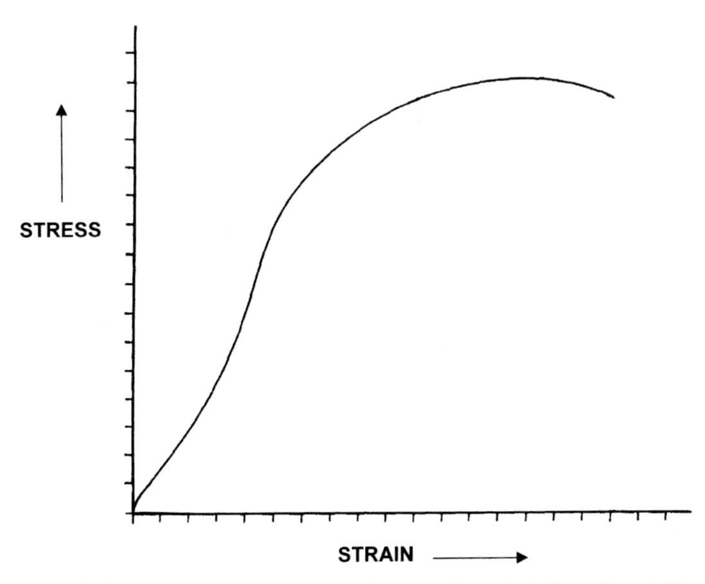

Figure 4.31 Typical room temperature stress-strain curve for oriented, semi-crystalline PET fibers.

Theoretical estimation of the ultimate tensile strength of oriented polymers is discussed in Sections 5.2.2 and 5.2.3. For fully extended and oriented chains, the tensile strength depends on intramolecular and intermolecular bond strength and on molecular weight. For less ordered structures, tensile strength is also influenced by chain extension, the degree of connectivity between crystals, the distribution of chain contour-lengths in the intercrystallite regions, the level of molecular orientation and the population and size of microvoids. When contour-length distribution of the tie chains is narrow and when chain extension, orientation and intercrystal connectivity are high, tensile strength is enhanced [248]. It can also be understood that a combination of narrow tie-chain length distribution and low orientation will result in a relatively high elongation at break.

The number of intercrystalline tie-segments per unit volume of noncrystalline fraction ν determines the degree of connectivity between crystals, and since this property plays a significant role in fiber properties, it useful to consider how polymer morphology influences ν. It has been shown by Buckley and Salem [18] that, assuming every molecule is incorporated into at least one crystal and all chain ends are located in the noncrystalline fraction:

$$\nu = \frac{1}{1-\chi}\left[\frac{\chi}{L_c A_o(\overline{n}_f + 1)} - \frac{\rho N_o}{\overline{M}_n}\right] \tag{4.19}$$

where A_o is the cross-sectional area of the unit cell perpendicular to the molecular axis, $\overline{n}_f$ is the mean number of chain folds per molecule per crystal, L_c is the crystal thickness (in the chain axis direction) and N_o is Avogadro's number. Equation 4.19 indicates the importance of crystal thickness and chain folding on crystal connectivity. Since thermal crystallization of a poorly oriented fiber results in a large fraction of chain re-entry (adjacent or otherwise) in a given crystal, the number of intercrystal tie-segments are low, producing poor intercrystal connectivity and brittle properties. Similarly, there is infrared evidence [250,251] (as well a inferences from SAXS data [252]) that thermal treatments of highly oriented PET at temperatures between 200 °C and the melting point cause a substantial increase in the fraction of adjacent crystal chain folds (even under constant length annealing conditions). The consequent reduction in ν causes a decrease in fiber strength and, as discussed in Section 4.6.3, contributes to an increase in the rate of penetrant diffusion.

Tensile strength, extensibility, and modulus not only depend on the test temperature but also on the strain rate imposed. Samuels [253] has shown that at low rates of deformation, the polymer chains in isotactic polypropylene have time to rearrange before failure, such that the final structure at failure is the same for all samples irrespective of the starting orientation, and all samples break at the same true stress. As deformation rate increases, samples break at lower stresses and lower extensions because the chains cannot redistribute stress fast enough, and at very high strain rates the breaking stress is directly related to the starting orientation.

4.6.2 Dimensional Stability

Thermal shrinkage of unrestrained fibers results from the relaxation of stresses imposed during fiber formation. Since the predominant shrinkage mechanism involves coiling of chains in the noncrystalline regions, higher noncrystalline orientation leads to higher shrinkage [248,251,254]. The degree of shrinkage is also related to the extent to which the initial molecular network is preserved during deformation. Entanglement slippage during deformation will fade the memory of the original network structure, reducing shrinkage. A higher shrinkage-force is expected when crystallinity is low because there are more chains in the mobile phase, although under some circumstances the presence of crystallites *prior* to deformation can aid subsequent shrinkage by enhancing network connectivity. When crystallization is induced in an oriented fiber in the process of deformation, or while held at constant length or under tension, the extended state of the most highly oriented chains is locked in, reducing the fiber's tendency to shrink in subsequent (unrestrained) thermal treatments below the melting point [255]. Relaxation of noncrystalline chains during high-temperature annealing at constant length also contributes to reduced shrinkage in subsequent treatments.

Creep is the time-dependent tendency for a polymer to elongate under a fixed load [256,257] and the creep behavior of a fiber at room temperature determines its suitability in load-bearing applications requiring high dimensional stability. Crosslinked polymers that are rubbery at room temperature show elastic deformation but do not flow, whereas uncrosslinked polymers generally display a combination of linear viscoelastic and non-linear flow behavior. The rate of creep is reduced by stronger interchain bonds, higher T_g, higher molecular weight and higher crystallinity. Polyethylene fibers, for example, have a strong tendency to creep at room temperature due to weak intermolecular interactions and a low glass transition temperature, whereas polyamide fibers have low creep rates due to the presence of hydrogen bonds. Woods, Busfield and Ward [258] have shown, however, that creep in polyethylene fibers can be dramatically reduced by using γ-radiation in an atmosphere of acetylene to crosslink the molecules.

In a number of commercial processes, fibers and fiber assemblies are heat-set at relatively low strain levels to permanently impart an imposed configuration or shape. The success of the heat-setting process depends on the ability of a thermal treatment to relax the internal stresses in the deformed fiber and erase its 'memory' of the undeformed state. In noncrystalline materials the molecular mechanisms involved in stress relaxation (and creep) have received considerable research attention and are fairly well understood. They involve time-dependent molecular motions which are accelerated in a predictable way by increasing temperature [256,257]. In semicrystalline materials where the presence of crystals can limit or prevent certain molecular motions in the noncrystalline regions, especially flow processes, the molecular origins of heat-setting were for a long time a matter of speculation. There is now strong evidence, however, that the stress relaxation processes giving rise to heat-setting in PET and polyamide fibers are the same as in amorphous polymers and that crystallinity simply slows down the relaxation rate [18,259,260]. On the other hand, in polymers of high crystallinity, such as polyethylene, it is probable that sufficient stress can only be relieved with the assistance of crystal relaxation processes and/or partial melting of crystals [199,261].

4.6.3 Penetrant Diffusion

For textile applications it is essential that the fiber be able to absorb small molecules, such as dyes and UV stabilizers. Since the rate of dye diffusion is sensitive to subtle changes in fiber structure, small differences in fiber processing conditions can result in highly nonuniform dyeing, and it has therefore been a long sought goal to quantify relationships between dye diffusion behavior and fiber structure. The free-volume concept of diffusive transport in amorphous polymers relates the diffusion coefficient D_a to the fractional free volume V_f through:

$$D_a = D_{ao} \exp - \left(\frac{1}{V_f} \right) \tag{4.20}$$

where D_{ao} characterizes the frequency and average direction of diffusive jumps [262]. Based on the work of Klute [263], Michaels et al. [264] and Peterlin [265], it can be argued that the diffusion coefficient of a penetrant molecule in an unoriented crystalline polymer D would lie below that of a purely amorphous polymer due to the increased path length imposed on the diffusing penetrant by crystalline objects (Detour Ratio ϕ) as well as to the constraining influence of crystallites on chain mobility (Immobilization Factor β). Thus:

$$D = \left(\frac{\phi}{\beta} \right) D_{ao} \exp - \left(\frac{1}{V_f} \right) \tag{4.21}$$

where $0 < \phi \geqslant 1$, and $\beta \geqslant 1$.

Extending this analysis to an oriented polymer, in which diffusion is perpendicular to the preferred axis of orientation, it can be argued that the diffusion coefficient would be further reduced by orientation of the molecules in the noncrystalline regions due to lower chain mobility (increasing β). Of course, in an oriented polymer, the diffusion coefficient and the pre-exponential terms become direction dependent.

The main microstructural features influencing ϕ and β are likely to be the degree of crystallinity, the number-density of tie segments in the noncrysalline regions (which has an inverse dependence on the degree of chain folding and crystal thickness – see Eq. 4.19), and the molecular orientation in the noncrystalline regions. An increase in any one of these parameters would decrease the penetrant diffusion rate, but the relative strength of their individual influence on diffusion rate is difficult to determine and the predictive capability of Eq. 4.21 is therefore highly limited. Some progress has been made in this area recently by preparing a range PET fibers in which only one key structural parameter varied at a time. It was found that an increase in noncrystalline orientation from 0.1 to 0.22 caused a seven fold decrease in dye diffusion coefficient [266], and that there is roughly a reciprocal square law relationship between D and the noncrystalline orientation [26]. The degree of crystallinity was found to have an even larger effect on diffusion coefficient than molecular orientation, and the impact of the number-density of tie segments remains to be assessed.

4.7 High Performance Fibers

4.7.1 Apolar Polymers

The formation of ultra-high modulus fibers from a flexible chain polymer was first achieved by Ward and coworkers in the 1970s [4–9,223]. They found that under optimized conditions of draw temperature and strain rate, the tensile modulus of linear polyethylene tapes and fibers is a single function of draw ratio for molecular weights M_w in the approximate range 65,000–300,000. When the absence of chain branching was ensured, a maximum draw ratio of about 35 and a tensile modulus of about 70 GPa was attained (Figure 4.32). While still only about 25% of its theoretical value, this represented an order of magnitude increase in the axial modulus of fibers and films from flexible chain polymers.

Although the morphology of the undrawn polyethylene starting material does not influence the maximum draw ratio attainable in the solid-state drawing operation, it does strongly influence the strain rate required to reach this draw ratio in samples where $M_w < 130{,}000$. By slow cooling the polymer to ensure that crystallization occurs at low degrees of super-cooling, more regularly folded lamellae with fewer tie chains are formed than in rapidly quenched samples which crystallize at high super-cooling. This results in greater ease of unfolding and more efficient unraveling of chains in the slowly cooled samples, which can therefore achieve the maximum draw ratio at faster rates of strain.

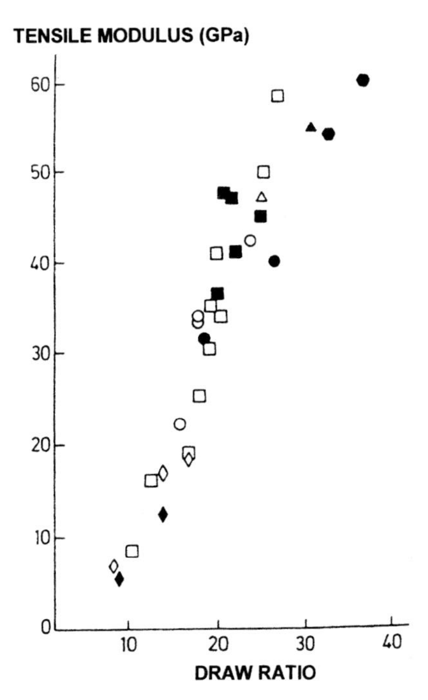

Figure 4.32 Tensile modulus versus draw ratio for quenched (open symbols) and slow-cooled (closed symbols) linear polyethylene samples drawn at 75 °C, with various molecular weights (M_w) in the range 68000 to 312000 (adapted from Ref. [223]).

For $M_w > 130{,}000$, however, the drawing behavior of quenched and slow cooled samples are indistinguishable, and for both samples a marked reduction in maximum draw ratio occurs as M_w is increased above about 270,000. By raising the draw temperature of the high molecular weight samples, high maximum draw ratios can be restored, but with increasing temperature the modulus values fall further and further below the unique modulus/draw ratio relationship identified at lower temperature (Figure 4.33). Eventually the slope of the modulus/draw-ratio curve becomes negative, indicating the highest temperature at which effective drawing can take place. The temperature position of the maximum in the modulus/draw-ratio curve depends on M_w, being in the range 75–80 °C for $M_w \sim 100{,}000$ and 125 °C for $M_w \sim 800{,}000$.

The reduction in the maximum draw ratio at high molecular weights appears to arise from the constraining effects of entanglements. This is discussed fully in Chapter 5, together with a detailed review of the gel-spinning method that overcomes this limitation. In brief, gel-spinning substantially reduces the entanglement density of the high molecular weight polymer by means of solvent-induced gelation of the polymer. The entanglements are not reformed after solvent removal, permitting dry-state drawing of the polyethylene filaments to draw ratios exceeding 50, with modulus values reaching 180 GPa and tensile strengths of 3–4 GPa [10–12].

Whereas similar ultra-drawing processes have been successfully applied to other apolar polymers, such as polypropylene [267] and polyoxymethylene, resulting in tensile moduli approaching the values determined for the crystal lattice, their application to polymers

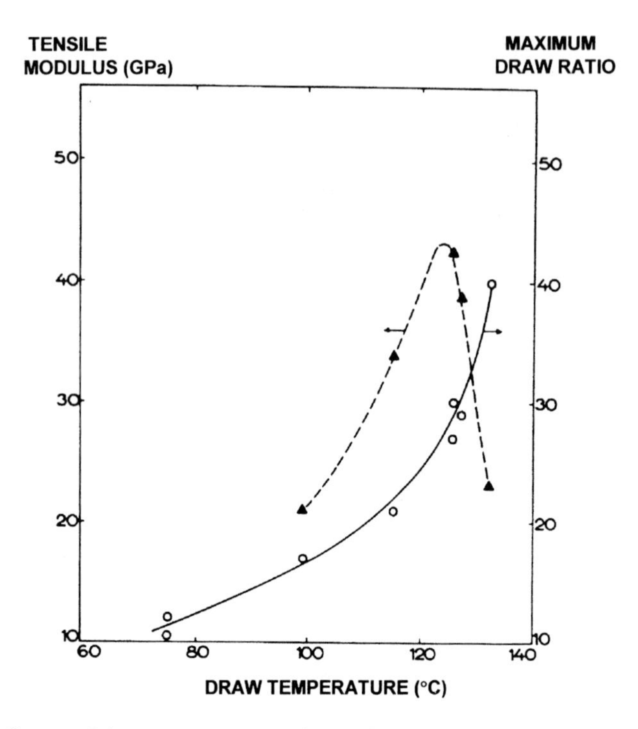

Figure 4.33 Influence of draw temperature on the maximum attainable draw ratio for high molecular weight LPE ($M_w \sim 8 \times 10^5$). The room temperature modulus of the samples drawn to the maximum draw ratio is shown at each draw temperature (broken line). (Adapted from Ref. [223]).

with significant polar interactions have failed to result in dramatic enhancement of mechanical properties.

4.7.2 Polar Polymers

The weak interchain interactions in apolar polymers seem to optimize the role of the crystals in the drawing process. On the one hand, these interactions provide sufficient coherence for the crystals to act as pinning points, impeding retraction of chains, and on the other hand they permit, under stress, sufficient chain-sliding in the crystals to promote efficient orientation of the intercrystalline tie-chains. Sliding of chains through the crystals prevents trapping of poorly oriented tie-chains when one, or a few, tie chains have become fully oriented. This is because, as drawing proceeds, fully oriented tie-chains will slip through the crystals and permit the slack in the less oriented tie-chains to be taken-up; simultaneously increasing the orientation and reducing the length dispersion of chains in the intercrystalline regions, and finally leading to crystal continuity. These mechanisms do not seem to be possible in polar polymers because the crystallites are essentially rigid and severely limit the molecular rearrangements that can occur in the intercrystalline regions during drawing (see, for example, Slutsker and Utevskii [268]).

Even when solvents are applied, the drawability of polyamides is not improved because the hydrogen bonds remain intact [269,270]. Smook et al. [271] suggest that, when there is a sufficiently reduced entanglement concentration, the maximum drawability of a polymer is directly related to its cohesion energy J_c, the energy required to break all the intermolecular contacts in the liquid state. For example, the J_c of polyamide 6 is about 80,000 kJ/kmol, mostly due to hydrogen bonds, and its maximum experimental draw ratio in the dilute state seems to be about 5, whereas the J_c of polyethylene is around 7,000 kJ/kmol and its maximum draw ratio is about 100. It may be noted that J_c for PET is estimated to be about 50,000 kJ/kmol, mostly from dipole-dipole interactions, but there is no reliable data on its maximum draw ratio in the dilute state.

This does not mean that the mechanical properties of polar polymers cannot be enhanced via innovative drawing sequences, most of which are based on the understanding that amorphous orientation should be as high as possible before high levels of crystallinity are induced [271–274]. As mentioned earlier, crystallization of unoriented or poorly oriented fibers leads to a low fraction of crystallite connectivity and low deformability, whereas partially oriented (as-spun) fibers containing no crystallinity or very low levels of crystallinity can be further oriented and crystallized in a subsequent drawing operation. Under suitable drawing conditions, usually involving draw temperatures in the range 150–250 °C, fibers with high intercrystallite connectivity together with high levels of crystallinity and amorphous orientation are obtained. This type of fiber forming sequence is used in the commercial production of so-called High Modulus Low Shrinkage (HMLS) polyester fibers for tire cord applications. According to one of the HMLS patents [273], PET with an intrinsic viscosity of about 0.9 dl/g is spun at speeds from approximately 2000–3000 m/min into filaments with partially oriented chains and a volume fraction crystallinity in the range 3–15%. The as-spun fibers are then drawn in three or four stages at temperatures between 90 °C and 230 °C. The resulting fibers possess an initial modulus of about 15 GPa, a tenacity of about 1 GPa and shrinkage of 4–10% at 177 °C.

In a similar vein, Kunugi and coworkers [275,276] have developed a zone-drawing/zone annealing method for producing PET and other polymers with relatively high tensile moduli. Here, filaments under tension are exposed to a narrow slit-shaped heater which is moved from the lower part of the sample to the upper part. The temperature applied is about 90 °C and this results in relatively high molecular orientation and relatvely low crystallinity. The filaments are then held under tension and zone-annealed at a temperature of about 200 °C. Variations of this method include cold drawing (below T_g) followed by high temperature drawing in one or more stages [247].

Although leading to significant improvements in mechanical properties, none of these multi-stage drawing techniques have produced PET fibers with a tensile modulus significantly higher than 20 GPa, which is less than 20% of its calculated modulus (110 GPa [277]). It is probable that the lack of dramatic improvement in the mechanical properties of PET fibers not only arises from limitations imposed by the rigidity of the crystals and the strength of the interchain interactions, but also from the fact that this polymer is difficult to produce in high molecular weights.

Ito, Takahashi and Kanamoto have succeeded in preparing high molecular weight PET, having an intrinsic viscosity of 2.4 dl/g, by solid state polymerization of solution-grown PET crystals that were prepared from commercially available polymer [278]. The high molecular weight PET was dissolved in a mixture of hexafluoro-2-propanol and dichloromethane to produce solution-spun fibers, which could be drawn up to a draw ratio of thirteen by a two stage drawing technique. The drawn fibers had a tensile modulus of 34 GPa and a tensile strength of 1.9 GPa, the highest values reported for PET so far.

Besides PET, there have been improvements in the tensile properties of fibers from other polymers of intermediate polarity, such as atactic polyvinyl alcohol (PVA) and polyketones, but reported tensile moduli are still less than one third of the theoretical values.

It is likely that the production of ultra-high modulus fibers from flexible-chain polar polymers, especially polyamides and other hydrogen-bonded polymers, will require a different approach from those established for polyolefins and other polymers with weak interchain bonding. One approach could be to block or decrease the interchain interactions (i.e dramatically reduce J_c) to permit ultra-high draw ratios and efficient chain orientation, and then restore them once the chains are almost fully oriented and extended. In polyamide, this could involve complexing the Lewis base (CONH) sites with a strong Lewis acid prior to deformation and decomplexing them after drawing. Candidate complexing agents are $AlCl_3$ and BCl_3, which are decomplexed in steam [279], and iodine, which is decomplexed in sodium thiosulphate. Drawing of a PA6-iodine complex resulted in a somewhat higher maximum draw ratio but did not so far provide significantly enhanced mechanical properties [280], and drawing with $AlCl_3$ and BCl_3 has not been reported so far. However, there is strong motivation for continued efforts in this direction because ultra-high modulus polyamide fibers would possess a high melting point, a high compressive strength and low creep, all of which are lacking in polyethylene fibers.

A way of blocking the dipole-dipole interactions in PET such that they can be restored after drawing is not apparent. There is, however, considerable interest in controlling the crystallization of PET for both fiber and non-fiber applications. As discussed in Section 4.2.2.1, when crystallization reaches a critical level, the drawing stress increases dramatically and the rate of development of molecular orientation (and the crystallization rate) decreases sharply. Since we attribute these changes to the constraints imposed by a well-developed crystallite network [3,23,27], it is apparent that 'too much' crystallinity in PET limits its drawability (maximum draw ratio) and the effectiveness of the orientation process (Figure 4.3). This, in turn, limits throughput rates in drawing because higher draw ratios permit higher take-up roll speeds. On the other hand the data of Salem and coworkers [3,27] indicate that low levels of crystallinity (below about 15%) can accelerate strain-induced orientation by providing a suitable distribution of permanent network junctions. There are therefore various motivations to optimize the crystallization rate and the crystallinity level for specific processes and properties. One approach is to disrupt the regularity of the polymer chain by, for example, replacing a fraction of the terephthalic acid units with isophthalic acid units [281]. Percentages of isophthalic acid in the range 10–30% restrict the crystallization of PET, but recent studies indicate that in high speed spinning the PET chain orientation is lower than for the PET homopolymer [282]. Fibers from the isophthalic-acid copolyester are more easily dyed (which is particularly useful for dye-resistant microfibers), have a much reduced softening temperature (which saves energy in nonwoven heat-bonding processes), and have high shrinkage levels above T_g (which can be used to provide bulk and texture) [283].

Another approach to controlling crystallization in PET, currently being pursued by Hermanutz et al., is by forming a copolymer of PET and poly(ethylene methyl terephthalate) PEMT [282]. The chemical structure of PEMT is very similar to PET, the only difference being that one of the hydrogen atoms on the benzene ring is replaced by a methyl group [170]. This is sufficient to completely suppress crystallization. Strain-induced and thermally induced crystallization are being investigated in a range of PET/PEMT copolymer ratios between 2% and 40%, and preliminary data from high-speed spun fibers show that, unlike the isophthalic-acid copolyester, suppression of crystallinity is not accompanied by lower amorphous orientation [282]. Drawing studies of PET/PEMT in these various copolymer ratios are in progress.

Although some recent, systematic studies have elucidated the effects of temperature, strain rate and molecular weight on structure development during drawing of crystallizable polymers at single temperatures, and under simple temperature sequences, there remains much to be learned about the effects of imposing more complex temperature, stress, and strain rate profiles on the deforming fiber. Such investgations are limited by the fact that it is not possible to sweep the temperature over a wide range, or to vary the strain rate in a controlled way, in the course of conventional drawing operations because the total deformation is completed in one or two very short zones. However, the Incremental Drawing Process (IDP), a commercial drawing technique in which continuous drawing occurs in up to 100 zones [284–287], would lend itself to the investigation of process-structure-property relationships under an array of conditions inaccessible to conventional drawing methods.

The IDP draws fiber in many small increments as it moves, in a helical path of increasing diameter, over a pair of skewed-axes rolls having steeply tapered, linear or non-linear profiles. The overall draw ratio is determined by the relative diameters at which the fiber enters and exits the rolls. Increasing the angle of skew between the roll-axes increases the helix pitch, thereby reducing the number of draw stages and increasing the rate of draw at each stage. In this way, the number of draw increments can be varied between 3 and 100. If linear-profile (conical) rolls are used, the draw rate at each stage is the same, but using non-linear profiles permits imposition of a strain rate gradient or profile. Thus, the imposed strain rate at each increment is decoupled from the throughput speed. A temperature gradient or profile can be set up by heating the fiber in each drawing increment by a separate heater.

Even under single-temperature and single draw-rate conditions, it has been found that IDP can produce polyamide and PET fibers with nearly constant moduli up to the breaking load, i.e both yield points that typically appear on the room-temperature stress-strain curves are eliminated. Fibers chemically-crosslinked by irradiation have shown similar yield-free behavior, implying that the crystallites and/or entanglements in IDP fibers provide more efficient crosslinking than in conventionally-drawn fibers. The reason that drawing in numerous small steps leads to these properties is not yet understood.

Since molecular mobility decreases with increasing molecular orientation and increases with temperature, it has been argued that the effectiveness of drawing would be enhanced by continiuously increasing the draw temperature to 'match' the state of orientation [285]. An IDP fiber has an orientation profile over the length of the fiber helix, and a temperature profile corresponding to the orientation profile may be imposed so that each draw increment occurs at a temperature best suited to the orientation state of the fiber at that increment. However, this type of temperature profiling has not been investigated so far, and neither have more complex temperature and strain-rate sequences.

Fiber forming processes frequently take place under conditions where equilibrium thermodynamics do not apply – a theme that is taken-up in Chapter 14 – and it is possible that benefit would be gained by pushing fiber formation conditions very far into the non-equilibrium regime. In Prigogine's approach to nonlinear thermodynamics, the rate of energy flow plays the determining role, and this implies that fluctuating energy conditions encourage the evolution of more and more complex structures by subdividing into systems and subsystems that minimize the dissipation of energy [288–290]. Unfortunately, Prigogine's ideas are difficult to apply to practical systems, and Lindenmeyer's attempts to develop a theoretical framework applicable to fiber formation and other processes are incomplete [291–293]. For example, it is not possible to extract from Lindenmeyer's analysis the size and frequencey of energy fluctuations required for a significant impact on structure evolution. However, the ideas are sufficiently intriguing to justify their exploration at the experimental level, and the IDP might be ideally suited to this type of study since it can take fibers through rapid changes of temperature, stress, strain-rate and other control factors.

4.8 References

1. Ziabicki A. and Jarecki, L. In High-Speed Fiber Spinning. Ziabicki A. and Kawai H. (Eds.) (1985) John Wiley and Sons, New York, p. 225
2. Yazdanian, M., Ward, I.M. and Brody, H. *Polymer* (1985) 26, p. 1779
3. Clauss, B and Salem, D.R. *Polymer* (1992) 33, p. 3193
4. Capaccio G. and Ward I.M. *Polymer* (1975) 16, p. 239
5. Capaccio G. and Ward I.M. *Polym. Eng. Sci.* (1975) 15, p. 219
6. Capaccio G., Crompton, T.A. and Ward I.M. *J. Polym. Sci. Phys. Edn* (1976) 14, p. 1641
7. Capaccio G., Crompton, T.A. and Ward, I.M. *Polym. Eng. Sci.* (1978) 18, p. 533
8. Capaccio G., Crompton, T.A. and Ward I.M. *J. Polym. Sci. Phys. Edn* (1980) 18, p. 301
9. Capaccio G and Ward I.M. *J. Polym. Sci. Phys. Edn* (1984) 22, p. 475
10. Smith, P. and Lemstra, P.J. U.K. Patent 2 040 414 (1979)
11. Smith, P. and Lemstra, P.J. *Polymer* (1980) 21, p. 31
12. Smith, P. and Lemstra, P.J. *J. Mater. Sci.* (1980) 15, p. 505
13. Kunugi, T., Oomori, S. Mikami, S. *Polymer* (1988) 29, p. 814
14. Matsuo, M, Inoue, K., Abumiya, N. Sen-I Gakkaishi (1984) 40, T-275
15. Kanamoto, T, Tsurata, A., Tanaka, M, Porter, R.S., *Polym. J.* (1983) 15, p. 327
16. Peters, R.H. In Diffusion in Polymers. Crank J. and Park G.S. (Eds.) (1968) Academic Press, London and New York, p. 315
17. Mark, J.E. In *Physical Properties of Polymers.* Mark J.E., Eisenberg A.,Graessley, W.W., Mandelkern, L and Koenig, J.L (Eds.) (1984) American Chemical Society, Washington D.C. p. 1
18. Buckley, C.P. and Salem, D.R., *Polymer*, (1987) 28, p. 69
19. Salem, D.R. and Ruetsch, S.B. In *Book of Abstracts of the Joint International Fiber Society Conference* Mulhouse, France (1997) p. 35
20. Falolle, R., Tassin, J.F., Sergot, P., Pambrun, C. and Monnerie, L. *Polymer* (1983) 24, p. 379
21. Thirion, P. and Tassin, J.F. *J. Polym. Sci. Phys. Edn* (1983) 21, p. 2097
22. Le Bourvellec, G., Monnerie, L. and Jarry, J.P. *Polymer* (1986) 27, p. 856
23. Salem, D.R. *Polymer* (1992) 33, p. 3182
24. Heffelfinger, C.J. and Burton, R.L., *J. Polym. Sci.* (1960) 47, p. 289

25. Lapersonne, P., Bower, D.I. and Ward, I.M. *Polymer* (1992) 33, p. 1266
26. Salem, D.R. Unpublished data.
27. Clauss, B. and Salem, D.R. *Macromolecules* (1995) 28, p. 8328
28. Pearce, R., Cole, K.C., Ajji, A. and Dumoulin, M.M. *Polym. Eng. Sci.* (1997) 37, p. 1795
29. Buckley, C.P., Jones, D.C. and Jones, D.P. *Polymer* (1996) 37, 2403
30. Vigny, M, Tassin, J.F., Gibaud, A. and Lorentz, G. *Polym. Eng. Sci.* (1997) 37, p. 1785
31. Salem, D.R. *Polymer* (1994) 35, p. 771
32. Spruiell, J.E., McCord, D.E. and Beuerlein, R.A. *Trans. Soc. Rheol.* (1972) 16, p. 535
33. Salem, D.R. In *Book of Extended Abstracts: Twelfth Annual Meeting of the Polymer Processing Society* Sorrento, Italy (1996) p. 5.6/1
34. Salem, D.R. In preparation
35. Salem, D.R. In *Book of Extended Abstracts: Ninth Annual Meeting of the Polymer Processing Society* Manchester, U.K. (1993) p. 83
36. Grime, D. and Ward I.M. *Trans. Faraday Soc.* (1958) 54, p. 959
37. Brody, H. *J. Macromol Sci. – Phys.* (1983) B22(1), p. 19
38. Long, S.D. and Ward, I.M. *J. Appl. Polym. Sci.* (1991) 42, p. 1911
39. Rebenfeld, L, Makarewicz, P.J., Weigmann, H.-D., and Wilkes, G.L. *J. Macromol. Sci. Chem.* (1976) C15 (2), p. 279
40. Peterlin, A. *J. Polym. Sci. Part C* (1960) 47, p. 289
41. Peterlin, A. *J. Polym. Sci. Part C* (1965) 9, p. 61
42. Peterlin, A. *J. Mater. Sci.* (1971) 6, p. 490
43. Peterlin, A. In *Structure and Properties of Oriented Polymers*, Ward I.M. (Ed) (1975) John Wiley and Sons, New York, p. 36
44. Stein, R.S. and Wilkes, G.L. In *Structure and Properties of Oriented Polymers*, Ward I.M. (Ed) (1975) John Wiley and Sons, New York, p 57
45. Keller, A. and Machin, *J. Macromol Sci.* (1967) B1, p. 41
46. Dees, J.R. and Spruiell, J.E. *J. Appl. Polym. Sci.* (1974) 18, p. 1053
47. Samuels, R.J. *J. Polym. Sci. Part A* (1965) 3, p. 1741
48. Samuels, R.J. *J. Polym. Sci. A-2* (1968) 6, p. 2021
49. Hay, I.L. and Keller, A. *Kolloid Z* (1965) 204, p. 43
50. Baranov, V.G. and Gasparyan, K.A. *J. Polym. Sci. A-2* (1970) 8, p. 1015
51. Peterlin, A.J. *J. Polym. Sci. Part C* (1966) 15, p. 427
52. Sadler, D.M. and Barham, P.J. *Polymer* (1990) 31, p. 36
53. Jambrich, M. and Diacik, I., *Faserforschung u. Textiltech* (1964), 15, p. 590
54. Sheehan, W.C. and Cole, T.B., *J. Appl. Polym. Sci.* (1964) 8, p. 2359
55. Hattori, H. Takagi, Y and Kawaguchi, T. *Bull. Chem. Soc. Japan* (1962) 35, p. 1163
56. Hattori, H. Takagi, Y and Kawaguchi, T. *Bull. Chem. Soc. Japan* (1963) 36, p. 675
57. Ahmed, M. *Polypropylene Science and Technology* (1982) Elsevier, Amsterdam, p. 262
58. Ziabicki, A. *Kolloid Z.* (1959) 167, p. 132
59. Sakaouku, K., Morosoff, N. and Peterlin, A. *J. Polym. Sci. Polym. Phys. Edn.* (1973) 11, p. 31
60. Heuvel, H.M. and Huisman, R. *J. Appl. Polym. Sci.* (1981) 26, p. 713
61. Salem, D.R. Moore, R.A.F. and Weigmann, H.D., *J. Polym. Sci. B: Polym. Phys.* (1987) 25, p. 567
62. Murthy, N.S., Bray, R.G., Correale, S.T., and Moore, R.A.F. *Polymer* (1995) 36, p. 3863
63. Babatope, B. and Isaac, D. H. *J. Mech. Behavior Mater.* (1992) 3, p. 195
64. Murase, S.; Hirami, M.; Nishio, Y.; Yamamoto, M. *Polymer* (1997) 38, p. 4577
65. Buckley, C.P. In preparation.
66. Buckley, C.P., Jones, D.C. and Jones, D.P. *Polymer* (1996) 37, p. 2403
67. Lazurkin, J.S., *J. Polym. Sci.* (1958) 30, p. 595
68. Vincent, P.I. *Polymer* (1960) 1, p. 7
69. Cavaille, J.Y., Perez, J. and Johari, G.P. *Physical Review B* (1989) 39, p. 2411
70. Buckley, C.P. Private communication.
71. Bechtel, S.E., Vohra, S. and Jacob, K I. *J. Appl. Mechanics* (2000). Submitted
72. Bechtel, S.E., Vohra, S. and Jacob, K I. *Text. Res. J.* (2000). Submitted
73. Bechtel, S.E., Vohra, S. and Jacob, K I. *Polymer*. (2000). Submitted

74. Bender, M.F. and Williams, M.L. *J. Appl. Phys.* (1963) 34, p. 3329
75. Bender, M.F. and Williams, M.L. *Text. Res. J.* (1963) 33, p. 1023
76. Bender, M.F. and Williams, M.L. *J. Appl. Phys.* (1965) 36, p. 3044
77. Dittman, H. *Acta. Polym.* (1979) 30, p. 365
78. Dittman, H. *Acta. Polym.* (1981) 32, p. 260
79. Matsuo, T. *Sen-I Gakkaishi.* (1968) 24(8), p. 366
80. F.S. Smith *J. Phys. D* (1975) 8, p. 759
81. Schultz-Gebhardt, F. *Acta. Polym.* (1986) 37, p. 247
82. Le Bourvellec, G. Beautemps, J. and Jarry, J.P. *J. Appl. Polym. Sci.* (1990) 39, p. 319
83. Desai, P. and Abhiraman, *J. Polym. Sci. B: Polym. Phys.* (1988) 26, p. 1657
84. Treleor, L.R.G. *Trans. Faraday Soc.* (1941) 37, p. 84
85. Treleor, L.R.G. *Trans. Faraday Soc.* (1947) 43, p. 284
86. Gent, A.N. *Trans. Faraday Soc.* (1954) 50, p.521
87. Alfrey, T. and Mark, H. *J. Phys. Chem.* (1942) 46, p. 112
88. Flory, P.J. *J. Chem. Phys.* (1947) 15, p. 397
89. Krigbaum, W.R. and Roe, R.J. *J. Polym. Sci. Part A* (1964) 2, p. 4391
90. Gaylord, R.J. *J. Polym. Sci. Polym. Phys. Edn* (1976) 14, p. 1827
91. Ziabicki, A. *Colloid Polym. Sci.* (1974) 252, p. 207
92. Ziabicki, A. *J. Chem Phys.* (1977) 66, p. 1638
93. Jarecki, L. and Ziabicki, A. *Polymer.* (1977) 18, p. 1015
94. Jarecki, L. *Colloid Polym. Sci.* (1979) 257, p. 711
95. Alfonso, G.C., Verdona, M.P. and Wasiak, A. *Polymer* (1978) 19, p. 711
96. Wasiak, A. *Colloid Polym. Sci.* (1981) 257, p.711
97. Yeh, G.S.Y. and Hong, K.Z. *Polym. Eng. Sci.* (1979) 19, p. 395
98. Yeh, G.S.Y., Hong, K.Z. and Krueger, D.L. *Polym. Eng. Sci.* (1979) 19, p. 401
99. Smith, F.S. and Steward, R.D. *Polymer* (1974) 15, p. 283
100. Sun, T. Pereira, J.R.C and Porter, R.S. *J. Polym. Sci. Polym. Phys. Edn* (1984) 22, p. 1163
101. Peszkin, P.N., Schultz, J.M., and Lin, J.S. *J. Polym. Sci. Polym. Phys. Edn* (1986) 24, p. 2591
102. Schultz, J.M In *Oriented Polymer Materials* Fakirov, S. (Ed.) Hüthig & Wepf, Heidleberg, p. 361
103. Hristov, H.A. and Schultz, J.M. *Polymer* (1988) 29, p.1211
104. Turnbull, D. and Fisher, J.C., *J. Chem Phys.* (1949) 17, p. 71
105. Frank, F.P. and Tosi, M. *Proc. Royal Soc. (London)* (1961) 263, p. 323
106. Hoffman, J.D. and Lauritzen, *J. Res. NBS* (1961) 65A, p. 297
107. Alexander, L.E. Ohlberg, S., and Taylor, G.R. *J. Appl. Phys.* (1955) 26, p. 1068
108. Mark, J.E. *Polym. Eng. Sci.* (1979) 19, p. 409
109. Pennings, A.J. and Kiel, A.M. *Kolloid Z.Z. Polymere* (1965) 205 p. 160
110. Keller, A. and Machin, M.J. *J. Macromol. Sci.* (1967) B1, p. 41
111. Keller, A. and Mackley, M.R. *Pure Appl. Chem.* (1974) 39, p. 195
112. Mackley, M.R. and Keller, A. *Phil. Trans. Royal Soc. London* (1975) 29, p.278
113. Mackley, M.R. *Colloid Polym. Sci.* (1975) 253, p. 393
114. Pennings, A.J. Schouteten, C.J.H. and Kiel, A.M., *J. Polym. Sci. C* (1972) 38, p. 167
115. Zwinjnenburg, A, and Pennings, A.J. *Colloid Polym. Sci.* (1976) 254, p. 868
116. Zwinjnenburg, A, and Pennings, A.J. *J. Polym. Sci. Letters Edn* (1976) 14, p. 339
117. Pennings, A.J. *J. Polym. Sci. C* (1977) 59, p. 55
118. Posthuma de Boer, A. and Pennings, A.J., *Macromolecules* (1977) 10, p. 981
119. Zwinjnenburg, A., van Hutten, P.F., Pennings, A.J. and Chanzy, H.D. *Colloid Polym. Sci.* (1978) 256 p. 729
120. Pennings, A.J. and Meihuizen, K.E. In *Ultra-High Modulus Polymers*, Ciferri, A and Ward, I.M. (Eds) (1977) Applied Science Publishers Ltd., Essex, U.K., p. 117
121. Keller, A. In *Ultra-High Modulus Polymers*, Ciferri, A and Ward, I.M. (Eds) (1977) Applied Science Publishers Ltd., Essex, U.K., p. 321
122. Yeh, G.S.Y. *Polym. Eng. Sci.* (1976) 16, p. 138
123. Le Bourvellec, G., Monnerie, L. and Jarry, J.P. *Polymer* (1987) 28, p. 1712
124. Le Bourvellec, G. and Beautemps, J. *J. Appl. Polym. Sci.* (1990) 39, p. 329

125. Salem, D.R. *Polymer* (1992) 33, p. 3189
126. Salem, D.R. *Polymer* (1995) 36, p. 3605
127. Salem, D.R. *Polymer* (1998) 39, p. 7067
128. Salem, D.R. *Polym. Eng. Sci.* (1999) 39, p. 2419
129. Lorentz, G. and Tassin, J.F., *Polymer* (1994) 35, p. 3200
130. Lapersonne, P. Tassin, J.F., Monnerie, L. and Beautemps, J. *Polymer* (1991) 32, p. 3331
131. Blundell, D.J., MacKerron, D.H., Fuller, W., Mahendrasingam, A., Martin, C., Oldman, R.J., Rule, R.J. and Rickel, C. *Polymer* (1996) 37, p. 3303
132. Ajji, A., Gievremont, J., Cole, K.C. and Dumoulin, M.M. *Polymer* (1996) 37, p. 3707
133. Tassin, J.F., Vigny, M., and Veyrat, D. In *Orientation in Polymers: SPE RETEC Conference Proceedings* (1999) p.281; Tassin, J.F. In *Solid State Processing of Polymers*, Ward, I.M., Coates, P.D. and Dunoulin, M.M. (Eds) (2000) Hanser (Munich), p. 214
134. Vigny, M., Aubert, A. Hiver, J.M. Abouflaraj, M. and G'Sell, C. *Polym. Eng. Sci.* (1999) 39, p.2366
135. Rietsch F., Duckett, R.A. and Ward, I.M. *Polymer* (1979) 20, p. 1133
136. Ward, I.M. *Proc. Phys Soc.* (1962) 80 p. 1176
137. Asano, T and Seto, T., *Polym. J.* (1973) 5, p. 72
138. Auriemma, F., Corradini, P., Rosa, C.D., Guerra, G., Petraccone, V., Bianchi, R. and Dino, G.D. *Macromolecules* (1992) 25, p. 2490
139. Tassin, J.F., Thirion, P. and Monnerie, L. *J. Polym. Sci. Polym. Phys. Edn*, (1983) 21, p. 2109
140. Dumbleton, J.H., Murayama, T., and Bell, J.P. *Kolloid Z.Z. Polymere* (1968) 228, p. 54
141. Dumbleton, J.H., Bell, J.P. and Murayama, T, *J. Appl. Polym. Sci.* (1968) 12, p. 2491
142. Dumbleton, J.H. *J. Polym. Sci. Part A-2* (1969) 7, p. 667
143. Brody, H. *J. Macromol Sci. – Phys.* (1983) B22(3), p. 407
144. Heuvel, H.M. and Huisman, R. *J. Appl. Polym. Sci.* (1978) 22 p. 2229
145. Hsiao, B. S.; Kennedy, A. D.; Leach, R. A.; Chu, B.; Harney, P. *J. Appl. Crystallogr.* (1998) 30 p. 1084
146. Jäckel, K. *Kolloid Z.* (1954) 137, p.130
147. Sakurada, I *Polyvinyl Alcohol Fibers* (1985) Marcel Dekker Inc., New York and Basel, p. 102
148. Guth, E. and Mark, H. *Lit. Chem.* (1934) 65, p. 93
149. Kuhn, W. and Grün, F. *Kolloid Z. (1942) 101, p. 242*
150. Flory, P.J. *Principles of Polymer Chemistry* (1953) Cornell University Press, Ithaca
151. Treloar, L.R.G. *The Physics of Rubber Elasticity 3rd ed* (1975) Clarendon Press, Oxford
152. Ward, I.M. *Mechanical Properties of Solid Polymers 2nd ed.* (1983) John Wiley and Sons, New York, p. 62
153. Hearle, J.W.S. *Polymers and their Properties* (1982) John Wiley and Sons, New York, p. 153
154. James, H.M. and Guth, E. *J. Chem. Phys.* (1947) 15, p. 669
155. Flory, P.J. *Proc. Royal Soc. London (A)* (1976) 351, p. 351
156. Erman, B. and Flory, P.J. *J. Chem. Phys.* (1978) 68, p. 5363
157. Flory, P.J. *Polymer* (1979) 20, p.1317
158. Flory, P.J. and Erman, B. *Macromolecules* (1982) 15, p. 800
159. Ball, R.C. Doi, M., Edwards, S.F. and Warner, M. *Polymer* (1981) 22, p.1011
160. Edwards. S.F. and Vilgis, Th. *Polymer* (1986) 27, p. 485 Equations 4.12 and 4.15 are not identical to those in this reference, because the latter appear to contain arithmetic (canceling) errors.
161. Thirion, P. and Weil, T. *Polymer* (1984) 25, p. 609
162. Mark. J.E., Kato, M. and Ko, J.H. *J. Polym. Sci. C* (1976) 54, p. 217
163. Andrady, A.L., Llorente. M.A. and Mark, J.E. *J. Chem. Phys.* (1980) 72, p.2282
164. Pinnock, P.R. and Ward, I.M. *Trans. Faraday Soc.* (1966) 62, p.1308
165. Nobbs, J.H., Bower, D.I. and Ward, I.M. *Polymer* (1976) 17, p. 25
166. Nobbs, J.H., Bower, D.I. and Ward, I.M. *J. Polym. Sci. Polym. Phys. Edn*, (1979) 17, p. 259
167. Matthews, R.G., Duckett, R.A. and Ward, I.M. *Polymer* (1997) 38, p. 4795
168. Muller, R. and Pesce, J. *Polymer* (1994) 35, p. 734
169. Ward, I.M., Bleackley, M, Taylor, D.J.R., Cail, J.I., and Stepto, R.F.T. *Polym. Eng. Sci.* (1999) 39, p. 2335
170. O'Neill, M.A., Duckett, R.A. and Ward, I.M. *Polymer* (1988) 29, p. 54

171. Ward, I.M. *Text. Res. J.* (1961) 31, p. 650
172. Doi, M and Edwards, S.F. *J. Chem. Soc Faraday Trans. 2* (1978) 74, pp. 1789, 1802, 1818
173. Doi, M. *J. Polym. Sci. Polym. Phys. Edn*, (1980) 18, p. 1005
174. Doi, M. *J. Polym. Sci. Polym. Phys. Edn*, (1983) 21, p. 667
175. Curtiss, C.F. and Bird, R. *J. Chem. Phys.* (1981) 74, pp. 2016, 2026
176. de Gennes, P.G. *J. Chem. Phys.* (1971) 55, p. 572
177. Rouse, P.E. *J. Chem. Phys.* (1953) 21, p. 1271
178. Tassin, J.F., Thirion, P. and Monnerie, L. *J. Polym. Sci. Phys. Edn* (1983) 21, p. 2109
179. Hasley, G. White, H.J. and Eyring, H. *Text. Res J.* (1945) 15, p. 295
180. Ward, I.M. *Polym. Eng. Sci.* (1984) 24, p. 724
181. Harward, R.N. and Thackray, G. *Proc. Royal. Soc. London* (1968) A302, p.453
182. Boyce, M.C., and Arruda, E.M. *Polym. Eng. Sci.* (1990) 30, p. 1288
183. Hasan, O.A. and Boyce, M.C., *Polym. Eng. Sci.* (1990) 35, p. 331
184. Dooling, P.J., Buckley, C.P and Hinduja, S. *Polym. Eng. Sci.* (1998) 38, p. 892
185. Bergstrom, J.S. and Boyce, M.C. *J. Mech. Phys. Solids* (1998) 46, p. 931
186. Buckley, C.P. and Jones, D.C. *Polymer* (1995) 36, p. 3301
187. Edwards, S.F. and Vilgis, Th. *Polymer* (1987) 28, p. 375
188. Kratky, O. *Kolloid Z.* (1933) 64, p. 213
189. Ward, I.M. In *Orientation in Solid Polymers*, G. Bodor (Ed) *J. Polym. Sci.: Polym. Symp.* (1977) 58, p. 1
190. Kahar, N., Duckett, R.A. and Ward, I.M. *Polymer* (1978) 19, p. 136
191. Andrews, R.D. and Hammack, T.J. *J. Polym. Sci. C* (1964) 5, p. 101
192. Jambrich, M. and Diacik, I., *Faserforschung u. Textiltech* (1961), 12, p. 522
193. Culpin, M.F. and Kemp, K.W. *Proc. Phys. Soc.* (1956) B69, p. 1301
194. Cannon, C.G. and Chappel, F.P. *Brit. J. Appl. Phys.* (1959) 10, p.68
195. Crawford, S.M. and Kolsky, H. *Proc. Phys. Soc.* (1951) B64, p. 119
196. Oya, S., Yasuda, A. and Nakai, *Sen-I Gakkaishi* (1961) 17, p. 1088
197. Amoedo, J. and Lee, D. *Polym. Eng. Sci.* (1992) 32, p. 1055
198. Duxbury, J. and Ward, I.M. *J. Mater. Sci.*, (1987) 22, p. 1215
199. Wilding, M.A. and Ward, I.M. *J. Mater. Sci.*, (1984) 19, p. 629
200. G'Sell, C. and Jonas, J.J. *J. Mater. Sci.*, (1981) 16, p. 1956
201. Liu, M.C.M. and Krempl, E. *J. Mech. Phys. Solids* (1979) 27, p. 377
202. Barham, P.J. and Arridge, R.G.C. *J. Polym. Sci. Phys. Edn* (1977) 15, p. 1177
203. Arridge, R.G.C. and Barham, P.J. *J. Polym. Sci. Phys. Edn* (1978) 16, p. 1297
204. Cox, H. *Brit. J. Appl. Phys.* (1952) 3, p.72
205. Reneker, D.H. *J. Polym. Sci.* (1962) 59, p. 539
206. Wunderlich, B. *Macromolecular Physics* (1976) Academic Press Inc., New York, p. 1
207. Sadler, D.M. *Polymer* (1987) 28, p. 1440
208. Hikosaka, M., *Polymer* (1990) 31, p. 458
209. Imai, M, Mori, K., Mizukami, T., Kaji, K. and Kanaya, T. *Polymer* (1992) 33, pp. 4451 and 4457
210. Imai, Kaji, K. and Kanaya, T. and Sakai, Y. *Phys. Rev. Condensed Matter* (1995) 52, p. 12696
211. Cahn, J.W. *J. Chem. Phys.* (1965) 42, p. 93
212. Furukawa, H. *Adv. Phys.* (1985) 34, p.703
213. Cakmak. M., Teitge, A., Zachmann, H.G. and White, J.L. *J. Polym. Sci: B: Phys.* (1993) 31, p. 371
214. Ryan, A.J., Terrill, N.J. and Fairclough, J.P.A. *Proceedings of the American Chemical Society, Div. Of Polymeric Materials: Science and Engineering* (1999) 81, p. 353
215. Terrill, N.J., Fairclough, J.P.A, Towns-Andrews, E., Komanschek, B.U., Young, R.J., Ryan, A.J. *Polymer* (1998) 39, p.2381
216. Olmsted, T.A., Poon, W.C.K, Macleish, T.C.B., Terrill, N.J. and Ryan, A.J. *Phys. Rev. Lett.* (1998) 81, p. 373
217. Wang, Z.G, Hsiao, B.S., Kopp, C., Sirota, E.B., Agarwal, P. and Srinivas, S. *Proceedings of the American Chemical Society, Div. Of Polymeric Materials: Science and Engineering* (1999) 81, p. 355
218. Wunderlich, B. *Macromolecular Physics* (1976) Academic Press Inc., New York, p. 115
219. Mandelkern, L. In *Physical Properties of Polymers.* Mark J.E., Eisenberg A., Graessley, W.W., Mandelkern, L and Koenig, J.L (Eds.) (1984) American Chemical Society, Washington D.C. p. 155

220. Abhiraman, A.S. *J. Polym. Sci. Phys. Edn* (1983) 21, p. 583
221. Jacob, K.I., Vohra, S., Dong, H., Hu, Q., Bechtel, S.E. and Salem, D.R. *NTC Annual Reports and Research Briefs* (1999) M98G5
222. Gibson, A.G., Davies, G.R. and Ward, I.M. *Polymer* (1978) 19, p. 683
223. Capaccio, G., Gibson, A.G., and Ward, I.M. In *Ultra-High Modulus Polymers*, Cifferi., A and Ward, I.M. (Eds) (1979) Applied Science Publishers, pp. 1–76
224. Hosemann, R., Loboda-Cackovic, J., Cackovic, J. *J. Polym. Sci. Symp.* (1973) 42, p. 563
225. Fischer, E.W. and Fakirov, S.F., *J Mater. Sci.* (1976) 11, p. 1041
226. Murthy, N.S. *J. Polym. Sci. B: Phys.* (1986) 24, p. 549
227. Prevorsek, D.C., Tirpak, G.A., Harget, P.J. and Reimschuessel, A.C. *J. Macromol. Sci. – Phys* (1974) B9(4), p. 733
228. Bonart, R. *Kolloid-Z.* (1968) 23, p. 438
229. Hinrichsen, G., Adam, H.G., Krebs, H., Springer, H. *Colloid Polym. Sci.* (1980) 258, p. 232
230. Suzuki, H. Grebowicz, J. and Wunderlich, B. *Polym. J.* (1985) 17, p.1
231. Havens, J.R. and VanderHart, D.L. *Macromolecules* (1985) 18, p. 1663
232. Schlosser, E. and Schonhals, A. *Colloid Polym. Sci.* (1989) 267, p. 963
233. Fu, Y., Annis, B., Boller, A., Jin, Y. and Wunderlich, B. *J. Polym. Sci. B: Phys.* (1994) 32, p. 2289
234. Fu, Y., Busing, B., Boller, Jin, Y., Affholter, K.A. and Wunderlich, B. *Makromol. Chem.* (1994) 195, p. 803
235. Flory, P.J., Yoon, D.Y. and Dill, K.A. *Macromolecules* (1984) 17, p.862
236. DiMarzio, E.A. and Guttman, C.M. *Polymer* (1981) 21, p. 733
237. Guttman, C.M. and DiMarzio E.A. *Macromolecules* (1982) 15, p. 525
238. Yoon, D.Y. and Flory, P.J. *Macromolecules* (1984) 17, p. 868
239. Sadler, D.M. In *Structure of Crystalline Polymers,* I.H. Hall (Ed) (1984) Elsevier, London and New York, p. 125
240. Holliday, L. In *Structure and Properties of Oriented Polymers*, Ward I.M. (Ed) (1975) John Wiley and Sons, New York, p 257
241. Hearle, J.W.S., Prakesh, R. and Wilding, M.A. *Polymer* (1987) 28, p. 441
242. Takayanagi, M. Imada, K. amd Kajiyama, T. *J. Polym. Sci. C* (1966) C15, p. 263
243. Uemura, S. and Takayanagi, M. *J. Appl. Polym. Sci.* (1966) 10, p. 113
244. Dulmage, W.J. and Contois, L.E. *J. Polym. Sci.* (1958) 28, p. 275
245. Biangardi, H.J. and Zachmann, H.G. *J. Polym. Sci. Polym. Symp.* (1977) 58, p. 169
246. Peterlin, A. *Polym. Eng. Sci.* (1979) 19, p. 118
247. Hofmann, D., Goschel, U., Walenta, E. Geiss, D. and Philip, B. *Polymer* (1989) 30, p. 242
248. Heuvel, H.M., Lucas, L.J., van den Heuvel, C.J.M and de Weijer, A.P. *J. Appl. Polym. Sci.* (1992) 45, p. 1649
249. Termonia, Y. and Smith, P. In *High Modulus Polymers*, Zachariades, A.E. and Porter, R.S. (Eds) (1988) Marcel Dekker Inc., New York, p. 321
250. Statton, W.O., Koenig, J.L. and Hannon, M.J. *J. Appl. Phys.* (1970) 41, p.4290
251. Wilson, M.P.W. *Polymer* (1974) 15, p. 277
252. Dismore, P.F. and Statton, W.O. *J. Polym. Sci. C* (1966) C13, p. 133
253. Samuels, R.J., *Polym. Eng. Sci.* (1979) 19, p. 66
254. Prevorsek, D. and Tobolsky, A.V. *Text. Res. J.* (1965) 33, p. 795
255. Cappaccio, G. and Ward, I.M. *Colloid Polym. Sci.* (1982) 260, p. 46
256. Ward, I.M. *Mechanical Properties of Solid Polymers 2nd ed.* (1983) John Wiley and Sons, New York
257. Ferry J.D. *Viscoelastic Properties of Solid Polymers 3rd ed* (1980) John Wiley and Sons, New York
258. Woods, D.W., Busfield, W.K. and Ward, I.M. *Polym. Comm.* (1984) 25, p. 298
259. Buckley, C.P. and Salem, D.R., *J. Appl. Polym. Sci.* (1990) 41, p. 1707
260. Hearle, J.W.S., Wilding, M.A., Auyeung, C. and Ihmayed, R. *J. Text. Inst.*, (1990) 81, p. 214
261. Castiff, E., Offenbach, J. and Tobolsky, A.V. *J. Colloid Sci.*, (1956) 11, p. 48
262. Cohen, M.H. and Turnbull, D. *J. Chem. Phys.* (1959) 31, p. 1164
263. Klute, C.H., *J. Appl. Polym. Sci.* (1959) 1, p. 340
264. Michaels, A.S. and Bixler, H.L. *J. Polym. Sci.* (1961) 50, p. 413
265. Peterlin, A., *J. Macromol. Sci. – Phys.* (1975) B11(1), p. 57

266. Salem, D.R., Weigmann, H.D. and Huang, X.X. In *AATCC Book of Papers* (1994) AATCC, Research Triangle Park, NC, p. 414
267. Peguy, A., Manley, R.S. st J, *Polym. Comm.* (1984) 25, p. 39
268. Slutsker, L.I. and Utevskii, L.E., *J. Polym. Sci. Phys. Edn* (1984) 22, p. 805
269. Postema, A.R., Smith, P. and English, A.D. *Polym. Comm.* (1990) 31, p. 444
270. Smook, J., Vos, G.J.H. and Doppert, H.L. *J. Appl. Polym. Sci.* (1990) 41, p. 105
271. Davis, H.L., Jaffe, M.L., Besso, M.M., U.S. Patent 3,946,100 (1976)
272. Davis, H.L., Jaffe, M.L., LaNieve, H.L., Powers, E.J. U.S. Patent 4,195,052 (1980)
273. Nelson, C.J., Bheda, J.H., Rim, P.B., Turner, J.M. U.S. Patent 5,630,976 (1997)
274. Abhiraman, A.S. and Song, J.W. *J. Polym. Sci. Letters Edn* (1985) 23, p. 613
275. Kunugi, T., Suzuki, A., Hashimoto, M. *J. Appl. Polym. Sci.* (1981) 26, p. 1951
276. Suzuki, A., Sato, Y., Kunugi, T *J. Polym. Sci. B: Polym. Phys.* (1998) 36, p. 473
277. Thistlewaite, T., Jakeways, R. and Ward, I.M., *Polymer* (1988) 29, p. 61
278. Ito, M., Takahashi, K., and Kanamoto, K. *Polymer* (1990) 31, p. 61
279. Roberts, M.F. and Jenekhe, S.A. *Macromolecules* (1991) 24, p. 3142
280. Hoe, H.C. and Porter, R.S. *Polymer* (1986) 27, p. 241
281. Jones, K.M., Kerr, G.P., Tindale, N. U.S. patent 5,409,983 (1994)
282. Hermanutz, F., Salem, D.R. and Oppermann, W. In preperation
283. Amoco Chemical Company Bulletin GTSR-113A, May 1994 and November 1995
284. Sussman, M.V. U.S. Patent Nos. 3,978,192 (1976), 4,891,872 (1990), 4,980,957 (1991)
285. Sussman, M.V. *Research and Development* (1989) p. 62
286. Sussman, M.V. *Fiber World* (1985) p. 58
287. Misra, A., Dutta, B and Prasad, V.K. *J. Appl. Polym. Sci.* (1986) 31, p. 441
288. Nicolis, G. and Prigogine, I. *Thermodynamic Theory of Structure, Stability and Fluctuations* (1971) John Wiley and Sons, New York
289. Nicolis, G. and Prigogine, I. *Self-Organization in Non-Equilibrium Systems* (1977) John Wiley and Sons, New York
290. Prigogine, I. *Order Out of Chaos* (1984) Bantam Books, New York
291. Lindenmeyer, P.H. *Polym. Eng. Sci.* (1994) 34, p. 252
292. Lindenmeyer, P.H. *Text. Res. J.* (1984) 54, p. 131
293. Lindenmeyer, P.H. *Polym. Eng. Sci.* (1981) 21, p. 958

5 Basic Aspects of Solution(Gel)-Spinning and Ultra-Drawing of Ultra-High Molecular Weight Polyethylene

Piet J. Lemstra, Cees W.M. Bastiaansen and Sanjay Rastogi
Dutch Polymer Institute/Eindhoven University of Technology, Eindhoven, The Netherlands

5.1 Introduction

The main characteristic of synthetic polymers is that the chemical bonding is strong along the main chain whereas usually relatively weak interactions (van der Waals bonds, hydrogen bonds) exist between the chains. As a consequence, the mechanical properties of (isotropic) polymers are usually not impressive (plastics) in comparison with inorganic materials such as metals, glass and ceramics, see Table 5.1.

In the past two decades, however, significant progress has been made in exploiting the intrinsic properties of the macromolecular chain, concerning ultimate mechanical properties, especially in the field of (almost) 1-dimensional objects such as fibers. Amongst the various developments in the area of high-performance fibers, two major routes can be discerned [1] which are completely different in respect to the starting (base) materials, namely intrinsically *rigid* as opposed to *flexible* macromolecules.

The prime examples of rigid chain polymers are the aromatic polyamides (aramids), notably poly(p-phenylene terephthalamide), PPTA, currently produced under the trade names Kevlar™ (Du Pont) and Twaron™ (Akzo Nobel). More recent developments include the PBO, poly-(phenylene benzobisoxazole), fiber from Toyobo (Zylon®) and the experimental fiber "M-5", from Akzo Nobel based on PIDP, poly{2,6-diimidazo [4,5-b:4′,5′-e]pyridinylene-1,4(2,5-dihydroxy)phenylene}. The latter fiber shows an enhanced compressive strength compared with aramids [2].

The *primus inter pares* of a high-performance fiber based on flexible macromolecules is polyethylene, currently produced by DSM (Dyneema™) and its licensee Allied Signal (Spectra™), see Table 5.1.

Table 5.1 Stiffness (Young Modulus) of Various Materials at Ambient Temperature

Material	Young's Modulus [GPa]
Rubbers	< 0.1
Amorphous thermoplasts, $T < T_g$	2–4
Semi-crystalline thermoplasts	0.1–3
Wood (fiber direction)	15
Bone	20
Aluminum	70
Glass	70
Steel	200
Ceramics	500
Carbon fiber	500–800
Diamond	1200
– – – – – – –	
Polyethylene Fibers (Dyneema™)	100–150
Aramid Fibers (Kevlar™, Twaron™)	80–130
PBO(ZylonTM/Toyobo)	180–280
"M-5" (Akzo Nobel)	300

High-performance polyethylene fibers are obtained via solution(gel)-spinning of ultra-high molar mass linear polyethylene, UHMW-PE. In this chapter, the basic aspects of solution-spinning and (ultra-)drawing of polyethylene and related flexible polymers will be discussed in detail. We will refer to the various techniques which have been developed to obtain oriented polyethylene structures possessing improved mechanical properties. An extensive discussion concerning all possible routes and methods to produce polyethylene fibers and other oriented fibrous structures based on flexible macromolecules is beyond the scope of this chapter. The route towards high-performance polyethylene fibers has been a tortuous path, see for example reference 1. In this chapter, we will refer, however, to the various key experiments and routes which have been developed on the way to high-performance polyethylene fibers. Figure 5.1 shows the overall development of the tensile modulus of oriented polyethylene structures in the 20th century. In the seventies, melt-spinning/drawing of linear polyethylene [3,4] and solid state extrusion/drawing processes [5] rendered oriented polyethylene structures possessing tensile (Young) moduli up to approximately 70 GPa. Around 1980, the solution(gel)-spinning technique was discovered [6,7,8] resulting in a major jump in properties: tensile moduli typically over 100 GPa could be obtained at that time. Since the early eighties optimized laboratory experiments resulted in tensile moduli over 200 GPa [9,10], i.e. close to the theoretical value, see below, Section 5.2. Figure 5.1, of course, shows only some highlights and for more historical details, the reader is referred to reference [1].

The outline of the present chapter is as follows. The driving force for pursuing chain orientation and chain extension was the theoretically estimated ultimate strength and stiffness values for single extended chains and 1-dimensional structures possessing chains in a parallel register to be discussed in Section 5.2. Some fundamental aspects concerning chain orientation and extension in solutions and polymer melts will be addressed in Section 5.3 whereas in Section 5.4. deformation in the solid-state will be discussed with the emphasis on polyethylene including modeling of the drawing behavior. A comparison

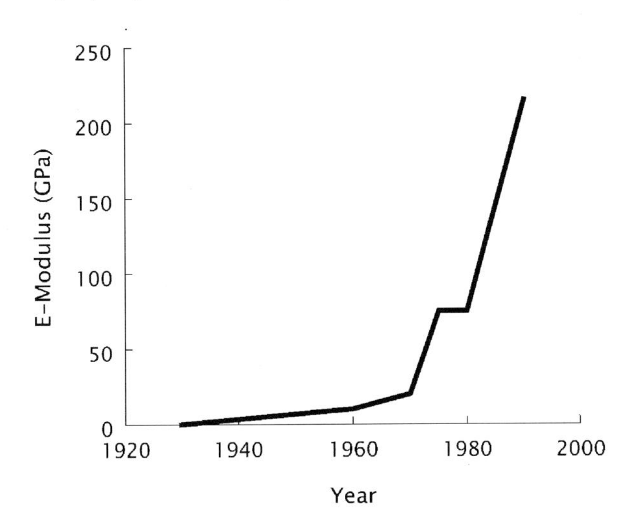

Figure 5.1 The development of the tensile modulus of polyethylene in this century.

will be made with other polymers concerning drawability. The limiting properties of high-performance polyethylene fibers will be discussed in Section 5.5.

5.2 The Ultimate Stiffness and Strength of Flexible Polymers

5.2.1 The Ultimate Tensile Modulus

Carothers and Hill [11] formulated more than 60 years ago the conditions requisite for the production of 'useful fibers', such as the necessity for long chain molecules which are perfectly ordered in an array with the chain axis parallel to the fiber direction. Meijer and Lotmar already mentioned in 1936 the high stiffness of an extended chain molecule [12]. An illustrative way, at least in the authors' opinion, to calculate the tensile (Young) modulus of a single extended chain was performed by Treloar [13] in 1960. Upon loading a single extended chain, both the C-C bonds and the bond angles in the main chain increase. Taking the force constants from low molar mass compounds, Treloar calculated the tensile modulus of a single extended polyethylene chain to be 182 GPa. In retrospect, this ultimate value is lower than the tensile modulus of present-day laboratory grade high-performance polyethylene fibers. With more modern approaches, using current force fields, the ultimate tensile moduli are nowadays estimated in the range of 180–340 GPa [14,15].

Estimates of the ultimate tensile moduli of polyethylene and other polymer systems, can also be obtained from X-ray diffraction measurements on oriented fibers during mechanical loading, assuming a homogeneous stress distribution [16–22]. Table 5.2 shows some representative data from the literature.

Generally, the tensile moduli derived from X-ray data are lower in comparison with data derived from theoretical calculations. Nevertheless, all literature data show that the tensile modulus of polyethylene in the chain direction is extremely high, viz. > 200 GPa. This is due to the small cross-sectional area of the chain, no side groups, and the planar zig-zag conformation in the orthorhombic crystal lattice. However, polyethylene is not unique. Poly(vinyl alcohol) comes close, the small -OH side group does not prevent standard atactic PVAL from crystallizing into an orthorhombic crystal structure with a

Table 5.2 Ultimate Tensile (Young) Moduli Derived from X-Ray Studies on Oriented Fibers

Material	X-Ray Modulus (GPa)	Reference
Polyethylene (PE)	235	[16]
Poly(vinyl alcohol) (PVAL)	230	[18]
Poly(ethylene terephthalate) (PETP)	110	[19]
Polyamide-6 (PAM-6)	175	[20]
Polypropylene (i-PP)	40	[21]
Polyoxymethylene (POM)	70	[22]

planar zig-zag chain conformation. Isotactic polypropylene, on the other hand, has a lower ultimate modulus in the chain direction related to the 3_1 helix conformation in the crystal lattice. In conclusion, the tensile modulus of an extended polymer chain is indeed very high, see also Table 5.1.

5.2.2 The Ultimate Tensile Strength

In the past, a variety of studies has been devoted to the theoretical tensile strength of oriented and chain extended structures, i.e the breaking of chains upon loading [23, 24]. The theoretical tensile strength of a single, extended, polymer chain can be calculated directly from the C-C bond energy. These calculations show that the theoretical tensile strength's are extremely high, in the order of 20-60 GPa! These values for the theoretical tensile strength are, in general, considered to represent the absolute upper limit of the theoretical tensile strength. In an array of chain extended polyethylene macromolecules, these theoretical values are approached only if all C-C bonds fracture simultaneously. This requires a defect-free, chain-extended structure and infinite polymer chains. In practice, however, we are dealing with finite chains and a completely different situation is encountered as will be addressed in the next section.

5.2.3 Infinite vs. Finite Chains

The theoretical estimates in Sections 5.2.1. and 5.2.2 concerning the ultimate stiffness and strength of (extended) polymer chains were based on loading *infinite* chains or, alternatively, infinite chains in perfect crystals. In practice, however, we are dealing with *finite* chains and, consequently, notably the tensile strength is determined not only by the primary bonds but equally well by the intermolecular secondary bonds. Upon loading an array of perfectly aligned and extended *finite* polymer chains, the stress transfer in the system occurs via secondary, intermolecular, bonds. Chain overlap is needed in order to be able to transfer the load through the system, see Figure 5.2.

Figure 5.2 The importance of chain overlap and chain alignment. a) perfect alignment, no chain overlap; b) perfect alignment with chain overlap; c) misalignment with chain overlap.

Qualitatively, one can easily envisage that the bonds in the main chains are only activated when the sum of the small secondary interactions, $\Sigma\ \varepsilon_i$, approaches E_i, the bond energy in the main chain. In this respect, one can distinguish between weak Van der Waals interactions, as is the case in polyethylene, or specific hydrogen bonds as encountered in the case of the polyamide or aramid fibers. Intuitively, one expects that in order to obtain high-strength structures in the case of polyethylene, a high molar mass is needed, in combination with a high degree of chain-extension, and sufficient chain overlap, to build up sufficient intermolecular interactions along the chains.

Termonia and Smith [25,26,27] used a kinetic model to simulate the fracture behavior of an array of aligned and extended polymer chains. Both chain slippage and chain rupture were considered by introducing a stress dependent activation barrier for rupture of both inter- and intramolecular bonds. It was found that the molecular weight (or number of chain ends) has a profound influence both on the fracture mechanism and on the theoretical tensile strength of these hypothetical structures. It was shown that chain slippage prevails at a low molecular weight, as expected. Figure 5.3 shows the calculated stress-strain behavior of polyethylene as a function of the molecular weight. In Figure 5.4, polyethylene is compared with PPTA. Figure 5.4 clearly demonstrates the influence of secondary interactions, viz. Van der Waals vs. hydrogen bonds. In order to obtain a strength level of 5 GPa, a molar mass of $> 10^5$ D(alton) is needed for polyethylene whereas 5×10^3 D is sufficient for PPTA! The conclusion is that polymers possessing strong secondary bonds require a smaller overlap length to obtain a high tenacity (in the case of perfectly aligned chains). This conclusion does not imply that any flexible polymer possessing hydrogen bonds, for example the conventional polyamides, is automatically an ideal candidate for obtaining high tenacity fibers. On the contrary, the hydrogen bonds also exist in the folded-chain crystals which are formed upon solidification of the melt. These hydrogen bonds provide a barrier for ultra-drawing, see also Section 5.5.3.

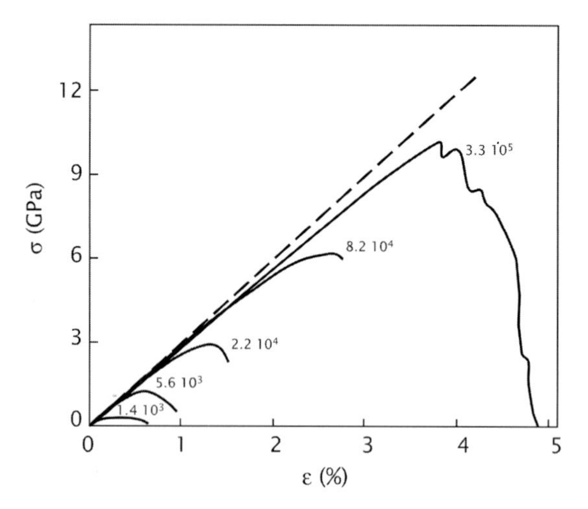

Figure 5.3 Stress-strain curves of polyethylene fibers as a function of molar mass. Adapted from Termonia et al. *Macromolecules* (1985) 2246.

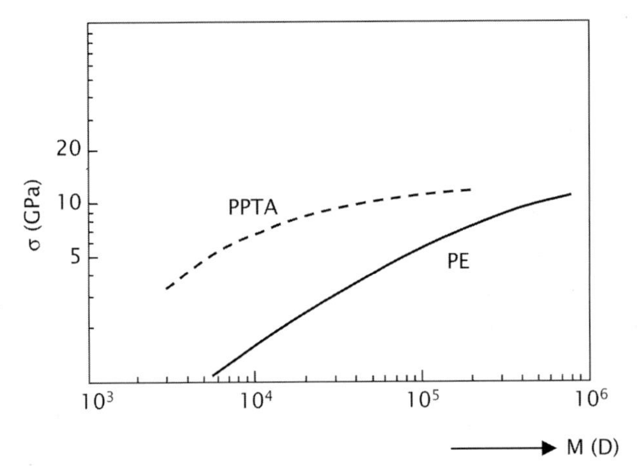

Figure 5.4 Theoretical strength of PPTA and PE fibers as a function of molecular weight. Adapted from Termonia Y., Smith P. *Polymer* (1986) 1845.

5.2.4 Chain Alignment, Orientation vs. Extension

Figure 5.2c shows a situation with a certain degree of misalignment of the extended polymer chains. Misalignment can have a profound effect on the stiffness of the structure since polymeric crystals possess highly anisotropic properties. The complete stiffness matrix C_{ij} and the compliance matrix S_{ij} of perfect polyethylene (single) crystals, were calculated by Tashiro et al. [28] and are presented below.

$$\mathbf{C_{ij}} = \begin{vmatrix} 7.99 & 3.28 & 1.13 & 0 & 0 & 0 \\ 3.28 & 9.92 & 2.14 & 0 & 0 & 0 \\ 1.13 & 2.14 & \mathbf{316} & 0 & 0 & 0 \\ 0 & 0 & 0 & 3.19 & 0 & 0 \\ 0 & 0 & 0 & 0 & 1.62 & 0 \\ 0 & 0 & 0 & 0 & 0 & 3.62 \end{vmatrix} \ \mathbf{GPa}$$

$$\mathbf{S_{ij}} = \begin{vmatrix} 14.5 & -4.78 & -0.019 & 0 & 0 & 0 \\ -4.78 & 11.7 & -0.062 & 0 & 0 & 0 \\ -0.019 & -0.062 & \mathbf{0.032} & 0 & 0 & 0 \\ 0 & 0 & 0 & 31.4 & 0 & 0 \\ 0 & 0 & 0 & 0 & 61.7 & 0 \\ 0 & 0 & 0 & 0 & 0 & 27.6 \end{vmatrix} \ (10^{-2})\ \mathbf{GPa}^{-1}$$

The compliance in the load direction can be calculated. For example, when the b-axis of the crystal is oriented perpendicular to the uniaxial drawing direction, the tensile modulus as a function of the angle θ between the test and the chain direction is given by [29,30,31]:

$$E(\theta) = S_{33}^{-1}$$

where

$$S_{33}(\theta) = S_{11}\sin^4\theta + S_{33}\cos^4\theta + 2(S_{13} + S_{55})\cos^2\theta\, \sin^2\theta - 2S_{35}\cos^3\theta\, \sin\theta - S_{15}\cos\theta\, \sin^3\theta \quad (5.1)$$

From Equation 5.1, the tensile (Young) modulus $E(0) = S_{33}^{-1}(0)$ can be calculated to be 312 GPa.

Figure 5.5 shows the tensile modulus as a function of θ and demonstrates the highly anisotropic character of the orthorhombic polyethylene crystal. The dramatic drop in the tensile modulus, even at small angles with the chain direction, is mainly caused by the low shear moduli of polyethylene, an advantage for ultra-drawing, see Section 5.4, but detrimental for off-axis properties of oriented polyethylene fibers, see Section 5.5. For comparison, the orientation dependence of graphite and PPTA is also plotted and last but not least of glass (isotropic).

The low transverse and shear moduli of polyethylene are due to the absence of specific interactions along the chain (only weak van der Waals bonding). Figure 5.5 clearly demonstrates the need for a low mismatch angle between chain-and test-direction to obtain structures with a high tensile modulus, see also Figure 5.2c, in the case of a perfect

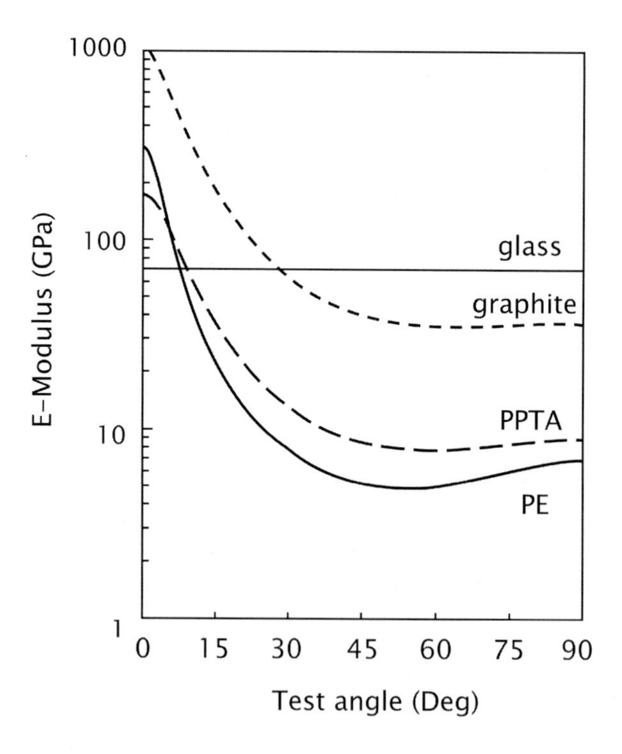

Figure 5.5 Theoretical moduli of various fibers as a function of the load (test) angle. Note the difference between polymeric 1-D (PPTA, PE) vs. inorganic (glass) 3-D fibers.

single crystal. In actual fibers we don't have a perfect single crystal, which actually makes the situation even more dramatic

Last but not least, a remark should be made concerning chain-extension vs. chain orientation. Figure 5.6 shows a situation of perfect orientation of molecules or molecular segments in a fibrous structure but with a strong variation in the degree of chain-extension. In all hypothetical structures, a high degree of orientation is present on a segmental molecular level but with a large difference in chain extension. In fact, X-ray (WAXS) studies would not differentiate between the various structures. It is obvious that notably the tensile strength will increase from situation a to c.

Summarizing, the previous discussion in Sections 5.2.3. and 5.2.4. were rather qualitative but served only the purpose to show that:

a. Calculations on the ultimate stiffness and strength of single extended-chain molecules is rather academic since chains in practice possess a finite length and secondary interactions are equally important. Nevertheless, calculations on the ultimate properties of extended (single) chains triggered many scientists to pursue chain orientation/extension from the 1960s.
b. The issue of chain overlap is rather trivial, but not often realized. For high-strength polyethylene fibers a very high molar mass is needed. In the case of aromatic polyamides, hydrogen bonding increases interchain interactions and a lower mass is sufficient to obtain high strength levels. Along these lines of thinking, one would expect that aliphatic polyamides, possessing hydrogen bonds as well, might be interesting candidates to obtain high-strength fibers at rather standard molecular weights. However, aliphatic polyamides form folded-chain crystals upon cooling from

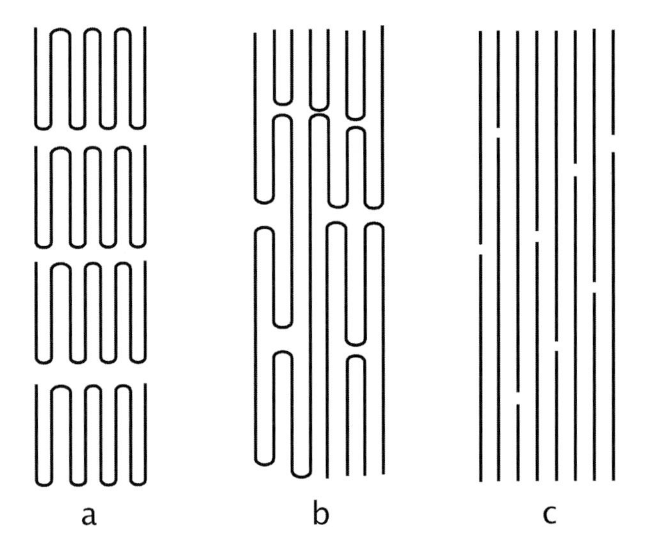

Figure 5.6 Hypothetical fibrous structures possessing perfect segmental (stem) orientation but a large variety of chain extension.

the melt and notably form solutions. Unfolding into extended chain structures, for example by drawing in the (semi)-solid state has not been successful up to now, despite many attempts made in industry and academia.

c. The high anisotropy of polymeric 1-D structures (fibers/tapes) is often overlooked. Polyethylene is an extreme example in this respect. The poor off-axis properties of polyethylene fibers will be noticed immediately when PE fibers are used to reinforce composites. Polyethylene fiber reinforced composites can not bear any load in compression, bending, shear etc. [30]. For the same reasons, biaxially drawn UHMW-PE films possess tensile moduli (in the plane of the film) < 10 GPa, to be compared with fibers, > 100 GPa!

In conclusion, long chain PE molecules can be put in an extended chain conformation with the molecules in a parallel array with the fiber axis (for the various processes, see below). Upon loading along the fiber axis, one will notice a high strength and stiffness (at short loading times, see Section 5.3). Perpendicular to the fiber axis, however, the properties are not different from standard polyethylene, since the crystals structure did not change, it remains orthorhombic.

5.3 Chain Orientation and Chain Extension

5.3.1 The Single Chain

Single or isolated chains are encountered only in dilute solutions below the overlap concentration φ^*. Stretching these isolated chains can only occur by frictional contacts between the chain and its surrounding medium, the solvent molecules. Simple shear flow is inadequate and in order to obtain full chain-extension, the flow has to possess elongational components. The effect of elongational flow fields on the transformation from a random coil into an extended chain conformation has been originally recognized by Frank [32], experimentally investigated by Peterlin [33] and addressed theoretically by de Gennes [34]. The general conclusion is that a chain will stretch out fully beyond a critical strain rate $\dot{\epsilon}_{cr}$. There is seemingly no intermediate region, below $\dot{\epsilon}_{cr}$ there is no chain extension and beyond $\dot{\epsilon}_{cr}$ the chain becomes practically fully extended as discussed recently by Keller and Kolnaar [35]. The critical strain rate $\dot{\epsilon}_{cr}$ is dependent on the molecular weight and scales with $M^{-1.5}$ as determined experimentally for monodisperse samples by Odell et al. [36]. This relationship implies that longer chains are more readily extensible.

Chain extension in dilute solutions can be made permanent if chain extension is followed by crystallization. Chain extension will promote crystallization since the melting point or dissolution temperature is raised due to lowering of the entropy of the chain. Consequently, a random coil in solution above the dissolution temperature can crystallize in an elongational flow field due to the increase in the conformation dependent dissolution temperature. Taking into account the foregoing discussions concerning the experimental observations that with increasing molar mass the chains become more readily extensible, and the fact that polymers such as polyethylene are usually polydisperse, one can easily envisage, in retrospect, that in an elongational flow field only the high molar mass fraction becomes extended and crystallizes into a fibrous structure. The remaining part will stay in solution as random coils and upon subsequent cooling crystallizes as folded-chain crystals, nucleating onto the fibrous structures. These composite structures are referred to in the literature as 'shish' (the central fibrous core)-'kebabs' (the lamellar overgrowth) (Figure 5.7). Mitsuhashi was probably the first to attempt inducing chain extension directly in solution using a simple Couette apparatus in the early 1960s [37]. He reported the formation of fibrous 'string-like' polyethylene structures upon stirring linear polyethylene in xylene. In their attempts to fractionate ultra-high molecular weight polyethylene using a Couette type apparatus, Pennings and Kiel [38] confirmed these 'string-like' structures and explained their occurrence as discussed above, stretching out of long chain molecules in an elongational flow field, provided by the Taylor vortices, followed by crystallization into fibrous structures and deposition of lower molar mass components in the form of lamellar type crystals.

The structure of shish-kebab type fibrous polyethylene is far from the ideal arrangement of PE macromolecules for optimum stiffness and strength. Due to the presence of lamellar overgrowth, the moduli of precipitated fibrous PE 'shish-kebabs' were limited to up to about 25 GPa [39], to be compared with > 50 GPa in the case of melt-spinning/drawing, as performed by Ward et al. [2]. The 'kebabs', i.e. the lamellar overgrowth, can be removed to some extent by selective dissolution. Keller and Wilmouth differentiate in this respect between macro and micro-shish kebabs [40]. The micro shish kebabs are obtained by removing the lamellar overgrowth, which is not structurally attached to the central fibrous core (shish), from the in-situ formed macro shish kebabs. Shish kebab structures obtained by stirring dilute solutions have never been exploited technologically, despite some efforts that have been made to use these structures for the

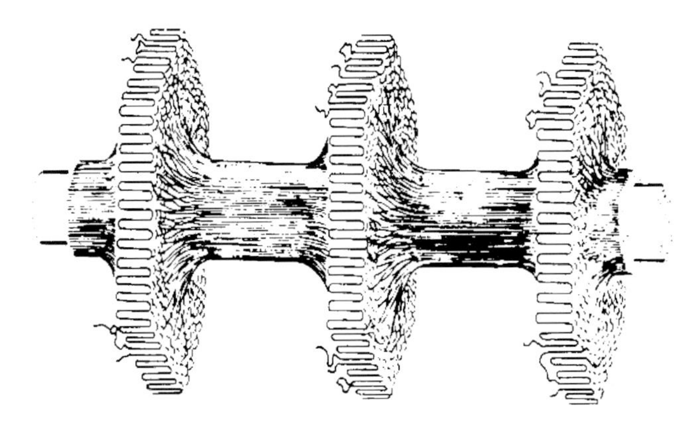

Figure 5.7 Schematic structure of shish-kebab fibers. *Courtesy Dr. M. Hill, Bristol.*

making of, for example, artificial paper. The observation, however, of shish kebab structures has been an important milestone on the route to high-performance fibers and, moreover, shish kebab structures are also encountered in polymer melt-processing such as film blowing, extrusion, injection-molding etc. In polymer melt-processing elongational flow fields are encountered and shish kebab formation is favored.

5.3.2 An Ensemble of Chains

In the previous section, the formation of shish-kebab structures was discussed with the empahsis on stretching single isolated polymer chains, i.e. at polymer concentrations below the overlap concentration φ^*. It is difficult, however, to make a clear distinction between the dilute regime, no chain overlap, and the regime where chain overlap starts in semi-dilute solutions. In elongational flows, the high molar mass fraction of the polymer will become elongated first. This high molar mass fraction is usually ill-defined in view of the difficulty to measure with high-temperature GPC the high molar mass components. Moreover, the amount of chain overlap required to stretch an ensemble of chains, rather than individual chains, is not well defined. Upon stretching chains, phase separation may occur, due to loss of entropy and an increase in melting temperature, and the degree of chain overlap and entanglement can be much higher on a local scale than calculated from the overall polymer concentration. These effects play certainly a role in the so-called surface growth technique which was introduced in the early seventies by Zwijnenburg and Pennings [41,42]. Figure 5.8 shows the experimental set up of the surface growth technique.

A seed fiber (polyethylene or even cotton) is immersed in a dilute solution of UHMW-PE and from the surface of the rotating inner-cylinder fibrous, tape-like, polyethylene structures could be withdrawn al low speeds. This pulling of fibers from the rotor is probably due to the formation of a thin gel-layer on the rotor surface. Upon stirring, the chains in the dilute solutions, especially near the wall of the rotor, become elongated to some extent, they phase separate and adhere to the surface of the rotor in the form of an entangled gel-layer. This view on the surface growth technique, became more apparent after the discovery of the solution(gel)-spinning process, see Section 5.4.2. Under optimized conditions, with respect to solution concentrations, temperatures and take up speeds, oriented UHMW-PE structures could be obtained possessing tensile moduli over 100 GPa and strength values above 3 GPa! In general, with increasing solution temperature, the lamellar overgrowth decreases and finally rather smooth oriented UHMW-PE structures could be obtained. The surface growth technique was another milestone on the route to high-performance UHMW-PE fibers and, in fact, the first experimental proof that high-modulus/high-strength structures could be made. The technique, however, possesses intrinsic draw backs such as very low production speed, a non-uniform thickness of the tape-like structures which were pulled of from the rotor and the problem of scaling up this process. Attempts have been made to develop technologies for continuous production of UHMW-PE tapes, such as the rotor technique by M. Mackley [43], see Figure 5.8. Supercooled UHMW-PE solutions were sheared and tape-

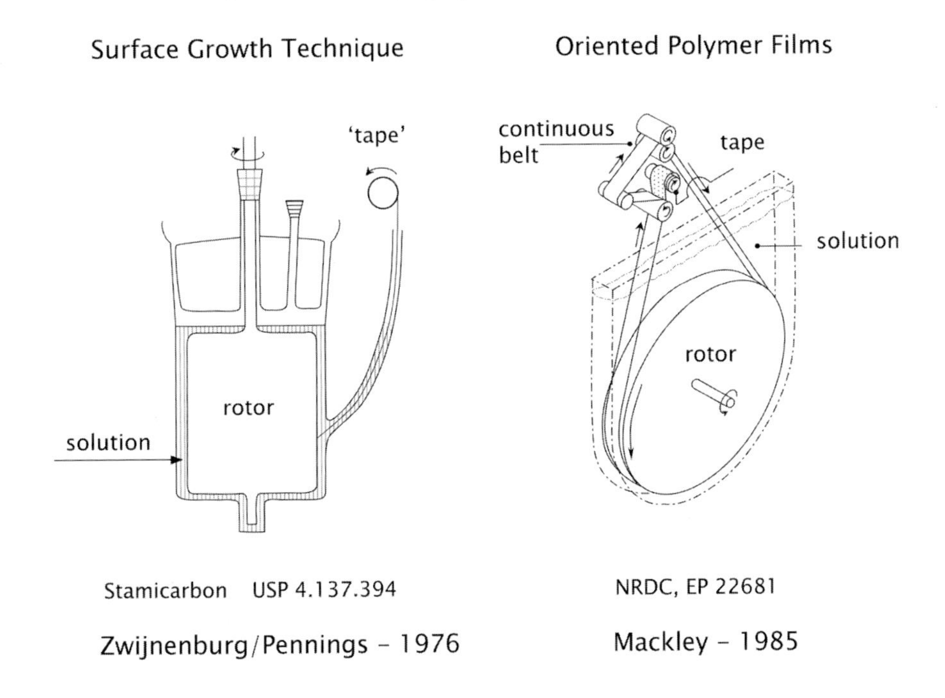

Figure 5.8 Schematic set-up of surface-growth technique and rotor technique.

like PE structures could be produced possessing stiffness values of approx. 60 GPa at take-up/roll-off speeds of several meters/min. The belt, see Figure 5.8, increases the local shear rate and drags the produced tape from the surface of the rotor. Although the linear take-up speed is still not impressive, the width of the tape is unlimited and hence the mass of tape per unit of time is much higher than for the surface growth technique. However, these and related attempts became of less interest after the discovery of the solution(gel)-spinning technique, see Section 5.4.2. In the solution(gel)-spinning technique, semi-dilute solutions are employed during spinning but the elongation of chains is performed by drawing in the semi-solid state, i.e. below the melting c.q. dissolution temperature. The process of spinning (shaping) and drawing (chain-extension) are thus separated, respectively above and below the melting c.q. dissolution temperature. In the literature various processes have been described to orient the chains directly in the molten state. The problem of chain-orientation and extension in the melt is that extensive relaxation processes occur, the chains resist deformation and retract back to a random coil conformation. Lowering the extrusion-spinning temperature is no real solution for this problem. It was shown already in 1967 by van der Vegt and Smit [44] that on lowering the extrusion temperature of polyethylene, and other crystallizable polymers, elongational flow-induced crystallization will occur and the solidified polymer will block the flow. This topic has been studied extensively by Porter et al. and they observed that the 'plugs' which are formed in the exit of the die, as a result of crystallization during blocked flow, are highly oriented and possess a high stiffness; tensile moduli of 70 GPA were reported [45].

For more details on capillary blockage etc., the reader is referred to reference 46. The conclusion is that the ultimate fate of experiments concerning chain extension directly in the melt is flow-induced crystallization in the processing equipment. Consequently, in order to obtain a high degree of chain-extension, drawing should be performed below the melting c.q. dissolution temperature to avoid relaxation processes, see Section 5.4.2.

5.4 Drawing of Polyethylene in the Solid-State

5.4.1 Solid-State Drawing of Polyethylenes

The requirements to produce high-modulus and high-strength polymeric fibers are presently well understood and documented as discussed in the previous sections. The polymeric chain should be preferably high molecular weight and fully aligned and packed into a parallel register with the chain-axis parallel to the macroscopic fiber axis. In quiescent polymer melts and solutions, the polymer chains adopt a random coil conformation and, upon cooling, the chains tend to adopt a folded-chain conformation in the case of crystallizable polymers. Traditionally, chain orientation and extension is generated in melt- and solution-spun fibers by two different methods: (i) applying a draw-down to the fibers during or immediately after spinning (in the molten state or super-cooled melt) and (ii) drawing of fibers at temperatures close to but below the melting-or dissolution temperature. A draw-down in the molten state or in solution is usually less effective, as discussed before, to generate chain-extension in view of extensive relaxation processes. Drawing in the (semi-)solid state, i.e. below the melting and/or dissolution temperature is usually much more effective since relaxation processes are restricted, due to reduced thermal motions and because the chains are trapped into crystals which act as physical network junctions.

Assuming that the molecular and macroscopic deformation are identical in solid state drawing i.e. assuming that the (pseudo-) affine deformation scheme is valid, the average orientation of statistical chain segments can be related to one single parameter: the macroscopic draw ratio λ The pseudo-affine deformation scheme was originally derived by Grün and Kuhn [47] for chemically crosslinked rubbers and relates the average orientation of statistical chain segments $<\cos^2\theta>$ to a single external parameter; the macroscopic draw ratio λ (for $\lambda > 5$):

$$\langle\cos^2\theta\rangle = \frac{\lambda^3}{\lambda^3 - 1} - \frac{\lambda^3}{(\lambda^3 - 1)^{3/2}}[\arctan((\lambda^3 - 1)^{1/2})] \qquad (5.2)$$

In Equation 5.2, the draw ratio is defined as the ratio of the length of a fiber after drawing (L) to the original length of the fiber (L_o). The pseudo-affine deformation scheme simplifies the quest for obtaining a high orientation and chain extension to a single objective i.e. the sole requirement for a high degree of chain orientation/extension is to obtain a high maximum draw ratio.

In the literature many reports [48,49,50,51] have been given concerning a unique relationship between the draw ratio, the degree of chain extension/orientation and the tensile modulus upon drawing polyethylenes. Moreover, it was shown that the birefringence and the tensile modulus of drawn polyethylene fibers solely depend on the macroscopic draw ratio λ, independent of the molar mass, the molar mass distributions and the type and content of branches and the initial morphology. Of course, this unique relationship only holds within a specific temperature and strain rate domain and deviations are usually related to unsuitable drawing conditions, for example a too high drawing temperature promoting chain slippage (chain relaxation processes are less important in solid-state drawing) instead of effective chain extension. The following requirements should be met:

(i) the drawing should be performed at temperatures below the melting temperature where effective drawing occurs (the effective draw ratio can be determined from simple shrinkage experiments after draw);
(ii) the drawing should be performed at optimized conditions with respect to drawing speed and temperature and
(iii) the specimen before draw should be isotropic.

Detailed neutron scattering experiments, using partly deuterated polyethylene samples, confirm the above view on affine deformation. By this technique, the influence of solid state drawing on the root mean square end-to-end distance of polyethylene macromolecules was investigated by Sadler and Barham [52]. It was found that the root mean square end-to-end distance of the macromolecular coil after drawing is directly related to the macroscopic draw ratio and it was concluded that the pseudo-affine deformation scheme also applies with respect to the degree of chain extension after drawing.

In this simplified scheme of relating the tensile modulus with the macroscopic draw ratio λ, the only remaining question is: what is the maximum attainable draw ratio λ_{max}? We focus on two extreme cases, respectively (a) single crystal mats consisting of well-defined stacked folded-chain polyethylene crystals and (b) solid polyethylene assuming a random-coil conformation.

In case (a), stacked lamellar crystals (see for example Figure 5.9) we can not expect a unique relationship between the tensile modulus and the draw ratio λ since the sample is not isotropic but the maximum draw ratio λ_{max} is given by the ratio of the fold-length L_f and the chain diameter δ:

$$\lambda_{max} = L_f/\delta \tag{5.3}$$

Taking typical values for the fold length L_f, 20–30 nm, and δ, 0.5–0.7 nm respectively, the maximum draw ratio for the case of well-stacked folded-chain lamellar crystals is between 30–60, independent of the molar mass [53].

The situation of well-stacked lamellar crystals is not representative for melt-crystallized polyethylene. Upon crystallization from the melt, in particular at higher molar mass, the chain is closer to a random-coil than to a folded-chain conformation [54]. Taking the extreme of no folding at all, viz. the chain remains in the random-coil conformation upon solidification, the maximum draw ratio is simply related to full chain-extension of individual molecules. Assuming that no chain-slippage occurs during draw, the maximum

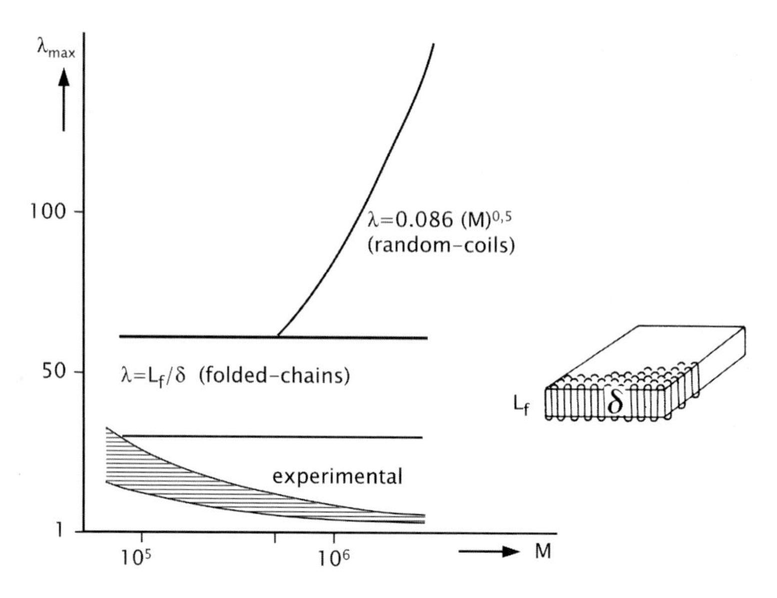

Figure 5.9 Maximum draw ratio of polyethylene based on two extreme models vs. the experimentally observed values (shaded area).

draw ratio λ_{max} is given by the ratio of the contour length L and the average unperturbed end-to-end distance $<r^2>_o$:

$$\lambda_{max} = \frac{Nl_b \sin(\theta/2)}{(C_\infty N l_b^2)^{0.5}} = \frac{N^{0.5}\sin(\theta/2)}{C_\infty^{0.5}} = 0.086(M)^{0.5} \tag{5.4}$$

In Equation 5.4, N is the number of C-C bonds (M/14), θ is the bond angle (112^0 in the case of polyethylene) and C_∞ is the characteristic ratio (6.7 for polyethylene). The 'theoretical' maximum attainable draw ratios, based on Peterlin's model and the random-coil approach are presented in Figure 5.9 as a function of the molecular weight together with the experimental data concerning the maximum drawability as obtained by numerous experiments by Ward et al. [2]. A large discrepancy is observed between the theoretical estimates of the maximum attainable draw ratio, especially with increasing molecular weights. The general trend from experiments is that with increasing molecular weight, the maximum draw ratio decreases monotonically. At very high molar mass ($M > 10^6$), the region of ultra-high molecular weight polyethylene, UHMW-PE, the maximum draw ratio levels off to approximately 4–5. A limited drawability in the solid-state is not unique for polyethylene. Many other polymers such as the polyamides show a limited drawability < 10. In the literature this is sometimes referred to as the natural draw ratio.

5.4.2 Solution(Gel)-Crystallized Polyethylene

In the late seventies, it was found that the gap between the theoretical and experimental maximum attainable draw ratio of high molecular weight polyethylene could be bridged, to a large extent, by spinning and casting from semi-dilute solutions [55–59]. For example, solution-spun UHMW-PE filaments could be ultra-drawn (draw ratios up to 50 or more) to high-strength/high-moduls fibers possessing tenacities > 3 GPa and tensile moduli of over 100 GPa. Figure 5.10 shows schematically this process, now often referred to as solution(gel)-spinning. A semi-dilute solution of UHMW-PE with a low polymer concentration, typically of 1–2%, was spun into water. Upon cooling, a gel filament is obtained consisting of a physical network, obtained by thermoreversible gelation, containing a large amount of solvent. The as-spun/quenched filaments are mechanically sufficient stable (gel-fibers) to be transported into an oven in which drawing is performed. At first glance, the ultra-drawability of these gel-fibers seems not too surprising in view of the large amount of solvent which could act as a plasticizer during draw. The remarkable feature, however, is that ultra-drawing is still possible after *complete removal* of the solvent *prior* to the drawing process. The solvent is necessary to facilitate processing of the rather intractable polymer UHMW-PE (melt-processing is impossible due to the excessive high melt-viscosity) and induces a favorable structure/morphology for ultra-drawing but the solvent is not essential *during* the drawing process.

In subsequent studies, the influence of initial polymer concentration in solution was systematically investigated [57,58] and it was found that the maximum attainable draw ratio scales with the square root of the inverse of the initial polymer concentration in solution ϕ as:

$$\lambda_{max} \propto \phi^{-0.5} \tag{5.5}$$

From the relationship between the maximum draw ratio and the initial polymer concentration in solution, a semi-empirical model to describe the drawing behavior of polyethylene in terms of topological constraints, viz. entanglements, was proposed by Smith, Lemstra and Booij [57]. In principle, this model is derived from classical rubber

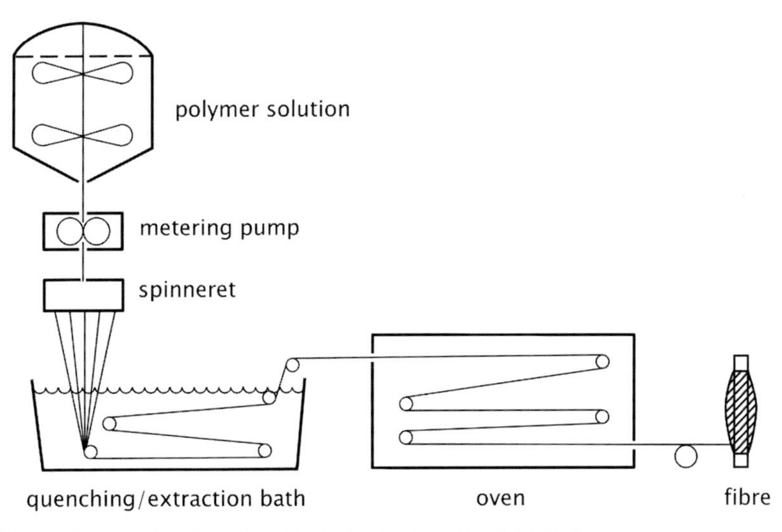

Figure 5.10 Schematic drawing of solution(gel)-spinning of UHMW-PE.

elasticity theory. It is assumed that entanglements are trapped in polyethylene upon crystallization and act as semi-permanent crosslinks upon solid state drawing. Upon dissolution, the entanglement density is reduced, about proportional to the inverse of the polymer volume fraction, and consequently the maximum attainable draw ratio in solution-crystallized samples is enhanced in comparison with melt-crystallized polyethylene. On this basis, the experimentally observed dependence of the maximum attainable draw ratio on the initial polymer concentration in solution could be theoretically predicted (Equation 5.5).

Based on Equation 5.5, one might be tempted to conclude that the maximum attainable draw ratio of solution-spun UHMW-PE fibers is solely determined by the polymer concentration in solution which is, of course, an oversimplification. Two factors play an important role during drawing in the solid state. First, during draw, the chains between entanglements will be stretched but chain slippage can occur as well, depending on the drawing temperature and the overall molar mass, since entanglements are physical rather than chemical (permanent) junction points. With increasing molar mass, chain slippage is likely to be reduced in view of an increased number of 'friction points' (entanglements) along the chain, scaling with the Bueche parameter: $\phi.M$ [31]. For example, the limiting value of the maximum drawability of melt-crystallized polyethylene with increasing molar mass is 4–5, see Figure 5.9. This value is to be expected from the proposed entanglement model. The molecular weight between entanglements in polyethylene melts, M_e, is approx. 2000 D and the maximum drawability λ_{max} is given by Equation 5.4 by substituting M for M_e with the result: $\lambda_{max} = 3.7$, in line with the experimental observations: $\lambda_{max} < 5$, see [57].

Second, during crystallization from the melt, disentangling may occur as a result of the 'reeling-in' of chains, especially at low supercoolings and at low and moderate molar masses. During slow (isothermal) crystallization from the melt, and from solution [59], chain-folding is promoted and the chains are 'reeled-in' onto the crystal surface from the melt, and consequently, become disentangled. Slow crystallization promotes drawability despite the fact that the crystallinity increases [59]. This observation is a strong indication that at least in the case of apolar polymers like polyethylene, topological constraints such as entanglements determine drawability rather than crystal structure, size and perfection. The crystals only seem to be necessary to fixate the disentangled state, but they do not resist deformation, see also Section 5.4.4.

5.4.3 Solvent-Free Processing of UHME-PE; Nascent Reactor Powders

The experimental results described in Section 5.4.2. show that the drawability of polyethylene in the solid-state can be controlled by adjusting the number of 'constraints', viz. 'entanglements' which limit draw. This simple model explains qualitatively the influence of the initial polymer concentration and also teaches that a relatively large amount of solvent is needed to remove entanglements prior to ultra-drawing. Especially in the beginning of the solution(gel)-spinning technique, only very low UHMW-PE concentrations could be handled, typically below 5%. Due to extensive development efforts and the use of efficient mixing equipment, such as twin screw extruders combined with temperature-gradient drawing processes, it is nowadays feasible to handle more concentrated solutions but nevertheless, solution(gel)-spinning requires a major amount of solvent which has to be recycled completely which is increasingly cumbersome in view of environmental legislation.

Solvent-free routes have been a challenge ever since the invention of the solution(gel)-spinning process and numerous attempts have been made to obtain disentangled precursors via different routes. One obvious route is to precipitate UHMW-PE from dilute solutions and to collect the crystals as the base material for subsequent spinning processes. Precipitated crystals are disentangled or partly disentangled, depending on the initial polymer concentration, and one could expect, in principle, an initial lower melt-viscosity during melt-spinning. The long chain molecules need a certain time to form an entangled melt again and this time slot could be used for melt-spinning and subsequent drawing. All these attempts have failed, however, since no memory effect could be observed related to re-entangling, see also Section 5.4.4. The disentangled UHMW-PE materials, obtained by casting or precipitating from solution, are immediately highly viscous upon heating above the melting point. In fact, no difference is observed between initial disentangled or standard melt-crystallized UHMW-PE with respect to the characteristic rheological properties upon heating into the melt [60,61]. This rather intriguing and disappointing result will be discussed further in Section 5.4.4.

A much more elegant route to obtain disentangled UHMW-PE is by direct polymerization in the reactor. In order to make UHMW-PE, a relatively low polymerization temperature is needed and a situation is easily encountered where the polymerization temperature T_p is below the dissolution temperature T_d of UHMW-PE in the surrounding medium in which the catalyst is suspended. In this situation, the growing chains on the catalyst surface tend to crystallize during the polymerization process. These UHMW-PE reactor powders, often referred to as nascent or virgin UHMW-PE, can be remarkably ductile. It was shown by Smith et al. [62] that reactor powders, after compacting below the melting temperature, could easily be drawn into high-modulus structures. Kanamoto et al developed a two-stage drawing process for reactor powders and tensile moduli over 100 GPa could be obtained [63]. Nippon Oil Company developed several solid state processing routes [64,65] for making strong and stiff PE tapes. A process consisting of three stages: compaction, roll drawing and tensile drawing, has been developed to pilot plant stage. The products obtained possess tensile moduli up to 120 GPa but moderate tensile strength's, up to 1.9 GPa.

The ductility of compacted UHMW-PE reactor powders can be understood in terms of the proposed entanglement model discussed in Section 5.4.2. The growing chains on the crystal surface can crystallize quite independently and consequently untangled UHMW-PE is obtained as a result of direct polymerization. Similar to solution-cast/precipitated UHMW-PE crystals, the nascent powder should not be heated above the melting temperature since the drawability is lost instantaneously, see also Section 5.4.4.

The morphology of these so-called nascent or virgin reactor powders has been studied in the past quite extensively. The high melting temperature, in the order of 142–144 °C, i.e. very close to the reported equilibrium melting temperature for polyethylene,

T_m^o = 145.5 °C*, and the absence of long range order (observed by SAXS), lead to the conclusion that nascent UHMW-PE reactor powders consist of 'extended-chain' crystals rather than the usual folded-chain crystals. The growing chains on the catalyst surface meet and form a crystalline structure in which the chains are aligned parallel to a large extent resulting in a high melting temperature [62].

In our view this model of rather extended-chain crystals in nascent UHMW-PE could be applicable in certain cases but the high melting can also be explained in a rather straightforward manner based on folded-chain crystals. During polymerization, the chains grow rather independently and form metastable folded-chain crystals. In the limit of a very low catalyst activity (low temperatures), one could envisage that ultimately one chain will form a monomolecular crystal. This implies for UHMW-PE, possessing a molar mass $M_w > 10^6$ D, that metastable crystals are formed possessing not only small dimensions, in the order of 10 nm, in the chain direction (the fold length) but equally well in the lateral dimensions. Taking as an example a molar mass of 10^6 D, the lateral dimensions are typically in the order of 10 nm. Consequently, the melting point depression is large with respect to the equilibrium melting temperature. The melting temperature T_m of folded-chain crystals is given by the well-known Gibss-Thomson equation, modified for lamellar crystals:

$$T_m = T_m^o[1-2\ \sigma_e/L_f\rho\Delta H - 2\ \sigma/A\rho\ \Delta H - 2\ \sigma/B\ \rho\Delta H] \qquad (5.6)$$

In Equation 5.6, T_m is the experimental melting point, T_m^o is the equilibrium melting point for infinite perfect crystals (145.5 °C in the case of polyethylene), σ_e is the surface free energy of the fold planes, σ the surface free energy of the lateral planes, ρ the crystalline density L_f the crystal thickness in the chain direction, and A and B the lateral crystal dimensions. L_f is usually in the order of 10–30 nanometers and A and B in the order of a few microns. Consequently, the last two terms in Equation 5.6 are usually ignored. However, when A and B are small with dimensions in the order of the crystal thickness L_f, the two last terms have to be taken into account and, consequently, contribute to the melting point depression.

In our view, nascent UHMW-PE reactor powders consist of highly metastable crystals which melt at a low temperatures. During heating in a DSC these crystals melt and reorganize continuously [66] during heating resulting in a high end-melting temperature (final melting). This high melting point has no relationship at all with the original morphology but is due to fast reorganizations upon heating, see further at Section 5.4.4.

Whether *monomolecular* crystals could be formed during polymerization is not known at this point of time and further investigations are needed. What is known, however, is that nascent UHMW-PE powders differ widely in drawing behavior after compacting (sintering) below the melting point as well as in sintering behavior above the melting temperature. These differences are partly related to chemistry in the reactor (type of catalyst, reaction temperature) and to the handling of these powders after polymerization, for example drying procedures in fluid beds at elevated temperatures which could destroy

*The equilibrium melting temperature of polyethylene is still a matter of debate. We took 145.5°C as an appropriate value

the chain-arrangement originally induced by chemistry. Some correlations have been made concerning the gross morphology of the UHMW-PE powders and sinterability [67]. These relationships, however, have been made at micrometer level, on not at nano-level, viz. molecular level and this level is needed to properly understand the phenomena. This will become more apparent in Section 5.4.4. where we discuss the annealing behavior of nascent reactor powders in more detail, in comparison with standard solution-crystallized UHMW-PE samples.

Processing of UHMW-PE reactor powders has been partly successful for making oriented tapes by sintering/compacting between rollers and subsequent drawing. All these operations have to be performed, as discussed above, below the melting temperature, viz. below approximately 140 °C. If premature melting occurs, the drawability of the (partly) re-crystallized samples is lost completely.

We will end this section on solvent-free routes by discussing one type of experiment which has been surrounded by quite some 'degree of magic' in the literature concerning processing UHMW-PE in the *hexagonal* phase. Up to now, we were often referring to melting of UHMW-PE and the observation that the favorable drawing characteristics are lost instantaneously upon melting and subsequent re-crystallization. In this case, we refer to melting of standard *orthorhombic* polyethylene crystals possessing an equilibrium melting temperature T_m^o of 145.5 °C. Solution-crystallized UHMW-PE samples show a melting temperature of approximately 135 °C, somewhat higher than the melting temperature of melt-crystallized UHMW-PE, approximately 133 °C. These experimental melting points are usually recorded in a Differential Scanning Calorimeter (DSC) at heating rates in the range of 10–20 °C/min. As mentioned before, the high melting point of nascent UHMW-PE reactor powders, 142–144 °C, is not related to the actual structure/ morphology but the result of extensive reorganization during the heating scan in a DSC apparatus which records only net heat effects.

Polyethylene, however, possesses different crystal structures and next to the standard orthorhombic structure, the *hexagonal* phase is important in processing UHMW-PE. In this respect we have to distinguish between two situations, respectively *isotropic* polyethylene and *oriented* polyethylene.

For *isotropic* polyethylene, the hexagonal phase is usually observed at elevated pressure and temperature, in fact above the triple point Q located at 3.4 Kbar and 220 °C according to the pioneering work of Bassett et al. and Wunderlich et al. [68,69]. Later, more detailed studies, involving in-situ light microscopy and X-ray studies [70,71,72], showed that the equilibrium point Q_0 is located at even higher temperatures and pressures, 250 °C and 5.2 Kbar respectively, see Figure 5.11. This temperature-pressure domain, above the triple point Q, is usually not encountered in standard processing techniques of polyethylene and high-pressure annealing has always been a rather academic topic up to now.

The triple point Q, however, is dependent on the crystal dimensions as shown recently in our laboratory by Rastogi et al. [73], rather similar to the melting point dependence of polyethylene on the crystal thickness (fold length). Especially for nascent UHMW-PE reactor powders, consisting of crystallites with very small dimensions, it was shown that a metastable hexagonal phase could be observed at pressures and temperatures as low as 1 Kbar and 200 °C, respectively. This hexagonal phase is transient in nature (metastable) and

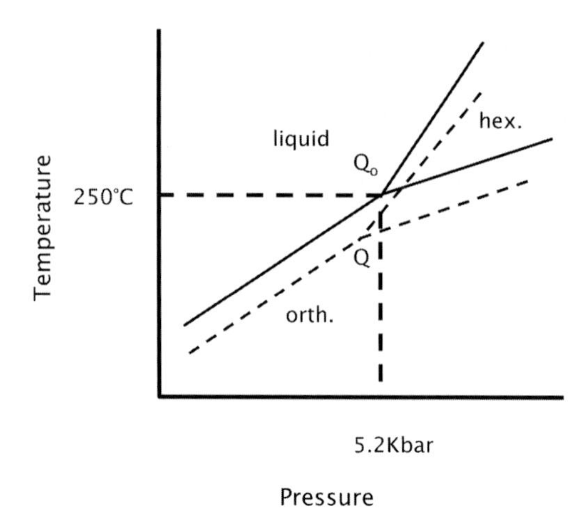

Figure 5.11 P, T diagram of PE.

upon annealing a transformation into the thermodynamically stable orthorhombic phase occurs. The observation that a thermodynamic stable crystal structure is reached via a metastable state of matter is not unique for polyethylene, nor for polymers, but has been invoked as early as in 1982 by Ostwald, commonly expressed as Ostwald's stage rule [74,75]. It has been shown [70–72] that PE crystals initially grow in the hexagonal phase. After a certain time, or crystal size these hexagonal crystals are transformed into thermodynamically stable orthorhombic crystals.

The occurrence of a transient, metastable hexagonal phase could have major implications for processing of the intractable polymer UHMW-PE possessing a molar mass typically $> 10^6$ D (according to ASTM definitions M $> 3.10^6$ D), see reference [73]. Processing of UHMW-PE is outside the scope of the present chapter but the hexagonal phase could be an important topic for future processes concerning the production of oriented PE structures. The hexagonal phase is a so-called 'mobile phase' with a high degree of chain mobility along the c-axis. Annealing in the hexagonal phase promotes chain-refolding and the formation of extended-chain crystals. The formation of extended-chain crystals invoked disentangling of chains, and consequently, promotes drawability. Annealing (UHMW)-PE samples, prior to drawing at elevated pressures (in the transient hexagonal phase region) will consequently facilitate the subsequent drawing operation as shown by Ward et al. [76].

For *oriented* UHMW-PE, the hexagonal phase can be observed at much lower temperatures and pressures, compared with isotropic polyethylenes. When a gel-spun/drawn UHMW-PE fiber like Dyneema^{TM} is heated under stress, for example embedded in a composite matrix or in a tensile tester under a constant load, the ultimate temperature at which the fiber fails is 155 °C. This seems rather surprising at first sight since the equilibrium melting temperature of a perfect polyethylene crystal is 145.5 °C. However, there is no apparent reason for an oriented 1-D polymeric structure to melt at all when heated under constrained conditions. For example, oriented gel-spun/drawn UHMW-polypropylene fibers can be heated for a prolonged time at 230 °C without any loss in mechanical properties [77]. In the case of UHMW-PE fibers, the ultimate temperature is 155 °C and this temperature is related to the onset of the hexagonal phase. The orthorhombic structure in the UHMW-PE fibers transforms in the hexagonal crystal structure at $T > 155$ °C and this so-called 'mobile phase' can not bear any load and, consequently, the fiber fails. This situation occurs only upon constrained heating. When the UHMW-PE fiber is chopped into small pieces, approx. 1 mm in length, and suspended in silicon oil in DSC pans, the recorded melting temperature is 145–146 °C, corresponding to the equilibrium melting temperature of polyethylene, see Section 5.5, see [1]. It is rather unfortunate that this solid-state transition from orthorhombic into the hexagonal crystal structure occurs since it sets an upper temperature limit to the use of UHMW-PE fibers in composite applications which would have been otherwise much higher, see also Section 5.5.5.

The occurrence of the hexagonal phase in oriented UHMW-PE plays also a role in melt-processing of this intractable material. As discussed before, UHMW-PE is an intractable polymer due to its excessive high melt-viscosity related to the high molar mass, typically $> 10^6$ D (according to ASTM definitions $M > 3.\ 10^6$ D). If one would attempt to process (extrude) UHMW-PE one would chose intuitively a processing temperature as high as possible within the limits of thermal decomposition. The result is that the extruded UHMW-PE strands show extensive melt-fracture. To the authors' surprise, they found in the early 80s that upon lowering the processing temperature, the extruded strands became rather homogeneous around temperatures as low as 150 °C! Figure 5.12 shows the extrusion characteristics of UHMW-PE in the three temperatures domains:

1. $T_{extr} < 135$ °C
2. 135 °C $< T_{extr} < 155$ °C
3. $T_{extr} > 155$ °C

At temperatures < 135 °C, region-1, extrusion is, of course, impossible below T_m and only some sintering of individual powder particles occurs. At high temperatures, region-3, extrusion is also not feasible due to extensive melt-fracture. In a narrow temperature range, region-2, strands could be extruded which look rather homogeneous upon visual inspection. This extrusion behavior was independent of the initial crystallization or polymerization history. Since the extruded strands in temperature region 2 showed no enhanced drawability, starting from nascent reactor powder or solution-crystallized flakes, the topic of melt-extrusion was not pursued by us. Recently, Keller and Kolnaar [78,79] revisited this topic and they were able to show that the hexagonal phase plays a role in this region 2 extrusion process. During extrusion, the UHMW-PE powder is in contact with the cylinder and die walls and orientation is induced, in particular at the interface polymer/metal. At this interface, the 'mobile' hexagonal phase could occur at temperatures from 155 °C, see discussions above, and this hexagonal interface lubricates the extrusion process of UHMW-PE strands. The core of the strands consists of compacted UHMW-PE powder particles which are just melted and poorly sintered/fused. The extruded strands, consequently, demonstrate a drawability in the solid state which is at most similar to standard melt-crystallized UHMW-PE samples, but usually the maximum drawability is lower due to poor fusion/welding of the individual UHMW-PE particles.

Summarizing, (1) solvent-free processing of UHMW-PE aiming at high-performance fibers has not been very successful up to now. (2) Processing of nascent UHMW-PE reactor powders

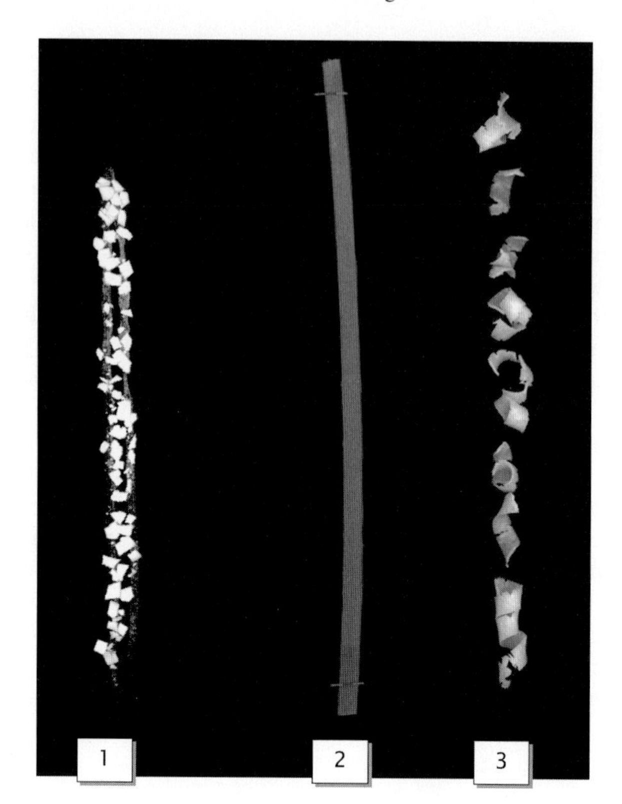

Figure 5.12 Extrusion behavior of UHMW-PE in various temperature domains.

is limited to solid-state operations, viz. below the melting temperature of approximately 135 °C, and oriented tapes etc. can be obtained. Solid state processing, however, has the disadvantage of powder handling and low-denier fibers can not be made straightforwardly. The advantage of UHMW-PE solutions is their low viscosity. (3) Processing of UHMW-PE via the melt, in a specific temperature domain (region 2), does not result in drawable precursors. Processing of nascent UHMW-PE with the aim of obtaining isotropic products with enhanced properties (boundary free materials) could be of interest.

Figure 5.13 summarizes the drawing characteristics of UHMW-PE in the three temperature domains discussed above, for standard melt-crystallized and solution-crystallized samples, respectively. The drawing characteristics of nascent reactor powders are not depicted since they are similar to the solution-crystallized samples. In region-1, below the melting temperature of folded-chain crystals T_m^1, a large difference is observed between melt-crystallized (M) vs. solution-crystallized (S) samples. Solution-crystallized samples become ultra-drawable and the drawability is dependent on the molar mass and initial polymer concentration in solution, λ_{max} scales with $\phi^{-0.5}$ for high molar mass polyethylenes. Figure 5.12a and b show the corresponding properties and stress-strain behavior, respectively. In region-2, $T_m^1 < T_{draw} < T_m^2$ (155 °C) there is often a noticeable difference in drawing behavior between melt-crystallized samples, obtained via compres-

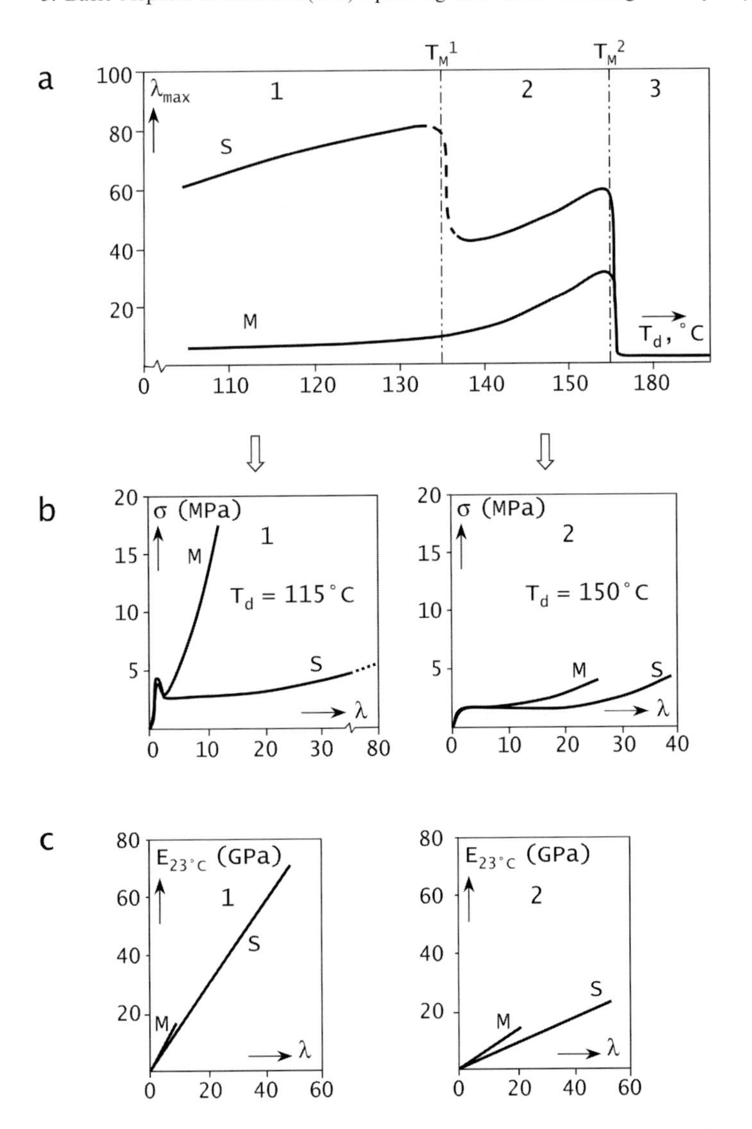

Figure 5.13 Drawing bahavior of UHMW-PE as a function of temperature T (a); the corresponding stress-strain behavior (b) and the resulting properties (c).

sion molding, and solution-crystallized samples. These differences are not related to a difference in entanglement network structures, see next section, but related to macroscopic effects like poor sintering in the case of melt-crystallized samples. Solution-crystallized samples are somewhat better drawable (no grain boundaries) but in both cases the drawing efficiency, in terms of the development of the tensile modulus as a function of the draw ratio λ, is low. In region-2 chain slippage and relaxation processes occurs and effective drawing is not feasible. In region-3, T_{draw} > 155 °C, drawing is not possible due to the onset of the hexagonal phase.

5.4.4 Modeling of the Drawing Behavior

The entanglement model as proposed by Smith et al. [57] could explain the overall effect of the dramatically enhanced drawability of solution-spun/cast polyethylene in comparison with melt-crystallized samples of chemically identical composition. The crystals are seemingly ignored in the drawing process, at least in the temperature region employed for ultra-drawing: above the alpha-relaxation temperature, $T_{\alpha lpha}$, approximately 80 °C for linear polyethylene, and below the melting temperature T_m. In this temperature region the polyethylene crystals are rather ductile and the chains within the crystals are rather mobile [58]. The crystals, however, play indirectly an important role because they fixate the disentangled state. In the entanglement model, crystals do not resist deformation upon draw, at least for polyethylene in the temperature range mentioned before. For polar polymers like the polyamides, however, hydrogen bonding between chains in the crystals plays a dominant role and solution(gel)-spinning is not feasible since the crystals can not be unfolded, they act as rather rigid bodies, see further Section 5.4.5.

After discussing the role of crystals, we now address the role of entanglements in the model for drawing of polyethylenes. Entanglements as such are ill-defined topological constraints which are usually visualized in text books as four strands leading away from a mutual contact, see Figure 5.14.

These type of entanglements are trapped during crystallization in the amorphous zones between the crystallites and act as physical crosslinks on the time-scale of the drawing experiment of the semi-crystalline solid. These entanglements can be removed effectively by dissolution in a good solvent. Another way for disentangling is slow (isothermal) crystallization from the melt or from solution. Slow crystallization promotes disentangling since the 'reeling-in' process of chains from the entangled melt onto the crystal surface is promoted [59]. Slow isothermal crystallization from dilute solution enhances the

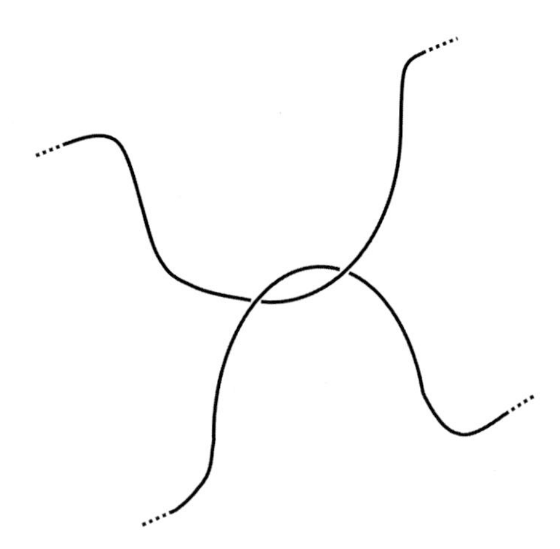

Figure 5.14 Classical model for an entanglement.

maximum drawability but also slow crystallization from the melt, at lower molar mass, promotes the drawability in the solid state. In the extreme limit of very slow isothermal crystallization, linear standard (not the UHMW-PE grades) polyethylenes can even loose their drawability completely and become brittle materials. These results strongly support the simple entanglement concept and the ductility of polyethylene crystals. Despite the fact that the crystallinity increases upon slow (isothermal) crystallization, the drawability is enhanced!

The way to remove entanglements, viz. the topological constraints limiting drawability, is seemingly well understood and crystallization from semi-dilute solutions is an effective and simple route to make disentangled precursors for subsequent drawing into fibers and tapes. A simple 2-D model visualing the entanglement density is shown in Figure 5.15.

The reverse route, from disentangled solid to the melt, is less well understood. As discussed in the previous Section, 5.4.3., many attempts have been made to use disentangled precursors for melt-processing UHMW-PE. The basic idea is that these long chain disentangled molecules need a certain time to (re-)establish an equilibrium entanglement network.

A model concerning molecular mobility in quiescent polymer melts was advanced originally by de Gennes [80]. Constitutive equations, based on this reptation concept, were derived by Doi and Edwards [81]. In these models it is assumed that the longest relaxation times in polymer melts correspond to the reptative motion of complete chains. These relaxation times can be very long in the case of high molar mass and branched polymers. Figure 5.16 shows the stress relaxation modulus of UHMW-PE which shows that relaxation times over 10^4 seconds are present, even at 180 °C.

To the author's knowledge, no effect of initial disentangling on the properties of linear polyethylene melts has ever been observed. There are a few reports [82,83,84] concerning 'shear-refining' of low-density polyethylenes, exhibiting a drop in the melt-viscosity related to 'disentangling' during repeated extrusion (shear), but these effects could not be

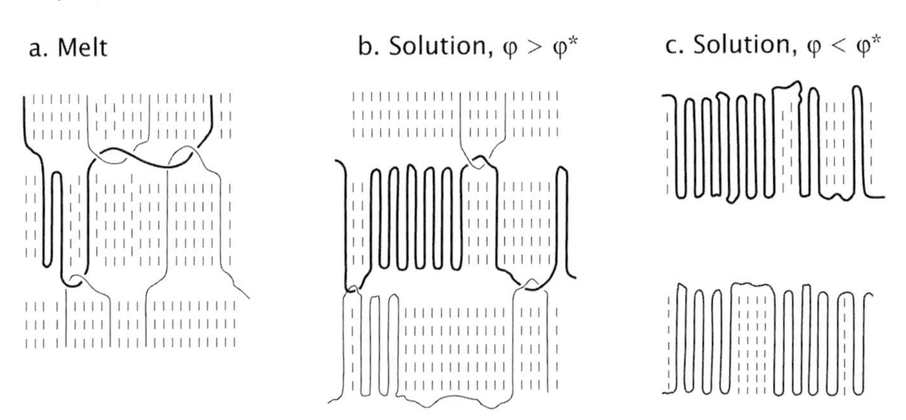

Figure 5.15 Chain topology in crystallized UHMW-PE showing the polymer test chain (bold) with entangled neighbours and non-entangled neighbours. Adapted from [1].

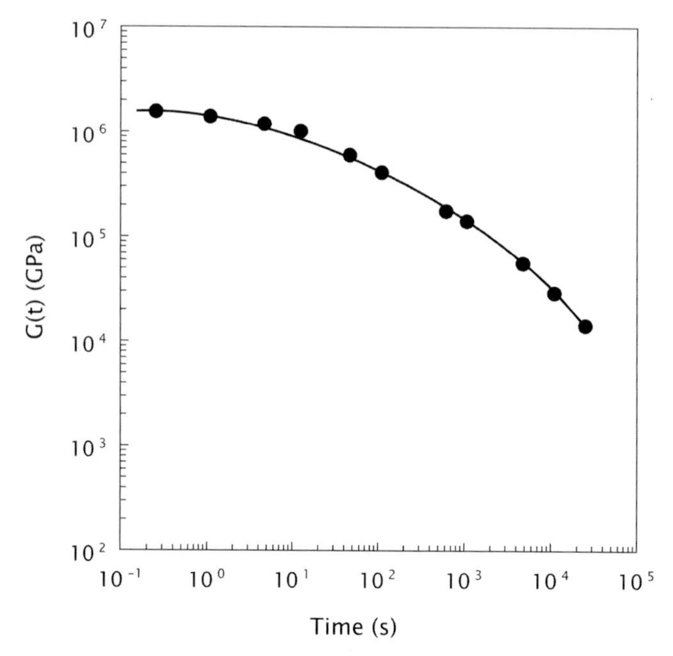

Figure 5.16 Stress relaxation modulus G(t) of UHMW-PE as a function of time (t) at 180 °C.

observed in the case of linear polyethylenes. In the case of UHMW-PE, a polymer possessing very long relaxation times, the following phenomena are observed on passing from the solid state to the melt. We distinguish between three characteristic UHMW-PE samples, respectively crystallized from the melt (M); crystallized from dilute solutions (S) and the so-called virgin or nascent reactor powder (N). In the solid state, below the melting temperature T_m, approximately 133 °C for UHMW-PE, the drawability differs remarkably as discussed above. The typical stress-strain curves were depicted in Figure 5.13 for respectively melt-crystallized (M) and solution-crystallized (S) samples. The drawing behavior of the nascent reactor powders (not shown in Figure 5.13) is similar to solution-crystallized UHMW-PE. To measure the drawability of nascent reactor powders the sample has to be compacted below T_m.

Upon heating all these samples, possessing widely different drawing characteristics in the solid state, above the melting temperature T_m, the rheological properties such as G', G'' and tan* are identical within experimental errors for all samples involved. No memory effect [86] with respect to polymerization nor crystallization history could be recorded. Moreover, upon re-crystallization by cooling from the melt, the enhanced drawability of the solution-crystallized (S) and nascent reactor powders (N) compared with melt-crystallized samples (M) is lost completely, all samples show a limited drawability, characteristic for melt-crystallized samples.

The experimental observations, described above, suggest according to the entanglement model that upon melting of disentangled structures such as solution-crystallized and nascent reactor powders an (equilibrium) entanglement network is restored instantaneously. In view of the long relaxation times, mentioned above, corresponding to the tube renewal time, and the entanglement model as depicted in Figure 5.14, the absence of any memory effect is rather puzzling.

This problem has been addressed experimentally by Barham and Saddler and theoretically by de Gennes. It was shown by Barham and Sadler using neutron scattering techniques and deuterated polyethylenes [85] that upon melting of solution-crystallized polyethylenes the radius of gyration, which is rather low in the case of folded-chain crystals, 'jumps' to its equilibrium value corresponding to a Gaussian chain (random-coil). The authors introduced the term 'coil explosion' for this instantaneous coil expansion upon melting, and the kinetics of this process is independent of the molecular weight. The coil expansion process upon melting implies that the chain will expand very rapidly taking no notice of its neighbors, in contrast with the concept of the 'reptation' theory where the neighboring chains play a dominant role by constituting a virtual tube that forces the chain to reptate along its own contour length.

In a recent note, de Gennes points to a way out of this dilemma [87]. He demonstrates that if a chain starts to melt, the free dangling end of the molten chain will create its own tube and moves much faster than anticipated from reptation theory, rather independent of the molar mass provided that the other end of the chain is still attached to the crystal.

Whatever is the cause for chain explosion upon melting, the effect is that the favorable drawability is lost completely, upon re-crystallization from the melt, but also in the molten state no advantage is obtained in processing initially disentangled UHMW-PE. Whether an entanglement network is formed, rather instantaneously, upon melting is still a matter of debate and partly a matter of semantics. The nature of entanglements is unknown and the model presented in Figure 5.14 is just one representation. The question remains, however, whether long chain molecules as present in UHMW-PE are capable of forming an (equilibrium) entanglement network on a short time scale based on interdiffusion of complete chains. Upon melting, there is an additional driving force, entropy, but to the experience of the authors and based on literature data, it is nearly impossible to form a homogeneous UHMW-PE melt. Even after compression molding/sintering for many hours, the grain boundaries between individual UHMW-PE powder particles can be observed upon crystallization afterwards and limit the long term applicability of UHMW-PE in high demanding products, for example in artificial hip-joints [73].

We have proposed in the past [88] an alternative model for ultra-drawing which is based on local diffusion processes rather than the movement of complete chains. In a simplified view one could compare the formation of an entangled homogeneous melt with 'weaving of complete molecules' (the molecules have to penetrate fully into each other in order to form entanglements as depicted in Figure 5.14). In our proposed alternative model, we compared melting of folded-chain crystals with 'knitting' of molecules, a localized process providing connectivity and loss of drawability as well. The entanglement model is based on topological constraints, entanglements, located *outside* the crystals in the amorphous zones, see Figure 5.15. The alternative model is based on stem arrangement *within* the crystals. This arrangement of molecular stems within the crystals determines the drawability.

As discussed in Section 5.2.2, the transverse moduli of polyethylene crystals, perpendicular to the chain direction, are low. Upon crystallization from solution, the

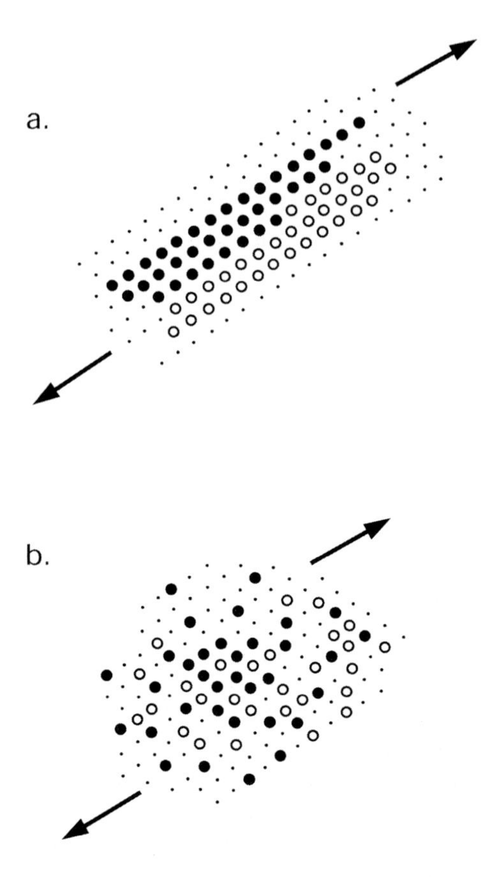

Figure 5.17 Arrangement of stems in lamellar crystals. The dots represent stems of molecules folding along the {110} plane, viewed along the c-direction. No folds are drawn for the sake of simplicity.

molecules fold usually along the 110 plane and the stems of a test chain (heavy dots) are shown in Figure 5.17 without indicating the folds. For the sake of simplicity, we assume that adjacent re-entry occurs during crystallization and that the chain is located within one crystal plane. Shearing and unfolding in the direction perpendicular to the chain and along the {110} plane is rather easy in view of the low shear moduli. Upon melting these crystals the chains will immediately adopt a random coil conformation as discussed before and stems of different molecules will interpenetrate in the 'coil explosion' process. Upon re-crystallization, the stems of the test chain are now crystallized in a more random order within the crystal and shearing (slip) is more difficult since the chains can not cross mutually during deformation.

The schematic representation of stems within the crystals is, of course, an oversimplification. In actual practice, superfolding will occur and crossover of stems belonging to one chain [89]. The presented model, however, only serves the purpose to demonstrate that adjacent re-entry and locality of molecules within a crystal facilitates the process of ultra-drawing which comprises fragmentation of lamellar crystals via shearing, tilting and subsequent unfolding of clusters. The instantaneous loss in drawability upon melting and re-crystallization is due to re-arrangement and intermixing of stems involving only *local* chain motions rather than movement of the *complete* chains chain as proposed for self-diffusion in polymer melts.

The driving force for the mixing of stems and coil-explosion upon melting is the gain in entropy (which is absent in the diffusion/reptation processes in quiescent polymer melts). The polymer chain, which is folded and rather compact in the lamellar crystal, gains entropy upon melting and its corresponding transformation into a random coil.

Arrangement of stems within crystals is difficult to measure and requires detailed neutron scattering experiments which often lack a model-independent interpretation. In the previous section, we discussed the concept of monomolecular crystals, which is also a matter of stem arrangement within one single chain. We assume that in the limit of low catalyst activity and at low polymerization temperatures that these monomolecular crystals could be formed. The experimental proof is, however, lacking for this concept. Nevertheless, we have obtained some evidence for the concept of monomolecular crystals by comparing the annealing behavior of solution-crystallized UHMW-PE vs. nascent UHMW-PE powders.

In the case of solution-crystallized UHMW-PE it is possible to make solution-cast films in which the lamellar crystals are stacked very well, see Figure 5.18a. Upon heating these (dry) solution-cast films above approximately 110 °C, it is observed that the lamellar thickness increases to twice its initial value, from 12.5 to 25 nm. In-situ synchrotron measurements confirmed the sudden quantum jump in lamellar thickness. The well-stacked lamellar arrangement is lost after the doubling process. Figure 5.18b shows the model for chain arrangement during heating. The drawability of annealed solution-crystallized films is not lost after lamellar doubling which is understandable in view of the

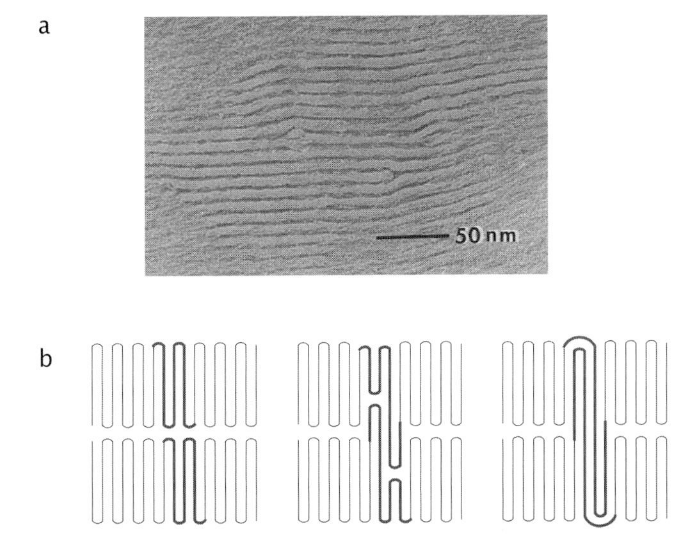

Figure 5.18 Transmisson electron-micrographs of solution-crystallized UHMW-PE showing regular stacking of lamellar crystals (a), and a model for lamellar doubling (b).

stem re-arrangement model which has been discussed before. The crystal stem increases twice in value but no stem intermixing occurs and the drawability remains high.

In contrast with solution-crystallized films, the behavior of nascent reactor powders upon annealing is completely different. No lamellar doubling is observed but the lamellar thickness increases continuously during in-situ synchrotron measurements, from approximately 8 nm up to over 100 nm (beyond the limit of the synchrotron apparatus). This fast increase and reorganization could be indicative of the fact that the nascent UHMW-PE consists of monomolecular crystals, as discussed in the previous section. In the case of monomolecular crystals, the molecule has no neighbors and reorganization/thickening processes are not hampered by neighbouring chains. Of course, these observations are not experimental proof for the concept of monomolecular crystals but the difference in annealing behavior between standard solution-crystallized UHMW-PE and nascent UHMW-PE powders is at least striking.

Summarizing, the drawing behavior of linear polyethylenes can be understood in terms of disentangling long chain molecules. The drawability is related to the degree of disentangling, either by crystallization from (dilute) solutions of by slow crystallization from the melt. The crystals fixate the disentangled state but do not resist deformation (unfolding). Whether entanglements, located *outside* the crystals, or the arrangement of stems *within* the crystals determine the (maximum) drawability is a matter of fine tuning which can not be confirmed by experiments. The entanglement model predicts, at least for high molar masses, the dependence of the maximum draw ratio on the initial polymer concentration in solution semi-quantitatively, λ_{max} scales with $\varphi^{-0.5}$ and has, consequently, the advantage of its simplicity. Stem arrangement within crystals could explain the fast decay in drawability upon melting and re-crystallization. Moreover, stem arrangement and their mutual interaction do play a dominant role for polar polymers as discussed below.

5.4.5 Drawing Behavior of Other Polymer Systems

The success of solution(gel)-spinning of ultra-high molecular weight polyethylene stimulated the research activities concerning the drawing behavior of other linear polymers, notably polypropylene [90,91,92], polyoxymethylene [93, 94] and the aliphatic polyamides, nylon 6 [95] and nylon 66. The prime motivation for using these polymers and to attempt to obtain high modulus and high strength fibers is their higher melting temperature in comparison with linear polyethylenes. In this respect, it has to be noted that a fundamental difference exists between the drawability of apolar polymers such as polyethylene and polypropylene on the one hand and polar polymers such as the polyamides, on the other hand. In Table 5.3, the tensile moduli of melt- and solution-spun, drawn polypropylene and polyoxymethylene are compared with the estimates of the maximum tensile modulus based on X-ray measurements. The experimental values for the tensile modulus approach the maximum tensile moduli, derived from X-ray studies, which illustrates that the concepts derived for linear polyethylenes can be used for other apolar polymers. The ultimate properties of solution(gel)-spun polypropylene fibers are, of

Table 5.3 The Experimental and X-Ray Tensile Modulus of Melt- and Solution-Spun Polypropylene and Polyoxymethylene Fibers

polymer	process	modulus [GPa] (exp.)	modulus [GPa] (X-ray)	Reference
i-PP	melt-spinning	20	40	[90, 91]
	solution-casting	36		[92]
POM	melt-spinning	40	73	[93, 94]

course, limited intrinsically due to the fact that the polypropylene chain possesses a 3_1 helix conformation in the solid state, see Section 5.2.1., and consequently the upper limit of the tensile modulus is below 40 GPa. Nevertheless, also in the case of i-polypropylene the theoretical limits are approached.

Much more interesting would be to produce fibers from polar polymers such as the polyamides, see Table 5.2. The high melting temperatures, compared with polyethylene, and the presence of hydrogen bonds, which could reduce the creep, see below, make the polyamides attractive candidates. An extensive research effort has been performed to produce high modulus and strength fibers based on aliphatic polyamides. These attempts have failed however despite major efforts in industry. It was demonstrated by Smith et al. [95] that the hydrogen bonds in lamellar, solution(gel)-crystallized polyamides, are essentially static up to the melting temperature and act as barriers prohibiting draw.

Another polymer of interest is poly(vinyl alcohol), PVAL, which is commercially available in its atactic form. The small -OH group does not prevent atactic PVAL from crystallizing and the combination of a small side group and a orthorhombic crystal structure, like polyethylene, renders a high theoretical stiffness value, see Table 5.2. The intermediate character in terms of polarity of PVAL, more polar than polyethylene but less directed hydrogen bonds (atactic) compared with polyamides, results in a drawability in between the both extremes, respectively polyethylene and polyamides. The major difference with drawing of polyethylene is that in the case of PVAL the alpha-relaxation temperature increases with the draw ratio. In the case of polyethylene, the α relaxation temperature remains constant upon draw, in other words the crystals remain ductile, even in a highly oriented/extended structure. This property is favorable for ultra-drawing but also is responsible for creep upon static loadings, see below. In the case of PVAL, the α-relaxation temperature increases upon draw and fiber fracture occurs as soon as the α-relaxation temperature approaches the melting and/or drawing temperature [31].

The tensile strength of PVAL fibers, at a fixed degree of polymerization, is high in comparison to linear polyethylene. Probably, this comparatively high tensile strength also originates from the presence of hydrogen bonds which, at a comparatively low molecular weight, induces a transition from a slippage dominated fracture mechanism to a C-C bond rupture type of fracture mechanism.

The above described observations diminish, to a certain extent, the need for a high maximum attainable draw ratio and a high molecular weight to obtain high strength fibers based on polymers with an intermediate polarity (cohesive energy density) such as

PVAL. Poly(vinyl alcohol) fibers with a tensile modulus and tensile strength of respectively $\sim$ 70 GPa and $\sim$ 2.5 GPa can be produced [31].

Similar observations were reported recently concerning another polymer with intermediate polarity, the polyketone fibers (PECO). Fibers, based on alternating copolymers of ethylene and carbon monoxide, possessing a tensile modulus and tensile strength of respectively $\sim$ 50 GPa and 3.5 GPa were produced by Lommerts [96].

Lommerts proposed that the maximum attainable draw ratio of semi-crystalline polymers is related to their cohesive energy density which, in principle, represents the total energy of all intermolecular interactions in a polymer. The experimentally observed relationship between maximum attainable draw ratio of semi-crystalline polymers and cohesive energy density further illustrates that enhanced intermolecular interactions in 'polar' polymers dominate their solid state drawing behavior. Research concerning drawing of polymers possessing an intermdeiate polarity is still going on. For example, the drawability of high molecular weight polyesters (PET) has been studied extensively by Ito and Kanamoto. Moduli up to appr. 35 GPa and tensile strength's up to about 2 GPa could be obtained [97].

5.5 Properties of Polyethylene Fibers, 1-D vs. 3-D

5.5.1 Tensile Strength (1-D)

Until the late seventies, the maximum tensile strength of textile and technical yarns based on flexible macromolecules was limited to approximately 1 GPa. This situation was changed with the discovery of solution(gel)-spinning of UHMW-PE fibers. Presently, UHMW-PE fibers possessing tensile strengths of 3–4 GPa are produced commercially, for example Dyneema™ by DSM and Spectra™ by its licencee Allied Signal. Figure 5.19 shows the properties of these high-performance polyethylene fibers in comparison with other advanced and classical (steel, glass) yarns. Due to its low density, the specific values for the stiffness and strength of polyethylene fibers are currently superior, at least at ambient temperatures. On a laboratory scale, fibers with a strength level up to 6–8 GPa can be made by optimized drawing procedures.

It is obvious that the experimental values for the maximum tensile strength of solution-spun, ultra-drawn UHMW-PE fibers (6-8 GPa) are still low in comparison with the theoretical values, viz. $>$ 20 GPa, for an extended polyethylene chain with an infinite molecular weight. In the past, different approaches were used to describe and to interpret the origin(s) of this discrepancy between experimental and theoretical values. This problem has already been addressed in Section 5.2.3., the difference between finite and infinite chains. In the case of finite chains, the overlap of and the secondary forces between the chains are of utmost importance. The molar mass distribution and in particular the number average molecular weight (chain ends) are important parameters.

The influence of the weight average molar mass on the tensile strength of melt-and solution-spun, ultra-drawn polyethylene fibers was systematically investigated by Smith

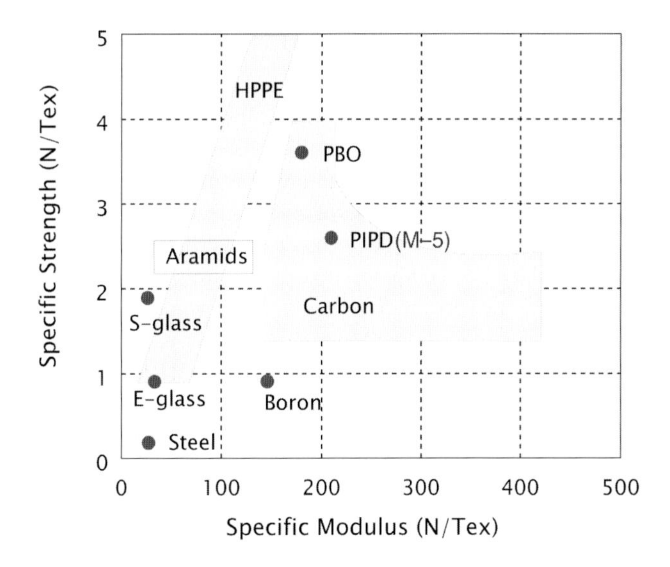

Figure 5.19 Specific strength vs. specific modulus of various fibers.

and Lemstra [98]. In these studies, the tensile strength of fibers was compared at a fixed tensile modulus to eliminate the influence of degree of orientation and chain extension on the tensile strength. It was shown that the tensile strength of drawn fibers increases with increasing molecular weight and an empirical relationship between the tensile strength, tensile modulus and the molecular weight was derived:

$$\sigma_t \propto E^n M_w{}^m \qquad n = 0.7;\ m = 0.4 \tag{5.7}$$

The influence of molecular weight distribution on the tensile strength of solution-spun, ultra-drawn UHMW-PE fibers was investigated as well [98] and it was found that a reduction in polydispersity ($Q = M_w/M_n$) enhances the tensile strength at a fixed tensile modulus and weight average molecular weight.

Smith and Termonia [25,26] have addressed the issue of finite chains theoretically and they have developed a kinetic model. The influence of the molar mass and the effect of chain-end segregation on the theoretical tensile strength of polyethylene and aromatic polyamides was investigated by Smith and Termonia using their kinetic model. For a molar mass in the order of 10^6 D, the theoretical tensile strength is estimated to be approximately 10 GPa.

Summarizing, the factors determining the tensile strength of polyethylene fibers are rather well understood. The tensile strength values of UHMW-PE fibers, obtained via optimized laboratory experiments, 6–8 GPa, are rather close to the theoretically predicted maximum values of approximately 10 GPa.

All experimental studies concerning the tensile strength of polyethylene fibers were, of course, focused on achieving the maximum values for strength and stiffness in the fiber direction, a typically 1-D(imensional) problem. Unfortunately, the world is 3-D(imensional) and this fact was noticed immediately when UHMW-PE fibers were used in composite applications as will be discussed in the Section 5.5.2.

5.5.2 Polyethylene Fibers in Composites (3-D)

The tenacity of UHMW-PE fibers is impressive, especially when the specific values for the stiffness and strength are compared with other fibers, see Figure 5.19. However, these UHMW-PE fibers also possess disadvantages, notably a relatively low melting temperature, creep under static loadings and a low compressive and shear strength. The limiting melting temperature of UHMW-PE fibers, 155 °C, has been discussed before in Section 5.4.3. When polyethylene fibers are embedded in a composite (matrix), melting of the fiber depends on the constraints put on the fiber by the surrounding matrix. In principle, the melting temperature can be increased to high values, provided that the matrix can survive the applied temperature. The melting point, in a first approximation, is given by: $T_m = \Delta H/\Delta S$ and the chains in the fiber can not melt when the fiber is constrained in a matrix, viz. adopt a random coil conformation. An increased melting temperature has been has been observed for solution(gel)-spun polypropylene [77] but in the case of UHMW-PE fibers the upper limit is 155 °C. Above this temperature, the orthorhombic crystals in the fiber transform into the hexagonal structure. In the hexagonal phase the chain mobility is high and the fiber can not sustain any load [1].

Next to a limiting melting temperature, another serious draw back of UHMW-PE fibers in structural composites is creep. UHMW-PE fibers are made by hot-drawing. There is no molecular 'lock-in' mechanism after the UHMW-PE fibers have been made. Drawing of UHMW-PE in the final stage of the fiber preparation, in a temperature range close to but below the melting point, or loading the fiber in actual use, notably at elevated temperatures, is principally the same.

The time/temperature dependence of the mechanical properties of UHMW-PE fibers was extensively investigated [99]. It was found that the properties of these fibers are strongly non-linear viscoelastic and can be mathematically described by a combination of a linear viscoelastic contribution and a non-linear plastic flow process. Improvement of the creep behavior of UHMW-PE fibers have been obtained by using UHMW-PE copolymers [31] and selective crosslinking, for example by UV crosslinking assisted by impregnation of benzophenone in carbon dioxide [100].

A major draw back of UHMW-PE in composite applications is the pronounced anisotropic character of the fiber, which is a consequence of the weak secondary interactions between the chains. This issue has already been addressed in Section 5.2.4., the stiffness matrix of PE crystals. In recent studies [30] it was shown by Peijs et al. that the low interlaminar shear strength, ILLS, obtained from a three point bending test of unidirectional composites containing UHMW-PE fibers, see Table 5.4, is inherently due

Table 5.4 Interlaminar Shear Strength (ILSS) of Unidirectional Composites; 50/50 Fiber/Epoxy Composites

fiber	ILSS [MPa]
UHMW-PE (untreated)	13
UHMW-PE (corona/plasma treated)	20–30
Aramid	45–70
E-glass	75–95
Carbon	80–120

to the poor shear and compressive properties of polyethylene fibers. Consequently, UHMW-PE fibers can not be used in structural composite applications, see Table 5.4.

5.5.3 Miscellaneous Properties of UHMW-PE Fibers

In the previous sections the properties of solution-spun, ultra-drawn UHMW-PE fibers were discussed in terms of their tensile modulus and tensile strength which show impressive values at short loading times but rather extensive creep at prolonged loading times. These properties are, of course, relevant in a large number of applications such as fishing nets and/or lines, ropes and composites. Solution-spun, ultra-drawn UHMW-PE fibers also possess a number of other favorable properties such as a high impact resistance, cutting-resistance, low dielectric constant, low dielectric loss, low stretch, high heat conductivity, and a high sonic modulus. Especially these 'secondary' properties are an important advantage in applications such as ballistic applications (bulletproof vests), helmets (impact), hybrid composites (impact), gloves (cutting-resistance), fishing lines (low stretch), loudspeaker cones (sonic modulus) and radomes/sonar domes (dielectric properties).

5.6 Concluding Remarks

It has been shown in this chapter that UHMW-PE fibers can be made possessing impressive strength and stiffness values, see Figure 5.19, especially when their specific values are taken into account. These short term values for strength and stiffness have not been challenged and matched by any other fiber based on flexible polymer molecules. Polyethylene is the 'primus inter pares' thanks to the availability of high molar mass base material, the enhanced drawability after removing the constraints limiting drawability, the absence of specific interactions such as hydrogen bonds and the small cross-sectional area of the PE chain. The penalty one has to pay for all these beneficial characteristics is that the oriented polyethylene fibers are prone to creep. Basically there is no difference between drawing and creep experiments ("easy draw – easy creep"), i.e. there is no lock-in mechanism after draw. Consequently, UHMW-PE fibers are less suitable for applications, such as reinforcing structural composites. The high strain-at-break of these fibers provide, however, unique possibilities for applications where impact resistance is important.

5.7 List of Symbols and Abbreviations

A	*cross-sectional area*
d	*diameter*
E	*tensile (Young) modulus*
F	*force*
L	*length*
L_o	*original length*
m	*constant*
M	*molecular weight*
M_n	*number average molecular mass*
Mw	*weight average molecular mass*
n	*constant*
Q	*polydispersity*
S	*compliance*
t	*thickness*
U	*internal energy*
Δl	*length change*
Δn	*birefringence*
θ	*angle*
λ	*draw ratio*
λ_{max}	*maximum attainable draw ratio*
ε	*strain*
σ	*stress*
σ_t	*tensile strength*
ϕ	*polymer concentration in solution*
PA	*polyamide*
PE	*polyethylene*
PECO	*polyketone*
PEEK	*polyetheretherketone*
PEO	*polyethyleneoxide*
PET	*polyethyleneterephthlate*
POM	*polyoxymethylene*
PP	*polypropylene*
PPTA	*poly(p-phenylene terephthalamide)*
PPX	*polyparaxylylene*
PTFE	*polytetrafluorethylene*
PVAL	*poly(vinyl alcohol)*
UHMW-PE	*ultra-high-molecular-weight polyethylene*

5.8 References

1. Lemstra, P.J., Kirschbaum, R., Ohta, T., Yasuda, H. *Developments in Oriented Polymers-2,* Ward, I.M. (Ed.), (1987) Elseviers Applied Science, London, pp. 39–79
2. Lammers, M. Ph.D. thesis ETH-Zuerich, 1998;
3. Capaccio, G., Gibson, A.G., Ward, I.M. *Ultra-High Modulus Polymers*, Ciferri, A. and Ward, I.M. (Eds.), (1979) Elseviers Applied Science, London, Ch. 1
4. Capaccio, G., Ward, I.M. *Polymer* (1974) 15, p. 233–238
5. Zachariades, A.E., Porter, R. S. *The Strength and Stiffness of Polymers*, Zachariades, A.E. and Porter, R.S. (Eds.), (1983) Marcel Dekker, New York, pp. 1–51
6. Smith, P., Lemstra P.J. U.K. Patent 2 040 414 (1979)
7. Smith, P., Lemstra, P.J. *Polymer* (1980) 21, p. 31–36
8. Smith, P., Lemstra, P.J. *J. Mater. Sci* (1980) 15, p. 505
9. Kanamoto, T., Tsuruta, A., Tanaka, M., Porter, R.S. *Polym. J.* (1983) 15 p. 327
10. Kunugi, T., Oomori, S., Mikami, S. *Polymer* (1988) 29, pp. 814–819
11. Carothers, W., Hill, J.W. *J. Am. Chem. Soc.* (1932) 54, p. 1586
12. Meijer, K.H. and Lotmar, W. *Helv. Chim. Acta*, (1936) 19, p. 68
13. Treloar, L.R.G. *Polymer* (1960) pp. 95–103
14. Tashiro, K., Kobayashi, M., Tadakoro, H. *Macromolecules* (1978) p. 914
15. Odajima, A., Madea, T. *J. Polym. Sci.* part C, (1966) 34, p. 55
16. Nakamae, K., Nishino, T., Ohkubo, H. *J. Macromol. Sci-Phys*, (1991) B30, pp. 1–23
17. Nishino, T., Ohkubo, H., Nakamae, K. *J. Macromol. Sci.-Phys* (1992) B31, pp. 191–214
18. Nakamae, K., Nishino, T., Ohkubo, H., Matsuzawa, S., Yamura, K., *Polymer* (1992) 33, p. 2281
19. Nakamae, K., Nishino, T., Yokoyama, F., Matsumoto, *T., J. Macrom. Sci. Phys.* (1988) B27(4), p. 404
20. Sakurada, I., Ito, T., Nakamae, K., Bull. *Inst. Chem. Res. Kyoto Univ.* (1966) 44, p. 77
21. Nakamae, K., Nishino, T., Hata, K., *Kobunshi Ronbonshu* (1985) 42, p. 241
22. Nakamae, K., Nishino, T., Shimiizu, Y., Hat, K., *Polymer* (1990) 31, p. 1909
23. He, T. *Polymer* (1986) 27, p. 253
24. Kelly, A. Macmillan, N.H. *Strong Solids* (1986) Clarendon Press, Oxford
25. Termonia, Y, Smith, P. *Macromolecules* (1987) 21, pp. 835–841
26. Termonia, Y., Smith, P. *Polymer* (1986) 27, pp. 1845–1851
27. Smith, P., Termonia, Y. *Polymer Com* m. (1989) 30, pp. 66–68
28. Tashiro, K., Kobayashi, M., Tadakoro, H. *Macromolecules* (1978) 11, p. 914
29. Ward, I.M. *Developments in Oriented Polymers* 2nd Ed. (1988) Elsevier New York
30. Peijs, A.A.J.M. *Ph.D. Thesis* (1993) Eindhoven University of Technology
31. Bastiaansen, C.W.M. *Ph.D. Thesis* (1991) Eindhoven University of Technology
32. Frank, F.C., *Proc. Royal Soc., London Ser. A* (1970) 319, p. 127
33. Peterlin, A., *J. Polym. Sci.* (1966) B4 p. 287
34. De Gennes, P.G. *J. Chem. Phys.* (1974) 60, p. 15
35. Keller, A., Kolnaar, H.W.H. *Materials Science and Technology*, Eds. Cahn, R.W., Haasen, P., Kramer, E.J. volume 18 (Ed. Meijer, H.E.H.) (1997), VCH, Weinheim, chap. 4
36. Keller, A., Odell, J.A. *Colloid Polym. Sci.* (1985) 263, p. 181
37. Mitsuhashi, S. *Bull. Text. Res. Inst.* (1963), 66, p. 1–*3*
38. Pennings, A., Kiel, A.M. *Kolloid Z.Z. Polymere* (1965) 205, p. 160
39. Pennings, A.J. *J. Polym. Sci. Polym. Symp.* (1977) 59, p. 55
40. Keller, A., Willmouth, F.M. *J. Macrom. Sci.* (1972) B6, p. 493
41. Zwijnenburg, A. *Ph.D. Thesis*, (1978) University of Groningen
42. DSM/Stamicarbon, U.S. Patent 4,137,394
43. Mackley, M. NRDC Eur. Patent 22681
44. van der Vegt, A.K., Smith, P.P.A. *Adv. Polym. Sci.* Monograph 26 (1967) p. 313
45. Southern, J.H., Porter, R.S. *J. Macrom. Sci.* (1970) B4, p. 541
46. Porter, R.S., Southern, J.H., Weeks, N.E. *Polym. Eng. Sci.* (1975) 15, p. 213
47. Grun, F., Kuhn, W. *Kolloid Z.* (1942) 101, p. 248
48. Ward, I.M. *Mechanical Properties of Solid Polymers* 2nd ed. (1985) John Wiley & Sons, New York,

49. Pazur, R.J., Ajii, A. Prud'homme, R.E. *Polymer*, (1993) 34, p. 4004
50. Dirix, Y., Tervoort, T.A., Bastiaansen, C.W.M., Lemstra, P.J. *J. Text. Inst.* (1995) 86/2 p. 314
51. Pinnock, P.R., Ward, I. *M., Brit J. Appli. Phys.* (1966) 17, p. 575
52. Sadler, D.M., Barham, P.J. *Polymer* (1990) 31, p. 36–46
53. Peterlin, A. *Polym. Eng. Sci* (1978) 18, p. 488
54. Stamm, M., Fischer, E.W., Dettenmaier, M. *Faraday Discussions of the Chemical Society* (1979) 68, pp. 263–278
55. Smith, P., Lemstra, P.J. UK Patent 2,051,661 (1979)
56. Smith, P., Lemstra, P.J., *Makrom. Chem.* (1979) 180, p. 2983
57. Smith, P., Lemstra, P.J., Booij, H.C. *J. Polym. Sci., Phys. Ed.* (1982) 20, p. 2229
58. Smith, P., Lemstra, P.J. *J. Mater. Sci.* (1980) 15, p. 505
59. Lemstra, P.J. Smith, P. *Brit. Polym. J.* (1980) 12, p. 212
60. Bastiaansen, C.W.M., Meijer, H.E.H., Lemstra, P.J. *Polymer* (1990) 31, p. 1435
61. Leblans, P.J.R., Bastiaansen, C.W.M. *Macromolecules* (1989) p. 3312
62. Smith, P., Chanzy, H.D., Rotzinger, B.P. *Polym. Comm.* (1985) 26, p. 258
63. Kanamoto, T., Ohama, T., Tanaka, K. Takeda, M. Porter, R.S. *Polymer* (1987) 28, pp. 1517–1520
64. Otsu, O., Yoshida, S., Kanamoto, T., Porter, R.S., *Proceedings PPS, Yokohama* (1988)
65. Eur. Patent EP 376 423/425 947
66. Engelen, Y. Ph.D. Thesis Eindhoven University of Technology (1991)
67. Egorove, V.M., Ivan'kova, E.M., Marikhin, V.A., Myasnakova, L.P. Baulin, A.A., *Vysokomol. Soedinnen* (in press)
68. Bassett, D.C., *Polymer* (1976) 17, pp. 460–470
69. Wunderlich, B., Grebowicz, J. *Adv. Polym. Sci.* (1984) 60/61, pp. 1–59
70. Hikosaka, M., Rastogi, S., Keller, A. and Kawabata, H., *J. Macrom. Sci. Phys.* (1992) B31, p. 87
71. Rastogi, S., Hikosaka, M., Kawabata, H. and Keller, A., *Macromolecules* (1991) 24, p. 6384
72. Hikosaka, M., Rastogi, S., Keller, A. and Kawabata, *H., J. Macromol. Sci. Phys. B.*, (1992) 31, p. 87
73. Rastogi, S., Kurelec, L., Lemstra, P.J., *Macromolecules* (1998) 22, p. 5022
74. Ostwald, W., *Z. Physik. Chem.* (1987) 22, p. 286
75. Keller, A., Hikosaka, M., Rastogi, S., Toda, A., Barham, P.J. and Goldbeck-Wood, G., *J. Materials Sci.* (1994) 29, p. 2579
76. Maxwell, A.S., Unwin, A.P., Ward I.M., *Polymer* (1996) 37, p. 3293
77. Bastiaansen, C.W.M., Lemstra, P.J., *Makrom. Chemie, Makrom. Symposia* (1989) 28, p.
78. Kolnaar, J.W.H., Ph. D. thesis Bristol (1993)
79. Kolnaar, J.W.H., Keller, A. *Polymer* (1995) 36, p. 821
80. De Gennes, P.G. *J. Chem. Phys.* (1971) 55, p. 572
81. Doi, M., Edwards, S.F. *J. Chem. Faraday Trans.* (1987) 2, p. 1789–1818
82. Munstedt, H. *Coll. Polym. Sci.* (1981) 259, p. 966
83. Schreiber, H.P., *J. Polym. Eng. Sci.* (1983) 23 p. 422
84. Leblans, J.R., Bastiaansen, C.W.M. *Macromolecules* (1989) 22, p. 3312
85. Barham, P., Sadler, D.M. *Polymer* (1991) 32, pp. 393–395
86. Bastiaansen, C.W.M., Meijer, H.E.H., Lemstra, P.J., *Polymer* (1990) 31, p. 1435
87. De Gennes, P.G. C. *R. Acad. Sci. Paris 321 series II* (1995) pp. 363–365
88. Lemstra, P.J., van Aerle, N.A.M.J., Bastiaansen, C.W.M. *Polymer Journal* (1987) 19, pp. 85–97
89. Keller, A. *Faraday Discussions of the Chemical Society* (1979) 68, pp. 145–166
90. Cansfield, D.L.M., Capaccio, G., Ward, I.M., *Polym. Eng. Sci.* (1976) 16, p. 721
91. Taylor, W.N., Clark, E.S., *Polym. Eng. Sci.* (1978) 18, p. 518
92. Peguy, A., Manley, R.S. st *J., Polym. Comm.* (1984) 25, p. 39
93. Capaccio, G., Ward, I.M., *Brit. Pat. Appl. 566626441/74* (1973)
94. Clark, E.S., Scott, L.S. *Polym. Eng. Sci.* (1974) 14, p. 682
95. Postma, A.R., Smith, P., English, A.D. *Polymer Comm.* (1990) 31, pp. 444–448
96. Lommerts, B.J. *Ph.D. thesi* s, University of Groningen (1989)
97. Ito, M., Takahashi, K., Kanamoto, T. *Journal of Appl. Polym. Sci.* (1990) 40, pp. 1257–1263
98. Smith, P., Lemstra, P.J. *J. Polym. Sci. Phys. Ed.* (1982) 20, p. 2229
99. Govaert, L.E. *Ph.D. thesis* Eindhoven University of Technology (1990)
100. Jacobs, M., *Ph.D. thesis*, Eindhoven University of Technology (1999)

6 Electrospinning and the Formation of Nanofibers

Hao Fong and Darrell H. Reneker
University of Akron, Akron, Ohio, USA

6.1 Introduction

Electrospinning is a straightforward method that produces polymer nanofibers. When the electrical force at the surface of a polymer solution or polymer melt overcomes the surface tension, a charged jet is ejected. The jet extends in a straight line for a certain distance, and then bends and follows a looping and spiraling path. The electrical forces elongate the jet thousands or even millions of times and the jet becomes very thin. Ultimately the solvent evaporates, or the melt solidifies. The resulting, very long, nanofiber collects on an electrically grounded metal sheet, a winder or some other object, often in the form of a non-woven fabric.

The electrospinning process makes fibers with diameters in a range one or two orders of magnitude smaller than those of conventional textile fibers. The small diameter provides a large surface area to mass ratio, in the range from 10 m^2/g (when the diameter is around 500 nm) to 1000 m^2/g (when the fiber diameter is around 50 nm). The equipment required for electrospinning is simple and only a small amount of polymer sample is needed to produce nanofibers.

Polymer nanofibers are being used, or finding uses, in filtration, protective clothing, and biomedical applications including wound dressings and drug delivery systems. Other possible uses include solar sails, light sails, and mirrors for use in space. Nanofibers offer advantages for the application of pesticides to plants, as structural elements in artificial organs, as supports for enzymes or catalysts that can promote chemical reactions, and in reinforced composites. Ceramic or carbon nanofibers made from polymeric precursors extend nanofiber applications to uses involving high temperature and high modulus. The electrospinning process can incorporate particles such as pigments or carbon black particles into nanofibers. Flexible fibers are needed on a scale commensurate with micro- or nano-electrical mechanical and optical systems. The use of electrical forces may lead to new ways to fabricate micro or nano scale devices.

The electrostatic spray literature contains many helpful insights into the electrospinning process. Lord Raleigh studied the instabilities that occur in electrically charged liquid droplets. He showed, over 100 years ago, that when the electrostatic force overcame the surface tension, a liquid jet was created [1]. Zeleny [2–5] considered the role of surface instability in electrical discharges from drops. He published a series of papers around 1910 on discharges from charged drops falling in electric fields, and showed that, when the discharge began, the theoretical relations for surface instability were satisfied. In 1952, Vonnegut and Neubauer [6] produced uniform streams of highly charged droplets with diameters of around 0.1 mm, by applying potentials of 5 to 10 kilovolts to liquids flowing from capillary tubes. Their experiment proved that monodisperse aerosols with a particle radius of a micron or less could be formed from the pendent droplet at the end of the pipette. The diameter of the droplet was sensitive to the applied potential. Wachtel and coworkers [7] prepared emulsion particles using an electrostatic method to make a monodisperse emulsion of oil in water. The diameters of the emulsion particles were from 0.5 to 1.6 microns. In the 1960's, Taylor [8–11] studied the disintegration of water droplets in an electrical field. His theoretical papers demonstrated that a conical interface, with a semi-angle close to 49.3°, was the limiting stable shape.

Electrospinning of solutions of macromolecules can be traced back to 1934, when Formhals invented a process for making polymer fibers by using electrostatic force. Fibers were formed from a solution of cellulose acetate. The potential difference required depended on properties of the spinning solution such as molecular weight and viscosity. Formhals obtained a series of patents on his electrospinning inventions [12–18].

Gladding [19] and Simons [20] improved the electrospinning apparatus and produced more stable fibers. They used movable devices such as a continuous belt for collecting the fibers. Later, Bornat [21–22] patented another electrospinning apparatus that produced a removable sheath on a rotating mandrel. The basic principles were similar to previous patents. He determined that the tubular product obtained by electrospinning polyurethane materials in this way could be used for synthetic blood vessels and urinary ducts.

In 1971, electrospinning of acrylic fibers was described by Baumgarten [23]. Acrylic polymers were electrospun from dimethylformamide solution into fibers with diameters less than 1 micron. A stainless steel capillary tube was used to suspend the drop of polymer solution and the electrospun fibers were collected on a grounded metal screen. Baumgarten observed relationships between fiber diameter, jet length, solution viscosity, feed rate of the solution and the composition of the surrounding gas.

In 1981, Manley and Larrondo [24–26] reported that continuous fibers of polyethylene and polypropylene could be electrospun from the melt, without mechanical forces. A drop of molten polymer was formed at the end of a capillary. A molten polymer jet was formed when a high electric field was established at the surface of the polymer. The jet became thinner and then solidified into a continuous fiber. The polymer molecules in the fiber were oriented by an amount similar to that found in conventional as-spun textile fibers before being drawn. The fiber diameter depended on the electric field, the operating temperature and the viscosity of the sample. The electrospun fibers were characterized by X-ray diffraction and mechanical testing. As either the applied electric field or the take-up velocity was increased, the diffraction rings became arcs, showing that the molecules were elongated along the fiber axis.

Reneker and coworkers [27–36] made further contributions to understanding the electrospinning process and characterizing the electrospun nanofibers in recent years. Doshi [27] made electrospun nanofibers, from water soluble poly(ethylene oxide), with diameters of 0.05 to 5 microns. He described the electrospinning process, the processing conditions, fiber morphology and some possible uses of electrospun fibers. Srinivasan [28] electrospun a liquid crystal polyaramid, poly (p-phenylene terephthalamide), and an electrically conducting polymer, poly (aniline), each from solution in sulfuric acid. He observed electron diffraction patterns of the polyaramid nanofibers, both as spun and after annealing at 400 °C. Chun [29] used transmission electron microscopy, scanning electron microscopy and atomic force microscopy to characterize electrospun fibers of poly (ethylene terephthalate). Fang [30] electrospun DNA into nanofibers, some of which were beaded. Polybenzimidazole nanofibers were electrospun by Kim [31, 32], who also studied the reinforcing effects of these nanofibers in an epoxy matrix and in a rubber matrix. Fong [33] described the electrospinning of beaded nanofibers. He also studied the morphology of phase separation in electrospun nanofibers of a styrene-butadiene-styrene tri-block copolymer [34]. Elastomeric nanofibers were made from this thermo-elastic copolymer. The smallest fibers of this tri-block copolymer had diameters around 3 nm. Reneker and coworkers [35, 36] also made carbon nanofibers, from polymers or pitch. The resulting carbon nanofibers had diameters of from 50 to 500 nm. The morphology ranged from highly oriented, crystalline, nanofibers to very porous ones with high values of surface area per unit mass.

6.2 Electrospinning Process

Electrospinning involves polymer science, applied physics, fluid mechanics, electrical, mechanical, chemical, material engineering and rheology. Many parameters, including the electric field, solution viscosity, resistivity, surface tension, charge carried by the jet and relaxation time can affect the process. A comprehensive mathematical model of this process was developed by Reneker, Yarin, et al. [37]

The electrospinning process has three stages: (1) jet initiation and the extension of the jet along a straight line; (2) the growth of a bending instability and the further elongation of the jet, which allows the jet to become very long and thin while it follows a looping and spiraling path; (3) solidification of the jet into nanofibers.

6.2.1 Jet Initiation and the Diameter of a Single Jet

In a typical experiment, a pendent droplet of polymer solution was supported by surface tension at the tip of the spinneret. When the electrical potential difference between the spinneret and the grounded collector was increased, the motion of ions through the liquid charged the surface of the liquid. The electrical forces at the surface overcame the forces associated with surface tension. A liquid jet emerged from a conical protrusion that formed on the surface of the pendant droplet. The jet was electrically charged. It carried away the ions that were attracted to the surface when the potential was applied. Increasing the potential increased both the charge density on the jet and the flow rate of the jet.

Figure 6.1 shows a sequence of droplet shapes taken at a frame rate of 500 frames per second and a shutter speed 2 ms. A 3.0% solution of polyethylene oxide in water was used. The solution flowed through a hole, with a diameter of 300 microns, in the bottom of the bowl of a metal spoon. The electric field along the axis of the jet was 0.5 kV/cm. The length of the horizontal edge of each of the images was 1.0 mm. When the semi-vertex angle of the droplet was around 22.5^0, the electric force was high enough to overcome both surface tension and viscoelastic forces, and a fluid jet was ejected. Jet diameters near the droplet were in the range from 20 μm to 100 μm. After the charged jet was ejected, the conical protrusion relaxed to a rounded shape, reaching a steady shape in a few

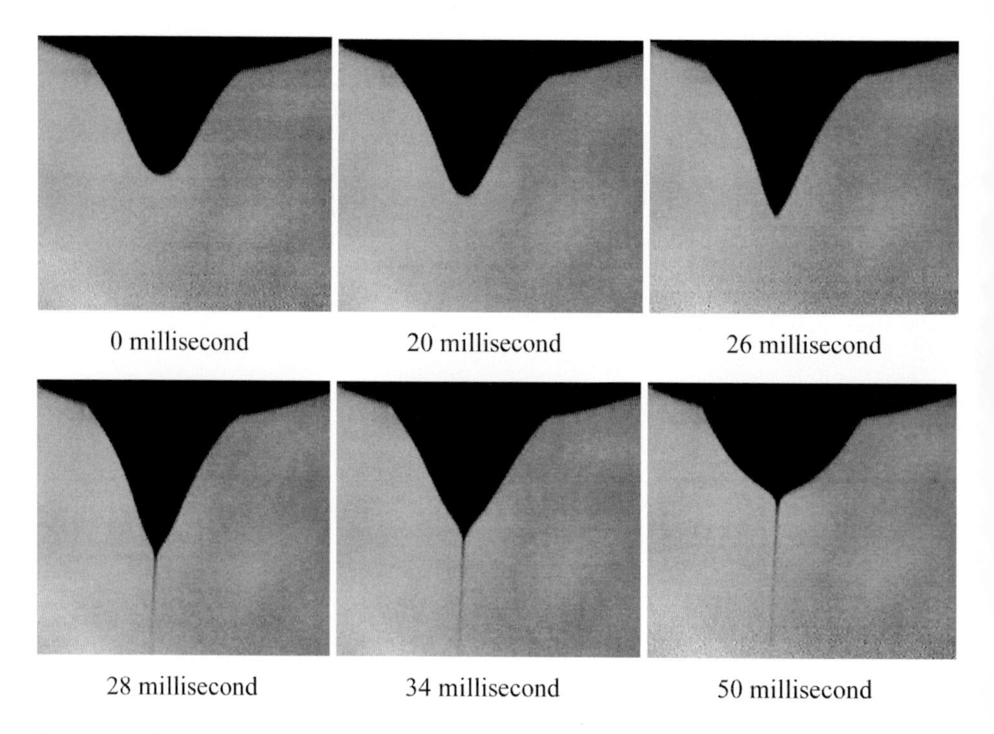

Figure 6.1 Photographs of the pendent droplet and jet, near the time the jet was ejected.

milliseconds. A jet can also be initiated, at a lower potential, by mechanically pulling a jet out of the pendent droplet, since the voltage required for initiation was higher than that required for maintaining the jet flow.

The observed semi-vertex angle as the jet emerged from the droplet was less than the 49.3^0 that Taylor calculated [8–11]. Sometimes the jet current and the shape of the droplet pulsated while the applied voltage was constant. A steady current was associated with a droplet that had a constant shape.

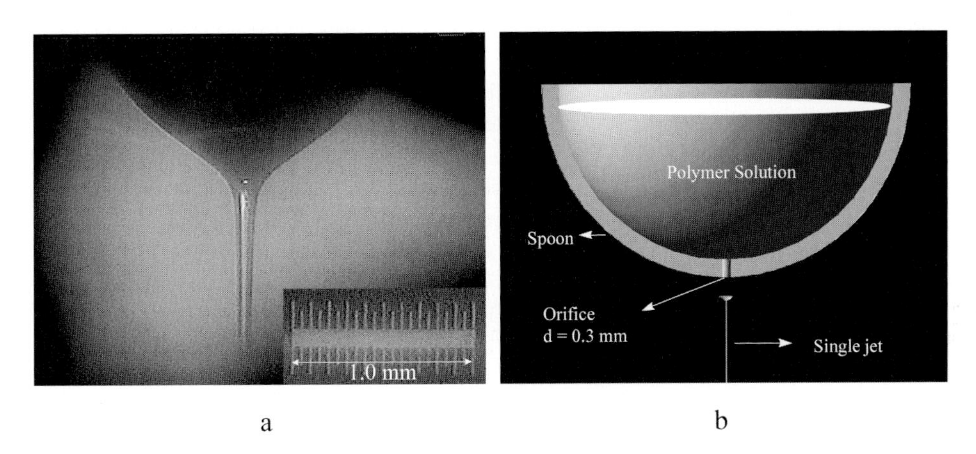

Figure 6.2 (a) Magnified image of the shape of the jet near the spoon. (b) Schematic drawing of a jet flowing from a hole in the bowl of a spoon. The jet is separated from the spoon for clarity.

The jet diameter decreased with the distance from the orifice. Higher electric fields and a lower surface tension coefficient favored the formation of a thicker jet. Addition of salt (NaCl) to the solution, with other parameters held constant, reduced the diameter of the jet. Increasing the viscosity of the solution did not always increase the jet diameter. The largest jet diameter occurred when the solution viscosity was in a medium range. Both higher and lower viscosity favored a thinner jet.

6.2.2 Bending Instability and Elongation of the Jet

After initiation, the path of the jet was straight for a certain distance. Then, an electrically driven bending instability grew at the bottom end of the straight segment. The bending allowed a large elongation to occur in a small region of space. The electrically driven bending instability occurred in self-similar cycles. Each cycle had three steps and was smaller in scale than the preceding cycle.

The three steps in each cycle were:

Step 1. A smooth segment that was straight or slightly curved suddenly developed an array of bends.

Step 2. As the segment of the jet in each bend elongated, the linear array of bends became a series of spiraling loops with growing diameters.

Step 3. As the perimeter of each loop increased, the cross-sectional diameter of the jet forming the loop grew smaller, and the conditions for Step 1 were established everywhere along the loop.

After the first cycle, the axis of a particular segment might lie in any direction. The continuous elongation of each segment was most strongly influenced by the repulsion between the charges carried by adjacent segments of the jet. The externally applied field, acting on the charged jet, caused the entire jet to drift towards the collector, which was maintained at an attractive potential.

Figure 6.3 shows two cycles of bending instability. The jet entered the image near the end of the straight segment, where the first electrically driven bending instability produced an array of helical bends. While the jet ran continuously, it shifted through a series of similar but changing paths. Most loops moved downward at a velocity of about 1 m/s, but some loops with larger diameters remained in the field of view for long time. The slightly curved thin segment that runs horizontally across the left image in Figure 6.3 is part of such a loop that remained in view for over 15 ms. This segment was smooth until, in a time interval of only one ms, the bends and loops shown in the right image of Figure 6.3 developed. During this 15 ms period, many bends and loops of the first cycle of bending instability formed and moved downward through the field of view. The diameter of every segment of the jet became smaller, and the length of every segment increased. The loops grew larger. Bending instabilities developed and grew.

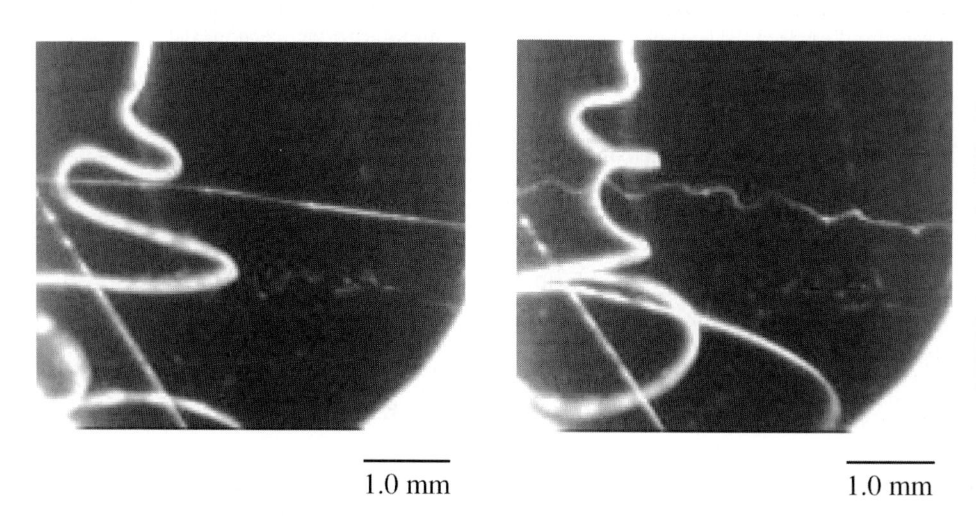

Figure 6.3 The development of the second cycle of bending instability. The time interval between these two images was 15 ms. The camera shutter speed was 0.25 ms.

It was often possible to follow the evolution of the shape of segments, such as those shown in the right part of Figure 6.3, back to the time at which they entered the upper left corner of the image, by stepping backwards in time through the image files created by the electronic camera. It was more difficult to follow the evolution of the jet into the third cycle of bending because the images of the path grew fainter as the jet became thinner, and soon were ambiguous. The elongation and the associated thinning of the jet continued as long as the charge on the jet supplied enough force to overcome the surface tension and viscoelastic forces. Meanwhile, the elongational viscosity increased as the solvent evaporated and eventually the elongation stopped. Details of the evolution of the solidification process remain to be investigated.

The "area reduction ratio", which was defined as the ratio of the cross-sectional area of the upper end of a segment to the cross-sectional area at the lower end of the same segment, was equal to the draw ratio if the volume of material in the segment was conserved. Consider a jet, with a 6% concentration of polymer in a volatile solvent, which went from a jet diameter of 50 μm to a dry nanofiber with a relatively large diameter of 0.5 μm. The area reduction ratio was 2500/0.25 = 10000. The drying process accounted for a factor of 16, and the elongation of the initial straight part of the jet contributed an additional factor of 5. The remaining area reduction ratio, 125, occurred in the parts of the jet affected by the bending instabilities.

Many nanofibers as thin as 0.05 μm (50 nm) in diameter were observed. The corresponding total area reduction ratio was 1,000,000, for an initial jet diameter of 50 μm. If, as above, evaporation of solvent contributed a factor of 16 and the elongation of the straight segment contributed a factor of 5, in this case, the bending and looping part of the jet provided the remaining the factor of 12,500 to the area reduction ratio.

While 12,500 is a high area reduction ratio, it occurred as many segments of the jet were drawn in different directions at the same time, in expanding loops. If the jet were drawn in a straight line to a ratio of 12,500, the velocity required at the nanofiber end of the jet would be much faster than the speed of sound in most solids. The actual path achieved very high elongation without such an unreasonably high velocity.

The longitudinal strain rate could be estimated as follows, by using the area reduction ratio and the time of flight. The time that a typical segment of the jet was in flight ($\delta t = 0.2$ s) could be estimated as the distance between the pendent droplet and the collector (20 cm) divided by the downward velocity of the jet (1 m/s). The longitudinal strain rate was $(\delta\zeta/(\delta t \cdot \zeta)$ where ζ is the initial segment length, and $\delta\zeta$ is the growth in length. The draw ratio $\delta\zeta/\zeta$ was, assuming no solvent loss, around 125 to 12,500, so that the longitudinal strain rate was around 625 to 62500 s^{-1} for the two cases described above.

Theory suggests that the transformation from a random coil to an elongated macromolecule occurs when the strain rate multiplied by the conformational relaxation time of the molecule is greater than 0.5 [38,39]. If the relaxation time of the polymer solution were longer than 0.01 s, which is a conservative estimate, the macromolecules are expected to be elongated and axially oriented during electrospinning.

Before high frame-rate, short exposure time cameras were available, visual observations and video images (30 frames per second) of electrically driven jets were interpreted as evidence for a process that splayed the primary jet into many smaller jets. The splaying jets were supposed to emerge from the region at the end of the straight segment. Figure 6.4a and 6.4b show images from a video frame with an exposure time of 16.7 ms. The jet path was illuminated with a single bright incandescent light that projected a narrow beam through the path of the jet. Figure 6.4c shows a jet similar to that shown in Figure 6.4a, illuminated with light from a broader source, and photographed at 30 frames per second. A shorter exposure time of 1 ms was used. Several loops were visible. The parts of the jet nearer the beginning of the bending instability appeared only as bright, short and unconnected lines. Specular reflections of the narrow beam of light, off nearly horizontal

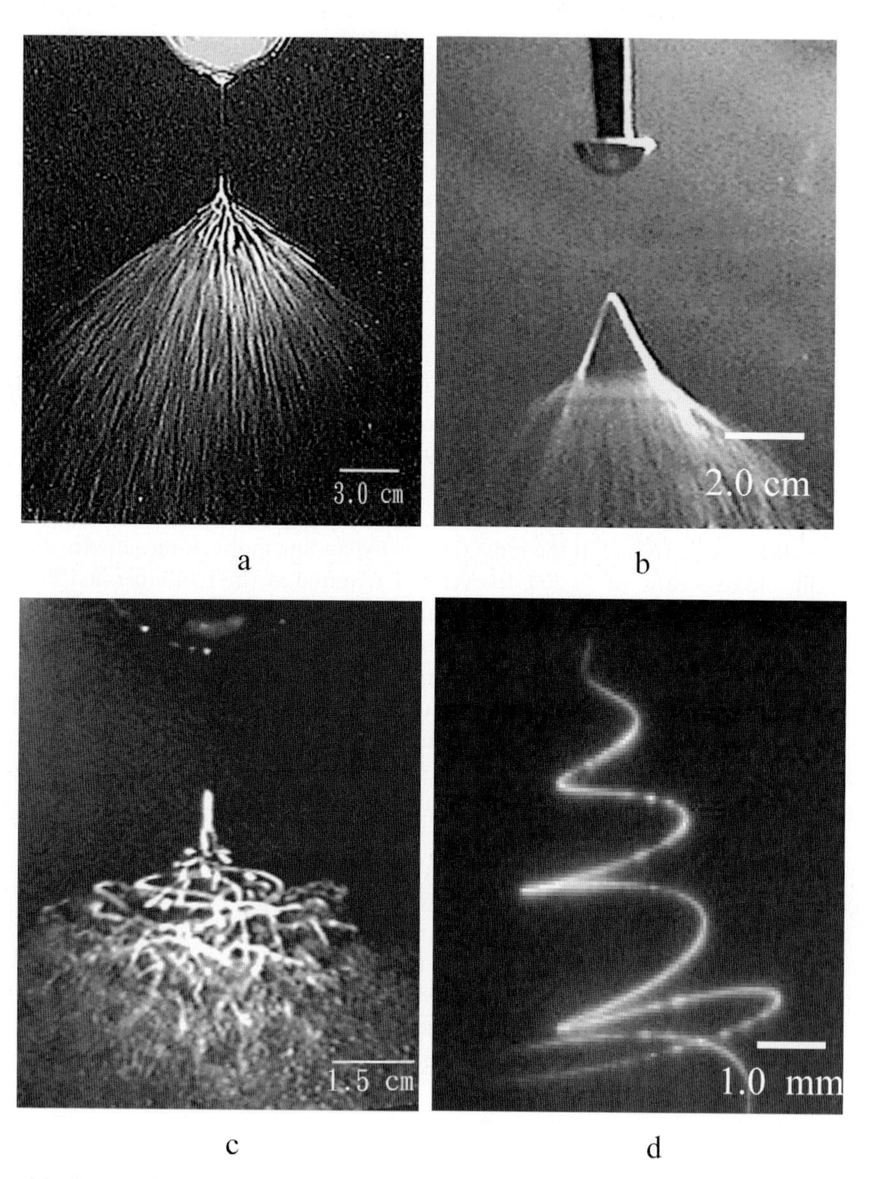

Figure 6.4 Images of electrospinning jet with different exposure times. (a), (b), and (c), shutter speed was 16.7 ms (d) shutter speed was 1.0 ms. The spots in (d) are artifacts produced by a faceted reflector used in the illuminating system.

segments of downward moving loops, were shown to be the cause of these bright spots. In Figure 6.4a, similar spots moved downward during the longer exposure and created the lines that are prominent.

At 30 frames per second, for any particular frame, the preceding and the following frames showed loops and spirals in completely different positions. After the illumination was broadened and made brighter, and a high frame rate electronic camera was used, it was obvious that the envelope cone was occupied by one long, looping, spiraling, continuous, and gradually thinner jet as shown in Figure 6.4d.

Although the elongation of the jet during the bending instability is considered as the major cause for the decrease in the diameter of the jet in the experiments described in this paper, splitting and splaying of the jet are seen in some experiments with other materials. These processes provide viable alternative mechanisms for the production of nanofibers.

6.2.3 Diameter of Nanofibers

Studies of the variation of nanofiber diameter as process parameters were changed, showed that when other parameters were held constant, the diameter of the polyethylene oxide nanofibers did not change much with electric field. The same result was found when the electric field was changed by either changing the voltage or by changing the distance between the pendent drop and the collector.

The diameter of the nanofibers did not change much with changes of the solution viscosity. When sodium chloride was added to the solution, the nanofibers were much thinner (Figure 6.5a). Substitution of ethanol for part of the water made the diameter of the nanofibers much larger (Figure 6.5b).

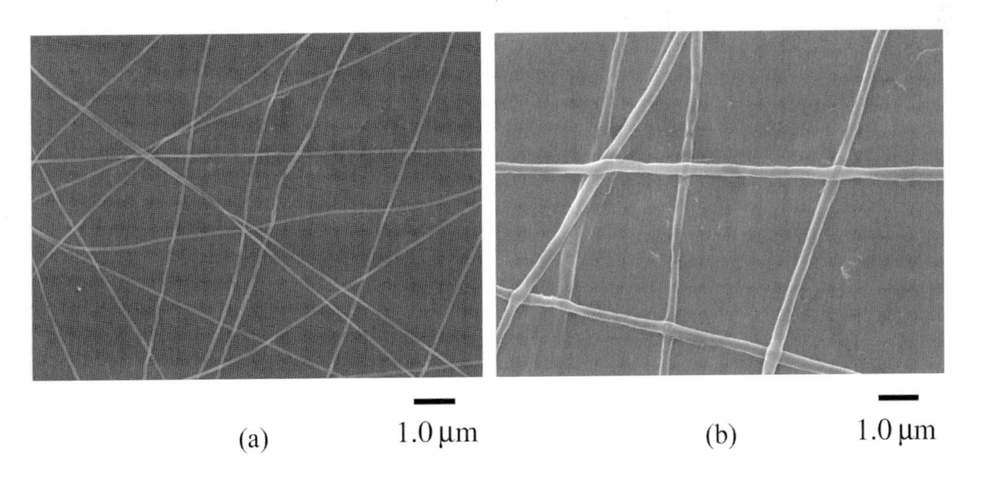

Figure 6.5 Variation of fiber diameter. (a) PEO, 3.0% in water, with NaCl. The ratio of NaCl to PEO was 0.5. (b) PEO, 3.0%, in water/ethanol mixture. The mass ratio of ethanol to water was 40% to 60%.

6.2.4 Observations of Electrospinning of Polyethylene oxide Solutions: Length of the Straight Segment, Flow Rate of the Solution, Current and Voltage

The electrical current carried by the jet increased with the square of the voltage, according to the empirical relationship,

$$I = 0.0144\ V^2 - 0.181V + 0.552$$

as shown in Figure 6.6a. The current was proportional both to the mass flow rate of the solution and to the excess charge density carried by the jet, each of which was found to be proportional to voltage, as shown in Figure 6.6b and 6.6c.

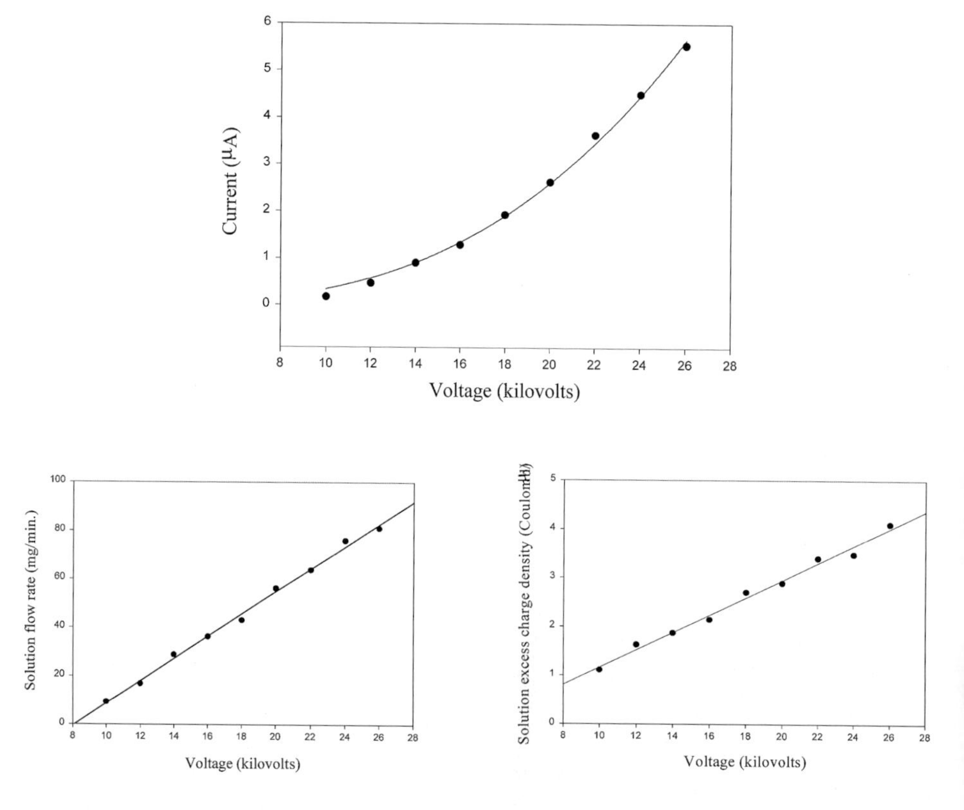

Figure 6.6 Relationship between jet current, voltage, solution flow rate, and excess charge density observed for a solution of 2.44% PEO in water. The distance from the droplet to the collector was 21 cm.

The length of the straight segment before the onset of the bending increased with the electric field and solution viscosity, but decreased as the coefficient of surface tension increased. Higher excess charge density was correlated with a shorter straight segment (Figure 6.7).

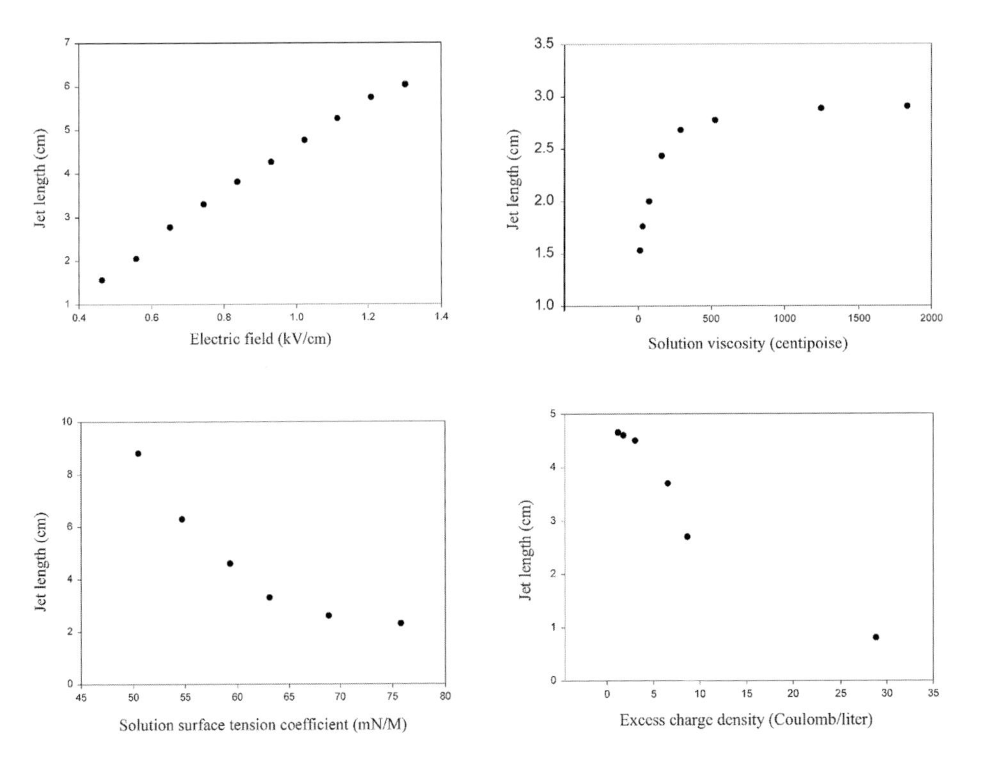

Figure 6.7 Dependence of the length of the straight segment on electric field, solution viscosity, coefficient of surface tension, and excess charge density.

The flow rate of the solution was calculated from the mass of the nanofibers collected, the concentration of the solution and the collecting time. A typical jet had a flow rate in the range from a few milligrams per minute to hundreds of milligrams per minute. As shown in Figure 6.8, the flow rate of the solution increased with the electric field, and decreased with the viscosity of the solution. Using the same concentration of PEO, but changing the solvent from water to a mixture of water and ethanol, caused the flow rate of the solution to increase. The addition of NaCl to an aqueous solution of PEO decreased the observed flow rate.

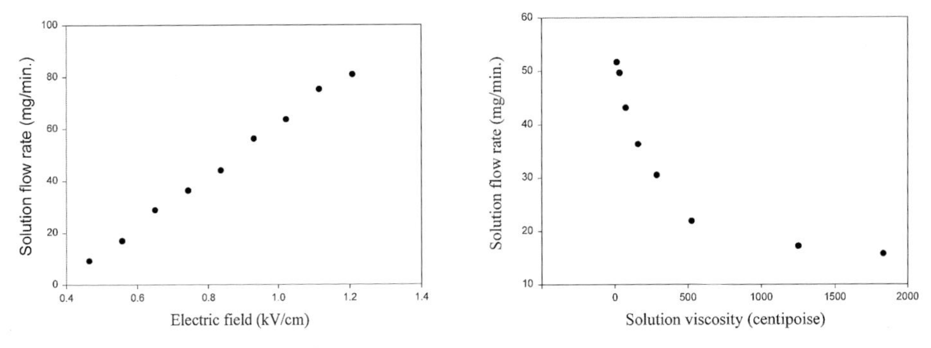

Figure 6.8 Variation of solution flow rate.

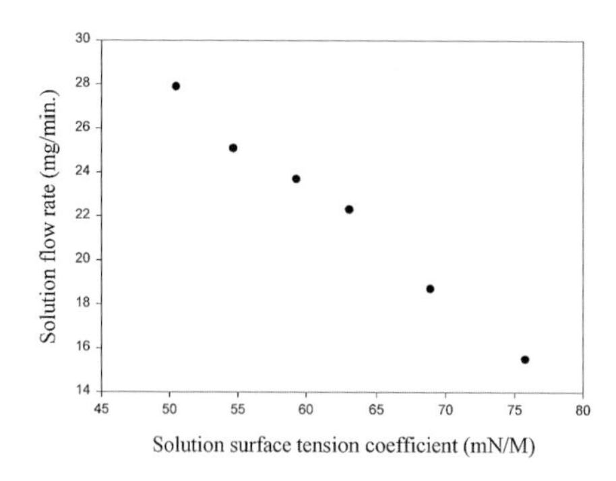

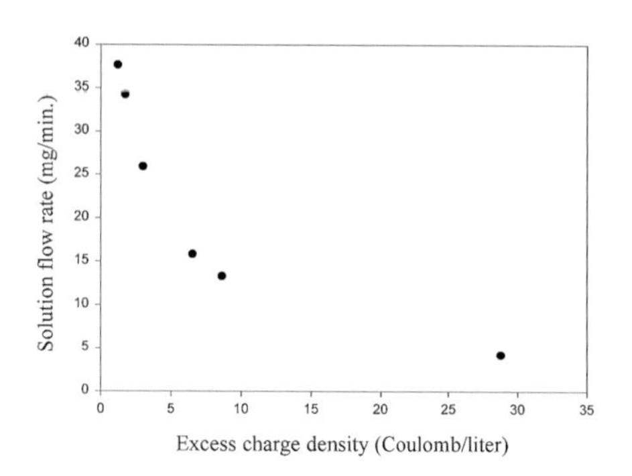

6.3 Nanofibers and Their Unique Properties

6.3.1 Beaded Nanofibers

Electrospun fibers sometimes have beads of polymer that formed during the process. An electrically driven jet of a low molecular weight liquid would form droplets, as occurs in electrospraying. The formation of such droplets is due to the capillary breakup of the jet by surface tension [40]. For polymer solutions, the pattern of the capillary breakup was changed radically. Instead of breaking completely in response to the capillary instability, the jets between the droplets formed nanofibers, and the contraction of the radius of the jet, which was driven by surface tension, caused the remaining solution to form beads.

As the viscosity of the solution was increased, the beads were larger, the average distances between beads was longer, the fiber diameter was larger and the shape of the beads changed from spherical to spindle-like. As the excess charge density increased, the beads became smaller and more spindle-like, while the diameter of fibers became smaller. Decreasing the surface tension by the addition of alcohol to the solution reduced the size and number of beads. Neutralization of the charge carried by the jet by ions from a corona discharge in air resulted in the formation of large numbers of beads. After the first bending instability, the tension along the axis of the fiber, which resisted the capillary instability, depended on the self-repulsion of the excess charge carried on the jet.

Jaeger et al [41] observed the surface of beads on polyethylene oxide nanofibers by atomic force microscopy. At the molecular level, the beads possessed a highly ordered surface.

Sometimes, the fibers between the beads were very thin and the beads were very small. A transmission electron micrograph of a beaded nanofiber fiber produced during the electrospinning of styrene-butadiene-styrene triblock copolymer (SBS) showed that the fibers between the beads had diameters as small as 3 nm. Since the diameter of a polyethylene oxide molecule or a polystyrene molecule is around 0.4 nm, there were fewer than 50 molecular chains in a typical cross-section of such a thin fiber.

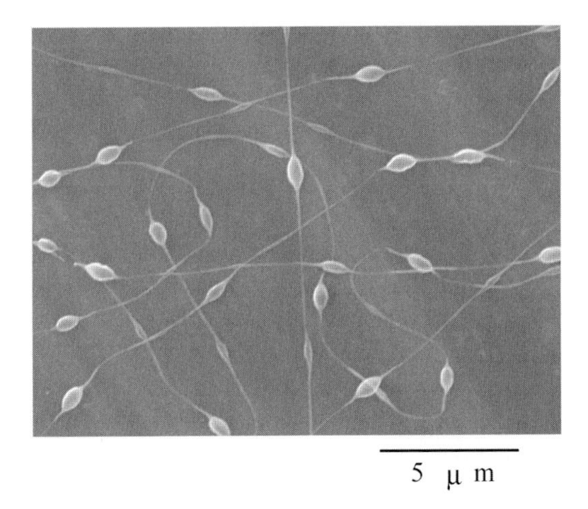

Figure 6.9 The morphology of beaded polyethylene oxide nanofibers.

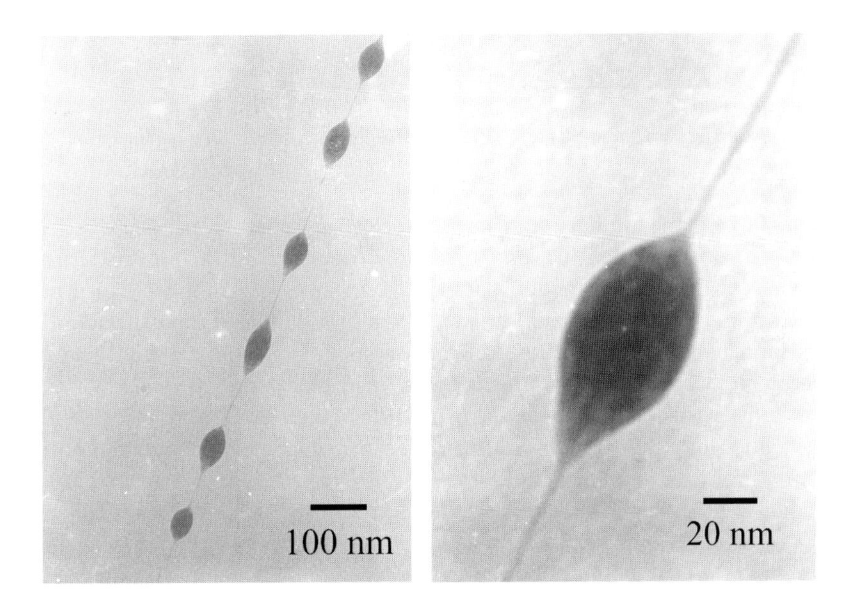

Figure 6.10 Beaded nanofiber of SBS copolymer, stained with osmium tetroxide.

6.3.2 Electrospun Poly (*p*-phenylene terephthalamide)

Fibers of poly (*p*-phenylene terephthalamide) (PPTA), commercialized by the Dupont company under the trade name Kevlar®, are well know for their excellent mechanical and thermal properties. Conventional highly oriented fibers are obtained when the liquid crystalline solutions are processed by a dry-jet-wet-spinning technique. PPTA nanofibers, with diameters around 100 nm, were electrospun from solution in sulfuric acid. The electrospun nanofibers are compared with a commercial fiber with a diameter of around 13 μm in Figure 6.11.

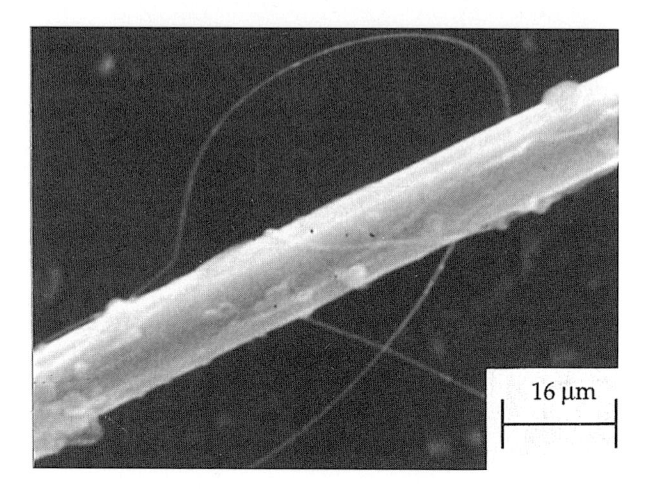

Figure 6.11 Scanning electron micrograph of an electrospun nanofiber with a commercial PPTA fiber.

Electrons from an ordinary transmission electron microscope penetrate polymer nanofibers, so it is straightforward to obtain electron diffraction patterns of the nanofibers. The electron diffraction patterns of an as-spun nanofiber and a nanofiber fiber annealed at 400 °C for 15 minutes are shown in Figure 6.12. The diffraction pattern of the as-spun nanofiber has diffuse diffraction spots along the equator. The annealed nanofiber has sharper spots on the equator, and sharp arcs on the meridian. The diffraction patterns show that the electrospun PPTA molecules were oriented along the fiber axis, and that the local crystallographic order of the as spun nanofibers was significantly improved by annealing.

The equatorial reflections correspond to spacings of 0.43 nm and 0.38 nm. The meridional arcs correspond to spacings of 0.21, 0.32 and 0.63 nm. These spacings are consistent with the crystal unit cell proposed by Northolt [42] and by Hasegawa et al. [43]. The equatorial spacings were indexed as 110 and 200, in order proceeding radially from the center of the diffraction pattern. The meridional spacings were indexed as 002, 004 and 006 planes.

An interesting feature of the crystallization behavior of the polymers in nanofibers is related to the confined morphology. In a nanofiber with a diameter of 100 nm or less, the crystals can grow to long dimensions only along the fiber axis. Since polymer crystals often grow to much longer dimensions from solutions or polymer melts, the small diameter of nanofibers imposes a size constraint on the crystallization process.

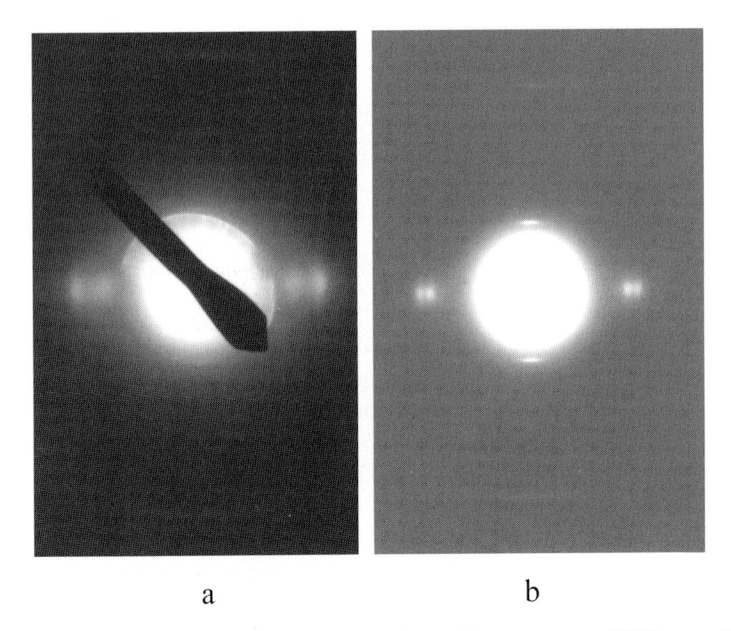

Figure 6.12 Electron diffraction pattern obtained from (a): an as-spun PPTA nanofiber, (b): an electrospun PPTA nanofiber annealed at 400 °C for 15 min.

6.3.3 Composites with Nanofiber Reinforcement

Polybenzimidazole (PBI) is a temperature resistant, linear polymer with a glass transition temperature of about 430 °C that is used in thermal protective, fire blocking and other aggressive temperature and environmental applications. Electrospun polybenzimidazole (PBI) nanofibers with diameters around 300 nm [31,32] were used as reinforcing fibers in polymer matrix composites.

After being reinforced with a non-woven fabric of PBI nanofibers, the epoxy matrix composites had higher tensile modulus (E), fracture toughness (K_{Ic}), and fracture energy (G_{Ic}). The higher the content of nanofibers, the higher were the values of these parameters. The fracture toughness and the fracture energy of the composites depended on the relative direction of the majority of the nanofibers in the fabric (the winding direction) and the crack. If the crack was transverse to the winding direction of the fabric in the composites, fracture toughness and the fracture energy were found to be higher than for those samples in which the crack is along the winding direction.

The PBI nanofibers, at a concentration near 10% by weight, provided dramatic reinforcement to a rubber matrix. Young's modulus of nanofiber reinforced SBR composite was ten times as large as for SBR containing an equal concentration of carbon black, and the tear strength was twice as large. These observations suggested that a combination of nanofibers and conventional rubber reinforcing particles could lead to rubbers with an improved balance of modulus and elongation to break.

These initial positive results indicate that further investigation of nanofiber composites is warranted. Both high strength graphite nanofibers and high toughness polymer nanofibers are expected to be useful as minor constituents of hybrid composites made with currently available carbon and glass fibers.

6.3.4 Elastomeric Poly(styrene-butadiene-styrene) Nanofibers, Phase Miscibility

Poly(styrene-butadiene-styrene) (SBS) nanofibers with diameters around 100 nm were made by electrospinning. Figure 6.13 shows the appearance of the nanofibers as they were observed between crossed polarizers, and then rotated. In the left image, fibers A-A and B-B were bright. Turning the sample clockwise by about 45° caused these nanofibers to become dark, as shown in the right image. The bright and dark segments in the looped nanofiber moved along the loops and remained approximately parallel to each other as the sample was rotated. These observations show that the electrospun nanofibers were birefringent. Both the elongational flow and the rubber-like strain of the molecular network contribute to the birefringence.

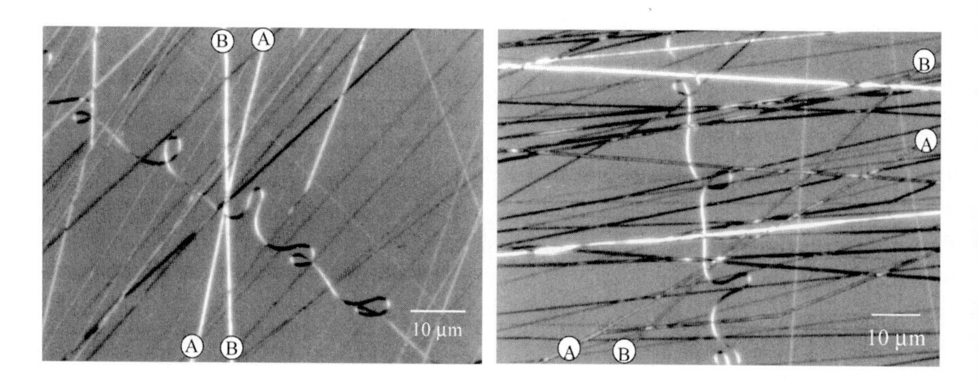

Figure 6.13 Birefringence of electrospun SBS fibers.

Figure 6.14 shows bright field transmission electron micrographs of the electrospun SBS nanofibers. The nanofibers on the evaporated graphite were stained with osmium tetroxide, which is known to preferentially stain the polybutadiene phase. The dark regions in the electron microphotographs were identified as polybutadiene domains. The polystyrene domains were lighter. After staining, the fibers were coated with a very thin layer of evaporated graphite to carry away electrical charge during examination in the transmission electron microscope.

In Figure 6.14, the left image shows an as-spun SBS nanofiber. The phase-separated domains were small and irregular. After the as-spun nanofiber was annealed at 70 °C for 30 minutes the right image showed that the transverse width of the polystyrene domains

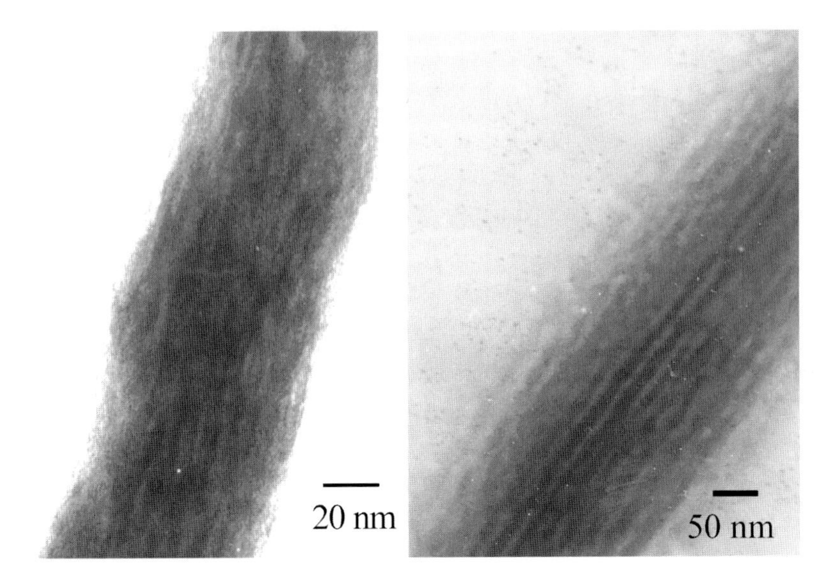

Figure 6.14 Transmission electron micrographs of electrospun SBS nanofibers.

increased from less than 10 nm in the as-spun fiber to over 20 nm, and a more ordered structure developed.

Figure 6.14 also shows that the single-phase domains were elongated along the fiber axis. The edges of the as-spun nanofibers were well defined. The more transparent edges of the nanofibers annealed at 70 °C indicate the nanofibers spread and flattened on the substrate during annealing. The attraction of the electrical charges carried by the fibers, to the electrically conducting layer of graphite is estimated to generate an electrical pressure of about one atmosphere, which acted to spread and flatten the nanofibers during annealing.

6.3.5 Carbon Nanofibers

The carbon fibers now widely used, with diameters around 7 μm, are usually made from polyacrylonitrile (PAN) or mesophase pitch (see Chapter 9). Melt spinning of PAN is impractical since polyacrylonitrile degrades below its melting temperature, so solution spinning is used. The as-spun PAN fibers are heated in air to 200 to 400 °C, under tension, to stabilize them for treatment at higher temperature. Stabilized fibers are converted to carbon fibers by heating to 1700 °C under inert gas. In this carbonization process, all chemical groups such as HCN, NH_3, CO_2, N_2, and hydrocarbons are removed. After carbonization, the fibers are heated to 2000 to 3000 °C under tension. This process, called graphitization, makes carbon fibers in which the c-axes of graphite crystallites are perpendicular to the fiber axis.

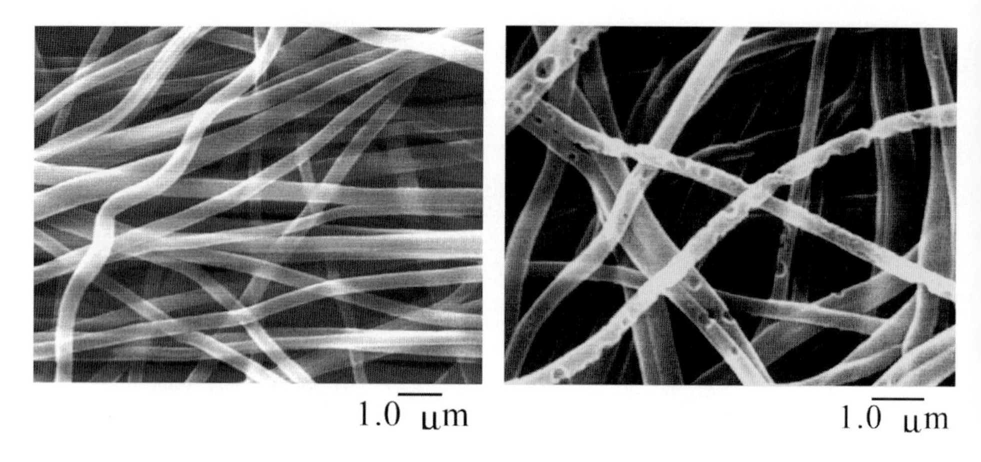

Figure 6.15 Carbon nanofibers from PAN before and after heating in water vapor.

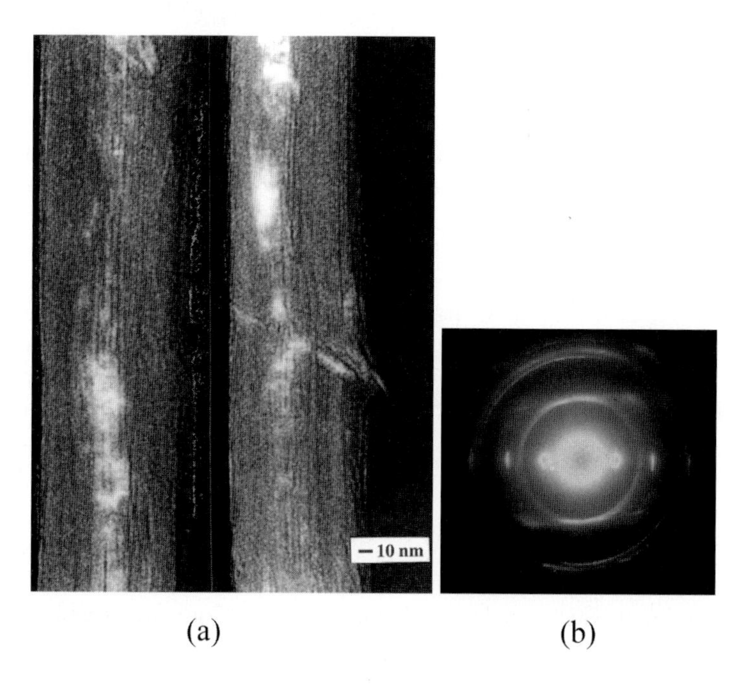

Figure 6.16 Bright field images of two segments of a carbonized mesophase pitch nanofiber and the electron diffraction pattern from one of the segments.

Stabilized carbon nanofibers, made from electrospun nanofibers of PAN by a simpler process, are shown in Figure 6.15a. Some of the carbon nanofibers were heated to a high temperature in water vapor (25) since this is known to increase the chemically accessible surface area of carbon. Pores that were 0.05 to 0.5 micron in diameter are shown in Figure

6.15b. Pores of this size would not have a large effect on the amount of chemically accessible surface area, but their presence indicated that the water vapor did react with the nanofibers.

Highly oriented carbon nanofibers were also made from mesophase pitch. Figure 6.16 shows two segments of the same carbonized pitch nanofiber. The spacing between the lines on the surface of the nanofiber in Figure 6.16a is somewhat greater than the spacing between the graphite planes. The diffraction pattern from this nanofiber is shown in Figure 6.16b. The diffraction pattern shows that both the lamellar structure and the graphite planes were oriented so that the axis of the fiber lies in these planes. The spacing

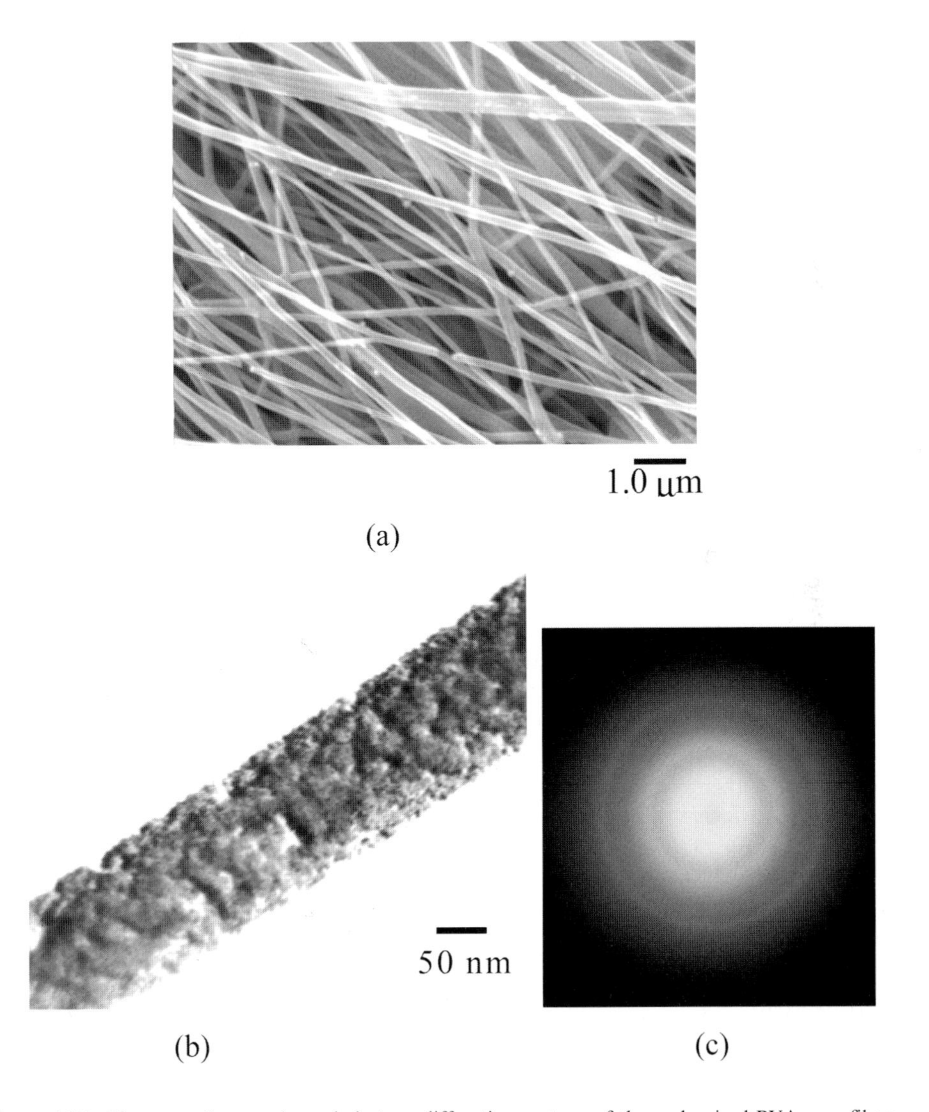

Figure 6.17 Electron micrographs and electron diffraction pattern of the carbonized PVA nanofibers.

of the (002) graphitic planes in the carbonized mesophase pitch nanofibers is the same as that observed in other carbon fibers. The height (along the arc) of the equatorial spots shows that the graphitic planes have waviness about the same as that of the lines on the surface of the nanofiber in Figure 6.16a.

Polyvinyl alcohol (PVA) is not widely used as a precursor for carbon fibers because PVA is not easy to stabilize. Recently, it was found that the presence of $(NH_4)_2HPO_4$ helped the stabilization process [44] and produced a porous form of carbon. Carbon nanofibers that were made from electrospun PVA nanofibers are shown in Figure 6.17a.

The nanofibers from PVA were strong enough to withstand the handling required to make samples for examination in the transmission electron microscope, but they broke and crumbled easily during ordinary handling. Figure 6.17b shows a magnified view of one nanofiber. Grains and pores with linear dimensions ranging around ten nanometers were present throughout the nanofiber. Figure 6.17c shows the electron diffraction pattern from a typical nanofiber derived from PVA. The electron diffraction patterns of these PVA carbon nanofibers indicated that the nanofibers contained small, imperfect and randomly oriented graphite crystallites. [36]. The nanoporous structure is consistent with a high specific surface area.

6.3.6 Nanofibers for Biomedical, Filtration, Agricultural, and Outer-Space Applications

Nanofibers made from many new synthetic polymers, and biologically derived polymers are being considered for use in tissue engineering, artificial organ applications, drug delivery, and for wound dressings. Nanofibers of DNA were made by Fang and Reneker [30]. The growth of cells on nanofibers was reported by Ko and Reneker [45].

Many kinds of bio-generated and bio-compatible materials are of interest. Conventional methods for making fibers require so much material that they are often impractical. Electrospinning provides a convenient way to make nanofibers of such materials. The amount of polymer required to produce useful quantities of nanofibers can be as little as a few hundred milligrams.

Nanofibers are finding use in filters to remove particles and droplets smaller than 100 nm from liquids or gases. They are also being considered for the absorption of noxious molecules, since their specific surface area is so large, and their surface chemistry can be tailored to be selective to many kinds of substances.

The application of pesticides to plants is another area where nanofibers may find large-scale applications. Nanofibers, spun in the field, and directed onto plants by a combination of electrical forces and air streams, will attach to plants with nearly 100% efficiency. This contrasts with the 3 to 5% sticking efficiency of conventional application methods for applying pesticides as dusts or sprays. The use of nanofibers to carry and attach pesticides could make the use of sophisticated but expensive pesticides cost effective. The burden placed on the environment by wasted pesticides would also be reduced.

Electrospinning of nanofibers from a polymer melt works better in a vacuum than in air because the dielectric breakdown strength of air is much less than the electric fields that can be applied to a polymer in a high vacuum. It is feasible to electrospin polymer nanofibers in space to create thin sheets with large areas, and as tensile members of other large scale structures. Solar sails to transport cargo between Earth and Mars need to be less than one micron thick, with an area of about five square kilometers. Electrospun polymer nanofibers could create such a structure, along with the shroud lines needed to attach the cargo. Proposals to explore interstellar space utilize light sails, similar to solar sails, driven by high power lasers, to carry instruments outside the solar system. The use of large mirrors to direct and focus sunlight, or microwaves generated by solar collectors in space, onto the earth, has been proposed.

Nanofibers, perhaps at the scale of single polymer molecules, can be expected to play a role in micro-electro-mechanical devices (MEMS). The possibility of making ceramic materials by chemical routes that use linear polymers as intermediates provides suggestions for ways to make ceramic nanofibers.

Commercialization of synthetic fibers with diameters in the range of nanometers has not been significantly developed, even though micron scale fibers are the basis of large industries. However, the future for nanofibers now looks brighter.

6.4 Acknowledgements

Support was provided by the National Science Foundation in Grants DMI-9813098 and CTS-9900949, by the U. S. Army Research Office, by the U. S. Army Soldier and Biological Systems Command, and by the Nonmetallic Materials Division of the U. S. Air Force Research Laboratory, Wright Patterson Air Force Base. The high-speed camera was acquired with support from the Hayes Investment Fund of the Ohio Board of Regents. A gift from the DuPont Company provided flexibility that would not have been otherwise available.

6.5 References

1. Rayleigh, F. R. S., *Phil., Mag.* (1882) 44, p. 184
2. Zeleny, J., *Physical Review*, (1914) 3, p. 69
3. Zeleny, J., *Proceedings of the Cambridge Philosophical Society* (1915) 18, p. 71
4. Zeleny, J., *Physical Review* (1917) 10, p. 1.
5. Zeleny, J., *Journal of the Franklin Institute* (1935) 219, p. 659
6. Vonnegut, B., and Neubauer, R. L., *Journal of Colloid Science* (1952) 7, p. 616
7. Watchel, R. E., and LaMer, V. K., *Journal of Colloid Science* (1962) 17, p. 531
8. Taylor, G. I., *Proceedings of the Royal Society of London* (1964) A280, p. 383
9. Taylor, G. I., *Journal of Fluid Mechanics* (1965) 22, p. 1
10. Taylor, G. I., *Proceedings of the Royal Society of London* (1966) A291, p. 145

11. Taylor, G. I., *Proceedings of the Royal Society of London* (1969) A313, p. 453
12. Formhals, A., U.S. Pat., 1,975,504 (1934)
13. Formhals, A., U.S. Pat., 2,077,373 (1937).
14. Formhals, A., U.S. Pat., 2,158,416 (1939).
15. Formhals, A., U.S. Pat., 2,160,962 (1939).
16. Formhals, A., U.S. Pat., 2,187,306 (1940).
17. Formhals, A., U.S. Pat., 2,323,025 (1940).
18. Formhals, A., U.S. Pat., 2,349,950 (1944).
19. Gladding, E. K., U.S. Pat., 2,168,027 (1939).
20. Simons, H. L., U.S. Pat., 3,280,229 (1966).
21. Bornat, A., U.S. Pat., 4,323,525 (1982).
22. Bornat, A., U.S. Pat., 4,689,186 (1987).
23. Baumgarten, P. K., *Journal of Colloid and Interface Science* (1971) 36, p. 71
24. Larrondo, L., and Manley, R. ST. J ., *Journal of Polymer Science, Polymer Physics Edition* (1981) 19, p. 909
25. Larrondo, L., and Manley, R. ST. J., *Journal of Polymer Science, Polymer Physics Edition* (1981) 19, p. 921
26. Larrondo, L., and Manley, R. ST. J., *Journal of Polymer Science, Polymer Physics Edition* (1981) 19, p. 933,
27. Doshi, J., and Reneker, D. H. (1995) *Journal of Electrostatics*, 35, pp. 151–160
28. Srinivasan, G., and Reneker, D. H ., *Polymer International* (1995) 36 pp. 195–201
29. Reneker, D. H., and Chun, I., *Nanotechnology* (1996) 7, pp. 216–223
30. Fang, X., and Reneker, D. H., (1997) Vol. B, 36, pp. 169–173
31. Kim, J., and Reneker, D. H. (1999) *Polym. Eng. Sci.*, 39 (5) pp. 849–854,
32. Kim, J., and Reneker, D. H., S. Korea *Polym. Compos.* (1999) 20 (1), pp. 124–131
33. Fong, H., Chun, I., and Reneker, D. H., *Polymer* (1999) 40, pp. 4585–4592
34. Fong, H., and Reneker, D. H., *Journal of Polymer Science, Part B: Polymer Physics*, (1999) 37, pp. 3488–3493,
35. Chun, I., Reneker, D. H., Fong, H., Fang, X., Deitzel, J., Tan N. B., Kearns, K., (1999) *Journal of Advanced Materials*, Vol. 31, No. 1, pp. 36–41
36. Fong, H., Chun, I, and Reneker, D. H., the *24th Biennial Conference on Carbon*, 380, (1999).
37. Reneker, D. H., Yarin, A. L., Fong, H., and Koomhongse, S. *J. Applied Physics* (2000) 87, pp. 4531–4547
38. de Gennes, P. G., *Journal of Chemical Physics* (1974) 60, p. 5030
39. Smith, D. E., and Chu, S. *Science* (1998) 281, p. 1335
40. Yarin, A. L., *Free Liquid Jets and Films: Hydrodynamics and Rheology* (1993) Longman, Harlow and Wiley, New York,
41. Jaeger, R., Schonherr, H., and Vancso, G. J., *Macromolecules* (1996) 29, pp. 7634-7636
42. Northolt, M. G., *Eur. Polym. J.* (1974) 10, p. 799
43. Hasegawa, R. K., and Chatani, Y., *Abstr. Meet. Soc. Cryst. Japan*, Osaka, Japan, (1973)
44. Tolkachev, A. V., Druzhinina, T. V., and Mosina, N. Y., *Fiber Chemistry* (1997) 29, No. 2,
45. Ko, F. K, Laurencin, C. T., Borden, M. D., and Reneker, D. H., 24th *Annual Meeting of the Society for Biomaterials*, San Diego, CA., April 22–26, 1998.

7 Fibers from Liquid Crystalline Polymers

Stephen Z. D. Cheng, Fuming Li, Christopher Y. Li, Kevin W. McCreight, Yeocheol Yoon and Frank W. Harris
University of Akron, Akron, Ohio, USA

7.1 Introduction: Liquid Crystalline Phases

The first example of a synthetic liquid crystal was *p*-azoxyanisole which was reported in 1889 by Gattermann and Ritschkle and investigated by Lehmann [1]. Since the discovery of the liquid crystalline phase, many liquid crystal molecules have been synthesized. Classification of the distinct liquid crystalline phases in small-molecule liquid crystals has been well-established [2–5]. As shown in Figure 7.1, the least ordered liquid crystalline phase is the nematic (N) phase which possesses molecular orientational order due to the anisotropy of the molecule's geometry. When a layer structure is introduced as another ordering level in

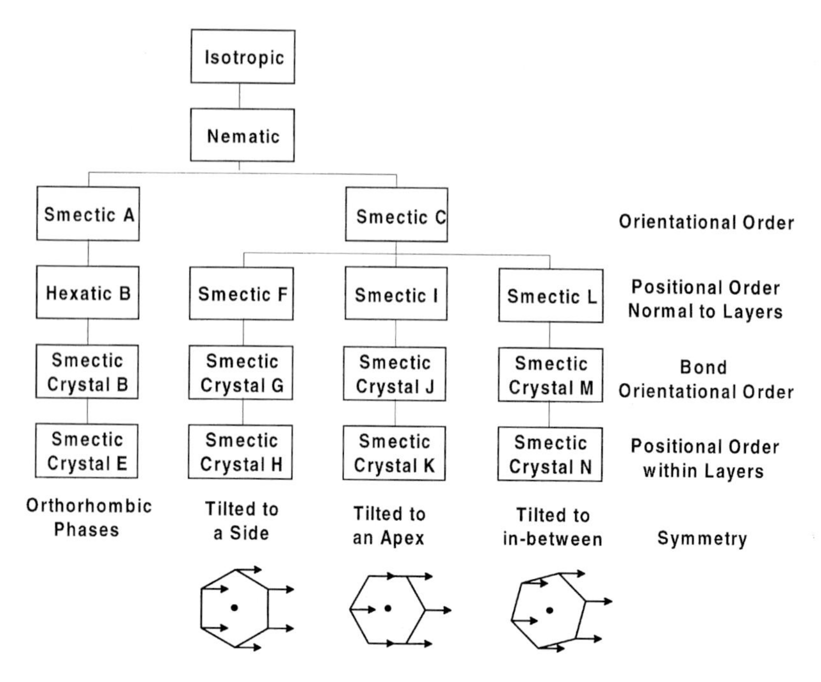

Figure 7.1 Classification of liquid crystalline phases.

addition to the molecular orientation, a smectic A (S_A) or a smectic C (S_C) phase may be formed. Three more ordered phases beyond the S_A phase may be observed. Listed in terms of increasing order, they are the hexatic B (H_B), smectic crystal B (S_B) and smectic crystal E (S_E) phases. In this series, the long axis of the molecules are oriented perpendicular to the layer surface while the increasing order develops from positional order normal to the layer in S_A, bond orientational order in H_B, positional order within the layers in S_B and, finally, asymmetric axial site symmetry in S_E [2–5]. Two separate series of highly ordered smectic phases are recognized after the S_C phase. Both of these series of phases possess long molecular axes that are tilted with respect to the layer surface normal. The development of order in both of these cases is identical to the first series which encompassed the phases from S_A to S_E. The difference between these two series lies in the tilt directions: the smectic F (S_F), smectic crystal G (S_G), and smectic crystal H (S_H) series possess a tilt in the long axis towards one side while the long axis directions in the smectic I (S_I), smectic crystal J (S_J) and smectic crystal K (S_K) series are tilted towards one apex (Figure 7.1). Furthermore, the S_F, S_G, S_I and S_J phases exhibit hexagonal (or pseudo-hexagonal) packing when viewed parallel to the long axis. The packing of the tilted long axes gives rise to a monoclinic lattice (for S_F and S_G phases $a>b$ while in S_I and S_J phases $a<b$). In contrast, the packing becomes an orthorhombic, herringbone type in the S_H and S_K phases [2–5]. Very recently, it has also been suggested that there is another series of tilted smectic phases in which the tilting direction is in between the above two series of phases (in between one side and one apex). De Gennes has proposed a broad concept in which he describes this state of matter as materials which have soft order [6].

Although the existence of liquid crystal polymers was postulated by Vorlander, the thermotropic poly(*p*-hydroxybenzoic acid) system he mentioned was infusible. The lyotropic system of poly(γ-benzyl-L-glutamate) (PBLG) in chloroform was, therefore, the first polymer to exhibit true liquid crystallinity. Great interest quickly developed in main-chain liquid crystal polymers from the point of view of both applications and academics. Fibers with greatly enhanced orientation, and consequently, strength and modulus could be spun from liquid crystalline solutions and melts [7].

The discovery of high molecular weight thermotropic and lyotropic liquid crystalline polymers led to developments in the design of molecular structure and architecture, phase behavior characterization as well as novel spinning technologies. The processing advantage is that in the liquid crystalline state, the rheological behavior of the system is very much anisotropic under an external shear force field. Along the direction of the force field, the solution and melt viscosity are substantially lower than in the other two directions. However, only a fraction of main-chain liquid crystalline polymers can form fibers with useful properties due to other requirements such as high thermal and thermo-oxidative stability. Therefore, liquid crystalline polymer fibers mainly consist of aromatic chemical structures in order to retain their linearity and rigidity. Although main-chain liquid crystalline polymers with flexible spacers have also been studied, it is evident that the mechanical properties and thermal stability are drastically reduced after the number of flexible units in the main-chain liquid crystalline polymers exceeds two.

All of these efforts have led to the commercialization of high tensile modulus, high strength and high temperature aromatic polymer fibers. These high performance products, in turn, stimulate research toward the understanding of phase transitions, structural formation during processing, and structure-property relationships of these liquid crystalline polymer fibers. The applications of these fibers cover almost every aspect of materials engineering from composites for structural materials in military and aerospace applications to tire cords and automobile parts. A substantial number of reviews, articles and patents for liquid crystalline polymer fibers have appeared during the past half century [8–12]. This chapter only covers a small portion of the developments and achievements which have been accomplished.

7.2 Rheology in Liquid Crystalline Polymers

Fibers made from liquid crystalline polymers are particularly attractive for two reasons. First, the fibers usually show excellent mechanical properties and thermal and thermo-oxidative stability. Second, the fibers can be spun in the low ordered liquid crystalline states where the substantially reduced viscosity leads to better processibility during fiber spinning. We focus on the anisotropic rheological properties of liquid crystalline polymers in this section, which are important features in the understanding and application of the unique molecular and supra-molecular structures of these materials. It is well-known that liquid crystalline polymers exhibit low viscosities compared with isotropic fluids. This has been explained as a result of very efficient molecular orientation in extensional fields, in particular, at high strain rates. However, the rheology of liquid crystalline polymers is complicated and a single viscosity coefficient is not sufficient to describe the flow behavior. As a result, several theories and molecular models have been proposed in order to explain their complex rheological behavior [2,13–15].

Phenomenologically, as first introduced by Asada et al. [16,17], the shear rate dependent viscosity can be separated into three regions (Figure 7.2). In region I, at low shear rates, the viscosity decreases with increasing shear rate. In region II, the viscosity plateaus, and in region III, the fluids become shear-thinning again. This is different from conventional polymers which do not have region I, i.e., their viscosities always approach the Newtonian viscosity as the shear rate goes to zero.

In lyotropic liquid crystalline polymers, the unusual concentration dependence of viscosity is well-known. As the concentration increases, the viscosity first increases. After the viscosity reaches a critical value, the viscosity begins to decrease. This can be found in the case of PBLG in *m*-cresol [18], a 50:50 copolymer of *n* -hexyl and *n* -propylisocyanate in toluene [19], poly(*p*-benzamide) (PBA) in hydrofluoric acid (HF) [20] and poly(*p*-phenylene terephthalamide) (PPTA) in sulfuric acid (H_2SO_4) (Figure 7.3) [21]. The viscosity-concentration relationships at different temperatures were also investigated [22–24]. Poly(*p*-phenylene benzobisoxazole) (PBZO) in H_2SO_4 has also been reported to show a similar concentration dependence on viscosity (Figure 7.4) [25]. This has been explained by the molecular development of liquid crystalline order. Below the critical concentration, the molecules are isotropic in solution. As soon as this concentration is exceeded, anisotropic lyotropic liquid crystalline behavior appears, resulting in a decrease in the solution viscosity. This explanation was supported by structural experimental observations such as polarized light microscopy (PLM). Thermodynamic theories and calculations of phase diagrams in lyotropic liquid crystalline polymers have also provided insight which illustrates the driving force for structural changes and order formation as well as macroscopic responses such as shear viscosity [26,27].

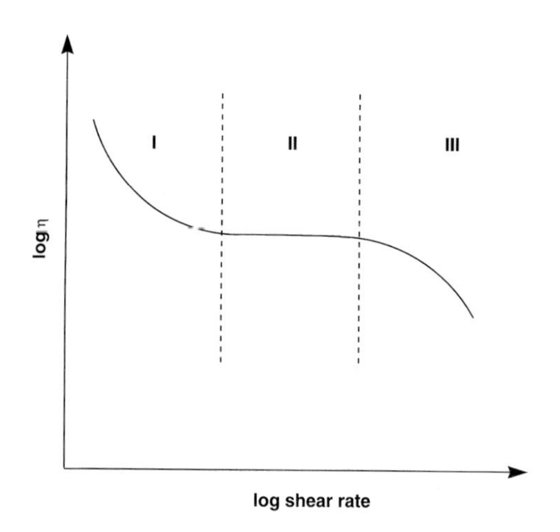

Figure 7.2 Three regions of rheological behavior in a plot of viscosity versus shear rate.

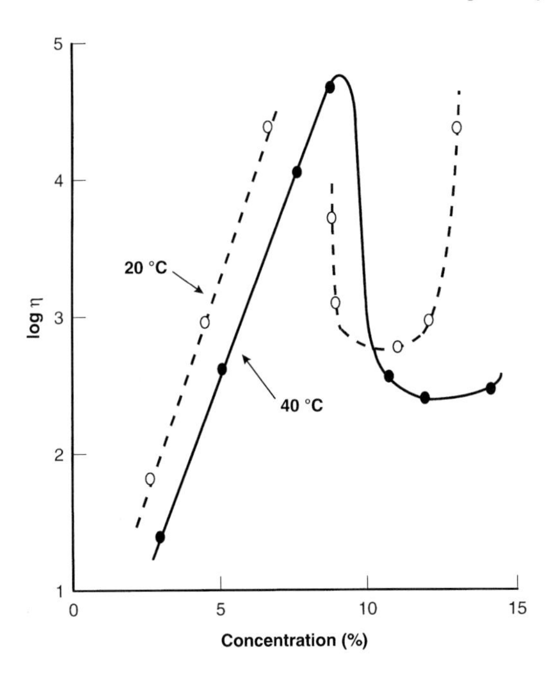

Figure 7.3 Relationship between viscosity and concentration for PPTA in H_2SO_4 [21].

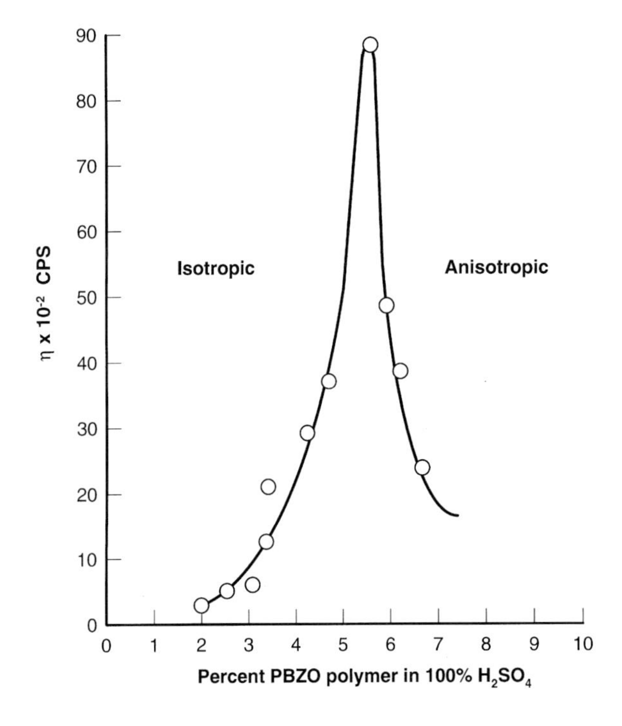

Figure 7.4 Relationship between viscosity and concentration for PBZO in H_2SO_4 [25].

In thermotropic liquid crystalline polymers, most of the rheological studies focus on the change in shear viscosity with composition, which determines the existence of low ordered liquid crystalline phases. When the liquid crystalline phase appears, the shear viscosity decreases. The steady-state shear viscosity also changes with temperature since at different temperatures, various phases appear. It is generally understood that in the low ordered liquid crystalline phases such as in N, S_A or S_C phases, the viscosity is usually lower than that of the isotropic melt. Research activity in this area began with Jackson and Kuhfuss [28] in 1976 on a liquid crystalline copolyester comprised of *p*-hydroxybenzoic acid (HBA) and poly(ethylene terephthalate) (PET). Since then, a number of research reports have been published [29–32]. When the composition of HBA exceeded 30%, a strong shear-thinning behavior was observed. Another series of copolyesters was developed by Celenese using a condensation of HBA and 6-hydroxy-2-naphthoic acid (HNA) [33]. The rheological behavior of this series was also extensively studied [34–36].

For thermotropic liquid crystalline polymers, a great deal of difficulty exists in resolving some conflicting rheological experimental observations due to the different thermal and mechanical histories involved in the sample preparations [29,37–41]. Han et al. recently reported their observations using a polyester with a relatively low isotropization temperature in order to overcome the thermal history difficulties [37–41]. One such example is poly[(phenylsulfonyl)-*p*-phenylene-1,10-decamethylene-bis(4-oxybenzoate)] [37]. Very recently, a series of polyesters was synthesized from 2-phenylsulfonyl-1,4-hydroquinone and 4,4′-dichloroformyl-1, *n*-diphenoxyalkane [41]. This series of liquid crystalline polymers exhibits a nematic phase before entering the isotropic melt. Figure 7.5 shows the reduced viscosity data *versus* reduced temperature for these polymers. Both reduced quantities are defined as the ratio of the measured values to the reference values. The reference values for the temperature scale are the isotropization temperatures for each polymer, while in viscosity, it is the maximum viscosity among these polymers. It is evident that the viscosities decrease when the temperature is in the N phase region. Before they approach the isotropization temperature at which the maxima of the viscosities can be found, the viscosities increase, and the system is assumed to be in a biphasic region. Above the isotropization temperature, the viscosities again decrease with increasing temperature. Mechanical history such as shear history also affects the rheological behavior [35,37]. It has been speculated that unless the unrelaxed stress is completely removed before the measurement starts, one may obtain erroneous values and signs on the first normal stress difference (N_1). Additional problems may also arise due to thermal degradation and post polymerization in these polyesters.

A striking observation in the flow of liquid crystalline polymers is the sign change of N_1, which was first reported for PBLG in *m*-cresol [42]. When the shear rate increases in the system in a concentration range between 14.0 and 22.1 weight percent, N_1 first increases and has a positive sign, and then decreases, giving rise to a negative value. Finally, it becomes positive again as the shear rate further increases. Recently, theoretical interpretations of the negative value of N_1 were suggested [43,44]. Marrucci and Maffettone found that over a range of shear rates, the sign of N_1 varied due to the tumbling of flow directors [45]. In the case of thermotropic liquid crystalline polymers, positive values of N_1 were observed in a copolyester of 73/27 HBA/HNA [35], while others reported negative values of N_1 [36]. In work reported by Cocchini et al. there was

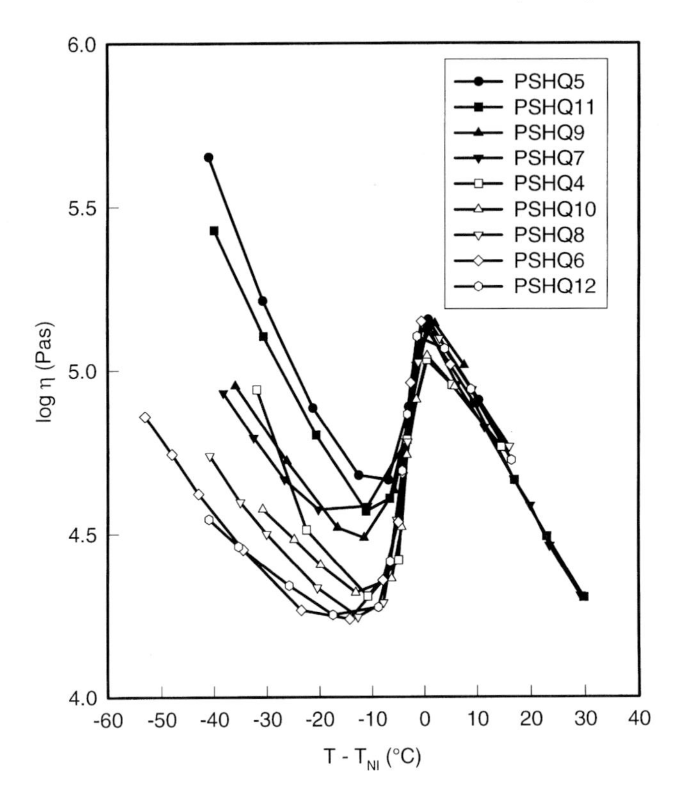

Figure 7.5 Relationship between viscosity and reduced temperature for a series of polyesters [41].

substantial normal stress introduced by the constricting flow during loading of the sample specimen. These unrelaxed normal stresses may affect the experimental measurements of N_1 in the samples. It is theoretically proposed that molecular weight has a greater influence on the rheological behavior of rigid-rod polymers than on the rheological properties of flexible polymers [45]. The zero-shear viscosity of flexible polymers and the first normal stress difference are proportional to the 3.4 and 6–7 powers of molecular weight, respectively [46,47]. For thermotropic liquid crystalline polymers, the zero-shear viscosity is proportional to the 6.5 power of molecular weight. The first normal stress difference does not have a significant relationship with molecular weight in the liquid crystalline state. In the isotropic state, the zero-shear viscosity and the first normal stress difference are proportional to the 6 and 6.7 powers of molecular weight, respectively [37].

7.3 Fibers from Liquid Crystalline Polymers

The strongest motivation for exploring liquid crystalline polymer fibers came from their applications in high-performance composite materials, specifically in the aerospace industry. The rapid development of this field generated a need for novel liquid crystalline fibers with improved properties and reduced weight. As shown in Figure 7.6, the superior tensile strength and modulus of these fibers, combined with their low density place them among the best materials when compared on the basis of specific strength and modulus. Technologists have long aspired to produce high-strength and high-modulus polymer fibers. The ideal structural model for these fibers was first suggested by Staudinger in the early 1930s. His model consisted entirely of extended chain molecules closely packed to each other [48].

The basic technology for forming synthetic fibers includes spinning and heat treatment using different drawing methods. To spin fibers from low ordered liquid crystalline states, it is important to distinguish lyotropic and thermotropic systems, although both of them have the advantage of relatively low viscosity, which improves processibility. Fibers spun from lyotropic liquid crystalline polymers usually employ a wet or dry-jet wet spinning method (sometimes, gel spinning for high molecular weight or rod-like polymers may also feasible), while those spun from thermotropic liquid crystalline polymers use a melt spinning method. The wet spinning process, so called because aqueous solutions are often used as coagulating agents, is employed mostly for lyotropic liquid crystalline polymers in nonvolatile solvents. The extrudate is immersed directly in a non-solvent bath, where fiber coagulation and solution extraction take place. A combination of wet and dry spinning methods, so called dry-jet wet spinning, is extremely useful in fiber formation from

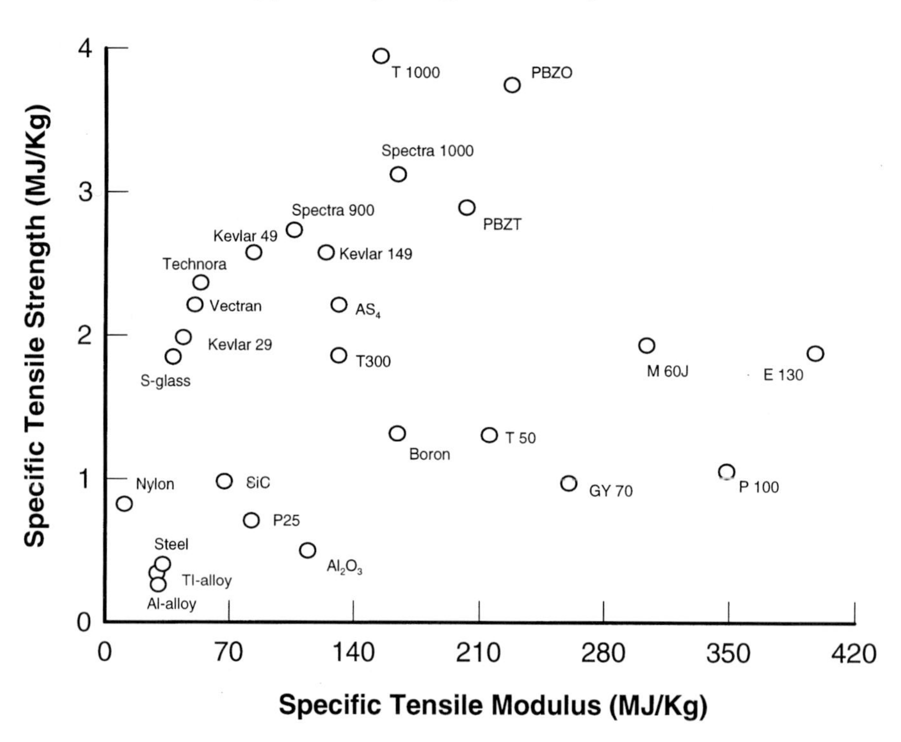

Figure 7.6 Relationships between specific modulus and strength of different polymer fibers compared with steel wires and inorganic fibers.

lyotropic liquid crystalline polymers. In this process, the polymer solution is passed through a short air gap prior to immersion in the coagulation bath (Figure 7.7). This prevents the solution from freezing inside the spinnerette. In fact, most high strength, high modulus synthetic fibers are spun using this method. To compare, the ordinary wet spinning process involves simultaneous solution spinning and coagulation, while in the dry jet-wet method, fiber spinning occurs in an air space and is followed by the coagulation process. As a result, the heat and mass transport phenomena which occur during fiber spinning in these two methods are quite different. It is generally understood that fibers produced by the dry-jet wet spinning method usually exhibit better mechanical properties. However, a quantitative analysis is not available at present.

Based on composition, fibers can also be spun from an isotropic solution with relatively dilute polymer concentrations. When the concentration increases, the solution enters the lyotropic liquid crystalline state. Further increasing the concentration may lead to gel formation, which requires gel spinning to produce fibers. It is evident that in the dry-jet wet spinning process, several factors are important in the determination of the ultimate fiber structure and properties. These include polymer composition, solution concentration, tractability during the fiber solidification and cost, in addition to fiber spinning speed and draw ratio.

For thermotropic liquid crystalline polymers, a melt spinning method is adopted to produce fibers. The anisotropic polymer melt is expelled from small capillary holes to form fibers which are attenuated by an external force, and then cooled into the solid state. The majority of the molecular alignment is developed by extensional flow during drawdown immediately after spinning from the low ordered liquid crystalline melt (N, S_A or S_C phases) [49]. At shear rates experienced during the fiber spinning, on the order of 10^3 to 10^4 s^{-1}, the viscosity of a thermotropic copolyester, for example, lies between 10^1 and 10^2 Nsm^{-2}. The marked shear thinning of the liquid crystalline melt can be seen as an

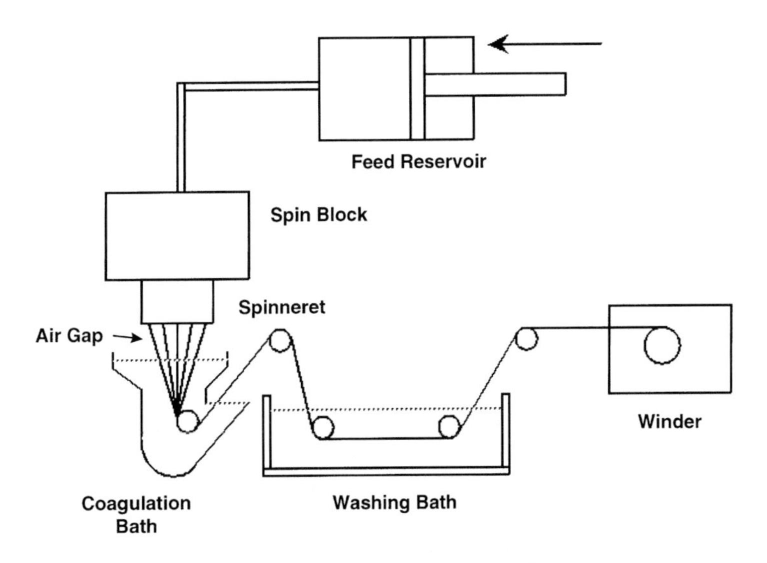

Figure 7.7 Schematic drawing of the dry-jet wet spinning process [8].

important contribution to the low viscosity at these high shear rates. Furthermore, the viscosity is also dependent upon molecular weight. Melt spinning is also possible for fibers spun from the isotropic melt at even higher temperatures. However, this process loses the advantage of the relatively low viscosity during fiber spinning. Therefore, in this fiber spinning process, the temperature and external force distribution along the spinning line are the two key factors which affect the performance of the fibers.

Following spinning, a fiber is heat treated by one of three methods: (1) free-end annealing, (2) fixed-end annealing, or (3) annealing under tension. Generally speaking, the heat treatment is applied to improve the fiber strength and/or modulus. Sometimes, a zone-annealing method is also used [50]. The method used varies from fiber to fiber, depending on the application. From a molecular point of view, the various spinning methods and heat treatment processes affect the chain orientation, crystallinity, crystal size and morphology in the fibers.

The molecular mechanism in the spinning and heat treatment that leads to chain alignment and crystallinity have several major components: rheological behavior of liquid crystalline polymers in the anisotropic melt or solution under shear flow in the spinnerette hole, elastic relaxation of the liquid at the capillary exit, elongational flow along the fiber axis before solidification, and crystallization upon cooling or coagulation of the fibers. The ideal case would be a two-step fiber formation process, namely, a polymer fluid which would allow the uniaxial alignment of molecules freely without the formation of crystals. After the perfect uniaxial orientation, the molecules could then crystallize extremely quickly in order to solidify the fibers. The design of fiber spinning for liquid crystalline polymers is an attempt to rationally approach this ideal case. One approach would be to utilize the concept of monotropic liquid crystalline behavior to form fibers. If one could design a liquid crystalline polymer which exhibits monotropic behavior and has an extremely slow crystallization rate or needs an external force field to induce crystallization, one might then spin fibers at relatively low temperatures (lower than the melting temperature) in the liquid crystalline state. The resulting fibers could then be crystallized during annealing under an external force field. All of the aspects associated with this process require a broad knowledge of polymer physics, chemistry, and processing technology. Basic structural parameters in fibers are usually characterized by various analytical methods during each step of the fiber formation. In addition, macroscopic thermal, chemical, and mechanical properties can be measured. However, the establishment of relationships between structure, properties, and processing conditions in high performance fibers obtained from liquid crystalline polymers is still not complete.

It should be pointed out that the properties of synthetic fibers spun from the liquid crystalline state (either lyotropic or thermotropic) are more critically dependent upon the fiber spinning process than those prepared using common fiber formation processes. This is because during fiber spinning, the anisotropic fluid is uniaxially orientated relatively easily and the fiber structure develops much earlier than in conventional fiber spinning. As soon as the fiber structure has formed, it is difficult to completely change the structure and develop new structure during the heat treatment process. Therefore, heat treatment can only improve the existing structure in order to enhance mechanical performance. This understanding can be supported by experimental observations that heat treatment causes an improvement in modulus but has little effect on strength in many high performance fibers spun from the liquid crystalline state.

7.4 Structure Formation and Properties of Liquid Crystalline Polymer Fibers from the Lyotropic Liquid Crystalline State

There are two fiber spinning processes commonly used in obtaining aromatic polymer fibers from solution: fibers may be spun from a lyotropic liquid crystalline state or from an isotropic solution. One of the best examples for the former case is Kevlar®. Other polyamide fibers developed by Teijin, Asahi and Unitika in Japan, Akzo in the Netherlands, Rhone-Poulenc of France and Russia, Korea and China are also in this category. Among the most notable products is Twaron® of Akzo. A recent report on aromatic polyimides serves a good example of a fiber spun from an isotropic solution. Technora® fibers, which were developed by Teijin are also in this category. Both spinning processes require very different fiber processing conditions and therefore, the structural formation mechanisms in these fibers are also drastically different.

A successful example in the commercialization of high performance organic fibers spun from lyotropic liquid crystalline polymers is poly(*p*-phenylene terephthalamide) (PPTA, Kevlar®), which was first produced by DuPont in 1972. Depending on the yarn size, mechanical performance and application, several different classes of PPTA fibers exist such as Kevlar® 29, 49 and 149. A less popular Kevlar® 68 is also available. Other examples along this line of development include lesser known aromatic polyamide hydrazides (X-500) which were first manufactured by Monsanto in the early 1970s.

There has also been a systematic investigation of aromatic heterocyclic polymers such as poly(*p* -phenylene benzobisthiazole) (PBZT) and PBZO, initially developed by U.S. Air Force researchers and subsequently produced by Dow and currently, Toyobo under the trade name of Zylon®. All of these fibers employ a common dry-jet wet fiber spinning process from the lyotropic liquid crystalline state.

Polymer molecular weight and chemical composition are two of the most important factors which affect ultimate fiber properties in addition to fiber processing technologies. Substantial effort was applied to change the chemical compositions and to try and correlate this change with the anisotropic polymer solution behavior and fiber properties. This was documented by a number of patents and articles which were issued and published during the last thirty years. In order to illustrate the topic of structural formation in fibers from lyotropic liquid crystalline states, we use the three most extensively studied fibers as examples: PPTA, PBZT and PBZO. General conclusions may be drawn based on observations of the formation of these fiber structures during processing. Some excellent reviews have been appeared recently and readers may refer to those references [51,52].

7.4.1 Molecular Parameters, Fiber Spinning, and Heat Treatments

As indicated in Section 7.2, in order to spin fibers from the lyotropic liquid crystalline state, it is necessary to know the critical concentration above which the viscosity begins to decrease. This critical concentration is generally a function of temperature and molecular weight in addition to the rigidity and linearity of the polymer conformation in solution. Furthermore, molecular weight is a key parameter in the production of high strength, high modulus fibers. Since these aromatic polymers are usually synthesized *via* a polycondensation reaction and the degree of polymerization is typically not high (about 100), a large amount of work has been conducted to increase the molecular weight of these polymers. To be useful, the intrinsic (or less precisely, inherent) viscosity of the polymers must exceed 4 dL/g. The intrinsic viscosity of PPTA polymer prepared by the low temperature condensation of *p*-phenylene diamine (PPD) and terephthaloyl chloride (TCl) in a dialkyl amide solvent is always higher than 5 dL/g [53–55]:

H_2N–⟨benzene⟩–NH_2 + ClOC–⟨benzene⟩–COCl ⟶

PPD **TCl**

[–HN–⟨benzene⟩–NH–CO–⟨benzene⟩–CO–]

PPD-T

A typical number-average molecular weight for PPTA is on the order of 20,000 (corresponding to a degree of polymerization of around 85 and a chain length of about 110 nm). The polydispersity is around 2 to 3. For PPTA in concentrated H_2SO_4, the intrinsic viscosity ($[\eta]$) correlates with the weight average molecular weight (M_W) as [56]:

$$[\eta] = 7.9 \times 10^{-5} (M_W)^{1.06} \qquad (\text{for } M_W > 12{,}000)$$

and

$$[\eta] = 2.8 \times 10^{-7} (M_W)^{1.7} \qquad (\text{for } M_W < 12{,}000)$$

In the cases of PBZT and PBZO, these polymers are synthesized by polycondensation of 2,5-diamino-1,4-benzenedithiazole dihydrochloride or 4,6-dihydroxy-*m* -phenylenediamine dihydrochloride with terephthalic acid in polyphosphoric acid (PPA). The polymers generally have high molecular weights with degrees of polymerization of 50–100 and intrinsic viscosities of 6–30 dL/g [57–59]:

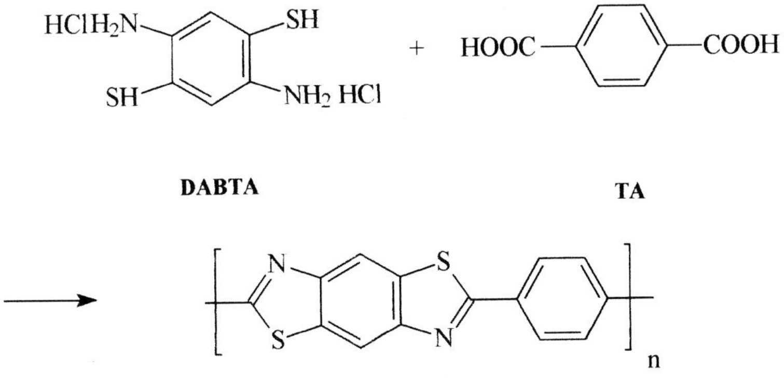

HClH₂N, SH, SH, NH₂ HCl + HOOC—COOH

DABTA **TA**

→ N, S, S, N, n

PBZT

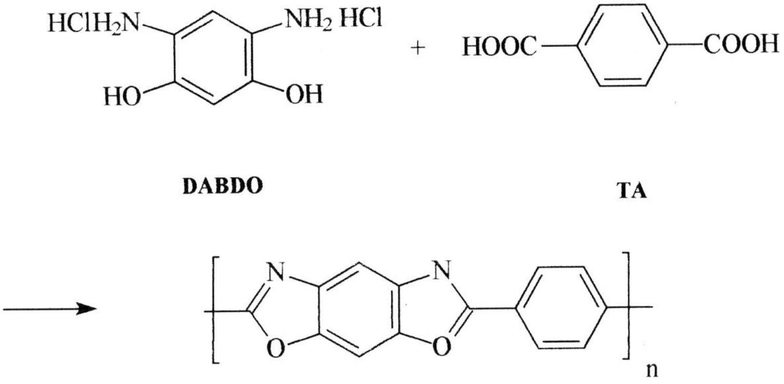

HClH₂N, NH₂ HCl, HO, OH + HOOC—COOH

DABDO **TA**

→ N, N, O, O, n

PBZO

For PBZT and PBZO in methanesulfonic acid (MSA), the relationships between intrinsic viscosity and weight average molecular weight are [60]:

$$[\eta] = 1.65\text{x}10^{-7}(M_W)^{1.8} \qquad \text{(for PBZT)}$$

and

$$[\eta] = 2.77\text{x}10^{-7}(M_W)^{1.8} \qquad \text{(for PBZO)}$$

Other polymers used to spin fibers from their lyotropic liquid crystalline states have similar molecular parameters as these aromatic polymers.

PPTA fibers are produced *via* a dry-jet wet spinning method. Based on available information, PPTA is usually dissolved in concentrated sulfuric acid with a solid content of around 15 to 20 wt%, which is in the anisotropic state. Fiber spinning can be carried out at a moderate temperature (around 80 to 95 °C). In the coagulation bath, an aqueous mixture is commonly used. The thermal treatment of as-spun, wet-PPTA fibers at high temperatures is efficient at improving the modulus but has very little effect on the strength. In fact, improper thermal treatment may lead to a decrease in the tensile strength of the fiber.

By dissolving isolated *trans* -PBZT in MSA (10% w/w) or a PBZT/polyphosphoric acid (PPA) solution (5–6% w/w), nematic lyotropic solution behavior can be found which may be used for dry jet-wet spinning. PPA solutions offer significant advantages of improved properties and processibility [60,61]. Allen et al. reported that their fibers were spun from solutions of 5–6% solid content in PPA or 10% solid content in a 97.5/2.5% mixture of MSA/chlorosulfonic acid [62,63]. The coagulation bath consisted of water or water/MSA. An advanced fiber spinning process was reported to further improve the fiber tensile properties. After high temperature treatment, the modulus of PBZT fibers exhibits a remarkable improvement of 0.8–16 times while the fiber strength increases approximately 50% [64]. PBZO fibers were also prepared by dry-jet wet spinning in a 13–17% solid content in PPA-polymer solution at 60–90 °C [65]. The as-spun fibers were heat treated at 450 °C–500 °C for one minute in air or nitrogen. The heat treatment was conducted under tension to further obtain a 1–7% fiber elongation compared to the as-spun PBZO fibers. Again, the heat treatment has a greater effect on the modulus (about an eight time improvement) than the strength (a 10–20% improvement).

7.4.2 PPTA Fiber Structure, Morphology, and Properties

PPTA fibers are the most commonly investigated fibers in terms of their ultimate structure, morphology and properties. In their final form, PPTA fibers are highly crystalline, uniaxially oriented materials. The birefringence of the fiber (overall orientation) can be as high as 0.65–0.75 [66]. Their crystallinity ranges from 65% to nearly 100%, depending upon the methods of experimental measurements. The degree of crystal orientation can reach 95% [67]. The crystal structure and unit cell lattice of PPTA fibers have been determined using wide angle X-ray diffraction (WAXD) experiments. Figure 7.8 shows a typical WAXD pattern along both the equatorial and meridian directions of a PPTA fiber. A two-chain centered pseudo-orthorhombic unit cell (monoclinic unit cell) has been reported [68–71]. The unit cell dimensions were $a = 0.787$ nm, $b = 0.518$ nm and $c = 1.29$ nm (fiber axis), and $\gamma = 90°$. The crystallographic density was calculated as 1.50 g/cm^3, compared with the observed value of 1.49 g/cm^3. The chain packing in the unit cell is shown in Figure 7.9a [68–71] with a space group of $P2_1/n\text{-}C_{2h}^5$. The angle between the *p*-diaminophenylene segment and the c-axis is about 6°, while the angle between the terephthaloyl segment and the c-axis is approximately 14°. The amide groups are parallel to the (100) planes. Polymorphism in PPTA fibers was also reported [72]. The polymorphism was obtained through fibers spun from the low concentration side of the anisotropic solution. The unit cell is still two chain with the same dimensions, but it possesses the different symmetry of a face-centered pseudo-orthorhombic structure (Figure 7.9b). When the fiber is spun from intermediate anisotropic concentrations, both polymorphs coexist.

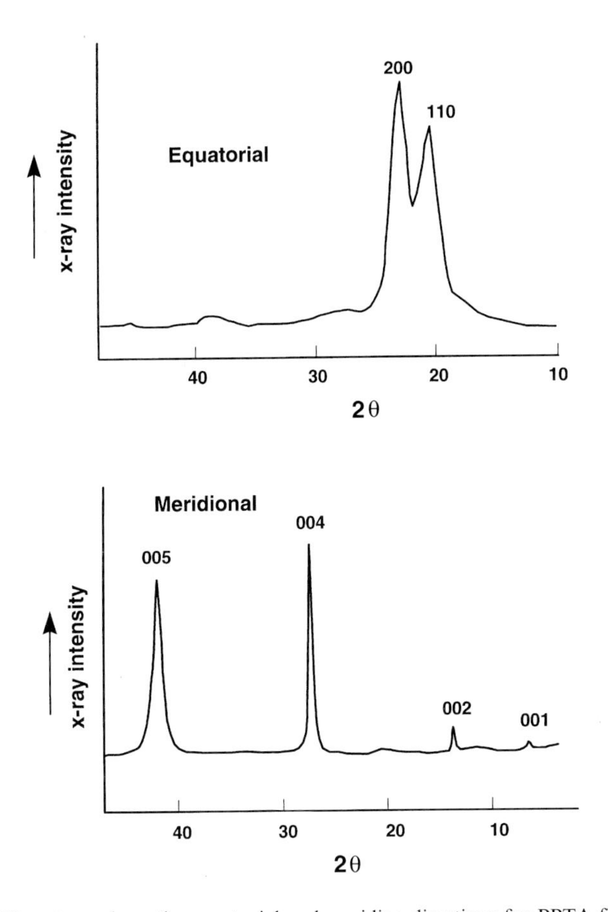

Figure 7.8 WAXD pattern along the equatorial and meridian directions for PPTA fibers [68–71].

Although PPTA fibers possess high crystallinity based on WAXD and density measurements, the type of crystal morphology in PPTA fibers is still in question. A paracrystalline concept originally used in metallurgy was proposed to describe the crystal structure and morphology of PPTA fibers, *i.e.*, a single-phase structure with nearly perfect crystallinity locally but long range lattice disorder. This was used to explain the transmission electron microscopy (TEM) observation of 30–40 nm spacings. Panar et al. suggested that these bands may be representative of stacks of crystalline layers perpendicular to the fiber axis, which are separated by defect layers [71]. Similarly, application of the Scherrer equation to WAXD experiments yields crystallite sizes of crystalline planes that are around 3.5–8 nm laterally and approximately 50 nm parallel to the fiber axis.

PPTA fibers contain highly oriented and ordered fibrillar/microfibrillar textures. The fibrils are oriented along the fiber axis, and are about 600 nm wide and up to several centimeters in length [71]. These fibrils also exhibit crystalline defect layers at spacings of 35 nm. They are joined by bundles of tie fibrils which pass from one fibril to another. The

(a)

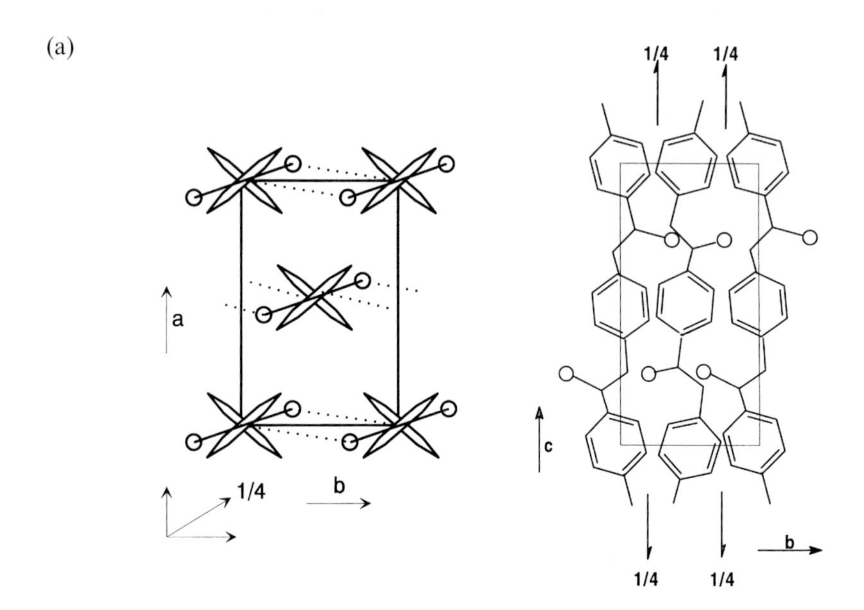

(b)

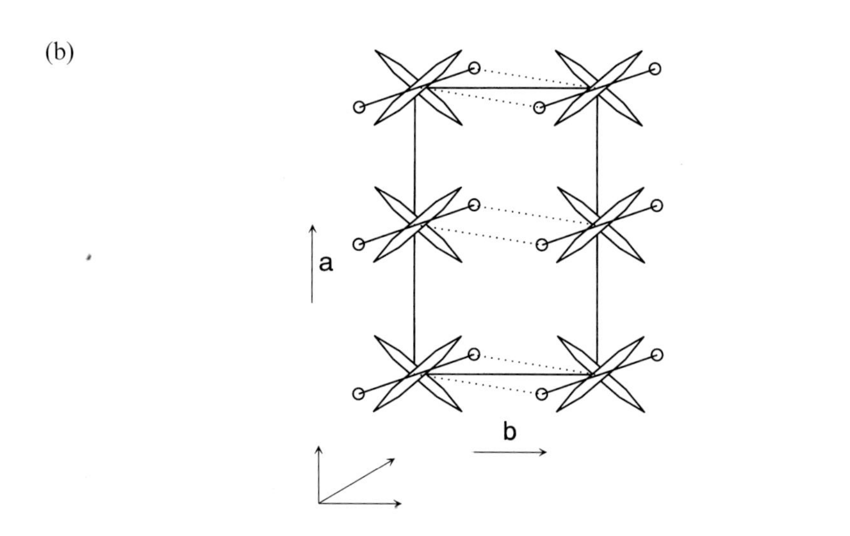

Figure 7.9 Molecular packing of (a) a two-chain C-centered pseudo-orthorhombic unit cell, and (b) a face-centered pseudo-orthorhombic unit cell [68-72].

axial disorder parameter is 1.7%. Figure 7.10 is a proposed structural model of the fibrillar texture which is superimposed on the crystalline structure [71]. The fibrillar texture is a key element in the load bearing ability of a fiber. This texture is not only important to the tensile properties but is also essential to the compressive strength of the fibers.

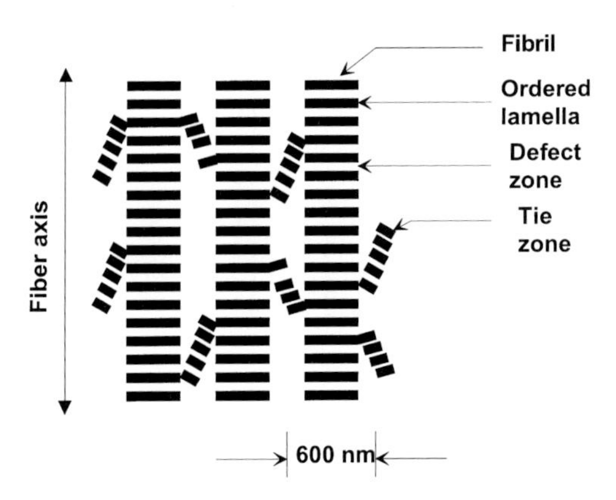

Figure 7.10 A proposed structural model for the fibrillar texture of PPTA [71].

Furthermore, it was also reported that for PPTA fibers, a series of transverse bands can be observed using polarized light microscopy (PLM). The spacing of these bands is 500–600 nm in what is referred to as a pleated structure. TEM dark-field imaging also reveals axial banding with spacings of 500 and 250 nm in longitudinal fiber sections. Dobb et al. explained that the 250 nm spacing, which is half of the 500 nm periodicity, results from a change in crystalline orientation. The (200) crystalline planes form two alternating bands along the fiber axis at a small angle to one another. A structural model was also proposed as shown in Figure 7.11 [73]. Recent reports indicate that this kind of banded texture many be a common feature in fibers spun from liquid crystalline polymers. A similar structural feature for PPTA fibers has also been reported from PLM observations and TEM experiments using the dark-field imaging technique [74,75]. It was concluded that in PPTA fibers, the periodicity in longitudinal fiber sections resulted from changes in crystalline orientation, and a structural model comprised of radial pleated sheets was proposed.

The largest scale of fiber texture includes skin-core observations. This phenomenon is largely dependent upon the fiber spinning process. This texture may also play a role in the tensile deformation process and may affect the tensile properties as a result.

When PPTA fibers are compressed, they first form kink bands at 55°–60° with respect to the fiber axis which can be observed under PLM. The critical compressive strain for the formation of kink bands is reported to be about 0.3% [76]. Further compression leads to fiber buckling. It has been reported that the kink band angle is related closely to the fiber structure and properties [77,78]. Dobb et al. observed that the propensity for and density of kink band formation in PPTA fibers increases with increasing tensile modulus and decreasing elongation at break [74]. There are several arguments about the mechanisms of kink band formation and compressive failure. Argon pointed out that for composite materials, kink band formation occurs when a compressive stress along the orientation direction causes locally misaligned elements to experience a relative shear stress [79,80]. DeTeresa et al. considered that the kink bands are associated with an elastic buckling instability in the fibers and correlated the compressive strength with the longitudinal shear modulus of the fiber [76,81–84]. It was suggested that the compressive strength of the

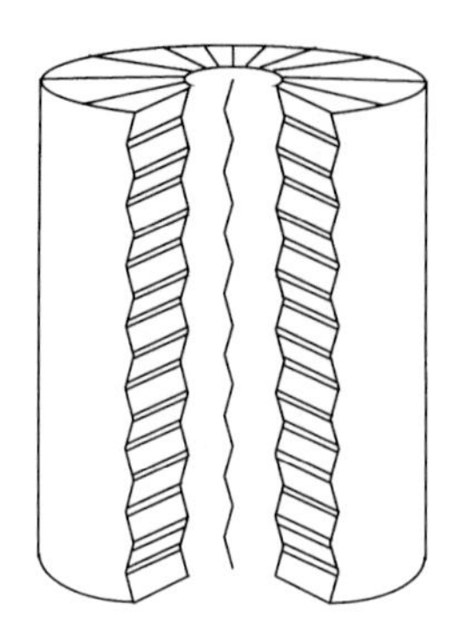

Figure 7.11 A proposed structural model for the pleated structure of PPTA [73].

fibers is limited by fibril buckling according to the Euler buckling model. Some researchers indicated that the kink bands may be attributed to a plastic deformation process [77,78], while others thought that the observation might be related to crystallographic twinning [85] or slipping [86,87].

The most important properties for fibers in general are their mechanical performance, dimensional stability, thermal and thermo-oxidative stability and moisture absorption. For PPTA fibers, these properties are listed in Table 7.1. This table clearly shows that at room temperature, PPTA fibers have a tensile strength at break of above 2.8 GPa. This is more than five times higher than that of steel wire and twice as high as the tensile strength of nylon, polyester and glass fibers. The initial modulus of 62 GPa to 143 GPa for PPTA is also unusually high. This is much higher than that of any steel wire, fiberglass, high tensile strength nylon or polyester. The elongation at break of PPTA fibers is relatively low (between 2.5% and 4%). This again reflects the extended PPTA chain conformations and the high crystallinity of the fibers. A semi-empirical quantum mechanical calculation has been used to predict the intrinsic axial molecular modulus of the polymer chain [88–91], as well as its strain and temperature dependence [92,93]. Although PPTA molecules possess rigid phenylene groups, they adopt a zigzag conformation in which the chain bonds are inclined at an angle to the molecular axis. Under tension, the strain energy is partially applied to the orientation of the phenylenes, which leads to an increase of the rigidity, and as a result, the calculated theoretical molecular modulus is not as great as in the cases of PBZT and PBZO molecules.

Crystals in PPTA fibers are infusible before their thermal decomposition and their glass transition temperature has not been clearly determined. Due to their highly crystalline nature, PPTA fibers do not show a sudden decrease in their tensile properties during heating. Only a gradual reduction of these properties can be found [94]. An extrapolated "zero-strength temperature" in PPTA fibers occurs at about 640 °C [95]. However, since the extrapolation extends over more than 300 °C, this temperature is questionable. The thermal resistance of PPTA fibers are significantly better than other conventional fibers. This is not because the rate of the loss of tensile properties in PPTA fibers is slower with increasing temperature. On the contrary, this rate is almost the same for PPTA fibers as other conventional fibers. The reason for the improved thermal resistance in PPTA fibers is simply due to their extremely high tensile properties at room temperature.

Table 7.1 Physical Properties of Kevlar® Fibers [92]

Properties	Kevlar® 29	Kevlar® 49	Kevlar® 68	Kevlar® 149
Density (g/cc)	1.44	1.44	1.44	1.47
Denier per filament	1.5	1.5	1.5-2.25	1.5
Elongation at break (%)	4.0	2.5	2.9	1.5
Strength (GPa)	2.8	2.8	3.0	2.4
Modulus (GPa)	62	124	91–99	143
Dry heat shrinkage (160 °C)		essentially zero		
Flammability		inherently flame resistant, does not melt		
Temperature resistance		useful properties from −251 °C to +160 °C		
Moisture retention (%)	7	4.5	6	1–1.2

Thermogravimetric analysis (TGA) of PPTA fibers shows a decomposition temperature of above 460 °C in nitrogen and about 380 °C in air (5% weight loss) [96]. The decomposition process can also be recognized by an exothermic peak in air and an endothermic peak in nitrogen using differential scanning calorimety (DSC) [97]. If the PPTA fiber is placed in an hot circulating air oven at 205 °C for 2500 hours, it is found that the tensile strength retention is less than 40% [98].

Other fiber properties which have not been listed in Table 7.1 include creep behavior, compressive properties and chemical resistance. It was found that the creep behavior of regular PPTA fibers (Kevlar® 29) is different than that of high modulus PPTA fibers (Kevlar® 49) [99–102]. Generally speaking, the creep is linearly proportional to the logarithm of time. When a regular PPTA fiber is tested with a load which is less than 50% of the break strength, creep is independent of temperature. At a load in excess of 70% of the strength at break, a secondary creep process occurs with an accelerated creep rate [102]. In high modulus PPTA fibers, stress relaxation is about 6–8% in the stress range of 0.14–1.0 GPa over the time scale of 0.1 to 300 s at room temperature [100]. The creep behavior is more obvious when moisture is present. The creep behavior was found to only have a small temperature dependence between room temperature and 150 °C. The loading dependence was also observed. In addition, the thermal shrinkage behavior of PPTA fibers was investigated. Using different heat treatment mathods, thermal shrinkage may be critically associated with structural parameters such as crystal orientation and apparent crystallite size [103].

The compressive strength of PPTA fibers range from 0.32 GPa to 0.48 GPa, depending upon the experimental testing technique [11,12]. An important problem in the study of compressive properties of a single fiber is that there is no satisfactory technique for measuring its axial compressive strength due to the ease with which it buckles. Several indirect methods have been developed during the past twenty years. These include bending beam [76,81,82], elastica loop [104,105], tensile recoil [106,107], single-fiber embedded composite [108], broken fiber pieces in a composite [109-111] and composite tests. Direct methods involve single-fiber compression with a micro-tensile testing machine (MTM) [112–114] and a nanocompressiometer [115]. These methods suffer from various deficiencies such as the compliance of the material holding the fiber in place for some methods or non-uniform stress, which can even alternate between compressive and tensile, in others [116]. Tensile properties of as received and compressed PPTA fibers show that the initial modulus of the compressed fiber is substantially reduced due to a recovery of the compressed dimension along the fiber axis [81]. The fatigue behavior of PPTA fibers under cyclic loading was also studied [117,118].

PPTA fibers resist most organic solvents and aqueous salt solutions. Their tensile properties are not affect by these solvents and solutions. However, strong acids and bases attack the fibers at elevated temperatures. PPTA fibers are also susceptible to hydrolysis under certain conditions. This leads to degradation of their tensile properties. The effect of ultraviolet (UV) light on the tensile properties of PPTA fibers depends on their shape and texture. However, it is known that PPTA fibers, like other polyamides, are not very stable under sunlight [94].

7.4.3 PBZT and PBO Fiber Structure, Morphology, and Properties

Both crystal structures and unit cell dimensions of PBZT and PBZO fibers have been studied using WAXD experiments. PBZT fibers possess a monoclinic unit cell having a = 1.22 nm, b = 0.365 nm, c = 1.25 nm and $\gamma = 106.2^\circ$ with two repeat units per cell [119]. A similar crystal unit cell for PBZO fibers was determined to have a = 1.12 nm, b = 0.354 nm c = 1.205 nm and $\gamma = 101.3^\circ$ [119]. The detailed molecular packing models of these two unit cells are shown in Figures 7.12a and 7.12b [119]. The b-axis lengths are roughly equal to the perpendicular distance between the faces of the heterocyclic rings in neighboring chains. The calculated PBZO fiber crystallographic density is 1.66 g/cm^3 compared to the measured value of 1.58 g/cm^3 for the heat-treated fibers [120,121].

It has been reported that in PBZO fibers, limited long range three-dimensionally ordered crystallites exist [120,121], although not to the extent seen in high modulus PPTA (Kevlar® 49) fibers. In PBZT fibers, only two-dimensional crystalline order with less long-range periodicity is found [122]. No quadrant reflections (hkl) are observed in PBZT WAXD fiber patterns due to the translational disorder along the fiber axis. For both types of fibers containing small crystallites, there exist many structural defects such as grain boundaries between crystallites and dislocations within crystallites [121,123]. These can be observed in synchrotron small angle X-ray scattering (SAXS) experiments on the fibers. Heat-treated PBZO fibers exhibit a four-point pattern as opposed to the equatorial streaks which are found in as-spun PBZO fibers. These indicate significant periodic density changes along the fiber direction [124] which are not found in PBZT fibers [120,121,124–126]. It has been suggested that the SAXS pattern may be caused by plate-like ordered regions with a tilt angle which deviates from the fiber axis in PBZO fibers, but not in the PBZT fibers [125,126]. The long period and lateral sizes of the PBZO crystallites calculated from the four-point SAXS pattern are 28 nm and 15 nm, respectively. The structural model of PBZO fibers is shown in Figure 7.13, in which crystal interfaces are inclined to the fiber axis, giving the crystals a rhombohedral shape [124]. Another possible reason for the four-point SAXS pattern may be preferential

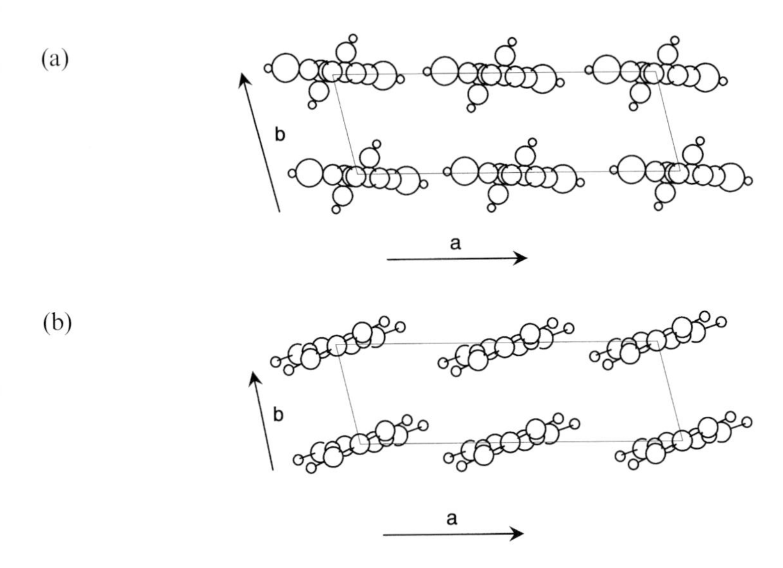

Figure 7.12 Molecular packing for (a) PBZT and (b) PBZO crystals [119].

arrangements of crystallites. In particular, the axial grain boundaries of the crystallites may have a propensity to orient at specific angles to the fiber axis [121,127]. High resolution TEM results show that in PBZO fibers, the grain boundary content is large. In addition, four different types of grain boundaries may exist between neighboring crystalline domains of lateral chain invariant, lateral chain rotation, axial chain invariant and axial chain rotation [127]. The heat treated PBZT and PBZO fibers exhibit significant crystallite growth along all three directions, but after heat treatment, PBZO crystallites are larger than those of PBZT along the lateral directions (*a*- and *b*-axes) [120].

The next morphological level to consider is the oriented fibrillar/microfibrillar structure in PBZT and PBZO fibers observed under scanning electron microscopy (SEM). The fibrils/microfibrils are predominantly lath shaped or ribbon-like [120,128], and they are entwined and connected *via* Y-shaped junctions [129]. Lamellar textures have also been reported in PBZT fibers [130]. The microfibrils in PBZT possess diameters of around 7–10 nm, while for PBZO fibers, the fibril size is about 0.2 μm. Fibrillar bundles are also found with diameters of about 5 μm [131]. The importance of these fibrils/microfibrils is that they consist of discrete, uniaxially oriented crystallites which are formed by highly oriented, extended chain molecules. The formation of these textures is a direct proof that these crystallites are connected by the chain molecules. The Hermans orientation function parameters are greater than 0.99. The axial disorder parameter is 0.53% for PBZT and 1.07% for PBZO [131]. However, certain instances of local disorientation have also been observed between adjacent crystallites on a 20 nm size scale in a high resolution TEM image of a PBZO fiber [132].

The skin-core structure in PBZT and PBZO fibers can be clearly observed under SEM. Their texture is similar to that of PPTA fibers [131,133]. In PBZO fibers, the skin thickness is around 10–100 nm [131]. The skin layers show higher molecular orientation

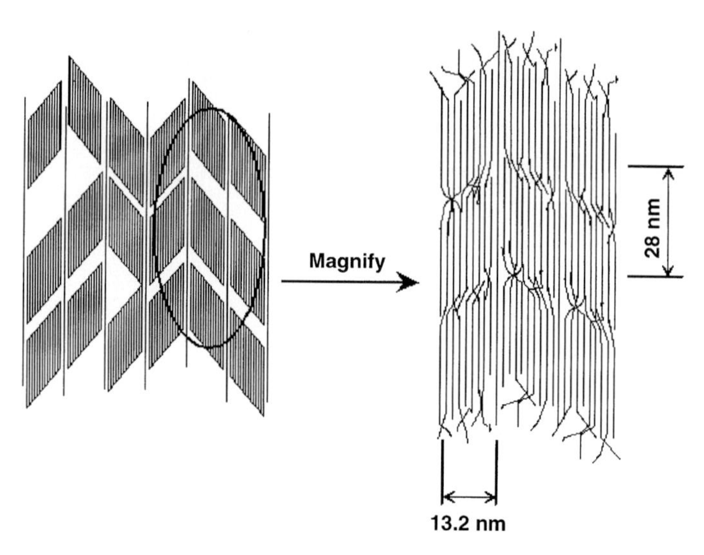

Figure 7.13 Structural model for PBZO fibers [124].

and a more well ordered structure compared with the core materials based on electron diffraction (ED) experiments. Furthermore, heat treatment enhances the crystallinity of the core materials more than it does for the skin layers [133–135]. Again, the formation mechanism of the skin-core texture is not fully understood. It appears to be closely associated with double diffusion processes of the solvent during the solidification process of the fibers in the coagulation bath.

Morphologies of compressed PBZO and PBZT fibers show that these fibers also illustrate the formation of kink bands. It is thought that kink bands are nucleated near the fiber surface position at which the compressive damage occurs. At a certain critical stress, the kink band starts to grow and continuously propagates toward the fiber center and the opposite side of the fiber surface. For PBZT fibers, the critical compressive strain at which the initiation of the kink bands appear under PLM is 0.1%. The density of kink bands increases linearly with the compressive strain up to a certain value. At larger compressive strain, the kink bands tend to saturate [136,137]. The kink bands seem to disappear when a tensile strain is applied after compression under PLM [136] as well as SEM [75,115]. However, the tensile properties of the compressed fibers must be affected. In particular, the fracture energies measured in tensile stress and strain are reduced. From TEM observations, PBZO fibers show kink band thicknesses from 45 nm to 1.8 μm. They are oriented at $69 \pm 3^\circ$ to the fiber axis. The thicker kink bands (> 1 μm in diameter) are more or less of uniform thickness and cross the entire fiber. Cracks and voids may also develop in the bands. The relatively thin bands are usually non-uniform and wavy. Buckling may occur in regions where there are high populations of kink bands and band intersections [138]. When the thickness of the kink bands is on the order of magnitude of 30–50 nm in PBZO fibers, one can judge that this thickness is smaller than the length of an average PBZO molecule (around 100 nm) and that the profile of molecular chains through the kink boundary appears to be constant. This suggests that in a kink band, there is local cooperative bending and/or reorientation of the covalently bonded polymer molecules, rather than chain scission [121,123,139]. Again, this kink band angle in the compressed fibers should be associated with the fiber structure and properties [77,78]. For instance, in PBZT fibers, the formation of kink bands is related to the (010) <001> crystalline slip mechanism [109,110,130,140,141]. Others have suggested that the extent of chain deformation facilitates "fine" shear [142] or both slippage and shear mechanisms [143].

Major properties of both PBZT and PBZO fibers are listed in Table 7.2. Tensile properties of these fibers are extraordinarily impressive. At room temperature, the tensile strengths are 4.1 GPa for PBZT fibers and 6.0 GPa for PBZO fibers, while their tensile moduli are 325 GPa and 360 GPa for the same fibers, respectively. The elongation at break is relatively low, about 1–2% and heat treated fibers are nearly linearly elastic.

Substantial effort has been applied in order to search for a molecular explanation for such high tensile properties, which had not previously been observed. Advances in computational materials science allow semiempirical quantum mechanical calculations of the intrinsic axial molecular modulus of polymer chain [88–91], as well as its strain and temperature dependence [92,93]. PBZT and PBZO molecules possess rigid rings which connect coaxially along the chain direction. Under tension, the molecules convert the strain energy directly into deformation of the stiff *para*-bonds and rings. PBZT and PBZO molecules thus have very high theoretical molecular moduli. In contrast to most

Table 7.2 Physical Structures and Properties of PBZT and PBZO Fibers

	PBZT		PBZO	
	AS	HT	AS	HT
Crystal unit cell		Monoclinic		Monoclinic
a (nm)		1.22		1.12
b (nm)		0.365		0.354
c (nm)		1.25		1.205
γ (°)		106.2		101.3
Lateral size L_a (nm)	1.3	8.6	9.4	16.7
Lateral size L_b (nm)	1.5	6.0	6.1	8.5
Measured density (g/cc)		1.58		1.6
Tensile strength (GPa)		4.1	4.3	6.0
Tensile modulus (GPa)		325	165	360
Elongation at break (%)		1–2	3	1–2
Recoil compressive strength (GPa)	0.2–0.3			0.2–0.35

AS: As spun
AT: After heat treatment

current fiber products, the tensile moduli of PBZT and PBZO fibers are close to their theoretical values, while only about 50% of the theoretical values for the tensile strength can be reached (for PBZO fibers, 20 GPa in actual observations versus a theoretical value of 60 GPa [91,144]). The difference is certainly associated with pre-existing strength-limiting defects which have to do with molecular parameters of the fibers on various morphological (size) levels such as impurities and inhomogeneities [145,146], molecular weight [147,148] and orientation [148], as well as testing conditions such as gauge length [145].

PBZT and PBZO fibers exhibit excellent thermal and thermo-oxidative stability. The degradation temperatures of these fibers in air are around 620 °C [150]. At 371 °C in air for 200 hours, the weight losses are reported to be 22% for PBZO and 30% for PBZT. At elevated temperatures, PBZT evolves gases of H_2S, HCN and CS_2, which constitute 95% of the total ionization flow [59,151]. For PBZT fibers, oxygen containing gases comprise the majority of released materials such as CO, CO_2, and H_2O. However, under 205 °C isothermal aging for 2500 hours, these fibers show tensile strength retention of 61% for PBZO fibers and 50% for PBZT fibers, which is much better than those of PPTA fibers [98]. Again, dimensional stability measurements on these fibers reveal a negative coefficient of thermal expansion along the fiber direction due to stress frozen in during the fiber spinning and heat treatment processes.

Compressive properties of PBZT and PBZO fibers have been extensively investigated. Several approaches have been proposed to improve the compressive strength of these materials. Attempts such as changing the fiber spinning and heat treatment conditions and/or introducing cross-linking moieties to enhance lateral packing have been reported. However, up to now, significant improvement of the compressive strength of these fibers has not been achieved. It seems that in the past, research about the morphological effects on compressive properties was not extensive. It may be possible to look in this direction for improving compressive properties of high-performance polymer fibers. For these fibers, the measured compressive strengths are 0.26–0.41 GPa for PBZT fibers and 0.2–0.4 GPa for PBZO fibers. These values vary slightly depending upon the method of measurement. Figure 7.14 shows a typical stress-strain curve for PBZO fibers under tension and compression [139,152]. On this plot, the yield stress in compression is only about one-tenth of that in tension. This value is much smaller than isotropic boron, alumina and silicon carbide fibers, which possess high compressive strengths. A nonlinearity in the compressive behavior was reported in PBZT fibers [75,115,153–155], and the compressive modulus decreases with increasing axial compressive stress. This may be referred to as "modulus softening". The fiber birefringence also decreases under compression, as explained by a crystal/fibril disorientation away from the fiber direction as detected by WAXD experiments [75].

Other fiber properties include chemical and solvent resistance. PBZT fibers exposed at 120 °C in a pH 5 buffer solution were found to be more stable than other organic fibers. Most organic solvents have no effect on PBZT fibers. These fibers do not undergo any significant hydrolysis. The moisture retention of PBZO fibers is about 0.5% after a prolong time period in a boiling water environment [145]. For PBZT fibers, the moisture retention is usually less than 1% after they are dried at 250–300 °C. Furthermore, the study of the radiation stability of these fibers shows good tensile property retention. The electron damage critical dose for PBZT fibers is 1.6 C/cm^2 [11]. Radiation damage

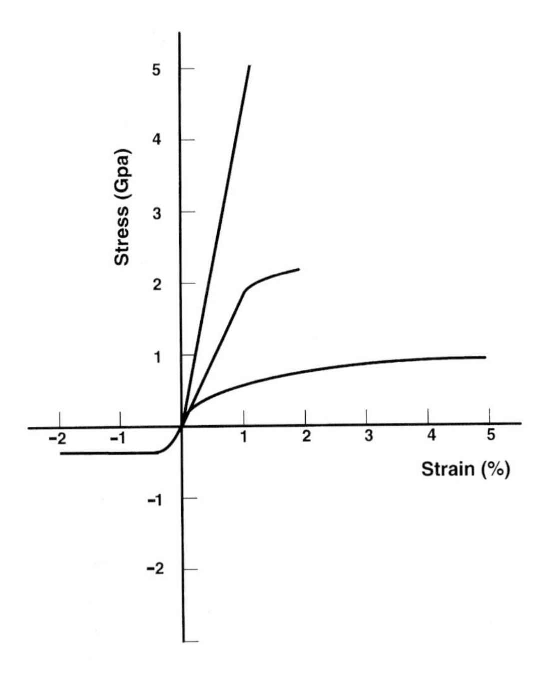

Figure 7.14 A typical stress-strain curve of PBZT fibers under tension and compression [139,152].

involves disruption of packing between molecules rather than scission or reorientation of bonds along the fiber axis [156]. Dielectric property measurements confirmed that these fibers possess high dielectric strength.

7.5 Structure Formation and Properties of Liquid Crystalline Polymer Fibers from the Isotropic Solution State

Fibers from isotropic solutions of aromatic polymers exhibit very different structures than those formed from anisotropic solutions. This is because in isotropic solutions only limited structure is formed during the fiber spinning process and heat treatment plays an important role in the structural development. As a result, as-spun fibers do not show impressive tensile properties, and high strength and high modulus can only be found after the fibers are drawn and annealed at elevated temperatures. A typical example of a commercialized fiber is Technora® (HM-50), which was developed by Teijin as a high strength, high modulus aromatic polyamide fiber in 1985. This fiber has different chemical compositions and spinning processes as compared to PPTA fibers and therefore, provides some unique fiber properties such as chemical and impact resistance. The tensile properties of Technora® fibers are in between those of regular and high modulus PPTA

(Kevlar® 29 and 49) fibers. This indicates that its chemical modification brings a loss in modulus due to the absence of lateral crystalline order and chain linearity.

Other examples include the organo-soluble aromatic polyimide fibers developed recently at the University of Akron [157–162]. It is well-known that aromatic polyimides are usually insoluble and infusible and that fibers from these polymers can only be spun from their soluble poly(amic acid) precursors. After fiber formation, thermal or chemical imidization takes place to form polyimides. This imidization reaction releases by-products such as water, etc. which substantially affect the ultimate fiber properties. In the past, a great deal of effort was devoted to developing aromatic polyimide fibers using this two-step polymerization approach [163,164]; none of which were successful. The special feature of this class of polyimides from the University of Akron is that they are synthesized using a one-step polymerization method and the poly(amic acids) precursors are not isolated. The key point is that these high molecular weight, aromatic polyimides are soluble in phenolic solvents and therefore, a dry-jet wet spinning method can be utilized to produce a new class of high strength, high modulus polymer fibers. A typical example is a polyimide synthesized from 3,3′,4,4′-biphenyltetra carboxylic dianhydride (BPDA) and 2,2′-dimethyl-4,4′-diaminobiphenyl (DMB). Some of the physical property data for this BPDA-DMB fiber is listed in Table 7.3. We will pay closest attention to the polyimide fibers in this section, but the general conclusions are still useful in other similar cases.

Table 7.3 Physical Properties of BPDA-DMB and BPDA-PFMB

	BPDA-DMB	BPDA-PFMB
Tensile strength (GPa)	3.3	3.0
Specific tensile strength (GPa)	2.44	2.0
Tensile modulus (GPa)	140	130
Specific modulus (GPa)	103	90
Elongation at break (%)	2–3	2–3
Density (g/cc)	1.35	1.44
Compressive strength (GPa)	0.66	0.70
Specific compressive strength (GPa)	0.49	0.47
Decomposition Temperature (5%, air, °C)	480	600
Decomposition Temperature (5%, N_2,°C)	520	600

7.5.1 Molecular Parameters, Fiber Spinning, and Heat Treatments

A series of organo-soluble, aromatic polyimides was synthesized using specific molecular designs and architectural control. This series of polyimides is based on BPDA with different 2,2′-disubstituted diaminobiphenyl diamines [157–159]. For example, two of the members of this family are fibers spun from polyimides synthesized from BPDA and DMB in *p*-chlorophenol and 2,2′-bis(trifluoromethyl)-4,4′-diaminobiphenyl (PFMB). The chemical reactions for their syntheses are

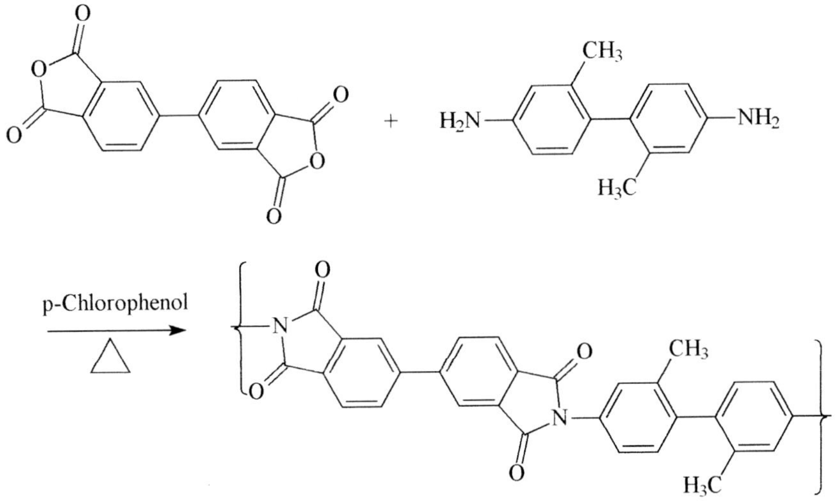

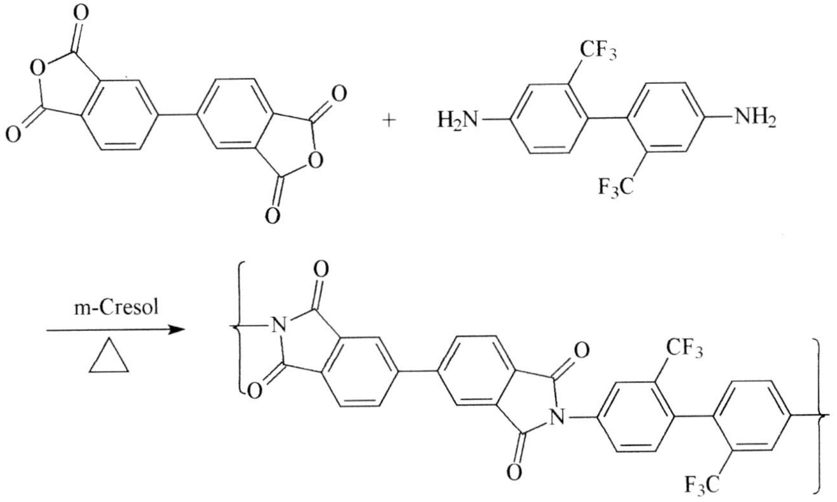

O O O O O O + H_2N CH$_3$ H_3C NH_2

p-Chlorophenol
Δ

O N O O N O CH$_3$ H_3C

O O O O O O + H_2N CF$_3$ F_3C NH_2

m-Cresol
Δ

O N O O N O CF$_3$ F_3C

10% (w/w) solutions of BPDA-DMB and BPDA-PFMB remained completely soluble during polymerization although the solutions formed gels at room temperature [165,166]. The intrinsic viscosity of BPDA-DMB in *p*-chlorophenol at 60 °C was 8–10 dL/g while that of BPDA-PFMB in *m*-cresol at 60 °C was about 5 dL/g. Both fibers were produced at the University of Akron. Fibers were spun from isotropic solutions of 10% (w/w) polymer in *p*-chlorophenol or *m*-cresol *via* dry-jet wet spinning into a coagulation bath of water/methanol (50/50) at room temperature. Although these polyimides enter an anisotropic solution state at even higher concentrations, testing indicates that the optimal spinning conditions require 10% (w/w) polymer solutions for the best ultimate fiber properties. The as-spun fibers were then drawn at an elevated temperature (above 400°C) up to a draw ratio of ten in air *via* a zone drawing method [157–159].

Technora® fibers, which are modified based on copoly(*p*-phenylene/3,4′-diphenyl ether terephthalamide), are spun from an isotropic solution instead of an anisotropic state. The polymerization is carried out at low temperature using *p*-phenylene diamine, 3,4′-diaminodiphenyl ether and terephthaloyl chloride in an amide solvent such as *n*-methyl-2-pyrrolidone (NMP), dimethyl acetamide (DMAc) or hexamethylphosphoramide (HMPA) *etc.* Small amounts of an alkali salt (such as calcium chloride or lithium chloride) are used during the polymerization [167,168].

$$mH_2N-C_6H_4-NH_2 + nH_2N-C_6H_4-O-C_6H_4-NH_2 + ClCO-C_6H_4-COCl \xrightarrow{\text{amide solvent/alkali salt}} \left[\left(HN-C_6H_4-NH \right)_m \left(NH-C_6H_4-O-C_6H_4-NH \right)_n \left(CO-C_6H_4-CO \right) \right] + 2HCl$$

The percentage of *meta*-linked phenylene is at least 15% and perhaps as high as 33%. As a result, kinks are introduced in the chain backbones and aramid linkages are replaced by ether linkages. The polymerization process was carried out at temperatures of 0–80 °C for 1–5 hours in a 6–12% solid-content solution. The inherent viscosity was about 2–4 dL/g. The reaction mixture was then filtered and spun from a spinnerette into an aqueous coagulation bath containing 35–50% $CaCl_2$ or $MgCl_2$ in order to control the coagulation rate. Next, the fibers were drawn 6–10 times and dried at high temperatures.

7.5.2 Organo-Soluble Aromatic Polyimide Fiber Structure, Morphology, and Properties

Traditionally, aromatic polyimide crystal structures were determined using WAXD fiber patterns. Since aromatic polyimide molecules usually have complicated chemical structures, the molecular weights of their chemical repeating units are often high. Several reviews have reported a number of aromatic polyimide crystal structures. Highly drawn BPDA-DMB and BPDA-PFMB fibers show crystallinities in excess of 60% and 50%, respectively. The birefringences (Δn), which are represented by overall degrees of orientation, are 0.24 and 0.25, while the degrees of crystal orientation for both fibers were about 85% based on WAXD data [157–159].

The crystal structures of these materials were determined using WAXD fiber patterns. The BPDA-PFMB unit cell is monoclinic with dimensions of a = 1.54 nm, b = 0.992 nm, c = 2.02 nm and $\gamma = 56.2^\circ$ [157]. The BPDA-DMB unit cell is triclinic with dimensions of a = 2.10 nm, b = 1.52 nm c = 4.23 nm, $\alpha = 61.3^\circ$, $\beta = 50.8^\circ$ and $\gamma = 79.0^\circ$ for highly drawn fibers. For heat treated fibers, the unit cell is still triclinic, but the dimensions change slightly to a = 2.05 nm, b = 1.53 nm c = 4.00 nm, $\alpha = 62.1^\circ$, $\beta = 52.2^\circ$ and $\gamma = 79.6^\circ$ [159]. Both unit cells have large volumes and high numbers of chain repeat units per cell. Although this phenomenon is not common, aromatic polyimides may exhibit large unit cell volumes. Recently, a method of crystallization during polymerization in aromatic polyimides was developed and it was shown that polyimide oligomers may form large lamellar single crystals and therefore, ED experiments can be used to determine their crystal structure and symmetry [169–171]. This method is very useful for providing a direct check of the crystal structure determination. On the other hand, WAXD experiments are not only used to determine crystal structure, but also provide useful information about crystallinity, crystal orientation and crystallite sizes.

One of the most prominent morphological characteristics of high performance polymeric fibers is their oriented fibrillar texture [159,160,162]. Figures 7.15a and 7.15b show SEM micrographs of BPDA-DMB fibers with microfibrillar textures. In Figure 7.15a, the fibrillar structure is shown after a portion of the skin was peeled off. The fibrils typically possess diameters in the sub-micrometer range and a length of tens of micrometers. Sometimes, the fibrils are observed to have a ribbon-like shape as shown in Figure 7.15b. The reason for this fibrillar formation is not yet fully understood. Generally speaking, it must be closely related not only to molecular structures and lateral chain interactions, but also to the fiber processing conditions. Polymer molecules in fibers are always anisotropic due to covalent bonding along the chain (and therefore, the fiber) direction and physical bonding perpendicular to the chain direction. The ratio of energies between these two kinds of bonding can be as high as ten to one. As a result, weak lateral interactions may lead to the formation of microfibrils. Fiber processing in a wet or dry-jet wet spinning method needs a substantial amount of mass transfer between the solvent in the polymer solution (*p*-chlorophenol here) and the solvent in the coagulation bath (ethanol-water) *via* diffusion processes. During polymer precipitation under tension in the ethanol-water mixture, microfibrils may be the favored texture. Quantitative research is necessary to obtain detailed understanding of the fibrillar formation mechanism. It is also worth noting that in Figure 7.15a, buckling is evident at the corner of the peeled half of the fiber surface. This reveals that on a large scale, the shear stress field introduced during peeling leads to a compressive banding deformation on the peeled fiber surface.

There is also a series of transverse bands with spacings of approximately 2.0 μm in the fibers as observed under PLM and TEM [161,162]. These bands are morphologically superimposed on the fibrillar structure. The band spacing of BPDA-DMB fibers is, however, much greater than that of PPTA fibers and the change of crystal orientation in BPDA-DMB fibers seems to be much larger than that of PPTA fibers. A possible origin of the banded texture in liquid crystalline polymers may be mechanical shearing, after which the characteristic morphology may be observed. It is known that both thermotropic and lyotropic liquid crystalline polymers exhibit banded textures after mechanical shearing. This banded texture, having a regular spacing of a few μm, is

(a)

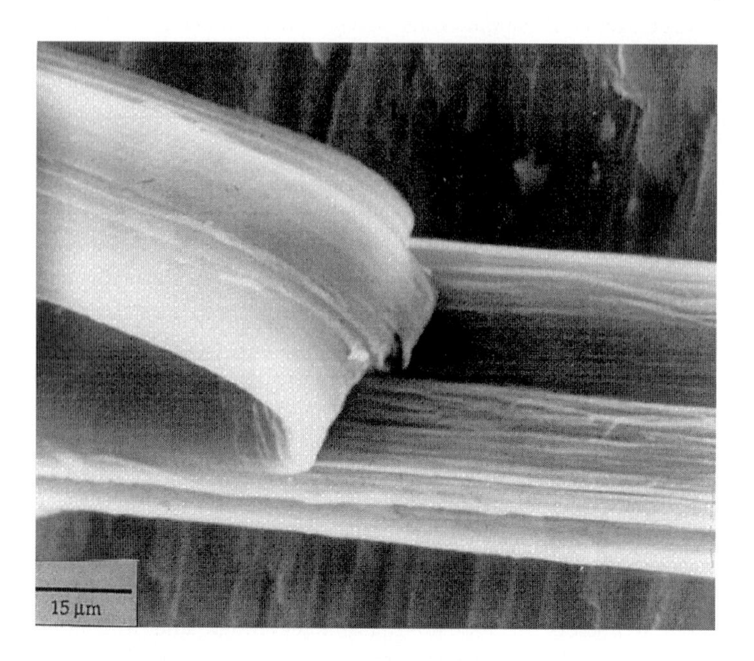

(b)

Figure 7.15 SEM microphotographs of (a) skin-core and (b) fibrillar textures in BPDA-DMB fibers [161].

perpendicular to the chain orientation direction and usually appears after some relaxation of the flow. Microscopic analysis has shown that in the banded texture, the orientational director continuously oscillates spatially about the direction imposed by the previous flow [172–174]. It is possible that the banded texture in BPDA-DMB fibers is formed through this mechanism since this polyimide does show mesophase behavior in solution.

The skin-core structure of a cross-sectional surface of an uncompressed BPDA-DMB fiber can also be observed under TEM. An average skin thickness of 100–150 nm was determined (indicated by the arrows). The existence of skin and core structural differences is believed to be critically associated with the processing conditions since *p*-chlorophenol diffuses away from the fiber surface to exchange with the solvent in the coagulation bath faster than it diffuses into the core of the fibers. Thus, the fiber surface quickly solidifies to form a thin layer. The skin-core structure observed from SEM (Figure 7.15a) shows that the skin separated from the core to become an independent entity. The thickness of the skin structure is in the submicrometer range, which is consistent with TEM observations. It is interesting to find that the skin of BPDA-DMB fibers is not as thick as that of other aromatic polymer fibers obtained from similar fiber spinning processes. This indicates that during the fiber spinning process, the solvent gradient built up along the transverse direction of the fiber in the coagulation bath is less severe than it is for other fibers.

Polyimide fibers show kink band formation after they are compressed beyond a critical compressive strain (0.5–0.54% for BPDA-DMB fibers). Figure 7.16 shows an example of

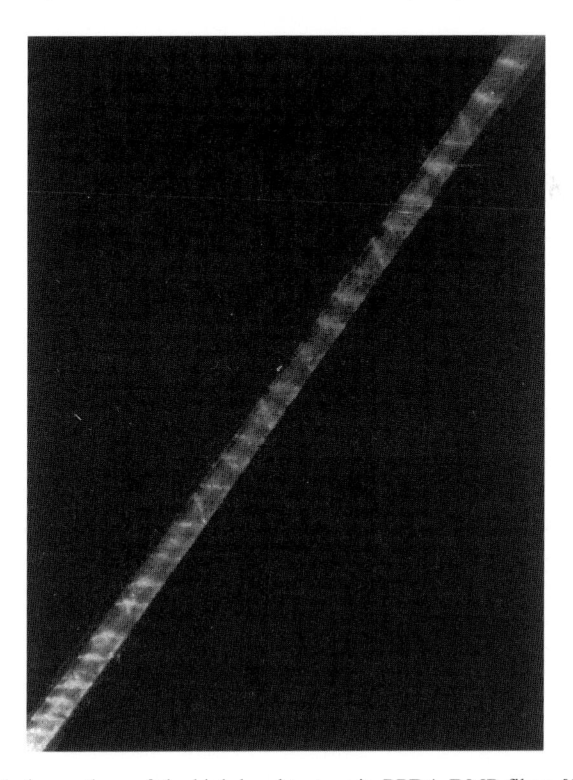

Figure 7.16 PLM observations of the kink band texture in BPDA-DMB fibers [161].

kink band textures under PLM which are elicited by the matrix shrinkage technique. It is clear that kink bands are formed with sharp boundaries and high birefringence at a consistent angle of $\pm 60^\circ(\pm 2^\circ)$ with respect to the fiber axis. The direction of the kink band, negative or positive, is more or less random. In some locations, a V-shaped band is observed which consists of two neighboring kink bands oriented in opposite directions with respect to the fiber axis ($\pm 60^\circ$). It has been proposed that the kink band angle is closely associated with the fiber crystal structure and morphology [77,78]. The initiation of kink bands may occur within a highly oriented non-crystalline region and/or within a crystalline region of the fibers. Mechanisms of uniform c -axis shear and inter-crystalline slippage are reportedly involved in the formation of kink bands [137,143]. The triclinic unit cell of BPDA-DMB crystals possesses an angle of $\alpha = 62.13^\circ$ [159] between the b- and c-axes. Assuming that the c-axis is parallel to the fiber direction and crystal retention of the lattice above and below the kink is permitted, it is possible that the shear and/or slippage along the b-direction may lead to a kink band having an angle of about $\pm 60^\circ$. In annealed PBZO fibers, the angle is around $\pm 67^\circ$ after compression, while for PPTA fibers, the kink band angle is between $\pm 50^\circ$ and $\pm 60^\circ$ with respect to the fiber axis. It should be cautioned that using the limited resolution of PLM to observe the kink band angle does not allow individual kinks to be distinguished, but rather shows an aggregation of kink bands [121]. Furthermore, the kink bands must be in the correct position (in-plane position) in order to observe their precise angles.

The processes of kink band formation and development which eventually lead to compressive failure are still being actively discussed. Earlier studies have shown that kink bands usually develop in groups localized around a sharp bend [115]. Kinking can be complicated by buckling and fibrillation, especially in the later stages of a relatively large scale compression. The fiber then fails in a catastrophic manner. Recently, based on our experimental observations, we have proposed a nucleation mechanism for the initiation of kink bands during compression which is followed by kink band propagation and fiber buckling [161,162].

Two major types of macroscopic kink bands are observed. One type of band possesses a uniform thickness of about 1 to 2 mm and appears across the whole fiber section. The second type of band has a thickness of 0.4 to 0.8 mm. These two types of kink bands may be representative of different aggregation processes. Both are oriented at around a $\pm 60^\circ$ angle with respect to the fiber axis. It is speculated that after the compressive strain exceeds the critical value (0.5–0.6%), the kink bands generate heterogeneous regions and the compressive stress is concentrated on these regions, which further enhances the compressive damage. However, it is uncertain what kind of structural changes may occur in the region between two macroscopic kink bands during compression. Microtomed thin fiber sections under TEM have shown that microkink bands exist in this region. The microkink bands possess uniform thicknesses of about 8–10 nm, and there is a 300 nm spacing between neighboring microkink bands. The boundaries separating the kinked and unkinked regions are sharp. This indicates that the heterogeneous structures in the fibers comprise multiple size scales. However, it is not clear whether the microkink bands form before or after the macroscopic kink bands. Further study is necessary to reach a definite conclusion.

Most of the polyimide fibers were spun from poly(amic acid) precursors which were later imidized in the fiber form. The imidization process produces by-products and usually generates microvoids, which substantially reduce the tensile properties of the fibers. In the past, the fiber tensile properties from this spinning procedure were not as impressive as those of PPTA fibers and other high-performance fibers. However, when the fibers are spun directly from a pre-imidized form, it is possible to achieve tensile properties that are as high or even higher than PPTA fibers. Two examples are BPDA-PFMB and BPDA-DMB after heat treatment. The tensile strengths of these fibers are around 3.0 GPa and 3.3 GPa, respectively, while their tensile moduli are 100 GPa and 130 GPa. Their elongations at break are about 3% [157,159].

BPDA-PFMB and BPDA-DMB fibers show excellent thermal and thermo-oxidative stability. From TGA experiments, 5% weight losses for BPDA-PFMB and BPDA-DMB occur at 600 °C and 530 °C in nitrogen and 580 °C and 500 °C in air, respectively [157,159]. TGA-MS data for BPDA-PFMB show that CO, HCN, NH_3, HF COF_2 and CF_3H are released from thermal cracking of the diimide and pendant loss at around 590 °C [175]. More importantly, after 2500 hours of isothermal aging at 205 °C, there is a 100% retention of tensile strength for BPDA-PFMB and 66% for BPDA-DMB fibers [98]. These values are substantially higher than those of PPTA fibers, and even slightly higher than those of PBZO fibers. Dynamic mechanical results indicate that for both fibers, two relaxation processes can be detected above room temperature: a β relaxation at relatively low temperature and an α relaxation which represents a glass transition temperature [158,159]. The β relaxation strength is dependent upon the crystallinity and has been assigned as a non-crystalline diamine rotational motion. The α relaxation is critically associated with the rigid part of the material in fibers above their glass transition temperatures [158,159].

In order to study dimensional stability of the fibers, the concept of a thermal shrinkage modulus was proposed [159]. This modulus is defined as the ratio between shrinkage stress and shrinkage strain at a constant temperature. It has been found that for these fibers the shrinkage modulus is roughly one-tenth of the tensile modulus [159]. This is consistent with PPTA fibers, where the shrinkage modulus is about one order of magnitude smaller than the tensile modulus [103].

The fibers exhibit surprisingly high compressive strengths. For example, BPDA-DMB fibers possess compressive strengths of 0.66 MPa from tensile recoil test measurements [161], which is about two to three times higher than any current high-performance fibers. Investigation reveals that the compressive strength is independent of the fiber diameter. Although kink bands can also be observed in BPDA-DMB fibers as found in other polymer fibers, the fibrillar/microfibrillar texture shows a sheet-like shape, indicating some morphological difference as compared with the other fibers. Furthermore, BPDA-DMB fibers show a β relaxation (sub-glass transition) around 120 °C to 200 °C below the glass transition temperature even though the fiber was heat treated at elevated temperature. At equivalent mechanical performance with PPTA fibers, BPDA-DMB fibers exhibit substantial sub-glass relaxation strength, which indicates that energy is absorbed in order to initiate non-cooperative motion [159,161]. As a result, less energy is applied to the formation of kink bands in these fibers as compared to other fibers. Tensile properties of before and after compressed fibers are shown in Figure 7.17 for BPDA-DMB [161]. Pre-compressed fibers exhibit a progressive loss in initial tensile strength with increasing compressive strain. This is because the initial tensile stress unfolds the kink

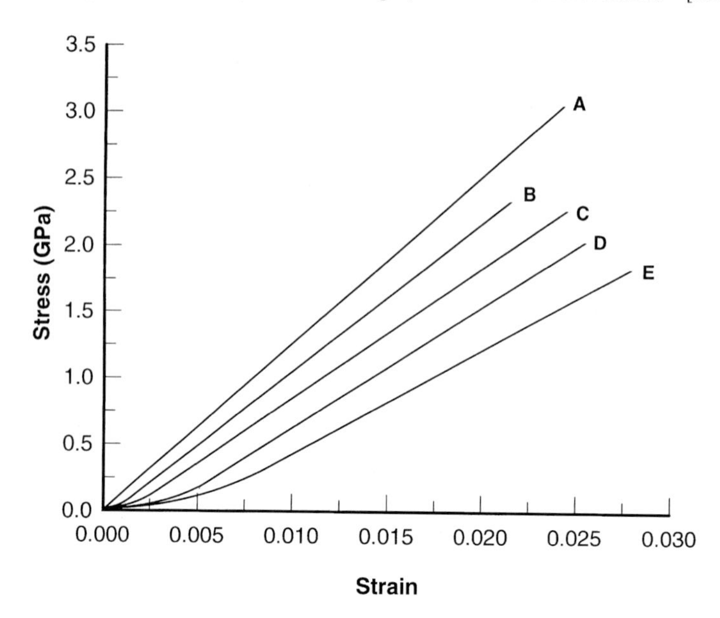

Figure 7.17 Stress-strain tensile relationships of the pre-compressed fibers at different compressive strains (A) as received, (B) 1%, (C) 1.5%, (D) 2% and (E) 3% [161].

bands in the fibers. As soon as the kink bands are stretched out, a modulus similar to that of the normal fibers without compression can be obtained. However, there is a substantial decrease in the fracture energy which increases with the compressive strain [161].

Aromatic polyimide fibers generally possess better UV resistance than PPTA fibers due to the absence of N-H bonds [176]. These materials are also highly resistant to nonoxidative radiation [177]. However, oxidative radiation quickly reduces their mechanical properties. The hydrolytic stability of polyimides is critically affected by the presence of poly(amic acids). Full imidization is thus necessary in order to enhance this stability. BPDA-DMB fibers usually exhibit better strong acid resistance, but weaker base resistance than PPTA fibers [178,179]. Substantial effort has also been applied toward the development of other high performance polyimides and copolyimide fibers [160,178–183].

7.5.3 Technora® Fiber Structure, Morphology, and Properties

Technora® fibers were introduced in 1985 by Teijin as HM-50. These are aromatic polyamide fibers, which differ from PPTA fibers in both their monomer composition and their fiber spinning method. It is believed that this chemical modification and fiber spinning method lead to some unique properties such as excellent chemical and impact resistance, while the expected tensile properties are maintained. Typical fiber properties are listed in Table 7.4.

When the chain conformations are investigated with both 3,4′-diphenyl ether terephthalamide and 4,4′-diphenyl ether terephthalamide, it is surprising to find that the

Table 7.4 Physical Properties of Technora® (HM-50) fibers

Tensile strength (GPa)	3.1
Tensile modulus (GPa)	70
Elongation at break (%)	4.4
Density (g/cm^3)	1.39
Thermal decomposition temperature (°C)	500
Equilibrium moisture absorption (%)	3.0
Heat of combustion (cal/g)	6800
Limited oxygen index (%)	25
Color	gold
Filament diameter (μm)	12

polymer containing 3,4′-diphenyl ether terephthalamide exhibits a nearly linear extended conformation with a slight shift of the chain axis [184]. On the other hand, the 4,4′-diphenyl ether terephthalamide containing polymer introduces a kink at the ether linkage and, as a result, there is a significant amount of nonlinearity. Since Technora® fibers are based on copoly(*p*-phenylene/3,4′-diphenyl ether terephthalamide), it is expected that these fibers can achieve a high degree of orientation by drawing and that the polymer chains are fully extended. WAXD results prove the existence of a high degree of molecular orientation and the presence of some three-dimensional order [185]. Blackwell *et al.* reported the occurrence of a series of aperiodic meridional diffractions in these fibers. This is an indication that the copolymers possess random compositional sequences. Both atomic positions were calculated and molecular models were proposed [185]. Based on the models, calculated WAXD maxima fit very well with experimental observations.

The inclusion of 3,4′-diphenyl ether terephthalamides disrupts the hydrogen bonding in the polymer and introduces crystal defects in extended chain crystals. These defects usually aggregate to form non-crystalline regions in the fibers. A fiber structure model was proposed by Imuro [184] to illustrate the molecular arrangements (Figure 7.18). The *p*-phenylene groups in 4,4′-diphenyl ether terephthalamides are linear, and form rigid segments having a periodicity of 11–13 nm. The 3,4′-diphenyl ether terephthalamides aggregate to form non-crystalline regions with periodicities of 7–9 nm. Therefore, the long spacing in the fibers is around 20 nm. From this model, it can be deduced that the highly oriented amorphous region causes a small apparent difference between the crystalline and amorphous regions which leads to the uniformly dense fiber morphology in Technora® fibers. However, this fiber does not offer the modulus of PPTA fibers due to the existence of large amounts of non-crystalline regions in the fibers.

As listed in Table 7.4, the tensile properties of Technora® fibers are impressive: a tensile strength of 3.1 GPa and a tensile modulus of 70 GPa. Its elongation at break is 4.4%. In fact, based on tensile properties, Technora® fibers are in between the regular PPTA (Kevlar® 29) and high modulus PPTA (Kevlar® 49) fibers. It is believed that the high draw ratio during heat treatment gives rise to these tensile properties.

Variable temperature testing of tensile properties shows a gradual loss of strength and modulus with increasing temperature. The rates of these losses with temperature are similar to those of PPTA fibers [186]. However, Technora® fibers exhibit a thermal

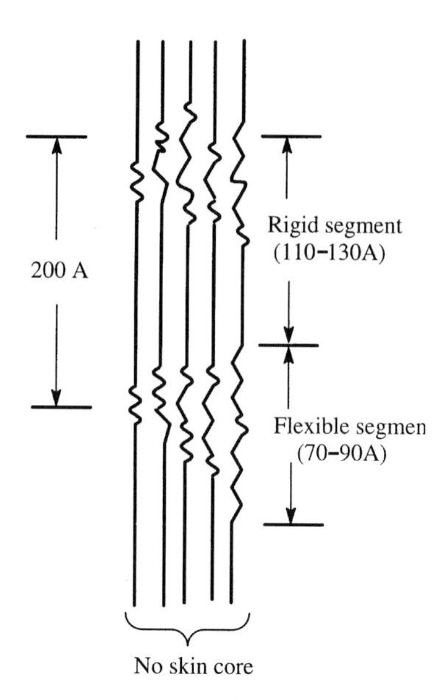

Figure 7.18 Molecular arrangement for Technora® fibers [184].

stability which is 50 °C lower than that of PPTA fibers. In addition, Technora® fibers have a glass transition temperature around 318 °C and an endothermic peak at about 485 °C. As a result, the fiber only exhibits dimensional stability below 200 °C. Above this temperature, a negative coefficient of thermal expansion (thermal shrinkage) is detected [186]. An equilibrium moisture content of 3% can be found in the fiber, which is only half that of PPTA fibers. Chemical resistance studies show that Technora® fibers are resistant to a number of common acids, alkalis and solvents. They are also hydrolytically stable in sea water and steam. Although the fiber loses its tensile properties after extended exposure to sunlight, its light stability can be improved by heat treatment in the presence of urea and thiourea [187].

7.6 Structure Formation and Properties of Liquid Crystalline Polymer Fibers from the Thermotropic State

Developmental efforts on fibers from thermotropic liquid crystalline polymers are mainly concentrated on aromatic copolyesters. Since these copolyesters show low ordered liquid crystalline phases (most are in the N, S_A or S_C phase), they exhibit anisotropic flow behavior. The fiber spinning process is almost exclusively a melt spinning process. During cooling and further heat treatment, the fibers undergo an ordering process which enhances mechanical performance. Three representative polyester fibers are HBA/PET fibers from Eastman Chemical, HBA/HNA fibers from Hoechst Celanese and Ekonol®, which was developed by Carborundum and is now produced by Sumitomo Chemical and Nippon Exlan. These fibers are typical examples of three different groups whose classifications are based on their transition temperatures and heat distortion temperatures.

7.6.1 Molecular Parameters, Fiber Spinning, and Heat Treatments

Aromatic polyesters were first reported almost ninety years ago in the case of a polyester synthesized from HBA by Fisher in 1910 [188]. The difficulties in synthesis kept the development of aromatic polyesters slow until the 1950s. Since then, successful preparations of high molecular weight aromatic polyesters started appearing and continuously added to the list of this class of polymers. A few examples are polyoxybenzoate [189,190], a series of polyesters containing bisphenols and dicarbonic acid [191–197], and random copolyesters containing carbonate and carboxylate groups [198]. However, the turning point came with two significant findings in the early 1970s. First was a fiber forming oxybenzoyl copolyester [199] which exhibited lower softening and melting temperatures than that of its homopolymer, and second was recognition of the liquid crystalline behavior of copolyesters containing HBA and PET blocks [200–225].

Following these findings, a number of new aromatic copolyesters were synthesized. Although a nearly infinite number of combinations of different monomers are available for research in this area, very few of these copolyesters are developed into fiber form. The focus of this research is, in fact, to obtain copolyesters which show high glass transition temperatures and high degrees of order. In this way, these materials have high temperature resistance, but relatively low melting temperatures and low ordered liquid crystalline behavior which provides a large processing window and low shear viscosity. To meet this goal, one cannot sacrifice the chain rigidity and linearity of the aromatic polyesters, but may only disturb the chain structure to a certain degree in order to lower the transition temperature. Thus, the series of copolyesters containing HBA and PET blocks does not fit into this category due to its relatively low transition temperatures and mediocre tensile properties caused by the introduction of flexible segments.

Three concepts to design chemical structures have been proposed [226]. First, one may introduce different lengths of comonomers to disturb the ordered structure along the fiber direction. This has been done in the case of HBA/HBP (4,4′-dihydroxybiphenyl). Second, kinks can be designed into the chemical structures such as in the case of HBA/HNA, where the HNA monomer is not linear. Third, a disturbance can be made in the lateral packing by introducing large side pendant groups such as in the case of *p*-benzenedicarboxylic acid (TPA)/phenylhydroquinone (PHQ)/(1-phenylethyl)hydroquinone (PEHQ) [227]. Combinations of these approaches may also be used.

Polyesterification reactions usually have very slow reaction rates and it is difficult to achieve high molecular weight samples. Reactions are thus carried out at elevated temperatures over long periods of time with the addition of catalysts in some cases. Four basic polymerization processes are usually used: melt, solution, interfacial and slurry polymerizations. For example, the copolymers of HBA/PET with the generalized chemical structure and synthesis shown below comprised one of the first series of thermotropic liquid crystalline polymers developed into fibers.

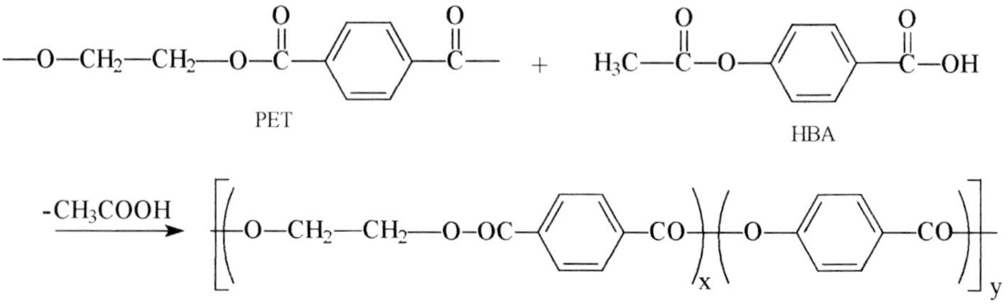

—O—CH_2—CH_2—O—C(=O)—C_6H_4—C(=O)— + H_3C—C(=O)—O—C_6H_4—C(=O)—OH

PET HBA

$\xrightarrow{-CH_3COOH}$ [(O—CH_2—CH_2—O-OC—C_6H_4—CO$)_x$(O—C_6H_4—CO$)]_y$

X7G

The compositions of x and y are varied and it was found that only when the HBA composition exceeds 30% does the liquid crystalline behavior appear. In this reaction *p*-acetoxybenzoic acid at 275 °C reacts with itself and with PET to give short acetoxy-terminated and carboxyl-terminated segments which are then condensed together to give a high molecular weight polyester by heating under reduced pressure.

Another example is HBA/HNA with different compositions. Comonomers of *p*-acetoxybenzoic acid and 6-acetoxy-2-naphthoic acid were synthesized by melt condensation around 300 °C. The degree of polymerization is approximately 100. Its generalized chemical structure is

H_3C—C(=O)—O—$C_{10}H_6$—COOH + H_3C—C(=O)—O—C_6H_4—C(=O)—OH

HNA HBA

$\xrightarrow{-CH_3COOH}$ [(OC—$C_{10}H_6$—O$)_x$(OC—C_6H_4—O$)_y]_n$

Vectran

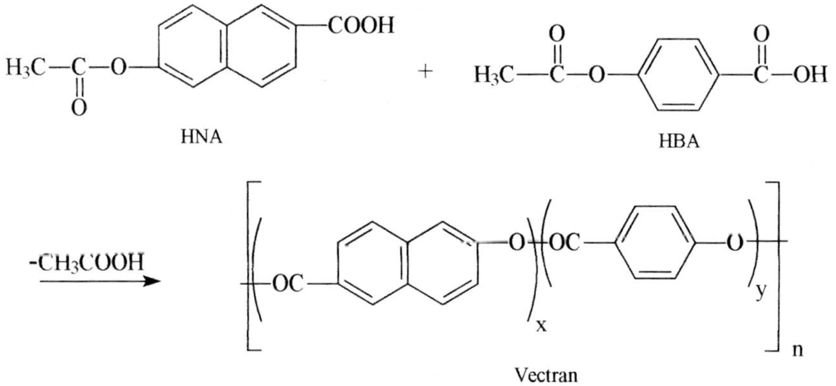

The third example is Ekonol® fibers. Its generalized chemical structure is

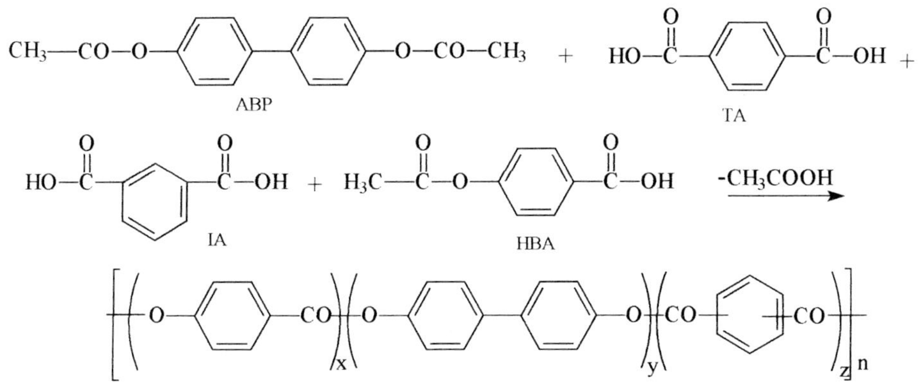

CH_3—CO—O—⟨ABP⟩—O—CO—CH_3 + HO—C(=O)—⟨TA⟩—C(=O)—OH + HO—C(=O)—⟨IA⟩—C(=O)—OH + H_3C—C(=O)—O—⟨HBA⟩—C(=O)—OH -CH_3COOH →

ABP TA IA HBA

Ekonol

The spinning process of these thermotropic liquid crystalline copolyesters is similar to the conventional PET melt spinning method. The polymers have a typical intrinsic viscosity of 1.5 to 3.0 dL/g. They are first washed in water, alcohol or acetone in order to remove impurities. Next, they are heated near the isotropization temperature and fibers are spun from their liquid crystalline state (usually in a temperature range of 300 °C to 400 °C). The liquid crystalline melt is then spun through spinnerettes to form fibers. The spinning speed is in the range of several hundred to several thousand meters per minute. The as-spun fibers are heat treated at elevated temperatures slightly below the crystal melting temperature for several hours in order to further build up the molecular weight and enhance the ordered structure. This heat treatment time is much longer than that of a conventional heat treatment process in PET fibers because in the melt, low molecular weight thermotropic liquid crystalline polymers are usually used in order to decrease the melt viscosity in order to ease processing. However, in the fiber form, higher molecular weight drastically increases fiber tensile properties, specifically, tensile strength. During heat treatment, the solid state polymerization which occurs during the long period of time at elevated temperatures can cause a significant increase in molecular weight and therefore, an improvement in the tensile properties. Some examples for aromatic copolyesters show that relatively short heat treatment times may also yield satisfactory tensile properties such as in the case of substituents attached to hydroquinone comonomers.

In contrast to the solution spun fibers from lyotropic liquid crystalline polymers, melt spun fibers from thermotropic liquid crystalline polymers do not require solvent recycling and disposal, they usually possess less skin-core structural differences and the fibers have a more regular cylindrical shape and less microvoids than those spun from liquid crystalline solutions.

7.6.2 HBA/PET Copolyester Fiber Structure, Morphology, and Properties

The copolymers of HBA/PET were first developed by Eastman Chemical Company under the trade name of X7G® [200–225]. Although this series of copolyesters was viewed to have random composition sequences, it was later determined that the two components are more or less phase separated by using several experimental methods. This biphasic morphology certainly affects the fiber structure and properties.

WAXD fiber patterns have been reported by several research groups for HBA/PET fibers with varying compositions. The most commonly reported compositions are 30/70, 60/40 and 80/20 for HBA/PET. For 30/70 HBA/PET, in WAXD fiber patterns, the strongest reflections are the d-spacings at 0.50 nm, 0.391 nm, and 0.348 nm. These reflections correspond closely to the 0.51 nm, 0.383 nm and 0.343 nm reflections from a pure PET WAXD fiber pattern [204]. After annealing, WAXD fiber patterns of 60/40 HBA/PET fibers show reflections of 0.448 nm, 0.424 nm and 0.373 nm, which are close to the reflections of 0.448 nm, 0.422 nm and 0.376 nm for a pure poly(*p*-hydroxybenzoate) [poly(HBA)] [200]. This has also been found in the case of 80/20 HBA/PET fibers in which the WAXD fiber patterns closely agree with poly(HBA) [207,224]. The crystal structures determined by these WAXD fiber patterns were also shown to be similar to those determined by the ED patterns of single crystals of poly(HBA). The reflection d-spacings and intensities were analyzed in detail in order to interpret the molecular packing in the fibers. The copolymers contain ordered regions rich in poly(HBA) with some ethylene terephthalate defects and disordered regions rich in ethylene terephthalate [211].

Morphological studies showed the existence of 30-40 μm ordered domains or lamellar blocks of HBA homopolymers embedded in an amorphous or mesophase medium [207,224]. The biphasic structure of 60/40 HBA/PET fibers was also confirmed by the study of transition behavior *via* DSC [205], TEM imaging and micro ED [225]. It was reported that the fiber matrix contained oriented, ordered regions rich in poly(HBA) and a separated phase of 0.3–1.5 μm in size which appeared to be amorphous and rich in PET [225]. The biphasic structure was also observed in 80/20 HBA/PET fibers under the influence of certain flow and temperature histories [228,229]. When the fibers were spun at relatively low temperatures, a domain structure of randomly oriented crystallites was observed, while in fibers spun at high temperatures a well-developed fibrillar structure was found. WAXD results showed that molecular orientation increased with increasing processing temperature [230]. Although it is generally agreed that the biphasic structure is an indication of microphase separation of the two components in HBA/PET copolyesters, there is still a great deal of discussion regarding the sequence of randomness or blockiness in the copolyesters. This must be associated with monomer compositions, polymerization conditions, as well as thermal and external force field histories of the polymers during fiber spinning.

Most HBA/PET copolyester fibers exhibit moderate tensile moduli. The highest modulus was around 32 GPa. However, the tensile strength at break was very low (less than 0.3 GPa) [204]. This is possibly due to the low molecular weight of the copolymers used in the investigation. These data were obtained for 60/40 HBA/PET fibers oriented at 225°C. The copolyester with 30/70 HBA/PET showed much lower tensile properties than 60/40 HBA/PET fibers. It was reported that the take-up speed of fiber spinning as well as the heat treatment were very important in determining the ultimate mechanical properties [204]. Recently, it was found that the tensile properties of these copolyester fibers could be improved by various annealing processes [231]. However, the thermal behavior of as-spun and heat treated copolyester fibers showed thermal transition behavior which was similar to that of the bulk samples [205].

7.6.3 HBA/HNA Copolyester Fiber Structure, Morphology, and Properties

HBA/HNA fibers have been named Vectran® fibers and are marketed by Hoechst Celanese and are under license to Kuraray in Japan. The most common compositions of these HBA/HNA copolyesters are 25/75, 30/70, 50/50, 58/42, and 75/25. The fibers exhibit very high molecular orientation which is developed during the fiber spinning process [232]. WAXD fiber patterns for these copolyesters were carefully analyzed [221,233,234]. The results showed relatively sharp reflections at d-spacings of about 0.45 nm and 0.26 nm along the equatorial direction and a reflection in quadrant at 0.33 nm. There was an amorphous scattering spot along the equatorial direction with a d-spacing which ranged from 0.3 to 0.6 nm. The number of reflections along the meridional direction differed with varying copolyester compositions. Weak diffuse intensity was also seen above and below the equator. These results indicate that the orientation of the molecular direction in the fibers is along the fiber axis. Laterally ordered structures exist in the fibers which lead to sharp reflections along the equatorial direction in the fiber patterns. The aperiodicity of the meridional reflections is attributed to the variation of copolymer compositions which disrupt the order along the fiber direction. Furthermore, these aperiodic reflections along the meridian were calculated from an atomic model for the polymers [234]. The agreement between the observed and calculated d-spacings as well as the intensities support the presence of random compositional sequences of HBA and HNA in the copolyesters. It was speculated that with increasing HNA composition, the chain linearity may decrease. This effect was studied *via* a concept of persistence length. It was found that the persistence length decreases from about 13 nm to 9 nm as the HNA composition increases [234].

Banded textures can be observed in sheared HBA/HNA copolyesters under PLM and dark field TEM if certain ranges of shear rate and viscosity are selected [235]. The bands always lie perpendicular to the shear direction. The bands are associated with a periodic variation in the director orientation about the flow axis. Copolyesters show a more gentle "serpentine" molecular trajectory [235]. Several levels of superimposed fibrillar structures were observed in the highly oriented copolyester fibers [236,237]. They included microfibrils about 50 nm in size, fibrils of about 500 nm and macrofibrils of around 5 μm. A defect structure with a periodicity of about 50 nm in size was also found [236, 237]. These features are similar to those observed in PPTA fibers. Skin-core morphology can also be found in the fibers although it is not as prominent as in the case of fibers from lyotropic liquid crystalline polymers. The formation mechanism is, however, quite different. In thermotropic liquid crystalline polymer fibers, the skin-core effect is mainly caused by the acceleration of elongational flow in the surface regions while there is deceleration in the central regions of the fibers [238].

Among the copolyester fibers with different HBA/HNA compositions, the 75/25 HBA/HNA copolyester fiber showed the best tensile properties. The tensile strength at break was 2.5 GPa and the tensile modulus was as high as 70 GPa. Copolyester fibers with other compositions showed only modest tensile properties [239]. A recent study reported by Nippon Exlan reported a 2.8 GPa tensile strength, a 71 GPa tensile modulus and 3.7% elongation at break for HBA/HNA yarn. Another report showed a tensile strength of 2.6 GPa and a modulus of 65 GPa [235]. These properties are close to regular PPTA fibers

(Kevlar® 29). Results for the coefficient of thermal expansion in the fibers along the fiber direction showed negative values with increasing temperature. These were accompanied by a decrease in the tensile modulus. Specifically, after the temperature exceeds the glass transition temperature (around 120°C), both properties exhibit a sharp decrease, indicating that much of the polymer is in the low ordered liquid crystalline state and the solid properties are maintained by a crystalline fraction of around 20% [235]. Finally, it should also be pointed out that the tensile strength is molecular weight dependent [232]. The compressive strength of the HBA/HNA fibers was reported to be around 0.45 GPa [240].

7.6.4 Ekonol® Copolyester Fiber Structure, Morphology, and Properties

Ekonol® fibers are a series of copolyesters composed of several different types of comonomers: *p*-acetoxybenzoic acid, 4,4′-diacetoxybiphenyl, and terephthalic or isophthalic acid. The copolyesters were synthesized by melt polymerization and enter an anisotropic melt at from 330 °C to 350 °C, depending upon the copolymer composition. Fibers can be spun in a temperature range between 360 °C and 400 °C [199,241].

Since the fiber spinning temperature is relatively high, the melt spinning was carried out in an inert atmosphere to minimize polymer degradation. It was reported that the fiber tensile properties were very sensitive to the heat treatment temperature [242]. The optimal heat treatment temperature range is a window from 300 °C to 320 °C. The tensile properties reported were 3.4 to 3.8 GPa for tensile strength (4.2 GPa was also reported), 2.4%–2.9% elongation at break and a tensile modulus of around 136 GPa. It seems that the introduction of biphenyl units is critical in the enhancement of the tensile properties for Ekonol® fibers. The fibers show good flame resistance and low moisture absorption. However, the fiber morphology and other properties are not known at this moment.

7.7 Concluding Remarks

High performance polymer fibers have excellent mechanical properties, thermal and thermo-oxidative stability, chemical resistance and electrical properties. These properties significantly exceed those of traditional textile fibers. The high strength and modulus of these fibers, combined with their low density make them very competitive among fibrous materials, especially when the comparison is based on specific properties. Liquid crystalline behavior provides an important route to polymer fiber formation. Fibers from liquid crystalline polymers include a major portion of high performance fibers, whether made from lyotropic or thermotropic liquid crystalline states. They generally show good axial strength and moduli. The natural extension of the molecules in the liquid crystalline state is maintained in the solid state. The crystallization process is an effective mechanism of self-organization which locks in the extended chain conformation of molecules and holds them in almost perfect parallel alignment with one other.

These fibers are not limited to be a reinforcing fiber component in structural applications, but may also be used in heavy duty ropes, belting and bullet-proof clothing. DuPont's Kevlar® fibers were the first series of commercially available high performance fibers on the market. It was almost twenty years before other high performance fibers were introduced. Along a similar line, fibers from lyotropic liquid crystalline polymers such as PBZO and PBZT fibers also exhibit impressive tensile properties. Polyamides and polyimides spun from their isotropic solutions are also based on dry-jet wet or wet spinning route. Fibers from thermotropic liquid crystalline polymers have the advantage of easy processing in conjunction with excellent moldability and the absence of solvent problems. However, crystallization in these fibers is commonly not perfect due to the chemical modifications which lower the transition temperature in order to make melt spinning of the polymer possible. As a result, tensile properties at elevated temperatures for fibers from thermotropic liquid crystalline polymers are usually not as high as those of lyotropic liquid crystalline polymers.

The last twenty-five years was a period of important development for high performance aromatic polymer fibers in research and industrial commercialization. After several generations of effort, the properties of these products have been improved to the point that the molecular and supra-molecular structural architectures may be designed to maximize macroscopic properties. On the other hand, we still only reach about half of the theoretical tensile modulus. In addition, quantitative structure, morphology and their relationships to fiber properties are still not fully established. Thus, this field will require continuous effort in order to further understand these relationships and develop new high performance fiber products.

7.8 References

1. Gatterman, L., Ritschke, A. *Ber Deutsch. Chem. Ges.* (1890) 23, p. 1738, from Kelker, H. *Mol. Cryst. Liq. Cryst.* (1973) 23, p. 1
2. de Gennes, P.-G., Prost, J. *The Physics of Liquid Crystals* (1993) Clarendon, New York
3. Chandrasekhar, S. *:Liquid Crystals* (1976) University Press, Cambridge
4. Gray, G. W., Goodby, J. W. C. *Smectic Liquid Crystals* (1984) Leonard Hill, London
5. Pershan, P. S. *Structure of Liquid Crystal Phases* (1988) World Scientific, New Jersey
6. de Gennes, P.-G. *Angew.-Chem. Int. Ed. Engl*, (1992) 31, p.842
7. Elliot, A., Ambrose, E. J. *Discuss. Faraday Soc.* (1950) 9, p.246
8. Young, H. H. *Aromatic High-Strength Fibers* (1989) J. Wiley & Sons, New York
9. Hongu, T., Phillips, G. O. *New Fibers* (1990) Ellis Horwood, New York
10. Cheng, S. Z. D., Harris, F. W. Polyimide Fibers, Aromatic, in "*International Encyclopedia of Composites*", (1992) VCH Publishers, New York, Vol. 6, pp. 293–309
11. Jiang, H., Adames, W. W., Eby, R. K. in "*Handbook of Fiber Science and Technology, Vol. III, High Technology Fibers, Part D*", Lewin, M., Preston, J. Eds. (1996) Chapt. 2, p. 171
12. Jiang, H., Adams, W. W., Eby, R. K., in "*Material Science and Technology, Vol. 12*", Thomas, E. L. Ed. VCH, Weinheim, (1993), Chapter 13, p. 597
13. Ericksen, J. L. *Arch. Ration. Mech. Anal.* (1960) 4, p. 231
14. Leslie, F. M . *Quart. J. Mech. Appl. Math.* (1966) 19, p. 357

15. Doi, M. *J. de Phys.* (1975) 36, p. 607
16. Asada, T., Muramatsu, H., Watanabe, R., Onogi, S. *Macromolecules* (1980) 13, p. 867
17. Asada, T. in *Polymer Liquid Crystals.* Ciferri, A., Krigbaum, W. R., Meyer, R. B. (Eds.) (1982) Academic Press, New York
18. Hermans, J. *J. Coll. Sci.* (1962) 17, p. 638
19. Aharoni, S. M. *Polymer* (1980) 21, p. 1413
20. Morgan, P. W. *Macromolecules* (1977) 10, p. 1381
21. Sokolova, T. S., Jefimova, S. C., Volokhina, A. V., Papkov, S. S., Kudrjacev, G. I. *Khim. Volokna* (1974) 1, p. 26
22. Vasileva, N. V., Platonov, V. A., Kulichikhin, V. G., Papkov, S. S., Efimova, S. G. *Khim. Volokna* (1975) 5, p. 71
23. Navard, P., Haudin, J. M. *J. Polym. Sci. Polym. Phys. Ed.* (1986) 24, p. 189
24. Mewis, J., Moldenaers, P. *Chem. Eng. Commun.* (1987) 53, p. 33
25. Won Choe, E. W., Kim, S. N. *Macromolecules* (1981) 14, p. 191
26. Onsager, L. *Ann. N. Y. Acad. Sci.* (1949) 51, p. 627
27. Flory, P. J. *Adv. Polym. Sci.* (1984) 59, p. 1
28. Jackson, W. J., Kuhfuss, H. F. *J. Polym. Sci. Polym. Chem. Ed.* (1976) 14, p. 2043
29. Wissbrum, K. F. *Brit. Polym. J.* (1980) 13, p. 163
30. Sun, T., Lin, Y. G., Winter, H. H., Porter, R. S. *Polymer* (1989) 30, p. 1257
31. Kalika, D. S., Giles, D. W., Denn, M. M. *J. Rheol.* (1990) 34, p. 139
32. Masuda, T., Fujiwara, K., Takahashi, M. *Int. Polym. Process.* (1991) 6, p. 225
33. Calundann, G. W. U.S. Patents (1978) 4,067,852; (1979) 4,161,470; (1980) 4,185,995
34. Wissbrun, K. F., Kiss, G., Cogswell, F. N. *Chem. Eng. Commun.* (1987) 53, p. 149
35. Cocchini, F., Nobile, M. R., Acierno, D. *J. Rheol.* (1991) 35, p. 1171
36. Guskey, S. M., Winter, H. H. *J. Rheol.* (1991) 35, p. 1191
37. Kim, S. S., Han, C. D. *Macromolecules* (1993) 26, p. 3176; p. 6633
38. Han, C. D., Kim, S. S. *J. Rheol.* (1994) 38, p. 13; p. 31
39. Kim, S. S., Han, C. D. *J. Polym. Sci. Polym. Phys. Ed.* (1994) 32, p. 371
40. Kim, S. S., Han, C. D. *Polymer* (1994) 35, p. 93
41. Chang, S. Ph.D dissertation, Department of Polymer Engineering, The University of Akron, Akron, Ohio 44325, 1996
42. Kiss, G., Porter, R. S. *J. Polym. Sci. Polym. Sympo.* (1978) 65, p. 193
43. Marrucci, G., Maffettone, P. L. *Macromolecules* (1989) 22, p. 4076
44. Larson, R. G. *Macromolecules* (1990) 23, p. 3983
45. Doi, M. *J. Polym. Sci. Polym. Phys. Ed.* (1981) 19, p. 229
46. Ferry, J. D. *Viscoelastic Properties of Polymers,* 3 rd Ed., (1980) Wiley, New York
47. Doi, M., Edward, S. F. *The Theory of Polymer Dynamics,* (1986) Clarendon Press, Oxford
48. Staudinger, H. in *Die Hochmolekularen Organischen Verbindunger,* (1932) Springer-Volag, Berlin, p. 111
49. Calundann, G., Jaff, M., Jones, R. S., Yoon, H. in *Fiber Reinforcements for Composite Materials.* Bunsell, A. R. Ed., Elsevier, Amsterdam, Chapt. 5
50. Kunugi, T., Suzuki, A., Hashimoto, M., *J. Appl. Polymn. Sci.* (1981) 26, p. 1961
51. Northolt, M. G., Sikkema, D. J. *Adv. Polym. Sci.* (1990) 98, p.115
52. Northolt, M. G., Sikkema, D. J. in *Liquid Crystal Polymers: from Structures to Applications.* Collyer, A. A. Ed., Elsevier Amsterdam, (1992) Chapt. 6
53. Kwolek, S. L. U.S. Pat. (1972) 3,671,542; (1974) 3,819,587
54. Bair, T. I., Morgan, P. W. U.S. Pat. (1972) 3,673,143; (1974) 3,817,941
55. Blades, H. U.S. Pat. (1973) 3,767,756, (1975) 3,869,429
56. Schaefgen, J. R., Folidi, V. S., Logillo, F. M., Good, V. H., Gulrich, L. W., Killian, F. L. *Polym. Prepr. (Am. Chem. Soc. Div. Polym. Chem.)* (1976) 17, p. 69
57. Wolfe, J. F., Loo, B. H. U.S. Pat. (1980) 4,225,700
58. Wolfe, J. F., Loo, B. H., Arnold, F. E. *Polym. Prepr. (Am. Chem. Soc. Div. Polym. Chem.)* (1978) 19, p. 1
59. Wolfe, J. F., Loo, B. H., Arnold, F. E. *Macromolecules* (1981) 14, p. 915
60. Wolfe, J. F. in *Encyclopedia of Polymer Science and Engineering*, Kroschwitz Ed. (1988), Wiley and Sons, New York, p. 601

61. Wolfe, J. F. in *The Material Science and Engineering of Rigid-Rod Polymers,* Adams, W. W., Eby, R. K., McLemore, D. E. Eds, (1989) Pittsburgh, 134, p. 83
62. Allen, S. R., Filippov, A. G., Farris, R. J., Thomas, E. L. *J. Appl. Polym. Sci.* (1981) 26, p. 291
63. Allen, S. R., Filippov, A. G., Farris, R. J., Thomas, E. L. Wang, C. P., Berry, G. C., Chenevey, E. C. *Macromolecules* (1981) 14, p. 1135
64. Uy, W. C., Mammone, J. F. *Am. Chem. Soc. Div. Cellulose, Paper and Textile Fibers* (1985) Miami
65. Wolfe, J. F. Sybert, P. D. U.S. Pat. (1985) 4,533,693; (1987) 4,703,103
66. Yang, H. H., Chouinard, M. P., Lingg, W. J. *J. Appl. Polym. Sci.* (1982) 20, p. 981; (1987) 34, p. 1399
67. Hindeleh, A. M., Halin, N. A., Zig, K. A. *J. Macromol. Sci. Phys.* (1984) B23, p. 289
68. Northolt, M. G., van Aartsen, J. J. *J. Polym. Sci. Polym. Lett. Ed.* (1973) 11, p. 333
69. Tashiro, K., Kobayashi, M., Tadokoro, H. *Macromolecules* (1977) 10, p. 413
70. Northolt, M. G. *Br. Polym. J.* (1981) 13, p. 64
71. Panar, M., Avakian, P., Blume, R. C., Gardner, K. H., Gierke, T. D., Yang, H. H. *J. Polym. Sci. Polym. Phys. Ed.* (1983) 21, p. 1955
72. Haraguchi, K., Kajiyama, T., Takayanagi, M. *J. Appl. Polym. Sci.* (1979) 23, p. 903; p. 915
73. Dobb, M. G., Johnson, D. J., Saville, B. P. *J. Polym. Sci. Polym. Phys. Ed.* (1977) 15, p. 2201
74. Dobb, M. G., Johnson, D. J., Saville, B. P. *Polymer* (1981) 22, p. 961
75. Socci, E. P., Thomas, D. A., Grubb, D. T., Adams, W. W., Eby, R. K. *Polymer* (1996) 37, p. 5005
76. DeTeresa, S. J., Allen, S. R., Farris, R. J. Porter, R. S. *J. Mat. Sci.* (1984) 19, p. 57
77. van der Zwagg, S., Kampschoer, G. in *Integration of Fundamental Polymer Science & Technology*, 2nd Ed., Vol. 2, (1988) Elsevier, London, p. 545
78. van der Zwagg, S., Picken, S. J., can Sluijs, C. P. in *Integration of Fundamental Polymer Science & Technology*, 2nd Ed., Vol. 3, (1989) Elsevier, London, p. 199
79. Argon, A. S. in *Glass Science & Technology*, (1980) Capt. 3, p. 5
80. Argon, A. S. *Proceeding of International Congress*, Gainesville, FL (1981) p. 393
81. DeTeresa, S. J., Farris, R. J. Porter, R. S. *Polym. Composites* (1982) 3, p. 57
82. DeTeresa, S. J., Porter, R. S., Farris, R. J. *J. Mat. Sci.* (1985) 20, p. 1645
83. DeTeresa, S. J., Porter, R. S., Farris, R. J. *J. Mat. Sci.* (1988) 23, p. 1886
84. DeTeresa, S. J., Farris, R. J. in *The Material Science and Engineering of Rigid-Rod Polymers,* Adams, W. W., Eby, R. K., McLemore, D. E. Eds, (1989) Pittsburgh, 134, p. 375
85. Robinson, I. M., Yeung, P. H. I., Caliotis, C., Young, R. J., Batchelder, D. J. *J. Mat. Sci.* (1986) 21, p. 3440
86. Hess, T. B., Barrett, C. S., *Trans. AIME J. Mat.* (1949) 185, p. 599
87. Gilman, J. J. *J. Mat.* (1954) 192, p. 621
88. Schemaker, J. R., Horn, T., Haaland, P., Pachter, R., Adams, W. W. *Polymer* (1992) 33, p. 3351
89. Welsh, W. J., Bhaumik, D., Jaffe, H. H., Mark, J. E. *Polym. Eng. Sci.* (1984) 24, p. 218
90. Wierschke, S. G. in *The Material Science and Engineering of Rigid-Rod Polymers,* Adams, W. W., Eby, R. K., McLemore, D. E. Eds, (1989) Pittsburgh, 134, p. 313
91. Wierschke, S. G., Shoemaker, J. R., Haaland, P., Pachter, R., Adams, W. W. *Polymer* (1992) 33, p. 3357
92. Klunzinger, P. E., Green, K. A., Eby, R. K., Farmer, B. L., Adams, W. W. Czornyj, G. *SPE Proceeding*, (1991) p. 1532
93. Klunzinger, P. E., Eby, R. K., Adams, W. W. in "*MRS Proceeding*", Aksay, I., Naer, E., Saeikaya, M., Tirrell, D. Eds., (1992) Pittsburgh, 255, p. 119
94. E. I. Du Pont de Nemours & Company, Inc. *Technical Bulletins* (1978) K-1 and K-2
95. Schaefgen, J. R., Bair, T. I., Ballou, J. W., Kwolek, S. L., Morgan, P. W., Panar, M., Zimmerman, J. in *Ultra-High Modulus Polymers*, Ciferri, A., Ward, I. M. Eds., (1979) Applied Science, London
96. Penn, L., Larson, F. *J. Appl. Polym. Sci.* (1979) 23, p. 59
97. Bair, T. I., Morgan, P. W., Killian, F. L. *Macromolecules* (1977) 10, p. 1396
98. Wu, Z.-Q., Yoon, Y., Harris, F. W., Cheng, S. Z. D., Chuang, K. C. *SPE Proceeding* (1996) Brookfield Connecticut, 54, p. 3038
99. Bunsell, A. R. *J. Mat. Sci.* (1975) 10, p. 1300
100. Cook, J. in *3rd Riso International Symposium on Metallurgy and Materials Science*, Lilholt, H., Talreja, R. Eds., (1982) pp. 193–198

101. Eriksen, R. H. *Polymer* (1985) 26, p. 733
102. Lafitte, M. H., Bunsell, A. R. *Polym. Eng. Sci.* (1982) 3, p. 57
103. Wu, Z.-Q., Zhang, A., Cheng, S. Z. D., Huang B., Qian, B. *J. Polym. Sci. Polym. Phys. Ed.,* (1990) 28, p. 2565
104. Sinclair, D. *J. Appl. Phys.* (1950) 21, p. 380
105. Greenwood, J. H., Rose, P. G. *J. Mat. Sci.* (1974) 9, p. 1809
106. Allen, S. R. *J. Mat. Sci.* (1987) 22, p. 853
107. Wang, C. S., Bai, S. J., Price, B. P. *Proc. Am. Chem. Soc. Polym. Mat. Sci. Eng.* (1989) 61, p. 550
108. Drzal, L. T. *AFWAL-TR-86-4003* (1986)
109. Miwa, M., Tsushima, E., Takayasu, J. *J. Appl. Polym. Sci.* (1991) 43, p. 1467
110. Miwa, M., Tsushima, E., Takayasu, J. *Sen-I Gakkaishi* (1991) 47, p. 171
111. Ohsawa, T., Miwa, M., Kawade, M. *J. Appl. Polym. Sci.* (1990) 39, p. 1733
112. Doyne, T. A., Palazotto, A. N., Schuppe, T., Lee, Y.-C., Wang, C. S. *Proc. Am. Chem. Soc. Polym. Mat. Sci. and Eng.* (1990) 63, p. 982
113. Fawaz, S. A., Palazotto, A. N., Wang, C. S. in *The Material Science and Engineering of Rigid-Rod Polymers,* Adams, W. W., Eby, R. K., McLemore, D. E. Eds, (1989) Pittsburgh, 134, p. 381
114. Fawaz, S. A., Palazotto, A. N., Wang, C. S. *Polymer* (1992) 33, p. 100
115. Macturk, K. S., Eby, R. K., Adams, W. W. *Polymer* (1991) 32, p. 1782
116. Jiang, H., Damodaran, S., Abhiraman, A. S., Desai, P., Kumar, S. in *High Performance Polymers and Polymer Matrix Composites,* Eby, R. K., Evers, R. C., Meador, M. A., Wilson, D. Eds, (1993) Pittsburgh, 305, p. 135
117. Bunsell, A. R. *J. Mat. Sci.* (1974) 9, p. 1804
118. Lafitte, M. H., Bunsell, A. R. *J. Mat. Sci.* (1982) 17, p. 2391
119. Frantini, A. V., Lenhert, P. G., Resch, T. J., Adams, W. W. in *The Material Science and Engineering of Rigid-Rod Polymers,* Adams, W. W., Eby, R. K., McLemore, D. E. Eds, (1989) Pittsburgh, 134, p. 134
120. Krause, S. J., Haddock, T. B., Vezie, D. L., Lenhert, P. G., Hwang, W.-F., Price, G. E., Helminiak, T. F., O'Brien, J. F., Adams, W. W. *Polymer* (1988) 29, p. 1354
121. Martin, D. C., Thomas, E. L. *Macromolecules* (1991) 24, p. 2224
122. Odell, J. A., Keller, A., Atkins, E. D. T., Miles, M. J. *J. Mat. Sci.* (1981) 16, p. 3309
123. Martin, D. C., Thomas, E. L. *J. Mat. Sci.* (1991) 26, p. 5171
124. Kunar, S., Warner, S., Grubb, D. T., Adams, W. W. *Polymer* (1994) 35, p. 5408
125. Hunsaker, M. E., Price, G. E., Bai, S. J. *Polymer* (1992) 33, p. 2128
126. Bai, S. J., Price, G. E. *Polymer* (1992) 33, p. 2136
127. Martin, D. C., Thomas, E. L. *Phil. Mag.* (1991) 64, p. 903
128. Allen, S. R., Farris, R. J, Thomas, E. L. *J. Mat. Sci.* (1985) 20, p. 2727
129. Rakas, M. A. Farris, R. J. *J. Appl. Polym. Sci.* (1990) 40, p. 823
130. Roche, E. J., Takahashi, T., Thomas, E. L. in *Am. Chem. Soc. Symposium Series,* (1980) 141, p. 303
131. Krause, S. J., Haddock, T. B., Vezie, D. L., Lenhert, P. G., Hwang, W.-F., Price, G. E., Helminiak, T. E., O''rien, J. F., Adams, W. W. *Polymer* (1988) 29, p. 1354
132. Kumar, S., Warner, S., Grubb, D. T., Adams, W. W. *Polymer* (1994) 35, p. 5408
133. Young, R. J., Lu, D., Day, R. J., Knoff, F., Davis, H. A. *J. Mat. Sci.* (1992) 27, p. 5431
134. Young, R. J., Day, R. J., Zakikhani, M. in *The Materials Science and Engineering of Rigid-Rod Polymers,* Adams, W. W., Eby, R. K., McLemore, D. E. Eds, (1989) Pittsburgh, 134, p. 351
135. Young, R. J., Day, R. J., Zakikhani, M. *J. Mat. Sci.* (1990) 25, p. 127
136. Huh, W., Kumar, S., Helminiak, T. E., Adams, W. W. *SPE Proceedings* (1990) p. 1245
137. Takahashi, T., Miwa, M., Susurai, S. *J. Appl. Polym. Sci.* (1983) 28, p. 589
138. Chau, C. C., Blackson, J., Im, J. *Polymer* (1995) 36, p. 2511
139. Martin, D. C., Thomas, E. L. in *The Materials Science and Engineering of Rigid-Rod Polymers,* Adams, W. W., Eby, R. K., McLemore, D. E. Eds, (1989) Pittsburgh, 134, p. 415
140. Takahashi, T., Xiao, C. F., Sakurai, K. *Sen-i Gakkaishi* (1991) 47, p. 397
141. Takahashi, T., Suzuki, K., Aoki, T., Sakurai, K. *J. Macromol. Sci. Phys.* (1991) B30, p. 101
142. Attenburrow, G. E., Bassett, D. C. *J. Mat. Sci.* (1979) 14, p. 2679
143. Shigematsu, K., Imada, K., Takayanagi, M. *J. Polym. Sci. Polym. Phys.* (1975) 13, p. 73

144. Ohta, T., Kunugi, T., Yobuki, K. *High Tenacity and High Modulus Fibers,* (1988) Kyo Ritsu, Tokyo
145. Im, J., Percha, P. A., Yeakle, D. S. in *The Materials Science and Engineering of Rigid-Rod Polymers,* Adams, W. W., Eby, R. K., McLemore, D. E. Eds, (1989) Pittsburgh, 134, p. 307
146. van der Zwagg, S. *J. Testing Evaluation* (1989) 17, p. 292
147. Uy, W. C., Mammone, J. F. *Can. Textile J.* (1988) 4, p. 54
148. Ledbetter, H. D., Rosenberg, S., Hurtig, C. W. in *The Materials Science and Engineering of Rigid-Rod Polymers,* Adams, W. W., Eby, R. K., McLemore, D. E. Eds, (1989) Pittsburgh, 134, p. 253
149. Jiang, H., Eby, R. K., Adams, W. W., Lenhert, G. in *The Materials Science and Engineering of Rigid-Rod Polymers,* Adams, W. W., Eby, R. K., McLemore, D. E. Eds, (1989) Pittsburgh, 134, p. 341
150. Denny, L. R., Goldfarb, I. J., Soloski, E. J. in *The Materials Science and Engineering of Rigid-Rod Polymers,* Adams, W. W., Eby, R. K., McLemore, D. E. Eds, (1989) Pittsburgh, 134, p. 345
151. Jones, E. G., Pedrick, D. L. in *The Materials Science and Engineering of Rigid-Rod Polymers,* Adams, W. W., Eby, R. K., McLemore, D. E. Eds, (1989) Pittsburgh, 134, p. 407
152. Kumar, S. in *International Encyclopedia of Composites* Vol. 4, (1990) VCH Publisher, Weinhein, Germany, p. 51
153. MaGarry, F. J., Moalli, J. E. *Polymer* (1991) 32, p. 1811
154. MaGarry, F. J., Moalli, J. E. *Polymer* (1991) 32, p. 1816
155. Klunzinger, P. E., Macturk, K. S., Eby, R. K., Adams, W. W. *Polym. Preprints (Am. Chem. Soc. Div. Polym. Chem.)* (1991) 32, p. 187
156. Kumar, S., Adams, W. W. *Polymer* (1990) 31, p. 15
157. Cheng, S. Z. D., Wu, Z.-Q., Eashoo, M., Hsu, S. L.-C., Harris, F. W. *Polymer,* (1991) 32, p. 1803
158. Eashoo, M., Shen, D.-X., Wu, Z.-Q., Lee, C. L., Harris, F. W., Cheng, S. Z. D. *Polymer,* (1993) 34, p. 3209
159. Eashoo, M., Wu, Z.-Q., Zhang, A., Shen, D.-X., Wu, C., Harris, F. W., Cheng, S. Z. D. Gardner, K. H., Hsiao, B. S. *Macromol. Chem. Phys.* (1994) 195, p. 2207
160. Shen, D.-X., Wu, Z.-Q., Liu, J., Wang, L., Lee, S., Harris, F. W., Cheng, S. Z. D., Blackwell, J., Wu, T. *Polym. & Polym. Composites* (1994) 2, p. 149
161. Li, W., Wu, Z.-Q., Jing, H., Harris F. W., Cheng, S. Z. D. *J. Mat. Sci.* (1996) 31, p. 4423
162. Li, W., Wu, Z.-Q., Leland, M., Park, J.-Y., Harris F. W., Cheng, S. Z. D. *J. Macromol. Sci. Phys.* (1997) B36, p. 3157
163. Harris, F. W. in *Polyimides,* Wilson, D., Stenzenberger, H. D., Hergenrother P. Eds., Chapman & Hall, New York (1990) Chapter 1, pp. 1–37
164. Harris F. W., Hsu, S.L.-C. *High Perform. Polym.* (1988) 1 p. 1
165. Lee, S.-K., Cheng, S. Z. D., Wu, Z.-Q., Lee, C. L., Harris, F. W., Kyu, T., and Yang, J. C. *Polym. Internl,* (1993) 30, p. 115
166. Park, J.-Y., Harris, F. W., Cheng, S. Z. D. *Polym. Internl,* (1995) 37, p. 207
167. Ozawa, S., Nakagawa, Y., Matsuda, K., Nishihara, T., Yunoki, H. U.S. Pat. (1978) 4,075,172
168. Shimada, K., Mera, H. M., Sasaki, N., Aoki, A. U.S. Pat. (1982) 44,355,151
169. Liu, J., Cheng, S. Z.D., Harris, F. W., Hsiao, B. S., Gardner, K. H. *Macromolecules* (1994) 27, p. 989
170. Liu, J., Kim, D., Harris, F. W., Cheng, S. Z. D. *Polymer* (1994) 36, p. 4048
171. Liu, J., Kim, D., Harris, F. W., Cheng, S. Z. D. *J. Polym. Sci. Polym. Phys. Ed.,* (1994) 32, p. 2705
172. Viney, C., Windle, A. H. *Polymer,* (1986) 27, p. 1325
173. Navard, P., Zachariades, A. E. *J. Polym. Sci. Poly. Phys. Ed.,* (1987) 25 , p . 1089
174. Chen, J., Zhang, A., Yandrasites, M. A., Cheng, S. Z.D., Percec, V. *Makromol. Chem.* (1993) 194, p. 3135
175. Arnold, Jr.,F. E., Cheng, S. Z.D., Hsu, S. L.-C., Lee, C. J., Harris, F. W., Lau, S.-F. *Polymer,* (1992) 33, p. 5179
176. Kaneda, Y., Katsura, T., Nakagawa, K. Jan. Pat. Appl. (1984) 59-69,079
177. Susuga, T. *Polymer* (1988) 29, p. 1562
178. Kaneda, T., Katsura, T., Nakagawa, K., Makino, J. *J. Appl. Polym. Sci.* (1986) 32, p. 3133
179. Kaneda, T., Katsura, T., Nakagawa, K., Makino, J. *J. Appl. Polym. Sci.* (1986) 32, p. 3151
180. Wu, T., Chvalun, S., Blackwell, J., Cheng, S. Z. D., Wu, Z.-Q., Harris, F. W., *Polymer,* (1995), 36, p. 2123
181. Wu, T., Chvalun, S., Blackwell, J., Cheng, S. Z. D., Wu, Z.-Q., Harris, F. W. *Acta Polymerica,* (1995) 36, p. 2123

182. Wu, Z.-Q., Zhang, A. Shen, D.-X., Harris, F. W., Cheng, S. Z. D. *J. Thermal Analysis* (1996) 42, p. 719
183. Kim, Y. H., Harris, F. W., Cheng, S. Z. D. *Thermochimica Acta* (1996) 282/283, p. 411
184. Imuro, H. *Proc. Int. Man-Made Fibers Conf.* (1986) Sept. p. 24
185. Blackwell, J., Cageao, R. A., Biswas, A. *Macromolecules* (1987) 20, p. 667
186. Teijin Ltd. *Technical Bulletin* (1986)
187. Minemura, N., Takabayashi, F., Yamada, S. (1986) U.S. Pat. 4,631,066
188. Fischer, E. *Ann. Chim.* (1910) 372, p. 32
189. Caldwell, J. R. U.S. Pat.(1952) 2,000,376
190. Gilkey, R., Caldwell, J. R. *J. Appl. Polym. Sci.* (1959) 5, p. 198
191. Levine, M., Temin, S. C. *J. Polym. Sci.* (1958) 28, p. 179
192. Conix, A. *Ind. Eng. Chem.* (1959) 51, p. 147
193. Eareckson, W. M. *J. Polym. Sci.* (1959) 40, p. 309
194. Korshak, V. V., Vinogradova, S. V. *Vysokomol. Soyed.* (1959) 1, p. 834; p. 1482
195. Vinogradova, S. V., Korshak, V. V. *Vysokomol. Soyed.* (1959) 1, p. 838
196. Korshak, V. V., Vinogradova, S. V., Lebyedyeva, A. S. *Vysokomol. Soyed.* (1960) 2, p. 61; p. 977; p. 1162
197. Korshak, V. V., Vinogradova, S. V., Valetskii, P. M., Salazkii, N. *Vysokomol. Soyed.* (1961) 3, p. 66; p. 72
198. Goldberg, E. P. U.S. Pat. (1962) 3,030,331; (1965) 3,169,121
199. Cottis, S. G., Economy, J. Nowak, B. E. U.S. Pat. (1972) 3,637,595
200. Wissbrun, K. F. *Br. Polym. J.* (1980) 12, p. 163
201. Manusz, J., Lenz, R. W., Macknight, W. J. *Polym. Eng. Sci.* (1981) 21, p. 1079
202. Jerman, R. E., Baird, D. G. *J. Rheology* (1981) 25, p. 275
203. Antoun, S., Lenz, R. W., Jin, J. I. *J. Polym. Sci. Polym. Chem. Ed.* (1981) 19, p. 1901
204. Acierno, D., LaMantia, F. P., Polizzotti, G., Ciferri, A., Valenti, B. *Macromolecules* (1982) 15, p. 1455
205. Meesiri, W., Menczel, J., Gaur, U., Wunderlich, B. *J. Polym. Sci. Polym. Phys. Ed.* (1982) 20, p. 719
206. Prasadarao, M., Pearce, E. M., Han, C. D. *J. Appl. Polym. Sci.* (1982) 27, p. 1343
207. Zachariades, A. E., Economy, J. Logan, J. A. *J. Appl. Polymn. Sci.* (1982) 27, p. 2009
208. Blumstien, A., Sivaramakrishnan, K. N., Blumstein, R. B., Clough, S. B. *Polymer* (1982) 23, p. 47
209. Blackwell, J., Gutierrez, G. A. *Polymer* (1982) 12, p. 671
210. Mitchell, G. R., Windle, A. H. *Polymer* (1982) 23, p. 1269
211. Blackwell, J., Liezer, G., Gutierrez, G. A. *Macromolecules* (1983) 16, p. 1418
212. Volksen, W., Lyerla, J. R., Economy, J., Dawson, B. *J. Polym. Sci. Polym. Chem. Ed.* (1983) 21, p. 2249
213. Dicke, H.-R., Lenz, R. W. *J. Polym. Sci. Polym. Chem. Sci. Ed.* (1983) 21, p. 2581
214. Higashi, F., Akiyama, N., Koyama, T. *J. Polym. Sci. Polym. Chem. Ed.* (1983) 21, p. 3233
215. Higashi, F., Hoshio, A., Kiyoshige, J. *J. Polym. Sci. Polym. Chem. Ed.* (1983) 21, p. 3241
216. Kyotana, M., Kanetsuna, H. *J. Polym. Sci. Polym. Chem. Ed.* (1983) 21, p. 379
217. Asrai, J., Torium, H., Watanabe, J., Krigbaum, W. R., Ciferri, A. *J. Polym. Sci. Polym. Phys. Ed.* (1983) 21, p. 1119
218. Liezer, G., Schwarz, G., Kricheldorf, H. R. *J. Polym. Sci. Polym. Phys. Ed.* (1983) 21, p. 1599
219. Acierno, D., LaMantia, F. P., Polizzotti, G., Ciferri, A., Krigbaum, W. R., Kotek, R. *J. Polym. Sci. Polym. Phys.* (1983) 21, p. 2027
220. Demartino, R. N. *J. Appl. Polym. Sci.* (1983) 28, p. 1805
221. Gutierrez, G. A., Chivers, R. A., Blackwell, J., Satamatoff, J. B., Yoon, H. *Polymer* (1983) 24, p. 937
222. Krigburm, W. R., Watanabe, J. *Polymer* (1983) 24, p. 1299
223. Ide, A., Ophir, Z. *Polym. Eng. Sci.* (1983) 23, p. 261
224. Zachariades, A. E., Logan, J. A. *Polym. Eng. Sci.* (1983) 23, p. 797
225. Sawyer, L. C. *J. Polym. Sci. Polym. Lett. Ed.* (1984) 22, p. 347
226. Cheng, S. Z. D., Zhang, A., Johnson, R. L., Wu Z.-Q., Wu, H. H. *Macromolecules* (1990) 23, p. 1196
227. Cheng, S. Z. D., Johnson, R. L., Wu, Z., Wu, H. H. *Macromolecules* (1991) 24, p. 150
228. Joseph, E., Wilkes, G. L., Baird, D. G. *Polymer* (1985) 26, p. 689
229. Joseph, E., Wilkes, G. L., Baird, D. G. *Polym. Eng. Sci.* (1985) 25 p. 377
230. Chen, G. Y., Cuculo, J. A., Tucker, P. A. *J. Polym. Sci. Polym. Phys. Ed.* (1988) 26, p. 1677

231. Itoyama, K. *J. Polym. Sci. Polym. Phys. Ed.* (1988) 26, p. 1945
232. Calundann, G., Jaffe, M., Jones, R. S., Yoon, H. in *Fiber Reinforcements for Composite Materials*, Bunsell, A. R. ed., (1988) Elsevier, New York, p. 1
233. Blackwell, J., Gutierrez, G. A., Chivers, R. A. *J. Polym. Sci. Polym. Phys. Ed.* (1984) 22, p. 1343
234. Chivers, R. A., Blackwell, J., Gutierrez, G. A. *Polymer* (1984) 25, p. 435
235. Donald, A. M., Windle, A. H. *Liquid Crystalline Polymers* (1992) Cambridge, London
236. Sawyer, L. C., Jaffe, M. *J. Mat. Sci.* (1986) 21, p. 1897
237. Sawyer, L. C., Grubb, D. T. *Polymer Microscopy* (1987) Chapman and Hall, London
238. Onogi, S., Asada, T. in *Rheology* Astarita G., Marrucci, G., Nicholais L. eds, (1908) Vol. 1, Plenum, New York
239. Calundann, G. W. U.S. Pat. (1979) 4,161,470; Br. Pat. Appl. (1979) 2,006,242A; Fr. Pat. (1980) 2,406,648; Ger. Pat. (1979) 052,844,817
240. Adames, W. W., Eby, R. K. *MRS Bulletin (Nov/Dec)* (1987) 22
241. *High Performance Plastics* (1986) 3, p. 1
242. Ueno, K., Sigimoto, H., Hayatsu, K. U.S. Pat. (1985) 44,503,005; EP. Pat. (1985) 92,843

8 Solvent Spun Cellulose Fibers

John A. Cuculo, Norman Aminuddin and Margaret W. Frey
North Carolina State University, North Carolina, USA

8.1 Structure and Properties

8.1.1 Introduction

Cellulose is a natural, high molecular weight polymer. It is the most abundant naturally occurring organic polymer as well as being renewable, and biodegradable. Due to complex crystalline and amorphous morphology, considerable hydrogen bonding and very high molecular weight, cellulose does not melt nor does it dissolve readily in many solvents. For this reason cellulose has not been exploited to its full potential. Hence, any process which effects, simplifies or hastens the dissolution represents a significant step forward in the development of cellulose as a viable, ecologically favorable, polymer source.

Numerous works have been published on cellulose, and readers are encouraged to refer to those by Hermans, Sisson, or Ott [1].

8.1.2 Cellulose Structure

The chemical structure of cellulose was first introduced by Haworth [2], see Figure 8.1, which was confirmed from the chemical evidence and X-ray diffraction by Meyer and Mark [3,4,5]. Cellulose consists of long chain β-glucopyranose units, joined through the 1,4-glucosidic linkages. The long cellulose chain is relatively rigid, with strong intermolecular forces and a highly symmetrically configured chain that can fit readily into a crystal lattice [6]. This results in the highly crystalline regions, with amorphous regions in between, which was first observed by Nishikawa and Ono, using X-ray diffraction diagrams [7]. Herzog and Jancke [8] observed that cellulose from widely different sources in nature gave identical X-ray diagrams. This 'native' cellulose crystal was later recognized as cellulose I. Meyer and Misch [9] proposed the lattice structure for cellulose, which was later modified by Gardner and Blackwell [10]. Presently, this modified crystal lattice has been generally accepted. Natural cellulose (cotton, ramie, wood cellulose, etc.) exists in the cellulose I crystalline state. The structure of cellulose I crystals has been well studied and conclusively shown to result from closely packed parallel cellulose chains.

Cellulose can conform to different crystal structures or polymorphs, attainable by means of chemical or heat treatments. Cellulose I can be modified by treatment with an alkali solution, such as sodium hydroxide. When the alkali is removed, the treated cellulose gives a distinct X-ray diffraction pattern different from that of cellulose I. This

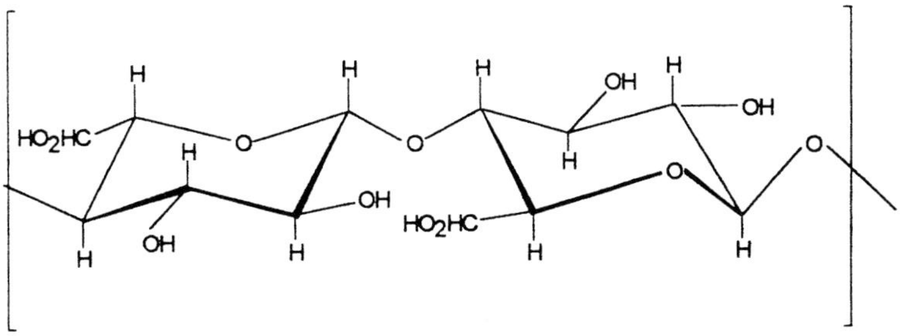

Figure 8.1 The repeat unit of cellulose.

modified cellulose is recognized as cellulose II. The cellulose II polymorph is found also in regenerated viscose rayon. Hess and Gundermann [11] observed that when cellulose I is treated with liquid ammonia and washed with water, the X-ray diffraction showed a pattern different from that of cellulose I. This "ammonia treated cellulose" was recognized as cellulose III. This observation was also reported by other investigators [12]. Meyer and Badenhuizen [13] found that cellulose II could be transformed to cellulose IV on heating at temperatures between 140 °C and 300 °C in glycerol or water (under pressure) or in formamide.

Regenerated and mercerized cellulose exist in the cellulose II crystalline structure. There has been extensive debate over the structure of this type of crystal, but recent work has shown that it can be formed in two different ways from either parallel or anti-parallel cellulose chains [9,14,15]. Regenerated cellulose, formed via coagulation of derivatized cellulose solutions, is known to have a cellulose II structure and is likely to contain both parallel and anti-parallel molecular chains. The anti-parallel configuration was found to be slightly energetically favored over the parallel configuration by early researchers [9]. Mercerization, however, involves swelling but not dissolution of cellulose I crystals and does not provide an opportunity for chains to move from the parallel to anti -parallel configuration. In this case the cellulose chains appear to remain in the parallel configuration and the change in crystal structure is achieved simply by rotation around C6 in the cellulose repeat unit [16,17]. The parallel structure for the cellulose II molecule has been confirmed by X-ray crystallography and NMR. The cellulose III crystal structure has been seen in fibers spun from NH_3/NH_4SCN solutions and in cellulose treated with ammonia [18]. This crystal structure, can also be achieved by rotating about the C6 carbon with parallel cellulose chains [16,17]. Cellulose I is the most tightly packed crystal structure and has the highest degree of intermolecular hydrogen bonding. Cellulose II is more loosely packed and has less hydrogen bonding than cellulose I and cellulose III is more loosely packed and has less intermolecular hydrogen bonding than cellulose II. It is possible to convert cellulose I to cellulose II or cellulose III and to interconvert within the cellulose I and cellulose II family by means of chemical or thermal treatment [19,20]. No route, however, has been found to convert cellulose II back to cellulose I. Hayashi et al. [19,20] observed that when cellulose I and II are transformed to cellulose III_I and IV_I (cellulose I family) and III_{II} and IV_{II} (cellulose II family), respectively, then cellulose III_I and IV_I can be transformed back to I and similarly III_{II} and IV_{II} to II. These differences in the cellulose polymorphs can be shown simply by x-ray diffraction patterns [21], see Figure 8.2.

The transformation between the polymorphs was investigated, using solid-state CP/MAS (cross polarization/magic angle spinning) ^{13}C NMR, by many, such as Horii et al. [22], Isogai et al. [23] and Cuculo et al. [16,17]. Observations on ^{13}C NMR spectra show that differences in the chemical shifts, especially C6 in the anhydroglucose units, are present between the polymorphs. Isogai et al. [21,23] proposed that the mechanism of cellulose polymorph transformation is due primarily to the changes in the conformation and/or in the environment of C4 and C6. Cuculo et al. [16,17] studied the dissolution of cellulose in the ammonia/ammonium thiocyanate solvent and observed the polymorph transformation. When cellulose I was dissolved, its polymorph would transform to II to III and finally to amorphous in solution. They found that these

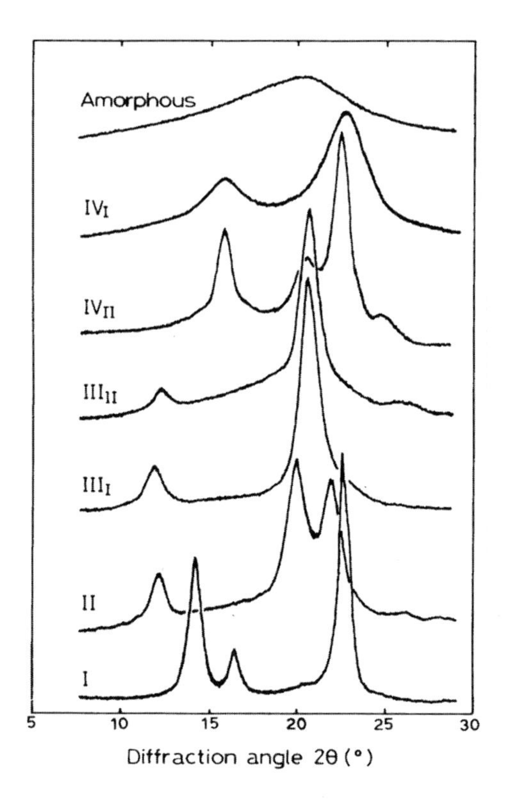

Figure 8.2 The various X-ray diffractogram of cellulose polymorphs. After ref. [21].

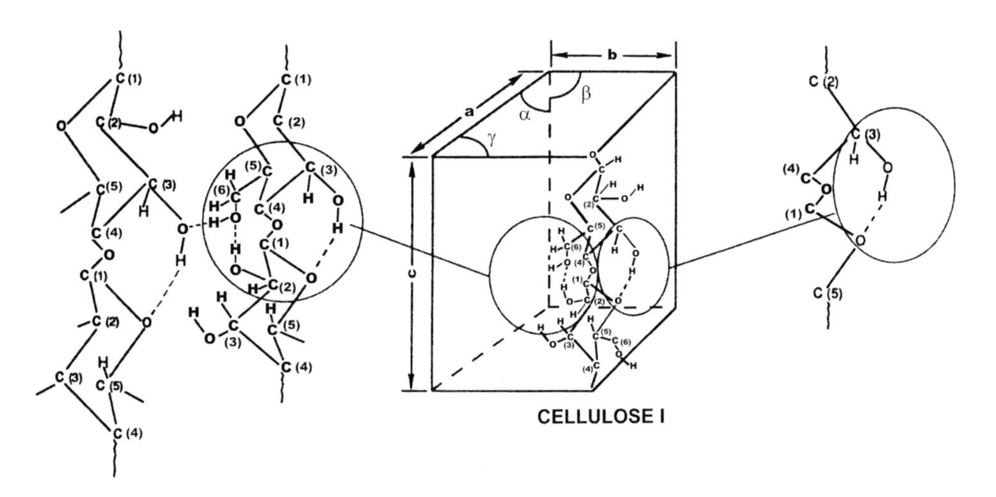

Figure 8.3 The intra-and intermolecular hydrogen bondings in cellulose I. After ref. [17]. Reprinted by permission John Wiley and Sons, Inc.

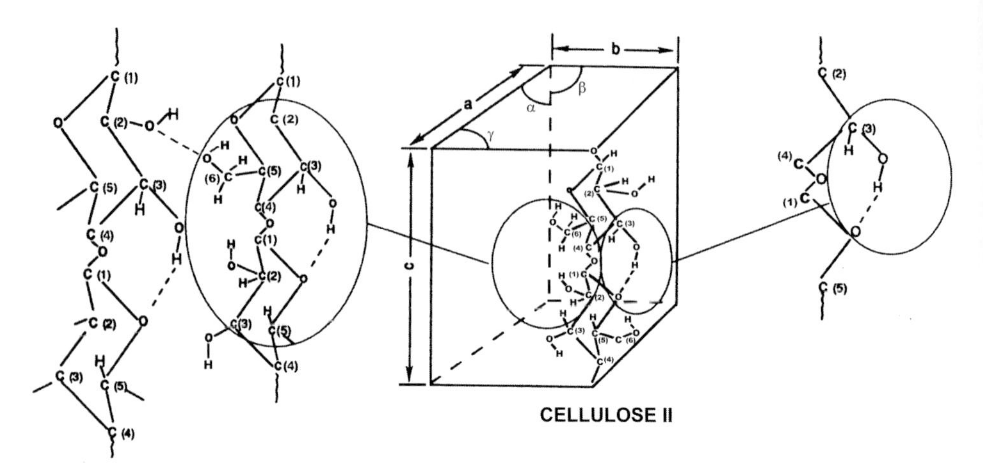

Figure 8.4 The intra-and intermolecular hydrogen bondings in cellulose II. After ref. [17]. Reprinted by permission John Wiley and Sons, Inc.

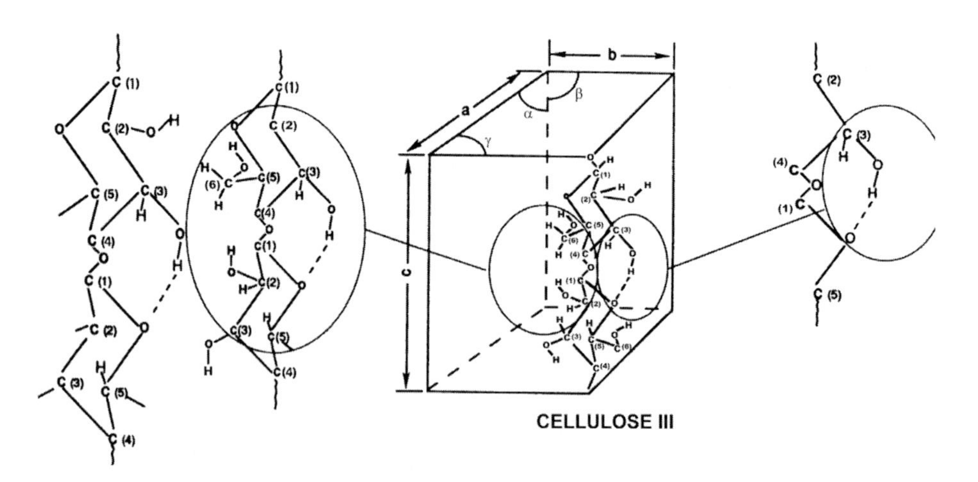

Figure 8.5 The intra-and intermolecular hydrogen bondings in cellulose III. After ref. [17]. Reprinted by permission John Wiley and Sons, Inc.

changes proceeded through the transformation of the CH_2OH conformations from trans-gauche, "tg", to gauche-trans, "gt", to gauche-gauche, "gg". These conformational changes were accompanied by the breaking of both inter- and intramolecular hydrogen bonds, and the changes in the polymorphs can be seen in Figures 8.3, 8.4 and 8.5. Owing to these unique characteristics of cellulose and to its abundance, a considerable amount of work has been done over the years on the processing of cellulose. Early on, because of the lack of adequate solvents, cellulose derivatives, mainly the xanthate and acetates, were studied. More recently, work on processing cellulose itself has been reported.

8.2 Fiber Formation

8.2.1 Introduction

Cellulosic fibers are among the first man made fibers ever produced starting with rayon fibers and the viscose process invented in the late 1800s. Since cellulose did not dissolve without degradation in any common solvent, processes for producing cellulosic fibers started by derivatizing the cellulose polymer chain. In the rayon process cellulose is first converted to a soluble form, extruded and finally regenerated to the original cellulose molecular structure in a different crystalline form. Cellulose acetates, however, are derivatized into any easily soluble form and that molecular structure is retained in the final fiber.

Several researchers have predicted that high modulus fibers could be formed from cellulose solutions [24,35,36] taking advantage of the semi-rigid nature of the cellulose molecule and the ability of cellulose to form liquid crystals. The semi-rigid nature of the cellulose molecule, a result of ring structures in the polymer chain backbone and intramolecular hydrogen bonding, makes cellulose a candidate for liquid crystal formation. Indeed, liquid crystalline phases have been found in most cellulose solvents under certain conditions of temperature, solvent composition and cellulose concentration [25].

8.2.1.1 Viscose Process

In 1892, Cross, Bevan and Beadle discovered the viscose process [26]. They found that treating cellulose with caustic soda and then carbon disulfide (CS_2) produced a cellulose derivative that would dissolve in water or dilute caustic soda. Thus, traditionally, cellulosic products, such as rayon, have been obtained indirectly by derivatizing the cellulose. Once the final shape has been obtained, the derivative can be converted (regenerated) by various means back to cellulose. These processes are rather cumbersome, expensive and generally have toxic by products and polluting effluents. In making viscose rayon, cellulose has to endure several rigorous processes: steeping, shredding, aging, dissolving, ripening and spinning. The industrial viscose process has been reviewed in great detail by Turbak [27].

The viscose process begins with a steeping step, where cellulose is converted into alkali cellulose prior to the addition of carbon disulfide (CS_2). Steeping consists of treating cellulose pulp with an 18–20 wt % solution of NaOH at a temperature of 25 °C. The cellulose is then pressed to remove most of the excess NaOH solution. The alkali cellulose is then shredded for ease of processing while the temperature is raised to ca. 30–32 °C. Next, the shredded alkali cellulose is aged to reduce and to control the cellulose DP.

When the desired DP is attained, CS_2 is added to the shredded alkali cellulose in a specially designed vessel called a Barette at 20–30 °C for 1–3 hours to carry out the xanthation [28]. This is a complicated process because in addition to the intended product, the xanthate, many reactions occur simultaneously upon addition of CS_2, one of which is Na_2CS_3, which can result in discoloration of the cellulose xanthate. The main reactions and the important side reactions may be shown as follows:

Main reactions:

$$\mathrm{CellOH + NaOH \rightarrow Cell-O^-Na^+ + H_2O}$$

$$\mathrm{Cell-O^-Na^+ + S{=}C{=}S \rightarrow CellO-\overset{\overset{S}{\|}}{C}-SNa}$$

$$\mathrm{CellO-\overset{\overset{S}{\|}}{C}-SNa + H_2SO_4 \rightarrow CellOH + CS_2 + NaHSO_4}$$

Side reactions:

$$\mathrm{CS_2 + H_2O \rightarrow H\overset{\overset{S}{\|}}{SC}OH \rightarrow H_2S + COS}$$

$$\mathrm{COS + H_2O \rightarrow H\overset{\overset{S}{\|}}{OC}OH \rightarrow H_2S + CO_2}$$

$$\mathrm{CO_2 + 2\ NaOH \rightarrow Na_2CO_3 + H_2O}$$

$$\mathrm{H_2S + CS_2 \rightarrow H\overset{\overset{S}{\|}}{SC}SH \xrightarrow{2NaOH} Na\overset{\overset{S}{\|}}{SC}SNa + 2H_2O}$$

$$\mathrm{Na\overset{\overset{S}{\|}}{SC}SNa + H_2SO_4 \rightarrow Na_2SO_4 + CS_2 + H_2S}$$

In the dissolving stage, the cellulose xanthate crumb is added to a cold dilute caustic soda solution. The dissolution is carried out at temperatures below 10 °C. The cellulose xanthate solution is then ripened for various reasons but mainly to redistribute the xanthate groups. This ripening stage is also carried out at low temperature over a period of several hours to several days. Once the cellulose xanthate is ripened, it is ready for spinning. The viscose is extruded into a coagulation bath, containing mainly acid and salt. It also generally contains other additives which can serve a multitude of purposes, one of which is to improve processability. The cellulose fiber is first coagulated and subsequently regenerated during the regeneration process. The rayon fiber may be modified during the processing by using additives either in the coagulating bath, as just mentioned, or in the viscose solution, prior to the spinning. The high wet modulus (HWM) is produced by adding various amines and polyglycols to the viscose and zinc to the acidic coagulation bath. It is important to realize that while the viscose process is obviously multistepped and complex, it still has its attributes in the relative ease with which wide variations in process and product may be effected.

Another solvent which has been used in dissolving cellulose is a mixture of ammonia and copper. This solvent has been used by the Bemberg rayon industry and is referred to as the Bemberg process. In 1857, Schweizer [29] found that certain natural fibers such as cotton, linen and silk would dissolve in a mixture of copper hydroxide and aqueous ammonia solution. Cuprammonium solutions may be prepared by dissolving freshly made copper hydroxide in aqueous ammonia. The concentration of copper must be greater than 25 g/liter and the ammonia concentration between 124 to 250 g/liter [30]. Some sodium hydroxide solution may be added to promote swelling of cellulose and interaction between the cellulose with the $Cu(NH_3)_4^{2+}$ ions. The cellulose molecules will interact with the cuprammonium ions to form a complex, as shown in Figure 8.6. Since cuprammonium will decompose when exposed to light and has a short shelf life, freshly made solutions are used to dissolve cellulose. Precaution must also be taken to limit the exposure of the cellulose solutions to light and oxygen.

To make cuprammonium cellulose solution, pretreated cellulose is added to a freshly made cuprammonium solvent. The cellulose may be pretreated by either ultrasonic dispersion or mechanical beating to "open" the structure which will improve the solubility of cellulose in the solvent. The cuprammonium cellulose solution is then spun into a coagulant/regenerating bath using a funnel as a processing aid to protect the delicate extrudate. The alkaline coagulant liquid will travel down the funnel and simultaneously attenuate and coagulate the extrudate. The velocity along the funnel will increase and causes the extrudate to stretch as much as 400%. Thereafter, the fibers are washed with 5% H_2SO_4 solution to remove any remnant copper or other chemical deposits.

The Bemberg process has several advantages over the viscose process. Cuprammonium can dissolve cellulose with higher DP, ca. 550 compared to DP ca. 400 which is used in the

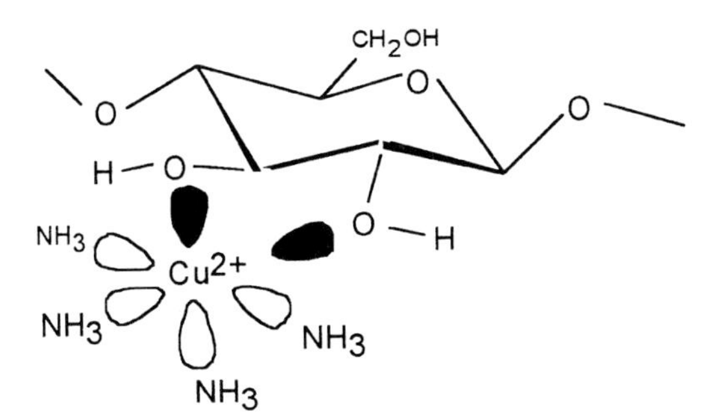

Figure 8.6 The interaction between cuprammonium and cellulose. After ref. [30]. Reprinted by permission John Wiley and Sons, Inc.

viscose process. The Bemberg process can spin cellulose solutions with concentration greater than 10% and the process itself is less complex, which could give it economic advantage. The properties of the cuprammonium fiber, however, are lower, if not similar, to those of the viscose rayon fiber. For this reason the Bemberg rayon process has not been able to completely replace the viscose process. Common obstacles that face both the Bemberg and the viscose processes today are the spate of restrictions imposed by the Environmental Protection Agency and how they impinge upon the recovery and recycling of toxic by-products.

Both the viscose and the cuprammonium rayon processes produce highly toxic by-product waste and pollutants. In the case of the Bemberg process, the recovery of both ammonia and copper poses a great challenge. In the viscose process, the recovery of CS_2 and the removal of H_2S from the air is very troublesome. In both processes the removal of these chemicals has proven to be very costly as well as being limited in success. In addition to these chemicals, other components from the solvents and coagulants are very difficult and costly to recover. Since these by-product chemicals are toxic and environmentally unfriendly, much effort has been directed toward recovering and recycling them. In general, speaking mainly of the viscose process, it is multistepped, complicated and polluting. Hence, it is natural that a considerable amount of work has been devoted to simplifying the process and to reducing the toxic by-product pollutants produced in the manufacture of cellulosic fibers.

Recent work on rayon, regenerated cellulose fibers, has focused on eliminating carbon disulfide from the derivatization and dissolution process [31,32]. Chegolya et. al. have evaluated various non-aqueous solvents as possible replacements for carbon disulfide in regenerated cellulose fiber production [33]. Solvents were evaluated for dissolution conditions (pressure, temperature, maximum cellulose concentration), solution viscosity, degradation of cellulose, solvent and solution stability, toxicity, availability, ability to recover reagents, and quality and processablility of fibers. Table 8.1 contains a summary of the solvents studied and their performance.

Although the fibers formed when using dimethylformamide–dinitrotetroxide (DMF-N_2O_4) showed improved properties and better dyeability than traditional rayons, the authors were not optimistic about implementation of this process since it would require retooling production facilities.

Along with the potential of improved properties and decreased pollution, producing cellulose fibers from direct dissolution and spinning provides opportunities for blending in a second polymer [33,34]. Chegolya, et. al. [33] blended polyacrylonitrile, polymethylmethacrylate, polyvinylacetate, polyvinylchloride, nitrile rubber, polyamide, etc. with cellulose. Blends containing 15% PAN showed improved crease recovery, higher knot strength and less water swelling than 100% regenerated cellulose fibers. Rabe, Collier and Collier [34] created composite fibers with a rayon sheath extruded on to a nylon core. Sorption and tensile properties were linearly related to the respective fraction of each polymer present. Properties of the nylon core dominated the mechanical properties of the composite fiber.

Technology for producing rayon and cellulose acetate fibers has been in place for decades. Recent advances in other fiber types leading to fibers with improved properties together with incentives to reduce chemical waste and pollution have motivated a recent

Table 8.1 Possible Non-Aqueous Solvent Systems for Regenerated Cellulose Fiber Production. After Ref. [33]. Reprinted by Permission Textile Research Institute.

	Dissolving Conditions		Solution Properties			
Cellulose Solvent	Temperature	Pressure	Time (hours)	Viscosity of 5% Solution, (Pa*s)	Maximum Concentration (wt%)	MPC, mg/m^3
H_3PO_4	0–5	normal	3–8	125 (DP = 740)	8 (DP = 740)	1
TFAA	50–70	normal	2–5	6.5 (DP = 740)	11 (DP = 740)	2
TEAO	55–75	normal	4–14	11 (DP = 740)	15 (DP = 740)	10
NMMO	80–150	normal	2–6	40 (DP = 500)	25 (DP = 500)	5
N_2H_4	160–200	0.3–0.7	0.5–1	jelly	30 (DP = 550)	0.1
DMF-N_2O_4	0–20	normal	1–3	7 (DP = 740)	14 (DP = 740)	2
DMSO-N_2O_4	0–20	normal	1–5	38 (DP = 740)	18 (DP = 740)	2
DMAA-N_2O_4	0–20	normal	2–6	—	—	1
AN-N_2O_4	0–20	normal	2–6	35 (DP = 740)	7 (DP = 740)	2
MA-N_2O_4	0–20	normal	6–12	—	6 (DP = 740)	2
EA-N_2O_4	0–20	normal	6–12	28 (DP = 740)	8 (DP = 740)	2
BA-N_2O_4	0–20	normal	10–20	—	—	2
NEPChl-DMSO	80–90	normal	3–5	—	—	5
DMSO-PF	90–130	normal	3–6	45 (DP = 480)	20 (DP = 950)	0.5
DMAA-LiCl	150,20–60	normal	6–10	114 (DP = 740)	12 (DP = 550)	1
NH_4SCN-NH_3-H_2O	–12–20	—	2–15	1300 (DP = 1015)	7.5 (DP = 1015)	20
DMSO-SO_2-DEA	–5–80	up to 0.3	0.1–24	300 (DP = 790)	10 (DP = 100)	10
DMF-Chl-P	20–40	normal	5–30	400 (DP = 1189)	8 (DP = 690)	5
$Ca(SCN)_2$-H_2O-CH_2O	130–140	—	0.5–1	—	10 (DP = 750)	0.5
DMSO-ETA-NaBr-NH_3	140–150	—	—	64 (DP = 260)	—	1

drive to produce cellulose fibers directly from solution without derivatization. Theory has shown that high modulus fibers were possible from cellulose which has chain stiffness between that of aramid liquid crystalline polymers and that of flexible polyethylene [35,36]. Several solvents were found for cellulose, three of which showed some promise for fiber formation [35,37–40]. Cellulose dissolves in LiCl/DMAc, NH_3/NH_4SCN, and $NMMO/H_2O$ solvents without degradation and has been shown to form liquid crystalline phases in each solvent under appropriate conditions of temperature and concentration [35–46].

A method that totally avoids the problems normally associated with derivatization/ regeneration that has shown promising results is the use of direct solvents for cellulose, which allows for a simplified, closed loop process, as well as providing better physical properties of the products.

8.2.1.2 Direct Solvents

The ideal requirement for a direct solvent for cellulose is that it dissolves cellulose without derivatization, preferably without degradation of the cellulose and with no toxic by-product and/or pollutant. The process should provide fibers with better physical properties than currently available and that permit a trouble free, reliable manufacturing process. All of the known viable direct solvents consist of two components. They are: N-methyl morpholine-N-oxide/water ($NMMO/H_2O$), lithium chloride/dimethyl acetamide (LiCl/DMAc), trifluoroacetic acid/dichloroethane (TFA/CH_2Cl_2), calcium thiocyanate/ water ($Ca(SCN)_2/H_2O$), ammonia/ammonium thiocyanate (NH_3/NH_4SCN), zinc chloride/water ($ZnCl_2/H_2O$) and sodium hydroxide/water ($NaOH/H_2O$).

To date, only $NMMO/H_2O$ has been commercialized, by Courtaulds to form a cellulosic fiber, Tencel®. Several of the solvents form cellulose mesophase solutions. These anisotropic solutions are considered potential precursors to improved cellulosic fibers. DuPont has confirmed this for non-cellulosic polymer by its industrial production of the high performance fibers, Kevlar® and Nomex® and also, for cellulose itself [47], albeit via a circuitous route. Kwolek et al. [48] showed that the system of aromatic polyamides, which comprises rigid polymers, exhibited the nematic phase. The high performance fiber, Kevlar®, which has a tenacity of 2.9 GPa (21.7 g/den) and modulus of 108 GPa (795 g/den) [49] is obtained from anisotropic solution.

Anisotropic solutions have also been obtained from semi-rigid polymers such as cellulose and cellulose derivatives. Werbowyj et al. [50] were the first to report the existence of a cholesteric mesophase structure of hydroxypropyl cellulose in water. They showed that 20–50% solutions of hydroxypropyl cellulose in water were birefringent.

A technique currently used to form high performance (high modulus and high tenacity) fibers is to eliminate chain folding, that is to spin fiber from liquid crystal (mesophase) systems [51]. Thus, one of the driving forces behind the study of mesophase solutions of cellulose and its derivatives is the potential of spinning anisotropic cellulosic solutions, or incipient-anisotropic solutions, which will produce high tenacity, high modulus cellulosic fibers. An increasing amount of research is now being done in the field of solvent spinning of cellulose solutions and, of course, in a continuing search for new solvents for cellulose.

8.2.2 Liquid Crystalline State

8.2.2.1 Formation of Liquid Crystalline Solution

Liquid crystal material was first observed by Reinitzer [52] and confirmed by Lehmann [53]. They noticed that on heating cholesteryl benzoate, partially ordered liquid phases were formed, which were then referred to as liquid crystal or mesophase behavior. Bawden and Pirie [54] also observed such behavior on a macromolecular solution, namely, the tobacco mosaic virus solution. Presently, there are many polymers that can form liquid crystalline solutions or melts. The former are referred to as lyotropic solutions, which consist of solvent and mesogenic molecules, the latter as thermotropic melts which contain the mesogenic molecules alone. In a lyotropic system, the liquid crystalline phase appears at a certain polymer solution concentration. This critical concentration will be discussed later in this chapter. In the thermotropic system, the mesophase can appear and exist only within a certain temperature range.

Many models have been developed to describe the behavior of liquid crystalline polymers. The lattice theory, developed by Flory [55,56], based on Onsager's theory [57], adequately describes and predicts the behavior of liquid crystalline polymers. This theory has been applied to many aspects of liquid crystalline polymer behavior. For a polymer to form liquid crystals, its chains must have the ability to conform to the most extended structure and to allow for efficient parallel packing in an ordered lattice. The onset for liquid crystal formation in solution or as a melt may be predicted based on its geometry [58]. In a concentrated system, the critical axial ratio (length, *L*, to diameter, *d*) may be expressed as,

$$x_{crit} = \frac{L}{d} = 6.417$$

Rigid polymers will have an axial ratio greater than 6.4, which is based on their molecular structure and their repulsive molecular forces. Any polymer with axial ratio greater than 6.4 tends to form a stable mesophase melt. The stability of the mesophase, however, is affected by the temperature of the systems. In thermotropic systems, the mesophase usually forms between the crystal melting temperature, T_m, (or T_g for non-crystallizing polymers) and the point where the mesophase becomes an isotropic liquid. This temperature is usually called the upper transition temperature, $T_{lc \rightarrow i}$.

In lyotropic systems, the stability of the mesophase system is governed by the respective properties of the solvents. Some solvents can induce polymers into their most extended conformation either by limiting rotation of the backbone chains or by favoring intramolecular hydrogen bonding [59]. The critical concentration (v_p^*) for formation of the liquid crystal phase may be estimated using the axial ratio of the polymer [56]:

$$v_p^* = \frac{8}{x}\left(1 - \frac{2}{x}\right)$$

The above expression has been used to estimate the minimum concentration at which a stable ordered phase begins to form. Although, Flory originally used the ideal rigid rod model in his lattice theory, in reality, only a few low molecular weight polymers are well represented by a rigid rod. High molecular weight semiflexible polymers are not strictly rigid rods, yet they are able to form mesophase solutions in suitable solvents. To apply the lattice theory to these polymer chains, a different chain model must be used. These polymer chains can be treated as a series of rigid segments connected together with flexible joints, see Figure 8.7. This model is known as the Kuhn chain. Flory has suggested that the Kuhn chain model is appropriate for lyotropic cellulose solutions where the cellulose chain appears to have sudden deviations from linearity [60]. The rigid rod length is replaced by the Kuhn segment length and the overall chain length in theoretical calculations. In applying the Kuhn chain model to the lattice theory, the axial ratio must be replaced by the axial ratio of the Kuhn segment x_k which is the persistence ratio of the chain. This persistence ratio of the chain is equivalent to the persistence length over the chain diameter. The persistence length of polymer is varied in different solvents and can be measured by various methods, such as light scattering, small angle X-ray scattering, etc.

The Kuhn chain model may be applied to cellulose, which is a high molecular weight polymer wherein its component molecules sometimes depart from the extended conformation.

8.2.2.2 Fibers from Liquid Crystalline Solution

Physical properties of fiber made from precursor liquid crystalline solutions are generally superior to those made from the corresponding isotropic solutions (see Chapter 7). The most well known, commercial fiber, derived from a lyotropic system is Kevlar^{TM}. A concentration of 20% of the aramid, poly(para phenylene terephthalamide) polymer is dissolved in 99.8% sulfuric acid. The solution is extruded using the dry-jet wet spinning method. Since the solution is solid at ambient temperature, it is extruded at ca. 90 °C into air. The extrudate then proceeds to the coagulant bath containing water at temperature ca. 5 °C. In the coagulation bath the solvent diffuses out and the fibers crystallize in highly oriented crystallites along the fiber axis. Finally, the fibers are washed, dried and post-treated according to the desired end uses.

Presently, cellulose fibers have not been produced commercially from mesophase solution, using the direct solvent route. Tencel®, the commercially produced cellulose fibers by Courtaulds, however, have shown marked improvement in physical properties compared to its predecessor, rayon. Tencel® has tensile strength of ca. 4.5 g/den (~0.6GPa). O'Brien et al., [61,62], however, have reported that cellulose fibers with tenacity ca. 20 g/den (~2.7GPa) have been produced from mesophase solutions of cellulose triacetate in trifluoroacetic acid

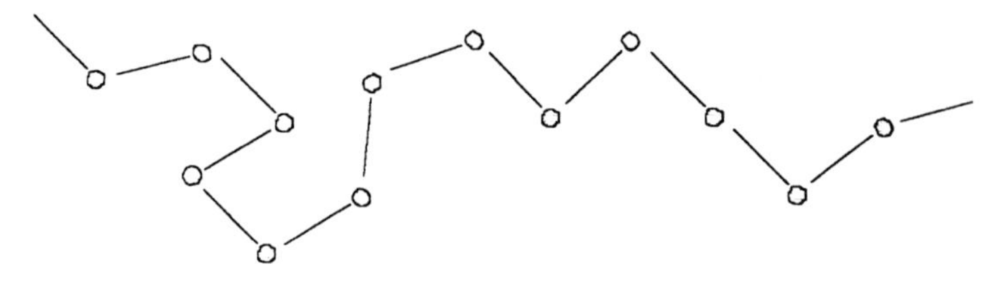

Figure 8.7 The Kuhn chain consists of rigid segments connected together with flexible joints.

and water. The resulting cellulose triacetate fibers were then transesterified with NaOMe/methanol mixture. Chen [63] has reported that cellulose fiber from cellulose/zinc chloride/water solutions have improved physical properties. He did not indicate whether the cellulose solutions were isotropic or anisotropic. The physical properties of the fibers were claimed to be in the order of 16 g/den ($\sim$2.2GPa).

The difficulty of extruding mesophase cellulose solutions to produce high strength fibers is a perplexing problem. Navard et al. [64] indicated that one of the problems of spinning liquid crystalline cellulose solutions in NMMO/water system was the instability of the solutions during extrusions. This instability resulted in uneven fiber dimensions with attendant poor physical properties.

8.2.3 Direct Dissolution of Cellulose

8.2.3.1 N-Methyl-Morpholine N-Oxide and Water

Chanzy et al. [37] reported that cellulose forms a lyotropic mesophase in a mixture of N-methyl-morpholine N-oxide (NMMO) and water. A phase diagram for this system is shown in Figure 8.8 [65]. The occurrence of the mesophase depends upon several parameters such as temperature of the solution, concentration of cellulose, water content of the solvent system and molecular weight (DP) of cellulose. Above 100 °C, the solution birefringence slowly disappears. Above 110 °C the solution is completely isotropic. The solution birefringence may be recovered by cooling to 90 °C. This transition temperature varies directly with the concentration of the solution, the DP of cellulose and indirectly with the molar ratio of water in the solvent. The water molar ratio is the number of moles of water per moles of anhydrous NMMO. For instance, a solution of 35% (w/w) with cellulose DP 600 and of water molar ratio of 0.3, has a transition temperature at 140 °C. A 25% solution with cellulose DP 130 and water molar ratio of 0.8 has a transition temperature of 90 °C. When the water molar ratio is above 1.0, no mesophase solution forms, regardless of the concentration of the solution.

In addition to the effects of DP and water content, the solution concentration also affects the occurrence of mesophase solution. Mesophase solutions occur at concentrations greater than 20% (w/w), with cellulose DP of 600. Below 20% (w/w) the solution is isotropic. When cellulose of lower DP is used, the mesophase solution is obtained at higher concentration. For instance, when the cellulose DP is 130, the mesophase solution occurs at a concentration of 25% (w/w).

The dissolution and spinning of cellulose in the NMMO system was developed commercially by Courtaulds. The preparation of the solution involves the addition of cellulose to a mixture of aqueous NMMO and n-propyl gallate. The n-propyl gallate is used as an anti-oxidant to stabilize the DP of the cellulose. The mixture, in an air tight vessel, is then stirred and heated at 130 °C in an oil bath. Much care is taken in heating the mixture since temperatures above 150 °C cause undesirably rapid decomposition of the solvent that could lead to explosions [65]. Complete dissolution normally occurs at 130 °C within 30 minutes. The temperature and time of complete dissolution varies, however,

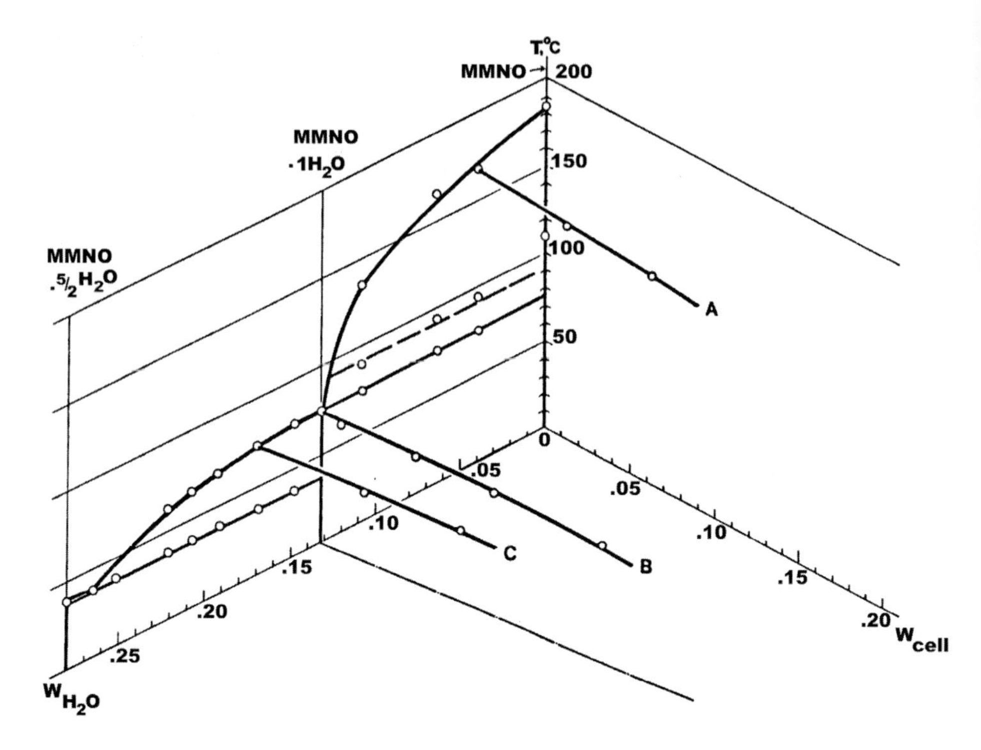

Figure 8.8 Phase diagram for cellulose in NMMO/water. Note: earlier publications referred to NMMO as MMNO. After ref. [65]. Reprinted by permission John Wiley and Sons, Inc.

with the composition of the solvent. The complete dissolution of cellulose at a given concentration depends directly on the DP of the cellulose.

8.2.3.2 Lithium Chloride and N,N-Dimethylacetamide

McCormick et al. [66] discovered that lithium chloride (LiCl)/N,N-dimethylacetamide (DMAc) solution would dissolve cellulose. He and his co-workers [42,67] also observed the formation of cholesteric lyotropic mesophase of cellulose in this solvent system. A phase diagram for this system is shown in Figure 8.9. The mesophase solution occurs at cellulose concentrations above 15% (w/w) in 9% LiCl/DMAc [42]. Ciferri et al. [38,41,43] also observed cholesteric lyotropic mesophases of cellulose in solutions of LiCl/DMAc. They found that LiCl/DMAc neither reacted with nor degraded cellulose. The mechanism by which the dissolution actually occurs is believed to proceed through the formation of complexes between the solvent and the cellulosic hydroxyl groups. The LiCl/DMAc complex has been referred to as a macrocation [42]. Incidentally, recurring literature references indicate that dissolution of cellulose in the solvent requires a pretreatment of the cellulose, an "opening" of the structure for reproducible ease of solution.

Terbojevich et al. [43] indicated that the best method of dissolving cellulose in the solvent involves prewetting the cellulose with DMAc. A known weight of DMAc is added to a weighed amount of dried cellulose. The mixture is refluxed at ca. 165 °C, in a nitrogen atmosphere, for 20–30 minutes. The mixture is then cooled to ca. 100 °C and a predetermined amount of LiCl is added while stirring. Continued stirring at 80 °C for a period of 10–40 minutes, ensures complete dissolution. Complete dissolution can be obtained with concentrations up to 15% (w/w) of cellulose (DP = 1130) [42]. Above this critical concentration, however, undissolved, swollen particles of cellulose are detected in the viscous solutions [42]. The authors suggested that the concentration of cellulose needed to form stable anisotropic solutions (ca. 24%) seems to exceed the solubility limit. Flory [56] derived and calculated the critical volume fraction for a stable anisotropic phase. In this solvent system, no anisotropic phases are observed with any solutions with concentrations up to 15% (w/w). A biphasic system (isotropic and anisotropic), however, is observed only after shearing a thin layer of sample between a microscope slide and the cover slip. A large area of birefringence is seen under a microscope equipped with cross-polars. Ciferri and McCormick concluded that the solution at the solubility limit of concentration was in a pretransitional stage. This means that the solution is mostly in the isotropic phase with some anisotropic occlusion. Therefore, shearing of the solution induces transition to the anisotropic phase.

The viscosity of the cellulose solution exhibits a rapid increase as the concentration of the cellulose increases. Unlike the cellulose derivative, cellulose triacetate in trifloroacetic acid [68], the expected decrease in viscosity of cellulose solution at a certain critical concentration was not observed. According to McCormick et al., this is due either to the

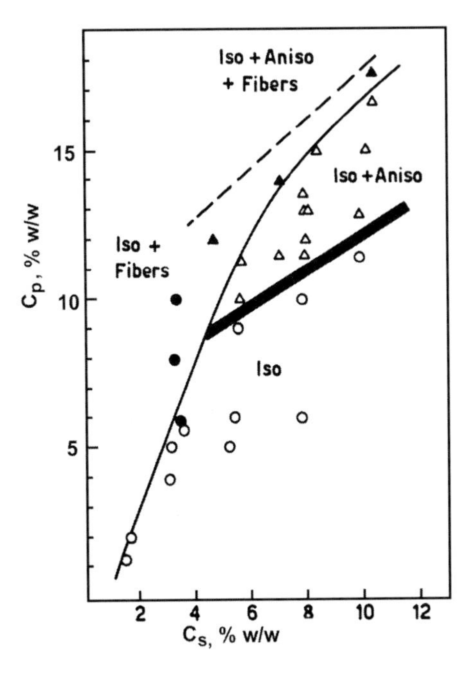

Figure 8.9 Phase diagram for cellulose in LiCl/DMAc. Note: "Fibers" refers to undissolved cellulose in the solution. After ref. [38]. Reprinted by permission John Wiley and Sons, Inc.

limits of solubility or the difficulties of adequate mixing at the high viscosities encountered.

Terbojevich et al. [43] found that the cellulose LiCl/DMAc system formed either stable aggregates composed of ca. seven cellulose chains, or a molecular dispersion. The seven cellulose molecules in the aggregate are thought to be in the conformation of fullest extension, and packed side by side. The aggregations of these cellulose molecules are stable even to a temperature close to degradation. The aggregation can occur even at low polymer concentration, especially with acid hydrolyzed cellulose. This aggregation corresponds to the low solubility of the polymer in the solvent, which will then prevent attainment of the critical concentration. Therefore, another factor that controls dissolution of cellulose, in addition to DP and polymer concentration, is the formation of aggregates. The stability of the aggregates, however, is affected by the amount of LiCl in the solvent. Terbojevich et al. [43], showed that an increase of LiCl, from 5% to 7%, induced the destabilization of the aggregates. The mechanism of the destabilization is thought to be through the formation of "pseudocomplexes involving DMAc", with the cellulose. This, however, occurs only with solvent whose LiCl composition is greater than 5 wt %.

Recently, Dave and Glasser [69] studied the processing and morphology of cellulose and cellulose hexanoate ester using LiCl/DMAc solvent. The solution preparations are similar to Terbojevich 's [43], as described above. Cellulose concentration ranges from 6.5–13.5% (w/w) in 7.8% LiCl/DMAc solvent. The concentration for cellulose hexanoate ranges from 5–40% (w/w). This specific cellulose derivative is stirred only with DMAc at ambient temperature. LiCl is not needed for dissolution in this case, due to the reduced hydrogen bonding propensity of cellulose hexanoate. Using an optical microscope equipped with cross-polars, the onset of liquid crystallinity for cellulose hexanoate solutions was indicated to occur at 24% (w/w) concentration. Biphasic solutions occurred at concentration greater than 10% (w/w) for the cellulose solutions. Using the cone and plate viscometer, however, the rheological behavior showed that the onset of crystallinity occurred at lower concentration, at 7% (w/w).

Dave and Glasser [69] also observed interesting rheological behavior of the cellulose solution. The viscosity of the cellulose solutions increased with concentration and reached a maximum at concentration ca. 7% (w/w), see Figure 8.10. Then the viscosity of the solution decreased with concentration to a minimum point. The author explained that the lowering of viscosity above the 7% (w/w) concentrations was due to the onset of liquid crystallinity or mesophase formation. The viscosity increased again above 12% (w/w) concentration. This behavior was not observed in cellulose solutions in other solvent system such as liquid ammonia/ammonium thiocyanate or TFA/CH_2Cl_2 solvent systems. Ciferri and McCormick also did not observe this behavior in cellulose/LiCl/DMAc solutions. In their findings, McCormick et al. [42] observed that the viscosity of the cellulose in LiCl/DMAc solution increased rapidly from 6% (w/w cellulose) concentration, see Figure 8.11. The viscosity drop, along with the formation of mesophase solution is very important in the processing of the solution.

The drop in solution viscosity also did not occur with cellulose hexanoate solutions as one might expect with cellulose derivatives. For instance, cellulose triacetate in TFA/CH_2Cl_2 solvent has shown the expected drop of viscosity as a function of concentration [68]. Cellulose in NMMO/water solutions, however, shows the drop in viscosity only as a function of temperature and not of concentration [65]. Dave and Glasser [69] explained

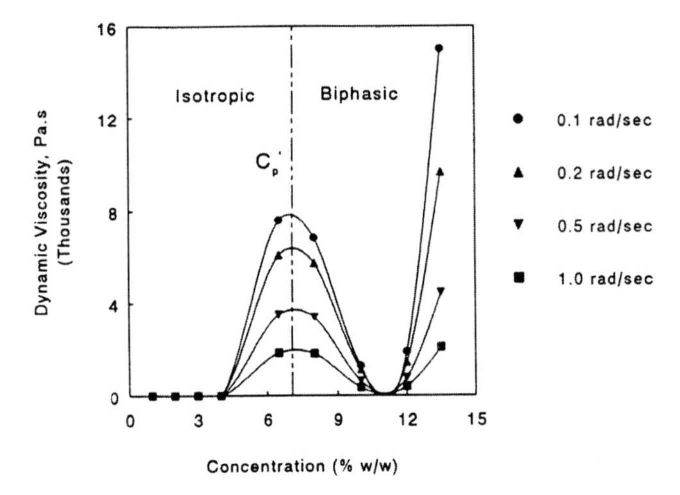

Figure 8.10 Dynamic viscosity vs. concentration of cellulose solution in DMAc/LiCl at different frequencies. After ref. [69]. Reprinted by permission John Wiley and Sons, Inc.

the absence of the viscosity drop with cellulose hexanoate in LiCl/DMAc solution possibly as due to the high side-chain association, which might prevent any mesophase order.

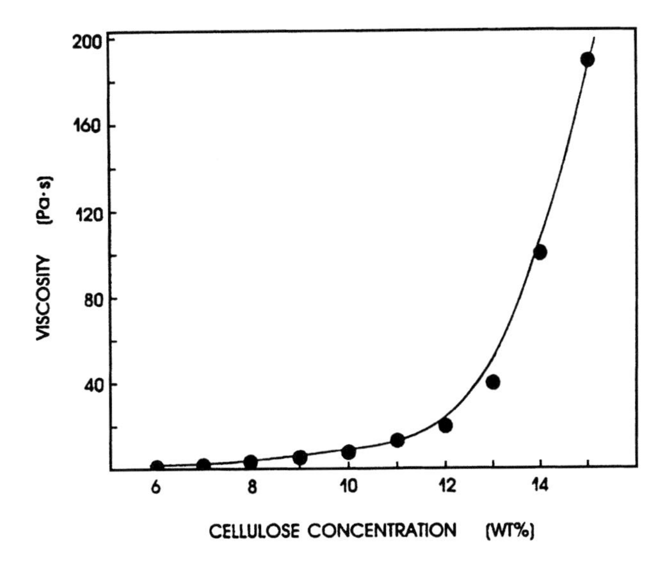

Figure 8.11 Relative viscosity vs. concentration (wt%) at low shear for cellulose in 9% LiCl/DMAc. After ref. [42]. Reprinted by permission American Chemical Society Publishing.

8.2.3.3 Trifluoroacetic Acid and Chlorinated Alkanes

Mixtures of trifluoroacetic acid (TFA) and chlorinated alkanes, such as 1,2-dichloroethane and methylene chloride, are also good solvents for cellulose and cellulose triacetate [70,71]. In this solvent system, the cellulosic lyotropic mesophases occur at 20% (w/w) concentration and they exhibit cholesteric characteristics. Degradation of cellulose, however, can occur in TFA/CH_2Cl_2 solvent [70,72]. The rate of degradation decreased with a decrease in the TFA/CH_2Cl_2 ratio. This solvent dissolves both cellulose and cellulose triacetate (CTA) [71]. The use of the latter was preferred because the viscosity of the CTA/TFA/CH_2Cl_2 solution decreased at some critical concentrations. However, degradation of CTA in the solvent with time was also observed.

The viscosity of the CTA solution reached a maximum at concentration ca. 20% (w/w) and reached a minimum at ca. 23% (w/w), then the viscosity increased again. The reduction of the viscosity is believed as a result of a reduction of polymer-polymer interaction in the ordered phase [68]. This behavior, however, has not been reported for cellulose, regardless of the solvent system, except for cellulose in LiCl/DMAc solvent as will be discussed below. This reduction of viscosity at a certain concentration may occur at a much higher concentration for cellulose than for the cellulose derivatives [42]. The impediment to reach this critical point is the dramatic increase of viscosity attending the formation of gel.

To make the solution, a mixture of known ratio of TFA and CH_2Cl_2 is thoroughly mixed to obtain homogeneity [68]. Then cellulose or CTA is added and stirred with the solvent. Interestingly, the authors introduced a third component to the cellulose or CTA in TFA/CH_2Cl_2 to form a ternary mixture. This ternary system consists of cellulose or a cellulose derivative and a synthetic polymer in the solvent. In this case, the components used were cellulose or CTA with polyethylene terephthalate (PET) or polymethyl methacrylate (PMMA) in the TFA/CH_2Cl_2 solvent system.

Flory [73] showed that for a ternary mixture of rod-like molecules and random coils in an isodiametrical solvent, the free energy of the nematic phase would increase, due to the presence of the random coils. Consequently, the nematic phase would contain no synthetic polymer (random coils). Upon phase separation of the ternary solution, Hong et al. [68] showed, in agreement with Flory's prediction, that the anisotropic phase was rich in cellulose or CTA (but not completely devoid, within experimental error, of PET or PMMA), and the isotropic phase was rich in PET or PMMA. The presence of the random coil molecules, such as PET, also lowers the critical concentration of the CTA for mesophase formation. This is because of the large excluded volume of PMMA or PET which decreases the lattice sites available for the CTA. Therefore, the CTA molecules are forced to pack more efficiently and are more aligned in the ordered phase. The presence of PET also lowers the viscosity of the solution.

8.2.3.4 Calcium Thiocyanate and Water

Dubose [74] found that solutions of calcium thiocyanate in water would dissolve cellulose. This was the first solvent system for cellulose. There is no evidence, however, of mesophase formation in these solutions. Little work has been done on this solvent system due to its high propensity toward thermal degradation of the cellulose. Despite this problem, the solvent demonstrates many properties of other direct solvent systems for cellulose.

In preparing the cellulose solution, dried cellulose may be added to a solvent comprising 34.4–68.3 wt % of $Ca(SCN)_2$ in water [79]. The mixture is then heated to 120 to 140 °C until a clear solution is obtained. The dissolution can be obtained within 10 to 40 minutes depending upon solution concentration and DP of cellulose. Cellulose solutions with concentrations in the range of 10–30 wt % are generally achievable with some evidence of some undissolved cellulose. Prolonged heating can lead to discoloration, from clear to yellow then finally, brown. This discoloration indicates the decomposition of the cellulose in solution. The dissolution is accompanied by cellulose degradation of up to 40% loss in DP [74]. The solution forms a clear gel which melts reversibly at 80 °C.

Hattori et al. studied the solubility of cellulose in aqueous calcium and sodium thiocyanate solutions and the structure of the dissolved cellulose [75,76,77]. X-ray diffraction, differential scanning calorimetry (DSC) and CP/MAS NMR were used to follow changes in crystalline structure and formation of cellulose/solvent complexes. In 55% aq. $Ca(SCN)_2$ at 100 °C interplanar spacing of (110), $(1\bar{1}0)$ increased and (200) decreased. Cellulose recovered from this system is a mixture of cellulose I and cellulose II if water is used as a coagulant. Methanol and acetone coagulants formed cellulose I only. After one day the structure of cellulose coagulated in water was all cellulose II, in methanol; cellulose I and in acetone; amorphous [75]. DSC and IR analysis indicate that dissolution of cellulose takes place first by association of $Ca(SCN)_2{*}H_2O$ with the ring oxygen in the cellulose back bone [76]. NMR studies on cellulose dissolved in 60% aq. NaSCN solution showed shifts in peaks associated with C(2), C(3) and C(6) indicating that Na ions formed complexes with hydroxide groups at these positions on the cellulose backbone [77].

8.2.3.5 Liquid Ammonia and Ammonium Thiocyanate

Hudson and Cuculo [40,81] discovered that ammonia/ammonium thiocyanate (NH_3)/(NH_4SCN) is an excellent solvent for cellulose. The solvent has several practical advantages including low cost and readily available components. The boiling point of the solvent is ca. 70 °C, so it can be handled easily. There seems to be no cellulose degradation nor reaction between the cellulose and the solvent [40,81]. Another advantage is that a mesophase solution can be obtained at reasonably low cellulose concentration.

To make these solutions, a known amount of NH_4SCN is dissolved in an appropriate volume of liquid ammonia (NH_3). A known amount of dried cellulose is then added to the NH_3/NH_4SCN solvent. The mixtures are then placed in dry ice for 24 hours. The frozen mixtures are thawed to room temperature and then warmed in a hot water bath at 50 °C for 30 minutes. If required, the process of cooling and thawing is repeated until the cellulose is completely dissolved. Once a complete dissolution is obtained, the solutions may be kept in a water bath at 25 °C until used.

Two new methods have been developed for the rapid dissolution of cellulose in the solvent, one involves the simple conversion of native cellulose I to cellulose III, the other involves the use of the so-called rapid temperature cycling method [18]. In the former method, the solvent is cooled to near its freezing point (ca. −30 °C) prior to the addition of cellulose III and the mixture of cellulose/NH_3/NH_4SCN is then stirred. The solution is then heated to 40 °C for 30 minutes, and then kept at 25 °C, until used. It was observed that cellulose III dissolved within minutes of its addition to the solvent. Solution concentration and DP of the cellulose affect the dissolution time. In the rapid temperature cycling method, cellulose of polymorph I or II is generally used. A cellulose solution may be prepared by adding and stirring dried cellulose to a cooled solvent. Next, the mixture is cooled to its freezing point (ca. −33 °C) for 30 minutes and then heated in a water bath at 40 °C for 30 minutes. The heating and cooling may be repeated, if necessary, until complete dissolution is achieved. Once a complete dissolution is obtained, the solution can be stored at room temperature until used. Cuculo and co-workers [16,17] proposed a working hypothesis for the mechanism of dissolution of cellulose in NH_3/NH_4SCN. They pointed out that cellulose I, in NH_3/NH_4SCN, undergoes conversions to II then III and finally to amorphous in the completely dissolved state. These conversions to the different polymorphs are aided by a sequential heating and cooling (cycling) process. The heating and cooling process leads to the breaking of inter- and intramolecular H-bonds.

The lyotropic mesophase in the cellulose/NH_3/NH_4SCN solutions has been reported at quite low concentrations with DP 210 cellulose [82]. The type of mesophase which developed depends, among other things, on the NH_3/NH_4SCN ratio and the concentration of cellulose in the solution. Both nematic and cholesteric mesophases have been observed in solutions of cellulose/NH_3/NH_4SCN with the solvent composition 24.5/75.5 (w/w). The cellulose solutions do not show any drop in viscosity at the isotropic to anisotropic transition, as do the cellulose derivatives in TFA/CH_2Cl_2 solvent. The viscosity increases continuously with concentration, similar to the cellulose/NMMO/H_2O solvent system, and the cellulose/NH_3/NH_4SCN solvent system shows a decrease in viscosity with temperature. In addition to the increase in the viscosity, the formation of gels with concentration introduces a challenging problem for processing. This gelation in the cellulose solutions has been observed to be thermoreversible [45]. An experimental phase diagram, Figure 8.12, was created from compiled data for DP210 cellulose in 24.5/75.5 w/w NH_3/NH_4SCN solvent. The diagram shows all of the phases predicted by theoretical phase diagrams based on Flory's lattice theory and a Kuhn chain model for semi-rigid polymers. All phases, however, form at lower volume fractions than predicted by the models. The models assume that only hard interactions involving excluded volume exist between polymer chains. In the cellulose/NH_3/NH_4SCN system soft interactions such as hydrogen bonding between polymer chains must be important and lead to anisotropy at volume fractions lower than predicted by hard interactions alone. The region of anisotropic non-gelled solutions is a likely region for the window of spinnability required for processing of such solutions for fiber formation [83].

Rheological studies were conducted to follow the gelation of solutions containing 8–16 g/100ml of DP210 cellulose in 24.5/75.5 (w/w) NH_3/NH_4SCN. Gelation times were measured at 15, 20, 25 and 30 °C. The increase in storage modulus, G', can be used as an indication for the gelation process. At lower temperatures the gel formed immediately and G' increased rapidly after the gel point. This effect can be shown in Figure 8.13 for 16 g/100ml cellulose solutions. At higher temperatures the gelation time was strongly dependent on the cellulose concentration and decreased with increasing cellulose

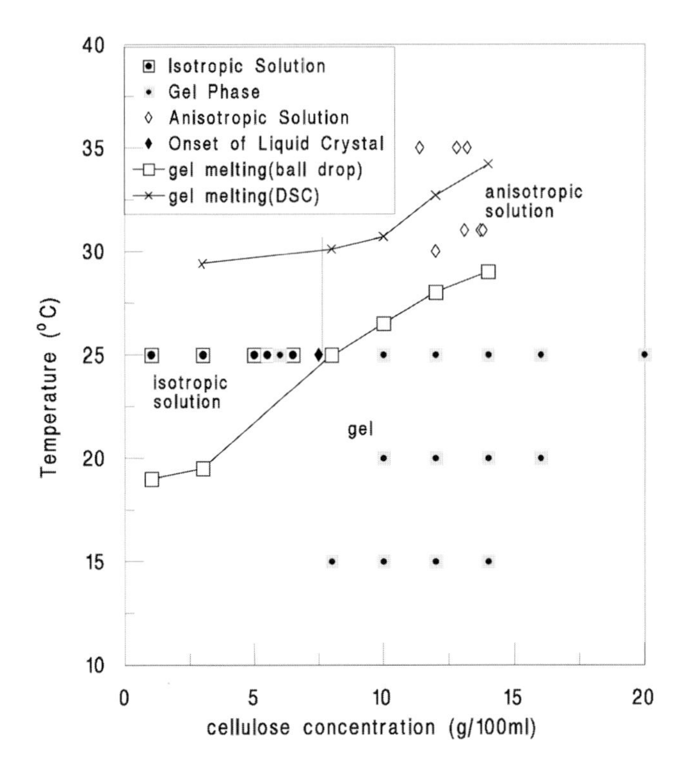

Figure 8.12 The experimental phase diagram for cellulose DP210 in 24.5/75.5 w/w NH_3/NH_4SCN. After ref. [83].

concentration. A distinct phase boundary occurred between 25 and 30 °C. At 30 °C none of the samples gelled and at 25 °C only the sample containing 8% w/v cellulose did not gel [84]. The others were all gelled. This region of anisotropic (non-gelled) solutions was confirmed in samples containing 11–13% (w/v) cellulose by polarized light microscopy. Samples stored at temperatures above the gel melting point were birefringent after 3 days at 31 °C and 1 week at 35 °C. Rheological experiments confirm that these samples were not gels at these temperatures. Yellowing of the samples increased with time and was most evident at high temperatures [85].

8.2.3.6 Steam Explosion

Another solvent that has been reported to dissolve cellulose is sodium hydroxide (NaOH) solution. The cellulose used in this case has been mechanically treated, by means of high pressure. The purpose of high pressure or steam explosion treatment on cellulose is to disrupt the intermolecular hydrogen bonding of cellulose. Thereafter, the cellulose can be dissolved in aqueous sodium hydroxide solution [86]. Kamide et al. [86] and Yamashiki et al. [87,88,89], dissolved regenerated cellulose with DP ca. 500 in 10 w % aq. NaOH at 4 °C to give a 5% (w/w) cellulose solution. The cellulose used had been pretreated by

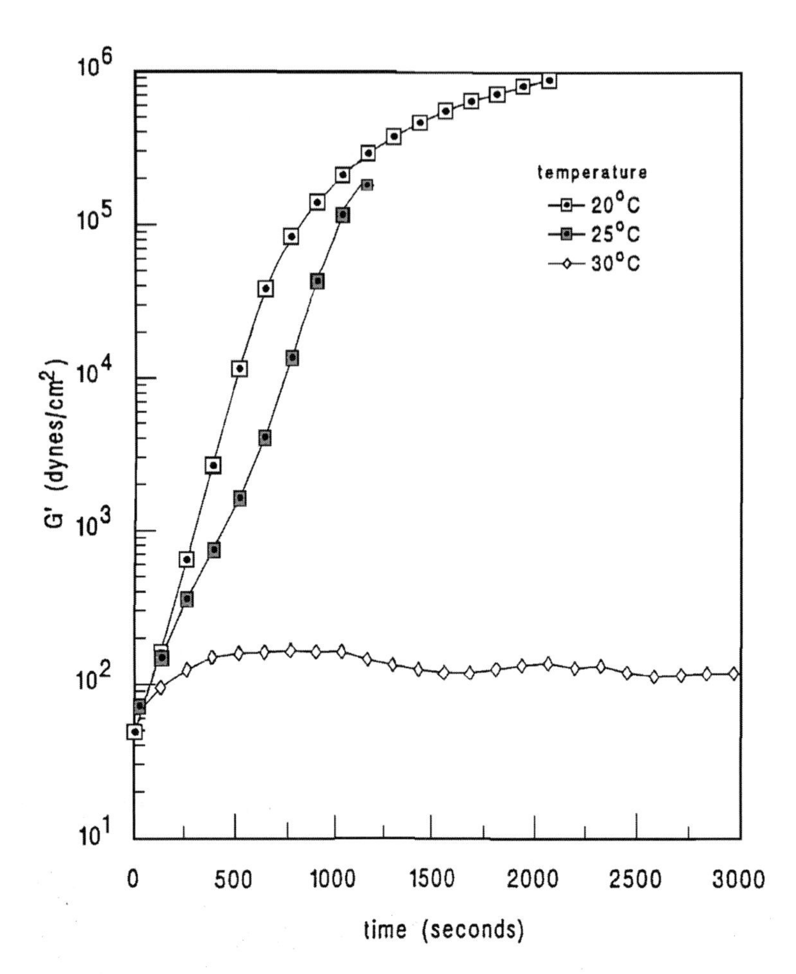

Figure 8.13 The effect of temperature on the storage modulus during the course of gelation for 16% DP210 cellulose in 24.5/75.5 NH_3/NH_4SCN solvent. After ref. [84].

steam explosion, prior to its dissolution in aqueous alkali solution. The steam explosion process causes breakdown of both inter- and intramolecular hydrogen bonds, which strongly affect the dissolution propensity of the cellulose [87]. The process of steam explosion is carried out at steam pressures in the range of 1.0–4.9 MPa, which correspond to steam temperature of 183–252 °C (456–525 °K). The cellulose contacts the steams for ca. 15–300 s. The cellulose is then cooled to room temperature, washed and dried prior to dissolution in the aq. NaOH.

Steam explosion brings about not only the breakdown of hydrogen bonds in cellulose, but also a reduction of DP and a change in the cellulose polymorph. The sensitivity of cellulose to mild steam explosion, in terms of the change in polymorph, reduction of DP, and the solubility is as follows: cellulose I > cellulose III > cellulose II.

Under severe steam explosion treatment, cellulose II has been observed to attain the lowest level-off DP, without a complete change in the polymorph [86]. This is not the case for cellulose III, however, under the same severe treatment its polymorph changes almost completely to cellulose I. The steam treatment of cellulose affects its degree of solubility in aq. NaOH and the effective solubility increases remarkably for cellulose I and cellulose III subjected to steam treatment, but increases only slightly for cellulose II. The authors indicated that the reason for the increase in solubility of cellulose I and III in aq. NaOH after the steam treatment was due to the increase in the degree of breakdown of intramolecular hydrogen bonds at the hydroxyl group at the carbon-3 position in the glucopyranose ring, estimated by solid-state cross-polar/magic-angle sample-spinning (CP/MAS) ^{13}C NMR.

To dissolve steam treated cellulose in aq. NaOH solution, steam treated cellulose (DP = 331), with water content ca. 8–12%, was added to 9.1 wt % aq. NaOH solution (pre-cooled at 4 °C) [90]. The mixture was left for 8 hr. with intermittent stirring. The resultant solution was then centrifuged at 10,000 rpm for one hour at 4 °C prior to spinning to remove remaining undissolved cellulose and air bubbles.

8.2.3.7 Zinc Chloride and Water

It has been known for many years that zinc chloride ($ZnCl_2$) in hydrochloric acid (HCl) dissolves cellulose [91]. This solvent, however, had been found to degrade the cellulose [33,92]. Chen [63] has observed that $ZnCl_2$ in water also dissolves cellulose. He did not mention, however, any degradation of the cellulose. Kasbekar [93] proposed a mechanism of degradation of cellulose with an oxonium salt, such as $ZnCl_2$. The degradation of cellulose occurs via the formation of a cellulose oxonium cation, which results in the hydrolysis of the cellulose [93].

Although there is no mention of mesophase formation, cellulose with DP between 100 to 3000 can be dissolved in $ZnCl_2/H_2O$ at concentrations up to 45% (w/v). The amount of salt in the solvent varies from 55–80% (w/w).

In dissolving the cellulose, a known amount of pre-wet cellulose is added to a known amount of solvent. The amount of water in the pre-wet cellulose is noted so that the final salt concentration may be calculated. The mixture is then stirred at a temperature of 70 °C. The cellulose will dissolve within 30 minutes, and the solutions may be spun at 70 °C.

8.2.3.8 Phosphoric Acid and Water

Processes have been developed to dissolve large quantities of cellulose in solutions of phosporic acid and water. Optically anisotropic solutions have been reported [94]. Several patents [95–101] have been filed on the process of dissolving cellulose in a solution containing 94–100% cellulose, phosporic acid (and/or its anhydrides) and water. Solutions are made in equipment that can provide intensive mixing via shearing and kneading. Both isotropic and anisotropic solutions have been spun into fibers and hollow tubes. Spinning may include an air gap prior to coagulation. During the coagulation step, pH of the fibers is adjusted to 7. Fibers with tenacity suitable for textile use and elongation greater than 7% have been claimed [94].

8.2.4 Fiber Extrusion and Properties

In 1980 Chanzy et al. [37] reported fiber formation from lyotropic cellulose solutions in NMMO/H_2O. Although it is possible to spin fibers from liquid crystalline solutions of cellulose in this solvent, the solutions can become unstable and explosive under conditions close to those for fiber spinning.

The system of isotropic solutions were studied extensively throughout the 1980s and early 1990s and eventually developed into a commercial process by Courtaulds Ltd., who led an effort to get a new generic name for cellulose fibers formed from direct dissolution. These fibers are currently known under the generic name lyocell, which serves at least two useful purposes. First, it emphasises a product different from rayon, with its characteristic chemical and physical properties. Second, it differentiates the lyocell fiber process, which is environmentally friendly, from the rayon process, which is polluting.

Currently, lyocell cellulose fibers are being produced commercially from isotropic NMMO/H_2O solutions for apparel and industrial markets including nonwovens [102–104,110–120].

The main reason that the lyocell fiber manufacturing process is environmentally friendly is that the fibers are produced by direct dissolution, and solvent can be more readily recovered and recycled [102,103,110,117]. This bypasses one of the major obstacles and costs of viscose production: waste solvent control. Another appeal of these fibers is that, as for all cellulose fibers, they are produced from a renewable resource and are biodegradable in both aerobic and anaerobic environments. This is not true of most other man-made fibers [111].

Lyocell fibers may be produced by two spinning methods, namely, wet spinning or dry-jet wet spinning. In the former, the solutions are forced through the spinneret directly into a coagulant, see Figure 8.14. The coagulant is a non solvent for the polymer. In the coagulant bath the solvents diffuse out of the extrudate. The fiber is further stretched and washed and post treated. In the dry-jet wet spinning method, the solution is extruded into air and immediately proceeds into the coagulant, see Figure 8.15. Post treatments are then applied to the fibers. The purpose of the air gap is to provide extensional flow to the extrudate to increase orientation across the fiber cross section. The air gap also allows the temperature of the polymer solution to be different, usually higher than the coagulant, which is the case in the production of Kevlar^{TM}. In the case of spinning cellulose in NMMO/water, the humidity and temperature within the air gap affects the fibrillation of the fiber [121].

8.2.4.1 N-Methyl-Morpholine N-Oxide and Water

Chanzy, et al. [120] spun cellulose/NMMO/water solutions. The characteristics of the fibers obtained were similar to those of the best viscose rayon with tenacity of 0.5GPa (3.7 g/den) and modulus of 21 GPa (153.3 g/den). Previous work with film indicated that the polymorph was cellulose II [37]. Certain additives in the solution, such as NH_4Cl, were found to boost the mechanical properties of the fibers. The addition of 2% NH_4Cl to 14% (w/w) solution of DP 600, increased the tenacity to 0.9 GPa (6.57 g/den) and modulus to 35 GPa (255.5 g/den). Chanzy et al. [120] noted that the fibers spun from

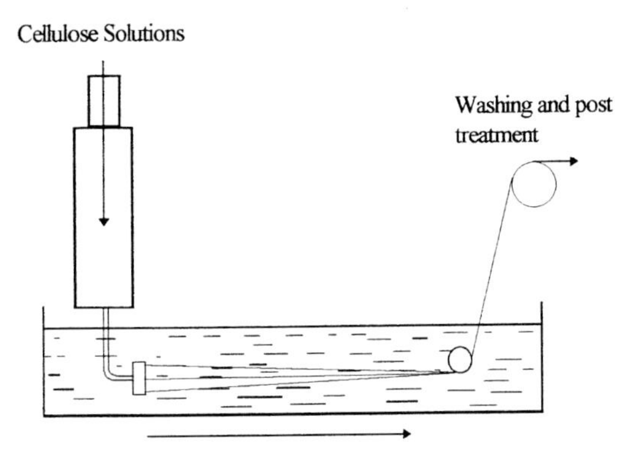

Figure 8.14 Wet spinning process.

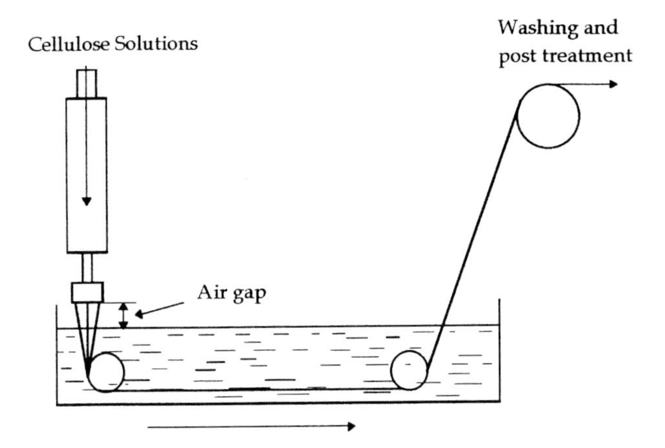

Figure 8.15 Dry-jet wet spinning process.

solutions with additives possessed a core structure consisting entirely of packets of microfibrils aligned parallel to the fiber axis. The addition of NH_4Cl to increase tenacity, however, decreased the lateral cohesion of the fibers and resulted in fibers which were easily fibrillated. Repeated handling, rubbing or bending would lead to delamination of the fibers. Mortimer and Peguy [122] reported that conditions inside the air gap during dry-jet wet spinning can affect the fibrillation tendency of the resulting fibers.

As lyocell fibers became commercially produced, methods were patented to make the spinning process more continuous by conducting dissolution in a series of pipes [105]. Numerous after treatments have also been patented, designed to increase or decrease the tendency of fibers to fibrillate [106]. Finishing fabrics and fibers with enzymes [107], crosslinking with multifunctional chemical treatments [108] or reactive dying [109] have been described to prevent fibrillation.

Tenacity of Tencel® fibers compare favorably with other commercial man-made fibers, such as polyester fibers while wet properties are intermediate between those of cotton (which actually gains strength when wet) and viscose fibers, see Table 8.2. The reduced disruption of cellulose molecular chains accounts for the improved properties over viscose although crystals within the fibers are not as well formed as in natural cotton [102,103,111,118,123].

Lyocell fibers have dyeing propensity similar to viscose rayon fibers with efficient uptake and are dye compatible for blending with rayon [118]. All standard dye types are applicable [102]. Fibrillation of the individual fibers can be controlled to give a range of fabric aesthetics from smooth to sueded [102], and the fabric properties are reputed to provide the breathability of cotton, the drape and softness of rayon and the shrinkage resistance of nylon or polyester [114].

Fibrillation of the fibers was seen as a challenge for apparel markets [102,103,110,114,115,116,117,118], but is actually touted as an advantage in nonwovens and technical textiles [102,103,111,115,119,123]. Fibrillation can be controlled to influence strength, absorbency, integrity and opacity of nonwovens. Fabrics will swell in the presence of moisture to provide a waterproof but breathable barrier [119]. Tencel® fibers have been processed into nonwovens by carding, wet laying, or airlaying from pulp or fiber. Fabrics have been successfully bonded with latexes, thermal bonding with polypropylene, needling and hydroentanglement [111,123]. Hydroentanglement in particular produces a completely biodegradable product [111]. The end use is very broad ranged: for example, absorbent wipes, fabric softener sheets, tampons, hospital drapes and gowns, filters, and coating substrates for materials such as artificial leather [111,115,119,123].

8.2.4.2 Lithium Chloride and N,N-Dimethylacetamide

Bianchi et al. spun fibers from solutions of regenerated cellulose (DP = 290, cellulose II) in Lithium Chloride/N,N-Dimethylacetamide (LiCl/DMAc) [38,41,124,125]. The cellulose

Table 8.2 Fiber Properties of Tencel® Fibers Compared to Other Commercially Available Fibers. After Ref. [102]. Reprinted by Permission Courtaulds Fibres, Inc.

	Tencel®	Viscose	HWM	Cotton	Polyester
Denier	1.5	1.5	1.5	—	1.5
Tenacity (g/den)	4.8–5.0	2.6–3.1	4.1–4.3	2.4–2.9	4.8–6.0
Elongation (%)	14–16	20–25	13–15	7–9	44–45
Wet Tenacity (g/den)	4.2–4.6	1.3–1.8	2.3–2.5	3.1–3.6	4.8–6.0
Wet Elongation (%)	16–18	25–30	13–15	12–14	44–45
Water Imbibition (%)	65	90	75	50	3

solutions were found to contain liquid crystalline phases at polymer concentrations greater than 11% by weight. True liquid crystalline phases were never achieved. Biphasic solutions containing liquid crystalline and isotropic phases were found. Anisotropic (biphasic) and isotropic solutions were stored for up to one year before spinning. Fibers obtained were between 20 and 50 μm in diameter (4–27 denier). Fibers spun from isotropic solutions had elastic moduli as high as 0.35 GPa (2.5 g/den) and from anisotropic solutions 22.0 GPa (161 g/den) [124]. Modulus increased with increasing cellulose concentration, sample storage or maturation time, and increased tension on fibers during spinning [124].

Dave and Glasser [69] also spun fibers from isotropic solutions and biphasic solutions containing liquid crystals of cellulose in LiCl/DMAc solvent, using the dry jet-wet spinning method. As in the earlier work by Bianchi, Ciferri et. al. [38,41,123,125] pure anisotropic solutions were not observed because of the limited solubility of cellulose in this solvent. Fiber tenacity and modulus increased as cellulose concentration increased and liquid crystalline phases formed. Maximum properties achieved for this system were tenacity of 0.15 GPa (1.1 g/den) and moduli of 20.8 GPa (151.9 g/den). Fibers spun from isotropic solutions had smooth uniform surfaces while fibers spun from biphasic solutions had rough surface morphology [69]. For cellulose hexanoate, however, the increase in fiber properties was not observed through the isotropic–anisotropic transition. The highest tenacity and modulus for cellulose hexanoate fibers were 0.096 GPa (0.7 g/den) and 3.11 GPa (22.7 g/den), respectively.

8.2.4.3 Trifluoroacetic Acid and Chlorinated Alkanes

Gilbert and co-workers spun fibers from CTA/PET/TFA/CH_2Cl_2 with composition of 29.1/0.3/70.6 (w/w/w). The solution was extruded using the dry-jet wet spinning technique, into a methanol/H_2O (70/30: v/v) coagulation bath. The resulting fiber had tenacity of 0.56 GPa (4.1 g/den) and modulus of 22.5 GPa (164 g/den), with CTA-I polymorph. O'Brien [61] spun anisotropic solutions of CTA/TFA/H_2O. The resulting triacetate fibers were transesterified with sodium methoxide in methanol and the regenerated cellulosic fibers had exceptionally good mechanical properties after they had been heat treated; tenacity of ca. 2.7 GPa (20 g/den) and modulus of ca. 49.3 GPa (360 g/den). O'Brien did not report the polymorph of the fiber. Roche et al. [126], however, suggested that cellulose IV polymorph is the preferred conformation. They found that a freshly made CTA-I film (coagulated by methanolic NaOH), transformed to cellulose IVwhen transesterified at room temperature.

8.2.4.4 Calcium Thiocyanate and Water

The fibers formed from cellulose/calcium thiocyanate/water solutions were very poor. The maximum tenacity obtained was 0.8 g/den. [127]. In addition to the poor physical properties, the fiber was thermally unstable. Discoloration of the fiber occurred on heating above 120 °C, due mostly to incomplete removal of salt.

8.2.4.5 Liquid Ammonia and Ammonium Thiocyanate

Cellulose fibers have also been spun from both isotropic and anisotropic solutions using the NH_3/NH_4SCN solvent system [128]. The spinning was done by means of the dry-jet wet spinning technique, using low molecular weight alcohols, such as methanol, isopropanol or n-propanol, as coagulants. The mechanical properties of the fibers from the isotropic solution were similar to those of normal viscose rayon. The anisotropic solutions produced fibers with slightly better mechanical properties than the isotropic solutions, as also reported by Bianchi et al. [123] and Dave and Glasser [69]. The fibers obtained from partially nematic solutions had tenacity of ca. 0.4 GPa (3 g/den.) and modulus of ca. 21.2 GPa (155 g/den). Liu, Cuculo and Smith [128] showed that the modulus of the fiber increased at lower coagulant temperatures. Lowering the coagulant temperature results in lower diffusion rate of solvent out of the cellulose solution. Cuculo and Aminuddin [78] observed that fiber properties can be improved by lowering the temperature of the coagulant and increasing the draw ratio during the spinning process. Cellulose fibers with tenacity and modulus higher than 0.68 GPa (5 g/den) and 21.9 GPa (160 g/den), respectively, have been obtained. Liu et al. [128] also reported that the cross-sectional shape of the fiber varied tremendously depending on the type of coagulant used and actually resulted in the reproducible production of a hollow fiber. The fiber spun from this system has cellulose III polymorph [18].

8.2.4.6 Steam Explosion

Fibers from isotropic solutions of cellulose in 9.1 wt % NaOH solutions have been spun by Yamashiki et. al. [89, 129]. The filaments obtained were between 50 and 85 denier and had a cellulose II crystal structure with higher percent crystallinity but less crystalline orientation than regenerated cellulose fibers. Tensile strengths of the fibers are still low, in the range between 0.21 GPa (1.53 g/den) and 0.25 GPa (1.82 g/den) with elongation between 4.5 and 7.5%. All of the fibers were spun using aqueous H_2SO_4 as the coagulant. The fibers were nearly circular, with a rough dense skin surrounding a more porous core which possibly contained some trapped solvent [129].

8.2.4.7 Zinc Chloride and Water

Cellulose fibers from isotropic cellulose/zinc chloride/water systems have been spun using the wet spinning method [63]. The cellulose solutions with concentration between 10–15 % (w/v) were spun using the wet spinning method. The cellulose solutions were extruded, using a 22 or a 27 gauge hypodermic needle, into a coagulant bath which contained respectively one or more alcohols or ketones. The fiber spinning and post treatment processes were not continuous. The fibers were collected and treated later. The collected fibers were further washed with fresh coagulant to remove any remnant of the solvent and dried at constant length. The dried fibers were then stretched and immersed in water for recrystallization to occur. Following this treatment the resulting fibers had much improved physical properties with tenacity as high as ca. 2.32 GPa (17 g/den). The initial modulus of the fibers was not reported.

Chen [63] has shown that the formation of crystals and their orientation in the cellulose fibers affect the physical properties of the fibers. The polymorph of the cellulose fibers may not hold a major role in the effect of the fiber physical properties. He suggested that the crystallization step needed to be separated from the coagulation step. Once the solvent has been removed completely from the fibers, the fibers are dried and stretched, to induce orientation along the fiber axis. The stretching of the fibers also inevitably induced crystallization. The dried and oriented fibers are then further crystallized by immersion in water.

8.3 Concluding Remarks

Cellulose is a renewable resource and the properties of cellulosic fibers are highly suited to a number of end-uses. It is therefore predictable that the demand for these fibers will increase in the future. The new breed of cellulose direct solvent systems has been shown to produce better cellulose fiber properties than those of rayon. Due to their simpler and potentially closed-loop processes as well as better fiber properties, the direct solvents systems promise a bright future, replacing viscose rayon and the Bemberg processes. Some of the direct solvents are able to produce anisotropic solutions, which will prove to be beneficial. Thus far, spinning anisotropic cellulose solutions has shown an improvement in properties over those derived from isotropic solutions. The fiber physical properties from anisotropic solutions, however, have not yet reached their theoretical predicted values. This is a result of some difficulties in spinning the anisotropic solutions, due to their high solution viscosity as well as the nature of the sometime accompanying gel formation.

8.3 References

1. For example: Hermans, P. H. in *Physics and Chemistry of Cellulose Fibres*, Elsevier Publishing Co., New York, NY, (1949); "Cellulose and Cellulose Derivatives" in *High Polymers* (1963) Vol. 5, pt. 1–3, Ott, E.; Spurlin, H. M. and Grafflin, M. W., Eds., Wiley-Interscience Publishers, Inc., New York, NY.
2. W. N. Haworth; in *The Constitution of Sugars* (1929) Edward Arnold, London p.90
3. Meyer, K. H.; Mark; H. *Ber.* (1928) 61B, 598
4. Meyer; K. H. *Z. angew. Chem.* (1928) 41, 936
5. Meyer, K. H.; Mark; H. in *Der Aufbau der hochpolymeren organischen Naturstoffe* Akadem. Verlagsgesellschaft. Leipzig., (1930) p. 130
6. Howsom, J. H. and Sisson; W. A. in *High Polymers* (1963) vol. 5, pt. 1., Ott, E.; Spurlin, H. M. and Grafflin, M. W., eds.; Wiley-Interscience Pub. New York, p. 233
7. Nishikawa, S.; Ono, S. *Proc. Math.-Phys. Soc. Tokyo* (1913) 7, p. 131
8. Herzog, R. O.; Jancke, W. *Z. Physik*, 1920, 3, p. 196; Ber. (1920) 53, p. 2162

9. Meyer, K. H.; Misch, L. *Helv. Chim. Acta.* (1937) 20, pp. 232–244
10. Gardner, K. H.; Blackwell, J. *Biopolymers* (1974) 13, pp. 1975–2001
11. Hess K. and Gundermann, J. *Ber.* (1937) 70B, p. 1788
12. For example: Hess, K.; Trogus, C. *Ber.* (1935) 68B, 1986; Barry, A. J., Peterson, F. C. and King, A. J. *J. Am. Chem. Soc.* (1936) 58, p. 333; Clark G. L. and Parker, E. A. *J. Phys. Chem.* (1937) 41, p. 777; Legrand, C. *J. Polym. Sci.* (1951) 7, p. 333
13. Meyer, K. H.; Badenhuizen, N. P. *Nature* (1937) 140, p. 281
14. Gardner, K. H. and Blackwell, J. *Biochom. Biophys. Acta* (1974) 343, p. 232
15. Sakthivel, A.; Turbak, A.F.; Young, R. A. in *New Structural Models for Cellulose II Derived from Packing Energy Minimization*, 10th International Cellulose Conference, Syracuse, NY, June 2nd 1988.
16. Cuculo, J.A.; Smith, B. and Sangwatanaroj, U. *J. Poly. Sci., Poly. Chem. Ed.* (1994) 32, p. 229
17. Cuculo, J.A.; Smith, B. and Sangwatanaroj, U. *J. Poly. Sci., Poly. Chem. Ed.* (1994) 32, p. 241
18. Aminuddin, N. Master of Science Thesis, North Carolina State University (1993)
19. Hayashi, J.; Kon, H. Takai, M.; Hatano, M.; Nozawa; T. In *The Structure of Cellulose*; ACS Symposium Series No 340 (1987) American Chemical Society: Washington DC, p. 134
20. Hayashi, J.; Sueoka, A.; Watanabe S. *J. Polym. Sci., Polym. Lett. Ed.* (1975) 13, p. 23
21. Isogai, A. In *Cellulosic Polymers, Blends and Composites* (1994) Gilbert, R. D., Ed.; Hanser/Gardner Publications Inc., Cincinnati, OH, p. 1–21
22. Horii, F.; Hirai, A. and Kitamura; R. *Polym. Bull.* (1983) 10, p. 357
23. Isogai, A.; Usuda, M.; Kato, T.; Uryu, T. and Atalla, R. *Macromol.* (1989) 22, p. 3168
24. Ciferri, A. *Polym. Eng. and Sci.* (1975) 15, pp. 191–198
25. Brandrup, J. and Immergut, E. in *Polymer Handbook* 3rd edition, V/139 (1989)
26. Cross, C. F.; Bevan, E. J.; Beadle, C. *Ber.* (1893) 26, p. 1090; U.S. Patent 520,770, (1894)
27. Turbak, A. in *Encyclopedia of Polymer Science and Engineering*, John Wiley and Sons, NY (1985) Vol. 14, pp. 45–72
28. Golben, M. U.S. Patent 2,492,421 (1949)
29. Schweizer, E. *J. Prakt. Chem.* 1857, 72, 109
30. Browning, B. L.; Sell L. O. and Abel, W. TAPPI 1954, 37, 273
31. Bisumic, A. *Chemistry in Australia* (1995) 52, p. 24
32. 'Rayon makers clean up image', *ATI: America's Textiles International* (1993) 22, 54
33. Chegolya, A.E. *Text. Res. J.* (1989) 59, p. 501
34. Rabe, R.L.; Collier, B.J.; Collier, J.P. *Text. Res. J.* (1989) 59, p. 735
35. *Encyclopedia of Polymer Science and Engineering*, Vol. 3, John Wiley and Sons, NY (1985) pp. 90–91
36. Chanzy, H.; Nawrot, S.; Perez, S. and Smith, P. *International Dissolving and Specialty Pulps*, TAPPI Proceedings (1983), p. 127
37. Chanzy, H.; Peguy, A.; Chaunis, S. and Monzie, P. *J. Polymer Sci.: Polymer Phys. Ed.* (1980) 18, p. 1137
38. Conio, G.; Corazzo, P.; Bianchi, E.; Tealdi, A. and Ciferri, A . *J. Polymer Sci.: Polymer Lett. Ed.* (1984) 22, p. 273
39. Mc Cormick, C.L. and Lichatowich, D.K. J . *Polymer Sci.: Polymer Lett. Ed.* (1984) 17, p. 479
40. Cuculo, J.A. and Hudson, S. M. U.S. Patent 4,367,191 (1983)
41 Bianchi, E.; Ciferri, A.; Conio, G.; Cosani, A. and Terbojevich, M. *Macromol.* (1985) 18, p. 646
42. McCormick, C.L.; Callais, P.A. and Hutchinson, B. H., *Macromol.* (1985) 18, p. 2394
43 Terbojevich, M. A.; Coseni, A.; Conio, G.; Ciferri, A. and Bianchi, E. *Macromol.* (1985) 18, p. 640
44. Chen, Y.-S. and Cuculo, J.A. *J. Polymer Sci.: Polymer Chem. Ed.* (1986) 24, p. 2075
45. Hudson, S.M. and Cuculo, J.A. *J. Polymer Sci.: Polymer Chem. Ed.* (1982) 20, p. 499
46. Yang, K.S.; Theil, M.H.; Chen, Y.S. and Cuculo, J. A. *Polymer* (1992) 33, p. 170.
47. O'Brien, J.P. U.S. Patent 4,501,886 (1985)
48. Kwolek, S.L. U.S. Patent 3,671,542 (1972)
49. Ihm, D. W.; Hiltner, A. and Baer, E. in *High Performance Polymers*; Baer, E. and Moet, A., Eds., Hanser Publishers, New York (1991) pp. 279–327
50. Werbowyj and Gray, D.G. *Mol. Cryst. Liq. Cryst.* (1976) 34, p. 97
51. Mark, H. in *High Performance Polymers: Their Origin & Development* (1986) Seymour, R. B.Kirshenbaum, G. S. Eds., Elsevier Sci. Pub. Co. Inc.
52. Reinitzer; F. *Monatcsh. Chem.* (1888) 9, p. 421

53. Lehmann; O. *Z. Phys. Chem., Stoechiom. Verwandschaftsl.* (1889) 4, p. 462
54. Bawden, F. C. and Pirie; N. W. *Proc. R. Soc. London, Ser. B* (1937) 123, p. 274
55. Flory; P. J. *Proc. R. Soc. London, Ser A* (1956) 234, p. 60
56. Flory; P. J. *Proc. R. Soc. London, Ser A.* (1956) 234, p. 73
57. Onsager; L. *Ann. N. Y. Acad. Sci.* (1949) 51, p. 627
58. Flory, P. J. and Ronca; G. *Mol. Cryst. Liq. Cryst.* (1979) 54, p. 289
59. Walton, A. G. and Blakcwell; J. in *Biopolymers* (1973) Academic Press, New York
60. Flory, P.J. *Advances in Polymer Science* (1983) 59, p. 1.
61. O'Brien, J.P. U.S. Patent 4,464,323 (1984)
62. O'Brien, J.P. U.S. Patent 4,501,886 (1985)
63. Chen, L. F. U.S. Patent 5,290,349 (1994)
64. Navard, P.; Haudin, J.-M. *Br. Polym. J.* (1980) 12, no. 4, p. 174
65. Chanzy, H.; Nawrot, S.; Peguy, A.; Smith, P. and Chevalier, J. *J. Polym. Sci., Polym. Phys. Ed.* (1982) 20, p. 1909
66. McCormick, C.L. U.S. Patent 4,278,790 (1981)
67. McCormick, C.L.; Callais, P.A.; Hutchison, Jr.; B.H. *Polym. Prepr* (1983) 24, no. 2, p. 271
68. Hong, Y.K.; Hawkinson, D.E.; Kohout, E.; Garrard, A.; Fornes, R.E.; Gilbert; R.D. in *Polymer Association Structures*, ACS Symposium Series No. 384; EL-Nokaly, M.A., Ed. (1989) p. 184
69. Dave V. and Glasser, W. G. *J. App. Polym. Sci.* (1993) 48, pp. 683–699.
70. Patel, D.L. and Gilbert, R.D. *J. Polym., Polym. Phys. Ed.* (1981) 19, p. 1231
71. Patel, D.L. and. Gilbert, R.D. *J. Polym., Polym. Phys. Ed.* (1981) 19, p. 1449
72. Hawkinson, D. Master of Science Thesis, North Carolina State University (1987)
73. Flory, P. J. *Macromolecules* (1981) 14, p. 1138
74. Dubose, J. Bull. Rouen (1905) 33, p. 318
75. Hattori, M, Shimaya, Y, Saito, M. *Polymer Journal* (1998) 30, p. 37
76. Hattori, M, Shimaya, Y, Saito, M. *Polymer Journal* (1998) 30, p. 43
77. Hattori, M, Shimaya, Y, Saito, M. *Polymer Journal* (1998) 30, p. 49
78. Aminuddin, N. Ph.D. Dissertation, North Carolina State University (1998)
79. Bechtold, H. and Werntz, J. U.S. Patent. 2,737,459 (1956)
80. Scherer, P. *J. Am. Chem. Soc.* (1931) 53, p. 4009
81. Degroot, A.W.; Carroll, F.I. and Cuculo, J.A. *J. Polym. Sci., Chem. Ed.* (1986) 24, p. 673.
82. Yang, K.S. Ph.D. Dissertation, North Carolina State University (1988)
83. Frey, M.W. Ph.D. Dissertation North Carolina State University (1995)
84. Frey, M.W.; Khan, S.A. and Cuculo, J.A. *J. Polym. Sci., Part B: Polym. Phys. Ed.* (1996) 34, p. 2375
85. Frey, M.W.; Spontak, R. and Cuculo, J.A. *J. Polym. Sci., Part B: Polym. Phys. Ed.* (1996) 34, p. 2049
86. Kamide K.; et al., U.S. Patent 4,634,470 (1987)
87. Yamashiki, T.; Matsui, T.; Saitoh, M.; Okajima, K.; Kamide, K. and Sawada, T. *British Polym. J.* (1990) 22, p. 73
88. Yamashiki, T.; Matsui, T.; Saitoh, M.; Okajima, K.; Kamide, K. and Sawada, T. *British Polym. J.* (1990) 22, p. 121
89. Yamashiki, T.; Matsui, T.; Saitoh, M.; Okajima, K.; Kamide, K. and Sawada, T. *British Polym. J.* (1990) 22, p. 201
90. Yamashiki, T.; Matsui, T.; Saitoh, M.; Okajima, K.; Kamide, K. and Sawada, T. *Cell. Chem. and Techcnol.* (1990) 24, pp. 237–249
91. For example: British Patent 1.6805, 1884; Letters, K. *Kolloid Z.*, (1932) 58 (2) p. 229
92. Hudson, S. M. and Cuculo, J. A. *J. Macromol. Sci.-Rev. Macromol. Chem.* (1980) C18 (1), pp. 1–82
93. Kasbekar; G. *Curr. Sci.* (1940) 9, p. 411
94. *Chemisch Weekblad* (1997) 93(15), p. 1.
95. Westerink, J.B. WO 9728298 (1997)
96. Rusticus, S.J. WO 9730196 (1997)
97. Westerink, J.B. WO 9730090 (1997)
98. Ypma. M. WO 9730198 (1997)
99. Boerstoel, H. US 5817801 (1998)
100. Boerstoel, H. US 5804120 (1998)

101. Boerstoel, H. US 5932158 (1999)
102. *Tencel Technical Overview*, product bulletin, Courtaulds Fibres Inc.
103. *Tencel, Natures Newest Luxury Fiber*, product bulletin, Courtaulds Fibres Inc.
104. Anson, R. *Textile Outlook International*, January (1989) p. 3–5.
105. Wykes, K. A. US 5401304 (1995)
106. Gannon, J.M. WO 9535399 and WO 9535400 (1995)
107. Watanabe, R. WO 9888432 (1998)
108. For Example: Potter, C.D. WO 9528516 (1995) Taylor, J.M. US 5580354 (1996)
109. Nicolai, M., Nechwatal, A., Mieck, K.P. *Angew. Mokromol. Chem.* (1998) 256, p. 21
110. Raven, G *International Fiber Journal*, April 1990, pp. 66-76
111. Woodings, C.R. *The New Nonwovens World*, Summer 1992, pp. 113–118
112. 'Ecology goes Mainstream', *Women's Wear Daily*, May 18, 1994
113. 'Are you Ecologically Influential', *International Textiles*, October 1994
114. Omori, Y. *Tradescope*, April 1993, pp. 2–4
115. Milbank, D. *Wall Street Journal*, January 9, 1995
116. 'Much Expected "Fifth" Fiber: "Tencel"', *JTN*, March 1994, p. 13
117. Rudie, R., *Bobbin*, February 1993, p. 10-12
118. Bullio, P. G. *Fiber World*, September 1992, pp. 16-18
119. 'Lyocell for Technical Textiles', *Chemical Fibers International* (1995) 45, p. 159
120. Chanzy, H.; Paillet, M. and Hagege, R. *Polymer* (1990) 31, p. 400
121. 'Lenzing: International cellulosic fiber investment', *Chemical Fibers International* (1995) 45, p. 162
122. Mortimer S. A. and Peguy, A. A. *J. Appl. Polym. Sci.* (1996) 60, p. 305
123. Bianchi, E.; Ciferri, A.; Conio, S. and Tealdi, A. *Journal of Polymer Science, Polymer Physics Ed.* (1989) 27, p. 1477
124. Woodings, C.R. *The New Nonwovens World*, Summer (1995), pp. 83–87
125. Conio, G.; Bruzzone, R.; Ciferri, A.; Bianchi, E. and Tealdi, A. *Polym. J.* (1987) 19, p. 757
126. Roche, E. J.; O'Brien, J. P.; Allen, S.R. *Polym. Commun.* 1986, 27, p. 138
127. Mac Donald, D. M. in *Solvent Spun Rayon, Modified Cellulose Fibers and Derivatives* (1977) Turbak, A. F., Ed.; ACS, Washington, D.C.
128. Liu, C. K.; Cuculo, J.A.; Smith, C. B. J . *Polym. Sci.: Part B: Polymer Physics* (1990) 28, p. 449.
129. Yamashiki, T.; Matsui, T.; Nawoaka, M. *J. Appl. Polym Sci.* (1992) 44, p. 691

9 Carbon Fibers

David J. Johnson
University of Leeds, Leeds, U.K.

9.1 Formation and Structure

9.1.1 Introduction

Carbon fibers have a high public profile following their successful exploitation in sporting end-uses such as fishing rods, golf club shafts, tennis racquets, racing and mountain bicycles, and Formula One racing cars. Less well known is their widespread use as a stiff, strong, lightweight reinforcement of composite materials in civil and military aircraft, the purpose for which they were originally conceived, and, indeed, for which tonnes of carbon fiber per aircraft are now utilised. Since their inception in the 1960s, carbon fibers have been prepared from a wide variety of precursors ranging from natural materials such as cellulose, lignin, wool, and pitch, to the semi-crystalline fibres of atactic polyacrylonitrile (PAN), and the high performance highly crystalline aramid fibers such as Kevlar, the primary requirement being for a material that does not melt on pyrolysis in an inert atmosphere. The first successful attempts to produce high strength, high modulus carbon fibres utilised viscose rayon precursor fibers (Bacon [1]), PAN fibers (Shindo [2]), and isotropic pitch (Otani [3]). Commercial carbon fibres were produced from viscose rayon by Union Carbide in the United States of America, and from petroleum pitch by Kureha Kagaku in Japan. Despite these early developments, it is carbon fibers from PAN and mesophase pitch (MP) precursors which now dominate the market, although there is still considerable interest in the production of low cost carbon fibers from isotropic pitch and from cellulosic materials.

The development of a successful commercial process for PAN-based carbon fibers was primarily a result of the fundamental work carried out by Watt and his colleagues at the Royal Aircraft Establishment in England during the 1960s (Mair and Mansfield [4]). Experiments with a specially prepared PAN fibre from Courtaulds (Courtelle) showed that these fibres could be oxidized at temperatures around 200 °C when maintained under tension, but care had to be taken to avoid a catastrophic runaway of the exothermic oxidation reaction. Subsequent carbonization at 1000 °C resulted in carbon fibers with tensile modulus in the range 155 to 190 GPa, a value which could be increased to between 350 and 420 GPa after heat treatment to 2500 °C. Four important conditions were shown in the British patent [5] for producing high modulus carbon fibers:

a. oxidation at a temperature below that at which thermal runaway occurs;
b. the necessity for oxidation to penetrate to the centre of the PAN fiber;
c. the necessity to restrict length shrinkage or even stretch the fibres during oxidation;
d. after oxidative stabilization there was no need for tension on the fibres during carbonizing and heat treatments.

It must be noted that two Japanese patents on heating PAN fibers at 200 °C before carbonizing, predate the British patent, but make no mention of tensile modulus or oxidation under tension. The first carbon fibers based on the British patent were produced by Morganite Ltd and Courtaulds Ltd. in 1966.

Concurrently, scientists and engineers at Rolls-Royce Ltd were developing a batch process for carbon fibers from PAN, and experimenting with carbon fiber reinforced plastic (CFRP) turbine blades for the then new RB 211 engine. Unfortunately, the bird strike problem, delamination of blades due to the low interlaminar shear strength of the composites, led to a major crisis for Rolls-Royce and resulted in a severe setback for the use of CFRP in aerospace applications. Fortunately, the demands of the sports enthusiasts led to the rapid development of many end uses for carbon fibers in sporting goods, which enabled the carbon fiber industry to progress during the years when aerospace development and test programmes involving the use of CFRP were being expedited with great caution [6].

9.1.2 Carbon Fibers from PAN Precursors

There are three important stages in the conversion of precursor polyacrylonitrile fiber to carbon fiber:

(a) Oxidative stabilization – carried out in an oxidizing atmosphere in the temperature range 200 °C–250 °C under tension, and which may take several hours for completion.
(b) Carbonization – carried out in an inert atmosphere at temperatures above 1000 °C. Tension is maintained on the fibers.
(c) Graphitization – carried out in an inert atmosphere at temperatures above 2000 °C. Again the tension is maintained. Some improvement in the graphitic nature of what is essentially a non-graphitizing material can be achieved.

9.1.2.1 Oxidative Stabilization

The key factor in producing good carbon fibers from PAN is the production of oxidized ladder polymer parallel to the fibre axis by the cyclization of the pendant nitrile groups and the incorporation of oxygen. It is now well known that the initiation of the ladder polymer formation (Figure 9.1) is enhanced in PAN precursors which contain small amounts of pendant acid groups. Thus Watt and Johnson [7] showed that Courtelle, which contains 4.6 mol% methacrylate and 0.4 mol% itaconic acid as copolymers, oxidizes at a much greater rate than Orlon which has 4.6 mol% methacrylate only as copolymer. In practice, oxidation is carried out at around 220 °C for several hours. Oxygen is incorporated into the ladder polymer in several ways; Watt and Johnson suggest a structure based on the conversion of naphthyridine rings and keto-formation to a number of tautomeric forms such as hydroxy pyridine and pyridone (Figure 9.2). Coleman and Petcavich [8] describe a similar mechanism based on the tautomerism of the enamine-imine conversion.

Although detailed descriptions of the chemical reactions involved in the oxidation process are available in the literature (see for example [9,10]), little attention is paid to the stereo-irregularity of the atactic PAN molecules. Recent attempts at molecular modelling of the structure of fibers based on atactic PAN, in conjunction with quantitative X-ray diffraction analysis, support the view that the ordered regions of PAN fibers consist of chains with irregular conformations arranged as a two dimensional hexagonal lattice [11]. It is has been suggested that the lengths of the conjugated sequences of cyclized goups which stabilize the system are of the order of 4 to 5 monomer units (n in Figure 9.2). Thus,

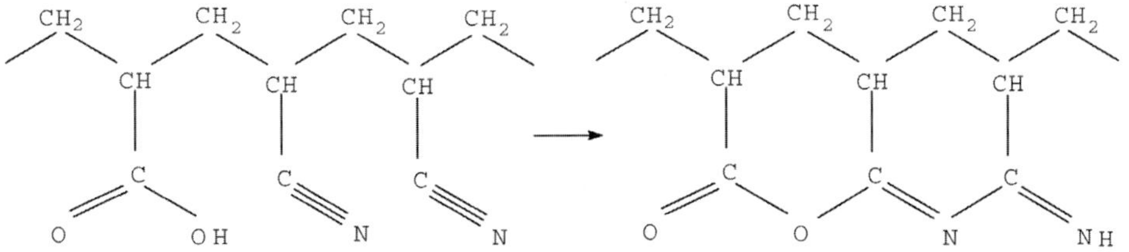

CH_2 CH_2 CH_2 CH_2 CH CH CH C C C O OH N N → CH_2 CH_2 CH_2 CH_2 CH CH CH C C C O O N NH

Figure 9.1 Initiation and formation of ladder polymer.

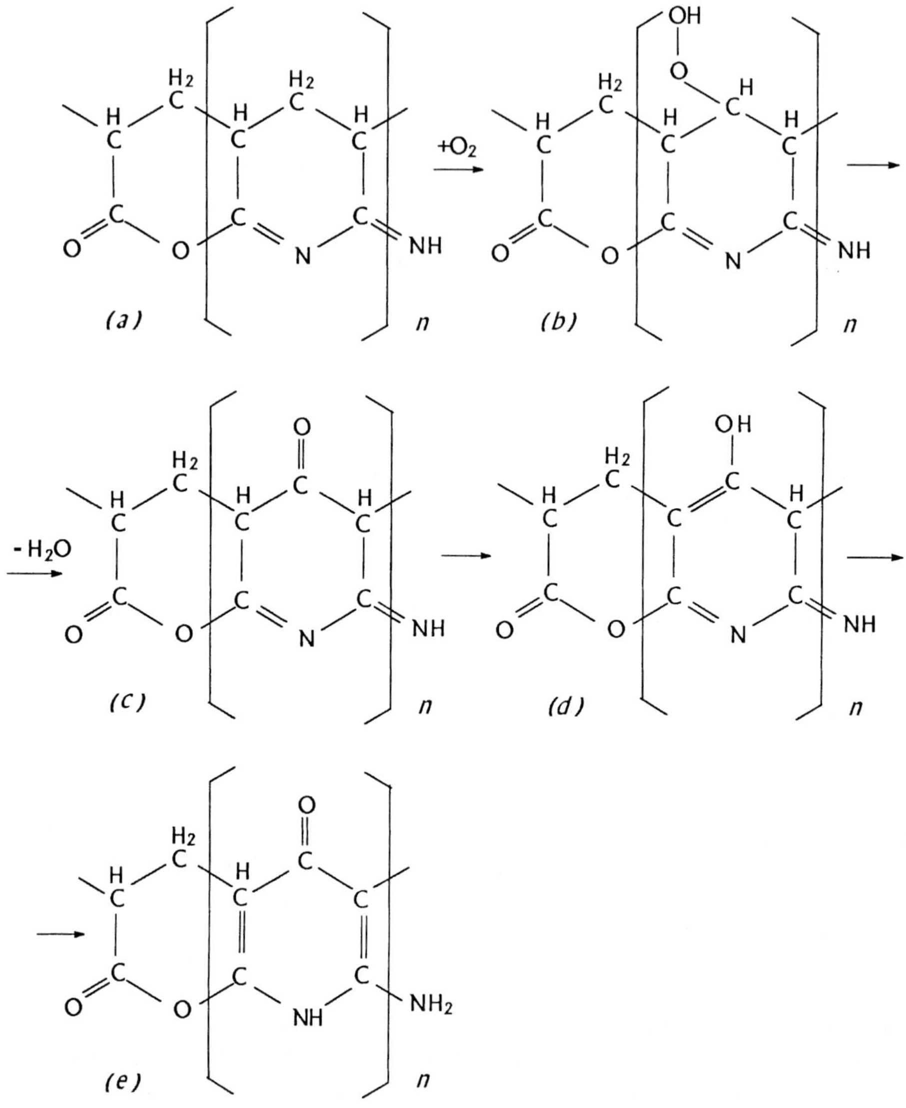

(a) Ladder Polymer
(b) Oxidation to hydro peroxide
(c) Keto-formation by loss of water
(d) Tautomeric change to hydroxy-pyridine structure
(e) Tautomeric change to pyridone structure

Figure 9.2 Oxidation of ladder polymer.

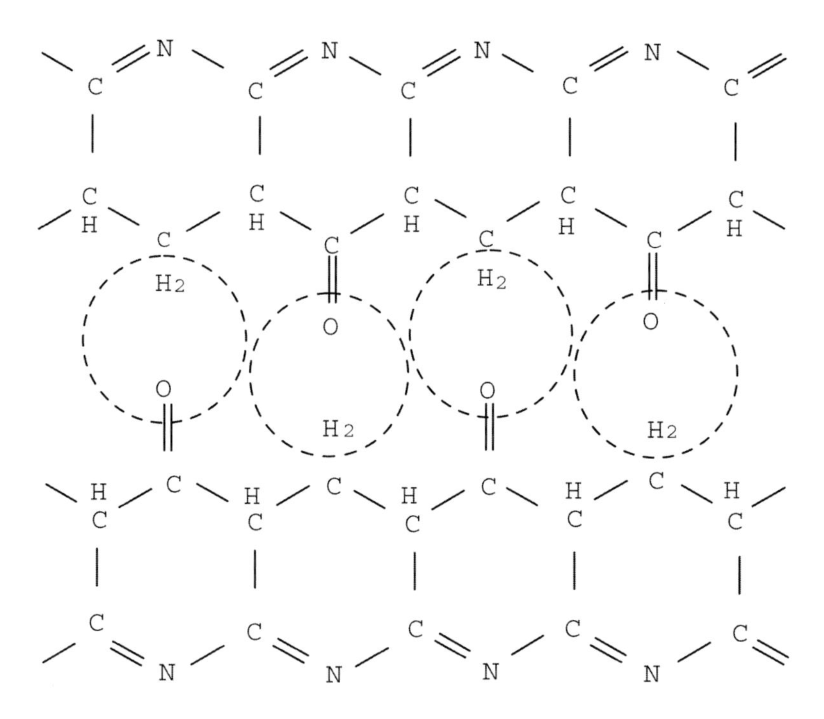

Figure 9.3 Cross-linking of oxidized ladder polymer with loss of H_2O.

after oxidation, lengths of cyclized polymer will be separated by lengths of uncyclized PAN with an irregular helical structure. Consequently, the alternating cyclized regions will tend to lie in different planes. Indeed, X-ray diffraction and transmission electron microscope studies of PAN fibres heat treated at 320 °C reveal both disordered PAN and ladder polymer with an approximate carbon layer plane spacing [12]. The lengths of these incipient building blocks for the carbonization process are approximately 1.0 to 1.5 nm, and are consistent with a length of 4 to 6 cyclized units in any one plane.

9.1.2.2 Carbonization

In practice, carbonization on a modern carbon fiber line takes place as a continuous process with the stabilized fiber passing through zones of increasing temperature up to around 1500 °C in an inert atmosphere, usually nitrogen. During this pyrolysis, non-carbon elements are removed as volatile products, the amounts varying with temperature, see for example Fitzer and Heyn [13]. The main gases evolved are HCN, NH_3, N_2, H_2O, CO_2, CH_4, and H_2, yielding carbon fibers with about 50% of the mass of the original PAN precursor. At temperatures up to 450 °C, HCN and various nitriles are evolved from the uncyclized regions of the chain molecules, and cross-linking of the cyclized regions begins with the evolution of H_2O (Figure 9.3).

Between 500 °C and 1000 °C, peaks in the evolution of NH_3 and HCN have been observed [4,9,12], involving end-to-end joining of cyclized regions, aromatization of non-cyclized regions and side-by-side condensation reactions between laddered structures to give broader heterocyclic regions (Figure 9.4). Dehydrogenation and denitrogenation take place (Figures. 9.5 and 9.6), giving rise to carbon ribbons containing approximately 5.8% N by mass at 1000 °C. X-ray diffraction and TEM studies show clearly the increased perfection of carbon layer plane packing as the temperature rises from 320 °C to 1000 °C [12].

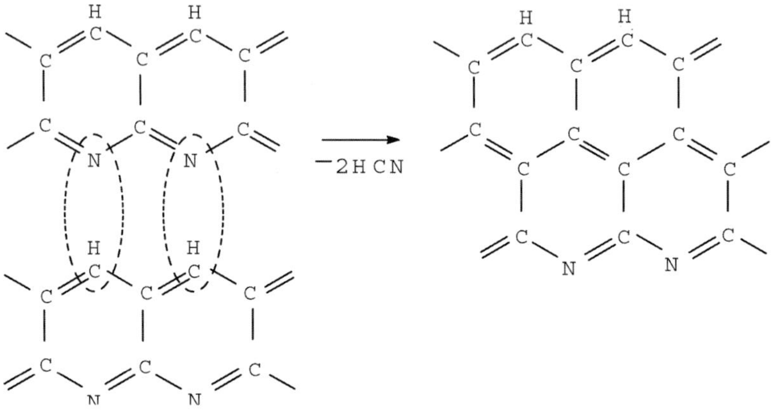

Figure 9.4 Cross-linking of unoxidized ladder polymer with loss of HCN.

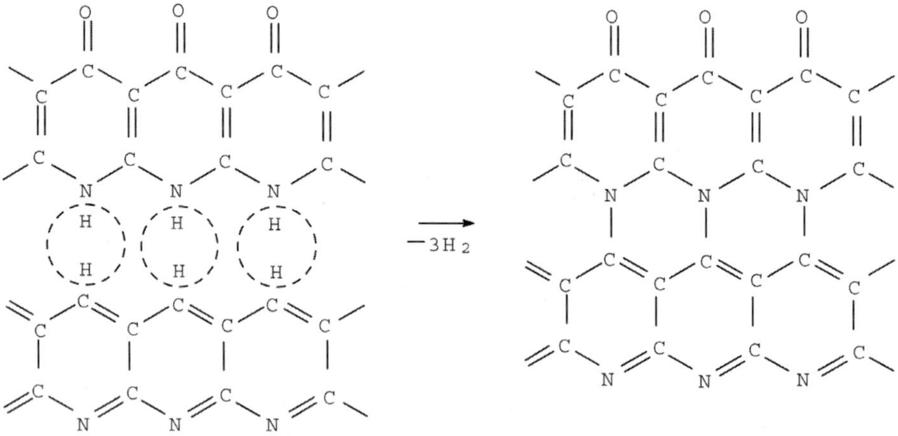

Figure 9.5 Cross-linking of ladder polymer by loss of H_2.

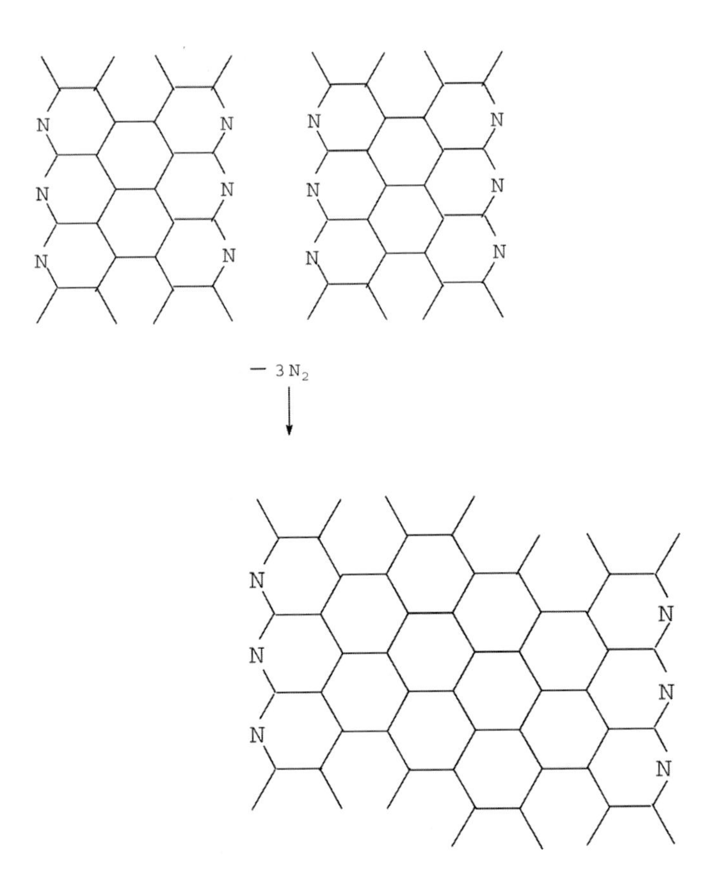

Figure 9.6 Cross-linking of layers by loss of N_2.

9.1.2.3 Graphitization

PAN-based carbon fibers are essentially non-graphitic, with no definite crystallographic arrangement of the layer planes in terms of a regular three-dimensional stacking. The layers are said to be 'turbostratic' and it is appropriate to refer to the structure of carbon fibers from PAN as 'turbostratic' graphite. Heat treatment of carbon fibers at temperatures up to and beyond 2500 °C in an atmosphere of argon will improve the packing of the layer planes and also their orientation with respect to the fibre axis, which effects an increase in the tensile modulus to around 500 GPa. Further increases in tensile modulus up to around 700 GPa have been obtained by hot stretching at 2800 °C [14]. However, despite considerable improvement in crystallite size and orientation, there are no indications in the X-ray diffraction pattern for the existence of a true graphitic structure in hot stretched PAN-based carbon fibres. Typical examples of the tensile properties of Toray's range of PAN-based carbon fibers are given in Table 9.1.

Table 9.1 Tensile Properties of Carbon Fibers

Precursor	Manufacturer	Fibre	Strength (GPa)	Modulus (GPa)	Strain (%)	Density ($g\,cm^{-3}$)
PAN	Toray	T300	3.53	230	1.5	1.76
		T800H	5.59	294	1.9	1.80
		T1000	7.06	294	2.4	1.82
PAN	Toray	M40	2.74	392	0.60	1.81
		M60J	4.40	392	1.0	1.77
		M50	2.45	490	0.50	1.91
MP Pitch	BP Amoco	P25	1.40	140	1.0	1.90
		P55	2.10	380	0.55	2.00
		P75	2.00	500	0.40	2.08
		P100	2.20	690	0.32	2.15
		P120	2.30	820	0.20	2.18
MP Pitch	Du Pont	E35	2.98	263	1.03	2.10
		E55	3.12	367	0.74	2.14
		E75	2.96	503	0.56	2.16
		E105	3.12	713	0.55	2.17
		E120	2.92	787	0.55	2.19
		E130	3.70	848	0.55	2.19

9.1.3 Carbon Fibers from Pitch Precursors

The low cost of raw pitch has persuaded several companies, particularly the major petroleum producers, that pitch-based precursors offer an attractive alternative to PAN for conversion to carbon fibres. Edie and Diefendorf [15] have listed some of the advantages of pitch as: a 40 to 50% cheaper raw material, less energy required to convert an essentially aromatic graphitizing material, and smaller percentages of hydrogen, nitrogen and other non-carbon elements, leading to around 75% yield for mesophase pitch precursor fibre compared with 40 to 45% for PAN precursor fibre. Figure 9.7 depicts a typical aromatic pitch molecule [9].

The steps required to convert raw pitch to carbon fibre are:

a. Refinement of raw pitch for isotropic pitch production, or
b. thermal polymerization of high aromatic content pitch to produce a mesophase or liquid crystal component.
c. Melt spinning of mesophase pitch to form precursor fibers.
d. Oxidative treatment of precursor fibers to stabilize them during carbonization and graphitization.
e. Carbonization and Graphitization at temperatures similar to those utilised in the PAN precursor conversion process.

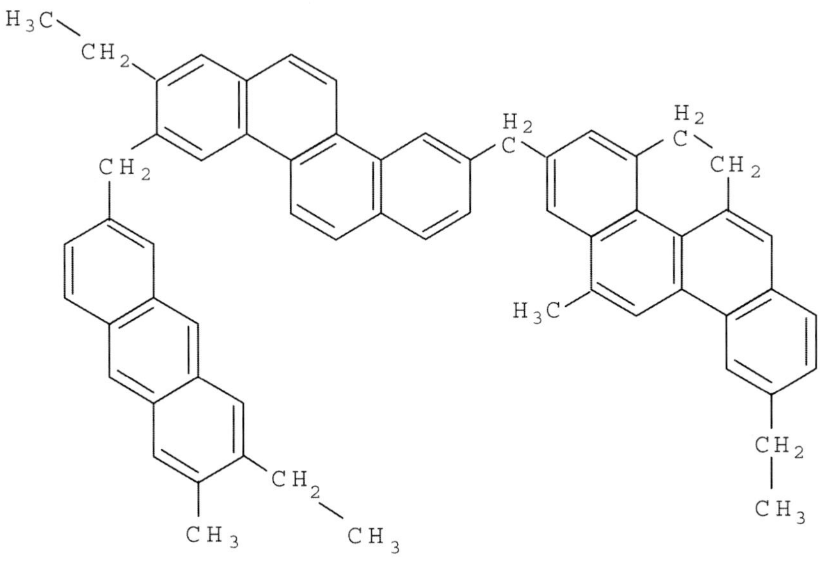

Figure 9.7 A typical pitch molecule.

9.1.3.1 Raw Pitch Refinement

Edie [16] has explained in detail why petroleum pitches are more suitable than coal-tar pitches as precursors for both isotropic and mesophase pitch fibers. In particular, he describes how a highly aromatic petroleum pitch such as Ashland 240, which is low in quinolene insolubles, can be converted to Aerocarb 60 and Aerocarb 75, both typical isotropic precursors for the melt spinning of pitch fibers. The pitch is refined in a process which allows the more volatile compounds to be removed effectively. The molten pitch is filtered to remove solid impurities, spun into pitch precursor fibers, then carbonized to produce a low quality but inexpensive carbon fiber for non-structural uses.

9.1.3.2 Thermal Polymerization

If a highly aromatic pitch is heated in an inert atmosphere to temperatures between 400 °C and 450 °C for an extended period, around 50% of the pitch will be transformed from an optically inactive to an optically active material, that is it will form an anisotropic liquid crystal phase. This is generally referred to as a mesophase. Edie also describes methods by which the thermal polymerization of the pitch can be controlled and by which a 100% mesophase fraction can be obtained prior to melt spinning.

9.1.3.3 Melt Spinning of Precursor Fibers

Conventional melt spinning through a spinnerette is the process usually used to form high quality continuous carbon fiber from both isotropic and mesophase pitches. An extruder melts the precursor and forces it through a spin pack containing a filter to remove solid particles. The melt is then extruded through the spinnerette, a plate containing a large number of small cylindrical holes. The threadline emerging from the spinnerette is quenched in air to solidify the filaments and wound up via a take-up roll. The process variables involved in melt spinning of pitch have been studied in detail, and the dependency of viscosity on temperature has been widely reported [16–18]. Edie [16] describes the difficulties inherent in producing a low diameter filament without exceeding the breaking stress of the fibre during draw down. Careful process control can be required, adding cost to this stage of the process.

It must be noted that, as with all liquid crystal melts, the aromatic molecular sheets which constitute mesophase pitch fibers are oriented initially by passage through the spinnerette, and further oriented by draw down in the winding process. No further drawing is necessary.

Although process control is slightly easier for isotropic pitch fiber production, it is generally necessary to hot stretch the fibers during carbonization and graphitization in order to achieve a satisfactory layer-plane orientation; this, unfortunately, is an expensive process. Some orientation at the precursor stage of isotropic pitch carbon fiber production may be achieved by different forms of melt spinning. Melt blowing and centrifugal spinning are two possibilities leading to low modulus mats of non-continuous material. Centrifugal spinning is of some interest in that the spinning bowl has a cusped rim from which fibres are formed and there are no spinnerette holes to become blocked with impurities in the melt.

9.1.3.4 Oxidative Stabilization

As is the case with PAN precursors, pitch fiber precursors must be stabilized by heat treatment in an oxidizing atmosphere, usually air. Isotropic pitch fibers have a lower melting temperature than mesophase pitch fibers, and complete thermosetting can take several hours (usually at around 250 °C). Mesophase pitch precursors are usually oxidized at temperatures around 300 °C for periods up to 2 hours. Crosslinking reactions take place between the aromatic units, forming layers which are considerably greater in extent than those produced at a similar stage with PAN. The mesophase pitch based precursor fibers are essentially graphitizable materials.

9.1.3.5 Carbonization and Graphitization

The processes of carbonization and graphitization of pitch precursor fibers are very similar to those carried out for PAN precursor fiber. The evolution of CH_4 and H_2 is allowed to take place without disruption of the structure in a precarbonization stage at 900 °C to 1000 °C. Heat treatment in the range 1000 °C to 3000 °C produces highly oriented graphitic carbon fibers with a tensile modulus directly related to the temperature of heat treatment. The most highly oriented MP-based fibers have moduli in the range 800 to 900 GPa, approaching the theoretical value for graphite (1000 GPa). Typical examples of the tensile properties of both BP Amoco and DuPont MP-based carbon fibers are given in Table 9.1; however, it may be noted that DuPont no longer produce the mesophase pitch based fibers discussed here.

9.2 Structure and Properties

9.2.1 Tensile Modulus

We have seen in Section 9.1, that it is possible to produce a variety of carbon fibers with different physical properties, particularly tensile modulus, from both PAN and pitch precursors. We should recall that the goal of the early research workers on carbon fibers, was to attain a tensile modulus well above the theoretical maximum of around 300 GPa for an organic fibre consisting of long chain molecules; indeed, to approach the theoretical modulus for graphite of about 1000 GPa. This goal has been more or less achieved with the high temperature treated mesophase pitch (MP)-based carbon fibers, and with all precursors it has been well established that the higher the heat-treatment temperature, the higher the tensile modulus, with stress graphitization giving the highest modulus of all. In effect, the input of either thermal or stress energy or both together, can produce that additional axial orientation of the layer planes which leads directly to high tensile modulus. In the case of the essentially non-graphitizing PAN-based carbon fibers, the carbon layers always remain in turbostratic arrangement, unless impurity particles induce catalytic graphitization. In the case of the essentially graphitizing MP-based carbon fibers, the layers can be aligned into three dimensional graphitic order at the highest heat treatment temperatures.

9.2.2 Tensile and Compressive Strength

The theoretical tensile strength of a solid is much more difficult to evaluate than the tensile modulus, but is generally estimated to lie in the range of 10% to 20% of the theoretical modulus. Thus in a perfect graphite, with modulus around 1000 GPa, a theoretical strength of 100 GPa might be anticipated. In fact, one of the highest reported values of tensile strength, 20 GPa, was achieved for a carbon whisker of almost perfect graphite, and it would seem that 10 GPa (1% of the tensile modulus) might be a more realistic goal for carbon fibers.

A complicating factor concerning the tensile strength of carbon fibers, is the effect of both surface and internal flaws. This flaw dependence has been widely studied in terms of statistical distributions showing how the tensile strength increases as the gauge length decreases, with extrapolations predicting strengths in flaw free fibres to be from 3.2 to 7.0 GPa.

The seminal work of Moreton and Watt [19] on clean-room spinning of PAN precursor fiber, and similar work on contaminated mesophase-pitch precursors by Jones et al [20], showed that contaminants are a major cause of low tensile strength, thus pointing the way to the production of commercial carbon fibers with tensile strengths at the top end of the range predicted by gauge-length extrapolations. Subsequently, with the introduction of clean methods of both precursor spinning and of carbon fibre processing, it was found possible to produce fibres with very low levels of flaw content. Structural features could then be perceived as strength determining factors.

It is well established that the compressive strength of most carbon fibers is inferior to their transverse strength, the ratios of compressive strength to tensile strength varying between 0.2 and 0.6. Since in many structural applications, carbon fiber reinforced polymer (CFRP) composites and carbon fiber reinforced carbon (CC) composites are subjected to compressive loading, and since the compressive strength of a composite is directly related to the compressive strength of the reinforcing fibers, an understanding of the behaviour of single fibres in axial compression is important. Indeed, an understanding of the relationships between fiber fine structure and both tensile and compressive strength are of paramount importance if achievement of the potential strength of carbon fibers is to be accomplished.

9.2.3 The Structure of Carbon Fibers

Information about the structure of carbon fibers has mainly been obtained by means of the indirect technique of X-ray diffraction, both wide-angle and small-angle, and the more direct methods of microscopy, particularly Scanning Electron Microscopy and Transmission Electron Microscopy, see for example the review by Johnson [21]. Wide-angle X-ray diffraction or less precisely Wide-angle X-ray scattering (WAXS), is a relatively simple technique for fiber characterisation; accounts specific to the analysis of carbon fiber patterns have been reported by Ruland et al. [22] and Johnson [23].

The WAXS patterns of PAN precursor fibres are particularly uninformative. Typical PAN fibers have only two reflections on the equator, no reflections on the meridian and no layer-line reflections, indicating a poorly oriented paracrystalline material. After stretching, the orientation improves, as indicated by sharper reflections on the equator and diffuse meridional and first layer line reflections. It is considered that the PAN fibres consist of molecular chains with irregular helical conformation arranged in a two-dimensional pseudo-hexagonal lattice [11]. We have described earlier (Section 9.1.2) the processes of carbonization and graphitization which convert the PAN molecules into regions of turbostratic graphite layers, i.e. carbon layers with no three dimensional regularity. This is particularly evident in the WAXS patterns of PAN-based carbon fibres which, after carbonization around 1000 °C, show a broad 002 reflection on the equator and a diffuse 100 ring, strongest on the meridian, and which, after graphitization around 2500 °C, show a much sharper 002 reflection on the equator, a sharper 100 ring, and evidence of a 004 reflection on the equator (Figure 9.8). These *hkl* indices are based on the typical graphite unit cell, but it must be emphasised that true graphitic reflections are not normally seen in PAN-based carbon fibres. In contrast, MP-based carbon fibers treated at high temperature have very sharp 00*l* and 100 reflections, with BP Amoco P120 fiber revealing a strong 101 reflection outside the normal 100 ring; indeed, the 101 reflection is stronger than the 100 reflection, a feature of a highly graphitic material (Figure 9.9).

Qualitative analysis of WAXS patterns may suffice for a cursory examination of structure, but serious characterisation necessitates a more rigorous analysis. By means of an X-ray diffractometer employed either in step-scan mode or with a Position Sensitive Detector (PSD), it is possible to obtain intensity data on the diffraction peaks in both

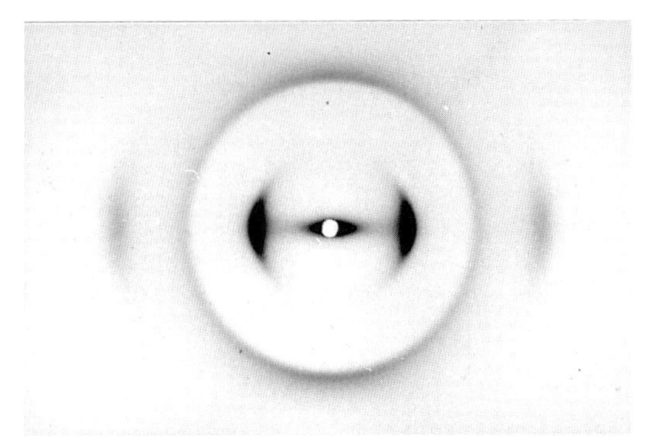

Figure 9.8 Wide-angle X-ray diffraction pattern of PAN-based carbon fiber (high modulus).

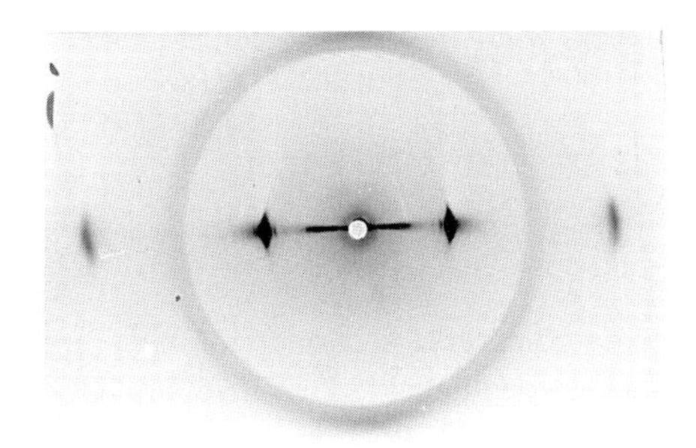

Figure 9.9 Wide-angle X-ray diffraction pattern of MP-based carbon fiber (high modulus).

equatorial and meridional directions. The data may be normalised, corrected, and overlapping peak profiles separated and resolved by means of computational methods [24]. The interlayer spacing $\frac{c}{2}$ and the crystallite size in the c-axis direction L_c may be evaluated from the position and width of the resolved 002 reflection. The crystallite sizes in the a-axis direction perpendicular and parallel to the fibre axis, $L_{a\perp}$ and $L_{a//}$, may be determined from the 100 reflections on the meridian and the equator, respectively. These measures may be considered to relate to short perfect lengths of crystal within the fibre. In reality, as revealed by TEM, the layer planes are very much more extensive, although imperfect. The preferred orientation Z may be measured from the azimuthal spread of the 002 reflection. Typical values of the key physical parameters for examples of both PAN and MP-based carbon fibres are included in Table 9.2.*

9.2.3.1 The Microstructure of PAN-Based Carbon Fibers

The Transmission Electron Microscope (TEM) is a very important tool for obtaining direct information concerning the structure of PAN-based carbon fibers, particularly

*σ_t and σ_c refer to laboratory measured tensile strength and compressive strength respectively.

Table 9.2 Physical and Structural Properties of Carbon Fibers

Precursor	Manufacturer	Fibre	σ_t (GPa)	σ_c (GPa)	$\frac{\sigma_c}{\sigma_t}$	$\frac{c}{2}$ (nm)	L_c (nm)	Orientation (Degrees)
PAN	Toray	T1000	5.7	2.2	0.39	0.348	1.7	31.5
	Experimental	Z	3.5	1.9	0.54	0.347	2.6	30.8
MP Pitch	DuPont	E35	2.4	0.82	0.34	0.342	3.7	28.4
		E55	3.0	0.78	0.26	0.342	8.5	20.1
		E75	2.9	0.67	0.23	0.342	12.0	16.3
		E105	3.1	0.61	0.20	0.340	16.8	13.8
		E120	3.2	0.57	0.18	0.340	19.2	11.4
		E130	3.6	0.63	0.17	0.338	23.8	12.5
MP Pitch	Amoco	P25	1.4	0.90	0.64	0.350	2.5	34.3
		P55	2.1	0.50	0.24	0.342	11.7	16.5
		P75	2.0	0.50	0.25	0.341	14.6	11.1
		P100	2.2	0.40	0.18	0.337	23.7	11.0
		P120	2.2	0.30	0.14	0.336	25.5	8.3

when ultrathin longitudinal or transverse sections can be prepared and lattice-fringe images of the 002 layers recorded. Bright-field, dark-field, and electron-diffraction modes of operation are also of significance. Bright-field is the normal mode of operation, but the dark-field mode allows a positive identification of those regions in a specimen which contribute to that reflection in the electron-diffraction pattern selected for dark-field operation. Dark-field images are particularly helpful in distinguishing the fine structure of fibers, for example sheath-core differentiation.

Although lattice-fringe images (Figures 9.10 and 9.11) have been extremely helpful in our understanding of the microstructure of carbon fibers, it must be emphasised that there is no exact one-to-one correspondence between layer planes in the specimen and lattice fringes in the image, see for example Johnson and Crawford [25] and Millward and Jefferson [26]. The lattice-fringe image is formed by the transfer of information from the object plane to the image plane via the objective lens, a process which is affected by the spherical aberration of the lens and its level of focus. The phase contrast transfer function, which takes into account spherical aberration and focus, reveals that small changes in focus can cause significant changes in the contrast of the lattice fringes. In some cases zero contrast can give the appearance of voids in the specimen, an artefact which must be avoided when interpreting the images of relatively disordered carbon fibers. Crawford and Marsh [27] in demonstrating this effect, show how valid and invalid images may be distinguished.

Some of the earliest TEM studies of carbon fibers were held to indicate the presence of a fibrillar structure, subsequent studies of bright-field and lattice-fringe images showed that the type of fibrils seen for example in natural fibres such as wool and cotton and in the high performance aramid fibre Kevlar, were not observed in carbon fibers; indeed, the

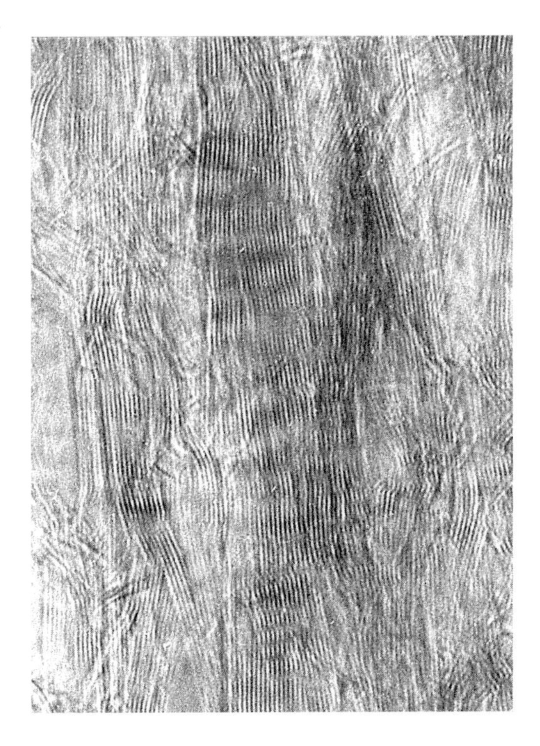

Figure 9.10 Lattice fringe image of longitudinal section of PAN-based carbon fiber (high modulus).

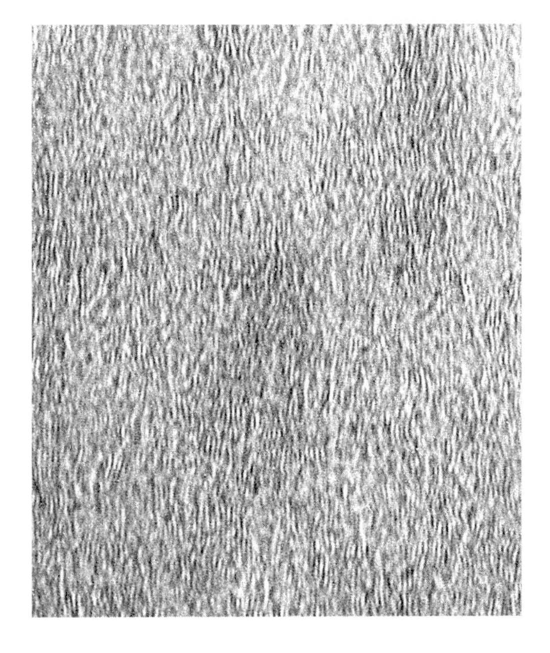

Figure 9.11 Lattice fringe image of longitudinal section of PAN-based carbon fiber (high strength).

structure of carbon fibers is essentially non-fibrillar. An early model, in which curvilinear layer planes are packed side by side enclosing voids of an approximately needle shape, was proposed by Ruland et al. [22] and has some merit for modelling physical behaviour. Studies of the complex nature of the lattice-fringe images in longitudinal sections (Figure 9.10 is a good example) suggested a more realistic model of structure indicating the three dimensional interlinking of the turbostratic layer planes as shown in the two dimensional schematic of Figure 9.12 [28,29].

Lattice-fringe images of transverse sections reveal a very complex structure. At the surface, and in the region close to the surface, the layer planes are essentially parallel to the surface; however, many layer planes can be seen to fold through angles of up to 180° in a 'hairpin' fashion (Figure 9.13), so that regions with layer planes at right angles to the surface are accessible if the outermost layers are removed. The ease with which carbon layers can deform into cylindrical shapes is demonstrated here; the ultimate perfection of distortion into a sphere being well recognised following extensive research carried out in the new field of fullerene carbons. In terms of a model for carbon fiber from PAN, a study of the lattice-fringe images from both transverse and longitudinal sections led to the schematic of Figure 9.14, which has been widely accepted as embodying most of the appropriate features of microstructure in high modulus PAN-based carbon fibers. If different degrees of layer-plane disorder are substituted in the schematic, then a general model for all types of PAN-based carbon fibers can be envisaged.

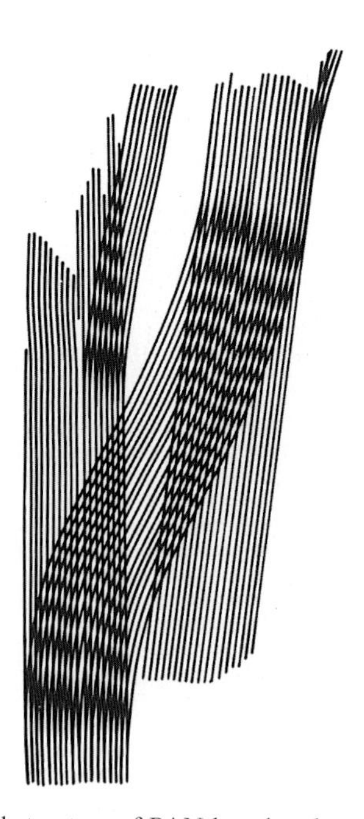

Figure 9.12 Model of longitudinal structure of PAN-based carbon fiber.

Figure 9.13 Lattice fringe image of transverse section of PAN-based carbon fiber.

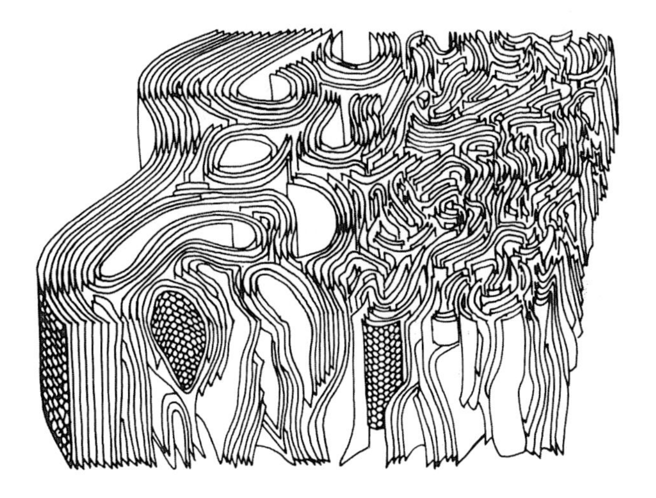

Figure 9.14 Model of transverse structure of PAN-based carbon fiber.

9.2.3.2 The Microstructure of MP-Based Carbon Fibers

Studies of the fracture faces and cut-face cross-sections of MP-based carbon fibers in the Scanning Electron Microscope (SEM) have shown that they have a variety of microstructures, and are rather more complex than PAN-based carbon fibers, which may be considered in general as having a random microstructure. Figure 9.15 depicts radial, flat-layer, onion-skin, radial-folded, line-origin and random structures. In effect these terms describe sheet-like entities which run parallel to the fibre axis with the named texture observed in transverse

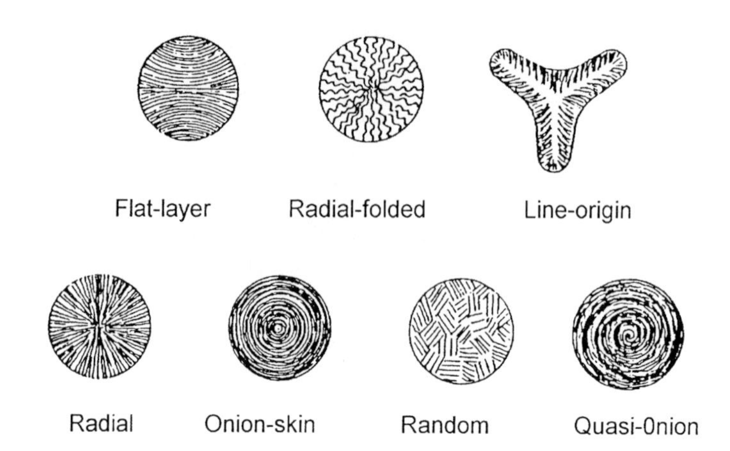

Figure 9.15 Microstructure in MP-based carbon fibers.

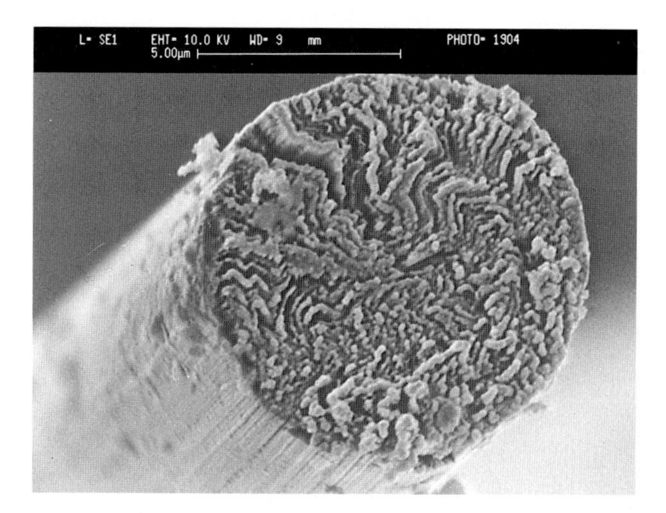

Figure 9.16 Microstructure of DuPont fiber.

section. An excellent example of radial-folded structure in DuPont MP-based fiber is shown in Figure 9.16. A typical flat-layer structure from a BP Amoco fibre is shown in Figure 9.17, and two radial-folded structures from the high thermal conductivity fibers from DuPont, E35C and C700, are depicted in Figure 9.18. These fibres also display the so-called 'missing-sector' appearance and have pseudo-radial microstructures.

It may be pointed out that the term 'microstructure' is not entirely appropriate in this context and that the term 'macrostructure' would have been more suitable. However, the use of 'microstructure' is already widespread in the literature and it would confuse the issue to be pedantic on the matter.

Singer [30] was the first to report that the precursor fibres for Union Carbide's Thornel MP-based fibers were characterized by onion-skin, radial or random textures, that they

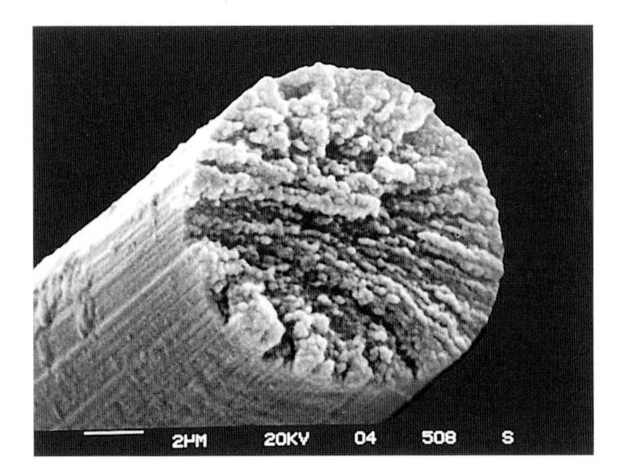

Figure 9.17 Microstructure of BP Amoco fiber.

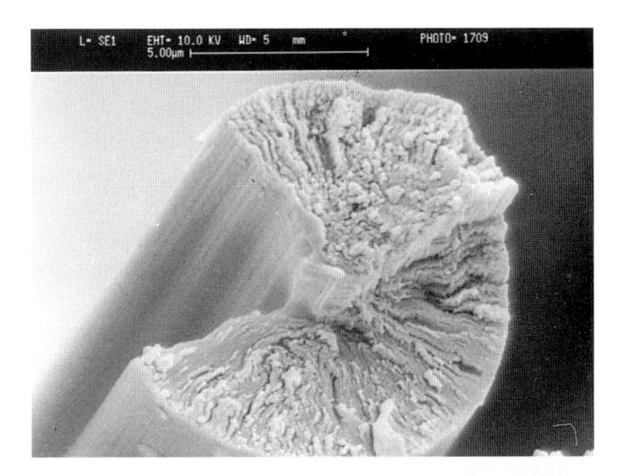

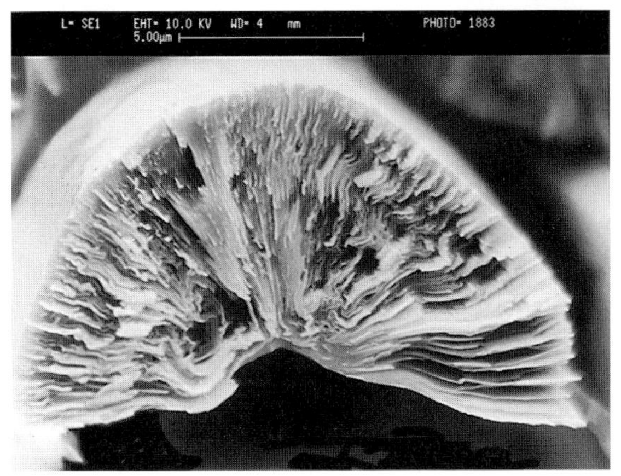

Figure 9.18 DuPont fibers with missing sectors (E35C above and C700 below).

may be varied by spinning conditions such as melt temperature and viscosity, and that the microstructures are maintained throughout carbonization and graphitization. More recently, Endo et al. [31] reported that extrusion without stirring tends to give a radial texture, whereas extrusion with stirring gives a random or onion-skin structure, depending upon spinning conditions. Edie [15] reports that newer varieties of Thornel fibers produced by BP Amoco Performance Products exhibit a flat-layer structure, whilst Endo [32] has shown that Kashima fibers consist of radial-folded layers.

White et al. [33] have described MP-based carbon fibers with radial and random structures; in particular the radial structure with a pie-shaped missing wedge, and a mixed structure with an outer radial section and an inner random section. They also demonstrated that variable structure along the length of a single fibre can be observed in many cases. Most importantly, the different structures and the changes between them were discussed in terms of disclinations.

Disclinations

Real crystals are never perfect; they contain imperfections which govern the properties of the material in which they are found. In metals and semiconductors, translations or linear displacements of one part of a crystal with respect to another give rise to dislocations. In liquid crystal or mesophase materials, if part of a solid is displaced by a rotation rather than a translation, the result is a disclination [34]. Zimmer and White [35] give examples of positive and negative wedge disclinations of the type found in mesophase carbonaceous systems. Deformation studies showed that a dense array of parallel wedge disclinations forms when plastic mesophase is drawn to a fine fiber. Whilst the mesophase remains sufficiently fluid for disclination motion, annihilation reactions will occur between disclinations of opposite sign; after normal cooling the bulk of the fiber consists of parallel wedge disclinations of opposite sign. Hence the outer sheath of more perfect radial structure surrounds a core of random, less perfect texture. Under certain processing conditions, there remains only a single positive disclination and a radial structure results, with a missing sector as a natural crystallization phenomenon.

Figure 9.18 shows how fibers with a missing sector possess layers which are typically unfolded and almost planar. This effect is more pronounced in the C700 fiber than in the E35C fiber; additionally, the larger the angle of the missing sector, the more planar the sheets. It would seem that when the wedge is formed by opening along a radial crack [36], the compressional stress which is stored within the fibre during initial formation is released and the layers within the fiber extend and become more planar [37].

Recently, Edie et al. [38] have spun fibres with trilobal and octalobal shapes to give MP carbon fibers with line origin macrostructures which possess higher tensile moduli and strengths than similar circular-section fibers. Mochida et al. [39] claim that fibers, only moderately stabilized and then carbonized under strain, exhibit a skin-core structure, and possess a higher orientation and hence a higher modulus than fully stabilized fibers. Clearly, changes in microstructure brought about by changes in processing can have a marked effect on tensile properties.

The early Thornel MP-based carbon fibers have been described as comprising bands of graphitic, microporous, and turbostratic carbon layers, the turbostratic phase being tightly folded parallel to the fibre axis. Subsequent work reported by FitzGerald, Pennock and Taylor [40] suggests that there are only two true phases in the more recent BP Amoco Thornel fibers, the microporous and the turbostratic. True graphitic layering is only found as a component of the dense turbostratic domains in the higher modulus fibres. A very detailed and comprehensive study of the range of DuPont MP-based carbon fibers has also been undertaken by Pennock, Taylor and FitzGerald [41] using the techniques of light microscopy, SEM, and TEM. Transverse sections reveal that an oriented core region is separated from a banded sheath region by a transition layer. The sheath has a radial-folded structure with a regular pattern of folding, which becomes much more irregular in the transition layer. There was no evidence of a microporous phase and it was considered that the fibers were spun from mesophase containing no isotropic phase, it being generally recognised that fibers prepared from mixed mesophase/isotropic pitches tend to have regions of mixed crystallization.. The fold regions have been very beautifully imaged in lattice-fringe TEM mode and disclinations of positive and negative sign can be seen clearly. It is suggested that fold separation is the feature which critically controls strength and that it is this feature which gives the DuPont range relatively high tensile strengths. Essentially this is a crack stopping microstructure.

9.2.4 Failure Mechanisms

9.2.4.1 Tensile Failure

Difficulties in describing the failure mechanism of PAN-based carbon fibers in terms of dislocation pile-up at grain boundaries, the unbending of curved ribbons, the presence of density fluctuations, or yield processes involving local shear deformation and slippage, have been thoroughly discussed in a review by Reynolds [42]. Despite the observation of a wide range of internal defects, no simple relationship could be found between flaw diameter, fiber strength, and surface free energy; consequently, Reynolds and Sharp [43] proposed a crystallite shear limit for fiber fracture based on the idea that crystallites are weakest in shear on the basal planes. It follows that, when tensile stress is applied to misoriented crystallites locked into the fiber structure, the shear stress cannot be relieved by cracking or yielding between basal planes. The shear strain energy may be sufficient to produce basal-plane rupture in the misoriented crystallite, and may initiate a crack which will propagate both across the basal plane and, by transference of shear stress, through adjacent layer planes.

An interpretation of the Reynolds and Sharp failure mechanism was first published by Johnson et al. [44]. A schematic diagram based on Figure 9.12 is reproduced as Figure 9.19, showing a misoriented crystallite well locked into the surrounding crystallites (Figure 9.19a) When stress is applied, basal plane failure takes place, (Figure 9.19b) and proceeds throughout the local region (Fig. 9.19c). However, before a crack can propagate throughout the fiber and cause failure, either one of two conditions must be fulfilled.

(i) The crystallite size in one of the directions of propagation of the crack, that is either L_c or $L_{a\perp}$, must be greater than the critical flaw size C for failure in tension.
(ii) The crystallite which initiates catastrophic failure must be sufficiently continuous with its neighbours for the crack to propagate.

The first condition is not normally fulfilled since both L_c and $L_{a\perp}$ are much less than C, although the effective value of $L_{a\perp}$ is considerably greater than the value measured by X-ray diffraction, as evidenced by the length of the curved layers seen in transverse sections. The second condition is most likely to be satisfied in the regions of enhanced crystallization and misorientation observed around defects.

It is now well accepted that internal and surface flaws which initiate failure often show evidence of misoriented crystallites within their walls, and continuity of crystallites is assured where catalytic graphitization has taken place around an impurity particle present in the precursor fiber or picked up during the spinning process. Continuity of crystallites then gives rise to a value of $L_{a\perp}$ which exceeds the critical size; thus any crack which is initiated by the Reynolds and Sharp mechanism may well propagate and cause catastrophic failure.

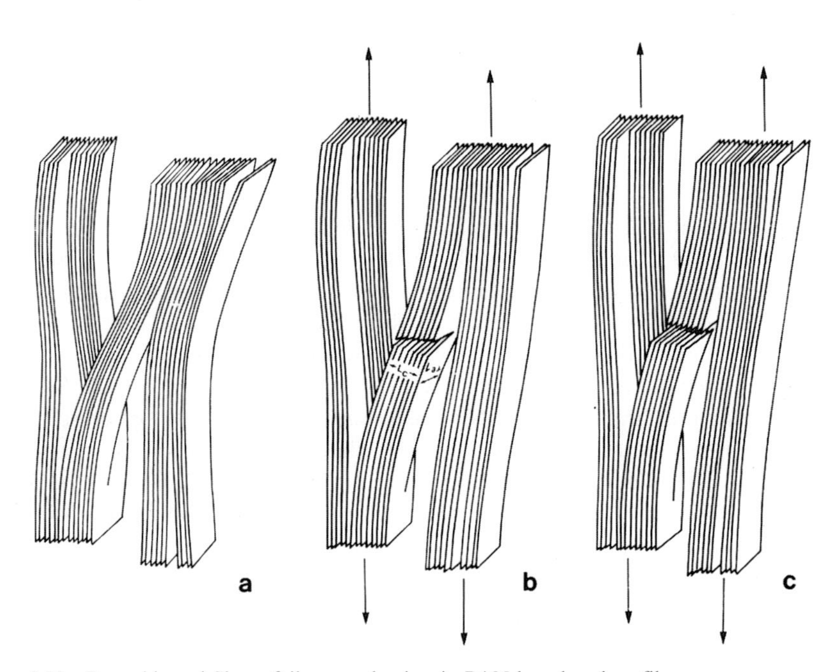

Figure 9.19 Reynolds and Sharp failure mechanism in PAN-based carbon fibers.
(a) Misoriented layers linking crystallites parallel to fibre axis
(b) Tensile stress causes rupture perpendicular to fibre axis
(c) Crystallite fails as crack propagates

Understanding of the mechanism of tensile failure indicates very strongly that high orientation and large crystallites, as found in HM PAN-based carbon fibers (Figure 9.10) will inevitably lead to a reduced tensile strength compared to the moderately well oriented, moderately sized crystallites found in the HS PAN-based carbon fibers (Figure 9.11). Furthermore, the surface skin itself may well be a cause of crack propagation and failure, so that treatments which remove some of the surface layers may well improve the tensile strength. In this respect, changes in layer-plane order are needed to prevent crack propagation; flaws or disorder in the layer-plane stacks are advantageous. Oberlin and her colleagues [45,46] have made a detailed study of the effect of disorder and layer plane radius of curvature on tensile strength of PAN-based carbon fibers. Increases in disorder correlate well with increased tensile strength, increases in layer-plane radius correlate well with decreased tensile strength. Layer-plane folding is also advantageous.

In terms of MP-based carbon fibers, it is most likely that the predominant sheet-like microstructure facilitates crack propagation, thus resulting in low tensile strength. Indeed Endo [47] has reported that Carbonic fibers produced by Kashima, which comprise a less graphitic, more folded structure than the Thornel fibers produced by BP Amoco, have higher tensile strengths; he has presented a model for crack propagation which agrees with the proposals put forward for PAN-based fibers by Oberlin. Further evidence that the spacing of regular folding is related to the critical crack length for failure, is found in Pennock's paper [41]; she concludes that a high density of folding and the presence of disclinations are significant factors in improving strength. Edie and Stoner [48] have detailed the effect of microstructure on the tensile properties of carbon fibres; they conclude that disturbance of the molten mesophase during spinning can disrupt the crystallization of the MP-based carbon fiber into regular sheets.

In essence it is clear that MP-based carbon fibers must have a microstructure which essentially mimics that of PAN-based carbon fibers in terms of crack stopping, if they are to achieve comparable strengths. At the same time measures must be taken at all steps in processing to reduce the possibility of contamination with particles which might induce catalytic graphitization and thus create crystallites of size greater than the critical length for failure. Flaws of this type are disadvantageous to tensile strength.

9.2.4.2 Compressive Failure

As indicated in Section 9.2.2, in many structural applications CFRP and CC composites are subjected to compressive loading; consequently, they are required to have good compressive properties, particularly strength. As the compressive strength of a composite is directly related to the compressive strength of the reinforcing fibers, an understanding of the behaviour of single carbon fibers in axial compression is important. If this behaviour is to be improved, it is imperative that the relationships between compressive strength and fine structure are investigated.

In a review of the properties of several high-performance fibres, Kumar and Helminiak [49] report that the ratios of compressive strength to tensile strength evaluated from composite data vary between 0.2 and 1.0. The main problem with testing composites is that application of the rule of mixtures to give a meaningful measure of compressive strength of the reinforcing carbon fibres is difficult. One way to overcome the problem is to investigate model composites comprising single carbon filaments embedded in epoxy resin.

The introduction of the fiber recoil method by Allen [50] gave a stimulus to research on the compressive properties of single carbon fibers. Subsequently, several improvements have been made to this method, particularly in terms of meaningful handling of data and symmetrical cutting of the fiber [51,52]. Other different approaches which enable compressive stress-strain curves to be plotted include piezoelectric microcompression of fibers [53] and a fiber/bending beam method monitored by a Laser Raman microprobe [54].

Boll et al. [55] suggested that changes in microstructure are needed for improved compressive strengths. Furthermore, they considered that, for a range of fibers, compressive failure occurred by shear with no evidence for microbuckling.

However, detailed SEM observations have revealed that, in general, PAN-based carbon fibers fail by a buckling mechanism, whereas high modulus MP-based carbon fibers fail by shear after tensile recoil [56]. A model for compressive failure in both PAN-based and some MP-based carbon fibers is presented in Figure 9.20. PAN-based fibers typically buckle on compression and form kink bands at the innermost surface of the fiber. A crack initiated at the side then under tension is propagated across the fiber leading to a stepped fracture face. Most BP Amoco-type MP-based fibers deform by a shear mechanism with

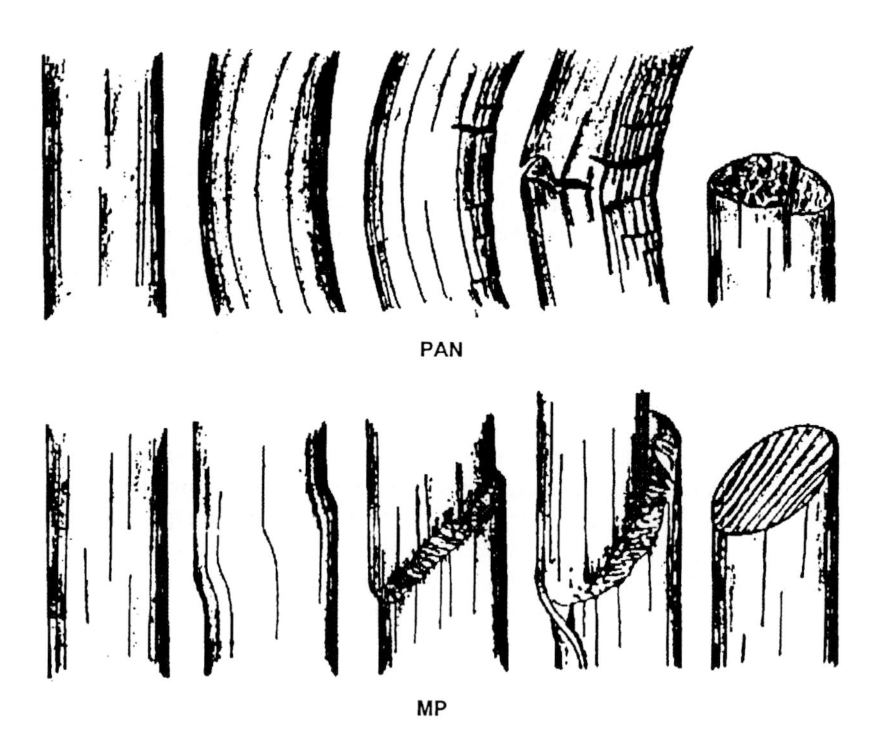

Figure 9.20 Model of compressive failure in PAN-and MP-based carbon fibers.

kink bands formed at 45° to the fiber axis; which initiate oblique rupture across the fiber on further deformation. Moreover, compressive strengths for various carbon fibers determined by recoil testing showed linear decreases with increasing tensile modulus for MP-based fibers. In the case of PAN-based fibers the compressive strength decreases more rapidly with relatively small increases in tensile modulus.

It has been suggested [56] that, under compression, the presence of disorder in carbon fibres may allow localized deformation, hence microbuckling, whereas the presence of highly ordered crystalline structures will promote shear. Thus, PAN-based carbon fibers, which have a predominantly random microstructure, tend to fail by microbuckling, and exhibit higher compressive strengths than those MP-based carbon fibers which have sheet-like microstructures, and tend to fail by shear. Edie and Stoner [48], reviewing the effect of microstructure and shape on properties, show that the compressive strength of MP-based fibers with a random structure is consistently higher than that of fibers with a layered microstructure. Clearly there is a need to understand the relationships between microstructure and compressive strength, and to find measures of disorder in the different types of carbon fiber.

9.2.5 Disorder in Carbon Fibers

There are two major sources of disorder in the microstructure of carbon fibers, the inter-crystallite disorder arising from the linking regions between the regions of relatively perfect layer-plane order, and the intra-crystallite disorder, arising from layer-plane imperfections and often referred to as 'lattice distortions'. The inter-crystallite disorder D_d can be evaluated from the proportion of scatter in the wide-angle X-ray diffraction pattern not contributing to the discrete carbon fibre reflections [56]. This scatter is often considered to arise from the so-called amorphous regions of the material, although it is extremely difficult to find examples of truly amorphous carbon in X-ray terms. The intra-crystallite disorder D_c may be evaluated, either from the increase in width of successive *00l* reflections [24], or from an empirical graphitization index based on interlayer spacings found in a perfect graphite, a very imperfect turbostratic carbon, and the material in question [56,57].

In the case of carbon fibers, selected-area electron-diffraction patterns from ultrathin longitudinal sections give more *00l* reflections than X-ray diffraction patterns, and have been utilised in comparative studies of the analytical methods employed to measure lattice distortion [24]. It is interesting to note that the only fibrous carbon studied in our laboratory which did not show increased broadening of successive *00l* reflections, and thus possessed zero lattice distortion, was a graphite whisker. Highly graphitic MP-based carbon fibers such as BP Amoco's P120 and DuPont's C700, which have very sharp *00l* reflections, nevertheless show broadening of the higher orders and thus possess measurable lattice distortion. Computational simulations of lattice distortion in carbon fibers have been attempted, in order to assess how the evaluations of both crystallite size and lattice distortion are affected [58]. A very detailed comparison of the mathematical methods employed for peak resolution, and the analysis of line broadening by size and distortion has also been reported [24].

Accounts of preliminary attempts to measure disorder in carbon fibers using methods of Image Analysis on both dark-field and lattice-fringe images from longitudinal fiber sections are now available [37,59]. Three new characterisation parameters have been defined; a crystallite imperfection factor (I_c) evaluated from the discrepancy between stacking size values L_c from X-ray diffraction measurements and direct measurements of L_c* from dark-field images, a direct lattice distortion or tortuosity (S) being a measure of lattice-fringe meandering, and a stacking imperfection factor (D_c*) corresponding to imperfection between the layer planes as measured from lattice spacings in local image areas.

Typical processed lattice-fringe images from DuPont E35 and C700 MP-based carbon fibers are shown in Figure 9.21, the tortuosity of E35 is 14.8% and of C700 is 2.1%. These values may be compared with tortuosities of 11.7% and 1.6% in the Toray T1000 PAN-based and the BP Amoco P120 MP-based carbon fibers respectively [59,60].

To illustrate the value of a direct measure of lattice disorder, a special PAN-based carbon fiber code-named 'Z' and the BP Amoco MP-based carbon fiber P25 can be exemplified from a study of structure-compressional property relations [60], the compressive strength of fiber Z being measured at 1.9 GPa, double the compressive strength of P25 (0.9 GPa). The two fibers have almost identical structural parameters, including the measures of inter-crystallite disorder D_d and intracrystallite disorder D_c. A qualitative appraisal of typical lattice-fringe images from the two fibers concerned shows an obvious difference (Fig. 9.22), the quantitative assessment by Image Analysis giving an average tortuosity of fiber Z as 12.1%, considerably greater than that of P25 at 9.2%.

9.2.6 Structure-Property Relations

We have observed that a highly oriented and highly crystalline microstructure will produce a high tensile modulus in both PAN-based and MP-based carbon fibers. Paradoxically, a well ordered system of layer planes will usually give rise to a lower than expected tensile strength, since crack propagation is enhanced by such organisation. Inclusions which have arisen by catalytic graphitization around impurity particles are particularly disadvantageous to good tensile strength. If relatively good orientation is maintained, a restricted size of the crystallites in both the *c* and *a* directions, as measured by L_c, $L_{a//}$, and $L_{a\perp}$, is advantageous to tensile strength, reducing the continuity which allows crack propagation.

A recent study [60] indicates that the ability of carbon fibers to withstand compressive stress is related to both the intra- and inter-crystallite disorder; relatively high disorder, particularly between the crystallites, enables the fiber to absorb compressive energy by buckling; relatively low disorder leads to shear failure of the fiber in compression. As with tensile failure, the compressive failure stress is also dependent on crystallite dimensions,

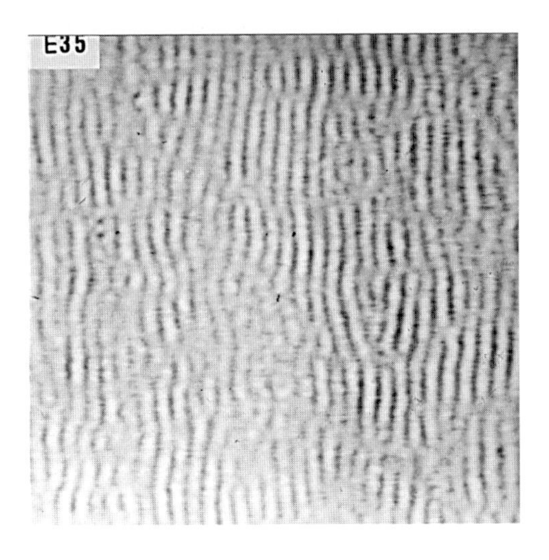

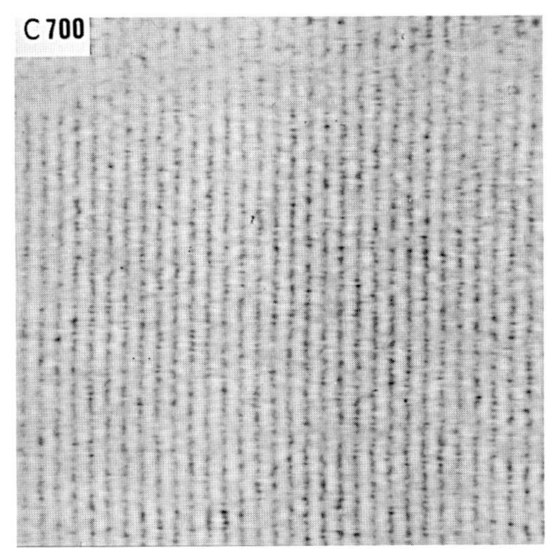

Figure 9.21 Image processed lattice fringe images of E35 and C700.

particularly L_c and $L_{a//}$; carbon fibers with these parameters below about 5nm tend to fail by buckling, whereas fibers with L_c and $L_{a//}$ greater than 5 nm tend to deform elastically before failing by shear.

The graphitizing nature of MP pitch precursors mitigates against the development of microstructures of around 5 nm in size, but appropriate attention to precursor formation can improve the ultimate tensile and compressive strengths of MP-based carbon fibers.

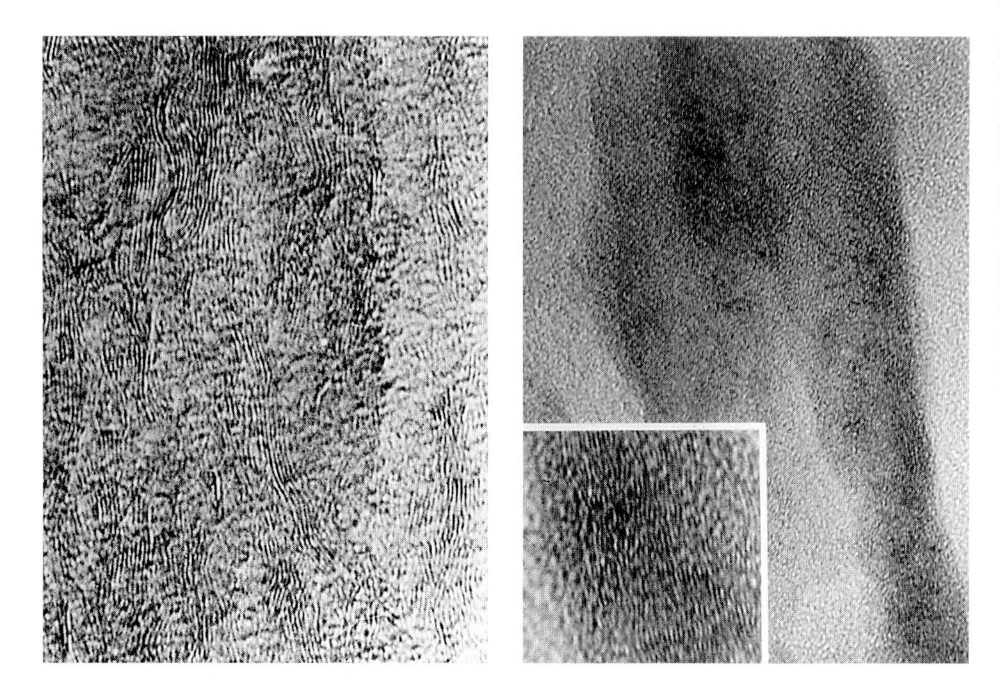

Figure 9.22 Comparison of lattice fringe images from Fiber Z (PAN-based) and Amoco P25 (MP-based).

9.3 References

1. Bacon, R., In *Chemistry and Physics of Carbon*, Walker, P. L., Thrower, P. A. (Eds.) (Vol. 9, 1973) Marcel Dekker, New York, pp. 1–102
2. Shindo, A. *Osaka Kogyo Gijitsu Shikenjo Koho* (1961) 12, pp. 110–119
3. Otani, S. *Carbon* (1965) 3, pp. 31–38
4. Mair, W. N., Mansfield, E. H. *Biographical Memoirs of Fellows of the Royal Society*, (1987) 33, pp. 643–667
5. Watt, W., Phillips, L. N., Johnson, W. U.K. Patent 110 791 (1968)
6. Anderson, B.W. *J. Phys. D: Appl. Phys.* (1987) 20, pp. 311–314
7. Watt, W., Johnson, W. *Nature* (1975) pp. 210–212
8. Coleman, M., Petcavich R., *J. Polym. Sci. Phys.* (1978) 16, p 821
9. Donnet, J. B., Bansal, R. C., *Carbon Fibers* (1984) Marcel Dekker, New York
10. Damoradan, S., Desai, P., Abhiraman, A. S. *J. Text. Inst.* (1990) 81, pp. 384–420
11. Hu, X., Johnson, D. J., Tomka, J. G. *J. Text. Inst.* (1995) 86, pp. 322–331
12. Watt, W., Johnson, D. J., Parker, E. *Int. Conf. on Carbon Fibres* (1974). The Plastics Institute, London, pp. 1/1–1/14
13. Fitzer, E., Heyn, M. *Chem. Ind.* (1976) 21, p 663
14. Johnson, W., *Proc. 3rd Conf. on Ind. Carbon and Graphite* (1971) Soc. Chem. Ind., London, pp. 447–452
15. Edie, D.D., Diefendorf, R.J. In *Carbon–Carbon Materials and Composites,* Buckley, J. D., Edie, D.D. (Eds.) (1992) NASA Reference Publication 1254, and Noyes Publications NJ, pp. 19–39

16. Edie, D.D. In *Carbon Filaments and Composites,* Figuerido, J.L., Bernado, C.A., Baker., R.T.K., Huttinger, K.J. (Eds.) (1990) NATO ASI Series E 177, Kluwer, Dordrecht/Boston/London, pp. 43–72
17. Rand, B. In *Strong Fibres*, Watt,W., Perov, B.V. (Eds.) (Vol.1, 1985) Elsevier, New York, pp. 495–575
18. Edie, D.D., Dunham, M.G. *Carbon* (1989) 27, pp. 647–655
19. Moreton, R., Watt, W. *Nature* (1974) 247, pp. 360–361
20. Jones, J.B., Barr, J.B., Smith, R.E. *J. Mater. Sci.* (1980) 15, pp. 2455–2465
21. Johnson, D.J. In *Chemistry and Physics of Carbon,* Thrower, P.A. (Ed.) (Vol. 20, 1987) Marcel Dekker, New York, pp. 1–58
22. Fourdeux, A., Perret, R., Ruland, W. *Proceedings of the First International Conference on Carbon Fibres* (1971) Plastics Institute, London, pp. 57–66
23. Johnson, D.J. *Proceedings of the First International Conference on Carbon Fibres* (1971) Plastics Institute, London, pp. 52–56
24. Hindeleh, A.M., Johnson, D.J., Montague, P.E. In *Fiber Diffraction Methods, ACS Symposium No. 141,* French, A.D., Gardner, K.H. (Eds) (1983), American Chemical Society, Washington, DC, pp. 149–182
25. Johnson, D.J.,Crawford, D. *J. Microsc.* (1971) 94, pp. 51–62
26. Millward, G.R., Jefferson, D.A. In *Chemistry and Physics of Carbon,* Walker, P.L., Thrower, P.A. (Eds.) (Vol. 14, 1978) Marcel Dekker, New York, pp. 1–82
27. Crawford, D., Marsh, H. *J. Microsc.* (1977) 109, pp. 145–152
28. Bennett, S.C., Johnson, D.J. *Carbon* (1979) 17, pp. 25–39
29. Johnson, D.J. J. Phys. D.: Appl. Phys. (1987) 20, pp. 286–291
30. Singer, L. S. *Carbon* (1978) 16, pp. 409–415
31. Hamada, T., Nishida, Y., Sajiki, Y., Matsumoto, M., Endo, M. *J. Mater. Res.* (1987) 2, pp. 850–857
32. Endo, M. *J. Mater. Sci.* (1988) 23, pp. 598–605
33. White, J. L., Ng, C. B., Buechler, M., Watts, E. J. *Proc. 15th Biennial Conf. Carbon*, Philadelphia, American Carbon Soc. and University of Pennsylvania, (1981) pp. 310–311
34. Harris, W. F. *Sci. Amer.* (1977) 237, p. 130
35. Zimmer, J. E., White, J. L. *Adv. Liq. Cryst.* 5, pp. 157–213
36. White, J. L., Buechler, M. in *Petroleum-Derived Carbons.* Bacha, J.D., Newman, J. W., White, J. L. (Eds.) American Chemical Society Symposium Series 303 (1986) p. 62
37. Guo, H. *Structure Property Relations in a Range of Mesophase Pitch-Based Carbon Fibres*, PhD Thesis, University of Leeds (1994)
38. Edie, D. D., Fox, N. K., Barnett, B. C., Fain, C.C. *Carbon* (1986) 24, p. 477–482
39. Mochida, I., Zeng, S. M., Korai, Y., Hino, T., Toshima, H. *J. Mater. Sci.* (1992) 27, p. 1960
40. FitzGerald, J. D., Pennock, G. M., Taylor, G. H. *Carbon* (1991) 29, pp. 139–164
41. Pennock, G. M., FitzGerald, J. D., Taylor, G. H. *Carbon* (1993) 31, pp.591–609
42. Reynolds, W. N. In *Chemistry and Physics of Carbon* Walker, A. C. and Thrower, P. A. (Eds.) Vol. 11 (1973) Marcel Dekker, New York pp. 1–67
43. Reynolds, W. N., Sharp, J. V. *Carbon* (1974) 12, pp. 103–110
44 Bennett, S. C., Johnson, D. J., Johnson, W. *J. Mater. Sci.* (1983) 18, pp. 3337–3347
45. Guigon, M., Oberlin, A., Desarmot, G. *Fibre Science and Technology* (1984) 20, pp. 55–72
46. Guigon, M., Oberlin, A., Desarmot, G. *Fibre Science and Technology* (1984) 20, pp. 177–198
47. Endo, M. *J. Mater. Sci.* (1988) 23, pp. 598–605
48. Edie, D. D., Stoner E. G. in *Carbon-Carbon Materials and Composites* Buckley, J. D. and Edie, D. D. (Eds.) Noyes Publications, Park Ridge, NJ (1993) pp. 41–69
49. Kumar, S., Helminiak, T. E. in *The Materials Science and Engineering of Rigid-Rod Polymers*, Adams, W. W., Eby, R. K., McLemore, D. E. (Eds.) MRS Symp. Proc., Pittsburgh, PA (1989) 134 p. 1
50. Allen, S. R. *J. Mater. Sci.* (1987) 22, p. 853
51. McGarry, F. J., Moalli, J. E. *Polymer* (1991) 32, p. 1811
52. Hayes, G., Edie, D. D., Durham, S. D. In *Proc. 20th Biennial Conf. Carbon*, Santa Barbara, American Carbon Society & University of California (1991) p. 326
53. Macturk, K. S., Eby, R. K., Adams, W. W. *Polymer* (1991) 32, p. 1782
54. Vlattas, C., Galiotis, C. *Polymer* (1991) 32, p. 1788

55. Boll, D. J., Jensen, R. M. M., Cordner, L., Bascom, W. D. *J. Comp. Mater.* (1990) 24, p. 208
56. Dobb, M. G., Johnson, D. J., Park, C. R. *J. Mater. Sci.* (1990) 25, pp. 829–834
57. Maire, J., Mering, J. *Proc. 1st Conf. of the Society of Chemical Ind. on Carbon and Graphite* (1958) p. 204
58. Johnson, D.J. *Proc. 15th Biennial Conf. Carbon*, Philadelphia, American Carbon Soc. and University of Pennsylvania, (1981) pp. 304–305
59. Dobb, M. G., Guo, H., Johnson, D. J. *Carbon.* (1995) 33, pp. 1115–1120
60. Dobb, M. G., Guo, H., Johnson, D. J., Park, C.R. *Carbon.* (1995) 33, pp. 1553–1559

10 Fibers from Electrically Conductive Polymers

Richard V. Gregory
Clemson University, Clemson, South Carolina, USA

10.1 Introduction

With the advent of the first industrially produced organic polymers in the early part of the twentieth century, materials have been fabricated that are strong, light weight, easily processed, and in many instances, resistant to harsh environments. Polymeric materials such as nylon and polyester have revolutionized the fiber and textile industries. Until quite recently however there was a significant area of applications that these organic polymers could not address. These applications were in the area of electrically conducting, electronic, and magnetic materials. Metals, inorganic crystalline structures, certain phases of carbon, and some ceramics were the only materials available for device construction in the electronics industry or in the case of metals, for wires to carry an electric current. In 1977 this changed dramatically with the work of Shirakawa, Heeger and MacDiarmid at the University of Pennsylvania, demonstrating that the organic polymer polyacetylene could be doped to high levels of electrical conductivity [1]. This polymer was the first truly organic conductor and consisted of only a carbon backbone whose energy levels had been altered to close the band gap between the HOMO (highest occupied molecular orbital, or valence band) and the LUMO (lowest unoccupied molecular orbital, or valence band) to a level where electrons could easily be promoted to the conduction band, delocalized over an extended distance, and the materials became more metallic in nature.

This work began an era of extensive scientific investigations into conductive polymers that were given the name *synthetic metals* or ICP's for *inherently conductive polymers*. Twenty years after the initial work at Pennsylvania thousands of papers and patents have been published or issued on many newly synthesized conductive polymers or older well known polymers investigated again to determine their inherent electrical properties. Each year national and international meetings devote many sessions, and in many cases entire meetings, to the subject of the electrical and electronic properties of these materials. Due to the unique electronic structure of the molecules involved, work has also begun in such areas as magnetic polymers, light emitting diodes, organic superconductors, polymeric actuators, and many other areas that have grown from the initial work [2]. Several small companies have been started devoted exclusively to the production of electronic synthetic polymers and many large corporations have established research groups to investigate these materials and bring them to a commercial reality [3]. These companies envision organic polymeric conductors for use in applications where the electrical or magnetic properties of a metal or semiconductor are needed and the light weight, strength, formability, and processing properties of polymers are desired. Applications range from polymeric batteries, electromagnetic shielding devices, polymer electronic devices such as light emitting diodes, photonic devices for optical computing based on non-linear optical properties, both second and third order, and if new ICP's are developed that are conductive enough, light weight motor windings and electrical transmission lines.

The purpose of the present chapter is to cover in some detail one of the many applications of these materials. The technology of fiber formation from intrinsically conductive polymers by spinning processes and its potential applications are addressed and contributions to the field will be discussed.. Areas of applications of magnetic polymers, non-linear optical materials, photonic applications, and organic superconductors are covered in depth in many recently published texts and a wealth of literature on these subjects has been published over the last decade [4]. Due to this extensive coverage in the existing literature these topics will not be addressed in this chapter. In order to review the effects of different ICP formation processes, particularly on fiber formation, it is necessary to initially discuss what energetic and morphological changes occur within the polymer structure when certain polymers are formed in such a way as to allow them to conduct charge in a metallic or semiconductor sense.

10.1.1 Polymer Conductivity

In general, most all of the organic conductive polymers have a conjugated polymer backbone that contains alternating double and single bonds. The model conductive polymer is polyacetylene and a schematic diagram of this material along with its pi (π) orbitals is shown in Figure 10.1. In its simplest form shown, the polymer can be considered to be a one dimensional material with two degenerate ground states. That is, the ground state energy of the polymer is not changed if we replace the single bonds with double bonds while at the same time replacing the double bonds with single bonds. It should be noted that this particular polymer can exist in either the trans or the cis form depending on the formation process. In the state shown in Figure 10.1, polyacetylene is not conductive in a metallic sense but does posses semiconductor characteristics.

The differences between a conductor, a semiconductor, and an insulator are generally the result of the differences between the HOMO and LUMO bands typically referred to as the energy gap. In metals, due to the close proximity of atoms in the metal lattice, these bands mix and form a continuous band that is half filled allowing the electrons at the top of the filled portion of the band (ie: at the Fermi level) to delocalize over the lattice when subjected to an energy field such as an applied voltage. In a semiconductor the bands do not quite overlap so that there is a small energy gap between the HOMO and LUMO levels. However due to the small gap, electrons can be excited either thermally or electrically over the gap where they are free to delocalize over the LUMO level or, in terms of the solid state physics, the conduction band. If enough atomic centers have these small band gaps so that a large delocalized band appears over the lattice and the electrons are free to "wander" over the structure a current will flow with electrons flowing in the conduction band and the vacant holes of positive charge "flowing" in the valence or HOMO band. This flow of charge is referred to as ohmic flow and generally obeys Ohms law.

Insulators normally have a large energy gap where it is difficult to promote electrons across the gap. As previously noted, metals normally do not posses an energy gap and electrons are free to delocalize in the half filled band over the entire metal structure. Semiconductors on the other hand have a small gap and the electrons must be promoted across it at room temperature by an applied electric field. This gap is usually under one electron volt (eV) in energy. Normal organic polymers such as nylon, rubber, polyester etc. have large energy gaps on the order of 5eV or greater and no significant promotion of electron density across this gap occurs.

The difference in energy between the HOMO and LUMO levels can be easily determined by ultraviolet/visible/near infrared spectroscopy. As the band gap decreases the observed spectrum is shifted to lower energies. The energy transition for small gaps on the order of 1eV is usually in the near infrared. For most insulators the band gap is wide and electrons are localized on atomic centers or a small group of atoms and once promoted are not free to delocalize over extended distances necessary for electrical conduction. Delocalization of electron density can be observed however for conductive polymeric materials by the development of a free carrier tail which is seen as a very broad highly absorbing band tailing off toward lower energy.

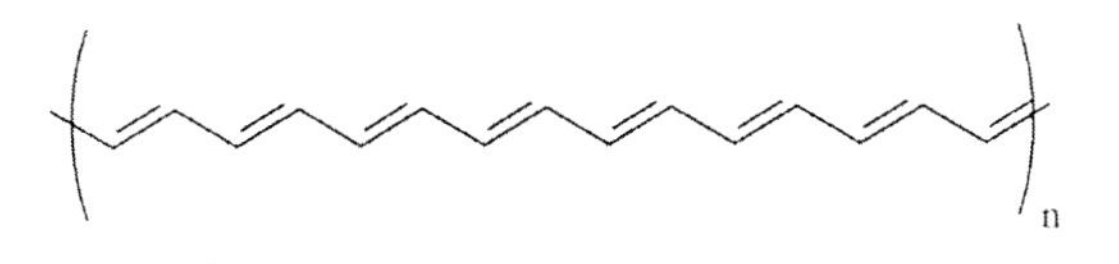

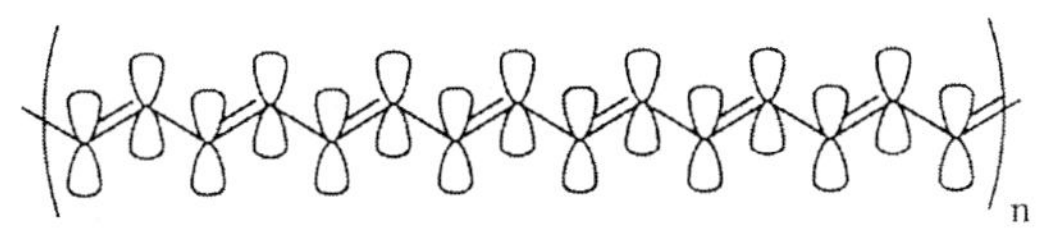

Figure 10.1 Ideal single polyacetylene chain showing π orbitals.

One might suspect that if enough thermal energy is supplied to an insulative polymer to create a large number of electrons in the conduction band that the polymer may indeed become conductive once the density of electrons is large enough. While this indeed may be true, one electron volt translates to over eleven thousand degrees Kelvin and most organic polymers known to the author dissociate long before an appropriate temperature is reached. Typically semiconductors, although intrinsically conductive, are doped with atoms which either supply an extra electron to the crystal or are electron deficient producing a "hole". This doping allows for enhanced electrical conductivity as charge can now flow within the structure due to the excess of electrons or the formation of "holes" due to the doping. In metals where doping is not necessary to produce extra electrons or holes due to the nature of the metallic lattice, the electrons are free to wander over the structure in a random manner. There is no directional flow of current until a field is applied since in the absence of an applied field the sum of the currents in all directions averages to zero.

In order for polymeric materials to become electrically conductive the energy gap between the valence and conduction bands must be significantly lowered or new bands must be formed in the gap into which electrons can be easily promoted or charge donated. In addition, the molecular orbitals must not be localized over single atomic centers or small groups of atoms such as benzene rings but must allow for significant delocalization of the electron density in order for a current to be carried. The question we now need to ask is how does one accomplish this in polymeric materials without destroying the polymer? The answer lies in forming defects along the polymer backbone in order to change the energy levels of the polymer and either reduce the band gaps or introduce new energy levels into the gap.

Again using polyacetylene as the model electrically conductive polymer we now introduce a defect into the polymer chain in one of several different ways. These differing ways may include chemical or photochemical oxidation, reduction, structural defects, and several others methods which produce a defect into the polymer chain. We can envision such defects in several different ways. For example if we consider only one isolated chain of polyacetylene and start at opposite ends of the chain and introduce the bond alteration we find that at some point we have two adjoining double bonds at the same carbon and if we also consider the attached hydrogen we have five bonds around the particular carbon in question. We find in practice that such a fifth bond does not actually bond to a neighboring carbon but is in fact unbonded and referred to in semiconductor physics as a "dangling bond". This dangling bond is at a different energy than the double and single bonds and many times this different energy level is located in the band gap. A dangling bond is depicted in Figure 10.2 (a). Likewise if an electron is removed from the polymer chain by an oxidative process as shown in Figure 10.2 (b) creating a radical cation, a structural defect occurs which results in a lattice distortion at that point. This distortion is at a different energy than an unoxidized portion of the chain and a different energy level associated with the lattice distortion results and this new energy state often appears between the HOMO and LUMO levels near the Fermi level in the gap.

Other types of defects will also introduce new energy levels in the polymers band gap. As more and more defects are introduced to the polymer chain, the density of the new energy states in the band gaps increases until such a point is reached that a significant delocalization of the electron density in the new energy states occurs and the electrical

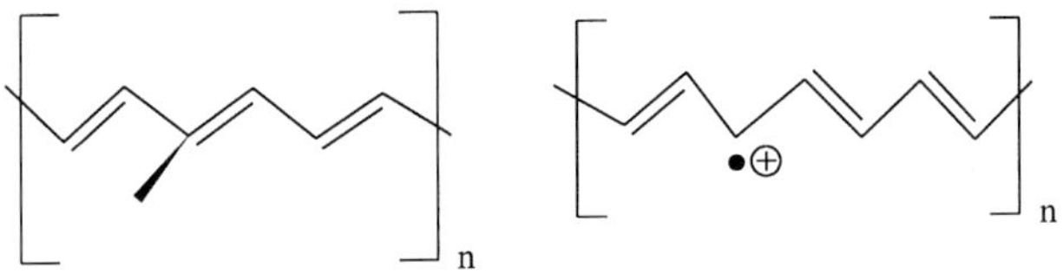

a) Dangling bond

b) Radical cation

Figure 10.2 Polyacetylene chain containing: a) dangling bond, b) radical cation defect formed by oxidation of the chain.

conductivity turns on. If the defects arise due to oxidative or reductive processes the charges introduced by this chemistry must be offset by appropriate counter ions. For example if polyacetylene is oxidized the process leaves radical cations on the polymer backbone. The positive charges may be offset by iodide ions and this process is known as doping.

The name doped polymer is loosely borrowed from semiconductor technology but is used somewhat differently. In conductive polymers the dopants lie between the polymer chains and are not incorporated into the polymer chains, whereas in semiconductors the dopant atoms form defects within the crystal structure by either forming a vacant hole, in the case of these atoms with fewer electrons than the majority of the those in the crystal, or the introduced defect atoms having additional electrons. Many workers in the area of conductive polymers do not like the use of the word dopant due to the above differences and the fact that doping of semiconductors is usually on the order of a few ppm, whereas in conductive polymers it is on the order of several percent. Another significant difference being that undoped semiconductors are intrinsically conductive whereas doped semiconductors are extrinsically conductive. By contrast, doped conductive polymers are referred to as intrinsically conductive polymers (ICP) and the dopants interact with the polymer to form a new three dimensional species with new molecular orbital energy levels. The undoped form of most conductive polymers are not conductive and in fact are insulators not semiconductors. One of the exceptions is polyacetylene, which does posses semiconductor characteristics in the undoped form, but polypyrrole, polythiophene, and polyaniline and most other conductive polymers are true insulators in their undoped form. In the most general form the use of the word dopant in conductive polymers refers to the introduction of a counter ion offsetting the positive or negative charge formed on the polymer back bone by oxidative or reductive processes. These counter ions may be the counter ion of the oxidizing or reducing agent such as the chloride ions in $FeCl_3$, a commonly used oxidizer, for the formation of polypyrrole. Polyaniline is somewhat unique in that the polymer can be doped in its emeraldine base form by simple protonation of the imine nitrogens in acid solutions. We normally refer to polyaniline as acid doped, and for the conductive form of polyaniline formed from the emeraldine base polymer, no electrons have either been removed or added to the polymer backbone. This is not the case for radical cation formation in other ICP's such as polypyrrole or poythiophene where the polymer backbones are in an oxidized state when doped to conductive levels. Dopant ions other than the counter ion of the oxidizing or reducing agent, or acid in the case of polyaniline, may also be introduced to displace other ions such as the previously mentioned chloride ions. In many cases these are organic molecules which may render a degree of stability to the formed polymer due to their size or reactivity.

The term *n* or *p* dopant has the same meaning as in semiconductor physics. An *n doped* conductive polymer is one in which the positive charge on the polymer backbone is offset by a negative counter ion and in a *p doped* a negative charge on the polymer backbone is offset by a positive counter ion. In both cases the dopants are associated with charged defects in the polymer creating new energy levels into which charge can be introduced and delocalized.

In polyacetylene these defects are referred to as solitons and can be either neutral or charged depending on the formation process. When enough of these defects are formed the HOMO and LUMO levels in the molecule may be bridged by the creation of these soliton energy states in the gap and polyacetylene under goes a transition to the metallic state. When this transition occurs one can no longer distinguish between the double and single bonds in polyacetylene since the electrons will delocalize over the structure as do the conduction electrons in true metals. Likewise, introduction of defects into polypyrrole and polythiophene by redox chemistries form lattice distortions and a significant change in the bonding structure occurs as seen in Figure 10.3 which shows the formation of polypyrrole from the monomer through the formed polymer. The polymer is formed in the conductive state with the chloride ions acting as the dopants, stabilizing the positive charges formed on the polymer backbone by the removal of electrons during the oxidative polymerization. In the case of polypyrrole the charge carriers are known as polarons and upon further oxidation become bipolarons.

Polarons have a structure containing a radical cation where the radical electron and the positive charge form a charge carrying species and move in unison. A bipolaron results from further oxidation of the polaron removing the radical electron, forming two positive charges which also is the charge carrying moiety on the polymer backbone. Creation of polarons and bipolarons form new energy levels in the band gap between the HOMO and LUMO. These energy levels can be populated with conduction electrons and provide for true ohmic conduction with the delocalization of electron densities in the new energy levels and the positive "holes" moving on the polymer chain.

The mechanism involved in conducting polyaniline is significantly different from that of polypyrrole, polythiophene, and most other known conductive polymers. Figure 10.4 shows the polymerization reaction for polyaniline (PANI). Once formed, polyaniline can exist in three possible oxidation states. These are the completely reduced or lecoemeraldine state, the partially oxidized or emeraldine base state, and the fully oxidized or pernigraniline state. Figure 10.5 shows these three different oxidation states. Only the emeraldine partially oxidized state can result in a conductive state for polyaniline and is shown in Figure 10.4 with the dopant chloride ions associated with the radical cations. In the emeraldine base form the imine nitrogens are protonated in acid solution. Upon protonation the molecule undergoes a transition such that aromaticity is introduced into the quinoid ring by reforming the benzene rings, leaving radical cations on the former imine nitrogens. The positive charges remaining on these sites are countered by the negatively charged ions in solution. This process results in polaron formation with electron density delocalized into the new energy bands formed from the polarons and the polarons acting as charge carriers on the backbone.

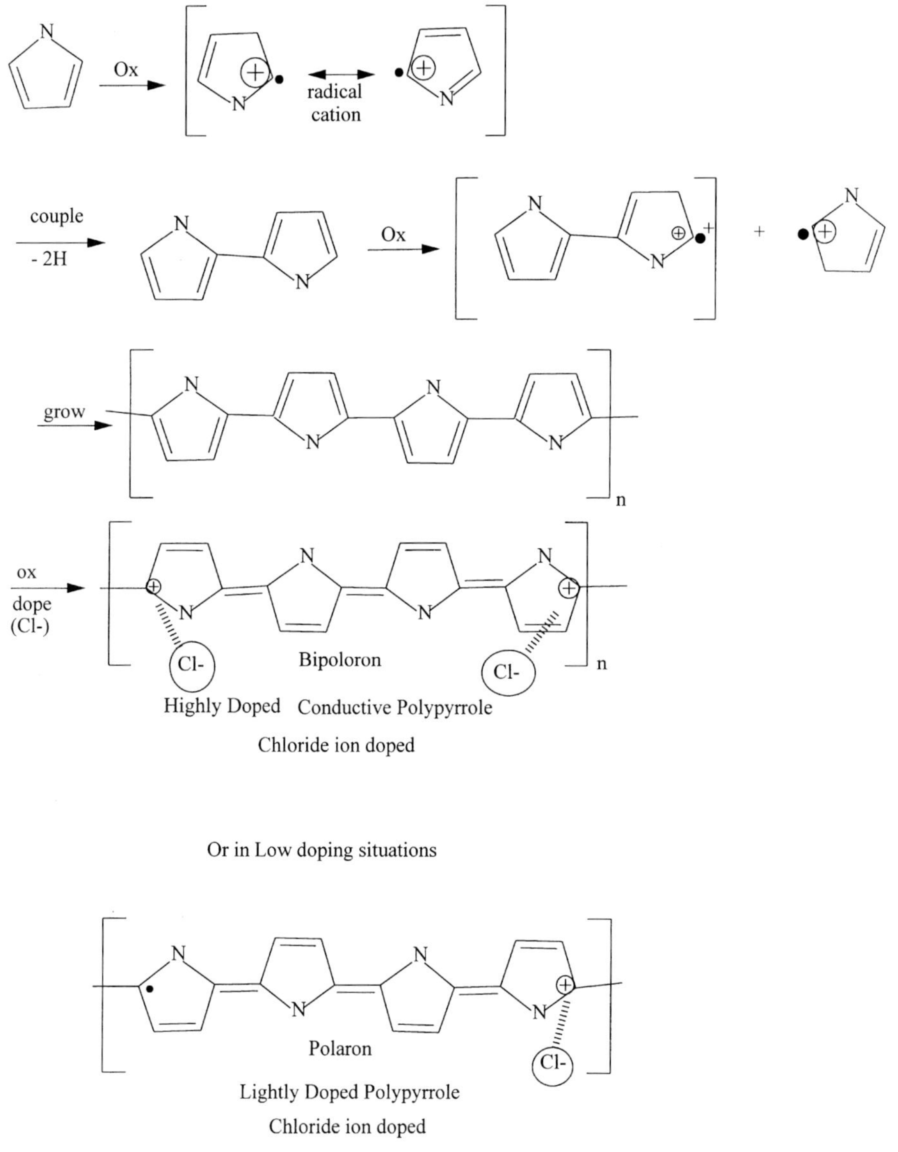

Figure 10.3 Synthesis of polypyrrole (PPY) from monomer to formed conductive polymer.

The difference between this polymer and most of the other conductive polymers is that the PANI backbone does not have electrons either removed or added to the original backbone structure and the structures conductivity was turned on by simple protonation in acid solutions of the imine nitrogens. The effect of this acid doping can be reversed for PANI by exposing the polymer to a base, and the conductivity can switched on and off by

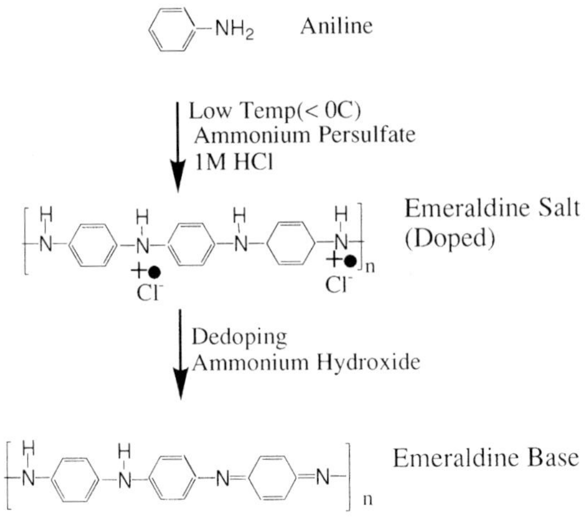

NH_2 Aniline

Low Temp(< 0C)
Ammonium Persulfate
1M HCl

Emeraldine Salt
(Doped)

Dedoping
Ammonium Hydroxide

Emeraldine Base

Figure 10.4 Synthesis of polyaniline (PANI).

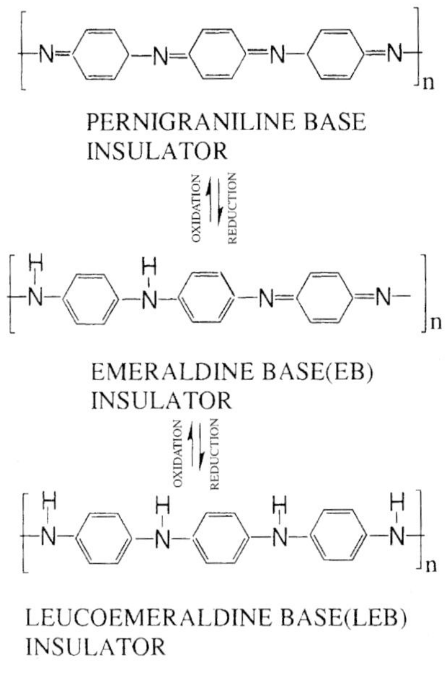

PERNIGRANILINE BASE
INSULATOR

OXIDATION REDUCTION

EMERALDINE BASE(EB)
INSULATOR

OXIDATION REDUCTION

LEUCOEMERALDINE BASE(LEB)
INSULATOR

Figure 10.5 Oxidation states of PANI; a; Leucoemeraldine base (reduced); b, Emeraldine base (partially oxidized); c, Pernigraniline (fully oxidized).

exposure to acid or base. This phenomenon can be used in applications of this polymer for sensor devices and in many other systems where this switching between the conductive and insulative states might prove useful.

The polymer morphology will also change depending on whether or not it is in the conductive state. In the conductive state, PANI is much more rigid than in the nonconductive emeraldine base state where there is significant flexibility in the polymer chain. This opens up some significant processing possibilities for this polymer as opposed to the other conductive polymers whose monomers are available on a large commercial scale. Polyacetylene, polythiophene, and polypyrrole are not stable in their undoped state and cannot be easily processed in this form. Their respective conductive forms are rigid rods with extensive delocalization making then intractable powders with little ability to process using existing fiber or film technologies.

10.1.2 Measurement of Polymer Conductivity

Conductivity (σ) is defined as the inverse of resistivity (ρ) and is usually reported in units of Siemens/cm (S cm^{-1}). The conductivity of a material is a bulk property and its measurement is dependent on the sample size and geometry. Generally, the conductivity is measured using a four point probe in which a steady known current (I_s) is applied through points one and four and the voltage drop (V_m) is determined between points two and three. Measuring the voltage drop between points other than the charge injection point eliminates a potential problem with contact resistance which may occur in typical two point measurements. A schematic diagram of a typical four point probe set up is depicted in Figure 10.6. Using the general relation for resistivity ρ (a bulk property) is determined by the following relationship:

$$\rho = \frac{\pi d}{\ln 2}\left(\frac{V_m}{I_s}\right) \tag{10.1}$$

where V_m is the measured voltage drop, I_s the known supplied current, and d the sample thickness. Fiber filament conductivities can be determined by applying four equidistant spots of silver paint or other conductive material and using the fiber cross sectional area. A device can be fabricated which has four platinum wires imbedded in an acrylic or other polymer matrix and arranged at known intervals. The fiber filament is

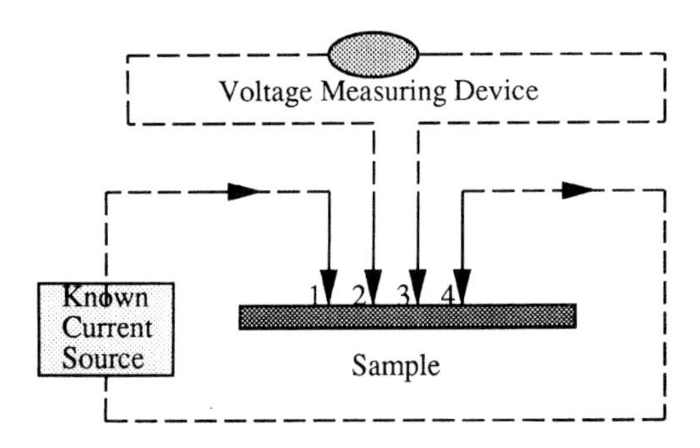

Figure 10.6 Typical four point probe for determination of electrical conductivity.

then placed across the wires and a covering plate affixed to insure uniform contact. A known current is applied as above and the conductivity measured as before. An accurate measurement of the fiber sample thickness is necessary to determination of the bulk property. Care must be taken when making these measurements as the geometry of the sample and the measurement electrode configuration is important and results will vary if not taken into account. A detailed discussion of the techniques for measurements of bulk electrical properties is given in Ref. [2]. The resistance or conductance, a surface effect measurement, can be obtained by typical two point multimeter determination but these, although useful, should not be confused with the bulk measurements. Normally, for surface resistance properties, the units are given in ohms/square. One can convert between the measurements, if the thickness is known, from the following relationships.

For the bulk conductivity, we have:

$$\sigma = \frac{\ln 2\ I_s}{\pi V_m d} \tag{10.2}$$

and knowing that the surface resistance can be related to the bulk conductivity by:

$$\sigma = \frac{1}{R_s} d^{-1} \tag{10.3}$$

we can set the two equations equal and solve for the surface resistance R_s.

$$R_s = \frac{\pi V_m}{\ln 2\ I_s} \tag{10.4}$$

The surface resistance can usually be easily measured and is useful for determination of materials such as fabrics that have been coated with electroactive polymers. If the thickness of the coating can be accurately determined then Equation 10.4 can be used to qualitatively estimate the bulk conductivity of the coating. Care must be taken in these measurements with regard to proper electrode configuration. For an isotropic material a square four inches on a side connected between two four inch electrodes on opposite sides will give the same resistance as a sample of the same isotropic material forming a square six inches on a side between opposite six inch electrodes. In order to obtain accurate results the electrodes must be the same size or larger than the sample size. Measurements of resistance or resistivities and their reciprocals, conductance and conductivity, may be sensitive to the environment of the measurement such as humidity, temperature or contaminants. One must also keep in mind that most polymeric conductors are anisotropic with regard to their electrical properties. For example in the case of fibers and films the conductivity may be quite different in the plane of the film or along the fiber axes than in a direction orthogonal to the plane of the film or the fiber axes. Several methods for determining the bulk conductivities and resistivities are given in Refs 2–4.

Application of four point probe techniques to fibers made from conductive polymers poses more of a problem. Inherent in the derivation of Equation 10.1 is the assumption that the spacing between the four points is very much smaller than that of the total surface area. A more fundamentally correct measurement technique would be to encapsulate the entire end of the fiber (short cut staple fiber) with a conductive graphite paste and to inject charge uniformly across the cross section of the fiber and to collect it at the opposite cross section face. Voltage drop can then be measured by two concentric rings around the fiber and the conductivity (σ) is determined by the more fundamental relationship which describes the conductivity as a function of the materials dimensions and resistance,

$$\sigma = \frac{L}{(R)(A)} \tag{10.5}$$

where R is the resistance, A, the cross-sectional area, and L, the fiber length. This method addresses the problem of probe spacing versus total surface area but does not account for variation in polymer thickness along the fiber axis. Another problem often encountered in the measurement of conductive polymer fibers is the presence of void spaces that may be introduced in the fiber forming process. These voids, if large, will significantly affect charge transport within the fiber matrix.

10.1.3 Processing of Conductive Polymers

Due to the formation of charge carrying species on the polymer backbone which rely on orbital overlap for charge delocalization, conductive polymers tend to form rigid rods and generally are intractable materials in their conductive form [5]. Efforts to develop novel polymers that are processable have been somewhat successful but the high cost of scale up to produce the unique monomers has delayed production of these materials. Major efforts to use monomers available in large quantities such as pyrrole, aniline, and thiophene has resulted in concentrated research efforts in industry and academia to utilize and process polymers formed from these monomers available in large commercial quantities. Polyaniline (PANI) is easily formed and is stable in its nonconducting emeraldine base form and has demonstrated processing potential for fiber formation on a large scale.

The early work on the processing of conductive polymers into useful articles such as fibers and fiber coatings was pioneered by workers at DuPont, Milliken and Co., and by Uniax. Uniax continues to be a world leader in conductive and semiconductive polymer applications and particularly in the development and application of polymer emissive displays [6]. Substantial exploratory work in fiber spinning was initiated by Cao, Andreatta and Smith at Uniax in the early 1990's.

Work by Cao, Smith, and Heeger reported that difficulties in the processing of PANI in the conductive form from reasonably high molecular weights could be overcome by the use of a functionalized protonic acid [7–9]. This functionalized system dopes PANI and simultaneously renders the conductive PANI complex soluble in common organic acids. In this study the authors defined a "functionalized protonic acid" as $H^+(M^-\text{-}R)$ in which the counter-ion anionic species, (M^{-}-R), contains an R functional group chosen to be compatible with non-polar or weakly polar organic solvents. An example of this is dodecylbenzenesulfonic acid (DBSA). These workers surmise that the long alkyl chains on the phenyl ring of this organic acid lead to solubility in common organic solvents such as toluene, xylenes, chloroform, etc. and the anionic part of the molecule (SO_3^{-}) dopes the PANI forming a complex that is conductive and soluble. Films prepared with DBSA and camphor sulfonic acid (CSA) exhibit initial conductivities of better than 100 S/cm. Smith and colleagues conclude that one can "design" the conducting PANI complex to be soluble in specific solvents and cite as an example the use of CSA as the functionalized protonic acid. After doping and complexation with CSA, the conducting PANI complex is soluble in meta-cresol. The CSA-PANI complex, due to the excellent solvation effects of meta-cresol, can be used to make films of high optical quality. Use of the soluble PANI complex in making appropriate blends with other non-conductive polymers soluble in the same solvent is also cited, as well as preparation of the neat conducting polymer. Since the polymer is processed in the conductive form, no post-processing chemical treatment is necessary.

Cao et al. prepared blends from PANI produced by counter-ion processes. Partial orientation of the PANI complex was accomplished upon drawing of the blend film [6]. An example cited by the authors is a PANI-DBSA polyethylene composite film which upon drawing at 105 °C to a draw ratio of 4 increases its electrical conductivity by two orders of magnitude at surprisingly low levels of PANI complex (less than 10%) in the blend. X-ray studies by these authors demonstrate that in the PANI complex there is a relatively high degree of orientation of the PANI chains, no doubt resulting in the increased conductivity. This is consistent with the results of MacDiarmid on polyaniline oriented films [10]. The mechanical properties of the polyblend material are also quite good with no substantial loss of mechanical properties at modest levels of the PANI complex that give conductivities of less than 10 S/cm.

Further work by Osterholm et al. in conjunction with Cao et. al. reported that a PANI complex of a proprietary nature, developed as a commercial product (POLARENE^{tm}) by Neste Oy in Finland, could be processed with various thermoplastic polymers such as polystyrene, PVC, and others including thermoplastic elastomers [7]. These blends could be processed by existing techniques such as injection and blow molding. Conductivities in the semiconductor range can be achieved at low loading levels in the 1 to 2% range. Due to the low loading levels of POLARENE^{tm} the mechanical properties of the blend are excellent, approaching those of the neat host polymer and in some cases exceeding them.

Several other methods have been attempted over the last several years with varying degrees of success in order to form processable electrically conductive polymers. These include the formation of precursor polymers that are processable, use of substituted derivatives, blending techniques, and several others. Some successes have been reported in the literature and a detailed review of various processing methods for these materials can be found in several published proceedings for different meetings of *The International Conference on the Science and Technology of Synthetic Metals* published in the journal *Synthetic Metals* by this and many other authors over the last several years.

One example is the use of precursor polymers. This method has been explored recently for making conductive polymers into useful end products of various geometries [11]. In this type of processing a precursor polymer is prepared and formed into the desired end product geometry and then converted to the final dopable ICP usually by the application of heat. This type of processing involves several steps including the preparation of a precursor polymer for subsequent processing. Although films and fibers can be produced by this process the synthesis of the precursor polymers can be difficult and the production of the precursor on a large scale may not be economically feasible. A major advantage of these materials is that the precursor polymer can be oriented by existing draw technologies prior to doping thus providing the formed materials relatively good mechanical strength and excellent electrical conductivity.

Adding solubilizing substituents to the monomer unit of the polymer structure is also a well established technique for rendering an intractable polymer soluble in certain solvents, and this technique has been used in solubilizing conductive polymers [12]. The addition of substituents onto the polymer chain may have stearic consequences that reduce the ability of the polymer to form delocalized π electron density over the chain thus reducing the level of achievable conductivity.

Another popular method for solubilization is the inclusion of solubilizing monomeric structures along the polymer chain rendering a modified chain which includes some percentage of different monomers used to enhance solubility. This method is not particularly useful for ICP's where reasonable levels of conductivity are needed since the inclusion of these structures at a level necessary for solubility may significantly reduce the conductivity of the polymer system due to these charge carrying "interruptions" along the chain. The morphology and the crystalline domains of the polymer may also be significantly affected by incorporation of these groups.

There is considerable work reported on methodologies employing co-polymerization. This polymerization technique is frequently used to combine diverse properties of differing polymers into one polymeric system. This process is widely used in the fiber industry to impart dyeability to non-reactive fiber forming polymers such as acrylics [13]. In this case the polymer is copolymerized with monomeric groups that form anionic moieties at the proper pH and can be readily dyed with cationic dyes. Using this approach one might anticipate that the solubility properties of one type of polymer might be incorporated with the electrical properties of an ICP by copolymerization to give a final soluble conductive polymer. Generally there are four different methods used in the copolymerization process: (i) in-situ polymerization of both monomers to form a random copolymer, (ii) grafting of the solubilizing polymer onto the backbone of the intractable polymer; (iii) grafting of the intractable polymer onto the backbone of the soluble polymer; (iv) combination of both soluble and intractable polymer to form a block co-polymer by various synthetic techniques. These techniques generally result in substantially lower electrical properties as previously mentioned. One notable exception to this is the counter ion induced processable conductive polyaniline, discussed earlier, where the dopant acts as a "pseudo graft" on the polymer back bone imparting to the PANI chain a degree of solubility in the doped form.

In addition to these forms of processing, utilization of a conductive polymer as a coating on other polymer substrates to induce a degree of processability has been explored. Work by Kuhn and co-workers at Milliken Research in the late 1980's resulted in polypyrrole coated textiles where the individual fibers were coated with the conductive polymer without any fiber to fiber bonding. This material available through the research division of Milliken & Co is called CONTEX^{tm} and may have widespread applications in composite structures and other materials [14]. Likewise, composites of polypyrrole and polythiophene prepared by in-situ polymerization in matrices such as poly(vinyl chloride), poly(vinyl alcohol), poly(vinylidine chloride)-co-(trifluoroethylene) and brominated poly(vinyl carbazole) have been reported. The conductivity of these composites can reach up to 60 S/cm when doped with appropriate species. Work reported by Liu and this author on alkyl substituted polythiophenes, synthesized with the addition of a urethane group on the ß substituted alkyl chain, shows excellent blending characteristics with hydrogen bonding polymers such as urethanes or polyamides [15]. These blends have significantly reduced percolation thresholds and are solution and melt spinnable.

10.1.4 Fiber Formation

Several groups around the world are currently engaged in the synthesis, production, and characterization of electroactive fibers. Large companies such as Neste Oy, DuPont, Monsanto, Milliken and Co. have pursued the production of electroactive polymers as either fibers or coatings for fibers and other materials. Many smaller companies have been started or spun off from universities or national laboratories and are primarily devoted to the production of electroactive fibers and films for a variety of applications.

This section will be devoted to polyaniline (PANI), one of the more promising conductive polymers for production of fibers, due to its facile processability in organic solvents in its base form. Many other conductive polymers are also being investigated for use as fibers but polyaniline and deriviatized polyaniline such as poly(o-toludine) (POT), and the previously mentioned counter ion induced processable polyaniline, have received the greatest attention in recent years [16]. One of the major reasons for this is the availability of the monomer in large quantities and the ease of formation of films and fibers in the base form which can be later doped to desired levels of conductivity.

Angelopoulos and co-workers reported that N-methylpyrrolidione (NMP) could dissolve PANI in the emeraldine base form. Subsequent work demonstrated that PANI solutions formed in NMP at solids concentrations above 6% formed gels after a short period of time [17]. The formation of these gels at concentrations necessary for the formation of fibers and films by solution spinning technologies ($\geqslant 10\%$) prohibited the use of this solvent. Work at DuPont by Hsu and co-workers reported that fibers could be spun from NMP solutions at fiber spinning concentrations that would normally gel by the addition of certain amines to the polymer solution [15]. These workers report that a PANI/1-4 diaminocyclohexane solution provides a stable spin dope and that fibers could be extruded by continuous dry-jet spinning with an initial tenacity of 0.8 grams per denier (gpd). The formed fiber could then be drawn at 215 °C with a final tenacity of 3.9 gpd, an elongation of 9.3%, and modulus of 83 gpd. (Assuming the density of the fiber is 1.2 g/cm^3, the increase in strength on drawing is from 80 to 400 MPa).

The spin dope for these fibers is prepared by the addition of gaseous ammonia or pyrrolidine to the NMP solution of PANI base of concentrations great enough for fiber spinning. The gelation of the PANI is reversed and a viscous solution is formed for use as the spin dope at concentrations of 20% solids. These authors speculate that the gelation process is related to the interaction of microcrystalline regions which the new binary solvent disrupts. They observed that when the solution is stirred at low speeds, on the order of 15–20 rpm, the solution remains smooth and stable but on increasing the stirring to 50–70 rpm the solution becomes lumpy. This phenomenon is suggested to be the result of shear induced crystallization. Doping of the hot-drawn fibers produced from the spin dopes in 1M aqueous HCl reduced the tensile strength by approximately 64%. This reduction was attributed to the microscopic incorporation of chain defects resulting from the exposure to HCl, such as chain scission due to acid hydrolysis, etc. The highest conductivity reported for the HCl doped fiber was $\approx$157 S/cm. When similarly prepared fibers were doped with sulfuric acid, the reported conductivity was $\approx$320 S/cm.

Hsu did not speculate on the difference in observed conductivity between HCl and H_2SO_4 or report if reduction in fiber strength was similar to the HCl doped fibers. These workers also produced fibers doped with camphor sulfonic acid (CSA) and reported that the results for the PANI-CSA fibers are much like those reported for PANI-CSA films which have been shown to have excellent mechanical and electrical properties. The stretched fibers are not as good as the oriented films and the authors speculate that this may be due to some areas of poorly conducting linkages. Fibers were also formed from the derivatized PANI, poly(o-toludine) and doped with CSA and show much higher electrical conductivity and weaker temperature dependence than similarly prepared HCl doped powders and films of this polymer. The tenacity reported for the PANI/CSA and POT/CSA as spun fibers was 0.2 gpd with the elongation to break and the modulus being 8.4%, 7.3 gpd, and 3.0%, 9.7 gpd respectively. Hsu did not speculate on the morphology of the as spun fibers nor the possibility of voids or other anomalies contributing to the very low mechanical properties.

10.2 Fiber Formation from Emeraldine Base Polyaniline

Since PANI is not thermoplastic in nature, it cannot be processed in the pure state by melt spinning technologies. However, if suitable solvents can be found that allow the formation of rheologically stable PANI base solutions of $\approx$15% wt./wt. without gel formation, then it is possible to spin PANI base into fiber form and subsequently dope the fibers. Although MacDiarmid and co-workers successfully produced small fibers by employing a syringe from NMP solution, long term solution stability studies were not carried out other than to point out the aforementioned gelation behavior of solutions whose concentrations were greater than 6% wt./wt. [18]. Studies at Drexel University by Wei and co-workers and others did show that solutions at higher concentrations demonstrate bimodal distributions of molecular weights in NMP [19]. The addition of

LiCl to the solution reduced the bimodal distribution with the removal of the higher molecular weight fraction indicating that the higher molecular weight fraction is most likely due to an agglomeration of polymer by interactive secondary forces between the polymer chains and not necessarily due to a very high molecular weight component. Salt addition, such as LiCl, is a well known technique used in spinning technologies for the reduction of "clustering" of polymers in solution [20]. Work at I.B.M. by Angelopoulos has quantified the interaction of the Li^+ ion in reducing the effect of hydrogen bonding interactions between the PANI chains, thus reducing the clusters of associated polymer [16]. This work is consistent with work done at Clemson in our laboratory, clearly demonstrating that PANI forms a thixotropic gel in reasonably concentrated solutions [21]. The clustering of PANI polymer in solution most likely continues to grow until the gel "sets".

Work in our laboratory on the rheology of NMP solutions of PANI and PANI/LiCl also demonstrated that the solution characteristics of the NMP and NMP/LiCl does not provide a stable window for spinning of PANI fiber since the rheological profile of the solution changes with time. NMP/LiCl is better than NMP alone but is still not satisfactory for fiber production over an extended period of time with fiber segment to fiber segment reproducibility. Figure 10.7 shows a plot of the normalized viscosities of a 3%, 6%, and 8% wt./wt. solution of PANI/NMP at 25 °C vs. time and clearly points out that at concentrations of 8% the viscosity quickly increases and the solution is not stable. A similar study of PANI/NMP/LiCl solutions (0.5 wt.% LiCl) at 6%, 8%, and 10% wt./wt. PANI demonstrated that although the solution viscosity is more stable it still increases at an undesirable rate at a concentration of 10%. For fiber production purposes a concentration of 10% is considered the lowest spinning concentration and fibers formed at this low concentration tend to have undesirable defects and reduced mechanical properties unless great care is taken.

Observing the viscosity changes over a period of time is a useful method to study polymer solution stability and determine which solvents may provide a stable solution

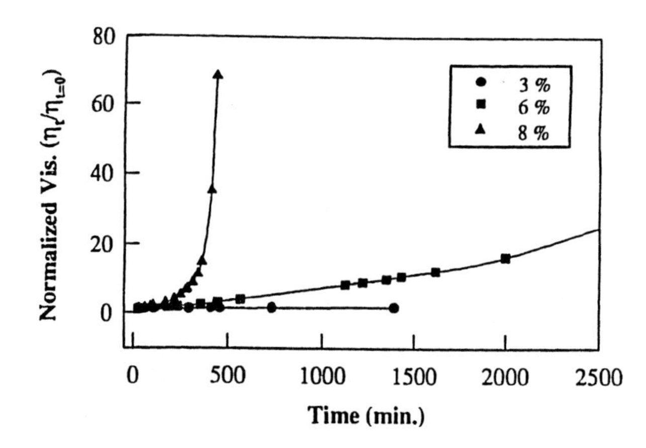

Figure 10.7 Normalized viscositites vs. time fo PANI/NMP solutions at 3 wt.%, 6 wt.%, and 8 wt.% at 25 °C; $\eta_{t=0}$ is the viscosity at time 0.

over extended periods of time from which to spin fiber. In order to reduce polymer/polymer interactions several solvents were studied for their ability to effectively solvate PANI in its base form at concentrations acceptable for wet spinning of emeraldine fibers. These solvents were selected from typical polymer solvents used for extraction and solvation of differing polymers and those with characteristics similar to NMP. Examples of solvents included derivatives of NMP, N,N' dimethylpropylene urea (DMPU), dimethyl formamide (DMF), triflouroethanol (TFE), tetramethyl urea (TMU), formic acid and other well known polymer solvent systems.

Of the solvents studied, N,N'dimethyl propylene urea (DMPU) showed promise for reducing the likelihood of gel formation for an extended period of time and providing solvent stability for fiber spinning under applied shear. Figure 10.8 shows the rheological stability of PANI emeraldine base in NMP, NMP/LiCl, and DMPU. It can be clearly seen that the PANI/DMPU solution is much more stable over an extended time even at concentrations high enough for fiber spinning. Figure 10.9 shows the dramatic change in solution viscosity of a 17.5% PANI/DMPU solution and a 8% PANI/NMP solution at a constant shear rate of 0.75 sec^{-1}, showing the superiority of DMPU as a solvent.

The importance of a good spinning solvent cannot be overstated. The nature of the solvent has a direct bearing on the degree of polymer solvation and the geometry of the polymer coil in solution. If the polymer/polymer interactions are strong enough for the polymer to fold in on itself resulting in areas of high polymer density ("bunched" coils) separated by areas mostly composed of solvent, the likelihood of polymer chain entanglement necessary for strength in the formed fiber is reduced. This chain entanglement may take the form of an actual mechanical entanglement between differing chains or secondary interactions between chains such as dispersion forces. In DMPU the PANI emeraldine chains are most likely more relaxed than in NMP based on the results of the viscometry and shear studies reported in Figures 10.8 and 10.9. This results in the polymer chains not being "bunched" or coiled in on themselves and the likelihood of different chains interacting due to their more relaxed nature is much greater. This chain

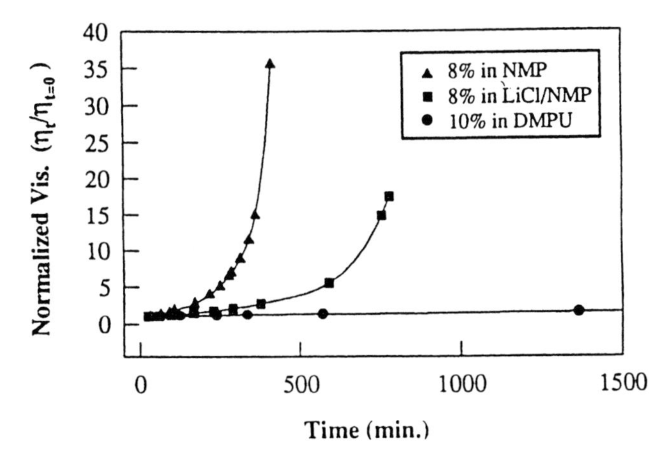

Figure 10.8 Normalized viscosities vs. time of PANI solutions in NMP (8 wt,%), 0.5 wt. % LiCl/NMP (8 wt.%), and DMPU (10 wt.%).

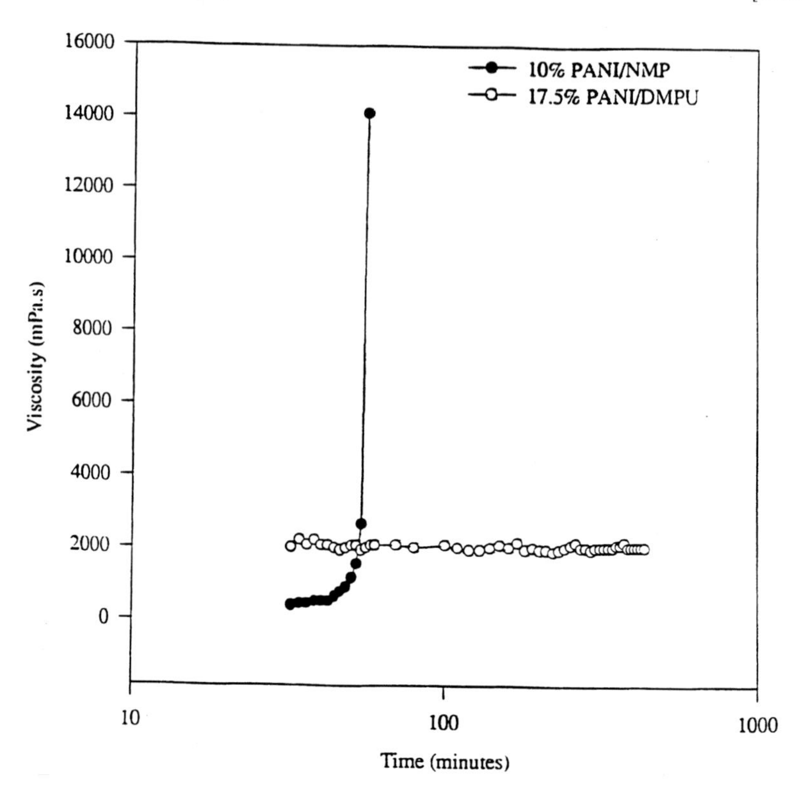

Figure 10.9 Rheological stability of solutions of PANI in NMP (8%) and DMPU (17.5%) at a shear rate of 0.75 sec^{-1}.

entanglement not only gives superior mechanical properties but should also facilitate interchain electron hopping thus giving a better overall macro conductivity. The improved chain to chain charge transport will result when different chains are in close proximity and energy necessary for charge "hopping" from chain to chain is reduced. The available number of sites for this interchain charge transport will be much higher if the chains are oriented and in a more random coil than if bunched in on themselves with little contact between different chains. Although the very nature of the extrusion process results in some chain orientation along the extrusion channel axis due to shear forces, extension will be substantially less for polymers that are initially bunched in on themselves, or self associated, in poorer solvents than for those polymers in good solvents that are more solvated and extended. Also polymers in poorer solvents will tend to rebunch during the die swell phenomena occurring at the spinnerette exit, thereby minimizing chain entanglement in the formed fiber [22].

The mechanical strength of fibers formed during extrusion processes may not necessarily be due only to entanglement of molecular chains. Polymers of sufficiently high molecular weight and that are highly crystalline in nature may interact via secondary bonding, such as dispersive forces, to form high strength materials. An example of this is the solution spun SPECTRA^{tm} polyethylene fiber. The molecular weight of the polymer in this fiber is on the order of several million and the crystallinity is around 85%. Although there is still considerable debate as to the actual molecular weight of PANI produced by differing methods it is reasonable to assume that the molecular weight and the degree of crystallinity are significantly less than that of SPECTRA^{tm} fiber and therefore the mechanical properties of the formed fiber to date results primarily from chain entanglement. PANI base films and fiber have been shown to be amorphous and have a molecular weight (M_w) below 100,000. Wide angle X-ray patterns for the PANI base fibers prior to and after doping further support the non-crystalline amorphous structure of the PANI base as-spun fiber. Based on these and other previously reported observations we believe the initial values for the tenacities, etc. result primarily from chain entanglement, although upon doping, dispersive interactions between microcrystallline regions may also play a role in the observed mechanical properties.

10.2.1 Solution Spinning of Emeraldine Base Fiber

Solutions, at concentrations high enough to allow for the production of PANI base fibers, have been prepared by this author using DMPU as the solvent. This "spin dope" was found to be stable for extended periods of time and if kept refrigerated at 10 °C the solution remains free of particulate matter indicating that gels have not formed after months of storage. The PANI base is made according to the methods described by MacDiarmid although special care is taken in the production of the PANI base by dedoping to insure all acid has been removed. Residual acid has been shown to catalyze gel formation [23]. The solution is filtered to remove any solid particles which may interfere with the spinning process due to clogging of the spin pack filters or spinnerette openings. After prolonged storage the spin dope may be again filtered to insure agglomeration has not occurred. In almost all cases we have seen little evidence of particulate formation upon prolonged storage provided adequate precautions are taken to remove and prevent exposure to gel catalyzing acids. The solution spinning equipment used at Clemson in the production of PANI emeraldine base fibers is designed to allow spinning of small amounts of fiber or prolonged runs producing thousands of meters of reproducible PANI base fibers. Interchangeable spinerettes allow for the formation of multi-filament or monofilament fiber. The linear density (i.e. mass/unit length, often expressed in denier, where 1 denier is 1 gram/9000 meters) can also be controlled by draw down (stretching) which induces high orientation, take up speed, and the initial spinnerette hole diameter. A gear pump can be used to deliver the spin dope to the spinnerette which provides a constant pressure that is easily monitored and can be controlled simply by adjusting the speed of the pump. Likewise, a commercial syringe pump can be employed. In all cases it is necessary to insure a constant pressure to the spinnerette and that the fluid flow rheology is well understood.

The polymer solution used for the production of the PANI filaments discussed in this chapter was a 15% wt./wt. DMPU/PANI spin dope having a viscosity of 21,000 cP at a shear rate of 1.5 sec^{-1} at a temperature of 25 °C. A Brookfield cone and plate rheometer was used for the viscosity determination. The polymer solution is placed in the spin dope tank and delivered to the spin pack which has an eighty hole circular spinnerette (L/D = 2.0) and into a coagulation bath consisting of NMP and water. The fiber is continuously taken up on a take-up roll system, equipped with a digital controller for precise take up speed control. Prior to take up, the yarn bundle is washed with copious amounts of water. Depending on the speed of the take roll, compared to the rate at which fiber is produced at the spinerette, some initial orientation can be achieved producing a partially oriented yarn (POY) prior to actual drawing.

The concentration of the coagulation bath is an important factor in the final mechanical and electrical properties of the formed fiber. Most of the fibers formed in our laboratories for initial studies are formed in a coagulation bath approximately 2 meters long with a NMP/H_20 concentration of 25% to 40%. The effects of changing coagulation bath concentrations and other process conditions will greatly affect the mechanical and electrical properties of initially formed PANI fibers. In an effort to completely characterize the electrical and mechanical properties of PANI fibers formed from emeraldine base, the initial properties and the effect of process conditions on these properties must be evaluated prior to annealing and drawing the fibers. Maximizing the initial mechanical and structural properties prior to draw should result in fibers with excellent mechanical and electrical properties.

Due to the many process steps involved in the formation of wet spun PANI fibers, the effect of these various stages must be studied to determine an optimal set of conditions to maximize desired mechanical and electrical properties. The development of constitutive equations for the process depends on an understanding of the effect of varying the spin conditions. Since the PANI fibers are initially undrawn, the molecular orientation and therefore the resultant properties are expected to be rather low. However, as previously stated it is generally assumed that in optimizing the initial stages of the fiber spinning process, the latter stages of processing (ie: drawing, annealing etc.) will be more efficient in approaching the intrinsic limits of the material.

PANI fibers formed according to the method previously described were found to have both the initial modulus and tenacity strongly dependent on both take up speed and on the coagulation bath concentration. The tenacity of the formed fiber increases by nearly 50% as the take up speed is increased from 2 to 10 m/min. Figure 10.10 shows this relationship quite clearly. Figure 10.11 shows increasing initial modulus as a function of take up speed. The increase in both the initial modulus and the tenacities is not surprising since the increased take up speed certainly imparts some degree of orientation to the formed fiber. Although tenacities of 0.3 gpd may be acceptable for some specialized applications they are not sufficient for most large scale industrial processes or textile processing. Subsequent drawing of the formed fiber increase tenacities to an acceptable level for some applications. Tenacities greater than 6 gpd can be obtained for the doped fibers after drawing and annealing the fiber.

Figure 10.12 shows the relationship between the conductivity and the take up speed of the formed fiber upon doping with methane sulfonic acid. As can be seen the conductivity increases with increasing take up speed demonstrating the effect of fiber orientation on conductivity. This observation is consistent with those of Epstein and MacDiarmid and other workers who have noted that drawing of PANI films and fibers increases conductivity [23,24]. It is unclear at the time of this writing whether doping and drawing

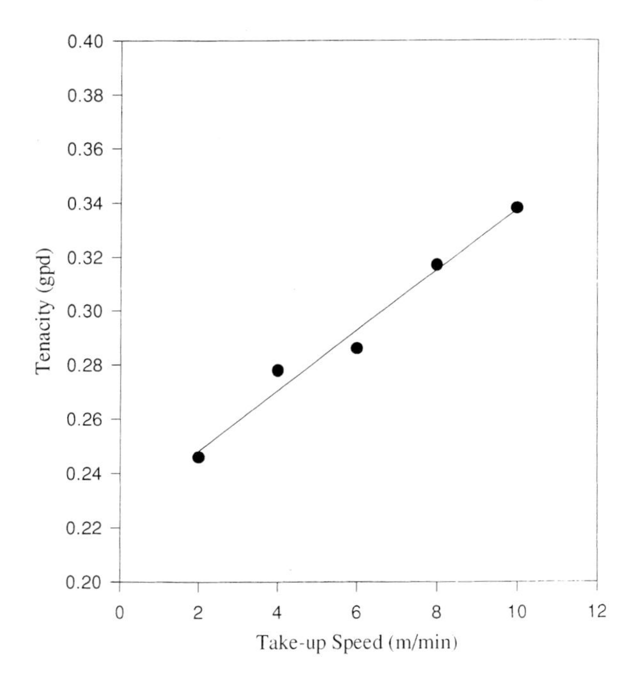

Figure 10.10 The effect of take up speed on the tenacity of as-spun PANI fibers. Coagulation bath concentration: 35 wt.% NMP/water.

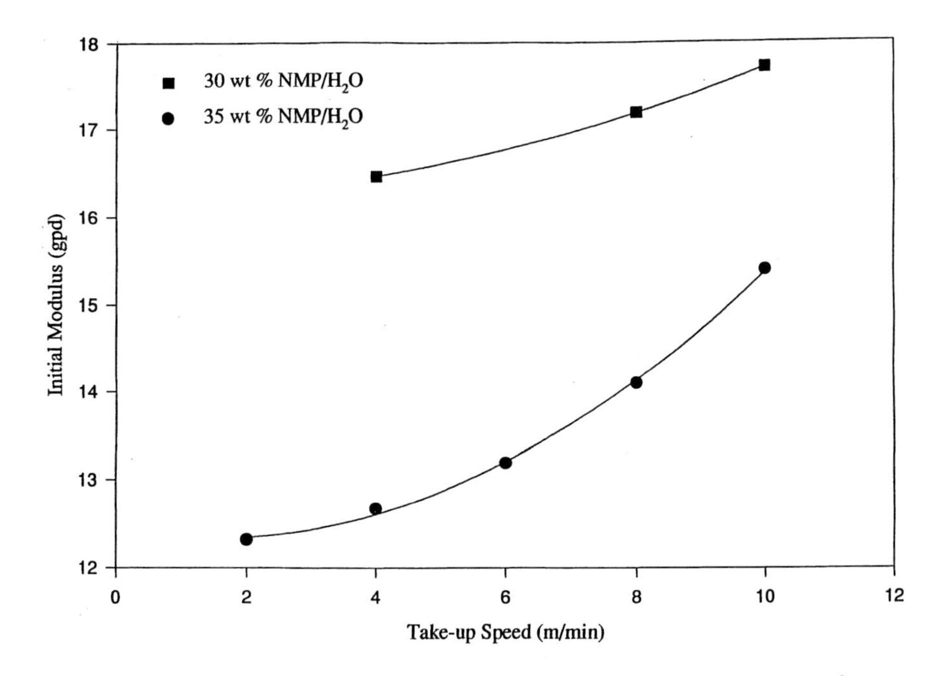

Figure 10.11 The effect of take up speed on the initial modulus of as-spun PANI fibers. Coagulation bath concentration: 35 wt.% NMP/water.

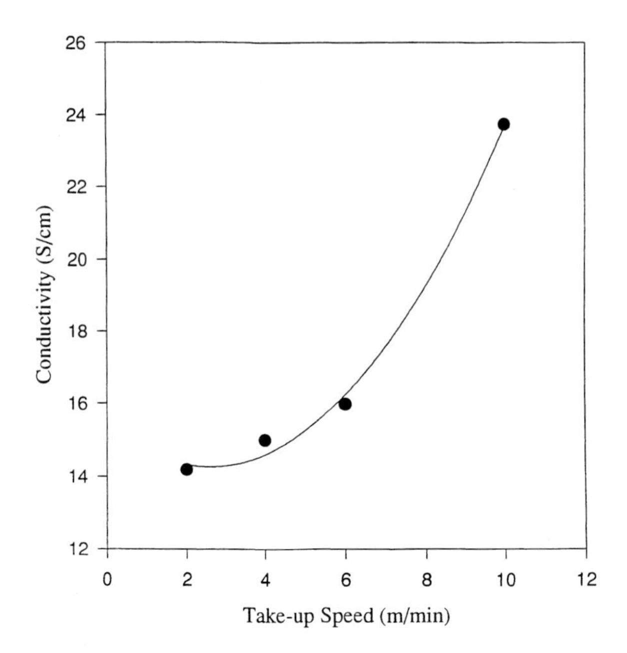

Figure 10.12 The effect of take up speed on the conductivity (1M $MeSO_3H$ doped) of as-spun PANI fibers. Coagulation bath concentration: 35 wt.% NMP/water.

in unison or in separate steps provides the highest and most reproducible conductivities. Again it is important to realize that the initially formed fiber must be as reproducible as possible on a segment by segment basis if subsequent drawing and/or doping are to be uniform.

From fiber diameter measurements and measured linear densities (denier) the fiber density can be calculated. Figure 10.13 shows the relationship between the calculated fiber density of the as spun fibers and the conductivity of the doped fiber using $MeSO_3H$ as the dopant. Fibers obtained at differing take up speeds had different calculated densities and the conductivity increases as a function of fiber density. The fiber densities range between 0.25 and 0.6 g/cm^3. These densities are much lower than expected based on the density of PANI itself which is approximately 1.3 g/cm^3. One infers from Figure 10.13 that if fibers are formed at higher densities the conductivity of the as spun fiber upon doping would be substantially increased.

The nature of the low densities was determined to be the result of large and small voids within the formed fiber. Figure 10.14 is a typical SEM photo of the undoped fiber cross section and is characterized by several interesting features. The most prominent of these is the large voids in the fiber which explains the low densities. At higher take up speed these voids are reduced in size due to the orientation (i.e. stretching) of the fibers and this results in higher fiber density. Several other important features are the circular fiber cross section and the grainy structure of the "solid" material. The circular cross-section is a confirmation that the spinnerette is properly machined and that the coagulation process is not too fast. More importantly however it indicates that the polymer solution (i.e. spin

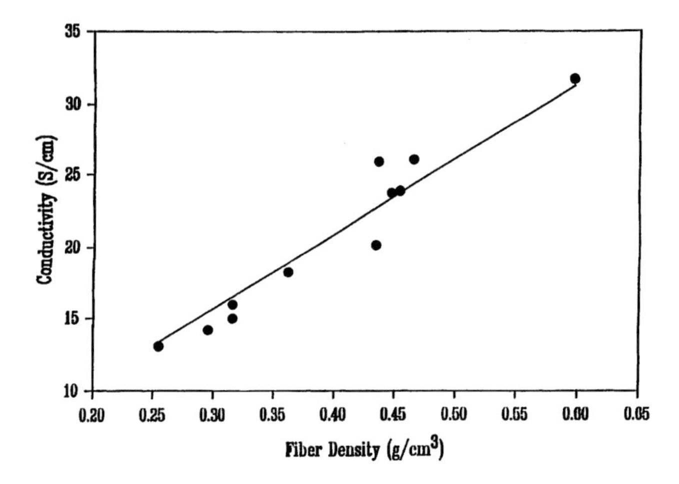

Figure 10.13 The effect of fiber density on the conductivity (1M $MeSO_3H$ doped) of as-spun PANI fibers.

Figure 10.14 Typical scanning electron micrograph of an as-spun PANI emeraldine base fiber (C1) cross section (900 X).

dope) is homogenous, giving further evidence that DMPU is a reasonable spin solvent. If the solution were not homogenous, shrinkage would not be uniform during fiber formation resulting in non-circular (crescent or dog-bone shaped) cross-sections. The grainy surface indicates that very small voids may be present as well as the larger ones.

Since the conductivity-density relationship is approximately linear as shown in Figure 10.13 the voids may be the major factor affecting both the mechanical and electrical properties of the as spun fiber. In order to more thoroughly investigate this possibility, the electron microscope images were digitally captured and analyzed using a binary threshold to distinguish void from solid. The void spaces were numbered and cross sectional areas calculated. This allows the formation of histograms showing the distribution of void sizes resulting from various processing conditions. Figure 10.15 is a series of histograms for fibers spun into a coagulation bath of 35% NMP in water. It clearly shows that the fiber spun at 4 m/min has larger voids than the one spun at 10 m/min. The largest void space in the 4 m/min fiber is approximately 430 μm^2 and the largest void size in the fiber spun at 10 m/min is 150 μm^2. Although the size may vary from cross section to cross-section, the fibers spun at higher take up speeds consistently have smaller voids when compared to those at lower speeds. When the total cross-sectional area of the voids within each fiber is related to both the fiber tenacity and the conductivity it is found that in fibers containing large void spaces (above 3000 μm^2 total void area) there is little dependence of the tenacity and conductivity on the total void area as the void area increases. However, as the total void area decreases the dependence of the measured conductivity and tenacity on the void area becomes very pronounced. This suggests that small decreases in total void area

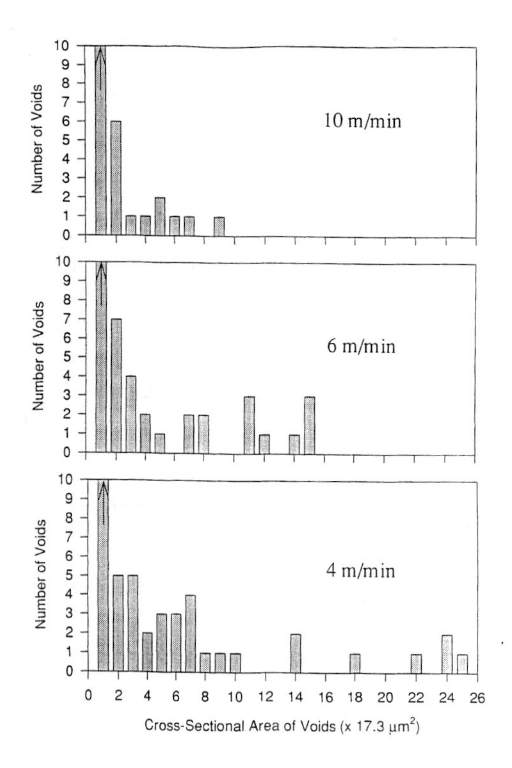

Figure 10.15 Histograms of void dimensions as a function of take up speed in as-spun PANI fibers. Coagulation bath concentration 35% NMP/water.

produce large changes in tenacity and conductivity upon doping. According to the Griffith failure criteria the material strength is proportional to $c^{1/2}$ where c is the defect size [25]. The increase in strength as the defect size decreases is clearly demonstrated by the formed PANI fibers.

As the data suggests, the conductivity of $MeSO_3H$ doped as spun fibers prior to any external drawing other than take up speed is dependent on the total void area. The highest conductivity achieved for these fibers prior to drawing was approximately 32 S/cm corresponding to a take up speed of 10 m/min and to the lowest total void area. When this fiber is drawn prior to doping the conductivity increases to better than 350 S/cm on average with good reproducibility on a segment by segment basis, representing a tenfold increase in the conductivity of the as spun fiber. Most likely the increase in conductivity of the as spun doped fiber results from minimizing the very small or micro voids, allowing better chain to chain electrical proximity. Table 10.1 lists some of the properties obtained for EB PANI fibers for the as spun fibers and after drawing to 400%. The reduced mechanical properties of the fiber labeled C1 is most likely a consequence of the large voids. Fiber C1a has smaller total void space area and higher conductivity and tenacity.

It is clear from the forgoing that in order to optimize the electrical as well as the mechanical properties of the as spun fibers, the void sizes must be reduced. The effect of two processing parameters, coagulation bath concentration and take up speed, on the developing fiber microstructure have recently been studied in our laboratory. Figure 10.16 shows the total void area, determined by the previously discussed method, vs. coagulation bath concentration for two different take up speeds. At low speeds, the total area of the voids goes through a minimum at a coagulation bath concentration of 30%. This behavior is not unexpected since it is common to observe a minimum in the diffusion coefficient and a maximum in fiber density verses concentration curves for various other well characterized wet spin systems. Figure 10.16 also shows that the total void area for a take up speed of 8 m/min is considerably less than for a 4 m/min take up speed, and that the dependence on coagulation bath concentration is significantly lower at the higher take-up speed. This seems to indicate that in order to reduce the void size significantly, thus increasing the mechanical properties of the PANI base fiber, higher take up speeds at lower coagulation bath concentrations of NMP in water are appropriate.

Figure 10.17 is an SEM of the drawn undoped fiber. The grainy fracture surface of the undrawn fiber in Figure 10.14 has been replaced by a smooth cross-section indicating the small microvoids have been substantially reduced or eliminated. Large voids are still

Table 10.1 Tenacity and Conductivity of Doped PANI Emeraldine (EB) Fibers

FIBER	Tenacity (gpd)	Conductivity (S/cm)	Doping Solution (5% Acetic Acid)
Cl as-spun	<0.2	10–32	1M $MeSO_3H$
Cl 4X drawn (voids $>3000\ \mu m^2$)	<1.5	235	1M $MeSO_3H$
Cla 4X drawn (voids $>1000\ \mu m^2$)	3.3	350	1M $MeSO_3H$

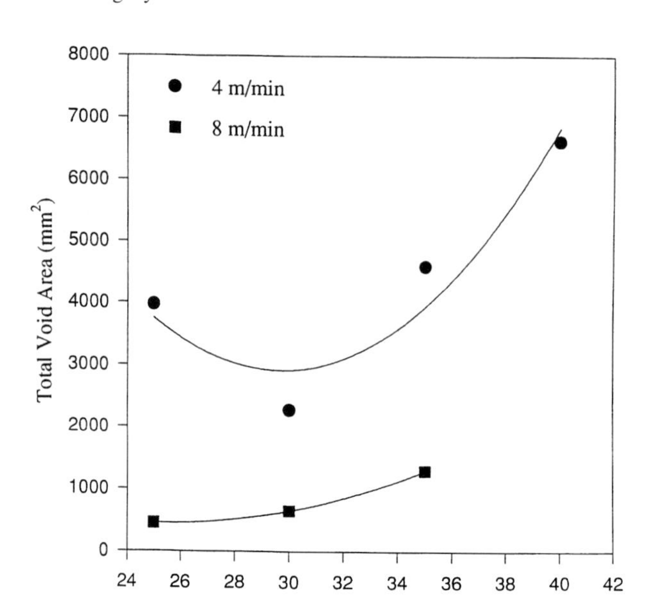

Figure 10.16 The effect of coagulation bath concentration and take up speed on the total void cross sectional area of PANI base fibers.

Figure 10.17 Scanning electron micrograph of a drawn undoped PANI Fiber (300X).

present in this fiber however, indicating that the intrinsic limit of conductivity for the doped fiber and the mechanical properties have probably not been reached. These studies indicate that the elimination of both the large and small scale voids is necessary to form a fiber which maximizes both the mechanical and electrical properties and approaches an intrinsic limit. Reduction of the voids should produce a fiber with a fiber density close to that of the polymer itself.

10.2.2 PANI Fiber Properties

Initial X-ray data obtained for the as spun fibers indicate them to be amorphous at slow take up speeds and some slight orientation is observed at higher speeds. This is consistent with the data obtained by MacDiarmid and Epstein on their films and fibers after orientation [26]. There is considerable scatter in the small angle X-ray photographs, further confirming the presence within the fiber of microvoids on the order of 60 nm across, which affect both the electrical and mechanical properties. The grainy fracture surface observed in the SEM scans (Figure 10.14) also is indicative of small microvoids. Our recent work and that of others suggests that to insure a more homogenous doping, the fiber should be doped immediately after or during the draw stage, and not after a prolonged post-draw time period [27].

Table 10.2 compares some of the properties of PANI/DMPU fibers after drawing with other synthetic fibers. These values are reproducible on a segment by segment basis. Much higher conductivities of 500 S/cm have been obtained on single filaments but it is difficult at this time to produce thousands of meters of this fiber with these conductivities on a segment by segment basis. After resolution of the void problem, by determination of the proper mass balance and thermodynamic coagulation bath parameters, production of uniform fibers at these higher conductivities, with level segment to segment reproducibility, should not prove difficult. Table 10.3 lists the processing parameters for formation of the PANI base as spun fibers. The spin dope solution of 15% wt/wt of PANI base in DMPU with a viscosity of 21,000 cP is stable and thousands of meters of fiber can be easily spun with negligible changes in the morphology of the as spun fiber. This is a direct result of the stability of the PANI/DMPU spin dope. This table also lists the take up

Table 10.2 Mechanical Properties of Some Commercial Fibers and PANI/DMPU

FIBER[a]	Tenacity (gpd)	Elongation (%)	Modulus (gpd)
PANI/DMPU (EB) (undoped drawn)	1.5–3.0	12–18	8–12
Nylon 6	4.0–7.2	17–45	18–23
PET[b]	2.8–5.6	24–42	10–30
Glass (e-type)	15.3	4.5	320
Rayon[c]	1.8–2.3	20–25	

(a) Physical Properties of Textile Fibers – Ciba-Geigy Inc.
(b) DuPont Dacron single filament regular tenacity
(c) North American Rayon

Table 10.3 As-Spun Fiber Processing Parameters

Spinning Solution	15 wt.% PANI/DMPU (either EB or LEB) $\eta = 21{,}000$ cP
Spinnerette	D = 0.004 in. L/D = 2.0
Coagulation Bath	25–40% NMP/H_2O
Take Up Speed	2–10 m/min (LEB solutions up to 100 m/min)

speed for the reduced PANI fibers formed from leucoemeraldine base (LEB). Fibers from reduced spin dopes are discussed in the following sections.

10.2.3 Viscoelastic Characterization of PANI Spin Dopes

To spin a fiber either by melt or solution spinning technologies, the viscoelastic properties of the fluid must be known in order to obtain reproducible mechanical properties of the formed fiber. In addition, the electrical properties are dependent on the final polymer structure in the condensed phase and will also be dependent on the viscoelastic characteristics of the fiber and the fiber spinning solution. The fiber forming solution (spin dope) must be stable and should not agglomerate or cross-link during spinning. Likewise the polymer spinning solution should not form cross-links, either chemical or physical during storage and no significant degree of interaction should occur between polymer chains during the spinning prior to extrusion at the spinnerette interface. Only certain specialized fibers employ cross-linking or gelation after extrusion from the spinnerette. As mentioned in previous sections, PANI base solutions tend to cross link and form thixotropic gels at higher solids concentrations and these interactions must be minimized if reproducible fiber is to be formed on an industrial scale. Although DMPU proved to be a good solvent for EB base, great care must be taken to dissolve 15% solids in this solvent to prevent any shear induced or acid catalyzed gelation due to the existence of the imine nitrogens. In order to study the changes in PANI base solutions with respect to time and shear, viscoelastic studies were carried out on PANI base solutions. As stated previously several laboratories including ours have recently characterized the nature of the interactions as being due to secondary bonding (i.e. hydrogen bonding) between the oxidized portion of the PANI chain (imine nitrogens) and amine protons on the same or on neighboring chains. In order to more throughly understand the nature and consequences of the chain to chain interactions, and their effect on fiber and film formation, we studied the viscoelastic properties of the emeraldine and the reduced (leucoemeraldene) form of polyaniline. In the lecoemeraldene (LEB) form, PANI contains significantly fewer (ideally no) imine nitrogens. It is well known that imine nitrogens show a much greater propensity for the formation of hydrogen bonds (secondary bonding) and this is thought to be responsible for gel formation in the emeraldine base spin dopes. Reduction of the imine nitrogens to amines, forming the leucoemeraldine base, should reduce the interactions between chains since the interaction between the amines is substantially weaker. These studies were carried out by viscoelastic oscillatory shear measurements. Oscillatory shear measurements can yield information on the storage modulus (G') which is a measure of the elasticity of the polymer. This elasticity normally results from a network formation between the polymer chains. Information is also gained on the viscous component (G") as well. This type of measurement on the polymer solutions can also give information on the complex viscosity, complex modulus and other data concerning polymer solutions. A review of oscillatory shear measurement techniques can be found in any standard text on the dynamics of polymeric liquids and solutions.

The viscoelastic characteristics of polyaniline (emeraldine base)/DMPU {PANI(EB)/DMPU} solution (14%wt/wt) and polyaniline (leucoemeraldine base)/DMPU {PANI (LEB)/DMPU} (14%wt/wt) were determined by oscillatory shear using a controlled stress rheometer in the frequency range of 1–100Hz. The polyaniline solution was subjected to an oscillatory shear as a function of frequency at various time intervals. Figure 10.18 shows the storage modulus G' as a function of frequency over time for PANI(EB)/DMPU. The storage modulus increases with frequency, however the frequency dependence decreases with increasing time suggesting a transformation from a viscoelastic liquid to a viscoelastic solid. The storage modulus also increased significantly with time indicating a development of elastic interaction between polyaniline base chains. Figure 10.19 shows the storage and loss moduli (G' and G") of a 14 wt% polyaniline(EB)/DMPU solution, which we will call PANI-C1 for brevity (this is the

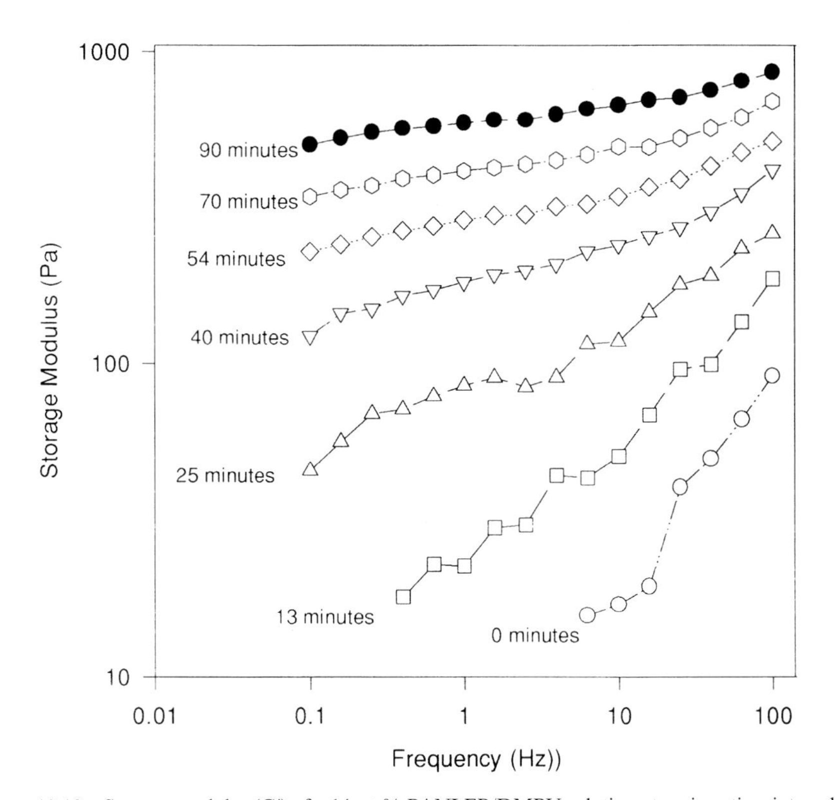

Figure 10.18 Storage modulus (G') of a 14 wt.% PANI EB/DMPU solution at various time intervals (min.).

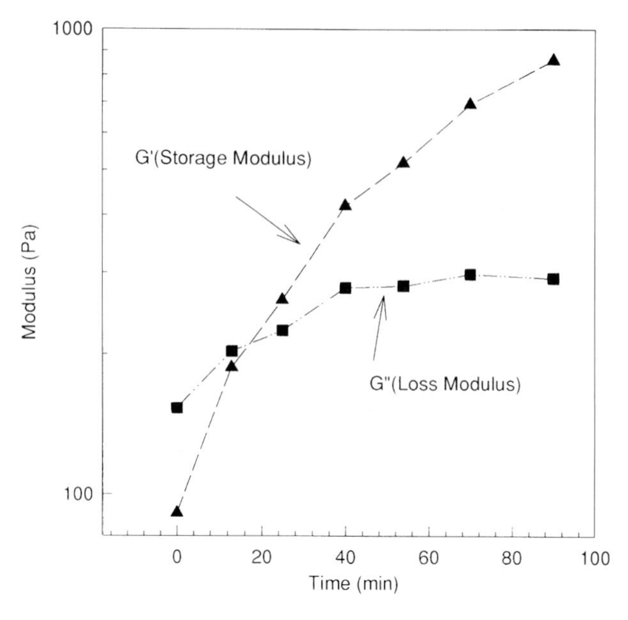

Figure 10.19 Time dependence of the evolution of the dynamic moduli of a 14 wt.% PANI(EB)/DMPU solution at 100 Hz.

solution used for spinning fibers from EB/DMPU), as a function of time at a frequency of 100 Hz. It can be seen that at times less than 20 minutes, the loss modulus (G") is greater than the storage modulus (G') indicating that the viscous contribution is greater than the elastic contribution. There is, however, a pronounced increase in the storage modulus with time suggesting that there is the formation of some type of network structure as a consequence of chain interactions resulting in the increased elasticity. Figure 10.20 shows the dynamic moduli of the PANI(LEB)/DMPU solution, which we now designate PANI-C2. In contrast to the C1 solution, the loss modulus remains greater than the storage modulus throughout the experiment indicating that formation of an elastic or gel structure is inhibited when PANI is in the reduced LEB form. The increased chain flexibility, due to the absence of cross links, is reflected in the order of magnitude difference in moduli between the EB and LEB solution. Dividing the loss modulus by the storage modulus gives the loss tangent ($\tan \delta = G''/G'$) which is the ratio of energy lost to energy stored in a cyclic deformation and is a measure of the fluidity of the solution. The loss tangent for both the C1 and C2 solutions is presented in Figure 10.21. The loss tangent is very low for a crosslinked material whereas it is very high for a dilute solution. The very low values of loss tangent for C1 solution compared to C2 solution and its decrease over time indicates the presence of elastic crosslinks in the C1 solution while few if any exist in the C2 solution. Scatter in the C2 data is due to the limited sensitivity of the rheometer in low elasticity solutions.

The dramatic changes in viscoelasticity when EB solution is converted to LEB solution motivated a study of the solution chain conformation since it is quite possible that some association is beginning to occur at an early time in the EB/DMPU solutions leading to

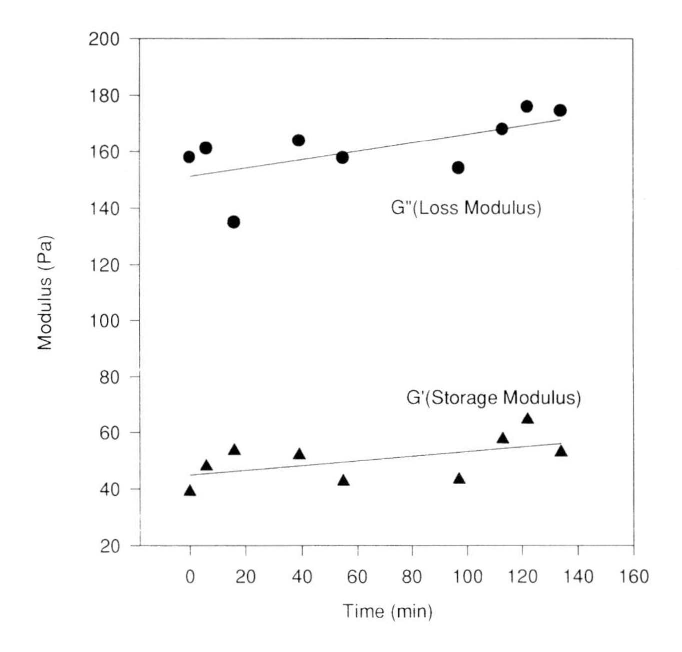

Figure 10.20 Time dependence of the evolution of the dynamic moduli of a 14 wt.% PANI(LEB)DMPU solution at 100 Hz.

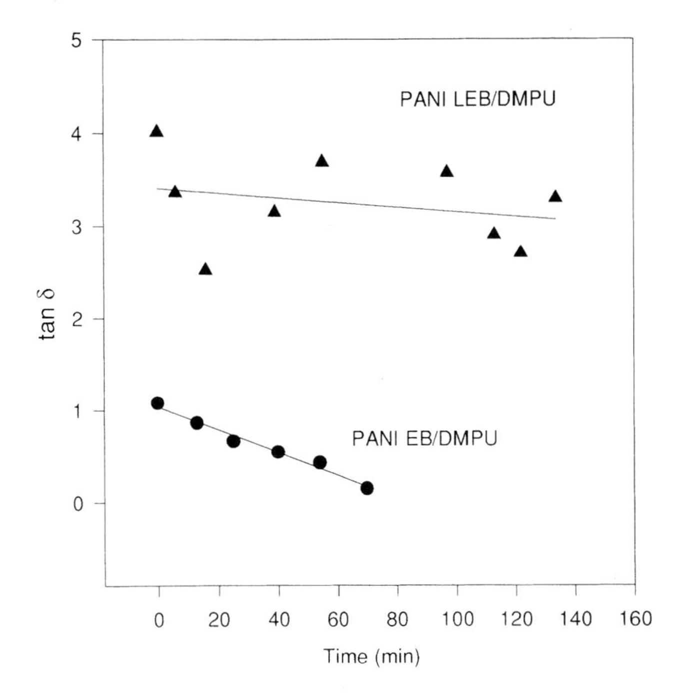

Figure 10.21 Comparison of the loss tangent (tan δ) of EB and LEB solutions of PANI in DMPU as a function of time at 100 Hz.

bunched polymer chains. This early association could be intrachain or interchain or both since the imine nitrogens would have a propensity to associate with amine nitrogen protons regardless of whether they are on the same or neighboring chains. DMPU being a better solvent most likely slows this association, as suggested by the viscosity data previously shown in Figure 10.8. Such agglomeration, if on a small scale early in the solution history, may not be seen in the studies on the normalized viscosities but may have significant effects on the fiber formation process and hence on the properties and morphology of the formed fiber.

If these early associations have not formed to a large extent, and those that have formed still have a significant solvent density in their immediate volume element, the shear in the capillary channel of the spinerette will cause a breakup of the forming network which is consistent with the thixotropic nature of the solution as stated earlier. If in fact there is some degree of network formation between neighboring chains the association may be reflected in differences in molecular weights determined by gel permeation chromatography (GPC). These differences would result from a higher molecular weight fraction occurring in the GPC for EB solution containing agglomerated particles than in the LEB solutions where these associations may be substantially reduced due to the absence of imine nitrogens. In addition, since molecular weight determination by GPC is based on chain size and shape (hydrodynamic volume), the tendency of the chains to bunch in on themselves due to poor solvation could be observed by comparison of the EB and LEB solutions made from the same stock solution.

Molecular weight of EB (C1 fiber forming) and LEB (C2 fiber forming) solution was obtained by GPC and compared. The EB-C1 material was estimated to be Mw = 85,000 with a polydispersity index of 2.4 while the molecular weight and polydispersity index of the LEB-C2 material was 25,000 and 1.6 respectively. When a portion of the reduced LEB-C2 solution was reoxidized and the molecular weight obtained again the by the same GPC method, the molecular weight value approached the 85,000 value obtained on the original solution. These preliminary studies indicate that the difference in the observed molecular weights is due primarily to a difference in hydrodynamic volume between the oxidized and reduced forms of the polymer and most likely not due to extensive chain degradation as reported in the literature.

This evidence supports the notion that there is some association due to bunching of the PANI chains resulting in a change in the hydrodynamic volume of the polymer chains in the spin dopes and that reduction of this association can be achieved by reduction of the EB to LEB. The decrease in the storage modulus seen in the C2 solutions suggest that this minimized chain interaction and changes in chain conformation might well be reflected in the processing flexibility.

In order to address this, both C1 and C2 solutions in DMPU were solution spun into fibers. Maximum attainable take-up speeds and draw ratios were greater for the C2 solutions than for the C1 solutions. As previously discussed, the C1 fibers exhibit a pronounced void structure. Fibers spun from leucoemeraldine base in DMPU had substantially fewer large area voids than did the C1 fibers, with better mechanical properties. Figure 10.22 is a cross section of a C2 fiber spun from the LEB spin dope. Since the fibers have substantially fewer imine nitrogens necessary for doping, the fibers must be reoxidized prior to doping. It is unclear as of this writing whether it is better to

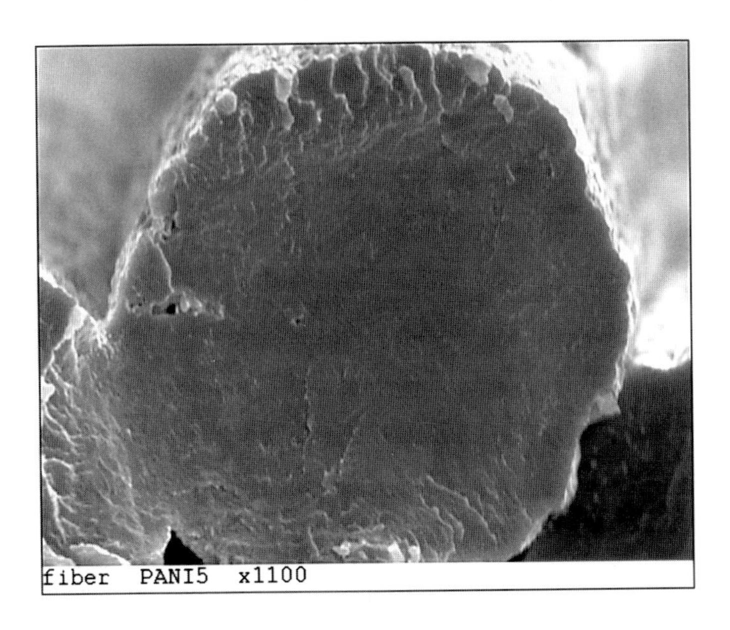

Figure 10.22 SEM of C2 fiber cross section (1100 X) spun from LEB spin dope.

oxidize and dope simultaneously or separately. The fibers reported here were oxidized and then doped in separate steps. Mechanical and electrical properties of the C2 fibers were measured and are presented in Table 10.4. These properties should be compared with those of the PANI C1 fibers (EB/DMPU) detailed in Table 10.1. The mechanical properties of the as spun C2 fibers are significantly higher than those of the as spun C1 fibers. This may well be due to much higher density of fiber obtained from the LEB solution. Even after doping with methane sulfonic acid (MSA) or HCl, these new fibers demonstrate mechanical properties strong enough for processing using existing textile technologies for yarn spinning [28]. Electrical conductivities for the C2 fibers after oxidation and doping gave reproducible values of 150 S/cm which is lower than the 350 S/cm observed for C1 fibers. The presence of voids in C1 fibers may facilitate easier diffusion of dopants throughout the fiber resulting in higher conductivities while smaller voids and non uniform oxidation give lower conductivity values for C2 fibers. The increased density of the LEB fiber may also prevent diffusion of the oxidizing agent and/

Table 10.4 Mechanical and Electrical Properties of PANI C-2 Fibers from Leucoemeraldine Base/DMPU Spin Dope

Property	As spun	Drawn (2X)	As spun doped	2Xdrawn doped
Tenacity (gpd)	1.1	3.6	0.8	1.9
Modulus (gpd)	57	89	15	41
Elongation (%)	51	15	75	23
Conductivity (S/cm)	–	–	20	150

or dopant throughout the fiber when this occurs under the same conditions used for the formation of the C1-EB fibers. With the development of a better procedure for oxidation and doping of the C2-LEB fibers and draw ratios in excess of 2, the conductivity should dramatically increase. Our laboratory is presently investigating methods to uniformly oxidize, dope, and draw the C2 fibers to attain maximum conductivity. However, these conductivity values already obtained are high enough for many applications where reasonably high mechanical properties are desired. Processing in reduced form also allows preparation of fibers from blends of polyaniline with other insulating and conducting polymers. In addition, variation of the oxidation state of the spin dope may well allow the fiber spinning process to be tailored to obtain a wide range of fiber properties (electrical, mechanical, morphological, etc.).

10.2.4 Thermal Characteristics of LEB PANI Fibers

Thermal characteristics of polyaniline fibers were determined in order to understand the effect of changes in oxidation state on morphology. In the first DSC scan of C2 fiber, the temperature was ramped to 400 °C. A broad glass transition was observed at $T_g \sim 190$ °C and a broad melting endotherm was observed around $T \sim 355$ °C, indicative of imperfect crystals with low sample crystallinity. The sample was quenched and measured under the same conditions. During the second scan, the T_g was observed, but the melting peak was absent. In order to understand this melting transition, a second sample was heated to 350 °C at 5 °C/min to anneal, followed by quenching. A DSC scan of the annealed sample revealed $T_g \sim 190$ °C and a sharper endotherm at $T \sim 365$ °C indicating more perfect crystals. A second scan revealed the same glass transition temperature, but no melting endotherm. The disappearance of the endotherm on the second heating may be due to either the crosslinking of chains preventing their recrystallization during a second heating, or possibly the recrystallization kinetics are sufficiently slow so that the melting temperature is reached before crystals can be formed. If the behavior is due to crosslinking, one would expect an exothermic reaction accompanied by an increase in the glass transition. The data indicates an exothermic trend for $T > 300$ °C, but the second scan shows no increase in the glass transition temperature. Figure 10.23 clearly shows the effect of temperature above 385 °C under nitrogen on thin pellets formed from LEB. The smooth surface which develops after exposure to temperatures greater than the apparent melt transition may well have application in small organic devices.

10.3 Conclusion

Fibers spun from polyaniline have a demonstrated promise for industrial applications where certain levels of conductivity are needed. Although fibers can be obtained from other ICP's, most of these are produced from novel monomers on a bench top scale. Scale

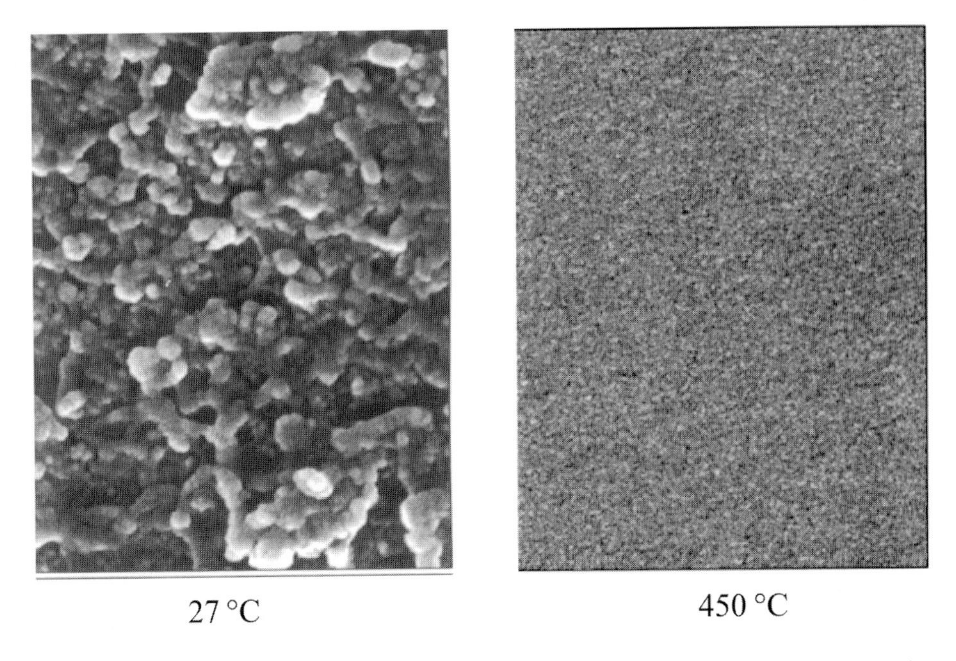

Figure 10.23 SEM micrographs of: a) formed pellet of LEB at low temperature, and b) after being held above the melt transition showing the formation of a smooth film. Both SEM's are at a magnification of 19,000.

up of these systems has not yet proved economically feasible and therefore ICP's derived from monomers available on a large scale such as aniline, pyrrole, or thiophene continue to receive wide attention in both industrial and academic laboratories. Of the polymers formed from these monomers only polyaniline is reasonably stable and processable in it's undoped state. This chapter has outlined some of the authors' work to date on applications of this polymer to fiber formation.

Fibers formed from polyaniline or other ICP's with very high levels of conductivity, on the order of several thousand Siemens and strengths of over 8 gpd are yet to be realized. Recent work reported by Monkman demonstrated fibers produced from the doped form of polyaniline that exceed 1000 S/cm [29]. Theoretical calculations on the upper level of conductivity for polyaniline show that conductivities on the order of tens of thousands are possible. Stable conductivities of several hundreds have been obtained and some recent work on higher molecular weight materials has shown promise of much higher levels of conductivity which are reproducible on a segment by segment basis. As we become better at limiting the defects in the polymer chain induced by various process routes, the levels of achievable conductivities will certainly rise substantially.

While there has been significant research effort in the synthetic chemistry of producing novel monomers in order to make very highly conductive materials, and also significant research into the solid state chemistry and physics of conduction mechanisms, there is precious little reported in the literature regarding the basic polymer science of these materials. Only recently have researchers begun to investigate the effect of processing, solvent interactions, structural morphological changes due to mechanical, electrical, or thermal stress on the formed electrical and electronic properties of the materials. Work on the effect of polymer orientation, structural morphology, etc, on the formation of charge carriers on the polymer, band gap energies, etc. is just now beginning to be reported in the literature. As this important work continues, conductivities and mechanical properties will greatly increase.

10.4 Future Trends

Electrical conductivities high enough for use as motor windings, or replacements for metallic wires, by fibers produced from electrically conductive polymers have not yet been commercially realized. However, uses for fibers formed from these materials may find many applications in the fiber, electronic, and composite areas. Of particular interest is the rapidly growing area of light emitting organic diode technologies. The use of conductive organic fibers in devices containing LED's may prove highly beneficial. The use of conductive polymeric fibers in actuators is also an area of investigation. All of these applications will rely on fibers with segment by segment reproducible conductivities. The ability to tune fibers to certain levels of conductivity will enable these materials to be used in such diverse areas as electromagnetic shielding, static dissipation, dielectric materials, corrosion inhibitors, bio-sensors and many other applications. Recently Mattes and coworkers at the Los Alamos national laboratory have reported that hollow fibers can be produced from polyaniline. These fibers may prove useful in many different applications and one reported is for ambient temperature gas separations [30]. Continuing research in the area of fiber production will address the issues of improved strength and conductivity, environmental stability, and large scale commercial production with segment to segment reproducibility of the electrical, electronic, and mechanical properties. Chemical modification of the basic monomeric structure to enhance the aforementioned properties will continue to be a strong area of research interest. Several new and innovative methods to produce electroactive fibers are presently being investigated. These include recent work by Reneker and co-workers on electro-spinning techniques [31] (see Chapter 6).

Whether or not conductivities and strengths can be obtained to replace metal fibers is at present uncertain. Conductive polymers, already beginning to find use in such unique applications as flexible light emitting diodes and other electronic, magnetic, and photonic device applications will certainly find a variety of usage in fiber form for use in the formation of tomorrow's high technology structures and smart materials [32].

10.5 References

1. H. Shirakawa, E. J. Lewis, A. G. MacDiarmid, C. K. Chiang, A. J. Heeger, *Journal of the Chemical Society*: Chemical Communications (1977) pp 578–580.

2. G. L. Baker, *Progress Toward Processable, Environmentally Stable Conducting Polymers*, in *Electronic and Photonic Applications of Polymers*, (M. J. Bowden, R. J. Turner ed.) *Advances in Chemistry Series*, (1989) Vol. 218, pp. 271–296; *Handbook of Conducting Polymers*, Vol 1 & 2, (T. A. Skotheim, ed.), Marcel Deckker, Inc. New York 1986
3. See for example: *Proceedings of the International Conference on the Science and Technology of Synthetic Metals*. Every other year 1988–2000, published in the journal *Synthetic Metals*
4. See for example: *Semiconducting Polymers; Applications, Properties, and Synthesis*, ACS symposium Series 735, B. Hsieh, Y. Wei, eds. ACS books (1999*); Handbook of Conducting Polymers*, 2nd Ed., (T. A. Skotheim, R. L. Elesenbaumer, J. R. Reynolds, eds.), Marcel Deckker, Inc. New York 1998
5. A. Monkman, P. Adams, P. Laughlin, and E. Holland, *Synthetic Metals* (1995) 69, p.183, 1995; A.P. Chacko, S. S. Hardaker, B. Huang, and R. V. Gregory, *Mat. Res. Soc. Symp. Proc. (1995)* 413, p. 503; S.S. Hardaker, A.P. Chacko, B. Huang, and R. V. Gregory, *SPE-ANTEC '96*, (1996) Vol. II, p. 1358.
6. See Uniax web site at: http://www.uniax.com/
7. Y. Cao, P. Smith, A.J. Heeger, *Synthetic Metals* (1992) 48, p. 91
8. O.T. Ikkala, J. Laakso, K. Vakiparta, E. Virtanen, H. Ruohonen, H. Jarvinen, T. Taka, P. Passiniemi, J.-E. Osterholm and Y. Cao, A. Andreatta, P. Smith, A. J. Heeger, *Synthetic Metals* (1995) 69, p. 97–100
9. R. Menon, Y. Cao, D. Moses, and A. Heeger; *Phys. Rev* (1993) B-47
10. J. E. Fisher, X. Tang, E. M. Scherr, V. B. Cajipe, A. G. MacDiarmid, *Synthetic Metals* (1991) pp. 661–664
11. T. Murase, T. Ohnishi, Noguchi, *Chemical Abstracts* (1990) 104, p. 169124; J. R. Reynolds, M. Pomereantz, Processable Electronically Conducting Polymers, in *Electroresponsive Molecular and Polymeric Systems*, (T. Skotheim, ed.) Marcell Dekker Inc., New York (1989) pp. 187–256; *Semiconducting Polymers; Applications, Properties, and Synthesis*, ACS symposium Series 735, B. Hsieh, Y. Wei, eds. (1999) ACS books
12. J. R. Reynolds, M. *Pomereantz, Processable Electronically Conducting Polymers* in *Electroresponsive Molecular and Polymeric Systems*, (T. Skotheim, ed.) Marcell Dekker Inc., New York (1989) pp. 187–256
13. R. Aspland, *Application of Basic Dye Cations to Anionic Fiber: Dyeing Acrylic & Other Fibers with Basic Dyes., A Series on Dyeing* (1993) Vol. 25 #6, Ch. 12, pp. 21–26; G. Odian, *Chain Copolymerization, Linear, Branched, and Crosslinked Polymers* in *Principles of Polymerization*, Wiley-Interscience John Wiley & Sons Inc. (1991) pp. 17–19
14. R. V. Gregory, W. C. Kimbrell, H. H. Kuhn, *Synthetic Metals* (1989) 28, pp. C.823–C835
15. M. J. Liu, *Synthesis and Characterization of a New Conducting Urethane β-substituted Polythipphene and its Blends*, PhD thesis Clemson University 1995
16. Y. Z. Wang, J. Joo, C.-H. Hsu, A. J. Epstein, *Synthetic Metals* (1995) 68, pp. 208–211
17. M. Angelopoulos, Y. H. Liao, B. Furman, D. Lewis, T. Graham, *Proceedings of the Soc. Plastics Eng.* (1995) Vol II, ANTEC '95, Vol 41, pp. 1678–1671; M. Angeloupolus, Y. Laio, B. Furman, and G. Teresits, *Macromolecules* (1996) 29
18. A. G. MacDiarmid, A. J. Epstein, *Science and Applications of Conducting Polymers*, W. R. Salaneck, D. C. Clark, and E. J. Samuelson, eds., IOP Publishing Limited Bristol (1990) pp. 117
19. Y. Wei, K. F. Hsueh, G. W. Jang, *Macromolecules* (1993) 27–2, p. 518
20. A. Ziabicki, *Fundamental of Fibre Formation, Wet and Dry-spinning from Solutions* (1976) Ch. 4, pp. 249–350
21. R. Jain, R. V. Gregory, *Synthetic Metals* (1995) 74, pp. 263–267
22. Z. Tadmore, *Principles of Polymer Processing*, Wiley-Interscience, John Wiley and Sons New York (1979)
23. A.G. MacDiarmid, A. J. Epstein, *Faraday Discuss. Chem. Soc.* 88, 1989, pp. 317
24. K. T. Tzou, R. V. Gregory, *Synthetic Metals* (1993) 55–57, p. 983
25. A. Kelly, N. H. MacMillan, *Strong Solids* (1985) 3rd. ed. Clarendon Press, Oxford, U.K
26. J. Pouget, M. Jozefowicz, A. Epstein, X. Tang, and A. MacDiarmid; *Macromolecules* (1991) 24, p. 779
27. National Science Foundation site visit report. Center for Advanced Fiber and Films, Clemson University, 1999
28. A. P. Chacko, S. S. Hardaker, R. J. Samuels, and R. V. Gregory; *Synthetic Metals* (1997) 84, pp.41–44

29. A. Monkman, *Proceedings of the Society of Plastics Engineers*, ANTEC, New York, 1999
30. B. Mattes,H. Wang, D. Yanga, Y. Zhu, W. Luenthal, M. Hundley, *Synthetic Metals* (1997) 84
31. G. Srinavasan, D. H. Reneker; *Polym. Int.* (1995) 36
32. Proceedings of the International Conference on the Science and Technology of Synthetic Metals, Snowbird Utah, *Synthetic Metals* 1996; Proceedings of the International Conference on the Science and Technology of Synthetic Metals, Monpilar France *Synthetic Metals* (1999) *Handbook of Conducting Polymers*, 2nd Ed., (T. A. Skotheim, R. L. Elesenbaumer, J. R. Reynolds, eds.), Marcel Deckker, Inc. New York 1998

11 Fibers from Polymer Blends and Copolymers

Rajendra K. Krishnaswamy and Donald G. Baird
Virginia Tech (VPI&SU), Blacksburg, Virginia, USA

11.1 Introduction

Several billion pounds of fibers are consumed every year in the United States, with a significant fraction of the amount constituting synthetic polymeric fibers. Traditional applications of polymeric fibers have been in textiles and furnishings. However, fiber applications have expanded dramatically with penetration of synthetic fibers into industries such as geotextiles, composites, insulation, filtration, aerospace components, biomedical products, and protective clothing. Thus, synthetic fibers usually need to satisfy a broad spectrum of performance criteria. For example, fibers employed in the textile industry need to possess a high softening temperature, adequate tensile strength over a fairly wide temperature range, solubility or meltability for spinning, high modulus, chemical and biological stability, dyeability, long-term stability, flame resistance, and good appearance. Enhancing the quality of fibers requires changes not only in the chemistry of the material to obtain desired properties, but also significant modifications in the spinning process. The search for higher performance materials has led researchers to either blend existing polymers or polymerize different monomers simultaneously. Blending and copolymerization approaches to obtain specific properties usually require relatively little investment in comparison to the development of new polymers or new processes. Hence, fibers from polymer blends and copolymers have assumed an important role in the synthetic fiber industry. This chapter will focus on the properties, morphology, and structure development during processing of fibers based on polymer blends and copolymers. Particular emphasis has been stressed on blend fibers wherein one of the components is a thermotropic liquid crystalline polymer.

11.2 Polymer Blends

11.2.1 Introduction

Incentives for polymer blending include the establishment of high performance materials from synergistic polymer combinations, as well as the economic advantages which can be achieved by the dilution of resins with low-cost polymer, and the potential recycling of industrial scrap. For engineering polymer blends, primary incentives for blending, as evident in the patent literature, include enhanced mechanical properties, processability, and thermal and chemical resistance. The factors that directly influence the performance of a polymeric blend are the properties of the blend constituents, phase behavior, blend composition, morphology and compounding technique. Commercial resins that are based on polymeric blends are listed by Utracki [1–2].

11.2.2 Miscibility

When the constituents of a polymeric blend exist in a completely homogeneous state wherein the polymer chains are physically mixed on a sizescale comparable to their segmental lengths such that the blend reveals a single glass transition that is intermediate to the blend constituent glass transition temperatures, they are said to be miscible. The basic requirement for miscibility is a negative free energy of mixing, which is given by:

$$\Delta G_{mix} = \Delta H_{mix} \; T\Delta S_{mix} \tag{11.1}$$

where, H, T, and S represent enthalpy, temperature, and entropy, respectively. In order to keep such a miscible blend from phase separating, the second partial derivative of ΔG_{mix} with respect to composition must be greater than zero:

$$\left(\frac{\partial^2 \Delta G_{mix}}{\partial \Phi_i^2}\right) > 0 \tag{11.2}$$

where Φ_i refers to the volume fraction of component i. Equations (11.1) and (11.2) are the necessary and sufficient conditions for miscibility [3].

The combinatorial entropy of mixing for a binary mixture is given by:

$$\Delta S_{mix} = k(N_1 \ln\Phi_1 + N_2 \ln\Phi_2) \quad (11.3)$$

where N_i represents the number of molecules of component i. From the above equation, it is evident that low molecular weight compounds have a high entropy of mixing due to a high degree of molecular activity. Hence, the contribution of ΔS_{mix} to a negative free energy of mixing is significant. However, for polymeric mixtures of high molecular weight, ΔS_{mix} approaches zero, thus hindering intimate mixing of the blend components. In such cases, the enthalpy of mixing has to be negative in order for the blend to be miscible. Therefore, some kind of specific interactions between the blend constituents must be present. Examples of specific interactions are hydrogen bonding, dipole-dipole, and acid-base interactions. Hydrogen bonding is the most commonly encountered interaction in miscible polymer blends [4–5].

Direct measurement of the heat of mixing is not possible due to strong contributions from viscous heating effects in polymers [4]. However, specific interactions in polymer blends have been observed experimentally via vapor sorption, inverse gas chromatography, infrared spectroscopy, light scattering, and osmotic pressure measurements. The interactions between blend components can be quantified through the Flory-Huggins interaction parameter, χ_{AB}, which represents specific interactions between blend components A and B. The Flory-Huggins interaction parameter is characteristic of the mixing segments, and is a function of temperature. In some blends, the interaction parameter was also observed to be a function of blend composition.

The miscibility of many blend systems has been observed to be a function of temperature. Some blends are miscible at high temperatures, but as they are cooled, the mixture may phase separate. The temperature above which they are miscible is referred to as the *upper critical solution temperature* (UCST). Low molecular weight oligomer mixtures which have a high ΔS_{mix} will tend to be more compatible at higher temperatures, thus exhibiting UCST behavior (see Equation 11.1). On the other hand, some high molecular weight polymeric mixtures which are known to be miscible at lower temperatures phase separate on heating, thus displaying a *lower critical solution temperature* (LCST) behavior. LCST behavior has been reported for several polymer blends [3, 5].

Optical clarity and a single glass transition are the two most widely used indicators of polymer blend miscibility. Optical clarity is not a conclusive criterion because it fails when the refractive indices of the blend components are very close. Also, if the sizescale of phase separation is smaller than the wavelength of light, light scattering fails to indicate immiscibility. On the other hand, turbidity is fairly good evidence of a phase separated blend.

The observation of a single glass transition is a more convincing indication of miscibility. However, this method is not very accurate when the glass transition temperatures of the blend components are not significantly different, or when one of the blend components crystallizes rapidly as the material becomes rubbery. The resolution of the experimental method employed is also important when determining miscibility. Several empirical relations have been developed to predict the T_g of miscible blends; the Fox equation is the most commonly used [5]. The Fox equation for a miscible binary blend of A and B is shown below:

$$\frac{1}{(T_g)_{AB}} = \frac{m_A}{(T_g)_A} + \frac{m_B}{(T_g)_B} \tag{11.4}$$

where m denotes the mass fraction of the blend constituents, A and B. Blends in which strong specific interactions are operative have been observed to display marked deviations from the Fox equation.

11.2.3 Multi-Phase Blends

The properties of miscible blends can be predicted with a fair degree of accuracy. Hence, reasonable control over microstructure and performance in such blends is possible. However, polymer mixtures that are incompatible far outnumber those that are fully miscible (thermodynamic considerations). The mechanical properties of immiscible blends are often poor, which precludes their commercial utilization. This is mainly due to poor transfer of force between the different phases. However, some immiscible blends have displayed spectacular improvement in mechanical properties in comparison to those of the blend constituents [1, 4]. This calls for further investigation into the morphology of such blends and how that morphology influences the ultimate properties of the blend-based materials.

The size, shape and stability of the different phases are important physical factors that can be controlled to improve the properties of incompatible blends. These factors are largely governed by interfacial tension between the phases and the rheological properties of the individual phases. On melting, the minor phase of a multiphase blend exists as spherical domains in order to minimize surface area to volume ratio. Enhanced mechanical mixing (shearing, like in the screw of an extruder) will lead to a breakup of the spherical domains into smaller droplets [6–7]. Deformation of the dispersed phase droplets within a continuous medium can be described by the *Weber Number* W_e (also referred to as capillary number), which is defined as the ratio of viscous stresses to that of interfacial stresses that exist between the two phases in the melt. It is expressed as:

$$W_e = \frac{\text{Deforming Forces}}{\text{Surface Tension Forces}} = \frac{\eta_m \gamma a}{\sigma} \tag{11.5}$$

where η_m, γ, a, and σ represent the matrix or continuous phase viscosity, shear rate, radius of the dispersed phase droplet, and interfacial tension, respectively. The deformation of a spherical droplet into an ellipsoid with a length L, and breadth B can be expressed by:

$$D = \frac{L - B}{L + B} = W_e \frac{19K + 16}{16K + 16} \tag{11.6}$$

where K (viscosity ratio) is defined as the ratio of the dispersed phase viscosity to the viscosity of the continuous phase ($K = \eta_d/\eta_m$).

The dispersed phase droplet will elongate into an ellipsoid and eventually break when the viscous normal stresses tending to elongate and disrupt the drop is greater than the interfacial capillary pressure which resists the deformation and breakup [8]. Thus, the lower the interfacial tension between the phases and smaller the viscosity ratio, the smaller will be the diameter of the dispersed phase droplets.

Interfacial adhesion is the most crucial non-mixing parameter that directly impacts the transfer of load between the phases, which in turn controls the ultimate performance of multiphase blends. Reactive compatibilization of multiphase blends has been used efficiently to enhance interfacial adhesion between the phases, thus imparting better properties to the material [4, 5]. The effective use of block and graft copolymers to enhance or induce compatibility in multiphase blends has been reviewed recently [9].

11.2.4 Blends of Thermoplastics and Thermotropic Liquid Crystalline Polymers

Thermotropic liquid crystalline polymers (TLCPs) are a relatively new class of materials that have received significant attention from both industrial and academic researchers in recent years. TLCPs are known for their excellent mechanical properties, good chemical and thermal resistance, low dielectric constant, excellent barrier properties and excellent dimensional stability. These polymers exhibit a phase intermediate to the isotropic melt state and the ordered crystalline state referred to as a mesophase or liquid crystalline phase. The mesophase displays order inherent to solid crystals, but can flow like isotropic liquids. Liquid crystallinity usually occurs in polymeric systems constituting highly elongated, rigid rod-like backbones. Excluded volume interactions between these stiff molecules tend to cause them to pack into oriented domains. The relaxation timescales of the molecular orientation are significantly longer than those of flexible chain molecules such that the orientation can be more readily preserved in the solid state. This preferred orientation of the TLCP molecules manifests itself in excellent mechanical properties along the orientation direction. However, orientational anisotropy results in relatively poor properties perpendicular to the orientation direction. An overview of the structure and properties of TLCPs can be found in Chapter 7.

Thermotropic liquid crystalline polymers (TLCPs) tend to form fibrillar morphologies in polymer processing operations that incorporate elongational flow fields. Hence, their applicability as reinforcements in polymeric composites has been explored since the mid to late eighties by Kiss [10], Baird and co-workers [11], Weiss and co-workers [12], and Isayev [13]. Such composites are often referred to as *in situ composites* as the reinforcing phase, namely the TLCP fibrils are formed during the melt-processing step. The properties of in situ composites have been reviewed recently [14]. In situ composites offer several advantages over traditional glass and carbon fiber reinforced composites. Some of the above mentioned advantages include lower weight, ease of processing, limited wear (and maintenance) on process equipment, efficient wetting of the reinforcing fibrils by the matrix, easily attainable enhancement of interfacial adhesion between the TLCP fibrils and matrix via compatibilization, and recycling potential.

It has been well established that strong elongational flow fields are necessary to maximize the properties of TLCPs. In fact, it has been experimentally shown that extensional deformation of TLCP melts is more effective in orienting the material compared to strong shear fields [11,12,15,16]. The extensional deformation imparted to the TLCP phase to obtain high aspect ratio fibrils with high degrees of molecular orientation results in highly anisotropic mechanical properties. Furthermore, TLCP fibrils generated in processes such as injection molding do not provide optimum reinforcement. Fiber spinning generates significantly higher levels of molecular orientation within the TLCP fibrils as compared to injection molding due to higher extensional forces involved. For example, Vectra B950, a commercial Ticona TLCP in highly drawn melt-spun fibers exhibits a tensile modulus of 75 GPa, while the highest attainable modulus along the machine/flow direction of injection molded plaques is less than 30 GPa. Thus, fiber spinning is superior to injection molding in the development of TLCP reinforcements within in situ composites. The properties and recent advances in the processing of in situ composite fibers are discussed in Section 11.4.

Basically, almost all combinations of thermoplastics and TLCPs are immiscible and incompatible. In fact, some degree of incompatibility is desirable for the formation of TLCP fibrils, which serve to reinforce the thermoplastic. Due to somewhat poor interfacial adhesion between the phases, the mechanical properties (especially strength and toughness along the transverse direction of injection molded parts) can be somewhat low in blends of thermoplastics and TLCPs. However, compatibilization of such blends have been considered to enhance both the reinforcing potential of TLCPs, as well as to increase the strength and toughness of such blends [17–22]. The compatibilizer was thought to lower the interfacial tension between the phases which leads to more efficient droplet deformation (see Equation 11.5). Also, the compatibilizer could prevent the TLCP phase (be it droplets or fibrils) from coalescing, resulting in more finely dispersed fibrils. Hence, the reinforcing potential of the TLCP is enhanced as fibrils with higher aspect ratios are obtained. The compatibilizer also serves to enhance interfacial adhesion between the phases, which in turn results in higher strength and toughness [17–22]. In samples with relatively high loadings of the compatibilizer, the dispersed TLCP droplets were too small to fibrillate subsequently. Therefore, optimization of the compatibilizer amount is critical to the overall performance of the in situ composite [18,21].

11.3 Bicomponent Fibers

A distinction between bicomponent fibers and fibers from polymer blends needs to be made prior to further discussion. When two or more polymers are physically melt-blended in an extruder prior to processing, the resulting fibers are said to be based on polymeric blends. However, when different polymers present within a fiber are bought in contact as separate streams just before the spinnerette (spinning pack), they are referred to as multicomponent fibers. The discussion on multicomponent fibers that follows will focus exclusively on bicomponent fibers for the sake of simplicity.

Typical phase arrangements sought in the bicomponent fiber industry are *side-by-side arrangement* and *sheath-core arrangement*. Figure 11.1a shows schematic representations of these arrangements as adapted from [23]. The side-by-side arrangement (Figure 11.1a) represents fibers wherein the blend components lie beside each other. The spinning pack design required to generate such an arrangement requires the addition of a conjugator plate just before the spinnerette. Thus the blend components are separated by a septum until just above the spinnerette. The two streams meet and are subsequently spun as they exit the spinnerette hole, with the two components lying beside each other. Adhesion between the different phases is critical to the integrity of such a fiber. Such fibers are desirable to generate self-crimping fibers. Self-crimping provides desirable aesthetic and functional features of stretch and bulk. Fibers based on two acrylic-type polymers which have very different crimping characteristics propelled the commercial utilization of side-by-side fibers [24]. One side of the above fibers shrinks more than the other on heating which produces a high degree of three-dimensional crimp. On immersion in water, the same side swells more than the other, which reduces the crimp. Subsequent drying of these fibers restores the crimp.

In fibers with a sheath-core arrangement (Figure 11.1a), one component (core) is entirely encapsulated by the other (sheath). In such fibers, the polymer that constitutes the sheath could provide characteristics such as appearance and chemical and thermal resistance properties, while that of the core could provide tensile properties such as modulus and strength or cost-savings via incorporation of a low-cost resin. The properties

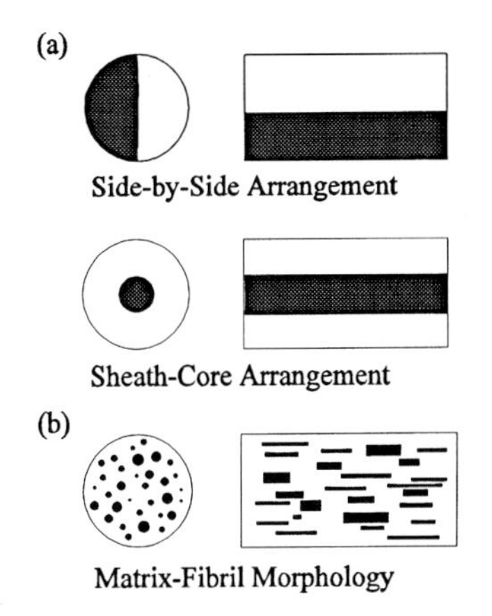

Figure 11.1 Schematic of typical multiphase fiber arrangements. (a) Bicomponent Fibers: side-by-side and sheath-core arrangements; (b) Polymer Blend Fibers: matrix-fibril morphology.

of fibers with a thermoplastic sheath (polypropylene) and a thermotropic liquid crystalline core (Vectra B) are discussed in Section 11.5.2. The spinning pack design to generate sheath-core fibers is somewhat different from that required for side-by-side fibers. The melt-spinning process involves extrusion of the core polymer through a pool of the sheath polymer in the conjugator plate. Thus the polymer melt that constitutes the sheath flows through the annular space created by the spinnerette hole and the core polymer. While adhesion between the sheath and the core is important, it is not as critical to the integrity of the fiber as is the case with the side-by-side fibers.

The stability of the above morphologies depend to a large extent on the rheological properties of the blend components. The lower viscosity component tends to relocate to areas of higher shear, leading to their desire to encapsulate the higher viscosity component in a configuration similar to a sheath-core arrangement. Therefore, a significantly good match between the component viscosities is important in order to generate fibers with side-by-side morphology. An instability will lead to a transformation from side-by-side to sheath-core arrangement.

Sheath-core fibers with the higher viscosity component in the core regions are inherently stable. However, if the lower viscosity component is located in the core region, it will try to relocate to the skin portions of the fiber. Therefore, the timescale of spinning (residence time in the spinnerette) is important. If the timescale of fiber spinning is lower than the time for rearrangement, the desired sheath-core structure can be preserved in the solid state [23].

11.4 Fibers from Polymer Blends

A good review on fibers from polymer blends can be found in Chapter 16 from the book on Polymer Blends [23]. This section will review some fundamental aspects regarding the morphology and processing of fibers from blends based on purely thermoplastic polymers. A subsequent section will focus more on thermoplastic fibers that are reinforced with thermotropic liquid crystalline polymers. Also, the following discussion will focus exclusively on the properties of *melt-spun* fibers formed from *multiphase blends*. Fibers based on miscible blends are omitted as they are less common. However, complete miscibility on a molecular scale is desirable in blend fibers when the additive or second component is added to impart color, or to enhance flame resistance, antistatic or chemical resistance characteristics.

Fibers based on polymer blends usually display a matrix-fibril morphology, wherein one phase is present as droplets or fibrils embedded within a continuous matrix phase (Figure 11.1b). The term *matrix-fibril* arrangement has been used to describe situations wherein the minor phase is present as both droplets and fibrils. Fibers with the matrix-fibril morphology (Figure 11.1b) do not require the use of specialized spinning equipment. The geometry and size distribution of the minor component within the matrix are dependent on the blend composition, rheological properties, and interfacial properties between the blend components. Adhesion between the dispersed phase and the matrix is not as critical as it is with the side-by-side or sheath-core fibers.

When different polymers are blended in an extruder and fibers spun thereof, a matrix-fibril morphology is typically obtained with the minor component present as droplets or fibrils dispersed within the continuous matrix phase. Such a structure development is largely random in nature. This approach is less expensive and easier in comparison to the generation of bicomponent fibers. As discussed earlier, the blend composition, rheological properties of the blend components, mixing history, and interfacial properties between the blend components control the ultimate morphology and performance of such fibers. A fine and uniform dispersion of the multiphase blend is usually desired prior to the melt-spinning step. Therefore, multiple extrusion steps prior to the generation of fibers are not uncommon. It follows from earlier discussions that when the viscosity of the minor component is lower than that of the major component, a fine dispersion will be obtained. Also, lowering of the interfacial tension between the blend components will lower the critical Weber number required for droplet deformation. Hence, appropriate use of interfacial agents such as graft or block co-polymers can lead to very fine dispersions.

The well-mixed melt is subsequently forced through the spinnerette to generate fibers. The dispersed droplets experience elongational stresses as they enter the spinnerette hole (typical diameter of spinnerette hole in melt-spinning $\sim$ 0.2–0.6 mm). These forces tend to elongate the dispersed phase droplet into a fibril. Again, the extent of elongation and the stability of the dispersed phase droplet depends on the rheological properties of the polymers, interfacial tension and spinnerette hole geometry. Under conditions of shear, droplet deformation to yield high aspect ratio fibrils is favored when the viscosity of the dispersed phase is comparable or lower than that of the matrix, and when the interfacial tension between the phases is low. However, droplet deformation in extensional flow occurs at a much lower critical Weber number and is less dependent on viscosity ratio in contrast to shear flow.

As the melt leaves the spinnerette, extensional forces exerted by the fiber take-up device can contribute further to structure formation within the blend fiber. The spinline forces enhance deformation of the dispersed phase and can also increase molecular orientation within the elongated droplet. The importance of molecular orientation is especially realized in thermoplastic fibers reinforced with thermotropic liquid crystalline polymers, where a small increase in the orientation within the TLCP fibrils results in a dramatic increase in the tensile properties of the blend fiber. Solidification of the fiber usually occurs within a few feet from the spinnerette and this locks in the morphology generated. The amount of fiber drawdown also plays a significant role on both the fibril geometry and distribution, as well as on tensile properties.

Blend fibers composed of poly (ethylene terephthalate) [PET] and nylon 6 have been studied in great detail [25,26]. When melt blended and subsequently spun, these fibers exhibit a matrix-fibril morphology with fibril diameters and lengths of the order of 0.1 microns and 100–200 microns, respectively. These fibers revealed higher tensile properties compared to those of the neat fibers [25,26]. Similar studies on blend fibers based on PET and poly (butylene terephthalate) [PBT] were also carried out [27]. PET fibers with tiny fibrils of PBT embedded within them revealed superior tensile and dyeing properties compared to neat PET fibers [27].

Synthetic polymeric fibers can build up significant static electric charges due to their low electrical conductivity. However, some polyethers such as poly (ethylene oxide) [PEO] possess electrical conductivity levels that are several orders of magnitude higher than those of common PET and nylon fibers. Hence, long fibrils of PEO generated within a nylon matrix during melt blending increases the electrical conductivity of nylon fibers [28,29]. The above approach thus reduces the extent of undesirable electrical discharge from nylon fibers.

The influence of interfacial agents on the properties of polypropylene-polyamide 6 matrix-fibril fibers have also been studied. The interfacial agent was found to have a strong impact on both the morphology and mechanical properties of these fibers, with smaller PA6 fibrils and superior mechanical properties evident in fibers with small amounts of the interfacial agent [30].

Blends of nylon 6 and nylon 66 have been commercialized recently for hosiery applications. These blend fibers reveal better transparency, higher tenacity, and easier finish compared to nylon 66 homopolymer fibers [31].

Polybenzimidazole (PBI) and polysulfone (PS) do not form miscible blends. However, in the presence of small amounts of lithium chloride, the blend fibers revealed tensile properties superior to those of PBI fibers [32,33].

It has also been shown that blend fibers with matrix-fibril morphologies are suitable in heat bonding fibers in nonwoven fabric applications compared to sheath-core fibers, which are very complex to produce [34,35]. Examples of blend fibers investigated for heat bonding applications include PET/PE [34] and PP/PVAc [35].

11.5 Thermoplastic Fibers Reinforced with Thermotropic Liquid Crystalline Polymers

11.5.1 Fibers Produced via Simultaneous Melting of Blend Components

Since the early realizations that TLCPs can be used to reinforce thermoplastics under appropriate processing conditions, several researchers have worked on the melt blending of thermoplastics with various TLCPs. It was also realized that higher degrees of extensional forces present in fiber spinning leads to a more efficient reinforcement as compared to processes such as injection molding. The tensile modulus of highly drawn melt-spun TLCP fibers ranges between 40-100 GPa, while those of thermoplastics are typically below 10 GPa. Hence, a vast majority of the tensile properties in thermoplastic/TLCP blend fibers is provided by the reinforcing TLCP phase. Such fibers are often referred to as in situ composite fibers. The properties of in situ composite fibers, wherein the thermoplastic and the TLCP are melt-blended in a single extruder followed by fiber spinning have been reported extensively over the last few years [10,13,36–60]. This more common approach requires the blend components to have overlapping processing temperatures.

When melt blended with thermoplastics, the TLCP melt is usually dispersed as droplets. Fibrillation of the TLCP droplets and hence useful reinforcement of the thermoplastic depends on the effective deformation of the dispersed TLCP droplets into elongated fibrils, thus resulting in a matrix-fibril morphology. This depends primarily on the extent of elongational strains imposed, relative melt viscosities of the blend components, interfacial tension between the phases, and the blend composition. The tensile properties of composite fibers produced thereof are a strong function of the aspect ratio of the reinforcing fibrils. Hence, fibrillation of the TLCP droplets is critical to the performance of in situ composite fibers. Generally, extensional flow fields are more effective in droplet deformation than shear flow. In shear flow, fibrillation of the TLCP droplets is favored when the ratio of the viscosity of the dispersed phase (TLCP) to that of the continuous phase is less than one and when interfacial tension between the blend components is minimal. The average diameter of the TLCP fibrils reported in the literature ranges between 1–5 microns, with the aspect ratio of the fibrils dependent on process history and TLCP composition.

In situ composite fibers reveal a distinct skin-core morphology, with the outer layer (skin) usually displaying higher orientation levels as well as finer and more continuous fibrils [45,50,59]. Also, in thermoplastic/TLCP blend fibers, the ratio of the area of the skin to that of the core region increases with increasing fiber draw ratio [59]. The skin-core morphology has been attributed to higher shear stresses acting on the skin as the melt converges into the die, which in turn promotes fibrillation and orients the TLCP phase to a greater extent [50]. However, the skin-core morphology is not evident in fibers of substantially high draw ratios.

The properties of melt-spun fibers, especially those of TLCP-based fibers are dependent to a great extent on the molecular orientation induced during melt spinning. In determination of molecular orientation by any means, it is essential to specify the orienting unit, be it a chemical bond, chain segment (represented by a vector joining specific atoms in the chain), or the complete chain (represented by the end-to-end vector). If φ is the angle between the vector representing the orienting unit and the fiber axis, the average orientation is characterized by the orientation parameter $<P_2>$, also known as Hermans' orientation factor (or order parameter in liquid crystalline polymers):

$$< P_2 >= \frac{3 < \cos^2\phi > -1}{2} \tag{11.7}$$

In the case of multiphase structures, orientation of the individual phases should be considered separately. TLCP fibers are believed to consist of a paracrystalline array of chains parallel to the fiber axis, with a mosaic of thin crystalline regions formed by matching sequences of the constituent units. Therefore, orientation of the crystalline planes in TLCPs can be used as an estimate of the mesophase orientation.

The tensile properties of in situ composite fibers are a strong function of fiber drawdown. The enhancement of mechanical properties with increasing draw ratio is usually attributed to higher molecular orientation of the reinforcing TLCP fibrils along the spinline direction. Molecular orientation has been reported for TLCP fibers as well as in situ composite fibers using techniques such as WAXS, sonic modulus, and birefringence. The orientation of the TLCP chain segments along the draw direction has been found to increase systematically with increasing drawdown ratio in both in situ composite fibers [46,47] as well as neat TLCP fibers [61–64]; this is consistent with the systematic increase in tensile modulus of these fibers with draw-down ratio. Both the tensile modulus and molecular orientation were observed to level off at very high draw ratios. However, some other studies on neat TLCP fibers [49,64] have shown that maximum orientation was achieved at relatively low draw ratios, while tensile modulus of the same fibers increased continuously up to much higher draw-down ratios. It was speculated that, while very small extensional strains are required for maximum orientation of the TLCP chain segments, elongation of the nematic domains (increase in aspect ratio) could be contributing to the increase in tensile modulus with draw ratio after maximum orientation was achieved [49]. Orientation of the reinforcing TLCP phase was also found to be a strong function of TLCP composition, with molecular orientation increasing with increasing TLCP content in the in situ composite fibers [42–44,59]. This was also a manifestation of the inability to obtain long TLCP fibrils at very low concentrations.

The tensile properties of thermoplastic fibers reinforced with TLCP fibrils can be treated similarly to those of fiber-reinforced composites. Therefore, the Halpin-Tsai equation can be used to estimate the tensile modulus of in situ composite fibers. The modulus of a uniaxially oriented fiber reinforced composite can be expressed as follows:

$$E_{comp} = E_m \frac{1 + 2\eta(L/D)\phi_f}{1 - \eta\phi_f} \tag{11.8}$$

$$\eta = \frac{(E_f/E_m) - 1}{(E_f/E_m) + 2(L/D)} \tag{11.9}$$

where, E represents tensile modulus, subscripts *comp*, *m*, and *f* represent composite, matrix, and reinforcing fiber, respectively. L and D represent the length and diameter of the reinforcing fiber, and φ represents volume fraction. The Halpin-Tsai equation assumes uniform aspect ratio (L/D) of the reinforcing fibers, and continuity of both stress and strain at the fiber-matrix interface. In the limiting case of very high aspect ratios, Equation (11.8) reduces to a rule of mixtures with the contributions of the components weighed on a volumetric basis:

$$E_{comp} = \phi_f E_f + (1 - \phi_f) E_m \tag{11.10}$$

Because the TLCP fibrils within the in situ composite fibers are largely oriented along the axial direction, the Halpin-Tsai equation has been used to predict the tensile modulus of the thermoplastic/TLCP blend fibers. A few models that account for the elastic properties of the TLCP phase as a function of fiber draw ratio have been incorporated into the Halpin-Tsai equations, and the tensile modulus predicted appears to be fairly consistent with experimentally measured values for in situ composite fibers composed of polycarbonate and Vectra B950 [54,59].

11.5.2 Fibers Generated Using the Dual Extrusion Process

Typical in situ composite fiber generation processes involve plastication of both the thermoplastic matrix and the reinforcing liquid crystalline polymer simultaneously in the same extruder [10,13,36–60]. This approach, however, presents a few drawbacks. TLCPs that exhibit good mechanical properties are typically processed at relatively high temperatures that are close to temperatures where degradation of commodity thermoplastics occur. This precludes the potential of reinforcing commodity resins with TLCPs. Also, if the blends are processed just above the melting point of the TLCP, the viscosity of the TLCP can be higher than that of the matrix which inhibits fibril formation and drawing along the spinline. Finally, TLCPs are known to supercool when cooled below their melting point, and the degree of supercooling is a strong function of the difference between the highest temperature it is exposed to relative to its melting point [65,66]. It is desirable to exploit the supercooling behavior of TLCPs to avoid early solidification within the spinline prior to optimal drawdown of the composite fiber. The simultaneous melting of the matrix and the TLCP in a single extruder will constrain utilization of the supercooling nature of TLCPs, as the matrix might degrade at the high temperatures that TLCPs need to be exposed to in order to obtain a reasonably large supercooling window.

A novel dual extrusion process was developed recently which addresses the above-mentioned concerns with simultaneous melting of the matrix and TLCP in a single extruder [67,68]. A schematic of the dual extrusion process is shown in Figure 11.2. In this process, the thermoplastic matrix and the TLCP are plasticated independently in separate extruders. The TLCP melt is supercooled below its melting point and subsequently introduced as continuous streams into the matrix through a phase distribution system (multiple port injection nozzle). The thermoplastic melt with continuous streams of TLCP is then forced through a series of static mixing elements which serve to divide the TLCP streams further. This results in a matrix-fibril morphology. The melt is then subjected to extensional forces as it enters the die opening and the melt is drawn further as it exits the spinnerette. The drawn fibers are quenched in a chimney that is approximately twelve feet in length and subsequently taken up on bobbins, with draw control provided by a godet. The detailed configuration of the process is described elsewhere [69,70].

The above process offers some significant advantages. First of all, it allows the combination of thermoplastics and TLCPs with non-overlapping processing temperatures, as it exploits the supercooling characteristics of TLCPs. This process thus provides for the imposition of independent thermal histories to the different blend streams. Also, injection of the TLCP as continuous streams into the matrix minimizes the dependence on flow-induced deformation (elongation and fibrillation of the dispersed droplets) to yield high aspect ratio fibrils. Most importantly, reinforcement of low melting matrices with high melting TLCPs provides a temperature window in which these fibers can be processed into usable parts by melting only the matrix and keeping the TLCP fibrils intact. Therefore, the high degrees of orientation imparted to the TLCP fibrils during melt spinning can be maintained in products fabricated using such in situ composite fibers. Also, it was found that this process yielded fibers with better tensile properties compared

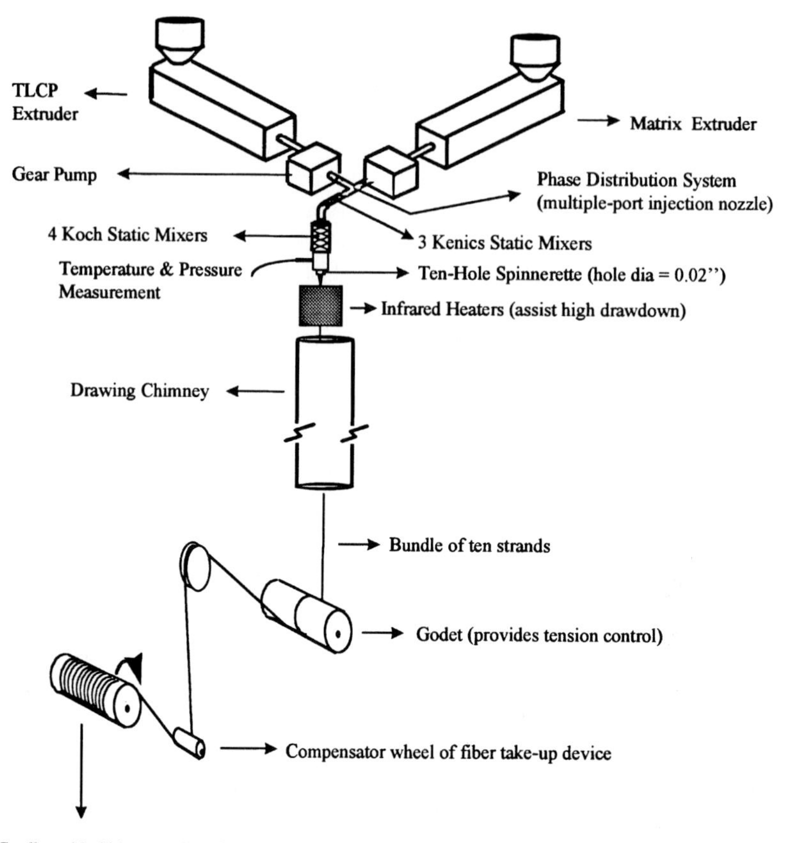

Figure 11.2 Schematic of the dual extrusion process.

to those generated via simultaneous melting of the blend components even when the constituents had overlapping processing temperatures [69].

In situ composite fibers based on nylon-11 and a hydroquinone-based liquid crystalline polyester (DuPont TLCP, HX8000) are chosen as an example to demonstrate the applicability and specific advantages of the dual extrusion process. Further details regarding the processing and properties of these fibers are reported elsewhere [71]. Figure 11.3 shows the cooling behavior of both nylon-11 and HX8000, plotted as complex viscosity versus temperature. Nylon-11 was cooled from 290 °C, while HX8000 was cooled from 310 °C. Both materials were cooled at the rate of 5 °C/min. The calorimetric melting temperature is indicated as dotted lines along the complex viscosity-temperature curve for both nylon-11 and HX8000. As the nylon-11 melt is cooled, it begins to solidify (crystallize) very close to its melting temperature. However, HX8000 supercools significantly below its melting point of 277 °C, as its complex viscosity begins to increase significantly only at temperatures close to 250 °C. This indicates that HX8000 is processable at temperatures well below its melting point. The extent of supercooling exhibited by HX8000 is significantly greater than that of nylon-11. This supercooling

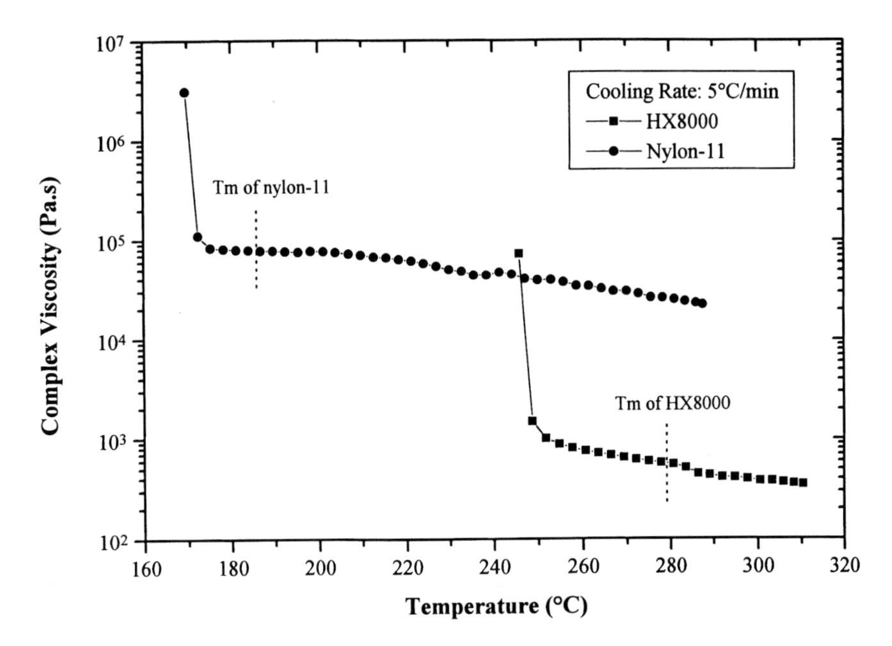

Figure 11.3 Complex viscosities of nylon-11 and HX8000 as a function of temperature, as they are cooled from the melt state.

behavior is consistent with previously reported literature for thermotropic liquid crystalline polymers [65,66]. It has also been shown that the extent of supercooling in TLCPs is a function of cooling rate, with larger supercooling observed for higher cooling rates. The extent of supercooling in TLCPs was found to be greater when the material was heated to temperatures well above its melting point. This behavior has been attributed to incomplete melting of residual crystallites at lower temperatures which in turn provide heterogenous nucleation sites for subsequent crystallization. When TLCPs are exposed to temperatures much above their calorimetric melting point, the residual crystallites melt out completely thus providing a larger supercooling window as compared to a specimen cooled from temperatures relatively close to its melting point [65,66].

Nylon-11 cannot be readily melt-spun at temperatures above 265 °C due to its poor melt strength. Therefore, the supercooling nature of HX8000 was exploited in the processing of composite fibers using the dual extrusion process. HX8000 was melted at 310 °C and supercooled to 260 °C before it was injected into the nylon-11 melt stream. The composite fibers were subsequently melt-spun at 260 °C. Also, without sufficient supercooling, HX8000 can solidify in the spinline prior to optimal drawdown of the composite fiber resulting in relatively low levels of molecular orientation and less than optimal tensile properties. Thus, the independent thermal histories imparted to nylon-11 and HX8000 streams in the dual extruder process could play a significant role in generating composite fibers thereof.

The tensile modulus of these composite fibers is plotted as a function of fiber draw ratio in Figure 11.4. The tensile modulus was found to increase systematically with increasing fiber drawdown, for all fiber compositions. The increase in tensile modulus with draw

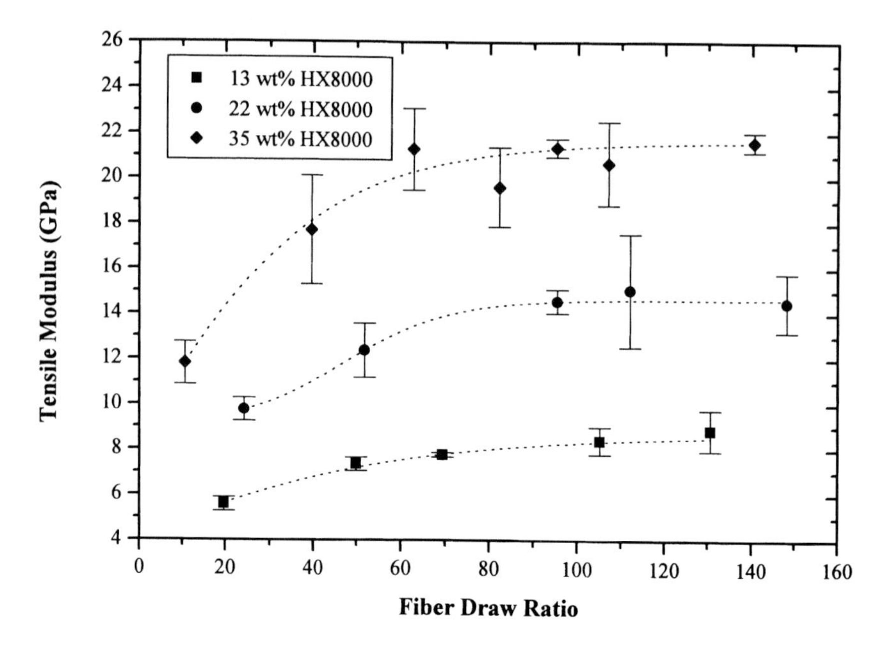

Figure 11.4 Tensile modulus of melt spun nylon-11/HX8000 blend fiber as a function of fiber draw ratio and blend composition. The dotted line is merely a best-fit line, drawn for purposes of clarity in trends.

ratio for the in situ composite fibers can be attributed to enhanced molecular orientation within the reinforcing TLCP phase as well as higher aspect ratios of the reinforcing TLCP fibrils. The tensile modulus of the composite fibers was observed to plateau at higher draw ratios (approximately 60). This is consistent with the trend observed for neat HX8000 fibers [71]. The tensile properties of the composite fibers were also found to depend strongly on the blend composition, with the modulus increasing with increasing TLCP content in the blend. This is a manifestation of the reinforcing potential of HX8000.

It has been shown that addition of small quantities of HX8000 to nylon-11 results in a significant enhancement of tensile properties. The morphology (size and shape) of HX8000 reinforcement was then observed using scanning electron microscopy. Cross-sectional views of the composite fibers (high draw ratio fibers compression molded uniaxially at 195 °C) are shown in Figure 11.5 as a function of blend composition. The reinforcing HX8000 phase was found to be largely fibrillar in nature. A distribution of HX8000 fibril sizes (diameters) is evident in these micrographs. The reinforcing fibril diameter was observed to be in the range between 0.1 micron to 3.0 microns in composite fibers with 20 wt% HX8000 and in the range between 0.1 micron to 4.5 microns in composite fibers with 35 wt% HX8000. The influence of draw-down ratio on the morphology of the composite fibers (20 wt% HX8000) is shown in Figure 11.6. These micrographs reveal HX8000 fibrils embedded in the nylon-11 matrix, and they clearly indicate that higher drawdown results in fibrils of smaller diameters. Also, high draw fibers reveal a narrower distribution of fibril diameters.

The aspect ratio of the reinforcing HX8000 fibrils in the composite fibers are inherently very high as the dual extrusion process basically injects continuous streams of the TLCP

(a)

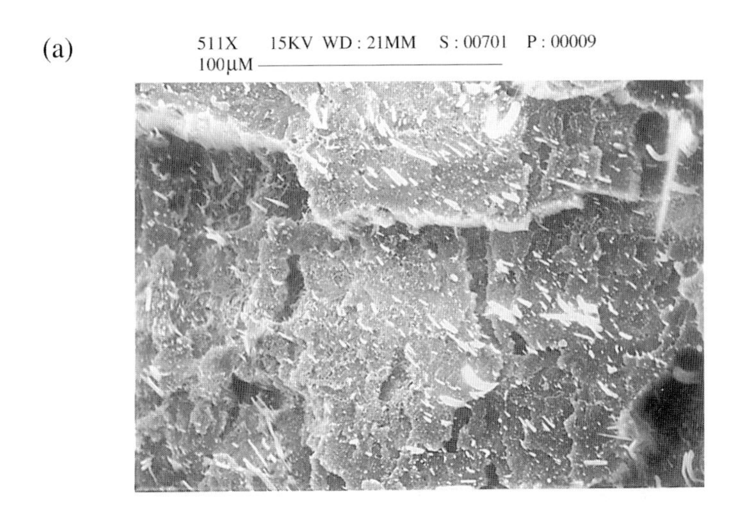

(b)

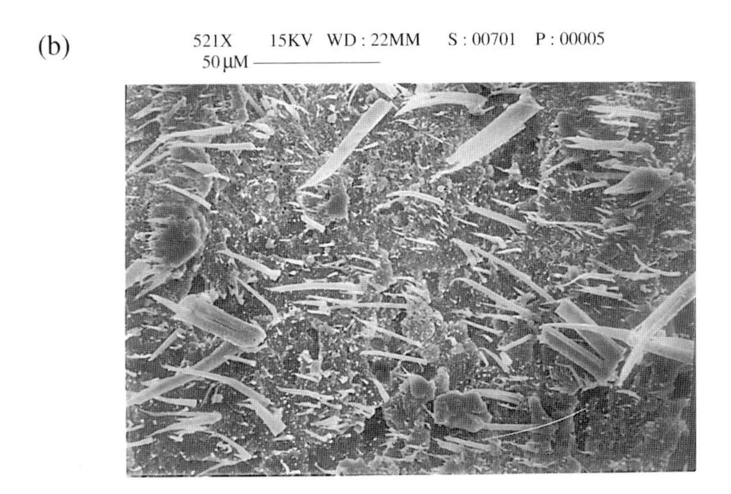

Figure 11.5 Scanning electron micrographs showing the cross section of uniaxially molded composite fibers. (a) Nylon-11/HX8000 80/20 wt%; (b) Nylon-11/HX8000 65/35 wt%.

into the matrix. Thus, the rule of mixtures (Equation 11.10) can be used to predict the modulus of composite fibers based on the tensile properties of the blend components (pure nylon-11 and pure HX8000 fibers). Likewise, it should be possible to back-calculate the average modulus of the reinforcing HX8000 fibrils in the composite fibers based on the measured tensile properties of neat nylon-11 and composite fibers and the blend composition. The contribution of HX8000 to the modulus of the composite fibers is shown in Figure 11.7, plotted as the back-calculated modulus of HX8000 fibrils in the composite fiber, as a function of fiber draw ratio and blend composition. The dotted line in the plot represents the tensile modulus measured for neat HX8000 fibers. It is clearly

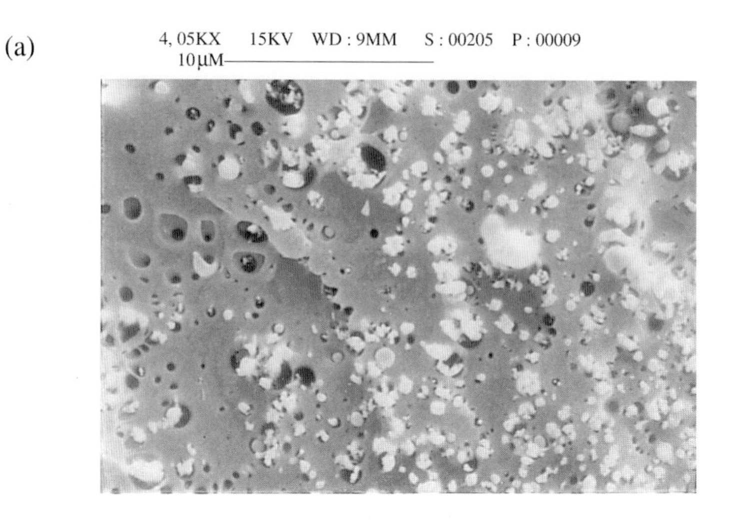

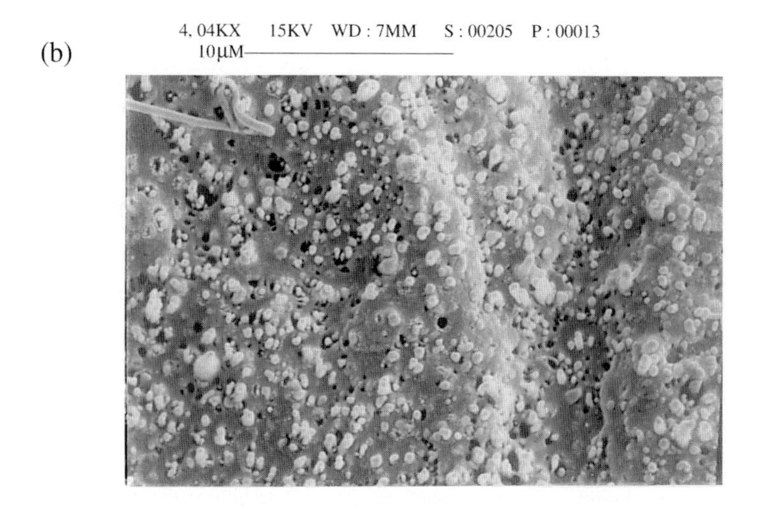

Figure 11.6 Scanning electron micrographs showing the cross section of a composite fiber (Nylon-11/HX8000 80/20 wt%) as a function of fiber draw ratio. (a) DDR = 20; (b) DDR = 140.

evident that the average modulus of the HX8000 fibrils in the composite fibers is significantly greater than that measured for the pure TLCP fiber. This indicates that the HX8000 fibrils are orienting more readily when drawn within the thermoplastic melt as compared to drawing of neat HX8000 fiber [75].

It has been shown that the tensile modulus of TLCP fibers is a strong function of molecular orientation, especially at very high orientation levels where the modulus increases exponentially with molecular orientation [72]. Therefore, small increases in molecular orientation within the HX8000 fibrils in the composite fiber relative to that of

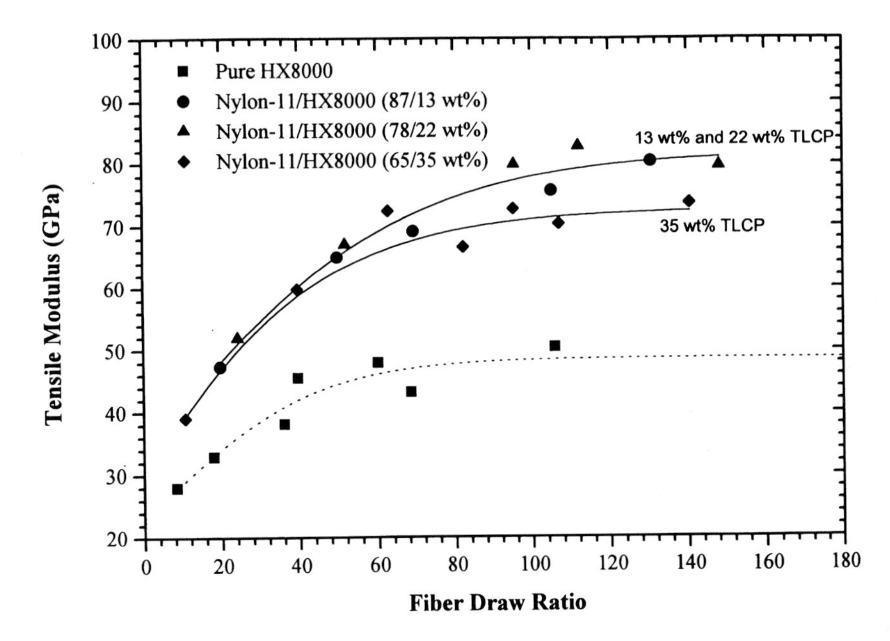

Figure 11.7 The average tensile modulus of HX8000 fibrils in composite fibers (back-calculated from Halpin-Tsai equation) as a function of draw ratio and blend composition. The dotted line is the tensile modulus measured for neat HX8000 melt-spun fibers.

neat HX8000 fiber can result in significant enhancements in tensile properties. The continuous TLCP streams are encapsulated by the matrix melt during melt-spinning. Therefore, the matrix can act as thermal insulation to the TLCP fibrils. This will prevent premature solidification of the fibrils during melt-spinning, thus enhancing molecular orientation within the fibrils as the distance over which extensional forces deform the TLCP fibrils is higher relative to the melt-spinning of neat TLCP fibers. Also, when fibrils within a neat TLCP fiber are orienting along the spinline direction in response to elongational forces, these fibrils have limited reorganizational freedom as they are constrained by neighboring fibrils. However, TLCP fibrils encapsulated by a thermoplastic melt do not encounter similar constraints. Hence, TLCP fibrils within an in situ composite fiber can orient more readily along the fiber drawing direction. The maximum contribution of the HX8000 fibrils in the composite fibers appears to decrease somewhat at higher TLCP concentrations. This is consistent with the above hypothesis as interactions between neighboring fibrils will be greater in fibers with higher TLCP concentration. This kind of synergism has only been observed in composite fibers generated using the dual extrusion process [71,73,74]. Wide-angle X-ray scattering results are consistent with tensile properties obtained, and they have indicated greater orientation levels within the TLCP fibrils in the composite fibers (generated using the dual extrusion process) relative to neat TLCP fibers of comparable draw ratios.

The properties of polypropylene fibers reinforced with Vectra B using the dual extrusion process have also been reported. These fibers displayed excellent tensile properties in fibers with both sheath-core and matrix-fibril morphologies [70,74]. The matrix-fibril fibers displayed slightly higher tensile modulus values compared to the sheath-core fibers. Numerical simulation of the spinning process suggested that some degree of Vectra B relaxation is possible at the center of the TLCP core in sheath-core fibers, thus resulting in slightly lower modulus values. However, both the matrix-fibril and sheath-core fibers displayed tensile properties that are significantly better than those predicted by composite theory [70,74]. These results are consistent with polypropylene fibers reinforced with other TLCPs [69,73] and nylon-11/HX8000 fibers [71] discussed previously.

Poly (phenylene sulfide) [PPS], a high performance engineering thermoplastic, was also reinforced with Vectra B using the dual extrusion process [75]. The tensile properties of PPS/Vectra B (50/50 wt%) fibers are shown in Figure 11.8. These values are again better than those predicted using composite theory. The tensile modulus of a PPS/Vectra A fiber (draw ratio ~ 30; ref [45], extrapolated to 50/50 wt% from four compositions with TLCP amounts lower than 50 wt%) based on simultaneous melting of both the matrix and the TLCP in a single extruder followed by fiber spinning is also shown in the figure (the tensile modulus of neat Vectra A fibers is very nearly equal to that of Vectra B fibers; hence, a direct comparison was possible). The tensile modulus of TLCP reinforced PPS fiber generated using the dual extrusion process is more than twice that of the fiber obtained via simultaneous melting of the blend components in the same extruder. This indicates that the dual extrusion process is capable of reinforcing thermoplastic matrices to a significantly greater extent than conventional means, and is consistent with the results obtained by Sabol [69].

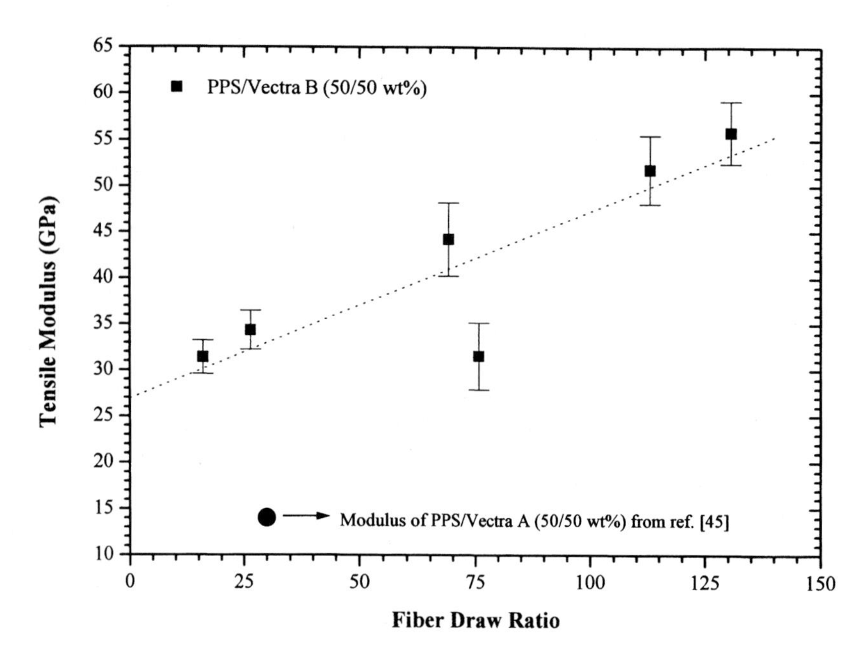

Figure 11.8 Tensile modulus of PPS/Vectra B (50/50 wt%) blend fibers as a function of draw ratio. The dotted line is merely a best-fit straight line, drawn for purposes of clarity in trends.

The properties of PET fibers reinforced with various TLCPs using the dual extrusion process can also be found [76]. The challenge of reinforcing PET with TLCPs is to find an appropriate TLCP which can accommodate the narrow processing window of PET. To accomplish this, two different TLCPs were melt-blended prior to the dual extrusion spinning process. The TLCPs (DuPont TLCPs HX8000 and HX6000) were melt-blended to overcome melt strength and supercooling window drawbacks that were imminent with the neat TLCPs. As the TLCP/TLCP blend was cooled from temperatures above their melting points, possible transesterification between the TLCPs yielded a single, composition-dependent solidification point that was intermediate to those of the homopolymers. Thus, appropriate TLCP reinforcement with desirable melting and cooling characteristics could be tailor-made via blending. Here again, the dual extrusion process allows one to achieve an appropriate TLCP/TLCP blend to reinforce various matrices. Such an approach would not be possible in situations where the matrix and the reinforcing TLCP/TLCP blend are melted simultaneously in a single extruder. All of the above results indicate that the dual extrusion process is far more effective in generating high modulus in situ composite fibers, than those obtainable via simultaneous melt-blending of the matrix and TLCP in a single extruder.

Other investigators have attempted to impose independent thermal histories to TLCPs and thermoplastic matrices prior to the generation of in situ composite strands [77,78]. The two melt streams were mixed only immediately prior to a section that contained several Ross static mixing elements. They found the approach to be particularly useful in reinforcing low-temperature matrices with high melting TLCPs. Their process, however, does not use a multiple port injection nozzle to introduce continuous streams of the TLCP into the matrix. This is the principal difference between their work and the process developed by Baird and co-workers [67,68,71,73,74]. Machiels and co-workers used the above-mentioned approach to reinforce thermoplastic elastomers [77] and polypropylene [78] with Vectra A. The tensile properties of PP/Vectra A fibers obtained by Machiels et. al. [78] are significantly inferior to those obtained by Baird and co-workers [73,74]. Machiels and co-workers [78] did not fully utilize the supercooling characteristics of Vectra A in their co-extrusion processing as the maximum temperature the TLCP was exposed to was 300 °C. Also, the utilization of a multiple-port injection nozzle to introduce continuous streams of TLCP into the matrix melt prior to fiber spinning can contribute significantly to the greater tensile properties obtained by Baird and co-workers [67,68,71,73,74].

11.5.3 Post Processing of Fibers Generated Using the Dual Extrusion Process

Post-processing of the in situ composite fibers is essential in order to translate the TLCP reinforcement imparted in the melt-spun fibers into useful products such as automobile floor panels, doors, hoods, and trunks. Because the dual extrusion process allows reinforcement of low-melting matrices with higher melting TLCPs, these fibers can be processed at temperatures above the melting point of the thermoplastic matrix, but well below the melting point of the TLCP. This allows retention of the high degrees of molecular orientation imparted to the TLCP phase during the melt-spinning step in the final product. Uniaxial compression molding of in situ reinforced fibers (based on polypropylene reinforced with various TLCPs) at temperatures just high enough for the matrix to flow yielded plaques whose tensile properties were very close to those of the individual strands [73,74]. Also, compression molding of short fibers in a random configuration revealed properties close to those predicted using composite theory [73]. These results highlight the post processing potential of in situ composite fibers generated by the dual extrusion process.

Pelletized in situ composite strands (generated by the dual extrusion process) based on polypropylene (PP) reinforced with various TLCPs were subsequently processed at temperatures below the melting point of the TLCP to generate plaques using injection molding, films via sheet extrusion, and bottles via extrusion blow molding [79–81]. While the TLCP fibrils were assumed to be intact during the post processing step, the mechanical properties of the plaques and films were significantly lower than predicted using composite theory. This loss in properties was attributed to a reduction in the aspect ratio of the reinforcing fibrils via fibril agglomeration and fracture of individual fibrils during post-processing. However, these specimens exhibited lower anisotropy compared to in situ composites generated in a single processing step. A more recent study [82] considered injection molding of polypropylene strands that were reinforced with HX8000 using the dual extrusion process. A modified injection nozzle with a larger opening was used to minimize shear effects at the nozzle tip. Injection molding was carried out at various temperatures and injection speeds. It was shown that lower processing temperatures and lower injection speeds yielded plaques with higher tensile properties. When pelletized strands of low draw ratio were injection molded at the lowest temperatures and moderate injection speeds, plaques with tensile properties comparable to those predicted by composite theory were obtained (based on random distribution of the TLCP fibrils along the plane of the plaque) [82]. The mechanical properties of these injection plaques were isotropic along the plane of the plaque and were superior to those generated under conditions wherein both the matrix and the TLCP were melt-blended during the injection molding process [82]. However, when strands of higher draw ratios were injection molded under similar conditions, the properties of the plaques obtained did not reveal noticeable improvements over those of the low draw strands; this is in contrast to the tensile properties of the individual strands themselves [82].

While post-processing of in situ composite fibers generated using the dual extrusion process is carried out at temperatures well below the melting point of the TLCP, the temperatures are above the glass transition of the TLCP. Hence, some degrees of orientation relaxations could be occurring that leads to less than optimum reinforcement. Also, at higher injection speeds in the injection molder, the shear in the nozzle tip and the sprue-runner assembly could destroy the TLCP fibrils to a reasonable degree. Thus, lower injection speeds yielded better retention of properties from strands to injection molded plaques. The mechanical properties of these injection-molded plaques approach those of glass-filled resins. However, injection molding of TLCP-reinforced resins inflicts less wear and tear on the equipment, and the resulting samples are lighter and have significantly smoother surfaces compared to those of glass-filled systems [82].

If the composite fibers are flexible enough, they can be woven into a continuous fabric which in turn will serve as a composite preform. These woven preforms can be consolidated into composite parts, with the consolidation step carried out at temperatures above the melting point of the thermoplastic matrix, but well below the melting point of the TLCP to minimize relaxation of the high degree of orientation imparted to the reinforcing TLCP fibrils during the melt spinning step. These woven preforms have been generated with fibers based on PP and Vectra B [83], as well as those based on nylon-11 and HX8000 [71]. Such composite plaques exhibited modulus (tensile and flexural) values ten times that of pure PP with approximately 31 wt% Vectra B [83], and five and one half times that of nylon-11 with approximately 35 wt% HX8000 [71]. The flexural properties of PP were also significantly enhanced when woven preforms were preferentially placed closer to the surface during the consolidation step [83]. The mechanical properties of representative composite plaques based on fabric preforms are included in Table 11.1. These properties are better than those based on continuous glass fiber mats, with comparable volume fraction of reinforcement [83].

Table 11.1 Mechanical (tensile and flexural) properties of composites from woven fabric preforms based on blend fibers generated using the dual extrusion process. Several layers of fabric preforms were stacked and consolidated at temperatures just high enough for the matrix to melt, but well below the melting point of the TLCP. Numbers in parenthesis indicate standard deviations of the measurements.

Fibers that constitute fabric preform	Tensile Modulus (GPa)	Tensile Strength (MPa)	Flexural Modulus (GPa)	Flexural Strength (MPa)
PP/Vectra B	10.3	63.8	10.0	65.6
(68/32 wt%) ref [83]	(2.2)	(4.5)	(1.7)	(7.3)
Nylon-11/HX8000	5.18	80.78	5.67	71.70
(65/35 wt%) ref [71]	(0.33)	(6.21)	(0.38)	(5.62)

11.6 Copolymers

Copolymers are polymers derived from more than one kind of monomer. A structure-based definition of copolymers indicates that the backbone structures of copolymers consist of more than one kind of repeat unit. Copolymers are usually classified based on the distribution of the different repeat units along the backbone. *Statistical copolymers* are copolymers in which the sequential distribution of the repeat units follow certain statistical laws. *Random copolymers* represent a fraction of statistical copolymers in which the sequential distribution of the repeat units is absolutely random. *Alternating copolymers* are made up of only two different kinds of repeat units along the backbone, and they are arranged alternately along the backbone. *Block copolymers* are linear chain copolymers in which each of the repeat units exist only in long sequences or blocks, as they are often referred to. *Graft copolymers* are branched polymers in which the side branches from the main chain are structurally different as compared to the main chain.

Statistical, random and alternating copolymers usually display properties intermediate to those of the corresponding homopolymers; for example the above copolymers display a single glass transition temperature which is a function of the copolymer composition. Therefore, these copolymers offer a means to combine the properties of different polymers in a single material. On the other hand, block and graft copolymers usually display properties characteristic of each of the constituent homopolymers. However, due to covalent bonding between the sequences in block and graft copolymers, the sequences do not behave independently of each other. Unlike multiphase blends, covalent bonding in block and graft copolymers prevents immiscibility that will lead to phase separation on a macroscopic scale. Therefore, some synergism is possible in block and graft copolymers.

Copolymerization has been used extensively to alter the properties of homopolymers in the synthetic fiber industry. Dyeing of PET fibers is an area where copolymerization has had an immense effect. Grafting acrylic acid onto PET was found to enhance the dyeing characteristics of polyester fibers substantially [84,85]. Various dyes can be incorporated into PET fibers depending on the choice of the graft segment [84,85]. Grafting acrylamide onto PET increases moisture regain of PET fibers [86]. The flame retardency of PET fibers was enhanced via copolymerization with halogenated co-monomers.

Aromatic polyimides are high performance materials known for their remarkable thermal and oxidative stabilities, excellent dimensional stability, excellent electrical and mechanical properties that can be sustained over long periods of time [87]. However, their backbone rigidity and interchain interactions makes them intractable. Therefore, flexible amide linkages are incorporated onto their backbone to make polyimides melt-processable. The properties of melt-spun aromatic polyamide-imide and other copolyimide fibers are well documented [88].

Random copolyesters based on p-hydroxybenzoic acid (p-HBA) and 2, 6-hydroxynaphthoic acid (2, 6-HNA) display liquid crystalline behavior and are marketed under the tradename *Vectra*. The HNA segment provides excellent melting point depression and because the interlocking naphthalene rings only offset the polymer chain linearity without changing its direction (crankshaft mechanism) there appears to be relatively little loss of liquid crystallinity or fiber properties [89, 90]. In addition, the wholly aromatic structure of this particular crankshaft formulation provides greater thermal stability compared to TLCPs with flexible spacers introduced in the backbone. The structure and properties of TLCP fibers are discussed in chapter 7.

11.7 Elastomeric Fibers

Elastomeric fibers are characterized by a very high elongation at break (several hundred percent), low modulus, and high recovery from large deformations [91,92]. Early elastomeric fibers were based largely on rubber-based materials. However, in recent years, elastomeric fibers based on polyurethane chemistry have dominated. The most important of these fibers are called *spandex* fibers which consist of at least 85% polyurethan.

Spandex fibers consist of segmented polyurethanes, which are basically alternating block copolymers of the type $(A\text{-}B)_n$ in which one of the chain segments (*soft segment*) provides the coilable-alignable material and the other segment (*hard segment*) gives long-range intermolecular attraction.

The soft segment domains are random-coiled aliphatic polyethers or copolyesters and they constitute up to 60–90 % by weight. They are relatively unoriented in the relaxed fiber, and they uncoil and align when extended. They tend to crystallize when extensional forces are applied, which provides stiffness to the chain and fiber. When the elongational forces are released, the oriented segments recover to their equilibrium state; i.e., the soft-segment crystallites melt and the chains recoil with a force derived largely from entropy change. The length of the soft segments dictates the extent of stretchability of the fiber.

Aromatic diisocyanates play a critical role in hard segment formation. They are responsible for the step-wise buildup of the blocky structure, allowing the hard segments to form between preformed soft segments. They provide an economical advantage due to their ability to condense rapidly with other intermediates (glycols and diamines) with little or no heat and with no elimination product. The hard segments are usually aromatic-aliphatic polyureas. Urethane linkages link the hard segments to the soft. During fiber formation, the hard segments from several chains associate into *tie-point* domains that act as crosslinks in a three-dimensional network. The nature of these associations are hydrogen bonds between the NH groups and carbonyls. Crystallization of the aromatic moieties can also provide additional integrity to the copolymer. The urea hard segments constitute less than 25% by weight. This ensures higher stretchability and low modulus. The copolymer structure can degrade at elevated temperatures and, therefore, these copolymers cannot be melt-spun. Solution spinning and reaction spinning are common. Important structure-property relationships include the influence of backbone make-up on thermal transitions, block segregation and resultant morphology, orientation and crystallization of the soft segments, and extent of interchain bonding between the hard segments.

In recent years, considerable interest has developed in thermotropic liquid-crystalline urethane polymers. During melt-processing of *thermoplastic* polyurethane elastomers, microphase separation between the hard and soft segments is often incomplete, leading to less-desirable morphologies. However, processing from a nematic (mesophasic) hard segment phase provides a high degree of phase separation between the hard and soft segments, resulting in better mechanical properties [93]. Liquid crystalline polyurethanes are usually synthesized by the polyaddition of a diisocyanate with a mesogenic diol in solution. The diol can be partially replaced by a polyether diol (flexible spacer). The polyaddition is carried out at a NCO/OH ratio of 1:1. Highly polar solvents are typically used. The product is isolated from the solution by pouring into water or precipitation with methanol.

The transition temperatures of the liquid crystalline polyurethanes are a strong function of the copolymer backbone structure [94]. Decrease in the amount of the mesogenic unit results in a reduction of the melting temperature, as well as the mesophase-isotropic transition temperature. Increase in the molecular weight of the flexible spacer at a constant content of the mesogenic unit exerts the same effect. Generally, the higher the molecular weight, the higher are the transition temperatures. However, the molecular weight of the liquid crystalline segment has a more pronounced effect on the transition temperature than the molecular weight of the flexible spacer. Other structure-property investigations on liquid crystalline polyurethanes based on various chemistries can be found [95–98]. The greatest advantage of these liquid crystalline polyurethanes over traditional elastomeric fibers is their melt-processability.

11.8 References

1. L. A. Utracki, *Polymer Alloys and Blends: Thermodynamics and Rheology*, Hanser Publishers, New York (1989)
2. L. A. Utracki, *Polymer Engineering and Science* (1995) 35, p. 2
3. D. R. Paul and S. Newman, *Polymer Blends Vol I,* (1978) Academic Press, New York
4. L. M. Robeson, "Advances in Polymer Blend Technology", from: *Contemporary Topics in Polymer Science, Vol. 6: Multiphase Macromolecular Systems* (1989) Plenum Press, New York
5. D. J. Walsh, "Polymer Blends", in *Comprehensive Polymer Science, Volume II: Polymer Properties* (1989) G. Allen, editor, Pergamon Press, New York.
6. C. E. Scott and C. W. Macosko, *Polymer* (1995) 36, p. 461
7. C. D. Han, *Multiphase Flow in Polymer Processing* (1981) Academic Press, New York
8. S. Wu, *Polymer Engineering and Science* (1987) 27, p. 335
9. S. Datta and D. J. Lohse, *Polymeric Compatibilizers: Uses and Benefits in Polymer Blends* (1996) Hanser Publishers, New York
10. G. Kiss, *Polymer Engineering and Science* (1987) 27, p. 410
11. K. G. Blizard and D. G. Baird, *Polymer Engineering and Science* (1987) 27, p. 653
12. R. A. Weiss, W. Huh, and L. Nicolais, *Polymer Engineering and Science* (1987) 27, p. 673
13. A. I. Isayev and M. Modic, *Polymer Composites* (1987) 8, p. 158
14. D. G. Baird, "In Situ Thermoplastic Composites", from *Polymeric Materials Encyclopedia* (1996) J. C. Salamone editor, v5, 3207, CRC Press, New York
15. G. G. Viola, D. G. Baird, and G. L. Wilkes, *Polymer Engineering and Science* (1985) 25, p. 888
16. A. Kohli, N. Chung, and R. A. Weiss, *Polymer Engineering and Science* (1989) 29, p. 573
17. A. Datta, H. H. Chen, and D. G. Baird, *Polymer* (1993) 34, p. 759
18. H. J. O'Donnell and D. G. Baird, *Polymer* (1995) 36, p. 3113
19. A. Datta and D. G. Baird, *Polymer* (1995) 36, p. 505
20. D. G. Baird and A. Datta, U.S. Patent 5,621,041 (1997)
21. R. K. Krishnaswamy, S. E. Bin Wadud, and D. G. Baird, *Polymer*, (1999) 40, p. 701
22. R. K. Krishnaswamy, D. G. Baird, and S. E. Bin Wadud, *U.S. Patent Pending*
23. D. R. Paul and S. Newman, *Polymer Blends Vol II* (1978) Academic Press, New York
24. E. M. Hicks, J. F. Ryan, R. B. Taylor, and R. L. Tichenor, *Textiles Journal* (1960) 30, p. 675
25. P. V. Papero, E. Kubu, and L. Roldan, *Textile Research Journal* (1967) 37, p. 823
26. B. T. Hayes, *Chemical Engineering Progress* (1969) 65 (10), p. 50
27. R. Gutmann and H. Herlinger, *Science and Technology of Fibers and Related Materials; Journal of Applied Polymer Science: Applied Polymer Symposium* (1991) 47, p. 199
28. E. E. Magat and D. Tanner, US Patent, 3,329,557 (DuPont) (1967)
29. L. W. Crovatt, US Patent, 3,388,104 (Monsanto) (1968)
30. I. Grof, O. Durcova, and M. Jambrich, *Colloid and Polymer Science* (1992) 270, p. 22
31. N. Koizumi and K. Okajima, *Polymer* (1996). *In Press*
32. T. Chung, M. Glick, and E. J. Powers, *Polymer Engineering and Science* (1993). 33, p. 1042
33. T. Chung, *Polymer Engineering and Science* (1994) 34, p. 428
34. W. Xiao and X. Zang, *Synthetic Fibers in China* (1993) 3, p. 17
35. W. Xiao, *Journal of Applied Polymer Science* (1994) 52, p. 1023

36. F. P. La Mantia, F. Cangialosi, U. Pedretti, and A. Roggero, *European Polymer Journal* (1993) 9, p. 671
37. A. M. Sukhadia, D. Done, and D. G. Baird, *Polymer Engineering and Science* (1990) 30, p. 519
38. R. A. Weiss, W. Huh, and L. Nicolais, *Polymer Engineering and Science* (1987) 27, p. 684
39. M. R. Nobile, E. Amendola, L. Nicolais, D. Acierno, and C. Carfagna, *Polymer Engineering and Science* (1989) 29, p. 244
40. K. G. Blizard, C. Federici, O. Federico, L. L. Chapoy, *Polymer Engineering and Science* (1990) 30, p. 1442
41. B. R. Bassett and A. F. Yee, *Polymer Composites* (1990) 11, p. 10
42. G. Crevecoeur and G. Groeninckx, *Polymer Engineering and Science* (1990) 30, p. 532
43. S. M. Hong, B. C. Kim, K. U. Kim, and I. J. Chung, *Polymer Journal* (1991) 23, p. 1347
44. C. Carfagna, E. Amendola, L. Nicolais, D. Acierno, O. Francescangeli, B. Yang, and F. Rustichelli, *Journal of Applied Polymer Science* (1991) 43, p. 839
45. M. T. Heino and J. V. Seppala, *Journal of Applied Polymer Science* (1992) 44, p. 2185
46. M. Kyotani, A. Kaito, and K. Nakayama, *Polymer* (1992) 33, p. 756
47. D. Dutta, R. A. Weiss, and K. Kristal, *Polymer Composites* (1992) 13, p. 394
48. J. X. Li, M. S. Silverstein, A. Hiltner, and E. Baer, *Journal of Applied Polymer Science* (1992) 4, p. 1531
49. W. C. Lee, A. T. Dibenedetto, J. M. Gromek, M. R. Nobile, and D. Acierno, *Polymer Engineering and Science* (1993) 33, p. 156
50. D. Beery, S. Kenig, A. Seigmann, and M. Narkis, *Polymer Engineering and Science* (1993) 33, p. 1548
51. G. Crevecoeur and G. Groeninckx, *Journal of Applied Polymer Science* (1993) 49, p. 839
52. H. Verhoogt, C. R. J. Williams, J. V. Dam, and A. P. De Boer, *Polymer Engineering and Science* (1994) 34, p. 453
53. S. L. Joslin, R. Giesa, and R. J. Farris, *Polymer* (1994) 35, p. 4303
54. Q. Lin and A. F. Yee, *Polymer Composites* (1994) 15, p. 156
55. S. Joslin, W. Jackson, and R. Farris, *Journal of Applied Polymer Science* (1994) 54, p. 289
56. J. He, W. Bu, and H. Zhang, *Polymer Engineering and Science* (1995) 35, p. 1695
57. Z. S. Petrovic and R. Farris, *Journal of Applied Polymer Science* (1995) 58, p. 1349
58. S. Mehta and B. L. Deopura, *Journal of Applied Polymer Science* (1995) 56, p. 169
59. C. L. Choy, K. W. E. Lau, Y. W. Wong, and A. F. Yee, *Polymer Engineering and Science* (1996) 36, p. 1256
60. H. S. Moon, J. Park, and J. Liu, *Journal of Applied Polymer Science* (1996) 59, p. 489
61. D. E. Turek and G. P. Simon, *Polymer* (1993) 34, p. 2763
62. A. Kaito, M. Kyotani, and K. Nakoyama, *Journal of Macromolecular Science–Phys. Ed.* (1995) B34, p. 105
63. T. S. Chung, *Journal of Polymer Science–Polym. Phys. Ed.* (1988) 26, p. 1549
64. K. Itoyama, *Journal of Polymer Science–Polym. Phys. Ed.* (1988) 26, p. 1845
65. D. Done and D. G. Baird, *Polymer Engineering and Science* (1987) 27, p. 816
66. D. Done and D. G. Baird, *Polymer Engineering and Science* (1990) 30, p. 989
67. D. G. Baird and A. M. Sukhadia, U. S. Patent 5,225,488 (1993)
68. A. M. Sukhadia, A. Datta, and D. G. Baird, *International Polymer Processing* (1992) 7, p. 218
69. Sabol, M. S. Thesis, Virginia Polytechnic Institute and State University (1994)
70. C. R. Robertson, M. S. Thesis, Virginia Polytechnic Institute and State University (1994)
71. R. K. Krishnaswamy and D. G. Baird, *Polymer Composites* (1997) *In Press*
72. T. S. Chung, *Journal of Polymer Science–Polym. Phys. Ed.* (1988) 26, p. 1549
73. E. Sabol, A. A. Handlos, and D. G. Baird, *Polymer Composites* (1995) 16, p. 330
74. C. G. Robertson and D. G. Baird, *International Polymer Processing* (1998) 12, p. 354
75. D. G. Baird, J. P. deSouza and C. G. Robertson, US Patent Application in Process (1997).
76. M. A. McLeod, Ph.D. Dissertation, Virginia Tech (1997)
77. A. G. C. Mahiels, K. F. J. Denys, J. Van Dam, and A. P. DeBoer, *Polymer Engineering and Science* (1996) 36, p. 2451
78. A. G. C. Mahiels, K. F. J. Denys, J. Van Dam, and A. P. DeBoer, *Polymer Engineering and Science* (1997) 37, p. 59
79. A. A. Handlos and D. G. Baird, *Journal of Macromolecular Science, Macromol. Chem. and Phys.* (1995) 135, p. 183
80. A. A. Handlos and D. G. Baird, *Polymer Composites* (1996) 17, p. 73
81. A. A. Handlos and D. G. Baird, *Polymer Engineering and Science* (1996) 36, p. 378

82. R. K. Krishnaswamy, J. Xue, and D. G. Baird, *J. Adv. Materials. To Appear*
83. C. G. Robertson, J. P. deSouza, and D. G. Baird, in *Recent Advances in Liquid Crystalline Polymers* (1996) A. I. Isayev, T. Kyu, and S. Z. D. Cheng eds., ACS Books, Washington D.C.
84. Z. Yao and B. J. Ranby, *Journal of Applied Polymer Science* (1990) 39, p. 1459
85. E. Oflaz and M. Sacak, *Journal of Applied Polymer Science* (1993) 42, p. 1909
86. E. Pulat and M. Sacak, *Journal of Applied Polymer Science* (1989) 38, p. 539
87. D. Wilson, H. D. Stenzenberger, and P. M. Hergenrother, *Polyimide* (1991) Blackie, Glasgow (UK)
88. M. G. Desitter and R. Cassat, *Synthetic Fibre Materials* (1994) H. Brody, editor, Longman, Harlow
89. G. W. Calundann and M. Jaffe, "Anisotropic Polymers: Their Synthesis and Properties", *Address to Welch Foundation Conference on Chemical Research*, Houston (TX) (1982).
90. W. J. Jackson, *Journal of Applied Polymer Science* (1985) 41, p. 25
91 M. Couper, "Polyurethane Elastomeric Fibers", from *Handbook of Fiber Science and Technology* (1985) vol VI, M. Levin and E. M. Pearce, editors, Marcel Dekker, New York
92. C. D. Eisenbach, K. Fischer, H. Hayer, H. Nefzger, A. Ribble, and E. Stadler, "Polymeric Elastomers, Segmented", in *Polymeric Materials Encyclopedia* (1996) Joseph C. Salamone editor, v9, p. 6957, CRC Press, New York
93. D. J. Gerbi, C. W. Macosko, W. Mormann, and S. Benadda, *Polymer Preprints*, ACS-Div. of Polym. Chem. (1992) 33, p. 1109
94. P. Penczek and B. Szczepaniak, "Liquid Crystalline Polyurethanes", in *Polymeric Materials Encyclopedia* (1996) Joseph C. Salamone editor, v5, p. 3779, CRC Press, New York
95. A. C. Griffin and Haven, *Journal of Polymer Science–Polym. Phys. Ed.* (1981) 19, p. 951
96. V. Frosini, S. dePetris, E. Chiellini, G. Galli, and R. W. Lenz, *Mol. Cryst. Liq. Cryst.* (1983).98, p. 223
97. P. J. Stenhouse, E. M. Valles, S. W. Kantor, and W. J. MacKnight, *Macromolecules* (1989) 22, p. 1467
98. S. K. Pollack, D. Y. Shen, S. L. Hsu, Q. Wang, and H. D. Stidham, *Macromolecules* (1989) 22, p. 551

12 Thermomechanical Processing: Structure and Properties

Jerold M. Schultz
University of Delaware, Newark, Delaware, USA

12.1 Introduction

In the production of fibers from flexible chain polymers, multiple processes of spinning, drawing and heat-setting may be used to produce a desired set of properties. These properties – e.g., thermal stability, dyeability, bulk, strength and stiffness – are defined by the structure of the material. "Structure" here refers to a broad spectrum: molecular structure and conformation, crystalline and noncrystalline arrangements of chains (and packings intermediate to these extremes), crystallite sizes and shapes, and the mutual arrangement of crystalline and noncrystalline regions.

It is to be understood from the outset that spinning and subsequent thermomechanical processing work together to define the final state of the fiber. For example, the details of the heat-setting operation and the degree to which it can be effective at imparting the desired structure, properties or form to the fibers depends upon the structural state of the fiber already established during the spinning, or spinning and drawing, process. In attempting to understand this area, one must be guided by developments in three distinct but closely related areas: (1) commercial heat-setting technology and its development, (2) relatively fundamental studies of structure, structure development and behavior in oriented polymers, and (3) research in which commercial treatments have been simulated in smaller-scale experiments aimed at understanding specific portions of the commercial process. To begin with, it is essential to have some understanding of commercial practice. This background defines for us what properties are to be controlled and provides a benchmark index of what has been found to be effective. The lab-scale structure development studies provide some understanding of the processes which occur on a fine scale within the material. And the simulated heat-treatment studies attempt to bridge the gap between practice and fundamental science. This chapter might be considered an abbreviated progress report on the degree to which the three areas begin to provide the detailed understanding we would like regarding the heat-setting process.

Some questions to be addressed are these:

1. What are commercial heat-treatments designed to do, how are they carried out, and what are their end products?
2. To what extent can we define and understand phenomenologically what transpires in heat-setting operations?
3. What is our understanding of the structural changes brought about in heat-setting, and how do these structural changes relate to behavior?
4. At what rates do the processes occur, and how are these rates related to processing variables?

12.2 Commercial Practice*

12.2.1 General Comments

The structure of a melt-spun fiber in its final form in a commercial product depends on its total history. The first influences occur during the continuous operations up to the first

*Section 12.2 was co-authored with Professor John W.S. Hearle.

break in industrial processing, when a solid fiber is wound up or otherwise collected. The effects involved are polymerization, molecular rearrangements in the melt, changes in extrusion and cooling, and the actions on the solid fiber until it is collected. This chapter is concerned with the molecular reorganization that occurs in the subsequent thermo-mechanical processing. For example, the hot-stretching of yarns to increase strength and reduce extensibility; the heat-setting of crimp in yarns to provide bulk and stretchability; the heat-setting of creases or pleats in fabrics. All these effects were discovered early in the development of synthetic fibers, and led respectively to nylon and polyester tire cords, to "textured" yarns, and to the "ease-of-care", "wash-and-wear" revolution in clothing. Optimization of these processes, both for improved properties and faster production speeds, is a continuing goal, requiring understanding of the underlying microstructural mechanisms taking place.

In order to understand current industrial thermo-mechanical processing, it is necessary to know the starting points. Whereas detailed discussions on structure formation in spinning and drawing are given in earlier chapters, a brief summary of many variants of industrial practice is as follows. Until about 1970, melt-spun yarns were wound up slowly at 1000 m/min or less in an unoriented form, which would extend plastically by about 4X. Nylon was about 50% crystalline but polyester was non-crystalline. Drawing to remove the plasticity, introduce molecular orientation and give useful textile fiber properties was carried out in a separate operation, which was combined with the slower process of inserting a small amount of twist to give coherence to the yarns. The drawing process caused crystallization of the polyester yarns. Thus, oriented, semi-crystalline yarns were the raw material supplied by the fiber producers to the textile industry. The situation was changed by two developments. Firstly, interlacing, to give coherence, could operate at a higher speed than twisting and could be introduced in the fiber formation process. Secondly, higher-speed wind-ups were developed. Spin-draw processes, in which undrawn spun-yarn was taken-up on a first set of rollers and the drawing stage followed through a second set of higher-speed rollers, were introduced. Alternatively, the extruded yarn can be taken directly to higher-speed rollers, which induces orientation in the molten thread-line. Wind-up at about 3000 m/min gives partially oriented yarns (POY), which require drawing by about 1.5X. In polyester, POY has a low or incipient crystallinity, which is sufficient to stabilize the yarn for transport and subsequent use, in contrast to less stable undrawn yarns, which needed to be drawn immediately after spinning. Wind-up at higher speeds, around 5000 m/min, gives oriented, crystalline yarns without appreciable plasticity, which can be used directly in textile fabrics for some uses, though they are low-strength, high-extension forms.

12.2.2 Consumer Textiles

In staple fiber production, large tows, which contain millions of filaments, are drawn, given some crimp in passing through stuffer boxes, and then cut and baled. This is the starting-point for the production of "spun" yarns and nonwoven fabrics composed of short fibers. Heat-setting may occur in later yarn or fabric processing. Other variants of the initial fiber formation stages occur in the production of spun-laid nonwovens composed of continuous filaments.

The production of textured yarns, which account for the largest part of the use of continuous filament yarns in consumer textiles, is a major industrial operation, which involves thermo-mechanical fiber processing. The first manufactured "synthetic" yarns were used in smooth, lustrous, silk-like fabrics, but this was a limited market. There was a need to mimic the bulkier and rougher fabrics made from natural fibers like cotton and wool. In addition to the improved feel and appearance of the fabrics [1], the bulk provides thermal insulation through entrapment of air. One route is to cut filaments into short fibers and make yarns on conventional cotton or wool spinning machinery. The alternative is to heat-set the filaments in contorted forms, so that the yarns have bulk, stretch and texture. Stuffer-box and edge-crimping were processes that became successful in the 1950s, but are no longer used. Knit-de-knit, which results in yarns set in the form of the knit loops, is still used to a small extent. The process, which became dominant, is twist texturing.

Although there was some early work on viscose and acetate yarns, it was the ability of nylon to be heat set, that led to the success of twist texturing. High-stretch nylon yarns, which were known as torque-stretch yarns and were useful in socks and other close-fitting garments, were produced. The first *Helanca* process involved the separate steps of imparting high twist to a yarn, giving twist angles close to the maximum stable angle of 70.5°, heat-setting in an autoclave, and untwisting. In fact, due to various technical problems, eight individual operations were needed, but, despite the cost, the market was strong. The application of false-twisting speeded up the process. The principle is illustrated in Figure 12.1. The twister effectively twists the yarn as it comes onto the pin and untwists it as it leaves the pin [2]. Between the input drive roller and the twister, the yarn is set in the twisted form by passing over a heater and then into a cooling zone. Between the twister and the output roller, the yarn is a set of parallel and untwisted, but potentially crimp-and-torque-lively, filaments.

The first false-twist machines provided yarn speeds of about 20 m/min. Today, friction spindles permit speeds of 1000 m/min, which gives about a tenth of a second on the heater. The other important development, which directly involves the thermo-mechanical actions, is the use of POY as the supply yarns. In the simultaneous process, the output rollers are run faster than the input rollers, so that the yarn is drawn by the residual draw ratio of about 1.5X as it enters the heater, as well as being twisted. In the alternative sequential process, the draw was effected by pairs of rollers before the entry to the twist zone.

For nylon 66, the process is effective with heater temperatures, or more strictly yarn temperatures, in the range 180 to 230 °C. Values over 210 °C are normally used to give a strong permanent set. This, in combination with cooling, gives a "permanent" set to the twisted yarn. Higher temperatures cause the filaments to stick together. For nylon 6, the temperatures are lower. The yarn is wound on a package under tension while it is still warm. Temporary set, on going through the glass transition temperature, means that when yarn is gently pulled off the package it is a flat continuous filament yarn without bulk or stretch. The yarn must be activated by the removal of temporary set, in order to develop its textured state. In textile manufacturing, subsequent mechanical treatments of loose fabric develop the bulk.

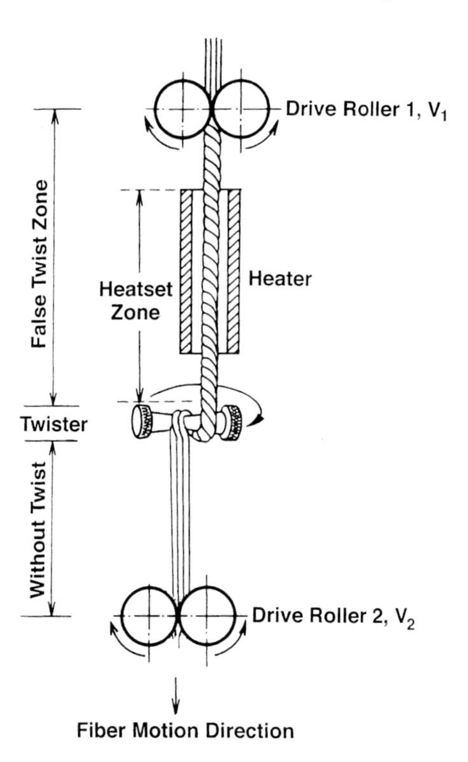

Figure 12.1 Schematic illustration of a false-twisting device. The yarn is twisted as it proceeds through the heater and is untwisted after passing the heater [2].

The forms which the filaments take up, when the temporary set is released, depend on the geometry in the yarn when the permanent set is imposed. Untwisting the yarn also reverse-twists the filaments from their set state, and consequently generates a torque in filaments held straight under tension. As is easily shown by twisting rubber rods, a torque-lively filament will buckle into 'pig-tail' snarls as tension is reduced and the ends are allowed to come closer together. Such pig-tails are found in torque-stretch yarns when they are released, and lead to large contractions from the extended length. The resulting yarns consequently have a stretch of up to 500%. However the filaments are not straight when they are set. They are following helical paths in the twisted yarn. Furthermore the helix radius alternately increases and decreases as the filaments migrate from the center to the surface of the yarn, in order to equalize path lengths. The mechanics problem is thus to determine how a helix of varying radius will buckle when it is untwisted and allowed to contract. The initial form is dominated by the attempt of the filaments to reach their bent forms. As occurs in helical telephone cords when they are untwisted and released, the form consists of alternating right-and left-handed helices. It is only when larger contractions occur that the pig-tails form.

The low modulus and good recovery from large strains make nylon particularly suitable for high-stretch uses, where it remains the preferred fiber. Polyester is not suitable because of its poorer recovery from large strains, which leads to bagging of stretch fabrics.

For other textile uses, high-bulk with lower stretch are needed, which is best provided by polyester. Double-heater machines [3] are used to produce these textured polyester yarns, in contrast to the single-heater machines used for nylon stretch yarns. The first stage has the same sequence of twist-heat-cool as in the single-stage process. In the second stage, beyond the false-twist spindle, the yarn passes over the second heater between two sets of rollers, with speeds allowing a contraction by 10 to 20%. The yarn buckles into the alternating helices, which are heat-set by the heater [3]. Pajgrt and Reichstädter [3] review the many other methods for on- and off-line stabilization.

Industrial experience shows an interesting difference in the behavior of nylon 66 and polyester yarns, though there is a lack of laboratory quantification of the effect. For nylon, it was found that, in order to stabilize the modified yarn, the second heater should be at an appreciably higher temperature than the first heater. In commercial polyester processing, this is not so. The second heater is typically run 20 to 30 °C cooler than the first heater.

Except for the use of flat filament yarns in smooth, lustrous fabrics, the fibers used in apparel textiles are subject to heat-setting in order to give the desired properties of texture, bulk, and stretch. This is accomplished by the stuffer-box setting of tows for staple fibers and by the twist-texturing of filament yarns. One other process, air-jet texturing, depends primarily on the mechanical action of locking-in projecting loops in the yarn, but is usually accompanied by heat-setting.

Many more heat-setting steps may occur in the subsequent production of apparel, some incidental and some to force the material into a required form. Dyeing, either of yarns on packages or of fabrics, is a high-temperature process. Finishing operations, such as calendering, which are intended to set overall fabric dimensions, to flatten fabrics and remove wrinkles, or to change surface appearance, involve heat-setting. Finally, garments are set to shape, and may require permanent pleats to be formed.

Nylon is a currently the preferred fiber for carpets and upholstery (although polytrimethylene terephthalate may start to compete in this area soon). The yarns are thicker than those used for apparel. Part of the production uses staple fibers spun on worsted machinery, when the fibers will have been set by crimping of tow. Bulked continuous filament (BCF) yarns are also extensively used. These yarns are processed by over-feeding the yarns in to a hot jet and then blasting them onto a rotating drum, where they cool and are set in crimped forms. The yarns, of either type, are then twisted to form the two-ply yarns used in carpets. The crimp may be temporarily lost. It is regenerated and stabilized by a subsequent thermal process, in which relaxed yarn is passed through machines applying either dry or wet heat. Heat-setting will also occur either in package-dyeing of yarn or piece-dyeing of carpet.

This account has covered the most important of the thermo-mechanical processes applied in the manufacture of consumer textiles. But the industry is highly diverse and other operations will be found in use for special purposes. Practically, it is necessary to take account of both temporary set and what is called permanent set, though more strictly described as a set that is retained at the severity level of subsequent operations. For nylon, it is necessary for subsequent treatments to be more severe, in terms of temperature, stress or moisture condition, in order to give a new permanent set, but for polyester, repeated setting can be achieved at the same or lower temperatures. The account has referred mainly to nylon 66 and poly(ethylene terephthalate), and they will continue to be the focus of this chapter. Other nylons and polyesters will differ in detail in their behavior in industrial thermo-mechanical processing, and there will be greater differences with polypropylene and other polymer fibers. However the commercial processes are generally similar to those described in this section.

12.2.3 Tire Cord and Other Industrial Applications

For industrial fibers and textiles the pattern is different. Major uses are in tires, conveyor belts, and ropes. For these applications, it is desirable to maximize strength and stiffness. Frequently, thermal shrinkage in subsequent processing should be minimized. These properties are easily measured and, within limits, controlled. Fatigue resistance, which must also depend on fiber structure, is also needed, but is not so easy to evaluate, except in use. The many other industrial applications may have other performance requirements. The right set of properties is mostly achieved by processes that involve combinations of hot-stretching and relaxing and are carried out in the fiber manufacturing plants. The patent literature is filled with examples of thermo-mechanical treatments, which are intended to impart improved properties to tire-cords and related yarns. Methods to continuously produce high strength fibers without high-speed spinning have been disclosed by several inventors [4–10]. The method of McClary [10] for PET fibers will be used here for illustration. The McClary method is sketched in Figure 12.2. In this method, filaments are extruded through a spinneret (6), air-quenched (12) and taken up on the take-up roller (24) for further on-line processing. The linear speed of the take-up roller is between 500 and 3,000 m/min, a range producing "partially oriented yarn" (POY). At this point, one has oriented filaments with no or very little crystallinity. Central to this processing stage is the quench. The utility of a quench, obviating massive crystallization and, consequently conferring easy drawability and the potential for crystallization treatment later, had been established early [11–14]. Continuing the description of the McClary process, roller (30) operates much faster than roller (24) and acts to stretch the filaments to a draw ratio up to 3:1 at room temperature (or any temperature below T_g). Rollers (30) and (28) operate at the same speed and serve to isolate the stress between them. Roller (34) operates faster than roller (28) and again causes drawing of the fiber between these two positions. In this section of the line, the fiber is heat-treated, via the steam jet (32). The temperature of the steam is between 90 degrees below the melting point to the softening point of the filaments. Roller (36) again isolates the stress and the fiber is finally taken up on roller (44). Thus the fibers are sequentially and continuously spun, cold-drawn, and heat-treated during a second draw. Similar heat-treatments, independent of the spinning step, have been patented in the U.S. [15] and Japan [16,17]. The McClary treatment produces PET fibers with initial elastic moduli in the range 11.8–13.6 GPa (100–115 g/den) and tenacities (tensile strengths) of about 1 GPa ($\sim$8.8 g/den).

Some variants of commercial practice have appeared in recent years. Abhiraman and coworkers [18,19] have described a process in which the orientational state of partially oriented PET fibers is first fixed through a low temperature crystallization (120–150 °C)

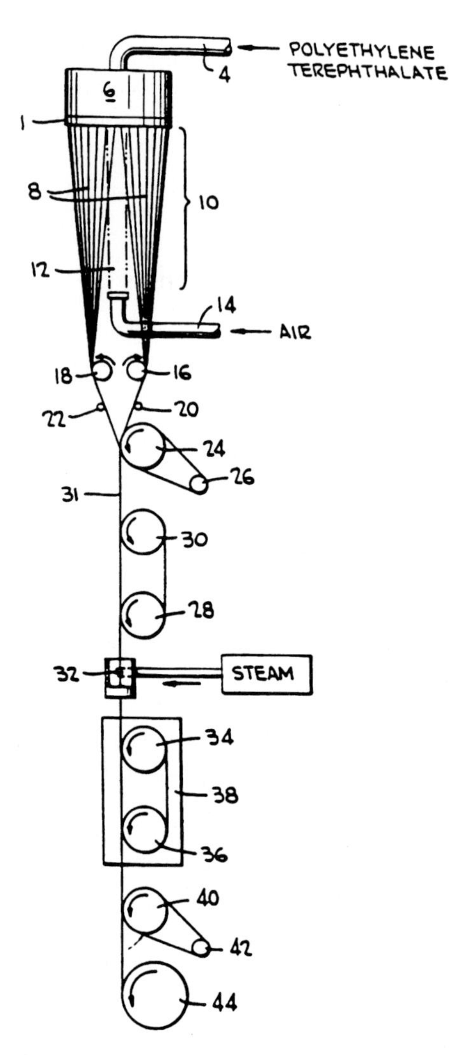

Figure 12.2 Schematic of McClary's method for heat-setting polyester fibers [10]. See text for description.

at constant length. The fibers are then subjected to a high temperature (230–250 °C) draw, in which the initial crystallites are melted and then recreated in a more perfect and more highly oriented state [18]. A high initial modulus, 15.3 GPa (130 g/den), and good tenacities, 940–1060 MPa (8–9 g/denier), are obtained by this process. As a standard of comparison, Itoyama has reported the highest modulus yet for PET [20]. He heat-treated drawn film under 1.25X extension for 15 days at 260 °C. The modulus of this material is 35.4 GPa.

After yarns have been supplied by the fiber manufacturers, there may be many subsequent thermo-mechanical processes involved in the production of tires and other products. Tire cords, either as single ends or in fabric, undergo a hot dipping process, which also functions as a heat-setting step. However, the most important thermo-mechanical processing, which incidentally but significantly affects the fibers, occurs during the application of molten rubber, its curing, and the forcing of the tire into the required size and shape. In the manufacture of conveyor belts, woven fabrics are stabilised in required dimensions by treatment in a stenter. Further thermo-mechanical processing may occur in coating or impregnation. Some synthetic fiber ropes are heat-set as a final manufacturing operation. These are just a few of the thermo-mechanical processes that are employed in the manufacture of industrial textiles and influence fiber structure and properties.

12.3 Phenomenology of Thermomechanical Treatments

12.3.1 Mechanical Behavior During Processing; Creation of Internal Stress; Heat-Setting

12.3.1.1 Conceptual Background

When a polymer fiber is subjected to elongational stress, chain segments in the noncrystalline regions become oriented preferentially toward the draw axis. This orientation lowers the entropy of these regions and creates an elastic force resisting the orientation. This is the usual physics associated with rubberlike elasticity [21]. If the applied stress is sufficiently high, plastic flow can take place locally throughout the fiber. Figure 12.3 provides a conceptual picture of this inhomogeneous deformation. The plastic flow produces the large macroscopic draw. However, even while the drawing is taking place, the stress is maintained and continues to be resisted by elastically strained noncrystalline regions. If the fiber is now unloaded, much of the drawn length is maintained by the plastically extended regions, which now hold the elastically deformed regions under strain, and consequently under stress. This, broadly, is the nature of the internal stress found in drawn fibers. If the temperature of the system is now elevated, the entropic restoring force (-TS) of the elastically oriented chains increases and can become high enough that the combination of this restoring force and the softening or destruction of the plastically deformed regions permit the extended chains to re-coil. What is described here is the process of shrinkage*. Two technological questions arise: (1) What is the stress level needed to produce plastic flow of fibers at temperatures of interest and (2) what is the level of internal stress following drawing? In the conceptual model just

* It is interesting to note that prominent polymer scientists had suggested in the 1960's that chain-folding during crystallization was responsible for shrinkage [22–24]. The basis for this suggestion was the observation that crystallization always accompanies shrinkage in heat-treated material. However, Wilson later showed that this observation was an effect of prolongation of the heat-treatment; most shrinkage occurs *before* chain-folded crystallization takes place [25].

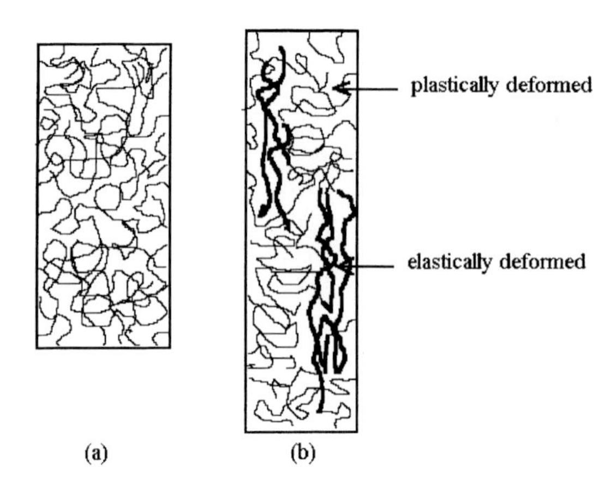

Figure 12.3 Polymer chains before and after drawing: (a) Initial state: homogeneous system of randomly coiled molecules. (b) After spinning: inhomogeneous system with more or less randomly coiled regions and more oriented regions.

described, the (plastic) flow stress of the fiber should be very nearly identical to the level of internal stress created in the drawing process. Is this really the case? And what is the effect of draw temperature on the flow stress and on residual stresses?

In commercial processing, it is the strain which is controlled, not the stress, and stress is normally not even monitored. Thus one looks for off-line measures of the yield stress and the flow stress. The picture becomes complicated for the measurement of mechanical behavior at higher temperatures. At higher temperatures some degree of stress relaxation can take place if the specimen resides at temperature before the load is applied. A striking example of the contrast in behavior is the 1959 experiment of Thompson [26]. He wrapped fibers around two rollers, the effective take-up roller rotating much faster than the feed roller. A hot plate was held against the fiber between the rollers to maintain a temperature of interest. Two stringing procedures were used. In one, the fiber was first wrapped on the feed roller, then passed over the hotplate and finally wrapped around the take-up roller. In the other, the fibers were first wrapped on the rollers and then the hotplate brought in contact. In both cases, the system was allowed to come to steady state and the stress measured. The stress generated in the second procedure is some 40 times that generated in first! In the first case, the rubberlike chains were able to re-coil without restraint before the temperature and strain were applied; the drawing process could then operate on relaxed material which had already been exposed to heat for a sufficient time. In the second case, re-coiling of the elastically extended chains was inhibited (but not eliminated) because macroscopic shrinkage could not occur. Thompson noted that for the first case drawing occurred gradually over the length of the hotplate, whereas in the second case drawing was concentrated near the beginning of the heating. Most standard methods of measurement (e.g., using a constant strain rate tensile tester) are similar to Thompson's first case, in that time is available for large-scale relaxation at temperature before deformation takes place. Commercial practice, on the other hand, conforms much more closely to Thompson's second method.

12.3.1.2 Stress-Strain Measurement at Temperature

In principle, one could measure the mechanical behavior of fibers using Thompson's second method directly. To obtain the equivalent of a stress-strain curve in this case, one would have to continuously vary the velocity difference between the rollers – and thereby the draw ratio – and continuously measure the stress response. Such measurements have not been reported. And the procedure would at any rate precondition the fiber at lower (or higher) draw ratio at temperature before measurement at any specific draw ratio, a condition similar to Thompson's first case.

To combat preconditioning and relaxation problems, Hristov et al have used the device shown in Figure 12.4 [27]. Here a weight insufficient to cause plastic deformation at room temperature is hung from a fiber. Liquid metal at the elevated temperature of interest is suddenly splashed over a section of the fiber. The specimen comes to temperature in 10 to 15 ms and maintains this temperature for some 300 ms. In this period, the length x of the fiber is continuously measured, using an LVDT. From Newton's second law,

$$M(d^2x/dt^2) = Mg - F \tag{12.1}$$

where M is the mass of the pendant weight, g is the acceleration due to gravity and F is the resistant force. Thus a stress-strain curve can be obtained using the second derivative of x at each extension x. Results for a PET yarn which had been spun at 1,450 m/min are shown in Figure 12.5. In Figure 12.5a we see stress-strain curves obtained using a specific

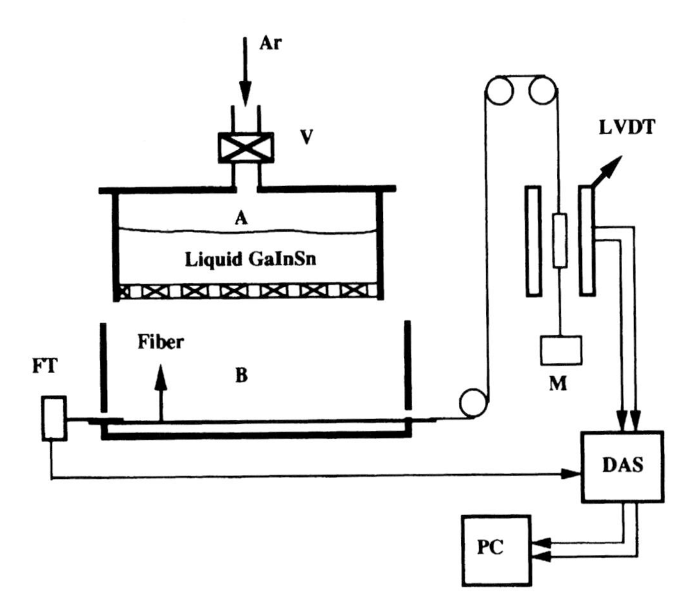

Figure 12.4 Schematic illustration of a device used for stress-strain studies of fibers at high temperatures [27]. Here Ar gas under pressure rapidly squirts a heated liquid metal alloy through holes in container A onto fiber in B. The fiber has been previously loaded at room temperature with the mass M and now extends at the higher temperature. The extension is monitored versus time as the solenoid to which the fiber is attached moves through a coil (a linear variable differential transformer–LVDT).

load at three different temperatures. The drawing stress is in the range of 5 to 15 MPa. Figure 12.5b shows the drawing force at 20% elongation as a function of temperature and load. Following a region of simple plastic flow, the material exhibits a region of extreme work-hardening, with stresses in the range of tens of MPa. These results are compared below to shrinkage stresses.

Other intriguing results emerge in the work of Hristov et al [27]:

1. The critical strain for the onset of work-hardening is in the range 100 to 150%.
2. This degree of plastic flow appears to be associated with necking or other localized deformation.
3. Necking is not necessarily associated with crystallization.

The last result is interesting of itself, in that other investigators had proposed that the large viscosity increase associated with crystallization would be necessary to stabilize the neck and prevent viscous failure [28–31]. It is certainly true, though, that large work-hardening must be requisite for the stabilization of a neck. Conversion of randomly coiled

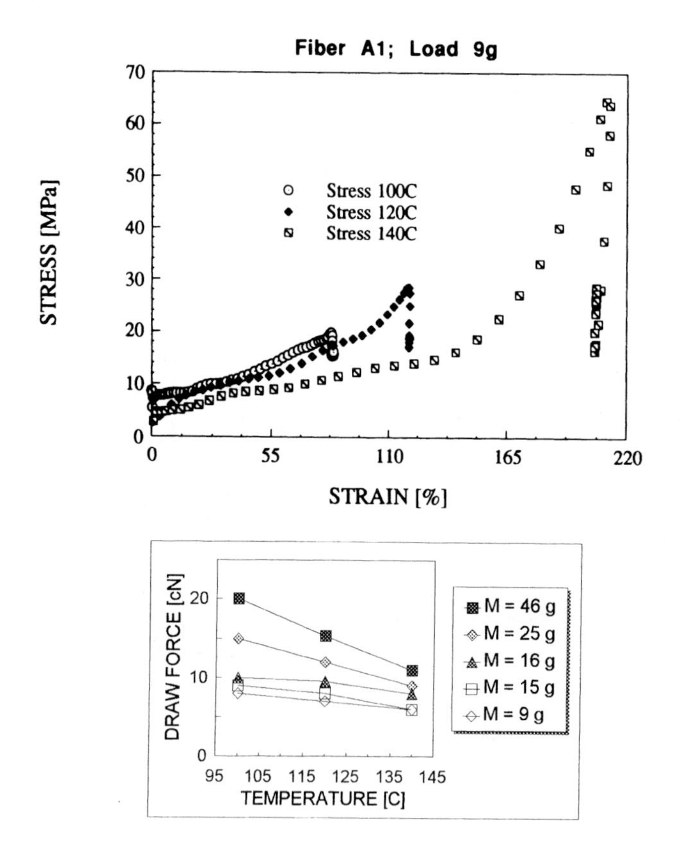

Figure 12.5 Results of fiber stress-strain experiments using the apparatus of Figure 12.5 [27]: (a) stress-strain curves taken using a 9 g load at temperatures 100, 120 and 140 °C, (b) draw force versus temperature for five different loads.

chains to locally oriented amorphous zones or to an oriented mesophase would work as well as crystallization to increase the modulus over the cross-section and thereby stabilize the neck.

Hristov et al propose a conceptual framework for strain localization. They point out that plastic flow normally demands the disentangling of mutually entangled chains. At sufficiently low stresses and sufficiently high temperatures chains can gradually disentangle and produce viscous flow. As the temperature decreases and/or the axial stress increases, such disentangling becomes increasingly more difficult, the result being that individual chains cannot fully disentangle. It is thus required that the chains neighboring any orienting chain likewise deform. Following the model proposed by Argon [32,33] and by Peterlin [34], one has then a zone in which orientation is maintained over some spatial range and the viscous flow associated with freely mobile chains is not possible. At high strain rates or low temperatures the flow is thus localized to clusters of chains which deform cooperatively. This localization leads to fine shear bands and these are the precursors of necks or crazes. We envision then that when the inverse of the deformation rate (a time) is smaller than the main relaxation time of the system disentanglement cannot become complete and localized deformation, leading ultimately to necking, must occur.

12.3.1.3 Shrinkage Stress

The level of internal stress in an oriented material is identified with the stress required to obviate shrinkage at a temperature at which shrinkage should occur. This shrinkage stress can be measured for a specimen held at constant length during rapid heating, using a thermomechanical analyzer. The shrinkage stress in an unset fiber which had been spun at 3500 m/min and subsequently drawn 1.364X is 40 MPa (0.34 g/den) [35]. This stress level is 2–3 times the drawing stress described in Section 12.3.1.2 and indicates that the deformation must have occurred in the region of high work-hardening (see Figure 12.5). Given the hypothesis that necking and work-hardening are correlated, one must conclude that the major shrinkage stresses are formed during the rapid deformation in the neck.

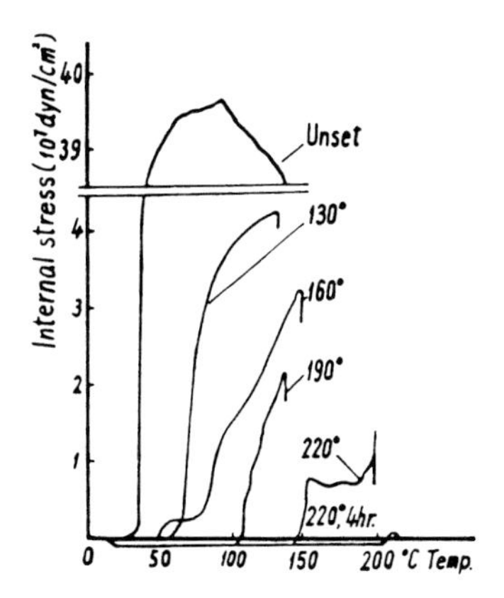

Figure 12.6 Shrinkage stress versus temperature for fibers which have been heat-set at various temperatures [35].

Figure 12.6 shows the results of Zhang et al on the effect of heat-setting temperature on the magnitude of the shrinkage stress present in PET fiber which had been spun at 3500 m/min, subsequently drawn 1.364X and finally heat-set [35]. We note the large decrease in shrinkage stress with increasing heat-setting temperature, an indication that whatever had held amorphous regions under strain has been released. Finally, the high level of stress in the amorphous regions is also experienced by the crystallites and is measured as lattice strain. Zhang et al report lattice strains along the three crystallographic axes for PET. The level of lattice strain along the chain axis is in the range 10^{-2}–10^{-3}, decreasing with increasing setting temperature. Simultaneous to the extension along the chain axis (c), the a and b axes also dilate, corresponding to a net unit cell dilation of some 14% in the drawn state (and decreasing toward zero with heat-treatment temperature and time).

12.3.1.4 Heat Setting and Recovery

Buckley and Salem [36], in an imaginative set of experiments and interpretations, characterize phenomena and suggest a detailed microstructural model for heat-setting.

In their experiments, PET monofilament is pretreated at fixed length in the untwisted state by heating at temperature T_p for time t_p. After this pretreatment, the monofilament is clamped at room temperature, and one clamp is twisted by N_i rotations. The fiber is held at constant length in this twisted condition and subjected to a heat-setting temperature T_s for time t_s. After quenching, the fiber is cooled back to room temperature and allowed to untwist. During recovery, the filament untwists N rotations. The fractional recovery, $f = N/N_i$ is reported. Figure 12.7 shows the take-off point of their results, the shape of the plot of fractional recovery versus heat-setting temperature. Shown here are two curves, one for a filament with no pretreatment and one for a filament pretreated at 200 °C for 35 minutes. The untreated filament exhibits a rapid drop

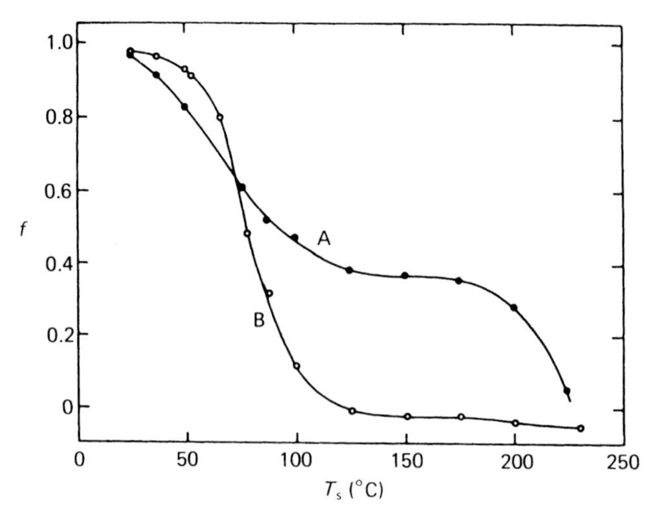

Figure 12.7 Fractional recovery f of PET fibers after heat-treatment at temperatures T_s [36]. Curve B is for as-drawn fiber; the fiber of curve A has been heat-treated for 35 min at 200 °C.

in f over a temperature range about the glass transition (ca. 70 °C). The pretreated filament exhibits a drop in f in this range, but plateaus before the fractional recovery reaches zero and then drops again over a higher temperature range. The lower temperature drop in f represents simply the response of those uncrystallized chains whose state of constraint permits them to relax. The higher temperature process represents the relaxation of noncrystalline chains that are constrained by the microstructure – presumably crystallites – which had prevented complete relaxation near T_g. It was also found that the high temperature heat-setting process is highly time-dependent, and that the rate of decay of f with time decreases with increasing pretreatment temperature.

Time of pretreatment at a specific T_p elevates the plateau level, while leaving the temperature range of the second recovery process unchanged The effect of pretreatment temperature T_p on f vs. T_s curves is shown in Figure 12.8. The effect of increasing the pretreatment temperature is to extend the plateau to higher temperature, while decreasing the level of the plateau. Over the range of pretreatment temperatures used, it is expected that the degree of crystallinity will be increased. The degree of constraint on the uncrystallized chains should then likewise increase. In this way, the observed increase in the temperature of the second recovery range (and the decrease in the rate of decay of f) can be qualitatively explained. The decrease in the plateau level with increasing T_p is, however, more difficult. Why should the *level* of heat-stabilization decrease with increasing crystallinity? Buckley and Salem suggest that a second effect of heat-treatment, namely the reduction in the number of intercrystallite tie-chains, is important. As the crystallinity increases during the pretreatment, the crystals thicken. During crystal thickening and crystal chain folding, the number of intercrystalline tie-chains are reduced

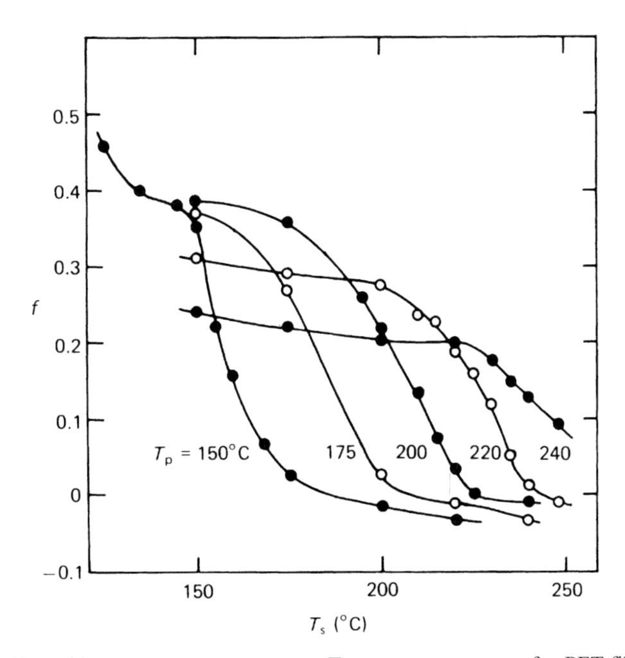

Figure 12.8 Effect of heat-treatment temperature T_p on recovery curves for PET fibers [36].

as these become incorporated into the thicker crystals. One is then left with a looser crystallite network with a lower modulus and a lower resistance to recovery in the plateau region. While this explanation is reasonable, there is no direct evidence for it.

12.3.2 The Development of Orientation of the Crystalline and Amorphous Substituents

In 1964 Krigbaum and Roe [37] studied the orientation in crosslinked polychloroprene which had been crystallized from the melt under different levels of uniaxial extension. They found that the crystalline orientation was always greater than that of the as-drawn, noncrystalline precursor, while the orientation of the amorphous material is lower than that of the as-drawn material. Further, the crystal orientation increased very rapidly initially with draw ratio, thereafter leveling off, while the amorphous orientation registered a low initial increase with draw ratio. This study was followed somewhat later by Krigbaum and Taga, in a linear polymer, isotactic polystyrene, with even more careful measurement of the development of orientation [38]. The same trend of results was reported. It was later shown that oriented PET fibers also crystallize such that the crystal orientation is greater than that of the precursor noncrystalline material [22–24].

Krigbaum and Roe [37] suggested that the very high orientation in the crystallites is due to a conformational selection during nucleation. They proposed that the ν segments of a chain strand entering a nucleus must all be oriented strongly toward the draw axis. If p_ψ is the probability that a segment is oriented at angle ψ to the draw axis, then $p_\psi\nu$ is the probability that ν segments in sequence all lie at angle ψ. Because of the high exponent in the probability that a chain strand enter a nucleus, the probability becomes strongly biased toward the most probable orientations, i.e., those lying closest to the draw axis. Since the most highly oriented chains are removed from the initial population to form the initial crystallites, the orientation of the remaining noncrystalline population must decrease slightly (in the absence of any simultaneous, orientation-inducing stress). As crystallization proceeds, highly oriented strands are preferentially removed from the noncrystalline population and one should continue to observe high crystal orientation and low amorphous orientation. Ziabicki and Jarecki, as part of a seminal analysis of crystal orientation in a flowing system, also looked at the effect of nucleation selectivity on crystal orientation [39]. They accepted the general concept of Kriegbaum and Roe and incorporated these physics into a general scheme which predicts quantitatively the development of crystal orientation. Abhiraman [40] modified the analysis to allow a nucleus to accept strands whose original orientation is somewhat lower than its own orientation. In this way, the overall orientation of the system becomes higher than that of the as-drawn system. Another consequence is that the orientation of chains in the uncrystallized matter near the growing crystal is reduced, through depletion of highly aligned chains. Abhiraman points out that this effect could lead to transverse alignment of chains in the regions adjacent to growing crystals [40,41].

The net result is generally that crystallization and amorphous relaxation are quite generally coupled. For some properties (such as small molecule penetration) correlations with crystallinity are misleading; the direct correlation is with the relaxation of the noncrystalline matter, and it is this which is coupled to crystallinity.

12.4 Fiber Structure

12.4.1 Basic Structure

It is useful at this point to provide a very brief overview of the structure of (heat-treated) PET and nylon fibers. While structural details depend upon processing detail, some features are always present in commercial textile, carpet or industrial fibers, as follows:

1. The fiber contains both crystal and noncrystalline phases, as would be expected for any heat-treated semicrystalline polymer.
2. The fiber contains microfibrils. These are reported in all electron microscope studies [42–47] and is also inferred from small-angle X-ray scattering [48–50]. Noncrystalline material separates the fibrils [51,52].
3. There is a periodic density modulation along the fiber axis, with a periodicity of the order of 10 nm. This periodicity makes itself known in small-angle X-ray scattering (SAXS) peaks, as first observed by Hess and Kiessig in 1944 [53] and has also been directly observed in transmission electron microscopy of PET fibers [41]. The SAXS peaks appear along the fiber axis in flat-plate X-ray photographs. This density modulation is normally interpreted as an alternation of crystalline and amorphous regions along the fiber axis.

These features combine to produce the basic model shown in Figure 12.9. What one has here is microfibrils which are separated from each other transverse to the fiber axis by amorphous matter and which are internally composed of axially alternating crystalline and amorphous regions.

This basic picture is the framework on which we shall paste detail as we proceed. We begin in the following subsections by examining two properties which are unusual and which have important implications in fiber structure and properties.

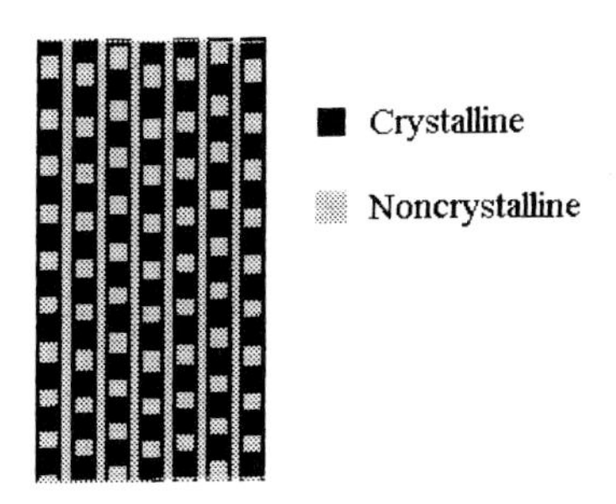

Figure 12.9 Schematic illustration of the basic structure of a polymer fiber.

12.4.2 Crystal Transformation During Heat-Treatment

A number of investigators (see Hearle [54] for a bibliography) have reported multiple melting points for polyester and polyamide fibers. Perhaps the most complete of these studies are those of Bell and coworkers [55–58] and Groeninckx and coworkers [59–61]. Typical of what is observed is the following sequence for nylon 66 fibers [54]. Spun yarn in the as-quenched state exhibits one DSC melting peak, at 256 °C. When the fiber is heat-treated at 220 °C for various times, a second melting peak develops. At short times, the new melting peak is near 230 °C and the original peak is retained at 256 °C, but at somewhat reduced amplitude. A portion of the original crystals has transformed into crystals which are somehow different. However, X-ray diffraction shows that the new crystals retain the crystal structure of the original crystals. Holding longer at 220 °C brings the new peak to progressively higher temperatures, while the 256 °C peak gradually vanishes. Finally, after very long heat-treatment, there is only one melting endotherm, at 260 °C. Groeninckx et al [59–61] show that what is happening is a reorganization of originally less perfect crystals into ones of increasingly better perfection.

This is, at first glance, very strange behavior. For any transformation to take place, the system must become more stable through the transformation – i.e., the system must go to a lower free energy state. The lower melting point of the new crystals, relative to the original crystals, indicates that the new crystals have a *higher* internal energy than do the original ones, whereas a higher internal energy should indicate a less stable form.

Hearle points out very clearly what this melting point conundrum indicates [54]. Suppose the initially formed crystals to be highly defective. Compared with more perfect crystals, these initial crystals have both higher internal energy and higher entropy. It is the higher entropy which can explain the transformation. Figure 12.10(a) schematically depicts free energy vs. temperature curves for melt, imperfect crystal and perfect crystal. What is significant here is that because of its higher entropy, the free energy curve for the imperfect crystal can intersect the melt curve at a higher temperature than would the more perfect crystal. Thus the free energy can be lowered by transforming imperfect crystals to perfect crystals at a temperature sufficiently below the melting point, but the melting point of the imperfect crystal can nonetheless be higher than that of the more perfect state. Now consider the more realistic case in which the original crystal transforms into a more perfect state, but in which the perfection of the more perfect crystal increases with time or temperature. This continuum of states is depicted in Figure 12.10(b). The model of a an imperfect crystalline state transforming to a kinetically ever more perfect crystalline state thus conceptually satisfies the requirement that heat-treatment produce a crystalline state with initially lower melting point than the original crystals, but in which the melting point continuously increases with time (or temperature), eventually exceeding the melting point of the initial state.

12.4.3 The Oriented Mesophase

The concept of stiff, oriented, noncrystalline phase is rooted in NMR studies of PET in the 1970s. It had been noted earlier by Farrow and Ward [62] that NMR measurements of

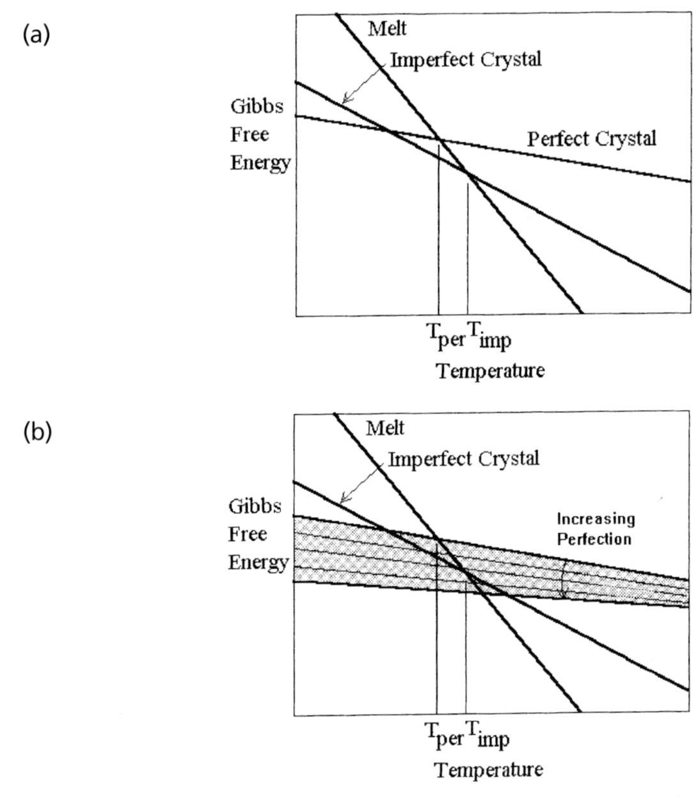

Figure 12.10 Depiction of free energy curves for the phases in a polymer fiber: (a) basic premise of melt, perfect crystal and imperfect crystal; (b) inclusion of a continuum of states for crystals becoming more perfect.

the degree of crystallinity are consistently higher than those based on WAXS measurements. The basis of the NMR crystallinity determination is that, in general, NMR spectra exhibit a broad component, representing immobile chain segments, and a narrow component, representing relatively mobile segments. Since only very limited chain motion is possible in the crystalline state, it had been assumed that the broad and narrow components characterized the amorphous and crystalline portions, respectively. Eichhoff and Zachmann found that when the oriented material is swollen, a portion of the broad NMR component is transferred to the narrow component [63]. At the same time, no change in the crystalline WAXS peak intensities is observed [63]. The portion of the broad NMR signal which became narrow upon swelling must therefore represent an immobile amorphous fraction. Adar and Noether [64] likewise reported a highly oriented, but noncrystalline portion in their Raman investigation of partially oriented PET filaments. Studies of oriented PET by Biangardi and Zachmann, in which orientation measurements were combined with NMR, SAXS and WAXS [65–69] showed that the rigid noncrystalline portion consists of relatively straight and highly oriented chain segments. It was separately noted in 1962 by Reichle and Prietzsch [70] and later by others [71–78] that the amorphous halo in oriented fibers can be separated into unoriented and oriented portions – UA and OA, respectively. The separation of the amorphous scattering into UA and OA portions has, in fact, recently become an important method of characterization for polymer fibers. A possible interpretation of these results is that the rigid portion of the noncrystalline material and the OA can be identified with a third phase [72–81] which has sometimes been called "oriented mesophase" [72] (see also Section 12.4.5.2) or "intermediate phase" [75,76]. In general, the fraction of the oriented amorphous material decreases with the temperature of heat-treatment of the fiber [72,73], although the "intermediate phase" identified by Fu et al. appears to show different behavior [76].

It is not yet clear what the effect of the oriented mesophase is on macroscopic properties. One exception to this statement is that the presence of OA material appears to inhibit small molecule diffusion and thereby lowers the dyeability of the material. This is inferred from two studies. Valk and Bunthoff [79,80] studied the dye uptake of drawn PET fibers, before and after heat-treatment at elevated temperatures. Comparing fibers of identical crystallinity but different portions of OA, they report that dye uptake is significantly greater in fibers with lower OA content. Likewise, Sharrow [81] has measured the fractions of unoriented and oriented noncrystalline material in nylon 66 fibers before and after dry and wet commercial heat-treatments. She reports low values of OA fraction and high dyeability in the wet-treated fiber and conversely for the dry-treated fiber. Both Valk and Sharrow note that when the OA content decreases, the crystalline phase shows a concomitant increase. Valk suggests that the conversion of highly oriented noncrystalline material to crystalline material produces a significant free volume increase and that it is this free volume which assists dye diffusion; dynamical mechanical evidence for increased chain mobility is cited in evidence [80].

Despite all the characterization of the oriented amorphous material, it is not yet clear whether it comprises a true mesophase or where it dominantly lies. On the basis of several property measurements, including dye uptake and mechanical behavior, Prevorsek [82] identifies the oriented noncrystalline material with taut tie molecules separating microfibrils radially, as sketched in Figure 12.11a. Marichin, on the other hand, uses packing considerations and spectroscopic results to demonstrate that the amorphous domains separating crystallites axially must be highly oriented [51]. It is suggested that the mesophase be identified with these regions, as sketched in Figure 12.11b. Finally, cylindrical distribution function analyses (based on careful X-ray scattering) by Sharrow

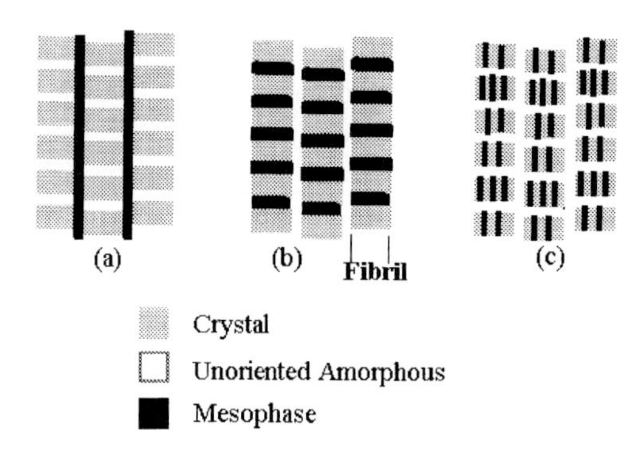

Figure 12.11 Possible sites for the oriented mesophase: (a) separating fibrils transversely, (b) separating crystallites within a fibril, (c) within the crystalline regions of a fibril.

[81] suggest that some portion of the mesophase be associated with defects within crystallites, as sketched in Figure 12.11c. Definitive characterization of the location and connection of the mesophase should be given high priority. It is likely that a combination of transmission electron microscope contrast methods could be effective in this regard.

12.5 Kinetics of Structure Development During Heat-Treatment

12.5.1 Effect of Orientation on the Crystallization Rate

The kinetics of structural transformations in highly oriented systems are very rapid. This is illustrated in Figure 12.12 [83], which shows the half-times of crystallization of PET from the melt [84] and from as-spun fibers [49,85] as a function of temperature. The as-spun fibers are oriented but noncrystalline. Different axial stresses were used in the heat-treatments, higher axial stresses producing higher degrees of orientation. The stress levels are indicated in the figure. We see here half-times of the order of tens or hundreds of milliseconds. Because of the rapidity of the transformation, direct, real-time measurements of structure development are extremely difficult.

Two methods have been used to track kinetics and structure development for these very rapid transformations. The methods are *in situ* studies and treat-and-quench investigations. These are reviewed in the following two subsections.

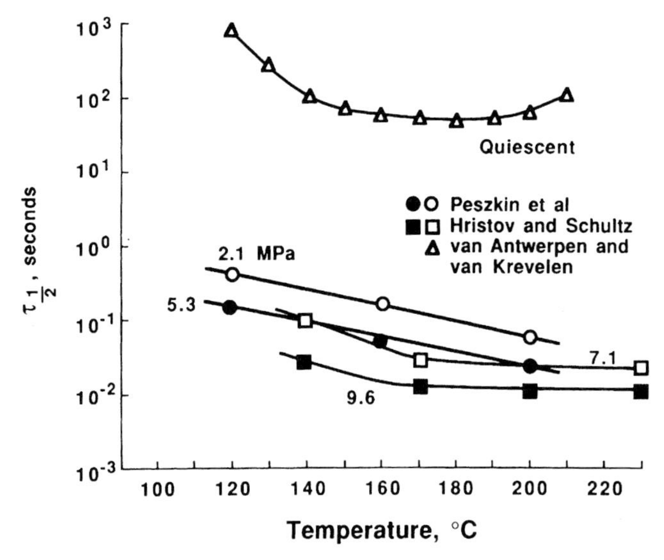

Figure 12.12 The dependence of PET crystallization half-times on temperature and axial stress [83]. The data for this figure are taken from van Antwerpen and Krevelan [84], Peszkin et al [49] and Hristov and Schultz [85].

12.5.2 In Situ Studies

In the past decade and a half, synchrotron sources of X-rays have become increasingly stable, intense and available, making these sources increasingly attractive for in situ studies of structure development in polymer fibers.

The earliest of such SAXS and WAXS studies were those of Zachmann and coworkers, on oriented PET film, using a beamline at the DORIS storage ring at DESY (Hamburg). In this work, an initial decrease (of some 30%) in SAXS long period with annealing time (in the 1–14 s range) was reported, for annealing temperatures between 96 and 108°C [86]. Simultaneously the WAXS azimuthal breadth decreased significantly [87,88]. No explanation for the long period decrease was suggested. It was suggested, however, that the orientation changes represent the straightening of initially bent crystal lamellae. Given the subsequent results of electron microscopy, it is more likely that the orientation results reflect a transformation from fibrillar to lamellar habit.

Soufffaché et al [89] later reported *in situ* synchrotron simultaneous SAXS and WAXS studies of virgin PET pellets and of POY PET fibers. For virgin PET and for fibers spun at 300 m/min a decrease in SAXS long spacing during the initial 100 s of heat-treatment at 130 or 160 °C with (apparently) free ends was observed. At the same time, the crystal thickness remained constant. The degree of long spacing decrease was a strong function of molecular weight, as was the asymptotic long spacing. Significantly, the relative long spacing decrement approximated the relative macroscopic shrinkage of the material. This correlation of long period change with macroscopic shrinkage suggests that the noncrystalline chains separating crystallites along the fiber direction relax with time – a form of physical aging [90]. Relaxation of the noncrystalline fraction with annealing time, toward an equilibrium state of low modulus, was also shown by Itoyama [20]. The dependence of long spacing on molecular weight, observed by Souffaché et al, suggests that the thickness of the amorphous layer separating crystallites depends at least partially on the degree of entanglement of the chains. The rate of crystallization would then depend on the rate of disentanglement. Strikingly, fibers which had been spun at 3300 m/min exhibited the same crystal thickness as the other materials, but much smaller long spacing and effectively no long spacing decrease upon heat-treatment. When heat-treated with fixed ends, fibers spun at 3300 m/min show a greater long spacing than do those annealed with free ends, but otherwise behave the same as the fixed-end system. Furthermore, these fibers showed no effect of molecular weight on long spacing. The authors conclude that two relaxation processes occur during annealing. The first of these is the re-coiling of the oriented chains; the slower process is the relaxation of entanglements. The authors suggest that the higher deformation experienced by the fibers spun at higher speed promotes rapid disentanglement of the chains, thereby enabling crystallization early in the annealing process. By implication, annealing with fixed ends apparently enables crystallization to occur without re-coiling of the noncrystalline chains. These results cohere with the two-stage kinetics described by Buckley and Salem [36] and offer a reasonable mechanistic interpretation of the stages.

Most recently, Hsiao et al [50] have studied the effect of drawing and temperature on nylon 66 fibers. Six draw ratios (1.01 to 1.37 X) and six temperatures (35 to 243 °C) were used. In this study, the fiber was continuously stretched between a feed roller and a take-up roller, as the fiber passed over a hot pin located 3.5 cm upstream of the X-ray beam. Two-dimensional SAXS and WAXS scans were made simultaneously. They report that at temperatures above the Brill transition (from triclinic to pseudohexagonal) the smallest, most defective crystals melt and are replaced, on cooling, by more perfect crystals. Such melting of small, defective crystals, they report, can also be effected by drawing alone. This relatively direct result agrees with the earlier suggestions, described in Section 12.4.2, that heat-treatment produces a transformation from less to more perfect crystals. It is not to be concluded, however, that the change in crystal perfection is directly responsible for important property changes. It is more likely that most property changes are related to changes in the noncrystalline material and the mesophase.

12.5.3 Treat and Quench Investigations

12.5.3.1 Background

In order to bring a larger number of characterization measurements to bear on the structure developing during the heat-treatment of polymer fibers, treat-and-quench experiments have proven to be of value. Such measurements begin with as-spun fiber at room temperature. The fiber is then very rapidly taken to a heat-treatment temperature T_a and held there for some time t_1 and very rapidly quenched to below the glass transition temperature. Another identical fiber is treated in the same way, except that the holding time is t_2. This procedure is repeated for many fibers. This set of heat-treatments is shown schematically in Figure 12.13. It is presumed that the quench is rapid enough to substantially retain the structure present at the end of the heat-treatment. After the preparations indicated in Figure 12.14, one has a set of fiber specimens representing heat-treatments for a series of times at temperature T_a. These specimens can then be examined by any characterization method and such data can be used to determine the change in any characteristic with time at T_a. The process can be repeated for other temperatures. In such experiments, the state of constraint or stress on the fiber can also be controlled and/or monitored.

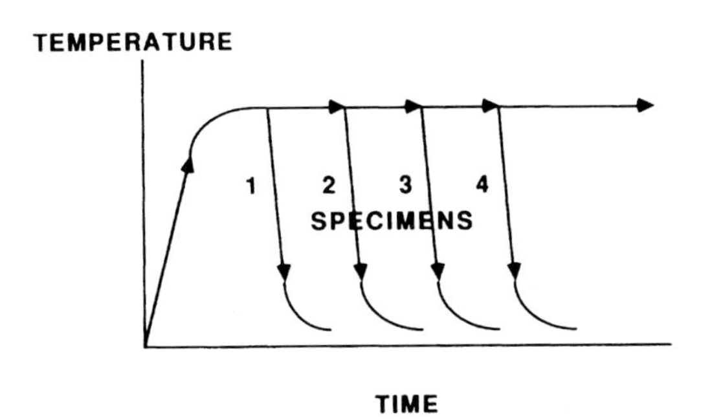

Figure 12.13 Schematic time and temperature sequence to create an isothermal set of treat-and-quench specimens [83].

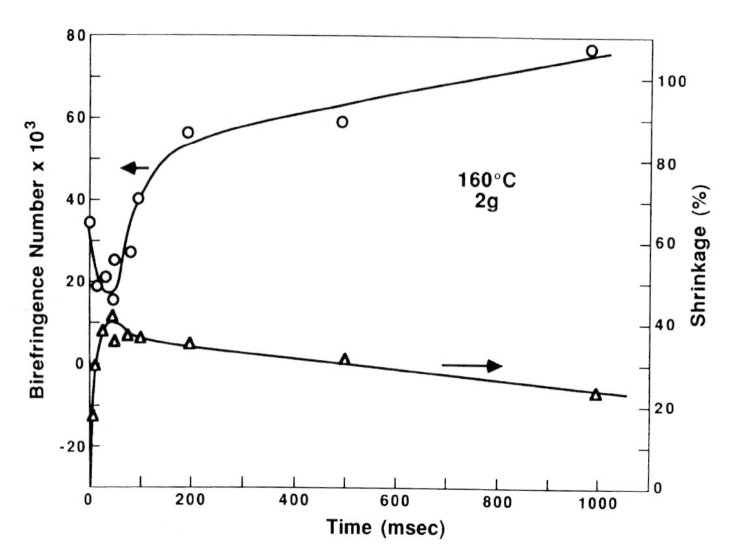

Figure 12.14 Birefringence and shrinkage development for a PET heat-treated under low load (2.1 MPa) at 160 °C [100].

12.5.3.2 Kinetics and Macroscopic Behavior

Treat-and-quench experiments were reported by Smith and Steward in 1974 [91]. They studied crystallization of PET fibers at 120 °C under conditions of constant length. Their characterization measurements were of birefringence and of shrinkage in boiling water. Their results showed that crystallization occurs very rapidly, being essentially complete at times less than 100 ms. They also reported an Avrami coefficient near 1, an indication of one-dimensional crystallite growth.

Following the lead of Smith and Steward, a series of studies by Ikeda [92], Desai and Abhiraman [93], and by several investigators in our laboratory [48,49,85,94–97] followed the development of structure in PET, nylon 66 and poly(butylene terephthalate) during heat-treatment. In all of these studies, the extensional stress and temperature were controllable.

Ikeda measured density and denier of PET POY, as well as shrinkage in 80 °C water [92]. He found that shrinkage was especially sensitive to the earliest stage of crystallization – e.g., what would occur during low temperature heat-treatment (80 to 100 °C). For such low temperature heat-treatment, changes in density and flat-film X-ray diffraction measurements are not perceptible, whereas shrinkage drops rapidly with time of heat-treatment. From the shrinkage measurements he inferred a half-time for crystallization τ. Using $1/\tau$ as an index of the crystallization rate, he reported an exponential increase in crystallization rate with elongational strain ($1/\tau = A \exp(B\varepsilon)$). This rate likewise increased with treatment temperature.

Desai and Abhiraman [93] were additionally able to measure the change in length of their yarn and the shrinkage force during heat-treatment in air under constant load at 150 °C. They found that shrinkage occurs rapidly at first and then levels off. At the same time, the sonic modulus undergoes a minimum and the birefringence undergoes rapid change (increasing sharply for fiber spun at 4500 m/min and decreasing for fiber spun at 2800 m/min). During constant-length annealing, the shrinkage force rises rapidly at first, followed by a much slower decline.

In the work described above, heat-treatment times down to 0.2 s were used. Since crystallization rate increases strongly with temperature and extensional stress, the ability to probe lower times is essential to studying heat-treatment under more severe conditions. A series of investigations in our laboratory have investigated crystallization under higher stresses and temperatures. In these investigations, it was necessary to improve heat-transfer to the specimen, so that the up- and down-quench become very rapid. In the earliest of these studies, Gupte et al [48] report heat-treatment times down to 0.1 s. Subsequent investigations by Peszkin et al [49,94–96] and by Lee and Schultz [97] lowered the earliest times measured to 30 and 5 ms, respectively. Selected aspects of these experiments are reviewed in the following paragraphs.

Before reviewing these results, it is useful to reiterate that in fiber heat-treatment there is a competition between relaxation (the re-coiling of extended chains) and crystallization (which acts to "lock" oriented chains to each other). Relaxation should result in shrinkage of a fiber, whereas crystallization preserves, or even increases, the local orientation of the system. The temperature and stress under which the transformation takes place define which of the competing processes dominates. Low stress levels and low temperatures favor the kinetics of relaxation; high stresses and high temperatures favor crystallization kinetics [49,98]. In another approach to this same question, Salem [99] has cast the competition in terms of the effect of draw rate and temperature on the rate of crystallization.

Figure 12.15 relates to a situation under which relaxation and crystallization are nearly balanced [49,100]. In this figure are plotted the relative fiber length ("shrinkage") and optical birefringence change with treatment time at several temperatures at 2.1 MPa threadline stress. At 120 °C, the fiber initially shrinks, then exhibits a maximum and finally increases in length. WAXS results show that the elongation parallels the development of crystallinity. In this case, relaxation initially occurs more rapidly than crystallization, but crystallization kinetics dominate after some 150 ms. At lower temperatures (e.g., 100 °C) crystallization never becomes effective. At higher temperatures (e.g., 160 and 200 °C), the crossover time becomes shorter and shorter. Orientation, manifest in birefringence measurements, exhibits the mirror image of the elongation result. Here an initial decrease in birefringence is due to relaxation. The later birefringence increases reflect a chain realignment due to oriented crystallization.

Several investigators have shown that crystallization dominates the process over all measurable time under higher stresses. In this case, the fibers always lengthen [49,97,98]. In Figure 12.15 fiber length change is plotted against time for heat-treatment under 10 MPa for several temperatures [97]. For temperatures above 120 °C, the half-times of crystallization are under 10 ms.

When heat-treatment is carried out under constant length, but with no controlled load, relaxation is the dominant feature for very long times. One way to measure this is through the crystallite orientation, measured, for instance, by the azimuthal breadth of WAXS hk0 peaks along their Debye arcs. Figure 12.16 demonstrates this behavior [48]. Eventually the orientation improves, through the axial shrinkage stress exerted by the re-coiling

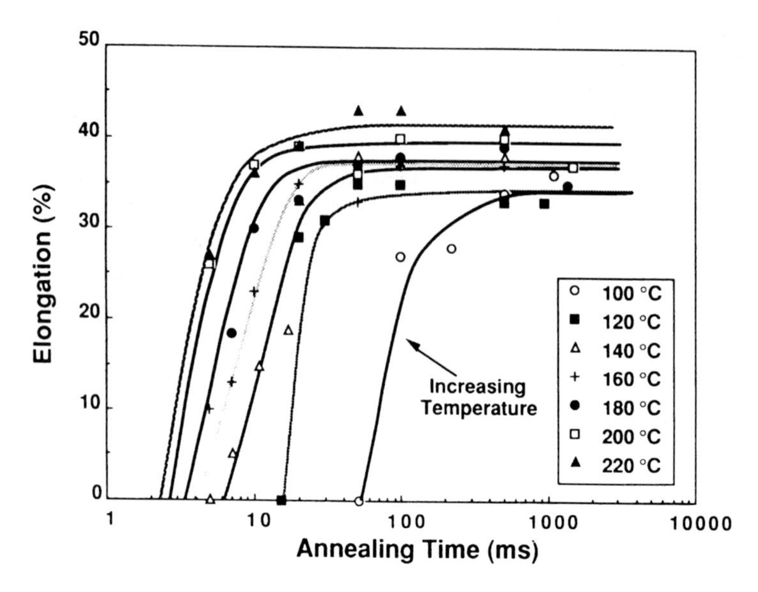

Figure 12.15 Fiber elongation versus heat-treatment time under 10 MPa tension for a number of temperatures [97].

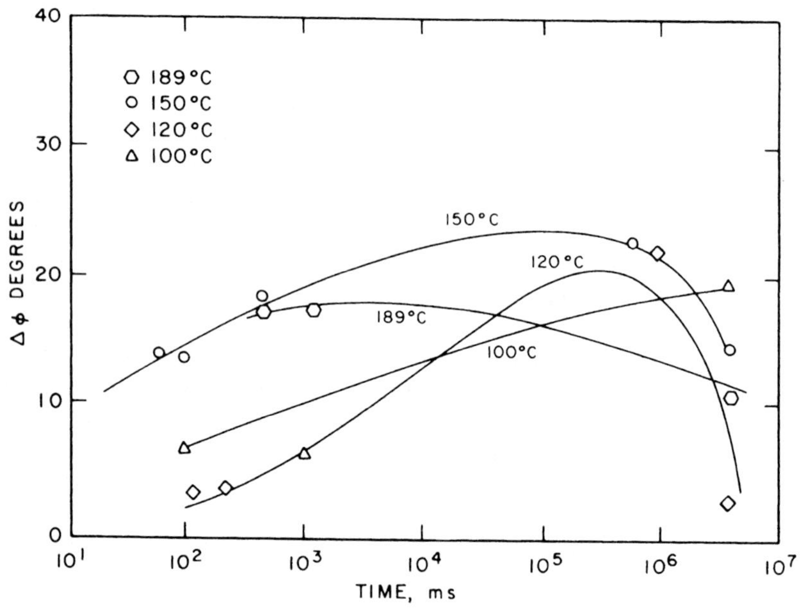

Figure 12.16 Development of the azimuthal breadth of (110) WAXS peaks with time at four temperatures of heat-treatment [48].

chains on the crystallites to which they are connected axially. Lee and Schultz [97] report two kinetic stages during heat-treatment at 10 MPa (0.083 g/den). The initial stage has half-times under 10 ms for temperatures above 120 °C. The later stage continues to nearly

100 ms at these temperatures. During the initial stage the fiber elongation due to crystallization goes to completion and the shrinkage in 80 °C water has stabilized. While crystallization is well underway in this period, it has not gone to completion. During the second stage, crystallization goes to completion and the mechanical properties (modulus and tenacity) are set.

12.5.3.3 Structure Development

Specimens created by the treat-and-quench method provide the opportunity for detailed structural investigation. Transmission electron microscopy (TEM), SAXS studies, and radial distribution function analysis of high quality WAXS scans are all possible and have provided evidence toward how structure develops during heat-treatment.

Phase contrast transmission electron microscopy was performed on PET POY fibers heat-treated at 10 MPa (0.083 g/den) at 140 and 160 °C [47]. The phase contrast method (defocus microscopy) operates with no staining of the specimen; ultramicrotomed thin films are viewed with no further preparation. The as-spun fibers exhibit no crystalline reflections in WAXS, but have a room temperature density greater than that of relaxed, amorphous PET.

At the initial stage of heat-treatment (10 ms), very fine microfibrils – approximately 1.5 nm in diameter – are observed for the material treated at 140 °C. Such narrow microfibrils are predicted to form as a means of effecting rapid removal of the heat of fusion during the very rapid crystallization of the fiber [101–103]. But these are ephemeral. Further heat-treatment causes a fibril-to-lamella transition, the lamellae being oriented normal to the fiber axis. The lamellae are a few tens of nm wide and form long, narrow, periodic stacks of such lamellae, as shown in the electron micrograph of Figure 12.17. (In this figure, the bright lines parallel to the fiber axis are microtome marks, not real structural features). During this entire process, the crystal c-axis (the chain direction) remains parallel to the fiber axis. At 160 °C, the initial fibrils are not observed; only lamellae are observed. Presumably the fibril-to-lamella transition has already taken place. It is likely that such long strings of stacked lamellae are the microfibrillar entities which have been observed in commercial fibers [42–46].

Small-angle X-ray scattering (SAXS) can provide information on crystallite dimensions and intercrystalline packing. SAXS studies on PET POY fibers yield values in the 3 to 4 nm range for the diameter, normal to the fiber axis, of structural features (presumably crystallites) [49]. These features appear early in the heat-treatment and persist at this dimension throughout heat-treatment.

hk0 WAXS Bragg peaks become increasingly narrow as heat-treatment progresses [48,49]. The breadth of hk0 peaks is determined by the transverse diameter of the diffracting entities (crystallites) and by the internal perfection of those entities. Increasing perfection and increasing transverse diameter of the crystallites both act to broaden Bragg peaks. The persistent constant value of crystallite diameter observed during crystallization by SAXS infers that the narrowing of WAXS peaks must derive from an increasing perfection of the crystallites during heat-treatment.

A model for structure development during heat-treatment emerges from these results [83]. A sketch of this development is given in Figure 12.18 and is described as follows:

Figure 12.17 TEM phase contrast image and electron diffraction pattern of a PET fiber heat-treated or 100 ms at 140 °C [47]. Arrow denotes fiber axis.

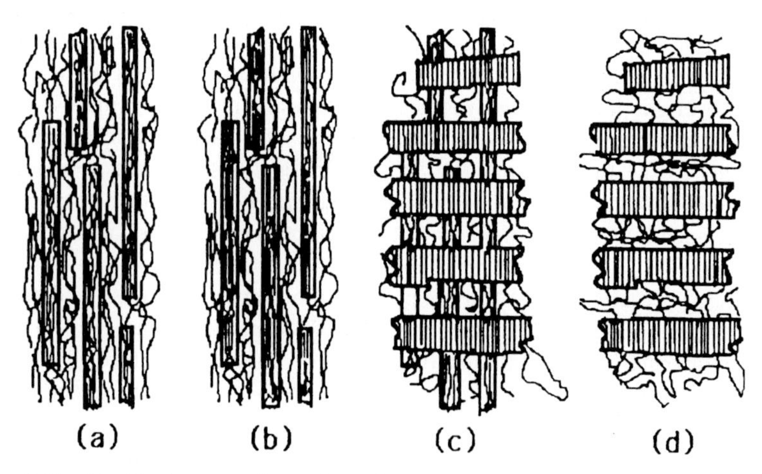

Figure 12.18 Model of stages of transformation: (a) formation of narrow, defective crystalline or mesophasic microfibrils; (b) axial segmentation of those microfibrils into crystalline and amorphous blocks; (c) transverse growth of the crystallites, and (d) dissolution of the remnants of the microfibrils [83].

1. Initially very fine fibrils – some 1.5 nm in diameter – form from the oriented melt. These microfibrils are highly defective and are metastable. These may be the oriented mesophase, but this is only a possibility; no direct identification of mesophase and defective microfibrils has yet been made. The fibrils very rapidly become periodically

segmented axially into alternating crystal-like and melt-like regions (likely by the transport of defects along the chain axis [104,105]).
2. The crystal-like regions then align with each other transverse to the fiber axis. Periodic stacks of lamellae are created as the crystallites merge and become more perfect. However, it is difficult for crystallites with different initial orientations about the c-axis to weld [106], and an imprint of the original crystal width remains, as evidenced by the continued 3 nm breadth derived from SAXS measurement [49].
3. The lamellar crystals gradually become more perfect.

This picture is in qualitative agreement with what is expected because of difficulties in dissipation of the heat of fusion at the advancing growth front. The generic prediction was for the formation of imperfect, rodlike crystals. Since such moieties can be only metastable, because of the energy stored in surfaces and internal defects, further changes with treatment time would be expected, and are manifested in the continuous increase in perfection and the fibrillar to lamellar transformation.

12.5.3.4 Property Development

The treat-and-quench study of Lee and Schultz [97] showed that transformation takes place during heat-setting in two kinetic stages. In the first, very rapid stage, a relatively small concentration of narrow, defective, fibrillar crystals is created [47,49,97]. These few defective fibrils are sufficient to fix the noncrystalline chains against large-scale shrinkage. As time goes on, a second stage occurs. In this stage a microfibrillar to lamellar transition takes place [47] and the crystallites become progressively more perfect, while the degree of crystallization increases [49,97]. In this second stage, under a given heat-setting tension, the tensile modulus increases nearly linearly with the density of the fiber [97]. The correlation between modulus and crystallinity can also be seen in the results of Itoyama [20]. And, most strikingly, very high modulus PET [20, 107,108] and nylon [109] fibers all have exceptionally high degrees of crystallinity. Kunugi points out, however, that the tension level is important, in that the modulus of the amorphous portion increases with applied tension [108].

12.6 Some Final Remarks

The heat-treatment required for fiber production depends upon the application for the fiber. Depending on the end use, some constellation of bulk, elastic modulus, dyeabilty, thermal stability and tensile strength is required. The heat-setting details depend on this constellation. What is required technologically, then, is an understanding of the role of heat-setting parameters in determining each of these properties. At a more fundamental level, we wish to understand how heat-setting parameters affect each element of fiber structure and how properties are defined by the structural detail. Some indicators relating to this goal are available.

Most heat-setting relates to simultaneous (or sequential) drawing and heat-treatment. We have seen that the drawing of fibers at elevated temperatures is at least sometimes done under stresses which produce a stable neck in the fiber. Such a stable neck infers the creation of microstructures which work-harden the material to the point at which further viscous deformation is not possible. This strain-resisting structure could be oriented noncrystalline regions, an oriented mesophase, or defective, microfibrillar crystals. In fact, these resisting structural features may be different levels of a continuum of strain-resisting structures. This is not known. Macroscopic shrinkage requires the "unlocking" of chains trapped in the locally oriented regions.

It has been established that the strain-resisting microsubstituent develops very rapidly during heat-setting and is important in defining the heat resistance and the dyeability of the material. The shrinkage resistance sets in with the occurrence of defective microfibrils. And we have seen that crystallization of oriented noncrystalline material coincides with improvements in dyeability. It is, however, not yet clear how this occurs, since even the location of the oriented noncrystalline material is not known.

The degree of crystallinity and the relaxation of chain strands in the noncrystalline regions are important in determining the fiber modulus. The degree of crystallinity develops gradually with time at temperature, long times at high temperature fostering high degrees of crystallinity. This process is associated first with a microfibrillar-to-lamellar transition and continues with the perfecting and growth of the lamellar crystallites. During these processes the oriented amorphous material may crystallize and the noncrystalline matter relaxes gradually. The imposition of high stresses during heat-treatment, coupled with short times at temperature, restrict the relaxation of chains in the noncrystalline regions and thereby cause a relatively high modulus for the noncrystalline matter to be retained.

12.7 References

1. Usenko, V. *Processing of Man-Made Fibres* (1979) Mir Publishers, Moscow
2. Gall, H. In *Synthesefasern*, von Falkai, B. (Ed.) (1981) Verlag Chemie, Weinheim
3. Pajgrt, O. and Reichstädter, B. *Processing of Polyester Fibers* (1979) Elsevier, Amsterdam
4. Chantry, W.A., Molini, A.E., U.S. Patent 3 216 187 (1965)
5. Martin, H.G., Glutz, B., U.S. Patent 3 715 421 (1973)
6. Davis, H.L., Jaffe, M.L., Besso, M.M., U.S. Patent 3 946 100 (1976)
7. Davis, H.L., Jaffe, M.L., LaNieve, H.L., Powers, E.J., U.S. Patent 4 101 525 (1978)
8. Davis, H.L., Jaffe, M.L., LaNieve, H.L., Powers, E.J., U.S. Patent 4 195 052 (1980)
9. Russell, W.N., Bruton, J.I., Twilley, C., U.S. Patent 4 113 821 (1978)
10. McClary, E.B., U.S. Patent 4 414 169 (1983)
11. Bull, T.H., Kinhear, A.M.G., U.S. Patent 3 259 681 (1962)
12. Bauer, G., Kramer, L., Kuhn, H., U.S. Patent 3 832 435 (1968)
13. Barbe, G.M., Curtillet, P.G., Lequay, C., U.S. Patent 3 840 633 (1974)
14. Kanetsuna, H., Kurita, T. U.S. Patent 3 946 094 (1976)
15. Vail, O.R., U.S. Patent 3 816 486 (1974)
16. Asada, Y., Tsujii, O. Japenese Patent Publication No. 21260 (1974)
17. Ohnuma, Fujimara, Y., Japanese Patent Publication No. 53019 (1976)
18. Abhiraman, A.S., Song, J.W. *J. Polym. Sci., Polym. Lett. Ed* 23 (1985) p. 613

19. Yoon, K.J., Desai, P., Abhiraman, A.S. *J. Polym. Sci., Polym. Phys. Ed* (1986) 24, p. 1665
20. Itoyama, K. *J. Polym. Sci., Polym. Lett. Ed.* (1987) 25, p. 331
21. Schultz, J.M. *Polymer Materials Science* (1974) Prentice Hall, Englewood Cliffs, NJ
22. Dismore, P.F., Statton, W.O. *J. Polym. Sci.* (1966) 13C, p. 133
23. Dumbleton, J.H. *J. Polym. Sci.* (1969) 7A-2, p. 667
24. Dumbleton, J.H. *Polymer* (1969) 10, p. 539
25. Wilson, M.P.W. *Polymer* (1974) 15, p. 277
26. Thompson, A.B. *J. Polym. Sci.* (1959) 34, p. 741
27. Hristov, H.A., Hearle, J.W.S., Schultz, J.M., Kennedy, A.D. *J. Polym. Sci., Polym. Phys. Ed.* (1995) 33, p. 125
28. Kase, S., Chang, M. In *Proc. Int. Symp. on Fiber Sci. and Techn., ISF-85, August, 1985, Hakone, Japan*
29. Ziabicki, A. In *High-Speed Fibre Spinning*. Ziabicki, A., Kawai, H. (Eds.) (1985) Wiley, New York, p. 21
30. Matsui, M. In *High-Speed Fibre Spinning*. Ziabicki, A., Kawai, H. (Eds.) (1985) Wiley, New York, p. 137
31. Perez, G. In *High-Speed Fibre Spinning*. Ziabicki, A., Kawai, H. (Eds.) (1985) Wiley, New York, p. 333
32. Argon, A.S. *Phil. Mag. (1973)* 28, p. 839
33. Argon, A.S. In *Polymeric Materials* (1975) ASM, p. 411
34. Peterlin, A. In *Polymeric Materials* (1975) ASM, p. 175
35. Zhang, A., Jiang, H., Wu, C., Zhou, L., Xuan, L., Qian, B. *Text. Res. J.* (1985) 55, p. 387
36. Buckley, C.P., Salem, D.R. *Polymer* (1987) 28, p. 69
37. Krigbaum, W.R., Roe, R.-J. *J. Polym. Sci.* (1964) 2A, p. 4391
38. Krigbaum, W.R., Taga, T. *J. Polym. Sci., Polym. Phys. Ed.* (1979) 17, p. 393
39. Ziabicki, A., Jarecki, L. *Colloid Polym. Sci.* (1978) 256, p. 332
40. Abhiraman, A.S. *J. Polym. Sci., Polym. Phys. Ed.* (1983) 21, p. 583
41. Desai, P., Abhiraman, A.S. *J. Polym. Sci., Polym. Phys. Ed.* (1985) 23, p. 213
42. Cobbold, A., Daubeny, R. de P., Deutsch, K., Markey, P. *Nature* (1953) 172, p. 806
43. Sikorski, J. In *Fibre Structure*. Hearle, J.W.S., Peters, R.H. (Eds.) (1963) Butterworths, London
44. Van Veld, R.D., Morris, G., Billica, H.R. *J. Appl. Polym. Sci.* (1968) 12, p. 2709
45. Prevorsek, D.C., Sibilia, I.F. *J. Macromol. Sci.–Phys.* (1971) B5, p. 119
46. Murray, R., Davis, H.A., Tucker, P. *J. Appl. Polym. Sci., Polym. Symp. Ed.* (1978) 33, p. 177
47. Chang, H., Lee, K.-G., Schultz, J.M. *J. Macromol. Sci.–Phys.* (1994) B33, p. 105
48. Gupte, K.M., Motz, H., Schultz, J.M. *J. Polym. Sci., Polym. Phys. Ed.* (1983) 21, p. 1927
49. Peszkin, P.N., Schultz, J.M., Lin, J.S. *J. Polym. Sci., Polym. Phys. Ed.* (1986) 24, p. 2591
50. Hsiao, B.S., Kennedy, A.D., Leach, R.A., Chu, B., Harney, P. *Proc. X Internat. Conf. on Small-Angle Scattering, July 21–25, 1996, Campinas, Brazil*, in press.
51. Marichin, V.A. *Acta Polymerica* (1979) 30, p. 507
52. Gupta, V.B., Ramesh, C., Gupta, A.K. *J. Appl. Polym. Sci.* (1984), 29, p. 3115
53. Hess, K., Kiessig, H. *Z. Phys. Chemie* (1944) 193, p. 196
54. Hearle, J.W.S. *J. Appl Polym. Sci., Appl. Polym. Symp.* (1977) 31, p. 137
55. Bell, J.P., Slade, P.E., Dumbleton, J.H. *J. Polym. Sci.* (1968) 6A-2, p. 1773
56. Bell, J.P., Dumbleton, J.H. *J. Polym. Sci.* (1969) 7A-2, p. 1033
57. Bell, J.P., Murayama, T. *J. Polym. Sci.* (1969) 7A-2, p. 1059
58. Sweet, G.E., Bell, J.P. *J. Polym. Sci.* (1972) 10A-2, p. 1273
59. Groeninckx, G., Reynaers, H., Berghmans, H., Smets, G. *J. Polym. Sci., Polym. Phys. Ed.* (1980) 18, p. 1311
60. Groeninckx, G., Reynaers, H. *J.. Polym. Sci., Polym. Phys. Ed.* (1980) 18, p. 1325
61. Fontaine, F., Ledent, J., Groeninckx, G., Reynaers, H. *Polymer* (1982) 23, p. 185
62. Farrow, G., Ward, I.M. *Brit. J. Appl. Phys.* (1960) 11, p. 543
63. Eichhoff, U., Zachmann, H.G. *Makromol. Chem.* (1971) 147, p. 41
64. Adar, F., Noether, H. *Polymer* (1985) 26, p. 1935
65. Zachmann, H.G. *Polym. Eng. Sci.* (1971) 19, p. 966
66. Biangardi, H.J., Zachmann, H.G. *Prog. Colloid Polym. Sci.* (1977) 62, p. 71
67. Biangardi, H.J., Zachmann, H.G. *J. Polymer Sci., Polym. Symp. Ed.* (1977) 58, p. 169
68. Rosenke, K., Zachmann, H.G. *Prog. Colloid Polym. Sci.* (1978) 64, p. 245

69. Biangardi, H.J. *Prog. Colloid Polym. Sci.* (1979) 69, p. 99
70. Reichle, A., Prietzsch, A. *Angew. Chem.* (1962) 74, p. 562
71. Jellinek, G., Ringens, W., Heidemann, G. *Ber. Bunsenges. Phys. Chem.* (1970) 74, p. 924
72. Lindner, W.L. *Polymer* (1973) 14, p. 9
73. Gupta, V.B., Kumar, S. *Polymer* (1978) 19, p. 953
74. Busing, W.R. *Macromolecules* (1990) 23, p. 4608
75. Fu, Y., Busing, W., Jin, Y., Affholter, K.A., Wunderlich, B. *Macromolecules* (1993) 26, p. 2187
76. Fu, Y., Busing, W., Jin, Y., Affholter, K.A., Wunderlich, B. *Macromol. Chem. Phys.* (1994) 195, p. 803
77. Fu, Y., Annis, B., Boller, A., Jin, Y., Wunderlich, B. *J. Polym. Sci., Polym. Phys. Ed.* (1994) 32, p. 2289
78. Murthy, N.S., Bray, R.G., Correale, S.T., Moore, R.A.F. *Polymer* (1995) 36, p. 3863
79. Valk, G., Bunthoff, K. *Chemiefasern/Textilind* (1979) 29/81, p. 334
80. Valk, G., Jellinek, G., Schröder, U. *Textile Res. J.* (1980) 50, p 46
81. Sharrow, J. R. *M. S. Thesis* (1996) University of Delaware
82. Prevorsek, D.C., Oswald, H.J. In *Solid State Behavior of Linear Polyesters and Polyamides* Schultz, J.M., Fakirov, S. (Eds.) (1990) Prentice-Hall, Englewood Cliffs, NJ, p. 131
83. Schultz, J.M. In *Oriented Polymer Materials* Fakirov, S. (Ed.) (1996) Hüthig & Wepf, Heidelberg, p. 361
84. van Antwerpen, F., van Krevelen, D.W. *J. Polym. Sci., Polym. Phys. Ed.* (1970) 11, p. 2423
85. Hristov, H.A., Schultz, J.M. *Polymer* (1988) 29, p. 1211
86. Elsner, G., Zachmann, H.G.,.Milch, J.R. *Makromol. Chem.* (1981) 182, p. 657
87. Gehrke, R., Zachmann, H.G. *Polymer* (1989) 30, p. 1582
88. Zachmann, H.G., Bark, M., Teitge, A., Röber, S. *Lecture on Chemiefasertagung,* Dornbirn, 1990.
89. Souffaché, E., Perez, G., Lecluse, C., Rault, J. *J. Macromol. Phys.–Phys.* (1988) B27, p. 337
90. Struik, L.C.E. *Physical Aging in Polymers and Other Materials* (1978) Elsevier, Amsterdam
91. Smith, F.S., Steward, R.D. *Polymer* (1974) 15, p. 283
92. Ikeda, R.M. *J. Polym. Sci., Polym. Lett. Ed.* (1980) 18, p. 325
93. Desai, P., Abhiraman, A.S. *J. Polym. Sci., Polym. Phys. Ed.* 23, p. 653
94. Elad, J., Schultz, J.M. *J. Polym. Sci., Polym. Phys. Ed.* (1984) 22, p. 781
95. Peszkin, P.N., Schultz, J.M. *J. Appl. Polym. Sci.* (1985) 30, p. 2689
96. Peszkin, P.N., Schultz, J.M. *J. Polym. Sci., Polym. Phys. Ed.* (1986) 24, p. 2617
97. Lee, K.-G., Schultz, J.M. *Polymer* (1993) 34, p. 4455
98. Warwicker, J.O., Vevers, B. *J Appl. Polym. Sci.* (1980) 25, p. 977
99. Salem, D.R. *Polymer* (1994) 35, p. 771
100. Schultz, J.M. In *Solid State Behavior of Linear Polyesters and Polycondensates* Schultz, J.M., Fakirov, S. (Eds.) (1990), p. 75
101. Tiller, W.A., Schultz, J.M. *J. Polym. Sci., Polym. Phys. Ed.* (1984) 22, p. 143
102. Schultz, J.M. *Polymer* (1991) 32, p. 3268
103. Schultz, J.M. *Polym. Eng. Sci.* (1991) 31, p. 661
104. Petermann, J. *Makromol. Chemie* (1981) 182, p. 613
105. Petermann, J., Gohil, R.M., Schultz, J.M., Hendricks, R.W., Lin, J.S. *J. Polym. Sci., Polym. Phys. Ed.* (1982) 20, p. 523.
106. Schultz J.M., Petermann, J. *Colloid Polym. Sci. (1984)* 262, p. 294
107. Kunugi, T., Suzuki, A., Hashimoto, M. *J. Appl. Polym. Sci.* (1981) 26, p. 213
108. Kunugi, T., Suzuki, A., Hashimoto, M. *J. Appl. Polym. Sci.* (1981) 26, p. 1951
109. Kunugi, T., Ikuta, T., Hashimoto, M., Matsuzaki, K. *Polymer* (1982) 23, p. 1983

13 Microstructure Characterization

13.1 Wide Angle X-Ray Diffraction Analysis of Fibers

John Blackwell
Case Western Reserve University, Cleveland, Ohio, USA

13.1.1 Introduction

Wide angle X-ray scattering (WAXS) analysis of fibers performed since the 1920s has given us the general structural models for the arrangement of the polymer molecules in the solid state. In most fibers, the polymer chains are arranged with their axes approximately parallel to the fiber axis, and frequently adjacent chains form ordered bundles or crystallites that are separated by less well ordered or amorphous regions. The mechanical properties can be highly dependent on the degree of orientation, degree of crystallinity, and the size of the crystalline regions, and methods have been established for the routine determination of these parameters from the X-ray diffraction data. In addition, these data provide information on the chain conformation, and its interaction with its neighbors in the ordered regions, giving us an understanding of the relationship between the chemical structure and the end properties. This chapter reviews the methods used to derive the above structural information.

Figure 13.1.1 shows the X-ray diffraction pattern of cellulose II: the specimen was a bundle of parallel fibers of Fortisan rayon, with their axes perpendicular to a collimated beam of CuKα X-rays. The data consist of a number of relatively sharp Bragg reflections that appear as arcs on "layer lines". It was recognized early on that these reflections originate from small crystallites within the fiber, and that the d-spacings can be indexed in terms of a unit cell. Data of this type are analogous to the rotation

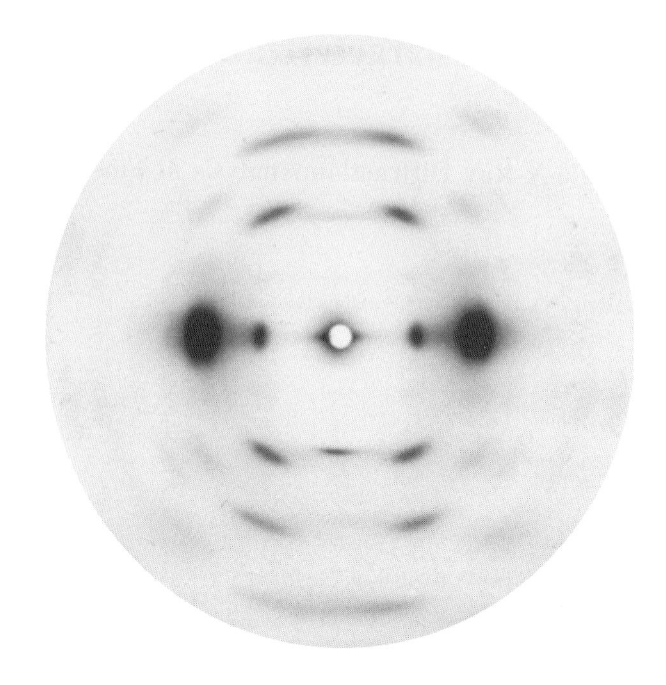

Figure 13.1.1 X-ray fiber diagram of Fortisan rayon fibers (cellulose II). The fiber axis direction is vertical.

diagram obtained when a single crystal mounted with a unit cell axis perpendicular to the beam is rotated about that axis, so that the different Bragg planes move in and out of the reflecting position. In an ideal fiber, the crystallites are arranged with their chain axes parallel to the fiber axis, but with cylindrical disorder, such that all possible orientations about that axis are present. Hence it is not necessary to rotate the specimen: rotation is effectively built into the structure. However there are important differences between the fiber diagram and the rotation photograph for a single crystal: the reflections are broadened in the radial direction due to the effect of small crystallite size; they are arced circumferentially due to imperfect orientation along the fiber direction; and there is a background of scattering due primarily to an amorphous component. These effects will be considered in turn in order to define the microstructure, after which we will review methods to analyze the chain conformation and packing in the ordered regions.

13.1.2 Degree of Crystallinity

Interpretation of wide angle X-ray data in terms of fiber morphology is best done in conjunction with analysis of the small angle scattering and direct visualization via electron and optical microscopy. Fiber structures range from completely non-crystalline, such as cellulose acetate, in which the irregular substitution prevents three dimensional ordering of the highly oriented chains, to structures like Kevlar that are essentially completely crystalline, with the only disorder occurring as defects and at the edges of the crystallites. The crystallites that give rise to Bragg reflections may be chain folded lamellae, as in many melt and solution spun fibers of synthetic polymers, or they may be due to lateral association of extended chains, as in cellulose or Kevlar fibers. The concept of a degree of crystallinity implies a two phase model: small polymer crystallites with a matrix of completely random chains between them. Such a model can only be an approximation: the edges to the crystallites must have a structure intermediate between crystalline and amorphous. The chains in the center of a crystallite are surrounded by identical chains, but this is not the case at the edges of the crystallite in contact with a disordered amorphous structure. Consequently there can be no unequivocal definition of the boundary between the crystalline and amorphous phase.

Determination of the degree of crystallinity by X-ray methods is based on the simple concept that the scattering can be divided into components due to the crystalline and amorphous structures. The usual procedure is to measure the scattered intensity as a diffractometer scan of an unoriented specimen, for which the arcs in Figure 13.1.1 would become isotropic rings. After correction for instrumental effects, such as air scattering, the contribution from the amorphous fraction is predicted from knowledge of the scattering from a completely amorphous sample, or based on the background intensity between the Bragg reflections from the crystalline regions. The degree of crystallinity is then given by

$$\%\chi = \frac{I_c}{I_a + I_c} \times 100 \tag{13.1.1}$$

where I_c and I_a are the total intensity scattered by the crystalline and amorphous regions, respectively. Typical scattering data for unoriented cellulose fibers are shown in Figure 13.1.2, recorded between $2\theta = 3$ and $60°$ [1]. The amorphous background was estimated based on the intensity between the peaks, but is similar to that for ball-milled celluloses. The areas under the two regions yield $\chi = 39\%$.

The above procedure assumes that the total intensity scattered is the same, regardless of the three dimensional arrangement. The scattered intensity is derived from the Fourier transform of the structure:

$$F(R) = \sum_j f_j \exp(-2\pi i R r_j) \tag{13.1.2}$$

The reciprocal space vector, R, is related to the scattering angle, θ, by $R = 2\sin\theta/\lambda$, where λ is the wavelength; f_j is the atomic scattering factor of the jth atom at real space coordinates defined by the vector r_j. The intensity depends on $|F(R)|^2$, which is the product of F(R) and its complex conjugate F*(R):

$$F(R)F^*(R) = \sum_j f_j \exp(-2\pi i R r_j) \sum_k f_k \exp(-2\pi i R r_k) \tag{13.1.3}$$

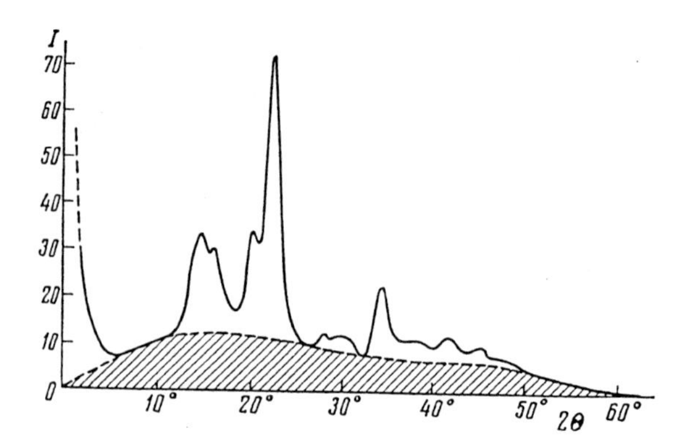

Figure 13.1.2 X-ray diffractometer scan of unoriented cotton cellulose showing separation into crystalline and amorphous contributions [1].

For convenience, this can be divided into summations when j = k and j ≠ k:

$$F(R)F^*(R) = \sum_j f_j^2 + \sum_j \sum_k f_j f_k \exp(2\pi i R(r_k - r_j)) \qquad (13.1.4)$$

The total intensity scattered is obtained as the integral of the terms in Equation 13.1.4 over all R. The integral of the second term is zero; integration of the first term yields a constant. This leads to the important conclusion that the total scattering is a constant, depending only on the atoms making up the structure, regardless of their coordinates.

We generally measure the scattered intensity only in a finite range, typically $2\theta = 5$–$50°$ using CuKα X-radiation. However, at higher scattering angles the intensity curve for a semi-crystalline polymer often becomes more or less featureless due to distortions, so that limiting the measurement to the region where there are actual differences in the scattering between the two components is not a major problem. Likewise, the diffractometer scan is only a one-dimensional scan through two dimensional data, but there too the approximation is not major. The primary source of error comes in the construction of the amorphous background, which frequently requires arbitrary assumptions. But in any case the limitations of the two phase model mean that the degree of crystallinity determined is more of an index than an absolute measurement. The important thing is to decide on a reasonable background and keep its shape constant, and the method is then extremely useful in comparing different preparations of the same polymer.

13.1.3 Degree of Orientation

The arcing of the reflections is due to the fact that there is imperfect orientation of the crystallites, so that the chain axes are only preferentially along the fiber axis, which can be described by a distribution function. When there is cylindrical symmetry to this distribution, the degree of orientation can be quantified in terms of the shape of the observed reflections in the X-ray fiber diagram, via a circumferential scan through one of the reflections. This can be done using scanning optical densitometer for the data recorded on film, or preferentially using a diffractometer or area detector to generate a plot of intensity, $I(\phi)$ versus azimuthal angle, ϕ. The degree of orientation, f_c, is given by the Hermans equation:

$$f_c = (3 < \cos^2\phi_c > -1)/2 \tag{13.1.5}$$

where ϕ_c is the azimuthal angle, and $\phi_c = 0$ corresponds to the fiber axis direction. The value of $<\cos^2\phi_c>$, the average value $\cos^2\phi_c$, is determined from the azimuthal distribution of intensity, $I(\phi_c)$, assuming cylindrical symmetry about the fiber axis:

$$< \cos^2\phi_c > = \frac{\int_0^{\pi/2} \cos^2\phi_c \sin\phi_c I(\phi_c) d\phi_c}{\int_0^{\pi/2} \sin\phi_c I(\phi_c) d\phi_c} \tag{13.1.6}$$

but can be approximated by taking ϕ_c as the half width of the azimuthal intensity distribution. Where deviations from cylindrical symmetry occur, a reflection may vary in both intensity and azimuthal profile as the specimen is rotated about the fiber axis. The orientation distribution can be quite complex and is best described using pole figures, which plot the intensity of a particular reflection for all orientations of the specimen.

13.1.4 Crystallite Size

The radial (2θ) broadening of the reflections is due to the small crystallite size. Bragg's Law ($\lambda = 2d\sin\theta$) requires that reflection occurs only at the angle θ, not at $\theta \pm \delta\theta$, but this condition was derived for an infinite crystal. For finite crystals this breaks down and the reflections have finite width. Crystals of low molecular weight compounds have a mosaic block structure, consisting of domains with dimensions of ~1,000–4,000Å, but the broadening is scarcely appreciable compared to instrumental effects, e.g. beam divergence. However, in polymers the ordered regions may extend only a few unit cells in each direction, and crystallite size becomes the dominant effect in the broadening of the reflections. The usual procedure is to determine the width at half height, B_{hkl}, from a plot of intensity versus 2θ for a particular hkl reflection, e.g. in the diffractometer scan in Figure 13.1.2, after correction by subtraction of the amorphous background. B_{hkl} is related approximately to the crystallite size, L_{hkl}, through the Scherrer equation:

$$L_{hkl} = \frac{0.9\lambda}{B_{hkl}\cos\theta_0} \tag{13.1.7}$$

where θ_0 is the Bragg angle of the reflection and B_{hkl} (corrected for instrumental broadening) is expressed in radians. The crystal dimensions are typically in the range 30–200Å. For example, for a crystallite width of 80Å, the width at half height is about 1° for a reflection at $2\theta = 40°$. L_{hkl} is the dimension perpendicular to that particular hkl plane. The dimensions are typically different in different directions. For example, in cellulose the perfection is much greater along than perpendicular to the fiber axis, and in chitin the lateral crystallite size is greatest along the direction of the intermolecular hydrogen bonds.

13.1.5 Crystal Structure Determination

The broadening, and also the arcing due to fiber disorientation, lead to significant overlap of the reflections. The small crystallites are also significantly distorted, with a result that one sees relatively few reflections at d < 2Å. Consequently most fiber patterns contain only 20–40 independent reflections, some of which will have multiple hkl indices due to overlap. With so few reflections it is not possible to apply the statistical methods that are used to solve the structures of crystals of low molecular weight compounds, for which several thousand Bragg reflections are frequently observed. In general, the only way to solve fibrous structures is by trial and error. The procedure is to determine the unit cell, and then construct a model which gives the best agreement between the observed and calculated intensities.

The unit cell for a crystalline polymer defines the packing at the monomer level. It is not meaningful to think in terms of the molecules, which are polydisperse and may participate in more than one crystallite as tie molecules. Rather we describe a unit cell containing a few monomers that just happen to be linked as polymer chains. For example the unit cell of lobster α-chitin is orthorhombic with dimensions $a = 4.76$Å, $b = 18.85$Å, $c = 10.32$Å, [2] and contains four monomer units consisting of the dimers of two independent chains of opposite sense. The dimensions of the unit cell are determined from the d-spacings of the observed reflections. Dimensions for a trial unit cell can be used to predict the d-spacings for the hkl reflections, and based on the agreement with the observed data, the actual dimensions are found by trial and error adjustment followed by least squares refinement.

The intensity of a reflection, I(hkl), is proportional to the squared modulus of the structure amplitude, F(hkl):

$$F(hkl) = \sum_{j=1}^{N} f_j \exp(2\pi i(hx_j + ky_j + lz_j)) \tag{13.1.8}$$

where f_j is the scattering factor of the jth of N atoms in the unit cell, and x_j, y_j, z_j are its coordinates. The procedure is to postulate a model for the chain conformation and packing within the unit cell and to compare the calculated and observed structure amplitudes, $F_{calc}(hkl)$ and $F_{obs}(hkl)$, respectively. The agreement will be improved by modifying the model in accordance with known stereochemical requirements, and computer software exist in order to do this in an automated routine [3]. The function minimized by least squares is R″:

$$R'' = \frac{\sum w(hkl)||F_{obs}(hkl)| - |F_{calc}(hkl)||^2}{\sum w(hkl)|F_{obs}(hkl)|^2} \quad (13.1.9)$$

where w(hkl) is a weight assigned to each $F_{obs}(hkl)$ based on the accuracy of measurement, and the summations are over all reflections. This is very similar to the traditional crystallographic R value:

$$R = \frac{\sum ||F_{obs}(hkl)| - |F_{calc}(hkl)||}{\sum |F_{obs}(hkl)|} \quad (13.1.10)$$

Typically R and R″ are in the range of 0.15–0.25 for correct structures, with higher values up to say 0.35 for hydrated structures, where the positions of the water molecules often cannot be specified. In general, it is essential to be sure that all reasonable possible structures have been considered and that the proposed solution is significantly better than all the others. Statistical tests exist to check the level of significance for a change in R″, based on the number of reflections and the number of structural parameters being refined [4].

The key step in the above determination is the definition of the trial model. The number of monomer units in the unit cell is easily derived from the cell dimensions and the measured density. Thereafter we need a model for the chain. Crystalline homopolymers usually have helical chains, in which the position of successive monomers are related by a screw operation. The mutual orientation of successive monomers in an isolated polymer chain depends mainly on the stereochemistry at the monomer linkage, so the twist of the chain on going from monomer 1 to monomer 2 is likely to be repeated at subsequent linkages, leading to a helical conformation. This conformation is described as N_M, i.e. N monomers repeating in M turns, with an axial repeat of *c*. In most fibers, *c* is the unit cell axis parallel to the fiber axis, so *a* and *b* define the intermolecular interactions. N and M can frequently be determined from the intensity distribution, as described below, and once this is known, the number of models to be considered in trial and error solution of the structure is considerably reduced.

To determine N and M it is useful to consider the intensity scattered by a single helical chain [5]. For a N_M helix of point monomers, the intensity at coordinate R on the lth layer line is given by

$$|F(R, l)|^2 \propto \sum J_n^2(2\pi Rr) \quad (13.1.11)$$

where $J_n(X)$ is a cylindrical Bessel function of order n and argument X. Bessel functions are tabulated, and their properties can be summarized as follows:

$J_0(X)$ is equal to 1 at $X = 0$, and oscillates about zero as X increases;

$J_n(X)$ is zero at $X = 0$ for $n \neq 0$ are and oscillates about zero as X increases, except that the first peak is at progressively higher X as n increases.

The orders (n) of the terms contributing to the summation for the lth layer line are given by the selection rule

$$l = Nm + Mn \tag{13.1.12}$$

where m is an integer.

This gives us a very simple way of determining the helix symmetry. Consider the meridian of the fiber diagram, i.e. the direction parallel to the fiber axis, which corresponds to $R = 0$. All the Bessel functions are zero when the argument is zero, except for $J_0(0) = 1$. Consequently, for there to be intensity at $R = 0$, there must be a J_0 contribution to that particular layer line. With $n = 0$, Equation 13.1.12 becomes $l = Nm$, and thus there can only be meridional intensity when $l = N, 2N, 3N, 4N,\ldots$.. So for a 6_1 helix, for example, there can be meridional intensity on the 6th, 12th, 18th,... layer lines but not on any others. For the polypeptide 18_5 α-helix, with c = 27Å, the first observed meridional reflection is on the 18th layer line at d = 1.5Å. For the 10_1 DNA-B double helix, which repeats in 34Å, the first meridional maximum is on the 10th layer line at d = 3.4Å. Consequently, we can determine N relatively easily from the meridional region of the fiber diagram.

M is determined from the intensity distribution on layer lines that do not have meridional intensity. The strongest such layer lines will be those with a J_1 (or J_{-1}) contribution: with $n = \pm 1$ the selection rule becomes $l = Nm \pm M$, which means that the strongest layer lines without meridional intensity should be the Mth, N-Mth, N+Mth, 2N-Mth, etc. Thus for the 18_5 α-helix we predict strong off-meridional intensity (close to the meridian) on the 5th and 13th layer lines.

The positions of the meridional reflections and the strong layer lines without meridional intensity are immediately apparent on reviewing the fiber diagram. Hence, if we know the chemical structure, the unit cell, the number of monomers therein, and the helix symmetry, we can soon produce reasonably plausible models for the crystal structure utilizing stereochemical criteria. For example, in the structure of β-chitin [6] the unit cell is monoclinic with dimensions a = 4.85Å, b = 9.25Å, c = 10.38Å and γ = 97.5° and contains two monomers of N-acetyl-**D**-glucosamine. Meridional reflections are seen on all the even layer lines, and the helix symmetry is 2_1. When we build a stereochemical model for the 2_1 chain we obtain a ribbon-like conformation which is elliptical in cross section. When stacked together such chains are separated by ~4.8Å in one direction, where they are linked by hydrogen bonds between the acetamido groups, and by ~9.2Å in the perpendicular direction. At this point the structure is basically solved. Intensity refinements modify the mutual inclination of the monomers and the orientation of the side groups, and lead to a final model with R = 0.23.

It should be stressed that most polymer structures are not solved so easily. Where there is more than one chain in the unit cell and the chain has a chemical "sense", the polarity of adjacent chains can be difficult to determine. In the structure of cellulose I, the X-ray data point to an approximate 2_1 helical conformation, and the unit cell contains dimer units of two chains [7]. The cellulose chain has a sense, and models with both parallel and antiparallel packing are stereochemically acceptable and for both the agreement between the observed and calculated intensities is reasonable. However, the refined parallel model had R = 0.179, compared to R = 0.208 for the best antiparallel model. The models had been constrained with rigid backbones, and the refinement was in terms of seven variables against 40 observed reflections. The significance tests were based on the 40 observed and 40 unobserved reflections (the latter are predicted but too weak to be detected) and these indicated that the parallel model is preferred at the 0.005 significance level, i.e. by 200:1.

Numerous caveats need to be borne in mind in applying the simple rules above for determining N_M. First, the rules are for a helix of point monomers, whereas in actual molecules the monomers are polyatomic, and we need also to consider interferences between the atoms within the monomer itself, which can affect the layer line intensities. For example, for the 18_5 α-helix we predict strong intensity on the 5th and 13th layer lines. We observe a strong 5th layer line, but frequently the 13th layer line is weak. Calculations of intensities for atomic models show that the latter effect is due to interferences within the amino acid monomer structure.

A more serious problem arises when multiple strand helices can be formed. This phenomena is relatively rare in synthetic polymers, but occurs frequently for nucleic acids, polysaccharides and fibrous proteins. A simple example of this is the structure thought to occur in B-starch, where two 6_1 amylose helices repeating in 20.8Å are coiled together [8]. The two chains have the same sense and are exactly out of phase, so that if the origin atom on the first chain is at x,y,0, the equivalent atom on the second chain is at x,y,c/2. The net effect is that the repeat is halved to c/2, and we only see layer lines at c/2, c/4 and c/6. The odd layer lines 1,3,5,... predicted for the repeat of the single strand helix are not seen. The 1st meridional intensity predicted for the single 6_1 strand at c/6 is still seen. Of course, when we look at the pattern we do not know that we have a double helix, and we see meridional intensity on what is the third layer line, so we begin to think of the structure as a 3_1 helix. (Actually this is not incorrect, because the double helix does have 3_1 symmetry relating the pairs of monomers (one for each chain) within the c/2 repeat.) However, it is not possible to build a stereochemically acceptable single helix with three monomers per turn repeating in 20.8Å. But the double helical model gives good agreement.

The above problem arises when the chains have the same sense. If we have chains of opposite sense, as in the DNA double helix, then the elimination of some of the layer lines predicted for the single chains does not occur. Even so, there are only small differences between up and down DNA chains, and although we see the odd layer lines, they are relatively weak.

13.1.6 Aperiodic Scatter from Fibers of Random Copolymers

In all the structures discussed above we have been dealing with repeating polymers, or at least those with repeating backbones as in the fibrous proteins and nucleic acids: generally random copolymers have amorphous structures. Exceptions are random copolymers where the chains are aligned and where it is possible for three dimensional order to occur despite the random sequence. The most well understood materials of this type are the wholly aromatic random copolyesters prepared from monomers such as p-hydroxybenzoic acid (HBA) and 2-hydroxy-6-naphthoic acid, which form liquid crystalline melts and can be processed as high strength high modulus fibers. X-ray fiber diagrams of copoly(HBA/HNA) contain Bragg reflections on the equator and lower layer lines, indicating the presence of three dimensional order akin to crystallinity [9]. Indeed, when these data were first obtained it was thought that the polymers must be blocky and that crystallites were formed by the segregated blocks, or that there was segregation of identical non-repeating sequences. But simulation of the scattering from distorted homopolymer crystallites shows that X-ray data are remarkably tolerant of defects, and that relatively large distortions are necessary before the Bragg reflections disappear, and arrays of parallel chains of different sequences of copoly(HBA/HNA) do predict the observed data.

The evidence for the random sequence comes directly from the X-ray data, in that the layer lines in the fiber diagrams are aperiodic, i.e. there is no obvious repeat along the chain axis. Figure 13.1.3(a) shows diffractometer scans along the fiber axis direction for copoly(HBA/HNA) preparations of different monomer ratio, where the aperiodic meridional maxima vary in number and position as a function of copolymer composition.

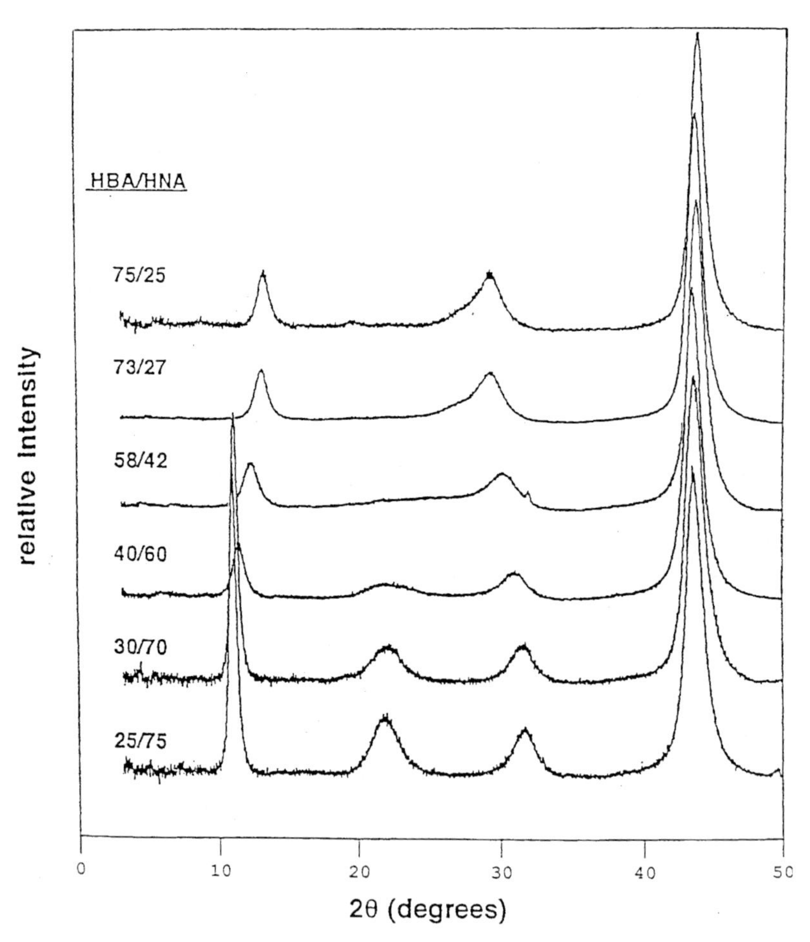

Figure 13.1.3 (a) Diffractometer scans along the fiber axis direction for preparations of copoly(HBA/HNA) for di erent monomer ratios.

The positions of these meridional intensities can be predicted with good accuracy for extended chains of completely random sequence, as will be explained below, and all but minimal non-randomness can be ruled out, as is confirmed by NMR analysis of low molecular weight preparations.

The aperiodic effect arises because the chains are extended, and the advance along the chain is different for each monomer type. The scattering by an extended polymer chain along the chain axis direction, I(Z), is calculated as the Fourier transform of the monomer autocorrelation function, Q(z), where z is the distance along the chain axis and Z is its reciprocal space coordinate [10]. Q(z) is a function defining the sequence statistics. For the case of copoly(HBA/HNA), we can generate Q(z) from the first nearest neighbor probability function, $Q_1(z)$, which depends on copolymer composition and combination probabilities. If $H_1(Z)$ is the Fourier transform of $Q_1(z)$, the z-axis scattering is given by

$$I(Z) = 1 + 2\mathrm{Re}\,[H_1(Z)/(1-H_1(Z))] \tag{13.1.13}$$

Re signifies the real component; $H_1(Z)$ is the product of three components which define respectively the monomer ratio, the combinatorial probabilities, and the phase terms arising from the different lengths of the projections of atomic structures of the monomers along the chain axis.

$$H_1(Z) = P \cdot M \cdot X \tag{13.1.14}$$

For copoly(HBA/HNA), P, M, and X can be written in matrix form:

$$P = \begin{bmatrix} p_3 & 0 \\ 0 & p_N \end{bmatrix} \quad M = \begin{bmatrix} M_{BB} & M_{BN} \\ M_{NB} & M_{NN} \end{bmatrix} \quad X = \begin{bmatrix} X_B & 0 \\ 0 & X_N \end{bmatrix} \tag{13.1.15}$$

where subscripts B and N designate HBA and HNA, respectively. p_B and p_N are the monomer mole ratios of HBA and HNA. The X matrix contains the phase terms for each monomer:

$$X_A = (\exp 2\pi i Z z_A) \tag{13.1.16}$$

where z_A is the length of monomer A projected along the chain axis. The M_{AB} terms define the probability of monomer A being followed by monomer B:

$$M_{AB} = \frac{r_{AB} p_B}{\sum_j r_{Aj} p_j} \tag{13.1.17}$$

The r_{AB} terms are unity in the random copolymer, and can be adjusted to generate non-random sequences. I(Z) is calculated from

$$I(Z) = \sum p_A F_{AA}(Z) + 2\mathrm{Re} \sum_A \sum_B p_A F_{AB}(Z) \frac{M_{AB} X_A(Z)}{1 - M_{AB} X_A(Z)} \tag{13.1.18}$$

where the F_{AB} is the Fourier transform of the cross-convolution of monomer A with monomer B:

$$F_{AB} = \sum_i \sum_j f_{A,i} f_{B,j} \exp 2\pi i Z(z_{B,j} - z_{A,i}) \tag{13.1.19}$$

f is the atomic scattering factor and z is the atomic coordinate in the chain axis direction; the subscripts A,i and B,j designate the ith atom of monomer A and jth atom of monomer B, respectively.

Figure 13.1.3(b) shows I(Z) data calculated for models of copoly(HBA/HNA) in the same monomer ratios as for the observed data in Figure 13.1.3a. The positional agreement is very good. The intensities do not match perfectly, but this simulation is only for the isolated chain. When the effect of chain packing is introduced, then this agreement is improved. Despite the apparent disorder, arrays of random copolymer chains predict

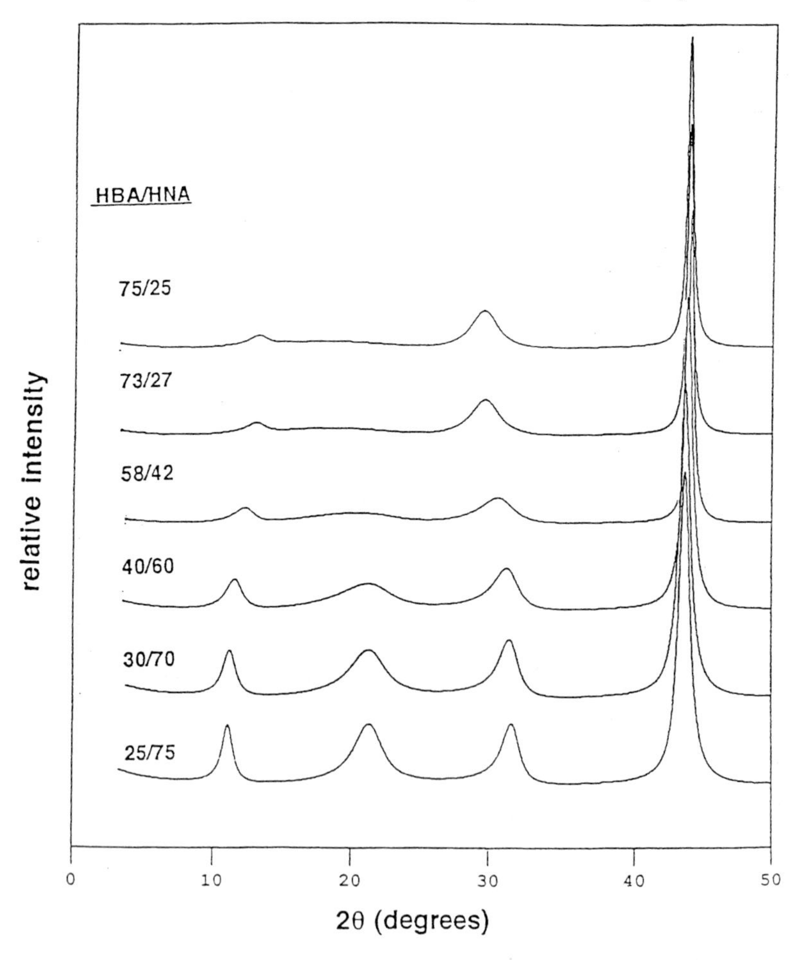

Figure 13.1.3 (b) Calculated I(Z) plotted against scattering angle (2θ) for copoly(HBA/HNA) with the same monomer ratios as in (a).

the observed strong Bragg reflections, and the potential energies are not very different from those for homopolyesters packed on the same lattice, indicating that despite the chemical difference of the two monomers there is not a major problem in packing the aligned chains.

13.1.7 Transesterification in Blends of Copoly(HBA/HNA)

The above analyses provide a unique method to study transesterification between copolyesters using X-ray diffraction [11]. Ester interchange reactions in melt blends of the different homopolyesters lead eventually to a completely random copolymer. The nonperiodic scattering of the wholly aromatic copolyesters allows us to follow this reaction in melt blends of preparations with different monomer ratios. The d-spacings of the meridional maxima are analytical, in that they are specific to a particular monomer ratio. Immediately after melt blending, mixtures of the 30/70 and 75/25 copolymers exhibit a single solid-nematic transition at a temperature well below the transitions of either the starting copolymers or those of intermediate compositions. Figure 13.1.4 (curve a) shows a diffractometer scan along the axis direction for fibers drawn from the melt blend of the 30/70 and 75/25 copolymers, in which the overall monomer ratio was 60/40. We observe the meridional maxima characteristic of the starting copolymers, which indicates the formation of an unreacted blend. Even though the monomer ratio is 60/40, the nearest neighbor statistics are those for the mixture of the two copolymers, not the

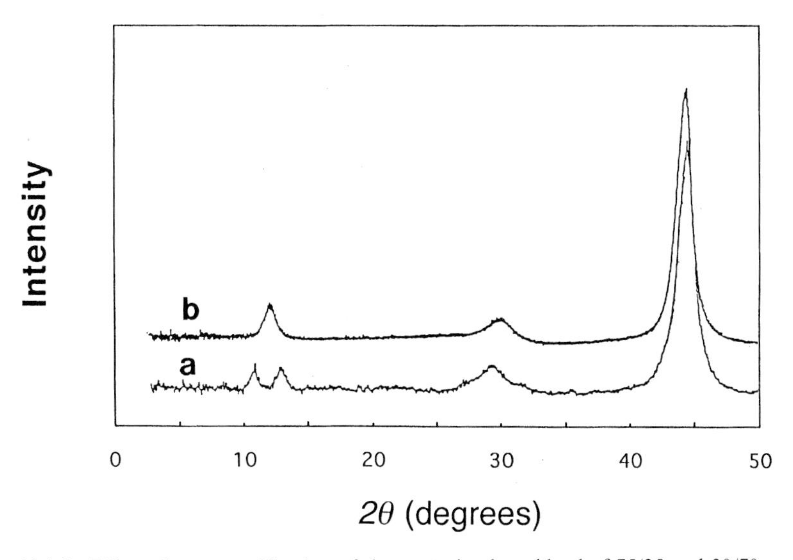

Figure 13.1.4 Effect of transesterification of the scattering by a blend of 75/25 and 30/70 copoly(HBA/HNA) with an overall 60/40 comonomer ratio.
(a) immediately after melt-blending;
(b) after compression molding at 315 °C for 1 h.

60/40 copolymer. Figure 13.1.4 (curve b) shows the data obtained for fibers drawn from the melt blend after compression molding for 1 h at 315°C. The peak position are now identical with those for the random copolymer prepared by reacting the monomers in the 60/40 mole ratio: most obviously, the first two peaks for the initial blend are replaced by an intermediate peak after the heat treatment. The progress of the reaction can followed by monitoring the shift in position of the meridional maxima as a function of time, by drawing fibers from samples held in the melt for different periods of time. Figure 13.1.5 shows the data recorder in the d = 6–8 Å region for such samples. The two peaks characteristic of the original copolymers move towards one another until they merge to give a single maximum at the d-spacing characteristic of the random copolymer of intermediate composition. During this process, the temperature of the solid-nematic

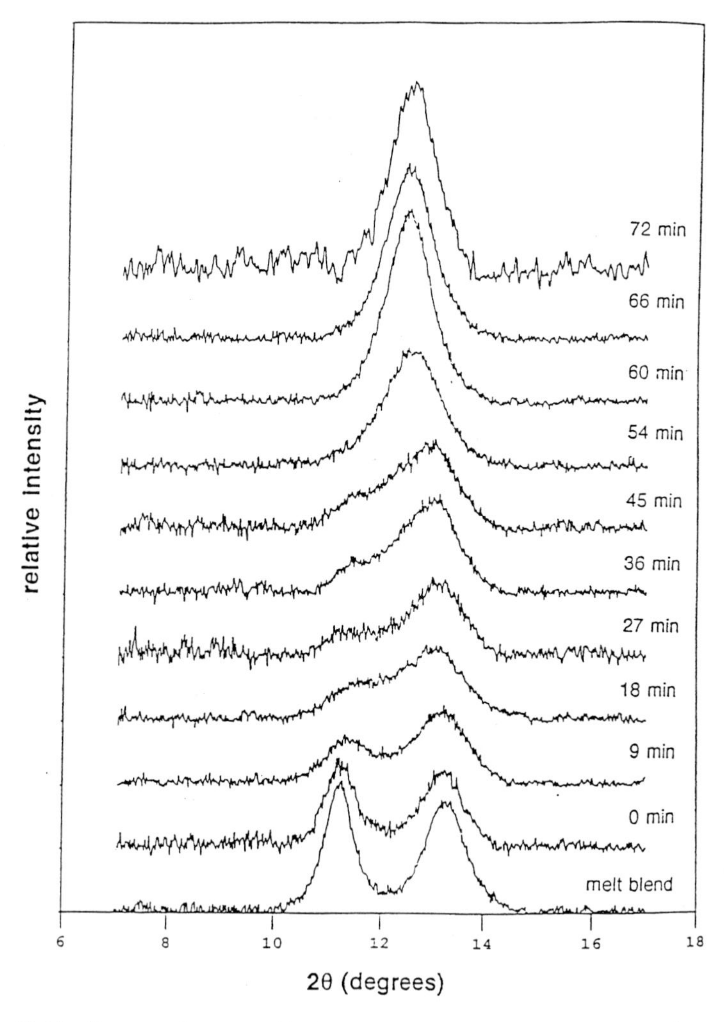

Figure 13.1.5 Progress of transesterification reaction. Diffraction in the d = 6-8 Å region along the fiber axis direction for fibers drawn from melt blends after compression molding at 315 °C for different times.

transition slowly rises until it reaches the value for the intermediate composition. The matrix calculation described above to predict I(Z) can be extended to model the changes in the statistics during the progress of the interchange reaction. This makes it possible to extract kinetic data, and the rate constants and activation energies prove to be comparable to those between PET and its deuterated analog.

13.1.8 Non-Linearity and Distortions

The above treatment of nonperiodic scattering was primarily for an infinite straight chain. However, the data suggest that the ordered regions are finite: the data for the polyesters are reproduced by copolymer chains with an average degree of polymerization of 10–12 that are sinuous rather than perfectly straight. These parameters, chain length and sinuosity, were examined in detail for dry-jet wet-spun fibers of a group of wholly aromatic copolyimides prepared by condensation reaction of 3,3′,4,4′-biphenyltetracarboxylic dianhydride (BPDA) with mixtures of o-tolidine (OTOL) and *p*-phenylene diamine (PPD):

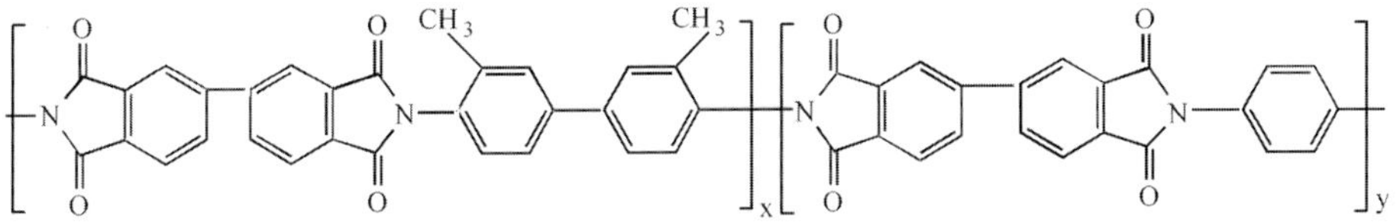

The copolymer fibers give rise to non-periodic layer lines, and the positions of these are reproduced by a random comonomer sequence [12].

To investigate the chain sinuosity, the distribution of monomer advances can be approximated using a truncated Gaussian function, so that X_A becomes

$$X_A(Z) = \exp(-2\sigma^2\pi^2Z^2)\exp(2\pi i Z z_A) \qquad (13.1.20)$$

The size of homopolymer crystallites can be obtained from the line broadening using the Scherrer equation, as described above. The effect of distortions on the breadth of Bragg reflections has been analyzed by Hosemann [13] in his treatment of paracrystallinity. Although there remain unsolved problems with this approach for a three dimensional structure, it can be used effectively to treat distortions in one dimension, and is based on the same approach as used to predict the non-periodic scattering. The analysis extracts information on the crystallite size and paracrystallinity for a homopolymer in terms of the breadth of a series of orders of hkl reflections. The breadth, d, is related to L_{hkl}:

$$d^2 = 1/L^2_{hkl} + (\pi g n)^4/d^2_{hkl} \qquad (13.1.21)$$

$g\ (= \sigma/z_A)$ is the index of paracrystallinity and n is the order of a reflection, so a plot of d^2 versus n^4 has intercept $1/L^2_{hkl}$ and slope $(\pi g)^4/d^2_{hkl}$ leading to estimates of L_{hkl} and g.

The random copolymers give rise to non-periodic peaks, and in general these cannot be treated as orders. The only exceptions are some of the peaks predicted for chains in which

z_A and z_B are simple multiples of a common length factor, *a*; for example, $z_A = 3a$ and $z_B = 4a$ [9]. In such a case, we predict very sharp periodic peaks (d-functions) at $Z = 1/a$, $2/a$, $3/a$.. etc., with broader, aperiodic peaks between them. The periodic peaks are invariant, i.e. their positions are independent of monomer ratio. For a finite and/or sinuous chain, these sharp maxima are broadened, and thus their observed widths can be used to evaluate the length and sinuosity of the scattering units.

Straight chain models for copoly(BPDA-OTOL/BPDA-PPD) have advances of 20.43 Å and 15.78 Å, respectively, for the two monomers. The calculated I(Z) data contain relatively strong, sharp maxima on the 4th, 8th and 12th layer lines, which are the first three orders of a repeat of ~5.2 Å. These peaks arise because of the approximate coincidence of the 4th order of 20.43 Å and the 3rd order of 15.78 Å. There is not an exact coincidence, and hence the predicted peaks are not d-function but have finite widths. Figure 13.1.6 (curve a) shows a plot of d^2 against n^4 for the calculated data for an infinite straight chain model for the 80/20 copolymer, which is approximately a straight line plot with positive slope. The intercept at $n^4 = 0$ is very close to the origin, consistent with the infinite chain model (i.e. analogous to infinite crystal size). An equivalent plot for the

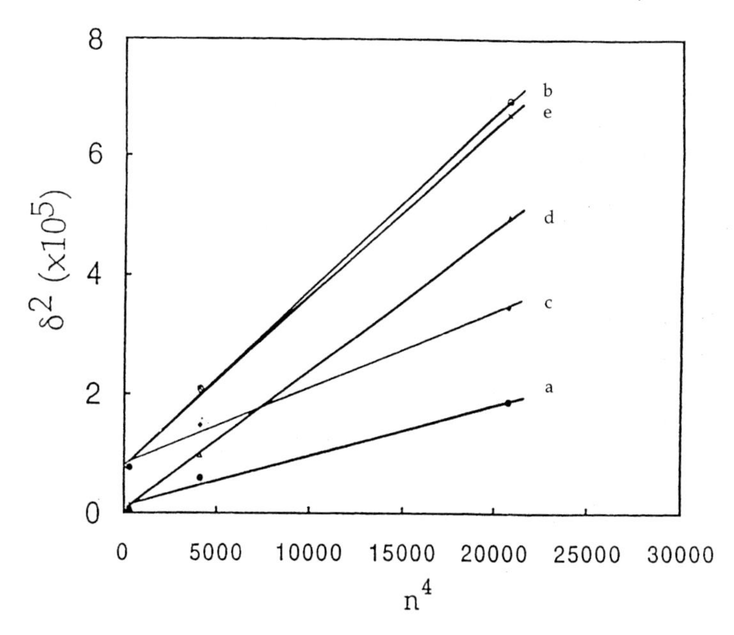

Figure 13.1.6 Plots of d^2 versus n^4 for 80/20 copoly(BPDA-OTOL/BPDA-PPD):
(a) predicted for infinite, straight chain;
(b) observed data for 4th, 8th and 12th layer lines;
(c) predicted data for a linear chain of length 390Å;
(d) predicted data for an infinite sinuous chain with $s = 0.166$Å ($g = 0.85\%$);
(e) predicted data for a finite sinuous chain of length 480 ± 20 Å and sinuosity $s = 0.164 \pm 0.002$Å ($g = 0.84\%$).

observed data for the copolymer fibers is shown as curve b, and is also approximately linear within experimental error. However, the latter plot has a positive intercept and is steeper than those predicted for the infinite straight chain models, suggesting that we need to consider a model with finite chain lengths and/or sinuous conformation.

Plots of d^2 against n^4 for finite linear chain models of the 80/20 copolymer with different chain lengths have approximately constant slope as we change the chain length, while the intercept moves to progressively higher positive values. We can match the intercept using a DP of 20 (curve c), which corresponds to a chain length of 390 Å. However, the slope is less than that observed, meaning that the predicted profiles for the maxima on the 8th and 12th layer lines are too narrow. Alternatively, if we change the sinuosity for the infinite chain, as defined by the s parameter, the primary effect is to increase the slope, with only a small increase in the intercept. In curve d we match the slope of the plot for the observed data by setting the chain sinuosity to s = 0.166Å (g = 0.85%). However, in this case the $n^4 = 0$ intercept is significantly below that for the observed data: it would correspond to a "crystallite size" of ~800 Å. But it is possible to match *both* the intercept and slope by trial and error refinement of the chain length and sinuosity parameters. The best fit (curve e) is achieved using DP = 25 ± 1 (chain length 480 ± 20 Å) and a sinuosity s = 0.164 ± 0.002Å (g = 0.84%).

The axial correlation lengths are much higher than the axial crystallite width of 183 Å determined for the homopolymer, for which the correction for paracrystalline distortion is negligible. In a separate work, we have measured the equivalent axial crystallite width for poly(*p*-phenylene terephthalamide) processed as Kevlar® 149 fibers, and obtained a crystallite dimension of 400 ± 12 Å (g = 1.71 ± 0.06 %). The correlation length for the 80/20 copolyimide exceeds this figure, which gives a qualitative picture of the high level of chain linearity in the copolyimide fibers, despite the random monomer sequence. We have also applied these analyses to the random sequence aromatic copolyamide Technora®, prepared from terephthaloyl chloride and equimolar proportions of p-phenylene diamine and 3,4′-diaminodiphenyl ether. The best agreement for unannealed Technora is obtained with an ordered segment length of 300 ± 15Å and sinuosity of g = 1.52 ± 0.05%. Consequently, although Technora is a random copolymer and the presence of the random diaminodiphenyl ether leads to a more distorted lateral packing of the chains than occurs in Kevlar, the fibers produced by the dry-jet wet spinning process have extended conformations that are very similar in linearity to those for Kevlar, which probably account for the analogous high tensile strengths and moduli.

13.1.9 References

1. Bonart, R., Hosemann, R, Motzkus, F. and Rush, H., *Norelco Reporter* (1960) 7, p. 3
2. Carlstrom, D., J. Biophys. *Biochem. Cytol.* (1957) 3, p. 669
3. Arnott, S., and Wonacott, A, *Polymer* (1966) 7, p. 157
4. Hamilton, W.C., Acta Crystallogr., *Acta Crystallogr.* (1967) 20, p. 473
5. Cochran, W., Crick, F.H.C., and Vand, V, *Acta Crystallogr.* (1952) 5, p. 581
6. Gardner, K.H., and Blackwell, J., *Biopolymers* (1975) 14, pp. 1581–1595
7. Gardner, K.H. and Blackwell, J., *Biopolymers* (1974) 13, p. 1975

8. Gardiner, E.S. and Sarko, A., *J.Appl. Polym. Sci., Appl. Polym. Symp.* (1983) 37, p. 303
9. Gutierrez, G.A, Chivers, R.A., Blackwell, J., Stamatoff, J.B. and Yoon, H., *Polymer* (1983) 24, pp. 937–944
10. Biswas, A., and Blackwell, J., *Macromolecules* (1987) 20, pp. 2997–3002
11. McCullagn, S.M., Blackwell, J., and Jamieson, A.M., Macromolecules (1994) 27, p. 2996
12. Wu, T.Z, Blackwell, J., and Chvalun, S.N., *Macromolecules* (1995) 28, pp. 7349–7354
13. Hosemann, R., *Z. Physik*, (1950) 128, p. 1

13.2 Small-Angle Scattering

N. Sanjeeva Murthy
Honeywell International Inc. Morristown, New Jersey, USA

13.2.1 Introduction

Small-angle X-ray scattering (SAXS) has been used since the 1950's [1] and is now routinely applied to studying structure in fibers. Small-angle neutron scattering (SANS), which is closely related to SAXS, is not as widely used because of limited access to instrumentation. In both SAXS and SANS, here collectively referred to as small-angle scattering (SAS), scattering of X-rays or neutrons within $\sim 5°$ from the direction of the incident beam is analyzed to obtain structural information over length scales of 1–100 nm. We will not discuss small-angle light scattering which provides information on much larger length scales ($\sim$ micrometers) and is therefore useful for studying spherulitic structures in unoriented polymers.

Fibers are oriented structures and are usually partially crystalline. A fiber in general could be regarded to consist of crystalline regions, noncrystalline (often referred to as amorphous) regions and voids. A commonly accepted fibrillar model for fibers is shown in Figure 13.2.1. Yarns are bundles of filaments (fibers) usually ~ 20 μm in diameter, and the figure shows ~ 10 nm fibrils within these filaments. Between the fibrils there are interfibrillar amorphous chain segments and elongated voids. The volume fraction of the voids vary greatly in fibers, and is typically around 1%. The fibrils some times form ordered aggregates which may be regarded as macrofibrils. Within the fibrils, there are crystalline lamellae separated by interlamellar amorphous chain segments. In some polymers such as polyamides (nylons), polyolefins and aliphatic polyesters, the crystals often have a folded chain structure (shown in Figure 13.2.1). In others, such as cotton,

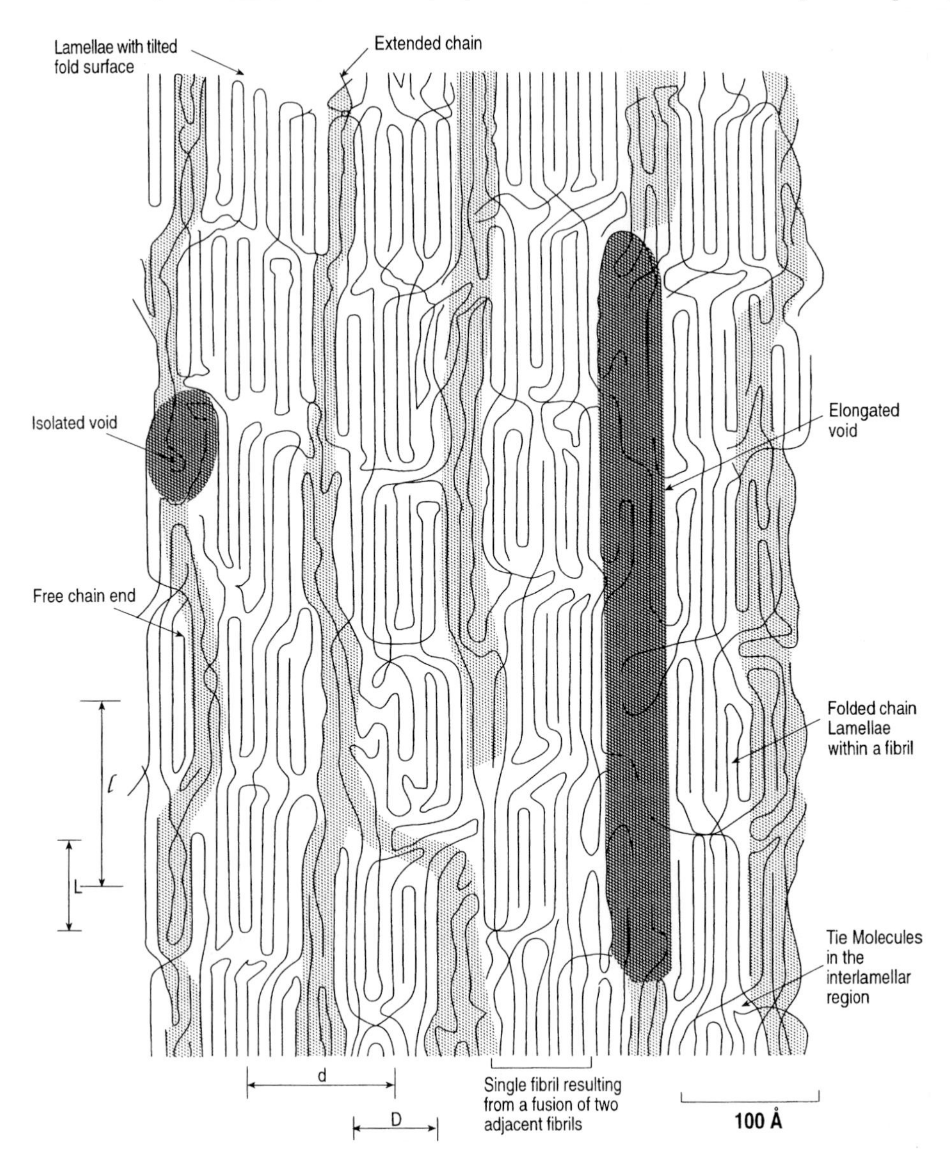

Figure 13.2.1 A structural model of a semicrystalline fiber [25].

wool, and liquid crystalline polymers, the crystals are always in the form of extended-chain micelles.

SAS is commonly used to study the distribution of interlamellar and interfibrillar amorphous regions in fibers by measuring the various dimensions shown in the figure. It has been observed, for instance during polymer deformation, that although the changes in the crystalline regions as seen in the wide-angle X-ray diffraction pattern may be small, the small-angle pattern can show large changes. The parameters commonly derived from

SAS data from fibers are listed in Table 13.2.1. In addition, SANS can be used study the difference in the distribution of solvent in the various amorphous regions.

13.2.2 Characteristics of SAS

SAS arises from the contrast between neighboring domains 5–50 nm in size. As can be seen in Figure 13.2.1, this contrast in SAXS arises from the difference in the electron densities between voids, crystalline regions and the amorphous domains; in SANS, the contrast is generated by selective deuteration of void spaces, amorphous domains and crystalline regions. In both SANS and SAXS, there are two distinct features, a diffuse equatorial streak and discrete reflections along or close to the meridian (Figure 13.2.2). The relation between the parameters derived from SAS data and the structural features are given in Table 13.2.1. The role of multiple scattering at very low angles, and surface scattering due to total internal reflection will not be discussed here.

The equatorial diffuse scattering (EDS) is due to elongated scattering objects which are few hundred nm long and ~10 nm in diameter. These scattering objects are attributed to both microvoids [2,3] and fibrils [4,5]. SANS data have shown that the scattering at least at $q < 0.01$ (q is the scattering vector related to the scattering angle 2θ by the relation

Table 13.2.1 Relation between SAS Data and Structural Features

Structural feature	Symbol	SAS parameter
Voids		
Diameter (or size distribution)	D_v	Guinier slope of the equatorial diffuse scattering (EDS)
Length	L_v	Width of the EDS along the meridian
Misorientation	ϕ_v	Variation in the width of the EDS with q_{eq}
Interfibrillar spacing	d	Position of the weak peak in the EDS
Lamellar stack		
Lamellar spacing	l	Position of the lamellar meridional peak (LMP)
Height	L	Height (along the fiber-axis) of the LMP
Diameter	D_f	Azimuthal width (perp. to the fiber-axis) of the LMP
Tilt angle	χ	Azimuthal separation of the LMP in 4-pt. pattern
Misorientation	ϕ_l	Variation in the meridional-width of the LMP with q_{eq}
Volume fractions		
Voids	I_v	Intensities of the three components
Fibrillar aggregates	I_f	
Lamellae	I_l	

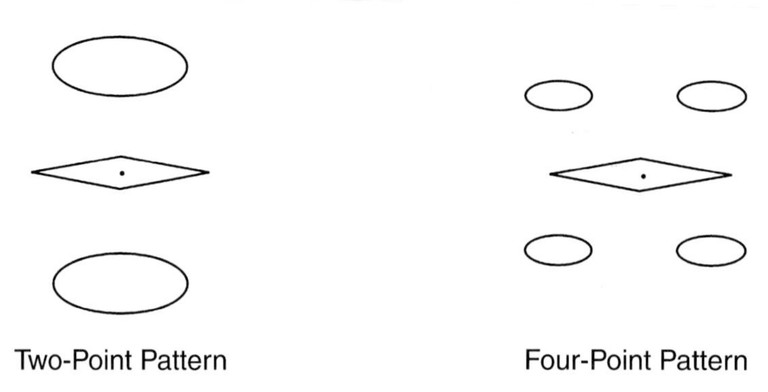

Figure 13.2.2 Typical SAS patterns from fibers.

$q = (4\pi \sin\theta)/\lambda$, where λ is the wavelength) is due to elongated voids [6]. The straight-line obtained in the Guinier plots even at $q > 0.01$ Å^{-1} (Figure 13.2.4) suggests that the scattering at these angles could be due to widely dispersed scattering voids, although the implied uniformity in size is somewhat disconcerting. Also, the similarity that is often observed between the diameter of the scattering objects determined from the Guinier slope at $q > 0.01$ Å^{-1} from EDS, and the crystallite dimensions from wide angle X-ray scattering WAXS [5,7–9] suggests that the scattering objects are fibrils. This interpretation is also supported by the presence of an interference peak in the EDS [5,8]. Even when a peak is not present, there could be a diffuse scattering due to liquid-like order among the fibrils [3].

Both these findings can be accommodated if we realize that there could be two components in EDS. An intense scattering due to widely dispersed elongated voids (at $q < 0.01$ Å^{-1}), and the other much weaker scattering due to interfibrillar interference (at $q > 0.01$ Å^{-1}). The q-range of the data and the relative intensities should be carefully considered in interpreting or apportioning the EDS to voids or fibrils. In either case, it is possible to analyze EDS and determine the volume fraction of the scattering objects, and their length, diameter and orientation, from the intensity and the shape of the EDS (Table 13.2.1).

The descrete small-angle scattering along or close to the fiber-axis is due to the lamellar structure in the fibers, and we will refer to these as Lamellar Meridional Peaks (LMP). This scattering is also called the lamellar reflection since it is somewhat similar to the Bragg reflection seen in highly ordered structures although orders higher than the first is rarely observed. The LMP seen in fibers fall into two categories, the two point pattern and the four-point pattern (Figure 13.2.2). The two point pattern is due to the lamellae whose

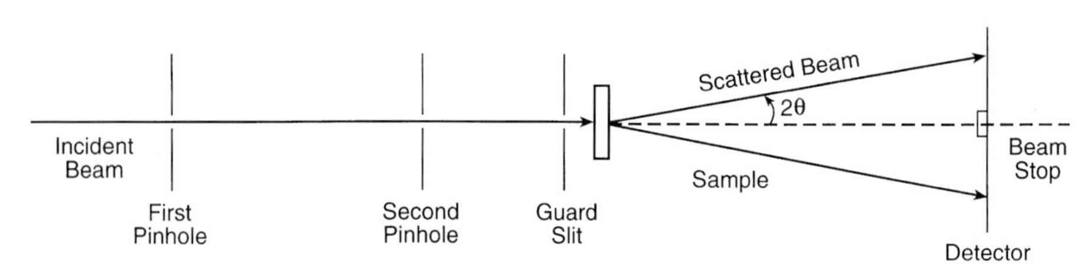

Figure 13.2.3 Schematic of the apparatus for SAXS measurement.

surface is perpendicular to the fiber-axis, or due to lamellar stacks whose position in adjacent fibrils is uncorrelated. The four point pattern also has two interpretations. In one interpretation, the off-meridional diffraction spots are from obliqueness of the fold-surface of the lamellar crystals within the stacks of lamellae, i.e., the normals to the lamellar surface are inclined with respect to the fiber-axis; and the basic model is one of uncorrelated fibrils [10,11]. Alternatively, these off-meridional spots have been interpreted by considering the lamellae in a paracrystalline lattice, i.e., by assuming that the longitudinal position of the neighboring fibrils are correlated [12–14]. Zheng et al. [15] have attempted to combine the two aspects, the inclined lamellar surface and the coherence between adjacent lamellar stacks (paracrystalline lattice).

13.2.3 Instrumentation

SAXS data from fibers are usually obtained using a pin-hole collimation (Figure 13.2.3) The collimation consists of a pair of pin-holes separated from each other by several tens of cm. to give a beam of about 0.1 mm in diameter. A third guard pin-hole is necessary to eliminate the parasitic scattering and obtain data at smaller angles. Franks' mirrors are also used in some X-ray cameras to obtain point collimation. Recently, capillaries have become available to focus the X-rays to a point; but, because of their limited efficiencies they are currently used to obtain small beam sizes (5–50 microns) for microbeam diffraction.

The data can be recorded on films, although they are now becoming widely replaced by image plates, multiwire area-detectors and CCD cameras. Image plates are the most economical substitute for photographic films. The exposed plates are read out by scanning with a laser beam. The laser beam causes the stored charge to fluoresce, which is detected by a photomultiplier. The image plates can be reused by erasing the previous image by illuminating the plate with a flood light of a different radiation. The image plates have a large dynamic range and low fog levels comparable to those of photographic film. Multiwire detectors have an array of wires and special read-out electronics keep track of the position and the number of X-ray photons entering the detector chamber and ionize the gas in the detector. In indirect CCD's, a phosphor plate converts the X-ray photons into visible light, a tapered fiber optic bundle demagnifies the image and projects it onto a CCD chip. X-rays can also be directly imaged on to CCD chip; although the area of operation is small, it is sufficient for SAXS work.

The three sources of X-rays are sealed tubes, rotating anodes and synchrotron. SANS measurements are done at central facilities using neutrons obtained either from a nuclear reactor (constant wavelength) or at a spallation source (varying wavelength).

13.2.4 Data Analysis

There are in principle two different routes for analyzing the SAS data. The data could be fitted to the scattering from a model whose parameters are refined by a least-squares procedure. Alternatively, the data could be parameterized by fitting the data to a few generic functions, and these parameters related to the structure. Both these approaches are used in analyzing simple systems. For instance, a Guinier plot is used in solution studies to calculate a radius of gyration to parameterize the data, and the data are also fitted to models such as an assembly of subunits to determine the subunit structures. Although there have been attempts to fit the data to a simple model [16], fibers are in general too complex to be modeled in sufficient detail. Hence, we will describe the analysis of the data in terms of a few parameters which are then related to the features of the structure (Table 13.2.1) [8].

The two-dimensional SAS pattern shown in Figure 13.2.2 is usually analyzed in several segments. A sector averaged equatorial scan (Figure 13.2.4a) can be used to determine the distribution of lateral sizes of the scattering objects (voids/fibrils). If it is assumed that the scattering objects are widely separated, elongated objects, then the radius r of the voids can be determined by Guinier slope in the plot of Ln (I) vs. q^2 (Figure 13.2.4b) according to the equation [1]

$$I(q) = I(0) \exp(-q^2 r^2/5) \tag{13.2.1}$$

The surface area per unit volume (S/V) of the scattering entities can be derived from Porod's law [17]

$$\lim_{q \to \infty} (Iq^4) = 2\pi (S/V)(\rho_1 - \rho_2)^2 \tag{13.2.2}$$

If the absolute intensities and the densities are not known, then one can use the relation

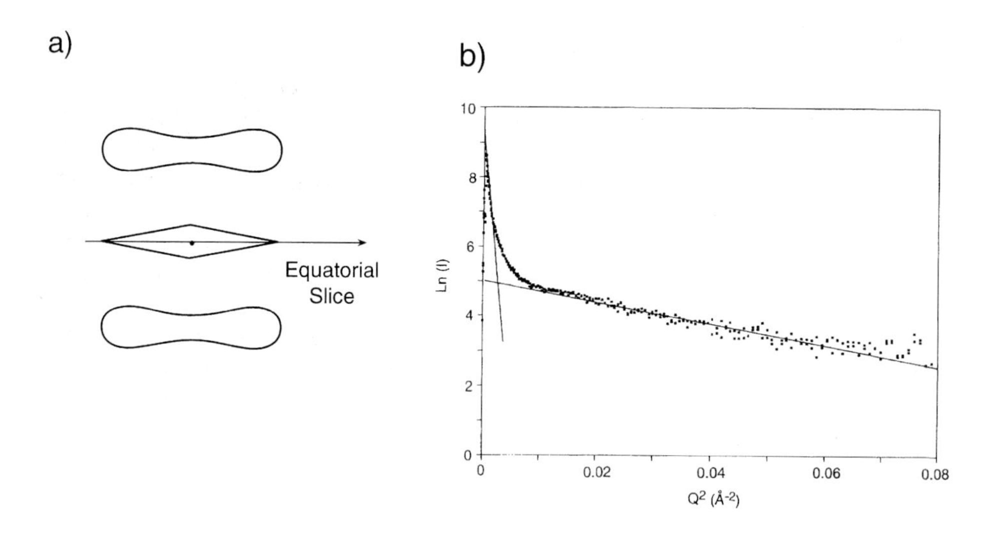

Figure 13.2.4 Analysis of the equatorial scans in SAS data. (a) One-dimensional slice through the equatorial steak. (b) Guinier plot of the intensity distribution in the equatorial streak for determining the diameter of the long scattering entities.

$$\frac{\lim_{q \to \infty} (Iq^4)}{Q} = \frac{(S/V)}{\pi\phi(1-\phi)} \tag{13.2.3}$$

where Q is the invariant given by

$$Q = \int_0^\infty q^2\ I(q)\ dq \tag{13.2.4}$$

In some instances it is possible to observe an interfibrillar interference peak in the EDS (Figure 13.2.4c). In such cases, the interfibrillar spacing d can be estimated from the Bragg's law

$$d = 2\pi/q \tag{13.2.5}$$

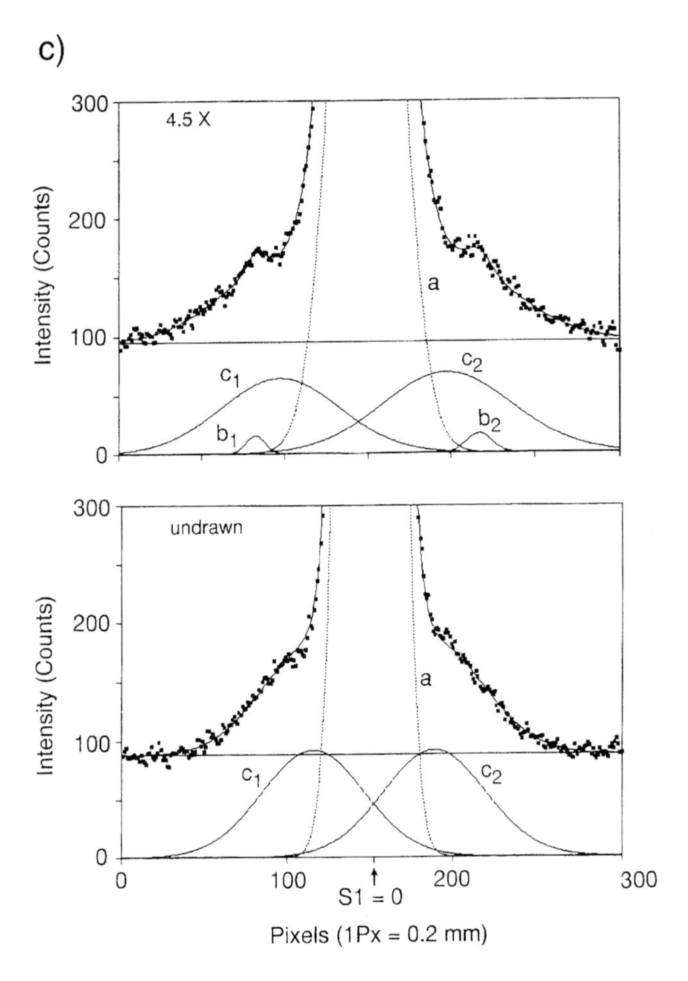

Figure 13.2.4 Analysis of the equatorial Scans in SAS data. (c) Profile fit to the equatorial streak showing the sharp reflection from ordered fibrillar aggregates in the highly drawn fibers and a diffuse interference peaks from the laterally uncorrelated fibrils in the undrawn fibers.

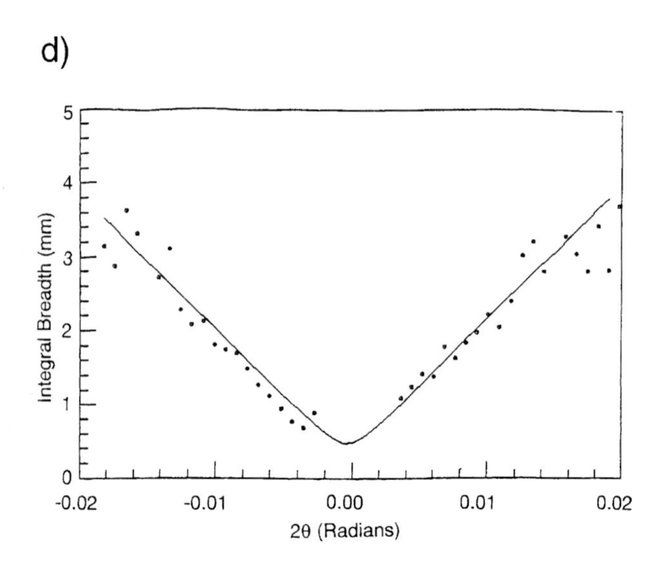

Figure 13.2.4 Analysis of equatorial scans in SAS data. (d) Plot of the width of the equatorial streak with scattering angle (along the equator) to determine the height and degree of orientation of the scattering objects oriented parallel to the fiber axis.

The width of the equatorial streak at zero angle, which can be determined by extrapolating the width of the streak measured at higher angles gives the length of the scattering objects. The changes in the width with q, gives the orientation of the fibril/void. These two parameters are respectively the intercept and the slope in a plot of width vs. q as shown in Figure 13.2.4d.

A longitudinal slice (Figure 13.2.5a) through the meridional reflection can be used to determine the position of the lamellar peak and using Equation 13.2.5 obtain the lamellar spacing (Figure 13.2.5b). It is preferable to calculate the long-period without any Lorentz correction, even though it may be somewhat higher, because the commonly used q^2 correction over corrects and gives too small a value [11]. The width of this lamellar peak corresponds to the height of the lamellar stacks. The lamellar structure can be analyzed in some detail, such as the determination of the thickness of the amorphous and crystalline domains using the correlation function $\gamma(x)$ which can be obtained from the relation

$$\gamma(x) = \left[\int_0^\infty I_1(q_2) \cos(xq_2) dq_2\right] / \left[\int_0^\infty I_1(q_2)\, dq_2\right] \qquad (13.2.6)$$

where

$$I_1(q_2) = (4\pi^2)^{-1} \int_0^\infty I_1(q_1, q_2)\, q_1\, dq_2 \qquad (13.2.7)$$

q_1 and q_2 are, respectively, the equatorial and the meridional axes in the diffraction pattern [8].

A slice through the meridional reflections perpendicular the fiber-axis (Figure 13.2.6a) is equivalent to an azimuthal scan. This transverse intensity profile can be used to calculate the diameter of the lamellae from the Guinier analysis as shown in Figure 13.2.6b [18]. This size (~60Å dia.) is similar to that obtained for the lateral crystallite size by

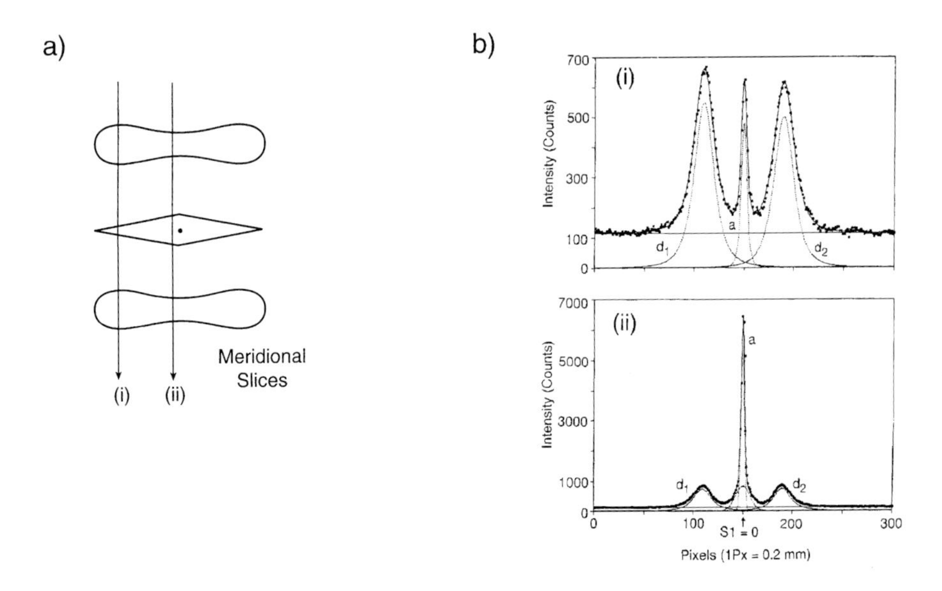

Figure 13.2.5 Analysis of the longitudinal scans. (a) One-dimensional slice through the lamellar reflections. (b) Profile fit to two typical longitudinal scans; (i) is away from the origin, and (ii) is near the origin.

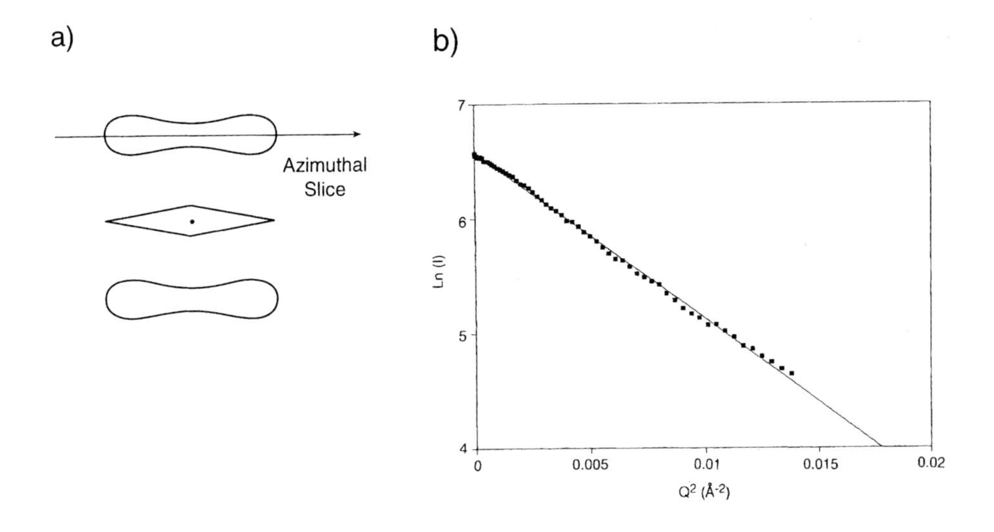

Figure 13.2.6 Analysis of the azimuthal scans. (a) One-dimensional slice through the top-pair of the lamellar peaks. (b) Guinier analysis of the azimuthal intensity distribution to determine the diameter of the lamellar stacks, i.e., fibrils.

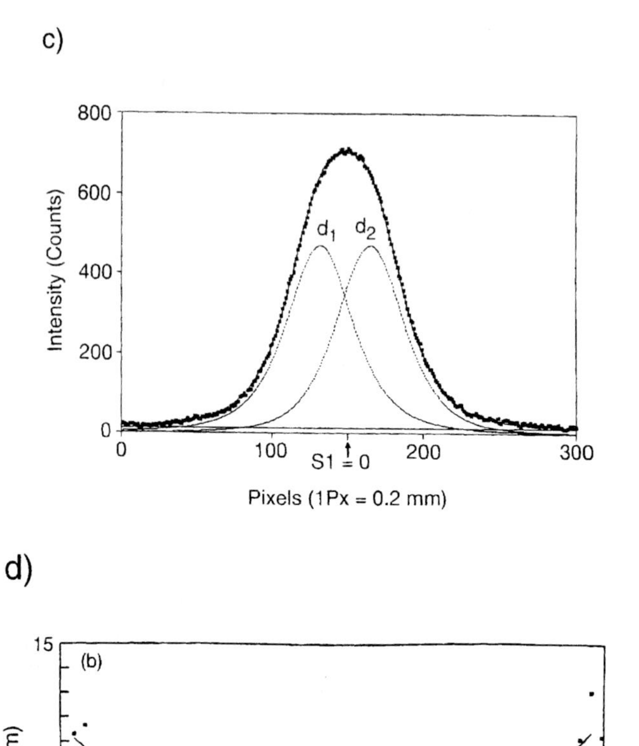

Figure 13.2.6 Analysis of the Azimuthal scans. (c) Profile fitting the azimuthal scan to determine the angular separation of the lamellar reflection, i.e., the tilt angle of the lamellar planes. (d) Plot of the width of the lamellar scattering with scattering angle (parallel to the equator) to determine the height and degree of orientation of the lamellar stacks.

WAXS. These two values do not always agree because the two methods give different moments of the size distribution. The χ angle between the lamellar reflection in the azimuthal scan (Figure 13.2.6c) is an important parameter, and relates to the tilt-angle of the lamellar plane. The orientation of the lamellar stack can be obtained from a plot of the variation in the width of the lamellar reflection with $q_{\perp}$ (Figure 13.2.6d). Other slices, such as diagonal slices are also occasionally used. Sometimes parameters such as longitudinal crystallinity and transverse crystallinity are calculated to quantify the fraction of the amorphous chains in the interlamellar and the interfibrillar regions.

The most efficient way is to analyze the SAS is to analyze the entire 2-D pattern in a single step where all the features of the EDS, the interfibrillar scattering and the LMP can be fitted simultaneously. It has been shown that such 2-D fit can be most effectively carried out in elliptical coordinates [19]. Such a fit can give a complete description of the structure of by a self consistent, least-squares determination of the parameters given in Table 13.2.1.

13.2.5 Applications

SAS has been extensively used to study the effect of drawing and heat setting and to study the effect of crystallization for instance under spinning conditions. Many aspects of the structure influence the performance, and here we will focus on those features accessible to SAS. We will discuss three following features commonly observed in fibers.

Void scattering is affected by the fiber processing. For instance, melt spinning shows the lowest EDS compared to dry, dispersion and wet spinning [2]. Drawing elongates the voids. The features of these elongated channels can be evaluated using the diffuse scattering along the equator. These voids behave as elongated channels through which liquids and gases can diffuse along the fiber axis. The voids affect the diffusion behavior as well as the mechanical properties (e.g., modulus, tenacity and elongation to break) of the fiber.

The fibrillar scattering most typically appears as a peak in the EDS whose position corresponds to the average distance between the fibrils in the aggregate. This peak can be used to study the changes in the aggregates due to drawing, heat setting and swelling. Whereas the WAXS measures the fibril diameter, SAS measures the diameter plus the thickness of the interfibrillar amorphous layer [19].

Lamellar spacings, tilt angle of the lamellar plane, intensity (lamellar and fibril), lamellar length, diameter and orientation have been shown to influence the performance of polymeric fibers. In general annealing increases the lamellae spacing, increases the diameter, decreases the stack orientation. The development of the four-point pattern from the two-point pattern upon drawing could be due to plastic deformation via interfibrillar shear.

13.2.5.1 Polyamide 6

Typical data from undrawn and drawn polyamide (nylon) 6 fibers, and control and heat-set fibers are shown in Figure 13.2.7 to illustrate the two common features, EDS and LMP. Some of the parameters reach a plateau at draw ratio of 3.5, others continue to increase even at higher draw ratios. The majority of the shrinkage occurs above an annealing temperature of 150 °C, and could be due to onset of new relaxation mechanisms such as the Brill transition [20] which perhaps lead to sliding of microfibrils [11].

Drawing decreases the misorientation of the voids from 24° to 15° and decreases their lengths from about 400 nm to 100 nm. (These scattering objects were identified as fibrils in the paper by Murthy et al. [8].) This shows that voids may break up at higher draw

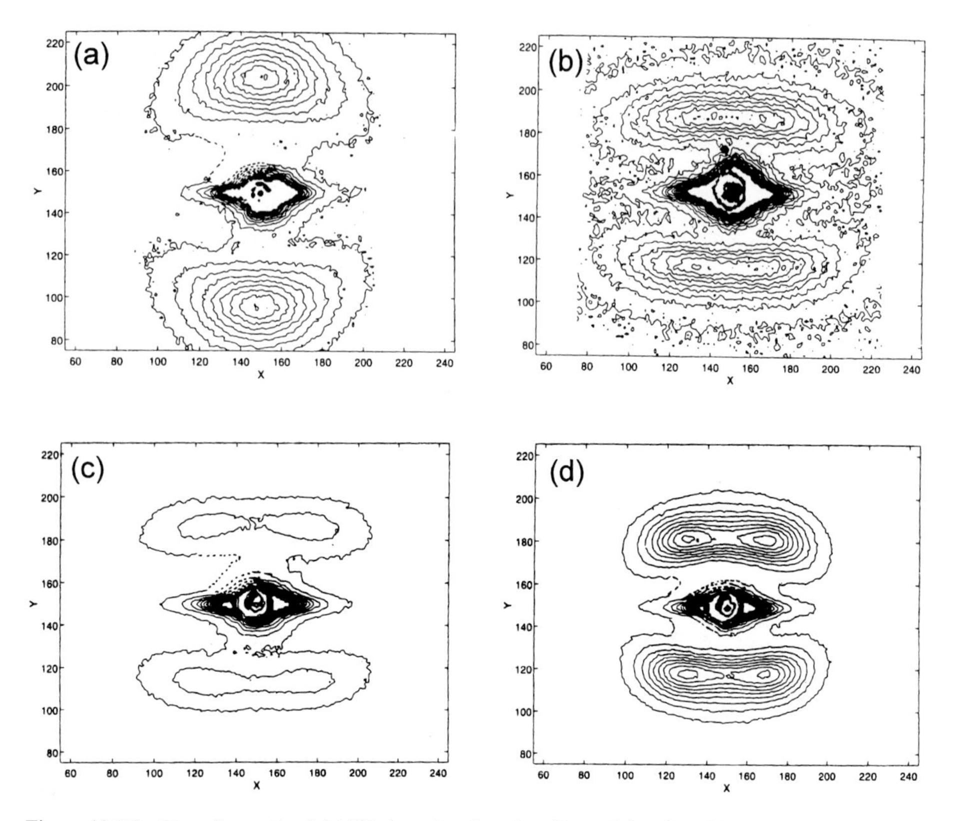

Figure 13.2.7 Two dimensional SAXS data showing the effect of drawing ((a) As spun and (b) 4.5 × drawn fibers) and the effect of annealing or heat-setting ((c) control and (d) heat set fibers).

ratios. Two fibers which are otherwise similar by characeristies such as crystallinity and orientation will dye differently mostly because of the difference in void content. The EDS increased after wet (autoclave or Superba process) heat-setting. No such change was observed in the dry (Suessen) process. In general, dry high temperature treatments greatly reduce the amount of diffuse scattering in fibers [2,7]. It is speculated that voids collapse in dry heat and that no such collapse occurs if the the fibers are heated wet or plasticized, or that voids are generated in wet fibers as a result of hydrolysis or vapor exiting from the fiber.

The interfibrillar scattering appears as a weak peak or a shoulder in EDS in many of the fibers [8]. The spacing between the fibrils is about 5 nm. The diameter of the the fibril is about 0.5 nm smaller than the interfibrillar spacing obtained by SAXS which is another 0.5 nm smaller than the value obtained by the SANS data.

Fibers with lower draw ratio give a two point pattern and this transfroms into a four-point pattern at higher draw ratios. The long-period (L) increases almost linearly from 6.1 to 9.2 nm with draw ratio (Figure 13.2.8). L first decreases with annealing temperature and then increases above 170 °C. The stack height first increases from 20–40 nm in undrawn fibers, to a maximum of 100 nm at a draw ratio of 3.5 and then begins to decrease at higher draw ratios. The stack height shows an inflection at 180 °C. The

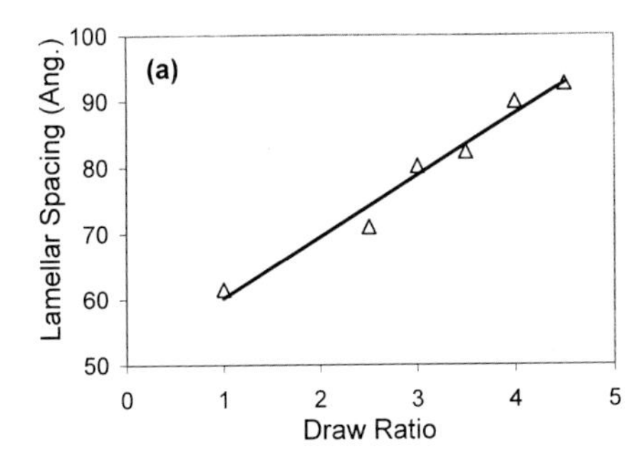

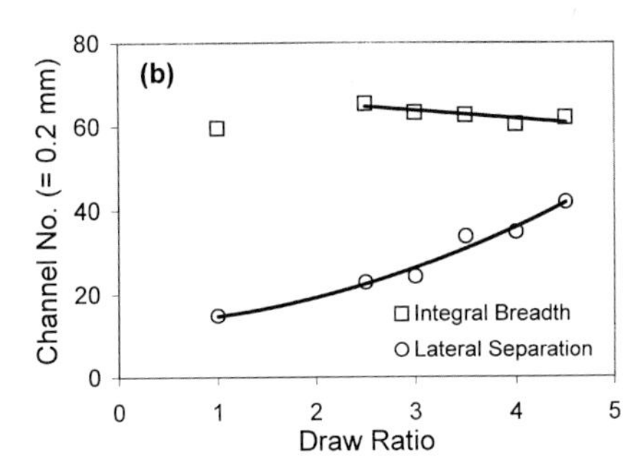

Figure 13.2.8 Variation in long-spacing (L) and the tilt-angle (χ, lateral separation) with draw ratio in nylon 6 fibers. Also shown is the variation in the integral breadth of the lamellar peak; this breadth is equivalent to the slope in the Guinier plot in Figure 13.6b.

diameter of the fibrils decreases upon drawing and increases upon annealing from 60 to 80 Å [8,11]. The orientation of the lamellar stack is similar to that obtained from WAXS.

The correlation between lamellae in the different lamellar stacks is weak compared to those between those within the same stack [15]. The lateral separation, the χ angle, increases from 15 to 60° upon drawing, The tilt-angle begins to decrease at 175 °C approaching 0 at 190 °C [11]. This is attributed to the structural rearrangement occurring during melting and recrystallization of the original crystal population and a result of partial crystallization of the interfibrillar amorphous chain segments which relax during moderate annealing.

13.2.5.2 Poly(ethylene terephthalate)

Extensive analysis of the SAXS data from PET have been carried out in many laboratories over the past 25 years. The equatorial scattering has been analyzed in terms of pores 2.5 to 100 nm diameter, and related to dye uptake [21]. However, as discussed earlier, the equatorial streak, cannot be unequivocally assigned to pore dimensions [22].

As-spun fibers show no evidence of a discrete SAXS pattern although the fibers have a significant fraction of oriented amorphous chains and fluctuation in amorphous density [23]. Typically, the SAXS appears before the appearance of crystalline peaks in the WAXS patterns. Upon further drawing discrete SAXS peaks appear. Annealing above 225 °C causes the SAXS pattern to be arced, implying disorientation. Lamellar spacing increases from 110 to 170 Å upon annealing [23,24] (Figure 13.2.9), although L might decrease in highly drawn fibers [25]. The intensity increases by about 100% upon annealing from 100 °C to 200 °C. Annealing above 225 °C followed by quenching reduces the SAXS intensity.

The stack height is between 180 to 300 Å and increases upon annealing. The diameter is between 50–100 Å [23,25], the higher values being prevalent in annealed fibers. This is due to the transverse growth of microfibrils at the expense of the oriented amorphous matrix. These values are consistent with those obtained by WAXS. The tilt angle (2χ) decreases with increase in annealing temperature from 100° to almost zero [24] (Figure 13.2.9b) suggesting less tilt of the lamellar plane in annealed fibers.

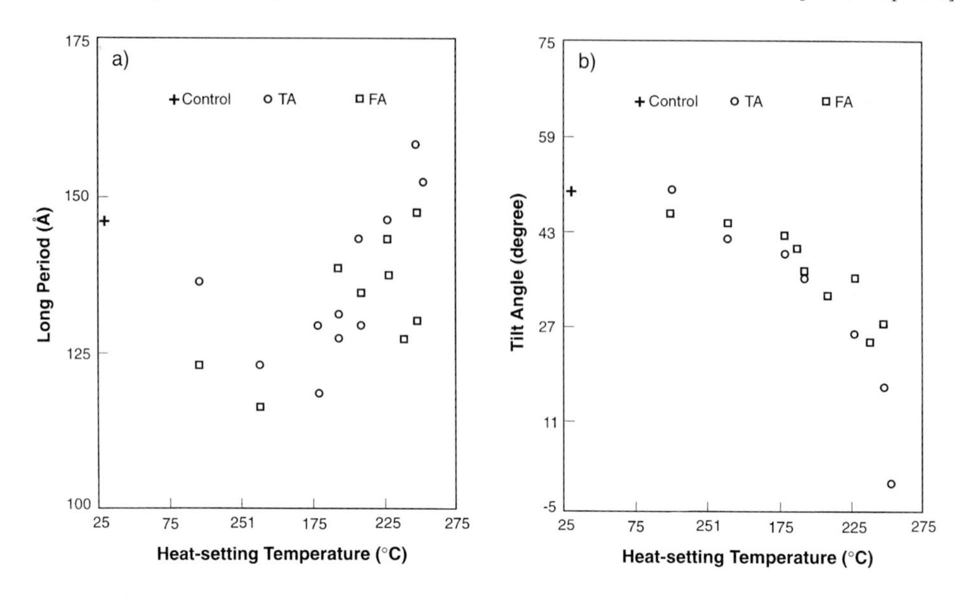

Figure 13.2.9 Variation in (a) long spacing (L) and (b) tilt angle (χ) with annealing in PET fibers [24].

13.2.5.3 Polyethylene

Polyethylene is one of the most studied fibers. In common with other fibers such as polyamides and polyesters, unoriented PE has a ring in SAXS corresponding to a lamellar structure on the scale 10 nm. Upon drawing, the ring transforms into a 2-point pattern. Further drawing and/or heating yields a 4-point pattern. This 4-point pattern has been attributed to obliquely oriented crystals in which the molecular chains pass through the lamellar crystals obliquely. The various parameters such as the one described for polyamide 6 and PET can be analyzed [26]. What is unique about PE, however, is that high-modulus fibers can be obtained by gel spinning. The intensity of the SAXS peak decreases with increase in draw ratio while the long-period remains essentially unchanged because of the unfolding of a higher proportion of lamellar crystals [27]. There is little fluctuation in the density in the axial direction on the scale of nanometers. The main structures in the extracted gel-spun PE fibers are shish-kebabs and the lamellae. Equatorial SAXS show that the former structure is very porous due to the presence of lamellar overgrowth that prevents close packing of the backbone fibrils [28]. Whereas the latter structure is relatively dense, both structures are transformed after hot-drawing into structures consisting of shish-kebabs or fibrils containing a void volume fraction of about 1%. Lamellar and shish-kebab structures were distinguished by filling the pores with paraffin oil. The elongated scattering objects, which are 150–200 nm long, have been attributed to fibrils, but these could be voids. The observed 5 to 10° misorientation is higher than that of the crystals, suggesting tapered scattering objects. These objects are typically 15–30 nm (in some cases 6 nm) in diameter, similar to that of crystals in WAXS. No lamellar reflections have been observed in extended chain PE fibers.

13.2.5.4 Liquid Crystalline Fibers

The major feature in these fibers is the equatorial streak, although in some instances a pair of faint off-axis streaks resembling a four point pattern, somewhat similar to that seen in semicrystalline polymers, has been observed (Figure 13.2.10). Although the equatorial streak has been attributed to voids, it is also possible that the fibril dimensions determine the intensity distribution within the streaks [9]. Grubb et al. [9], suggest that the "void scattering" in Kevlar arises from crystals in the fibers and not from voids, because the scattering objects have the same length, width and orientation distribution as crystals. The main feature is that the fibrils are longer and less oriented (200 nm and 20° misorientation) in Kevlar 149 than in Kevlar 49 (>70 nm and 15.3° misorientation). There are other features in Kevlar 149, such as off-axes reflections, the source of which are not easy to interpret. Another liquid crystalline fiber, PBO, has also been studied by SAXS [29]. As-spun PBO has only an equatorial streak from needle shaped voids. Heat treatment brings out the off-axis features in the form of a four-point pattern with radial streaking which are attributed to rhombohedrally shaped pseudo crystals [30].

13.2.5.5 Carbon fibers

SAXS is used to study the development of microstructure in fibers as well as in carbon-carbon composites. Carbon fibers are made mostly from pitch and PAN (see Chapter 9),

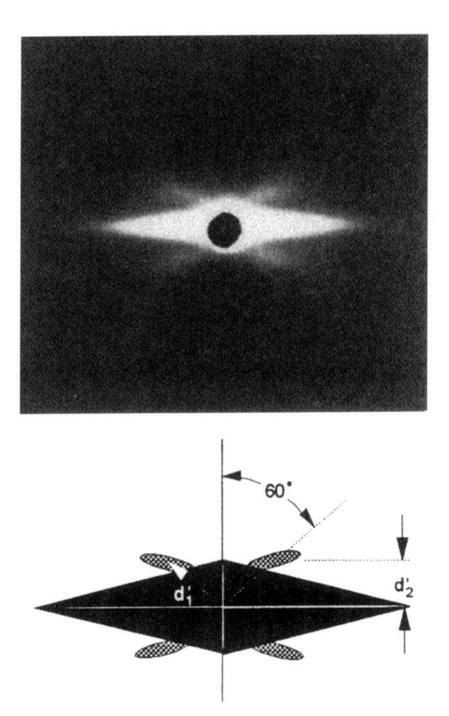

Figure 13.2.10 SAXS pattern from a liquid crystalline fiber (PBO). Also shown at the bottom is a schematic of this pattern. Provided by S. Kumar [29].

and to a small extent from cellulosic fibers. The data are analyzed to determine the volume fraction and the size distribution of the microvoids, and to follow the local density variation on the 1–10nm scale [31]. Orientation of the needle like pores are also analyzed by SAXS [32]. The length of the voids is about 10–30 nm. Other than the difference in the volume fraction of the voids (e.g., lower in PAN than from rayon), the microstructure in fibers from different precursors may not be that much different [33]. Two types of boundary regions are observed. First, voids separating the adjacent packs of ribbons parallel to the layer planes; these follow the general orientation of the ribbons and their lengths are much larger than the average lengths of the straight parts of the ribbon. Second, the regions formed by the atoms located at the lateral boundaries of the ribbon. In one study on C-C composites it was found that the fibers are only slightly influenced by the heat treatment, showing a small increase of the pore size and a weakly preferred orientation in the fiber-length direction [34]. The matrix however shows a pronounced increase in pore size. Large pores develop in the material as a consequence of the stress-induced graphitization of the matrix.

13.2.5.6 Cellulosic Fibers

Native cellulose fibers (e.g., Ramie) have only diffuse scattering. Such equatorial streaks have been is attributed to Lozenge shaped voids of diameters 20–250 Å [35]. Although microvoids could be the principal scatterers in dry cellulosic fibers, the nature of the diffuse scattering in swollen fibers is rather ambiguous [3]. An interfibrillar interference peak appears upon swelling, and there should be a similar, but more diffuse peak at higher angles even in dry fibers. The fibril diameters obtained by analyzing the EDS from swollen fibers are 46 Å for Fortisan and 37 to 44 Å in various rayons [4]. Regenerated cellulose fibers give merdional reflections and the meridional spacing is typically 170–220 Å in Fortisan and rayon fibers [35,36]. The diameter of the fibrils in Fortisan fiber derived from the layer line (merdional peak) in deuterated fiber is 34 Å. These elementary fibrils may form aggregates of $\sim$100 Å diameter [35].

13.2.5.7 SANS and the Diffusion of Solvents into Fibers

Although deuterium labeling of polymers is not always practical, SANS can be used to examine the voids and amorphous structures by diffusing deuterated solvents in to these spaces [35]. A similar study in SAXS will provide complementary results. Murthy et al. [6,37] have done work on polyamide fibers to investigate the diffusion of water (D_2O) through the amorphous regions in polyamide 6 fibers. The data show that the water diffuses into the fiber through a series of interconnected channels that run parallel to the fiber-axis (interfibrillar spaces) and perpendicular the fiber-axis (interlamellar spaces). The presence of an equatorial streak and meridional reflections show that as the water diffuses into polyamide 6, it is partitioned into interfibrillar and interlamellar spaces respectively. The longitudinal channels are 150 to 200 nm long and the misorientation of the voids increases from 5 to 15° upon drawing. Diffusion of water into the interfibrillar voids is different from that into interlamellar and interfibrillar amorphous domains. The decrease in the orientation and the increase in the length of the voids is opposite to that observed by SAXS suggesting that SAXS (at $q > 0.01$) sees the fibrils and SANS (at $q < 0.01$) sees the voids. Lamellar intensity in SANS patterns decreases, whereas the equatorial intensity remains essentially unchanged upon drawing. Because the intensity is proportional to the amount of water in the amorphous domains, this indicates that the fraction of interlamellar amorphous domain decreases whereas the interfibrillar amorphous domain remains unchanged. This suggests that the increase in crystallinity observed during drawing occurs at the expense of the interlamellar amorphous domains. In general, SANS can be used to understand the difference in the diffusion characteristics of the amorphous regions between the lamellae and between the fibrils.

13.2.6 Concluding Remark

With the availability of new detectors and new methods of rigorous quantitative analysis, SAXS can now be used to better follow the structural effects of processing parameters, and relate structure to performance.

13.2.7 References

1. A. Guinier and G. Fournet *Small-Angle Scattering of X-Rays*, Wiley, New York (1955)
2. W.O. Statton, *J. Polym. Sci.* (1962) 58, p. 205
3. P.H. Hermans, D. Hekins and A. Weidinger, *J. Polym. Sci.* (1959) 35, p. 145
4. A.N.J. Heyn, *J. Am. Chem. Soc.* (1953) 23, p. 782
5. A.N.J. Heyn, *Textile Res. J.* (1950) 72, p. 5768
6. N.S. Murthy and W.J. Orts, *J. Polym. Sci. Polym. Phys. Ed.* (1994) 32, p. 2695.
7. N.S. Murthy, A.C. Reimschuessel and V. Kramer, *J. Appl. Polym. Sci.* (1990) 40, p. 249
8. N. S. Murthy,, C. Bednarczyk, R.A.F. Moore, D.T. Grubb, *J. Polym. Sci. Polym. Phys. Ed.*, (1996) 34, p. 821
9. D.T. Grubb, K. Prasad and W. Adams, *Polymer* (1991) 32, p. 1167
10. D.P. Pope and A. Keller, *J. Polym. Sci. Polym. Phys. Ed.* (1975) 13, p. 533
11. R.J. Matyi and B. Crist, Jr., *J. Polym. Sci. Polym. Phys.* (1978) 16, p. 1329
12. W. O. Statton, *J. Polym. Sci.* (1959) 41, p. 143
13. C.G. Vonk, *Colloid and Polymer Sci.* (1979) 257, p. 1021
14. W. Wilke and M. Brarrich, *J. Appl. Cryst.* (1991) 24, p. 645
15. Z. Zheng, S..Nojima, T. Yamane and T. Ashida, *Macromolecules* (1989) 22, p. 4362
16. R.J. Rule, D.H. MacKerron, A. Mahendrasingam, C. Martin and T.M.W. Nye, *Macromolecules* (1995) 28, p. 8517
17. C.G. Vonk, Synthetic Polymers in the Solid State, in Small Angle X-ray Scattering Eds. O. Glatter and O. Kratky, Academic Press, New York (1982) pp. 456-457
18. B. Crist, *J. Appl. Cryst.* (1979) 12, p. 27
19. N.S. Murthy, K. Zero and D.T. Grubb, *Polymer* (1997) 38, p. 1021
20. N.S. Murthy, S.A. Curran, S.M. Aharoni and H. Minor, *Macromolecules* (1991) 24, p. 3215
21 A.K. Kulshreshtha, M.V.S. Rao and N.E. Dweltz, *J. Appl. Polym. Sci.* (1985) 30, p. 3423
22. K. -G. Lee and J.M. Schultz, *Polymer* (1993) 34, p. 4455
23. R.J. Matyi and B. Crist, Jr., J. *Macromol. Phys.* (1979) B16, p. 15

24. C. Ramesh, V.B. Gupta and J. Radhakrishan, *J. Macromol. Phys.* (1997) B36, p. 281
25. N.S. Murthy, D.T. Grubb. K. Zero, C.J. Nelson and G. Chen *J. Appl. Polymer Sci.* (1998).70, p. 2527
26. D.T. Grubb and K. Prasad, *Macromolecules* (1992) 25, p. 4575
27. Y. Fu, W. Chen, M. Pyda, D. Londono, B. Annis, A. Boller, A. Habenschuss, J, Cheng and B. Wunderlich, J. Macromol. Phys. (1996) B35, p. 37.
28. W. Hoostein, G. TenBrinks and J. Pennings, *J. Mat. Sci.* (1990). 25 p. 1551
29. S. Kumar, S. Warner, D.T. Grubb and W.W. Adams, *Polymer* (1994) 35, p. 5408
30. B.A. Asherov and B.M. Ginsburg, *J. Macromol. Sci. Phys.* (1997) B36, p. 689
31. A. Gupta and I.R. Harrison, *Carbon* (1994) 32, p. 953
32. W. Ruland, *J. Polym. Sci.* (1969) 28, p. 143
33. R. Perret and W. Ruland, *J. Appl. Cryst.* (1970) 3, p. 525
34. H. Peterlik, P. Fratzi, and K. Kromp, *Carbon* (1994) 32, p. 939
35 E.W. Fischer, P. Herchenroder, R.St. J. Manley and M. Stamm, *Macromolecules* (1978) 11, p. 213
36. W.O. Statton, *J. Polym. Sci.,.* (1956) 22, p. 385
37. N.S. Murthy, *Text. Res. J.* (1997) 67, p. 511
38. N.S. Murthy, K. Zero and D.T. Grubb, *Macromolecules* (2000) 33, p. 1012

13.3 Density, Birefringence, and Polarized Fluorescence

Bernd Clauss* and David R. Salem**
*Institut für Chemiefasern, Denkendorf, Germany and
**TRI/Princeton, Princeton, New Jersey, USA

13.3.1 Crystallinity from Density

An increase in the measured density ρ of a polymer reflects a decrease in interchain distances associated with crystallization. From knowledge of the polymer density in the fully crystalline state ρ_c and in the fully amorphous polymer ρ_a, the volume fraction crystallinity χ can be calculated from:

$$\chi = \left(\frac{\rho - \rho_a}{\rho_c - \rho_a} \right) \tag{13.3.1}$$

Most polymers can be obtained in the fully amorphous state (although this sometimes requires extremely rapid quenching from the melt) and ρ_a can therefore be determined by measuring the density of the amorphous polymer. So far, it has not been possible to produce fully crystalline flexible-chain polymers and ρ_c is calculated from the unit cell structure (determined from WAXS – see Section 13.1.5) and the monomer molecular weight. If two crystal forms are present, Equation 13.3.1 can of course be adapted to provide the volume fraction of both crystalline phases provided the phase ratio has been obtained by some other method [1].

It is usually assumed that ρ_c and ρ_a are constant for a given polymer, but variations in these parameters can occur. Processing conditions, especially thermal history, can influence the packing density of chains in the unit cell, without necessarily changing the crystal form. However, by WAXS, it is often possible to check the unit cell parameters of a sample with sufficient accuracy to determine its ρ_c and use this value in the calculation of χ from Equation 13.3.1 (although in cases where the crystals are very small, the apparent lattice spacings measured may need to be corrected to account for the Lorentz-polarization factor and the so-called Wallner effect [2,3]).

It is generally supposed that increasing chain orientation in the noncrystalline phase would tend to decrease the average interchain distance in this phase and increase ρ_a. Surprisingly, however, reliable experimental documentation of this effect is lacking. It is very clear from studies on above-T_g drawing of amorphous PET film that molecular orientation can reach values of about 0.2 without any significant effect on polymer density [4,5], but the onset of crystallization prevents investigation of the relationship at higher orientations. A simple model derived by Nobbs et al. [6] predicts a smoothly increasing rate of change of ρ_a with noncrystalline orientation $<P_2(\cos\theta)>_a$, such that there is a negligible change in ρ_a up to $<P_2(\cos\theta)>_a$ values of about 0.2, followed by a region of gradually increasing ρ_a, and finally a rapidly increasing region starting at a $<P_2(\cos\theta)>_a$ value of about 0.6. However, the authors make clear that this model involves various assumptions and unavoidable approximations. Heuvel and Huisman's experimental results indicate a small linear increase in ρ_a with noncrystalline orientation [7] in "amorphous" PET fibers, but they do not describe how they assessed the (absence of) crystallinity in their samples, which may not have been truly noncrystalline.

It is worth emphasizing that an increase in both polymer density and $<P_2(\cos\theta)>_a$ in the absence of the development of *detectable* crystalline WAXS reflections is not necessarily acceptable evidence for a dependence of ρ_a on $<P_2(\cos\theta)>_a$. At a volume fraction crystallinity less than about 0.1 combined with crystal dimensions of the order of 3nm (which may be on the high side for the early stages of stress-induced crystallization – see Section 4.3.2.6), crystalline WAXS reflections would usually be too broad and of too low intensity to be distinguishable from the diffuse amorphous scattering. It is also important to realize that when adjacent, fully oriented chains are parallel to each other but not in crystalline register, the interactions and steric hindrances between mismatched units may *prevent* the chains from packing significantly more closely than in a random arrangement. In other words, high orientation and even high orientation with parallelization, does not *necessarily* mean high density.

It is evidently very difficult to obtain a functional relationship between $<P_2(\cos\theta)>_a$ and ρ_a in polymers that crystallize in the process of orientation. Unfortunately, orienting the polymer by drawing at temperatures below T_g, where there is insufficient molecular mobility for crystallization, is not a way around this problem. In below T_g-drawing, the uncrystallized chains can be forced into close packing arrangements (including the formation of a mesophase) that they would resist under conditions of higher mobility. Therefore, $<P_2(\cos\theta)>_a$ vs ρ_a relationships obtained from cold drawing cannot be applied to hot-drawn samples.

Disparate pieces of evidence indicate that, for fibers drawn above T_g, the effect of amorphous orientation on ρ_a is very small for values of $<P_2(\cos\theta)>_a$ up to about 0.45. But there is a need to investigate this relationship for specific polymers in a more rigorous and systematic fashion than has been attempted so far.

There are issues of interpretation and definition with all methods of determining crystallinity in oriented polymers, and the assumptions involved in the density method are arguably less problematic than most, especially if any significant changes in ρ_c are taken into account. A major advantage of density measurements, when the density gradient column technique is employed, is their high level of accuracy. Careful experimental practice of this method permits density values to be reliably reported to four significant figures and allows very small changes in crystallinity to be detected.

A density gradient column is made by filling a glass, temperature-controlled column (about 1 meter tall) with two miscible liquids of high and low density. A liquid column of linear density gradient is formed covering the range required for the polymer under investigation. It is important, of course, to select column liquids that do not interact with the polymer and induce crystallization. Once the column has been calibrated using floatation balls of known density, the specimens are dropped into the liquid. After 12-16 hours each specimen will have reached its equilibrium height in the column, corresponding to its density.

13.3.2 Molecular Orientation from Birefringence

A polymer has optical birefringence Δ if its refractive index n is anisotropic, as evidenced by its ability to rotate the plane of polarized light. Thus, for uniaxially oriented polymers:

$$\Delta = n_{\parallel} - n_{\perp} \tag{13.3.2}$$

where $n_{\parallel}$ and $n_{\perp}$ are, respectively, the refractive indeces parallel and perpendicular to the fiber axis (or draw direction). The value of Δ depends on the optical anisotropy of the monomeric unit and on the degree of molecular orientation. Thus, for a fully crystalline polymer, the orientation function $<P_2(\cos\theta)>_c$ is related to Δ through:

$$\Delta = <P_2(\cos\theta)>_c \Delta_{oc} \tag{13.3.3}$$

where Δ_{oc} is the intrinsic birefringence of the crystalline phase. For full orientation parallel to the fiber axis, $<P_2(\cos\theta)>_c = 1$ and for random (isotropic) orientation $<P_2(\cos\theta)>_c = 0$.

Similarly, for a fully amorphous polymer:

$$\Delta = <P_2(\cos\theta)>_a \Delta_{oa} \tag{13.3.4}$$

where Δ_{oa} is the intrinsic birefringence of the crystalline phase.

It follows that for a partially crystalline polymer [8]:

$$\Delta = <P_2(\cos\theta)>_c \Delta_{oc}\, \chi + <P_2(\cos\theta)>_a \Delta_{oa}\, (1-\chi) \tag{13.3.5}$$

where χ is the volume fraction crystallinity. This equation neglects any contribution from form birefringence that can result from the geometric anisotropy of crystals but which is generally very small compared with the other two terms. Since $<P_2(\cos\theta)>_c$ can readily be obtained from WAXS measurements, it is possible to calculate $<P_2(\cos\theta)>_a$ – a key determinant of fiber properties – from knowledge of χ (using density, WAXS or another method), Δ (using polarized optical microscopy techniques), and intrinsic birefringence values (discussed below).

Δ_{oc} and Δ_{oa}, representing the limiting birefringence of a perfectly oriented chain in the crystalline or amorphous environment of a given polymer, do not usually have equal values. This is because the polarizability of a polymer chain in an electric field is influenced by the induced dipolar field of neighboring chains, which depends on the orientation and configuration of those chains [9]. Since molecular arrangements are usually quite different in the crystalline and noncrystalline state, Δ_{oc} and Δ_{oa} may differ substantially from each other.

By rewriting Equation 13.3.5 as

$$\frac{\Delta}{\chi\langle P_2(\cos\theta)\rangle_c} = \left(\frac{1-\chi}{\chi}\right)\left(\frac{\langle P_2(\cos\theta)\rangle_a}{\langle P_2(\cos\theta)\rangle_c}\right)\Delta_{oa} + \Delta_{oc} \tag{13.3.6}$$

Δ_{oc} and Δ_{oa} can be obtained from the intercept and slope, respectively, of the linear plot predicted by Equation 13.3.6 [10]. This requires the determination of volume fraction crystallinity and noncrystalline orientation from samples with a range of microstructures. For this purpose, crystallinity has most frequently been determined by density (see Section 13.3.1) or WAXS (see Section 13.1), and noncrystalline orientation has commonly been obtained from the sonic modulus technique involving sonic velocity measurements along a specimen held under a very small load [10,11]. By this method, the noncrystalline orientation is given by [10]:

$$\langle P_2(\cos\theta)\rangle_a = \left(\frac{3}{2(E_u - E)} - \frac{\chi\langle P_2(\cos\theta)\rangle_c}{E_{oc,t}}\right)\frac{E_{oa,t}}{(1-\chi)} \tag{13.3.7}$$

where E is the sonic modulus or the oriented specimen, E_u is the sonic modulus of an unoriented reference specimen and $E_{oc,t}$ and $E_{oa,t}$ are the intrinsic transverse moduli of the crystalline and amorphous regions respectively. It is generally acknowledged, however, that aspects of this technique cast doubt on its reliability. It assumes that for noncrystalline chains, the linear relationship between sonic modulus and birefringence at low orientations [11,12] extends to higher orientations where the relationship is likely to be more complex [13,14], and where its experimental determination is precluded by the presence of crystallinity. Also uncertain is the assumption that $E_{oc,t}$ and $E_{oa,t}$ are material constants, unaffected by processing history. Finally, it is arguable that sound waves propagate preferentially along the taut tie molecules, resulting in values of noncrystalline orientation that are higher than the true average.

For some polymers, especially PET, there are wide variations in the reported values of intrinsic birefringence, which have often been attributed to inadequacies in the experimental methods used. However, the analysis of Gupta and Ramesh [15] and of Clauss and Salem [4] indicate that the predominant cause of these discrepancies comes from the assumption that Δ_{oc} and Δ_{oa} are independent of the state of order in the crystalline and noncrystalline regions. Since the optical anisotropy of a chain depends on the configuration of surrounding chains, Δ_{oc} may be influenced by lattice distortions and defects in the crystalline phase, and Δ_{oa} may be influenced by the overall level of noncrystalline orientation and/or the *trans/gauche* content in the noncrystalline phase. If Δ_{oc} and Δ_{oa} are not material constants, but depend on the processing history of the polymer, it is clear that this would sometimes lead to significant errors in the evaluation of $<P_2(\cos\theta)>_a$ from the birefringence method (Equation 13.3.4).

The intrinsic fluorescence method of determining noncrystalline orientation in PET has been developed relatively recently (see Section 13.3.3) and, being a *direct* method providing highly reliable values of $<P_2(\cos\theta)>_a$, it has been used to check the constancy of Δ_{oc} and Δ_{oa} in PET fibers. Plots for determining intrinsic birefringence from Equation 13.3.6 were constructed for a series of high speed spun fibers (with $<P_2(\cos\theta)>_a$ values from 0.22 – 0.32) and for some spun-drawn fibers (with $<P_2(\cos\theta)>_a$ values from 0.45 – 0.55) [4]. Different linear fits were obtained for the two sets of fibers (Figure 13.3.1) yielding $\Delta_{oc} = 0.28$ and $\Delta_{oa} = 0.12$ for the set of high speed spun fibers, and $\Delta_{oc} = 0.24$ and $\Delta_{oa} = 0.27$ for the spun-drawn fibers.

13.3.3 Molecular Orientation from Polarized Fluorescence

While the structure and orientation of the crystallites in oriented polymers can be accurately determined by X-ray diffraction, attempts to develop a reliable method for the characterization of the orientation in the noncrystalline (amorphous) phase of polymers have been more problematic. The indirect and time-consuming method described in Section 13.3.2 (involving a combination of birefringence, X-ray and density measurements) is most commonly employed for this purpose.

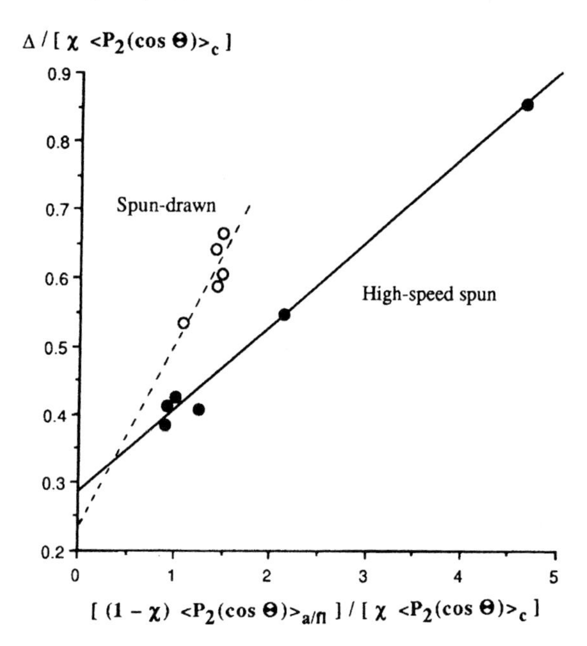

Figure 13.3.1 Determination of the intrinsic birefringences of the crystalline and noncrystalline phases of spun-drawn fibers (with $<P_2(\cos\theta)>_a$ values from 0.45 – 0.55) and high speed spun PET fibers (with $<P_2(\cos\theta)>_a$ values from 0.22 – 0.32).

This section will briefly describe a *direct* technique for measuring noncrystalline orientation in fibers and other uniaxially oriented polymers, which has been established relatively recently. It is based on the polarized fluorescence of extrinsic molecules incorporated in the polymer or of chain-intrinsic moieties which arise from inherent features of the polymer's chemical structure. We will refer to both these types of fluorescing units as fluorescent probes. The polarized fluorescence technique is rapid, reliable and straightforward, and the chain-intrinsic fluorescence method is particularly suited to online measurement of noncrystalline orientation in fiber spinning and drawing operations.

13.3.3.1 The Fluorescence Phenomenon

The first terms of the expansion of the orientation distribution function can give valuable information about the average orientation of chain segments in polymers. The task is therefore to find tools which permit detection of these orientation averages experimentally. In the case of the measurement of orientation in the noncrystalline phase of polymers, fluorescent radiation has been used with notable success.

A schematic description of the fluorescence process is given in Figure 13.3.2. If a fluorescent molecule is illuminated with radiation in a suitable region of the spectrum, light absorption leads to a promotion of the molecule from the ground singulet electronic state S_0 to vibrational levels of the first excited electronic state S_1. This is followed by a fast vibrational relaxation to the lowest energy state of S_1. The molecule in the S_1 state can then drop to some vibrational levels of the S_0 state by emitting energy in the form of light. This process, which generally happens within 10^{-9} to 10^{-8} seconds, is called fluorescence, and the fluorescence emission spectrum is usually the approximate mirror image of the absorption spectrum.

Depending on the symmetry of the electronic structures of the ground state and the excited state, the transition between these states can be, to a greater or lesser extent, anisotropic. If polarized light is used for the excitation of molecules with anisotropic transitions, the excitation probability will depend on the angle between the polarization direction and the so-called transition moment of the molecule. This property can therefore be used to detect the orientation of anisotropic fluorescent molecules, since there exists a correlation between the orientation of the molecule with respect to the polarization direction and the fluorescence intensity.

Three different techniques have been utilized in order to exploit this phenomenon for the measurement of polymer chain orientation in the noncrystalline phase of oriented polymers. First, highly anisotropic fluorescent molecules can be used as probes, if they can be

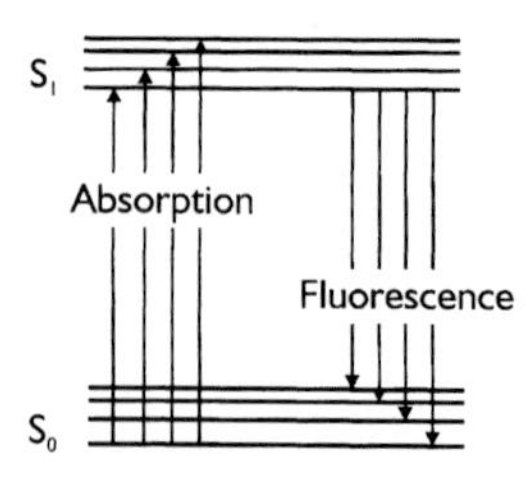

Figure 13.3.2 Absorption (excitation) and fluorescence (emission).

incorporated in a polymer such that they align with the polymer chains. Second, fluorescent probes can be incorporated in the polymer chains themselves by chemical modification of the polymer backbone. And finally, some polymers possess intrinsic, optically anisotropic moieties, which can serve as probes for the detection of chain orientation.

The basic principle of these techniques is shown in Figure 13.3.3, illustrating that the probability of the excitation process, and therefore also the intensity of the fluorescent light, depends on the orientation of the optically anisotropic probes with respect to the polarization direction of the incident light.

The fluorescence technique gives more information on orientation than many other methods such as birefringence or dye-molecule dichroism. This is due to the fact that fluorescence involves the absorption and the emission process, which leads to a fourth moment dependence between intensity values and the orientation function, i.e. not only P_2 but also P_4 can be determined in principle. This additional information can be useful for distinguishing features of the orientation distribution function, which are not available from the second moments. In practice, however, small errors in measurement can lead to large errors in the calculated value of P_4, and caution should be applied in attempts to determine this parameter.

13.3.3.2 Orientation Measurement by Polarized Fluorescence

Several research groups have been studying the polarized fluorescence phenomenon [4,16-43], and the mathematical background of the correlation between measured intensity values and the moments of the orientation distribution function have been derived for different cases. There are three main topics that must be considered, which finally determine the complexity of the mathematics that has to be applied.

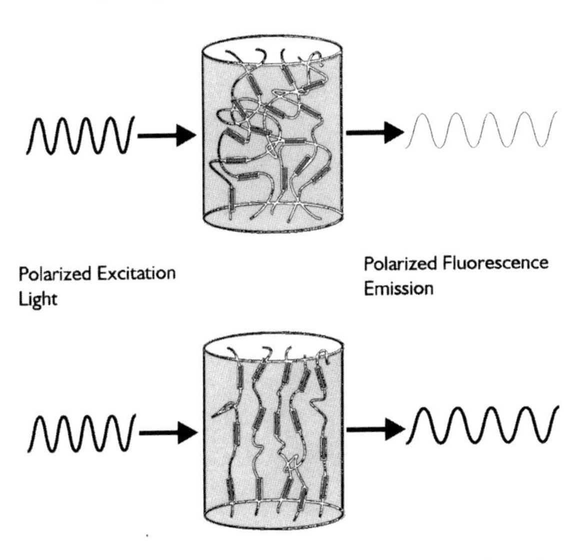

Figure 13.3.3 Schematic illustration of the dependence of fluorescence intensity on the orientation of fluorescent probes, if polarized excitation light is used. The probes shall be anisotropic, with the highest excitation probability along the long probe axis.

The first topic is the symmetry of the orientation distribution of chain segments that may be imparted to the polymer by a particular deformation mode. It is impossible to obtain enough data by the fluorescence measurement or other techniques to describe the distribution function in the amorphous phase exactly, if no restrictions can be made in terms of the expected symmetry of the distribution. However, in most polymer fibers a uniaxial, cylindrical distribution of the orientation of polymer chain segments can be assumed, in which case a relatively simple correlation can be derived.

Second, if extrinsic fluorescent probes are used for the detection of the chain orientation, it cannot be assumed that the probes are perfectly aligned with the polymer chain. In other words, the task is to correlate the detected orientation distribution of the fluorescent probes with the actual orientation average of the polymer chain segments, which is not trivial.

Third, one has to be aware that the measured intensity values reflect the orientation of transition moments of the fluorescent probes, which are in most cases not identical with a symmetry axis of the fluorescent probe. Correction procedures have to be derived, to take this into account.

Evidently there are different levels of complexity, ranging from the situation where the transition moments of excitation and emission are not identical in direction and do not coincide with the molecular axis of the probe molecule, to the simplest situation, where the two transition moments are identical and do coincide with the probe axis. A very clear and extensive overview about different approaches to this topic is given by Nobbs and Ward [42].

A comprehensive review of the mathematical treatments of the different cases mentioned above cannot be given here, but it has been shown by Desper and Kimura [43] that for uniaxial symmetry the orientation averages $<\cos^2\theta>$ and $<\cos^4\theta>$ are given by:

$$\langle\cos^2\theta\rangle = \frac{I_{33} + 2I_{31}}{I_{33} + 4I_{31} + \frac{8}{3}I_{11}} \tag{13.3.8}$$

$$\langle\cos^4\theta\rangle = \frac{I_{33}}{I_{33} + 4I_{31} + \frac{8}{3}I_{11}} \tag{13.3.9}$$

where I_{ij} is the fluorescent emission intensity measured with the polarizer along the direction i and the analyzer along the direction j, such that directions 3 and 1 are, respectively, parallel and perpendicular to fiber axis (Figure 13.3.4). These averages can then be used to calculate the second and fourth moment of the orientation distribution function:

$$\langle P_2(\cos\theta)\rangle = \frac{1}{2}\left(3\langle\cos^2\theta\rangle - 1\right) \tag{13.3.10}$$

$$\langle P_4(\cos\theta)\rangle = \frac{1}{8}\left(35\langle\cos^4\theta\rangle - 30\langle\cos^2\theta\rangle + 3\right) \tag{13.3.11}$$

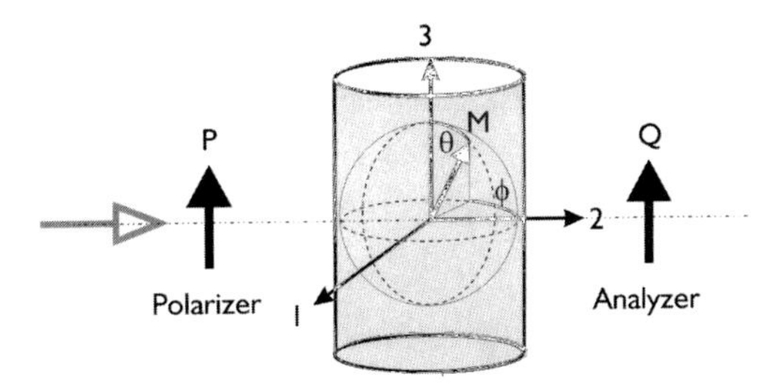

Figure 13.3.4 Definition of coordinate axes with regard to the fiber symmetry.

Equations 13.3.8 and 13.3.9 assume that the transition moments of fluorescent absorption and emission of the probe are identical and perfectly anisotropic, and that their direction coincides with the molecular axis of the polymer chain segments. However, assuming, the coincidence of absorption and emission moments and making some additional assumptions concerning the tensor symmetries, a correction can be made for incomplete emission anisotropy [41]. The resulting formulae for $<\cos^2\theta>$ and $<\cos^4\theta>$ are relatively cumbersome and are given elsewhere [4]. Nevertheless, they involve only the readily measured intensities I_{33}, I_{11}, I_{31}, I_{13} (and note that for uniaxial deformation, I_{31} and I_{13} are theoretically equal).

If the specimen has significant opacity, depolarization from light scattering may occur. This can be corrected for, by a procedure involving measurement of scattering factors for the excitation and fluorescence wavelengths [44].

13.3.3.3 Instrumentation

The preferred measurement set-up for performing orientation measurements by the polarized fluorescence technique is comparatively simple (Figure 13.3.5) and has been described in Ref. [4].

The four intensities I_{33}, I_{11}, I_{31} and I_{13} can either be measured by changing the polarizer and analyzer position (to 0° and 90°) or by rotating the polarizer and the sample. The second procedure has the advantage that no corrections have to be made for the different

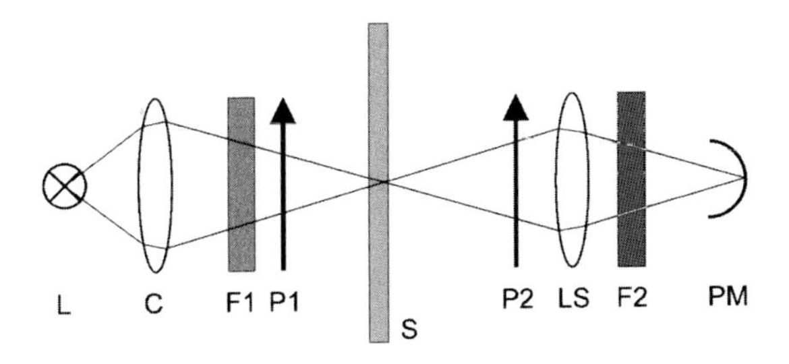

Figure 13.3.5 Experimental set-up for measurement of polarized fluorescence intensities. L: mercury arc lamp, C: condenser, F1: filter for excitation wavelength, F2: filter for fluorescence wavelength, P1, P2: UV polarizers, S: sample (film or aligned fibers), LS: lens system, PM: photomultiplier.

sensitivity of the photomultiplier for light of different polarization direction, but this is obviously restricted to off-line measurements. For on-line measurements of fiber or film formation, these corrections are essential.

13.3.3.4 Studies Based on Extrinsic Fluorescence

Intensive work concerning orientation development in uniaxially drawn poly(ethylene terephthalate) PET film has been conducted by Bower and Ward and their collaborators. Different techniques like polarized Raman, infrared dichroism and polarized fluorescence have been applied and compared. In the polarized fluorescence studies [40,45,46] the stilbene derivate 2,4'-(dibenzoxazolyl)stilbene (VBPO) was used as a fluorescent probe molecule. The compound was added to the material during the polycondensation in concentrations between 50 ppm and 200 ppm. One of the results, deduced from measurements of unoriented isotropic film, showed that the emission and absorption axes of the probe molecule were non-coincident. This was taken into account in the calculation of the second and fourth moments of the orientation distribution function. The calculations were based on the measurement of polarized fluorescence intensities parallel and perpendicular to the draw direction of the films. It was stressed that these intensities are not affected by birefringence effects, but that a correction for dichroic absorption is necessary. By comparing the fluorescence method with other techniques it was shown that the unique axes of the fluorescent probe molecules (VPBO) are more highly oriented than the PET chains in the amorphous regions, but the two orientation distributions still appear to be uniquely related over the applied range of temperatures (65 °C to 90 °C) and draw ratios (1 to 5.85). It was concluded that the probes are preferentially aligned parallel to the chain axis direction of the polymers in the *trans*-conformation, which was consistent with the accompanying birefringence and infrared studies. The above-T_g deformation behavior at low draw ratios was modeled by a rubber network model, in which the PET sample appears to behave like a crosslinked rubber with 5.6 freely jointed links between crosslink points at the temperature and deformation rates studied (see Section 4.4.1.2). It was also shown that shrinkage of the oriented films was always associated with considerable disorientation in the amorphous regions, confirming that this is the primary shrinkage mechanism.

A polarized fluorescence study of polystyrene films, in which diphenylanthracene was used as probe molecule, was published by Fuhrmann and Hennecke [47]. In a subsequent study by the same authors the behavior of a variety of fluorescent probes were examined during uniaxial drawing of low density polyethylene film [41]. It was shown that the properties of the probes influence the detected orientation averages because the probe orientation is controlled by the interaction of the probe with the polymer chain. For example, the measured molecular orientation is effectively the average orientation of a number of polymer chain segments spanned by the rodlike fluorescent probe, and the number of segments that are averaged depends on the probe's length. It was also remarked that the quantum yield of certain probe molecules can be influenced by local deformation forces. A further paper by Hennecke and Fuhrmann [48] deals with the accuracy of the fluorescence polarization experiment and the discrimination of uniaxial deformation models.

Other studies using extrinsic fluorescence probes have been published by Monnerie and coworkers. All-*trans* 1,8-diphenyloctatetraene was employed as a fluorescent probe in an investigation of the orientation of the noncrystalline phase in polypropylene [44]. The second and fourth moment of the orientation distribution function was measured from the necking zone of cold drawn samples and during stretching at various temperatures. The results show that noncrystalline orientation in PP depends mainly on the crystalline orientation and morphology, which suggests that the "amorphous chains" are strongly embedded or attached to the crystalline phase. Results of studies of uniaxially deformed PET films by using VPBO as probe have also been reported by Monnerie et al., in which the role of noncrystalline orientation on the development of draw-induced crystallization was investigated [49-51].

13.3.3.5 Studies Based on Intrinsic Fluorescence

The first polarized fluorescence studies based on chemically incorporated probe units were conducted by Nishijima [32]. This technique was later revived by Hibi et al. [52-58], who reported on orientation development in PVC films. Polyene and carboxylic groups produced by heat treatment of the PVC served as chain intrinsic probes.

Monnerie and his colleagues investigated segmental orientation of labeled polyisoprene networks [49] and of polystyrene [56] by incorporating anthracene in the polymer chains during the anionic polymerization. Deviations from the stress-orientation relationship of the rubber network theory were found for the polyisoprene networks.

The most recent development in polarized fluorescence analysis is based on the chain-intrinsic fluorescence found in unmodified PET. It has been shown that an associated ground state dimer, first identified by Allen and Mc Kellar in 1978 [57], can be used as an intrinsic fluorescent probe of noncrystalline orientation in PET. Fundamental studies of the nature of the fluorescing site suggest that it arises from the association of two terephthalic moieties in adjacent PET chains [58-61]. A precise side by side alignment of the phenylene rings with a 0.35 nm distance between them is needed to form the dimer which, after excitation at 340 nm, emits fluorescent radiation with a maximum intensity at 390 nm. This arrangement of the rings is randomly realized in the noncrystalline phase, but in the crystalline phase fluorescence excitation is prevented due to the displaced arrangement of the phenylene rings and a ring separation that is too large (0.4 nm to 0.45 nm). The transition moment of absorption and emission is highly anisotropic (its emission anisotropy value is in the range 0.3 – 0.35, being close to the value for fully anisotropic emission of 0.4.), which is essential for the polarized fluorescence experiment. Also important is the fact that the fluorescent probes do not change their optical properties (fluorescence spectrum and time decay law) during the extension of the samples. By using polarized incident light and detecting the polarized components of the fluorescent light, one can directly calculate averages of the orientation distribution function as in the case of extrinsic probes, but with the advantage that no deviation of the probe distribution from the polymer chain distribution need be considered.

It should be mentioned that Quian et al. proposed another explanation for fluorescent emission from amorphous PET at 390 nm [62]. They believe that the absorption at 340 nm has its origin in the excitation of carbonyl groups, and that the fluorescent emission results from an interaction of carbonyl groups and the phenylene π -electrons. In any case, this view (which does not appear to have been adopted by other authors in subsequent publications) does not call into question the usefulness of the fluorescent properties of PET for orientation measurements.

Hennecke, Fuhrmann et al. used intrinsic fluorescence to study the development of molecular orientation during drawing of PET films that had been crystallized in the undeformed, isotropic state [53]. A deformation model was fitted to the results, in which a certain amount of the amorphous phase orients cooperatively with the crystallites, i.e. according to a pseudo-affine model, whereas another "true amorphous" part is deformed according to the affine network model. Another conclusion of the investigations was that the detected increase of fluorescence intensity with increasing crystallinity most likely results from an increase in the excited volume due to light scattering by the crystallites, and additional evidence for this interpretation was later provided by Clauss and Salem [4].

An interesting finding in subsequent papers from Hennecke and Fuhrmann's group was that similar fluorescent moieties are formed in other aromatic polyesters like poly(ethylene terephthalate co-p-oxybenzoate) [63] and poly(butylene terephthalate) [64]. In the case of PBT, the spectrum and the fluorescence decay law were almost identical to those of PET. As in their PET study, the deformation behavior of the noncrystralline regions of crystalline PBT could be described by a superposition of crystallite-like and true-amorphous orientation.

The power of the chain-intrinsic fluorescence technique was further demonstrated by Clauss and Salem [4,5,65] in detailed studies of the development of noncrystalline orientation during drawing of (initially noncrystalline) PET films and fibers above T_g. Effects of strain rate, draw temperature, molecular weight and deformation mode were reported (see Sections 4.2.2.1 and 4.3.2) and a clearer understanding of the interrelations between orientation development and orientation-induced crystallization was obtained.

Gohil and Salem [66] have shown that the intrinsic fluorescence method can also be used to follow the in-plane redistribution of noncrystalline orientation during biaxial deformation of PET films, and they were able to predict certain mechanical properties from the data.

A set-up for on-line measurements of noncrysatalline orientation during PET fiber spinning has been designed and implemented, as described by Clauss, Oppermann et al. [67]. It has been shown that changes in noncrystalline orientation with varying spinning speeds can be monitored on-line with high accuracy. A suitable technique for online measurements of axial and biaxial (in-plane) orientation of PET film has also been developed recently [68].

13.3.3.6 Summary

The reliable measurement of orientation in the noncrystalline phase of polymers is highly important for the optimization and modification of melt spinning and of fiber and film drawing processes. The polarized intrinsic-fluorescence technique, which has been developed during the last three decades, has proven to be a powerful direct method for these orientation measurements in some commercially important polymers, and its potential as an on-line method is considered very promising. Polarized fluorescence measurements can ideally supplement other methods like infra-red, Raman and NMR spectroscopy as well as optical birefringence and X-ray diffraction.

13.3.4 References

1. Salem, D.R., Moore, R.A.F. and Weigmann, H.-D. *J. Polym. Sci., Polym. Phys. Ed.* (1987) 25, p. 567
2. Northolt, M.G. and Stuut, H.A. *J. Polym. Sci., Polym. Phys. Ed.* (1987) 25, p. 2561
3. Salem, D.R. *J. Polym. Sci., Polym. Phys. Ed.* (1987) 25, p. 567
4. Clauss, B. and Salem, D.R. *Polymer* (1992) 33, p. 3193
5. Salem, D.R. *Polymer* (1998) 39, p. 7067
6. Nobbs, J.H., Bower, D.I. and Ward, I.M. *Polymer* (1976) 17, p. 25
7. Huisman, R. and Heuvel, H.M. *J. Appl. Polym Sci.* (1978) 22, p. 943
8. Stein, R.S. and Norris, F.H. *J. Polym. Sci.* (1956) 21, p. 381
9. Stein, R.S. *J. Polym. Sci. C* (1966) 15, p. 185
10. Samuels, R.J. *J. Polym. Sci.* (1965) A3, p. 1741
11. Moseley, W.W., *J. Appl. Polym Sci.* (1960) 3, p. 266
12. Dumbleton, J.H. *J. Polym. Sci. A2* (1968) 6, p. 795
13. Ward, I.M. in *Mechanical Properties of Solid Polymers 2 nd Edition* (1983) p. 291
14. Ward, I.M. *Text Res. J.* (1964) 34, p. 806
15. Gupta, V.B and Ramesh, C. *Polym. Commun.* (1987) 28, p. 43
16. Morey, P.E. *Text Res. J.* (1933) 3, p. 325
17. Morey, P.E. *Text Res. J.* (1934) 4, p. 491
18. Morey, P.E . *Text Res. J.* (1935) 5, p. 483
19. Nishijima, Y., Onogi, Y., Asai, T. *J. Polym. Sci. C* (1966) 15, p. 237
20. Nishijima, Y., Fujimoto,T., Onogi, Y. *Rep. Prog. Polym. Phys. Jap.* (1966) 9, p. 457
21. Nishijima, Y., Onogi, Y., Asai, T. *Int. Symp. Macromol. Chem.* (1966) 7, p. 161
22. Nishijima, Y., Onogi, Y., Asai, T. *Rep. Prog. Polym. Phys. Jap.* (1967) 10, p. 461
23. Nishijima, Y., Onogi, Y., Asai, T., Yamazaki, R. *Rep. Prog. Polym. Phys. Jap.* (1967) 10, p. 465
24. Nishijima, Y., Onogi, Y. *Rep. Prog. Polym. Phys. Jap.* (1968) 11, p. 395
25. Nishijima, Y., Onogi, Y., Asai, T., Yamazaki, R. *Rep. Prog. Polym. Phys. Jap.* (1968) 11, p. 399 and 403
26. Nishijima, Y., Asai, T. *Rep. Prog. Polym. Phys. Jap.* (1968) 11, p. 419 and 423
27. Onogi, Y., Nishijima, Y. *Rep. Prog. Polym. Phys. Jap.* (1971)14, p. 533 and 537 and 541
28. Nishijima, Y., Onogi, Y., Yamazaki, R., Kawakami, K. *Rep. Prog. Polym. Phys Jap.* (1968) 11, p. 407
29. Nishijima, Y., Onogi, Y., Yamazaki, R. *Rep. Prog. Polym. Phys. Jap.* (1968)11, p. 415
30. Yamazaki, R., Onogi, Y., Nishijima, Y. *Rep. Prog. Polym. Phys. Jap.* (1969)12, p. 439
31. Asai, T., Onogi, Y., Nishijima, Y. *Rep. Prog. Polym. Phys. Jap.* (1969)12, p. 433
32. Nishijima, Y. *J. Polym. Sci* C (1970) 31, p. 353
33. Seki, J. *Sen-I Gakkaishi* (1969) 25, p. 16
34. Seki, J. *Sen-I Gakkaishi* (1969) 25, p. 24
35. Mc Graw, G.E . *J. Polym. Sci. A-2* (1970) 8, p. 1323
36. Roe, R.-J. *J. Polym. Sci. A-2* (1970) 8, p. 1187
37. Bower, D.I. *J. Polym. Sci., Polym. Phys. Ed.* (1972) 10, p. 2135
38. Kimura, I., Kagiyama, M., Nomura, S., Kawai, H. *J. Polym. Sci. A-2* (1969) 7, p. 709
39. Nobbs, J.H., Bower, D.I., Ward, I.M., Patterson, D. *Polymer* (1974) 15, p. 287
40. Jarry, J.P., Monnerie, L. *J. Polym. Sci., Polym. Phys. Ed.* (1978) 16, p. 443
41. Hennecke, M., Fuhrmann, J. *J. Colloid Polym. Sci.* (1980) 258, p. 219
42. Nobbs, J.H., Ward, I.M. In *Polymer Photophysics* Phillips, D.C. (Ed.)(1985) Chapman & Hall, London, pp. 159
43. Desper, C.R., Kimura, I. *J. Appl. Phys.* (1967) 38, p. 4225
44. Pinaud, F., Jarry, J.P., Sergot, Ph., Monnerie, L. *Polymer* (1982) 23, p. 1575

45. Nobbs, J.H., Bower, D.I., Ward, I. *Polymer* (1976) 17, p. 25
46. Nobbs, J.H., Bower, D.I., Ward, I. *J. Polym. Sci., Polym. Phys. Ed.* (1979) 17, p. 259
47. Fuhrmann, J., Hennecke, M. *J. Colloid Polym. Sci.* (1976) 254, p. 6
48. Hennecke, M., Fuhrmann, J. *Polymer* (1982) 23, p. 797
49. Lapersonne, P., Tassin, J.F., Sergot, P., Monnerie, L., Le Bourvellec, G. *Polymer* (1989) 30, p. 1558
50. Le Bourvellec, G., Monnerie, L., Jarry, J.P. *Polymer* (1986) 27, p. 856
51. Le Bourvellec, G., Monnerie, L., Jarry, J.P. *Polymer* (1987) 28, p. 1712
52. Hibi, S., Maeda, M., Kubota, H., Miura, T. *Polymer* (1977) 18, p. 137
53. Hibi, S., Maeda, M., Kubota, H., Miura, T. *Polymer* (1977) 18, p. 143
54. Hibi, S., Maeda, M., Kubota, H., Miura, T. *Polymer* (1977) 18, p. 801
55. Jarry, J.P., Monnerie, L. *J. Polym. Sci., Polym. Phys. Ed.* (1980) 18, p. 1879
56. Fajolle, R., Tassin, J.F., Sergot, P., Pambrun, C., Monnerie, L. *Polymer* (1983) 24, p. 379
57. Allen, N.S., Mc Kellar, J.F. *Makromol. Chem.* (1978) 179, p. 523
58. Hennecke, M., Fuhrmann, J. *Makromol. Chem., Macromol. Symp.* (1986) 5, p. 181
59. Hennecke, M., Kud, A., Kurz, K., Fuhrmann, J. *J. Colloid Polym. Sci.* (1987) 265, p. 674
60. Hemker, D.J., Frank, C.W., Thomas, J.W. *Polymer* (1988) 29, p. 437
61. Sonnenschein, M.F., Roland, C.M. *Polymer* (1990) 31, p. 2023
62. Chen, L., Jiu, X., Du, J., Qian, R. *Makromol. Chem.* (1991) 192, p. 1399
63. Hennecke, M., Kurz, K., Fuhrmann, J. *Polymer* (1989) 30, p. 1072
64. Hennecke, M., Kurz, K., Fuhrmann, J. *Macromolecules* (1995) 25, p. 6190
65. Clauss, B., Salem, D.R. *Macromolecules* (1995) 28, p. 8328
66. Gohil, R.M., Salem, D.R. *J. Appl. Polym. Sci.* (1993) 47, p. 1989
67. Clauss, B., Bayer, A., Oppermann, W., Hartrumpf, M. *Chem. Fib. Internat.* (1996) 46, p. 175
68. Hartrumpf, M., Clauss, B. and Salem, D.R. Unpublished report.

13.4 Spectroscopic Methods: Infrared, Raman, and Nuclear Magnetic Resonance

David R. Salem and Nadarajah Vasanthan
TRI/Princeton, Princeton New Jersey, USA

13.3.1 Introduction

Vibrational spectroscopy has been used more than fifty years to characterize polymeric materials [1,2]. Macromolecules are formed by connecting atoms through chemical bonds. All atoms in the macromolecule possess kinetic energy and they vibrate. Infrared radiation is absorbed by macromolecules and converted into energy of molecular vibrations. These vibrations can be separated as normal vibrational modes and appear as bands rather than as lines because of multiple rotational energy levels. The normal modes are classified into several groups depending on the symmetry of the molecule. Molecular vibrations in the infrared (IR) result from electric dipole moment changes and only an oscillating dipole will interact with the electromagnetic field with absorption energy. In a centro-symmetrical valence vibration, the dipole moment is not changed and the vibration is IR inactive. Highly polar molecules produce particularly intense IR bands because the corresponding vibrations cause large changes in the dipole moment.

The Raman effect does not involve absorption, but is a scattering process. The inelastically scattered Stokes lines are measured, and it is not the change in dipole moment that determines the intensity of a Raman band but the periodic change in polarizability resulting from the vibration. Thus IR and Raman are sensitive to different vibrations and can provide complimentary information [3].

The introduction of fourier transform infrared (FTIR) spectrometry has permitted spectroscopists to develop new sampling techniques such as attenuated total reflectance (ATR), diffuse reflectance, and photoacoustic (PA) methods for characterization of polymeric materials [4]. The development of microscope attachments for infrared and Raman spectrometers has allowed the study of very small samples, such as single fibers. Resolution can be attained to 10μ for infrared spectroscopy and 1μ for charge-coupled device (CCD) Raman spectrometers with a confocal microscope.

Nuclear magnetic resonance (NMR) spectroscopy is another form of absorption spectroscopy that can be applied to polymers [5]. In strong magnetic fields at excitation frequencies in the radio frequency region, materials can absorb electromagnetic radiation at frequencies governed by the characteristics of the atomic nuclei and their chemical or structural environment. This effect is restricted to nuclei with nuclear spin not equal to zero. Since the nuclear spin differs from zero only if the number of protons and/or neutrons are odd, nuclei having an equal number of protons and neutrons (e.g., ^{12}C and ^{14}N) cannot be measured and nuclei with spin ½ are favored (e.g. ^{13}C, ^{1}H, ^{15}N and ^{19}F). Nuclei with higher spin possess a quadrapole moment, which leads to unfavorable line broadening. The NMR spectra of spin ½ nuclei are commonly used to analyze polymer structure [6].

It is relatively straightforward to obtain high resolution NMR spectra from polymer solutions because of rapid local motion of polymer chains. Of course, fibers must be characterized in the solid state and in the last two decades there has been substantial development of improved methods for structural studies of polymers by solid-state NMR. 'Magic angle' spinning of the sample must be applied to remove line broadening due to the chemical-shift anisotropy [7]. Decoupling of dipolar interaction between nuclear spins is a prerequisite to obtaining high-resolution spectra, and cross polarization from protons to rare nuclei is also a technique widely used for signal enhancement. The introduction of two-dimensional and three-dimensional NMR has greatly enhanced the power and versatility of the NMR technique. By systematically incrementing the evolution time t_1 and acquisition of the NMR signal during the detection period t_2, a two dimensional time signal is obtained. In a 2D exchange spectrum detailed information is provided about mixing processes, which may be chemical exchange, molecular reorientation, or dipolar exchange [8]. In 2D separation spectra, one dimension separates the signals by their well-resolved isotropic shifts, while a broadline spectrum of the same site, reflecting dipolar couplings, chemical shift anisotropy etc., is displayed along the other dimension. 3D NMR can be used to separate overlapping 2D spectra, to correlate a further anisotropic interaction, or to study details of a complex exchange process. The main chemical and structural information that can be obtained from NMR analysis of polymer fibers and other oriented polymers are: Chemical groups (from magic angle spinning experiments); intermolecular distances (from dipolar splitting); orientation distributions in crystalline and amorphous regions (from anisotropic interaction measurements); chain packing in the crystal structure (from chemical-shift differences); molecular mobility in various phases (from relaxation times and line shapes); the size of supermolecular domains in semicrystalline polymers, polymer blends, and block copolmers (from spin and noble-gas diffusion studies).

The purpose of this short review is to demonstrate the sort of measurements that can be carried out and the type of microstructural information that can be obtained for polymeric fibers using infrared, Raman and NMR spectroscopy. Examples of recent work on characterization of polyolefins, polyamides, polyesters and liquid crystalline fibers by these methods are briefly summarized.

13.4.2 Infrared

13.4.2.1 Polyolefins

Infrared spectroscopy has been used to determine the crystallinity and molecular orientation of polyethylene [9,10]. 'Real' crystallinity bands, which arise from intermolecular interaction of adjacent chains in the crystal, have been observed only for a few semicrystalline polymers. In polyethylene, for example, the two chains in the crystallographic unit cell are very close to each other and intermolecular interaction between them causes band splitting in the IR spectrum (720 and 730 cm^{-1}). For many semicrystalline polymers, crystallinity changes are assessed from bands originating from conformational arrangements that are required for a specific crystal morphology.

The orientation of polyethylene samples has been measured recently using the bands at 1080 and 1378 cm^{-1}, assigned to the crystalline phase, and at 1894 cm^{-1}, assigned to the amorphous phase [11]. These bands are overtones and relatively week compared to the other vibrations. Three samples with different levels of crystallinity were studied and quite similar orientational behavior was observed during drawing in spite of the large differences in crystallinity (Figure 13.4.1). It is apparent that, under the drawing conditions used, both crystalline and amorphous orientation functions increase as a function of draw ratio up to $\lambda = 5$ and then level off.

Crystalline orientation of polypropylene has been investigated by IR spectroscopy [12]. The crystalline orientation was determined using the band at 1220 cm^{-1}. The crystalline orientation function obtained by X-ray diffraction was plotted against the dichroic ratio of the band at 1220 cm^{-1} (Figure 13.4.2) and the transition moment angle for the band at 1220 cm^{-1} with respect to c-axis was found to be 72^0. The band at 1256 cm^{-1} has contributions from crystalline and amorphous components and, therefore, the dichroism of this band was used to obtain average overall orientation of the sample. After determining the phase fractions, the amorphous orientation was determined by subtracting the crystalline orientation from the average orientation values obtained by IR spectroscopy.

13.4.2.2 Polyamides

Having found much ambiguity in the infrared band assignments for polyamide 66, Vasanthan and Salem [13] recently revisited some of these assignments. They showed that, whereas the 1144 cm^{-1} and 1180 cm^{-1} bands are not appropriate for characterizing the noncrystalline phase, the bands at 924 cm^{-1} and 1136 cm^{-1} can be attributed unambiguously to this phase. They also confirmed that the bands at 936 and 1200 cm^{-1}

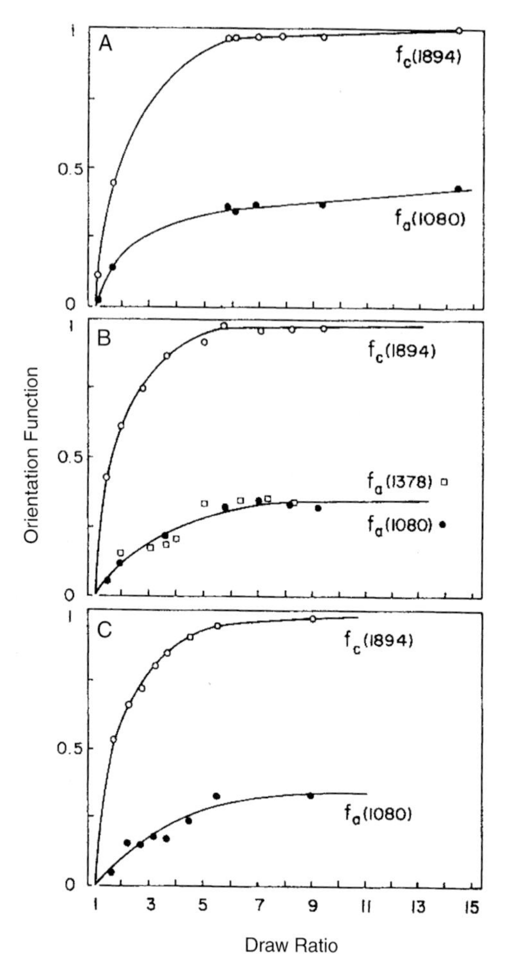

Figure 13.4.1 Crystalline and amorphous orientation functions of polyethylene containing different crystallinities versus draw ratio A) 65% crystallinity B) 50% crystallinity C) 32% crystallinity. After Ref. [11].

are crystalline bands. These results were obtained by plotting the normalized absorbance of each band against fiber density and determining whether the extinction of the crystalline and amorphous sensitive bands corresponds to the reported amorphous and crystalline density values. Determination of PA66 crystallinity can be carried out by one of two infrared methods. In one method the 1200 cm^{-1}/1630 cm^{-1} band ratio is calibrated against crystallinity measured by density, whereas the other method is an independent IR approach based on changes in the 936 and 924 cm^{-1} bands. The independent method showed good agreement with crystallnity values obtained from density measurements.

The band at 1224 cm^{-1} has been assigned to the regular chain fold conformation in single crystals of PA66 [14]. The normalized absorbance of this band in PA66 fibers appears to increase with heat treatment temperature in the range 150 – 200 °C in a similar

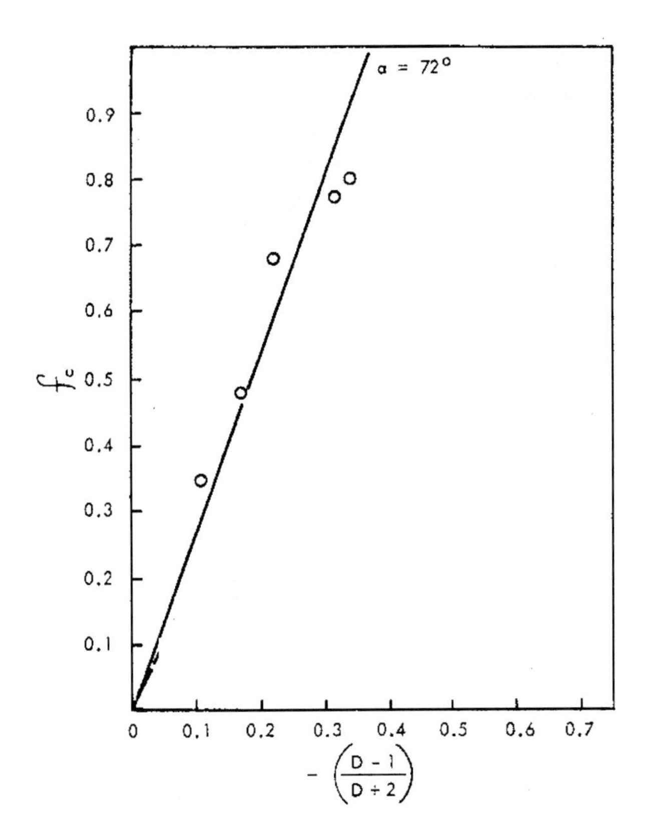

Figure 13.4.2 Determination of transition moment angle for the band at 1220 cm^{-1} in isotactic polypropylene. After Ref. [12].

way to the crystalline bands. This indicates that chain folding is the predominant mechanism of thermally induced crystallization in oriented PA66 [13].

The 936 cm^{-1} crystalline band has been used to determine crystal orientation in PA66 fibers as a function of spinning speed and molecular weight and as a function of draw ratio. The transition moment angle of the 936 cm^{-1} vibration has been reported as 39°, 41° [15] and 48° [16].

Measurements of the molecular orientation in PA66 films has been made by selective deuteration [17]. Deuteration was carried out by immersing the film in the D_20 medium. Diffusion of D_20 in the crystalline phase is extremely slow compared to the amorphous phase, permitting selective deuteration of the amorphous phase. The vibration associated with N-D stretch was used to obtain molecular orientation. A similar approach could be used to study the crystalline and amorphous orientation of polyamide fibers.

In polyamide 6, the bands at 930 and 960 cm^{-1} have been assigned to the α phase, the bands at 919 and 977 cm^{-1} have been assigned to the γ phase and the 983 cm^{-1} band has been attributed to the amorphous phase [18,19]. Unpolarized spectra taken for drawn fibers having α, γ and amorphous phases are shown in Figure 13.4.3. Depending upon the thermal and mechanical treatment, the fractions of the α, γ and amorphous phase can change dramatically and these band assignments have been used to determine the fractions of individual phases under a range of fiber formation conditions. The bands at 930 and 977 cm^{-1} have been used to measure the crystalline orientation of α and γ crystalline phases, respectively, using the measured transition moment angle of 39°. Since the direction of the transition moment is not known for the amorphous phase, only the dichroic ratio of the 983 cm^{-1} band was reported in the fiber drawing study of Sibilia et al. [18]. Other problems with the 983 cm^{-1} band are that it is of low intensity and it is highly submerged by strong overlapping bands, so that reliable curve fitting is often impossible.

13.4.2.3 Polyesters

Investigations of the IR spectra of poly(ethylene terephthalate) PET samples that were subjected to various thermal and mechanical treatments have shown that several infrared bands vary in intensity as a function of crystallinity [20-22]. However, it has been pointed out that changes in the band intensity can be interpreted in terms of conformational changes in the ethylene glycol unit rather than directly to crystallinity changes. The bands at 1453, 1370, 1040 and 895 cm^{-1} have been assigned to *gauche* conformation of the ethylene glycol unit, which occurs only in the noncrystalline phase, and the bands at 1470, 1340, 973 and 846 cm^{-1} have been assigned to the *trans* conformation, which can occur in both the crystalline and noncrystalline phases. The best bands to monitor the conformer content quantitatively are the *trans* band at 973 cm^{-1} and the *gauche* band at 898 cm^{-1}. The band at 793 cm^{-1} is generally used as the internal

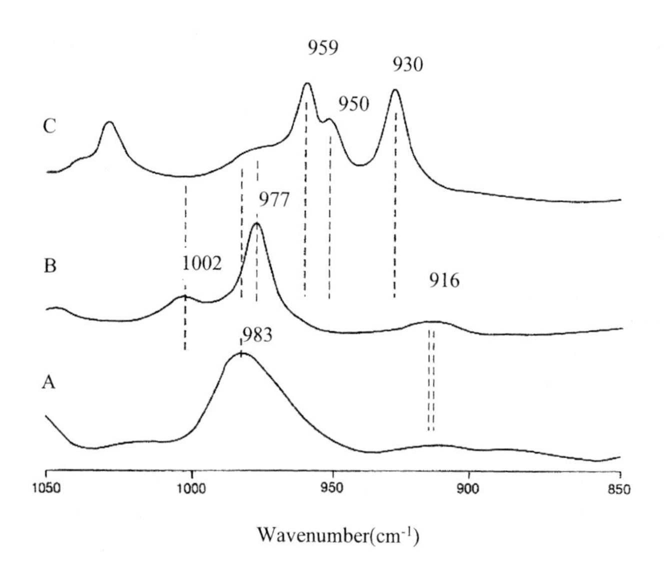

Figure 13.4.3 Unpolarized IR Spectra of polyamide 6 A) amorphous, B) high γ content, C) high α content. After Ref. [19].

reference band. Changes in the *trans/gauche* ratio have been studied as a function of draw ratio in single stage and two-stage drawing of PET and after thermal shrinkage of high tenacity fibers [23]. Polarized infrared spectroscopy has been used to obtain orientation changes of the *gauche* and *trans* conformers during drawing [24], and the overall orientation was obtained from bands associated with the phenyl ring.

The 988 cm^{-1} band in PET has been assigned to adjacent chain folding at the surface of the crystals [25]. The band appears to be absent in semicrystalline PET fibers drawn in the region of the glass transition temperature, but is present in semicrystalline, high-speed-spun fibers [26]. Thermal annealing of PET fibers in the range 200 – 240 °C causes the absorption at 988 cm^{-1} to increase quite strongly, indicating that chain folding is the predominant mode of crystallization at these temperatures.

The crystal to crystal transition in poly(butylene terephthalate) PBT fibers has been followed using infrared spectroscopy [27]. The α crystal form is obtained at zero stress condition while the β crystal form is obtained during fiber extension. The changes in the crystal form produced dramatic changes in the infrared spectra. For example, the bands at 972 cm^{-1} and 752 cm^{-1}, assigned to the C-H rocking vibration in the α phase, disappear during extension and new bands appear at 962 and 844 cm^{-1}. Changes in the spectrum were attributed to conformational changes occurring during the extension, and it was suggested that butylene glycol adopts a *gauche-trans-gauche* conformation at zero stress, which tranforms to a nearly all *trans* conformation under high stress.

13.4.2.4 Liquid Crystalline Polymers

The two crystal modifications in poly(p-phenylene terephthalamide) (Kevlar 49) fibers have been identified by infrared spectroscopy. Normal cordinate analysis was carried out for fuller understanding of the spectrum and the major differences between the two crystal forms were found to be in the vibrations associated with amide bond. FTIR photoacoustic spectroscopy in combination with deuterium exchange has been used to determine the accessibility of the N-H groups of Kevlar 49 fibers [28].

13.4.3 Raman

The theories and procedures for determining polymer orientation by Raman spectroscopy have been developed by Bower [29], who showed that for samples with uniaxial symmetry it is necessary to measure five independent components of the scattered radiation. Unlike infrared dichroism and birefringence, Raman spectroscopy can yield the fourth moment of the orientation distribution in addition the second moment. An experimental problem with orientation analysis by Raman specroscopy is polarization scrambling that arises from multiple reflections and refractions of the incident and scattered radiation near the surface of the sample, and it is necessary to employ special sampling techniques to reduce this effect.

Stress-induced changes in the vibrational frequences of infrared and Raman bands, arising from the anharmonicity of the vibration [30], can provide the means to analyze the molecular deformation mechanisms resulting from tensile deformation of a fiber. Changes in bond lengths, internal rotation angles, and the associated force constants, cause changes in the frequency, intensity and width of the vibrational bands, which can be precisely measured. The application of Raman spectroscopy to study the deformation processes in high performance fibers, such as liquid-crystalline polymer fibers [31–36], gel-spun polyethylene fibers [37–43] and carbon fibers [44–49], has been particularly fruitful, and some success has also been obtained in its application to molecular deformation analysis of poly(ethylene terephthalate) fibers [50–52].

13.4.3.1 Liquid Crystalline Polymers

Extensive investigations by Young and coworkers have established the power of Raman microscopy as a tool for monitoring (room temperature) molecular deformation processes in aramid [31–34], PBO [36] and PBT [35] fibers. When such fibers are deformed, the frequencies of Raman active bands tend to decrease by an amount dependent upon the material, the nature of the band and the fiber modulus. Shifts have been reported of $-12\,cm^{-1}$ per 1% strain for high modulus PBT fibers, $-8\,cm^{-1}/\%$ for the 1280 cm^{-1} band of PBO fibers and $-5\,cm^{-1}/\%$ for the $1610\,cm^{-1}/\%$ band of Kevlar. Deformation appears to take place by a combination of crystal rotation and crystal stretching [31], and since the shifts of the Raman bands arise directly from the stretching of covalent bonds along the polymer backbone, they reflect only the crystal stretching component, which is higher in high modulus fibers. Raman spectra of these fibers also display significant peak broadening, which indicates that the molecules are experiencing different levels of stress, some being overstressed and some under stressed relative to the average stress level. Band shifting in Raman microscopy can also be used to follow the micromechanics of deformation in high-performance fibers within composites, and to serve as a highly effective molecular strain gauge for these materials [53–56].

13.4.3.2 Polyolefins

Rigorous quantitative measurements of orientation in high density polyethylene HDPE fibers have been made by Citra et al. [57]. Using the 1418 cm^{-1} band, which is characteristic of the orthorhombic crystal structure, they determined a $\langle P_2(\cos\theta)\rangle_c$ value of 0.82 (in excellent agreement with their WAXS data) and a $\langle P_4(\cos\theta)\rangle_c$ value of 0.53. From the 1130 cm^{-1} band, which has been attributed to C-C symmetric stretching modes contributed from both the crystalline phase and the *trans* conformers of the noncrystalline phase, a $\langle P_2(\cos\theta)\rangle$ of 0.76 and a $\langle P_4(\cos\theta)\rangle$ of 0.49 were determined. These results are consistent with the high crystallinity of the sample (80 – 85%); the contribution from the noncrystalline fraction being small.

The shape and position of Raman bands in ultra-high modulus (gel-spun) PE fibers under stress have been studied by various groups [37–43]. The stress-sensitive bands (at $\sim 1063\,cm^{-1}$ and $1128\,cm^{-1}$) shift to lower wavenumber with increasing strain and the line width increases with stress. The line shape remains symmetric at relatively low deformations, but at higher deformations a broad tail emerges at the low wavenumber end of the peak and the main peak stops shifting. This assymetric tail, which in some highly-strained samples has been observed to develop into a separate peak [39], has been attributed to a fraction of highly stressed load-bearing chains. In some studies involving strains of around 2% (at room temperature), 40% of the crystalline material was in the form of the high load-bearing chains [39,42]. A current interpretation is that there are two, differently-stressed types of molecular chains within the crystalline phase [38,40,43], but the mechanism of this type of stress concentration is not well understood.

13.4.3.3 Polyesters

Purvis, Bower and Ward [58] have shown that polarized Raman analysis can provide information on the overall molecular orientation in PET, since good correlations were obtained between orientation functions from Raman and optical birefringence. Melveger [59] related the half-width of the carbonyl band at ~1725 cm^{-1} to the degree of crystallinity in PET, attributing the sharpening of the band in the crystalline phase to the coplanarity of the carbonyl and aromatic groups. Raman bands at 1096 and 1000 cm^{-1} in PET were originally assigned to the crystalline phase [60], but Adar and Noether [61] have argued that they actually represent the *trans* conformation of the glycol group.

An informative comparison of the IR and Raman spectra of poly(ethylene terephthalate), poly(trimethylene terephthalate) PTT, and poly(tetramethylene terephthalate) PBT was made by Ward and Wilding [27]. They concluded that the differences between the spectra of these polymers can be explained by changes in the conformation of the molecular chains involving both *trans/gauche* isomerism in the glycol unit and the planarity of the terephthalate unit.

Yeh and Young [52] have found that the shift of the 1616 cm^{-1} C-C ring stretching band shows a linear relationship with stress and has a Raman band shift factor of –4.0 cm^{-1}/GPa in all the highly oriented PET fibers they studied. The Raman bandwidth provided information about the distribution of stresses in the molecules, and the authors interpreted the (room temperature) molecular deformation processes in terms of a modified series aggregate model.

It has been shown that for PET fibers of moderate orientation, the Raman band shifts depend on the polarization direction [50,51]. This is because observations with different polarization directions are selectively sensitive to chains oriented in different directions, and these chains are in different states of stress.

A useful application of confocal Raman microscopy is in the study of bicomponent fibers, as demonstrated by Natarajah and Michielsen who examined the structure and orientation of the PET interior of PET/PP core/shell fibers [62].

13.4.4 Nuclear Magnetic Resonance

13.4.4.1 Polyolefins

Gel-spun polyethylene has been investigated at room temperature using CP/MAS spectra [36]. Three components – crystalline, amorphous and oriented amorphous – were identified. The chemical shifts values for the crystalline and the oriented amorphous components are 34.06 ppm but the relaxation times are different (28 s for the crystalline component and 1.8 s for the oriented amorphous component). The chemical shift value for the amorphous phase is 31.7 ppm and the relaxation time is 0.3 s. The observed chemical shift differences between crystalline and amorphous carbons can be interpreted using the γ-gauche effect [64]. Two dimentional rotor-synchronized solid state NMR spectroscopy has been used to investigate the orientational order of several moderate and ultra-high molecular weight polyethylene fibers [65]. Three orientational orders were identified (two crystalline and one amorphous). One of the crystalline components was found to be much less oriented than the other, although the orientational order of the more highly oriented component varied substantially from sample to sample. The different crystal components could not be related to crystal polymorphism, since no evidence to support this was found from WAXS analysis.

13.4.4.2 Polyamides

Solid state ^{13}C and ^{15}N NMR spectra of the α, γ and amorphous phases of polyamide 6 has been obtained [66]. The observed chemical shift differences were discussed in terms of hydrogen bonding and chain packing. It was shown that hydrogen bonding interaction is stronger in the γ phase than in the α phase. ^{15}N NMR spectra showed a chemical shift difference of approximately 5 ppm for the α and γ forms of PA6 (Figure 13.4.4). Two dimensional solid state NMR spectroscopy has been used to investigate the molecular orientation of hot-drawn PA6 fibers at draw ratios in the range 1–5.5 [67].

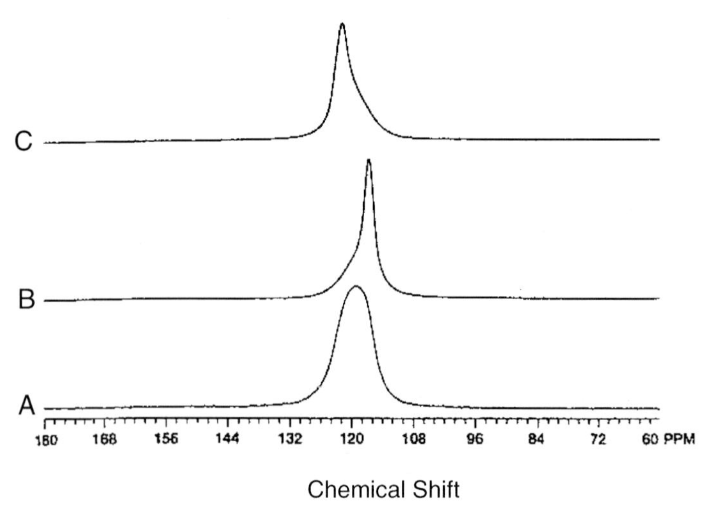

Figure 13.4.4 ^{15}N CP/MAS NMR spectra of polyamide 6 A) amorphous, B) high α content, C) high γ content. After Ref. [39].

13.4.4.3 Polyesters

NMR has been applied to the structural analysis of PET for some time. Two methods have been used in the past: line shape analysis and relaxation time measurements [68,69]. The chemical shift of the ethylene carbon associated with the ethylene glycol unit has been resolved into two components. The narrow component is associated with the crystalline fraction and the broad component is associated with amorphous fraction. Integrated peak intensities can be used to calculate crystallinity values. Roland et al [70] used solid state proton NMR spectroscopy recently to determine the crystallinity of PET fibers. NMR spectra of amorphous material were fitted with bi-exponential functions, associated with amorphous and constrained amorphous material. Drawn fibers were fitted with tri-exponential functions, associated with amorphous, constrained amorphous and crystalline material. The degree of crystallinity obtained using proton NMR showed a linear relationship with the crystallinity obtained from density measurements.

English has studied PET fibers using various solid state ^{1}H and ^{13}C variable-temperature NMR experiments [71]. The oriented fibers used for this study were either completely amorphous or 49% crystalline. The results showed four separate motional processes: very rapid reorientaion, benzene ring motion, *trans-gauche* motion and isotropic motion. It was also shown that results depend on the crystallinity of the sample. The molecular orientation of a series of PET fibers was studied by C-13 2D CP-MAS NMR [72]. As would be expected, significant differences in the orientation were apparent as a function of processing history, and it was shown that overall orientation values obtained by NMR are linear related to birefringence data.

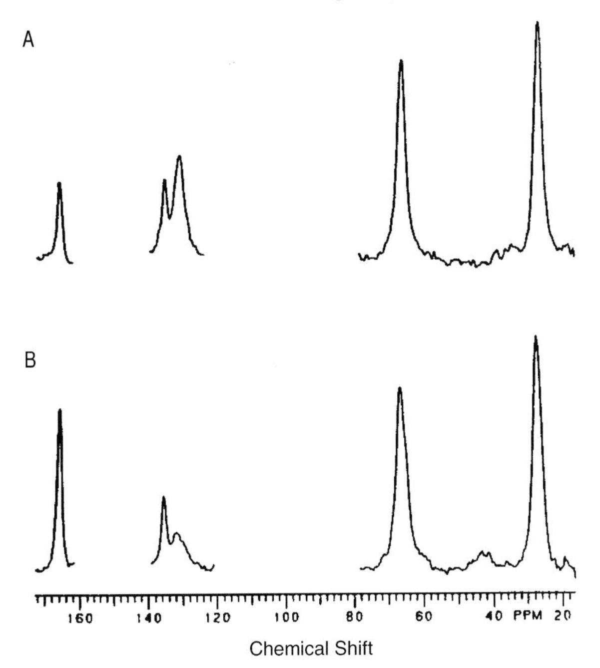

Figure 13.4.5 13 C CP/MAS/DD NMR spectra of polybutylene terephthalate A) high α content, B) high γ content. After Ref. [46].

The reversible crystal to crystal transition of PBT during uniaxial extension was characterized by solid state NMR spectroscopy [73]. The CPMAS/DD C-13 NMR spectra were obtained at elevated temperature to remove the contribution of the glassy amorphous component (Figure 13.4.5). Identical chemical shifts were observed for the central methylene carbons in both α and β crystal forms, suggesting that conformation of the butylene glycol is the same (all *trans*) for both crystal forms. If the glycol unit adopts different conformations, as discussed in the infrared section, one would expect to see a chemical shift difference for the central methylene carbon due to the γ-gauche effect. The authors concluded that conformational changes are not associated with differences in the α and β crystal structure.

13.4.4.4 Liquid Crystalline Polymers

Poly(p-phenylene terephthalamide) (Kevlar 49) has been studied by solution and solid state NMR spectroscopy [74]. It has been shown that the chain structure is highly ordered in the solid state and is of lower symmetry than in solution. High resolution solution state NMR results suggest that the polymer adopts an all *trans* conformation in a dilute isotropic solution. Spin lattice relaxation times have been obtained for Kevlar 49 fibers under dry and wet conditions at room temperature and at 80 °C [75]. Water sorption did not affect the NMR spectrum significantly, which suggests that the water molecules sorbed by Kevlar 49 do not interact so strongly with the polar amide group. However, two components with different relaxation times (t_1) were observed. The longer t_1 is not effected by hydration, whereas a slight decrease was observed for the shorter t_1. These results indicate that Kevlar 49 fibers are not completely crystalline (uniphase), as suggested previously, but have a small noncrystalline fraction.

13.4.5 References

1. Koenig, J. L. *Spectroscopy of Polymers, American Chemical Society* (1992).
2. Siesler, H. W and Holland-Moritz, K. *Infrared and Raman Spectroscopy of Polymers* (1980) Practical Spectroscopy Series
3. Urban, M. W. and Provder, T. *Multidimensional Spectroscopy of Polymers* (1994). American Chemical Society
4. Urban, M. W, *Vibrational Spectoscopy of Molecules and Macromolecules on Surfaces* (1993) Wiley & Sons, New York
5. Bovey, F. A. *Nuclear Magnetic Resonance* Academic Press (1988), New York
6. McBrierty, V. J.; Packer, K. J. *Nuclear Magnetic Resonance in Solid Polymers* (1993) Cambridge University Press, Cambridge
7. Tonelli, A. E. *NMR Spectroscopy and Polymer Microstructure: The Conformational Connection* (1989) VCH Publishers, New York
8. Scmidt-Rohr, K.; Spiess, H. W. *Multidimensional Solid State NMR and Polymers* (1994) Academic Press, San Diego
9. Read, B. E and Stein, R. S . *Macromolecules* (1968) 2, p. 116
10. Ward, I. M. *Development of Oriented Polymers* (1982) Applied Science Publishers, London
11. Rossingnol, J. M.; Seguela, R and Rietsch, F. *J. Poly Sc., Lett. Ed.* (1989) 27, p. 527

12. Samuels, R. H. *J. Polym Sci, Part A* (1965) 3, p. 1741
13. Vasanthan, N and Salem, D. R. *J. Poly. Sci., Phys. Edn* (2000) 38, p. 516
14. Koenig, J. L and Agboatwalla, M. C. *J. Macromol. Sci.* (1968) B2, p. 391
15. Samanta, S. R.; Lanier, W. W.; Miller, R. W and Gibbson, JR, M. E. *Applied Spectroscopy* (1990) 44, p. 1139
16. Vasanthan, N and Salem, D. R. *Material Research Innovations* (2000). In press
17. Garton, A and Phibbs, M. K. *Makromol. Chem.,Rapid Commun.* (1982) 3, p. 569
18. Sibilia, J. P. *J. Polym Sci. Polym. Phys. Edn* (1971) 27, p. 19
19. Murthy, N. S.; Bray, R. G.; Correale and Moore, R. A. F. *Polymer* (1995) 20, p. 3863
20. Miyake, A. *J. Polym. Sci* (1959) 38, p. 479
21. D'Esposito, L and Koenig, J. L. *J. Polym. Sci, Polym. Phys. Edn* (1976) 14, p. 1731
22. Rodriquez-Cabello.; Satos, J.; Merino, J. C and Pastor, J. M. *J. Polym. Sci, Polym. Phys. Edn*, (1996) 34, p. 1243
23. Padibjo, S. R and Ward, I. M. *Polymer* (1983) 24, p. 1103
24. Cunningam, A.; Davis, G. R and Ward, I. M. *Polymer* (1974) 15, p. 743
25. Koenig, J. L and Hannon, M. J. *J. Macromol. Sci. Phys.* (1967) B1, p. 119
26. Cuculo, J. A. *J. Polym. Sci, Polym. Phys. Edn* (1995) 33, p. 909
27. Ward, I. M and Wilding, M. A. *Polymer* (1977) 18, p. 327
28. Chatzi, E. G.; Urban, M. W.; Ishida, H and Koenig, J. L. *Macromolecules* (1986) 19, p. 3861
29. Bower, D. I. *J. Polym. Sci, Polym. Phys Edn* (1972) 10, p. 2135
30. Tashiro, K. *Prog. in Polym. Sci.* (1993) 18, p. 337
31. Galiotis, C., Robinson, I.M., Young, R.J., Smith, B.J.E. and Batchelder, D.N. *Polym. Commun.* (1985) 26, p. 354
32. Van der Zwaag, S., Northolt, M.G., Young, R.J., Robinson, I.M., Galiotis, C, Batchelder, D.N. *Polym. Commun.* (1987) 28, p. 276
33. Young, R. J, Lu, D., Day, R. J. *Polym. International* (1991) 24, p. 71
34. Young, R. J, Lu, D., Day, R. J., Knoff, W. F., Davis, H. A. *J.Mater. Sci.* (1992) 27, p. 5431
35. Day, R. J., Robinson, I. M., Zakikhani, M. and Young, R. J. *Polymer* (1987) 28, p. 1833
36. Young, R. J., Day, R.J., Zakikhani, M. *J. Mater. Sci.* (1990) 25, p. 127
37. Prasad, K and Grubb, D. T . *J. Polym. Sci, Polym. Phys. Edn* (1989) 27, p. 381
38. Grubb, D. T. and Li, Z. *Polymer* 30 (1992) 2587
39. Kip, B. J., Van Eijk, M. C. P. and Meier, R. J. *J. Polym. Sci, Polym. Phys. Edn* (1991) 29, p. 99
40. Monnen, J. A. H. M., Roovers, W. A. C., Meier, R. J., Kip, B. J. *J. Polym. Sci, Polym. Phys. Edn* (1992) 20, p. 361
41. Tashiro, K., Wu, G. and Kobayashi, M. *Polymer* (1988) 29, p. 1768
42. Wong, W.F. and Young, R. J. *J. Mater. Sci.* (1993) 28, p. 510
43. Wong, W.F. and Young, R. J. *J. Mater. Sci.* (1994) 29, p. 520
44. Morita, K., Murata, Y., Ishitani, K., Murayama, T., Ono, T. and Nakajima, A. *Pure Appl. Chem.* (1986) 58, p. 455
45. Robinson, I. M., Zakikhani, M., Day, R. J., Young, R. J. and Galiotis, C. *J. Mater. Sci. Lett.* (1987) 6, p. 1212
46. Huang, Y. and Young, R. J. *J. Mater. Sci. Lett.* (1993) 12, p. 92
47. Sakata, H., Dresselhaus, G., Dresselhaus, M.S., and Endo, M. J. Appl. Phys. (1988) 63, p. 2769
48. Galiotis, C., Batchelder, D. N. *J. Mater. Sci. Lett.* (1988) 7, p. 545
49. Huang, Y. and Young, R. J. *Carbon* (1995) 33, p. 97
50. Fina, L. J., Bower, D. I. and Ward, I. M. *Polymer* (1988) 29, p. 2146
51. Bower, D. I., Lewis, E. L. V. and Ward, I. M. *Polymer* (1995) 36, p. 3473
52. Yeh. W. Y. and Young. R. J. *J. Macromol. Sci.–Phys.* (1998) B37(1), p. 83
53. Fan, C. F. and Hsu, S. L. *Macromolecules* (1989) 22, p. 1474
54. Melantis, N. and Galiotis, C. *Proc. Roy. Soc. London* (1993) A440, p. 379
55. Andrews, M. C. and Young, R. J. *J. Raman Spectroscopy* (1993) 24, p. 539
56. Huang, Y. and Young, R. J. *Composites Sci. Tech.*, (1994) 52, p. 505
57. Citra, M. J.; Chase, D. B.; Ikeda, R. M and Gardner, K. H. *Macromolecules* (1995) 28, p. 4007
58. Purvis, J.; Bower, D. I and Ward, I. M. *Polymer* (1973) 14, p. 398

59. Melveger, A. J. *J. Polym Sci. A2* (1972) 10, p. 317
60. Boerio, F. J.; Bahl, S. K and McGraw, G. E. *J. Polym Sci, Polym Phys. Edn* (1976) 14, p. 1029
61. Adar, F and Noether, H. *Polymer* (1985) 26, p. 1935
62. Natarajah, S. and Michielsen, S., *Text. Res. J.* (1999) 69, p. 903
63. Cheng, J.; Fone, M.; Reddy, V. N.; Schwartz.; Fisher, H. P and Wunderlich, B . *J. Polym. Sci, Polym. Phys. Edn* (1994) 32, p. 2683
64. Tonelli, A. E and Schilling, F. C. *Accts. Chem. Res* (1981) 14, p. 233
65. Tzou, D. L.; Huang, T. H.; Desai, P and Abhiraman, A. S. *J. Poly Sci, Phys. Edn* (1993) 31, p. 1005
66. Hatfield, G. R.; Glans, J. H and Hammond, W. B. *Macromolecules* (1990) 23, p. 1654
67. Tzou, D. L.; Spiess, H. W and Curran, S. *J. Poly Sci, Phys. Edn* (1994) 32, p. 1521
68. Sefcik, M. D.; Schaefer, E. O.; Stejskal, E. O and McKay. *Macromolecules* (1980) 13, p. 1132
69. Tzou, D. L.; Desai, P.; Abhiraman, A. S and Huang, T. H. *J. Poly Sci. Phys. Edn* (1991) 29, p. 49
70. Roland, C. M.; Walton, J. H and Miller, J. B. *Magnetic. Resonance in Chemistry* (1994) p. S36
71. English, A. D. *Macromolecules* (1984) 17, p. 2182
72. Gabrielse, W.; Gaur, H. A.; Feyen, F. C and Veeman, W. S. *Macromolecules* (1994) 27, p. 5811
73. Gomez, M. A.; Cozine, M. H and Tonelli, A. E. *Macromolecules* (1988) 21, p. 388
74. English, A. E . *J. Poly Sci, Phys. Edn* (1986) 24, p. 805
75. Fukuda, M.; Kawai, H.; Horii, F and Kitamaru, R. *Polym. Commun.* (1988) 29, p. 97

14 Fiber Formation and the Science of Complexity

John W. S. Hearle

14.1 Introduction

The structures of manufactured polymeric fibers, from which their properties and performance result, are formed on the fly as the materials move through industrial processes. For melt-spun fibers, yarns are commonly moving through changing thermal and mechanical fields at speeds between about 10 and 100 m/s (35 to 350 km/hr). The complete sequence from extrusion to wind-up covers about 10 meters, which gives times of about 1 second or less. Critical developments, which lock in structures, may occur over distances of the order of millimeters, which means that structure formation is occurring on a microsecond/millisecond time-scale. This is very different from laboratory studies of polymer crystallization in minutes, hours or days. Inevitably, the structures formed reflect the complexity of a hurried packing together of the molecules.

Because of the complexity, there will be little that is specific in this chapter. However, in order to avoid the need to refer to particular technological features of the formation of different fiber types, it is convenient to concentrate on the melt-spinning of polyethylene terephthalate, abbreviated to polyester or PET, as an example running through the account. The polyamides, nylon 6 and 66, will be referred to from time to time. Most of the comments will apply to the structure formation processes in melt spinning of all the partly aliphatic polyesters and aliphatic polyamides, which have molecules consisting of inert, flexible segments and more interactive segments. Some detailed differences will result from differences in chain stiffness and inter-molecular forces. Melt-spun polyolefins will differ more; and the differences in detail become greater with other polymers and other spinning systems.

Wet-spinning of cellulosic, acrylic and other polymers is slower than melt-spinning, but adds the complication of solvent movement, evaporation or coagulation reactions, and sometimes chemical changes in the polymer. When liquid crystals are formed in solution, for example in aramids and PBO, or in the melt, as in fully aromatic polyester copolymers, the more ordered molecular packing prior to fiber formation leads to less complex structures, which are highly crystalline and highly oriented. In the alternative route to high performance fibers from linear polymers, by super-drawing of polyethylene in gel form, the processes are much slower, so that the molecules can be disentangled into fully extended forms. For example, a patent issued to Dunbar et al [1] refers to post-stretching at take-up speeds of 20 m/min ($\sim$300 mm/s), which is two orders of magnitude slower than polyester melt-spinning.

There is little doubt that the rapid transition from a more-or-less random polyester melt to an oriented semi-crystalline fiber is a far-from-equilibrium process, which involves the developing advances in the science of complexity. This would be true to some degree for the crystallization of small molecules in such conditions, but is greatly intensified in polymer systems, where the movement of one unit of the chain has effects over thousands of units back along the molecule. One consequence of the structures of polyester and similar fibers being far from equilibrium, as initially formed and indeed after further processing, is that the structures are easily changed by thermo-mechanical treatments. This is the technologically important but, in scientific terms, poorly understood phenomenon of heat setting. For the other fiber-spinning routes, where structure formation is easier, the systems are not so far from equilibrium, but there is still considerable complexity in the fiber structures.

In contrast to manufacturing operations, the natural fibers are formed more slowly, over weeks for cotton and months for wool. Under genetic control, enzymes bring the monomers to the growing polymer molecules as they are laid down in well-defined structures, which are in "equilibrium" states that are not easily changed.

Carbon fibers are a special case. The formation processes are slow, but complexity comes from the diversity of structures that might occur. In the commonest manufacturing process, a collection of linear polyacrylonitrile molecules is converted to a partially disordered assembly of graphitic sheets, which are probably linked by molecular defects. The degree of perfection of the sheets and the extent to which they are distorted from a planar form have led researchers to draw many different sketches of possible structures. Although the geometry is different, the uncertainty is similar to that found in drawings of the fine structure of fibers from linear polymers, which are discussed below. Although, when used at ordinary temperatures, carbon fibers will be firmly locked in stable structural states, the formation processes at high temperature will be far from equilibrium.

14.2 A Historical Account of Structural Complexity

14.2.1 The Early Fringed Micelle Models

Until about 1930, a classical definitive approach to the organic and physical chemistry of the structure of the natural and regenerated fibers was dominant. At this time, synthetic fibers had not been commercialised. The naturally occurring molecules were assumed to have a specific size and crystals were well defined entities. The X-ray diffraction studies of Meyer and Mark [2] showed that the fibers contained a mixture of crystalline and disordered material, but the picture, Figure 14.1, shown by Seifriz [3] and Meyer [4] still followed a mid-19th century view, as proposed by Nageli [5], in showing separate crystallites packed in an assembly with what Nodder [6] described as an inter-micellar mortar. Indeterminism was only present in a lack of knowledge of the nature of the disordered intermicellar matrix.

The work of Staudinger [7], summarised in 1932, introduced molecular complexity, which was reinforced by the clarifications from Carothers' investigations of polycondensation, which have been described by Hermes [8]. Although conflict continued until the Faraday Society meeting in Manchester in 1933, it became clear that macromolecular systems had distributions of chain lengths and that these lengths were greater than was indicated for the size of crystallites by the diffraction data. The fringed micelle theory, in which long polymer molecules meander between crystallites through the intervening amorphous matrix, was proposed by Gerngross and Herrmann [9].

From the start, there was great uncertainty in the models due to the complexity of the structure. The diagrams, which were published in papers and books, were imperfect representations of views of the structures, which were only imperfectly visualised in the minds of the different scientists. To a considerable degree, they reflected the artistic skills, patience and temperaments of those making the drawings. They were two-dimensional, or, at best, quasi-three-dimensional, views of a 3D structure, and the molecular complexity was reduced to lines. A collection of these pictures, from Hearle [10] is shown in Figures 14.2 and 14.3. In passing, it may be noted that, as late as the mid-1950's, the fringed micelle structure was seen as applicable to the natural plant fibers and animal hairs, although, as noted below, explicit fibrillar forms were soon to be recognised.

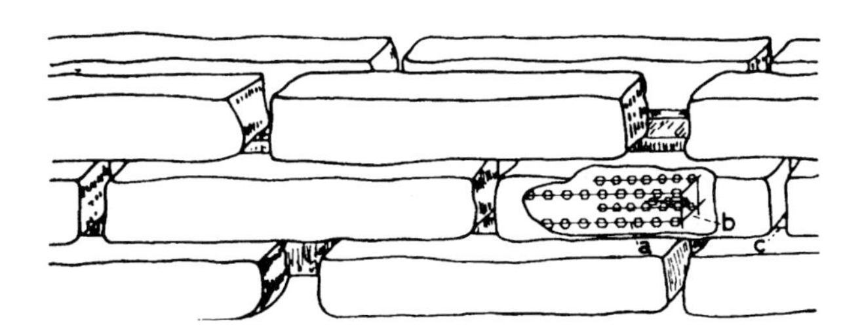

Figure 14.1 Micellar structure as shown by Seifriz [3] and Meyer [4].

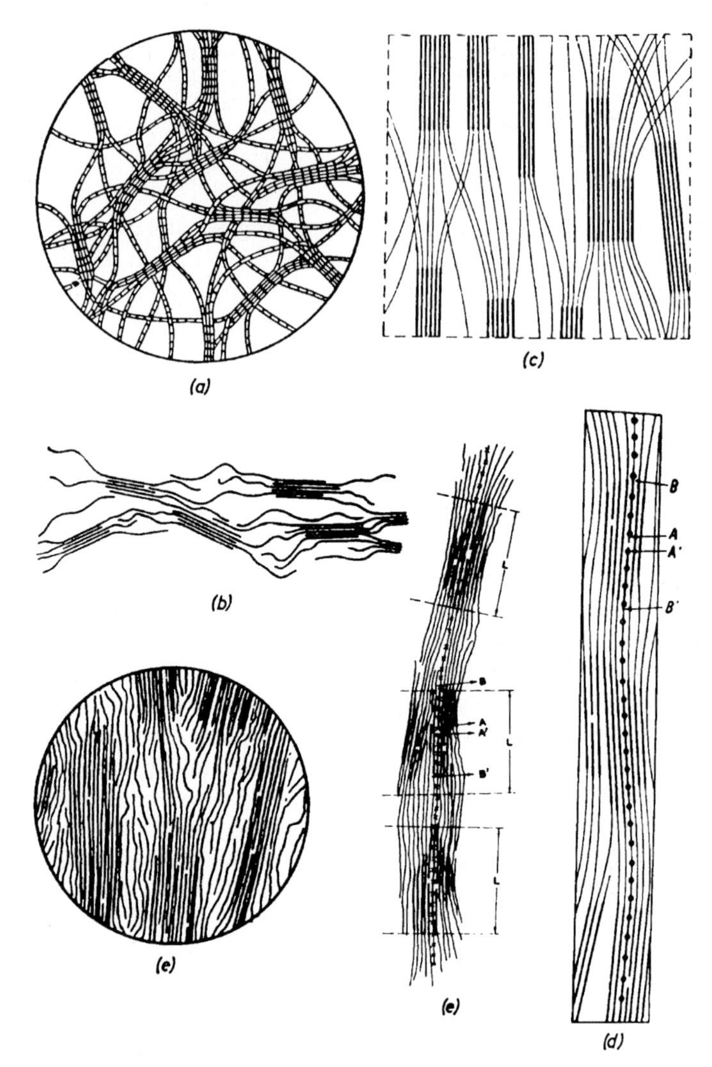

Figure 14.2 Fine structure of fibers and crystalline polymers, 1932–1958, from Hearle [10]. (a) Herrmann and Gerngross [11]. (b) Kratky and Mark [12]. (c) Frey-Wyssling [13]. (d) Kratky [14]. (e) Mark [15].

Despite their historical inaccuracies, it is worth looking at these pictures in order to understand the problems of complexity. A major divergence of view, which is relevant today, is apparent. There is a spectrum of opinion, which runs from clearly defined crystallites separated by a continuous amorphous matrix through more blurred divisions to a continuous mixture of order and disorder. The latter view was clearest in Staudinger's [7] picture, Figure 14.4, which was later reproduced by Hearle [33] as a first approximation to the structure of highly oriented linear polymer fibers, such as para-aramid and HMPE.

In 1963, when Peters and I edited the book on Fibre Structure [34], the fringed micelle model, which, however imperfectly visualised, had been the working model for both natural and manufactured fibers for 30 years, was being challenged from two directions.

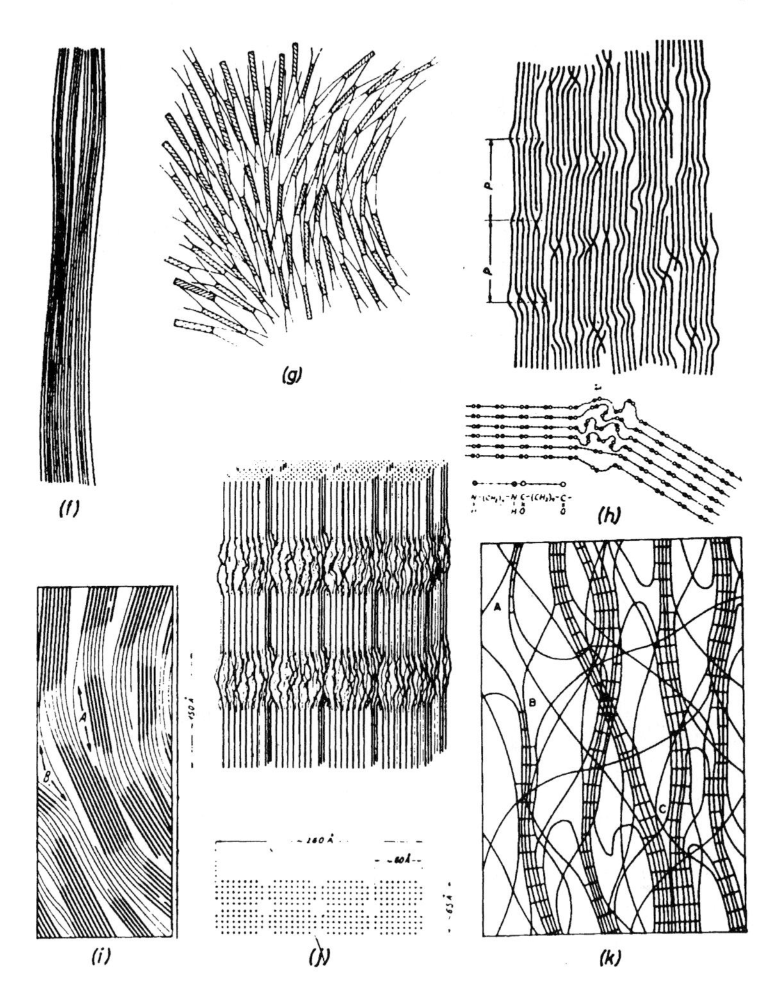

Figure 14.2 Fine structure of fibers and crystalline polymers, 1932–1958, from Hearle [10]. (f) Meyer and van der Wyk [16]. (g) Hermans [17]. (h) Hess and Kiessig [18]. (i) Götze [19]. (j) Hess, Mahl and Gütter [20]. (k) Hearle [21].

In my contribution to [34], I described the growing electron microscope evidence for the presence of fibrils or fibrillar structures in cotton, wool and other natural fibers, which led to the well-defined structures mentioned above. For manufactured fibers, less well-defined fibrillar forms are thought to occur in the newer high-modulus, high-tenacity fibers, in high-wet-modulus viscose rayon, and in some other fibers spun from solution. The fringed fibril model, Figure 14.2(k), was an attempt by Hearle [21] to combine features of the fringed micelle model with fibrillar forms. The earlier forms of viscose rayon and the melt-spun polyester, polyamide and other fibers are thought to have a more micellar texture.

In [34], Keller described the emerging fundamental studies of polymer crystallization. On a larger scale, this introduced spherulites, which are a complicating factor in the

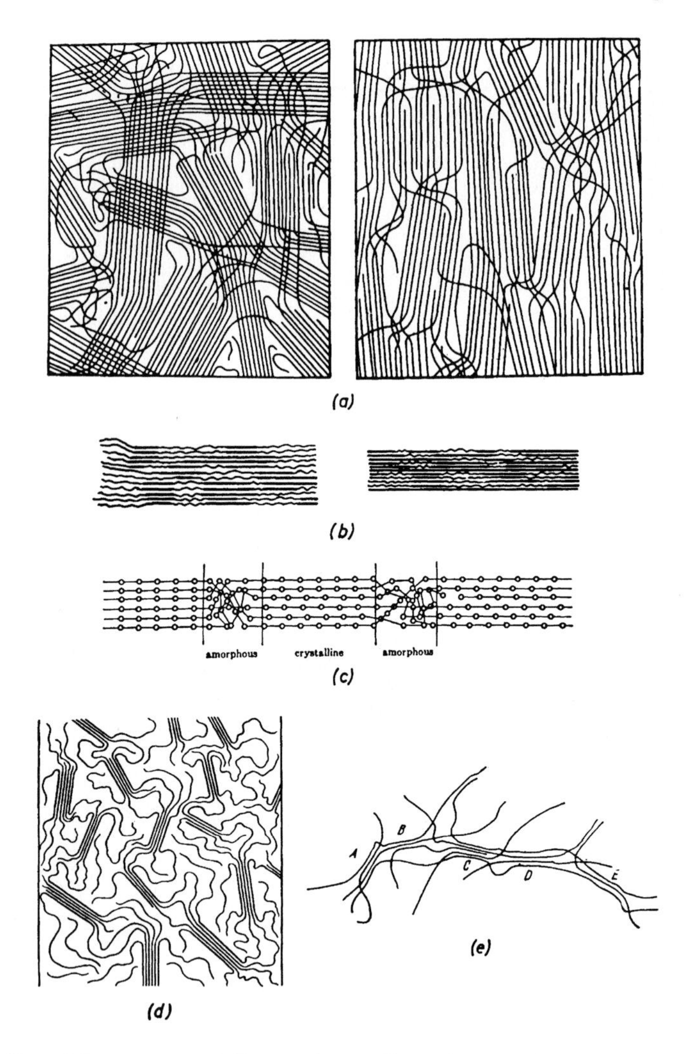

Figure 14.3 View of fine structure of fibers and crystalline polymers in books and articles, 1953-1959, from Hearle [10]. (a) Bunn [22] in *Fibres from Synthetic Polymers*. (b) Mark [23] in *Cellulose and Cellulose Derivatives*. (c) Sherer [24] in *Matthews' Textile Fibers*. (d) Alexander and Hudson [25] in *Wool: its Chemistry and Physics*. (e) Stuart [26] in *Die Physik der Hochpolymeren*.

formation of fiber structures. If the nucleation conditions are right and there is enough time before the material cools, occluded spherulites will develop in melt-spun nylon fibers. In PET, the crystallization rates would be too slow in any normal spinning conditions. Generally, it is better to avoid conditions giving rise to the formation of spherulites, but, there may be special circumstances, for example to change the optical properties of the fibers, where it is desirable.

A more important development was the recognition that chain folding occurred in single polymer crystals. None of the pictures in Figures 14.2 and 14.3 show chain folds at

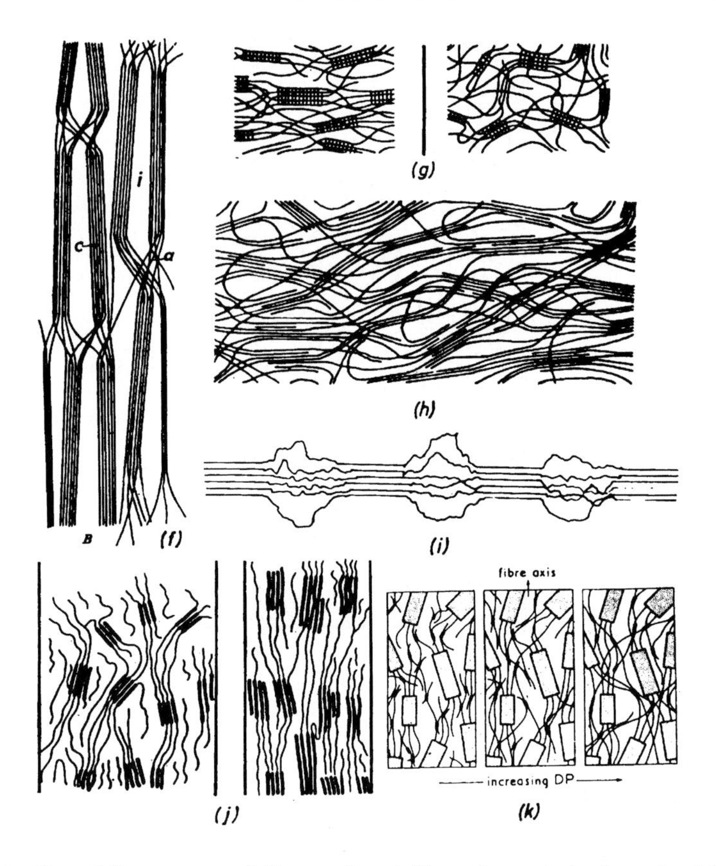

Figure 14.3 View of fine structure of fibers and crystalline polymers in books and articles, 1953-1959, from Hearle [10]. (f) Jane [27] in *The Structure of Wood.* (g) Hearle [28]. (h) Boulton [29]. (i) Battista [30] in *Fundamentals of High Polymers.* (j) Hohenstein and Ullman [31] in *Unit Processes in Organic Synthesis.* (k) Cumberbirch and Harland [32].

Figure 14.4 Staudinger's [7] view of fine structure of cellulose.

the ends of crystallites. The molecules all fringe off into the amorphous regions -and in most cases they continue in the same general direction from one end of the picture to the other, with no reversals of direction even between crystallites. Once the mind-block was removed, Hearle drew a modified fringed-micelle structure, which was used in Morton and Hearle [35] and is shown in Figure 14.5.

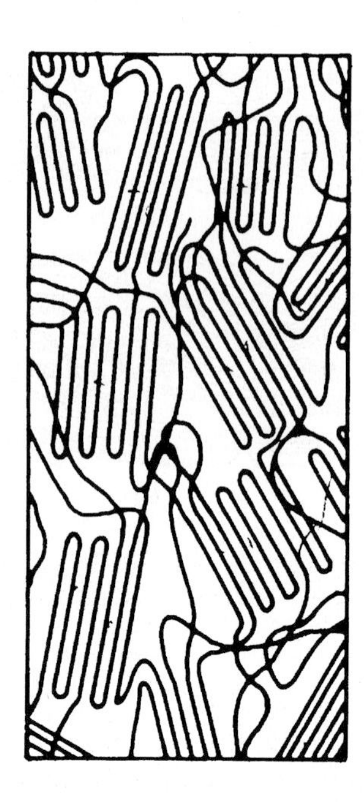

Figure 14.5 Modified fringed micelle structure from Morton and Hearle [35].

There is one other major problem, which comes from a lack of quantitative considerations in drawing the structural pictures. When converted into three dimensions, any model which has crystallites randomly located within the amorphous matrix will have low crystallinity. Figure 14.5, which is only partly random, would have a crystallinity of about 25%. In order to reach the typical crystallinities of 45 to 50%, the crystallites must be more regularly stacked.

Once these lessons had been appreciated, a consensus developed around a "common working model" of the structure of melt-spun polyester and nylon fibers – but with some countervailing views which reflected complexity through differences dependent on particular formation processes. The current view is mostly contained in an account by Hearle [33], which is largely followed in the remainder of this chapter.

14.2.2 The Common Working Model

In order to set the scene and contrast an academic approach with industrial reality, it is useful to quote from Schultz [36]. In relation to polyesters and polyamides, it is stated (with my added emphasis):

> These materials are generally semi-crystalline: that is, they are composed of crystalline and noncrystalline regions. The crystal volume fraction, size, shape, orientation and perfection are all *controllable variables,* as are the size, orientation, homogeneity, and

density of the noncrystalline regions. It is just these variables which determine the properties of the end product.

This may be valid, in principle, as a statement of ultimate Utopian potential, but the reality is that we cannot measure many of these variables with confidence, let alone set out to control them operationally.

The practical industrial approach strives to improve the structural measurement techniques. But then academic thought shows that the problem is deeper rooted: the words used impose views of structure that may not be true of reality. Theoretically, the definitions may be questionable; practically, in addition to possible experimental errors and artefacts, the conversion of data as directly obtained into named parameters may mislead. Even the statement that the manifest state of intermediate order, called semicrystalline, implies the existence of crystalline and noncrystalline regions may not be true. The alternative consists of regions of varying levels of intermediate order or even uniform partial order. We need to heed the wise words of Thoreau:

How can he remember well his ignorance – which his growth requires – who has so often to use his knowledge?

We must use the information that is available, we must make models to help our thinking, but we must maintain a skeptical approach and not delude ourselves, and, more importantly our students, into confusing our aids to understanding improperly with reality. Some features can be accepted with a high degree of confidence, but many others may be uncertain or obscured by our mind set.

But even if the problems are difficult, the opportunities are great. There can be no doubt that a great diversity of arrangements of polyamide or polyester molecules can form as stable states of partial order, nor that these structural states will be determined by the processing history in going from the melt to the final product, and nor that structure determines properties and properties determine performance. Can we advance from the empirical adjustment of process conditions shown on the left of Figure 14.6 to the fundamental understanding shown on the right? Which poses the greater difficulty: the enormous expense of vast parametric experiments or the intellectual obstacles to understanding the complexity of fiber formation?

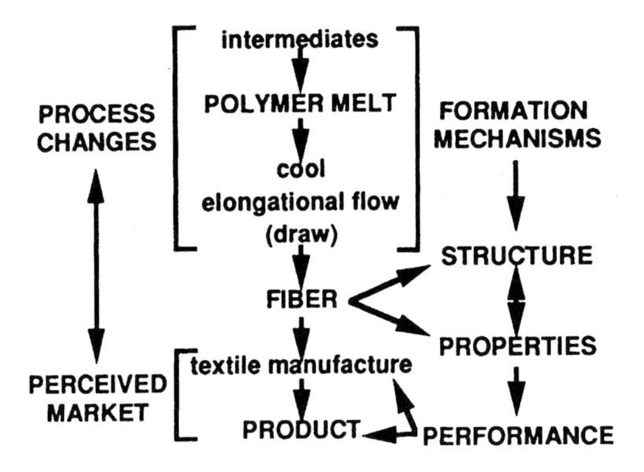

Figure 14.6 Melt-spun fiber production process.

The common working model, around which a consensus has developed over the last 30 years, is illustrated by the examples shown in Figure 14.7. The general intended features of these models are:

- a clear distinction between crystalline and noncrystalline regions
- brick-shaped crystallites stacked in pseudofibrils
- perfect parallel packing of chains along the length of the crystallite
- crystallite ends composed of a mixture of chain folds and chains leaving the crystallite in a fringe
- crystallites linked by tie molecules, which form the intrafibrillar and interfibrillar amorphous regions

It must be noted that these pictures are drawn not with an intent to be a true representation of the total structure but to illustrate particular features. Figure 14.7(a) was drawn to show the influence of the angle of the *ab* plane on the small-angle diffraction pattern of nylon, and Figure 14.7(b) contrasted the proposed alternation of location of crystallites in neighboring fibrils of polyester with planes of crystallites in register across the structure in nylon. At best the drawings must be regarded as approximations to a structure which is less regular, and may vary from place to place in the fiber. They have the following in-built limitations:

- two-dimensional representations of a three-dimensional reality
- simplification of molecular strings of atoms to lines
- perfect crystal orientation over the small region shown
- static form of a system having intense thermal motions

Only rarely has there been any attempt to match the pictures of supposed structures to quantitative details from experimental structural investigations. A notable exception is the picture of nylon 6 in Figure 14.8, which was created by Murthy et al [39]. This is a more complex structure with less regular shapes and sizes for the crystallites – but it is still

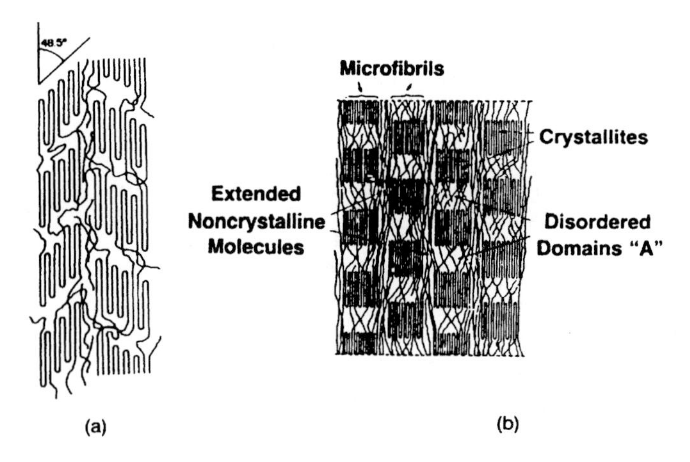

Figure 14.7 Two views of the common working model of melt-spun fiber structure. (a) Nylon according to Hearle and Greer [37]. (b) Polyester according to Prevorsek et al [38].

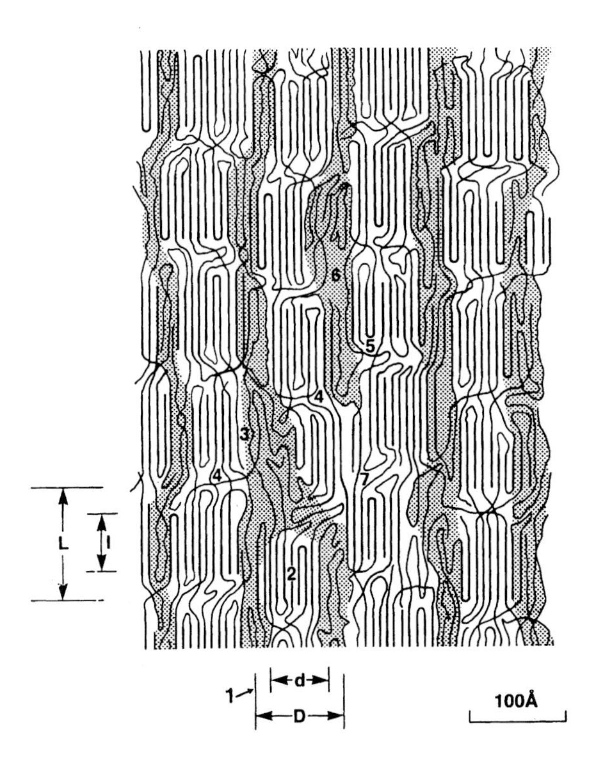

Figure 14.8 Fine structure of nylon 6, as interpreted by Murthy et al [39] from SAXS and WAXD data.

almost two-dimensional. The molecules do cross over one another in noncrystalline regions, thus dividing locally and infrequently into two layers, but they return to the same single plane in all the crystallites.

Scientists working on the structural mechanics of such complex systems are pulled in two opposing directions. On one hand, they want the models to reflect the complex reality; on the other, mechanical analysis requires simplification. Consequently, in contrast to cotton and wool fibers, there have been few attempts to model the mechanics. Simple composite models were shown by Hearle et al [40] to be unsuitable.

Two structural parameters – crystallinity and a series/parallel indicator – are not enough. Hearle [33] outlined a treatment which remains to be published in full. This reduced the complexity to the simple representation of the most important features of the common working model, as shown in Figure 14.9. The crystalline blocks are assumed to be undeformable. The mechanics is computed by the minimisation of two energy terms, which depend on (1) the volume of the system and (2) the extended chain length of the tie-molecules. Despite the simplifications, the model still requires the selection of values for a large number of independent variables. A typical set is listed in Table 14.1. Calculations with reasonable values of these parameters show the right sort of shape for the stress-strain curve of nylon at a temperature around 150 °C, above the glass transition, but the values of stress are about an order of magnitude too low. This shows the need for further refinement of the approach.

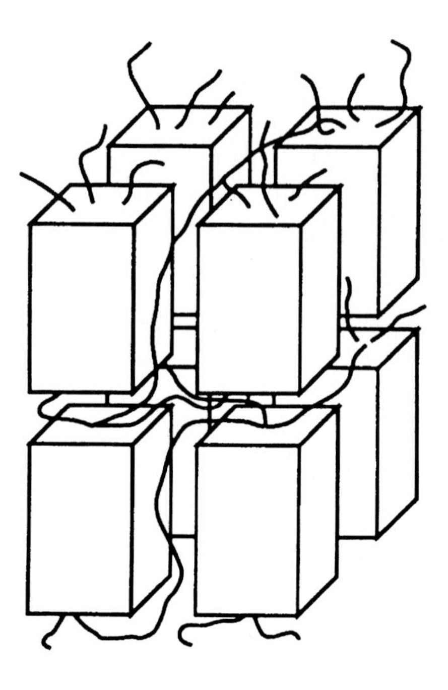

Figure 14.9 Simplified model of crystallites linked by tie-molecules, which was the basis for a mechanical analysis outlined by Hearle [33].

One important qualitative conclusion comes out of the study. When the fiber is under zero stress, the tie-molecules are not, as may be thought from the diagrams, under zero stress. They are under considerable tension as a result of their extension from the random coil state, exhibiting the classical features of rubber elasticity generated by the entropic forces due to thermal vibrations. The system acts like a set of wooden blocks held together by taut elastic bands. The tensions in the tie-molecules act against the resistance of the material to shrink to a smaller volume.

14.2.3 Alternative Continuous Models

Are we right to simplify the real structure by idealising the "two-phase" form of crystallites in an amorphous matrix? The alternative view is a more uniformly continuous structure.

A dominant feature of the common working model is a sharp distinction between crystalline and noncrystalline regions. I believe that this is right for fibers that have been annealed to some extent. In metals or other materials composed of small molecules, thermal treatments allow defects to migrate, eventually disappearing at the edges of the crystals. In this or other ways, more perfect packing can be achieved, and large crystals

Table 14.1 Set of Parameters Used in Analysis of Mechanics of a Simplified Model

Features of the polymer
- # molar mass of the repeat unit
- # length of repeat unit in crystal
- # crystal density
- # amorphous density, stress-free
- * number of equivalent free links per repeat
- | degree of polymerisation

Features of fine structure
- # fractional mass crystallinity
- # number of repeats in crystallite length
- # number of repeats across crystallite
- # series fraction of amorphite
- | fraction of sites with crystallographic folds
- | fraction of sites with loose folds
- | length factor for free ends
- | length factor for loose folds
- | relative probability of connector types

Other parameters
- * bulk modulus of amorphous materials
- * stress at which chains break
- * temperature
- * mass of proton
- * Boltzmann's constant

required to characterise two-phase structure
| required to characterise connectivity
* required to analyse mechanics

can devour small ones. This is not easily possible in polymers because of the linkage along chains. Removing a disturbance in one place pushes it to another: It is not easy to reach the edge of the system. Nevertheless, I believe that a lower free energy state is achieved when the order and disorder are segregated into separate regions. Although the system may be driven in this direction, there will be many barriers to overcome and much hindrance to the rearrangement, which involves the kinetics of motion of molecular segments over considerable distances. This is one of the complexities of fiber formation processes, which needs to be explored more.

In crystallization in a rapid quenching from the melt, as in nylon or oriented polyester melts, or in a short exposure to a temperature which allows crystallization from what was an amorphous polyester glass, a different type of structure may be formed. Consider the situation in the liquid polymer above the melting point. At any instant, many segments will be locally associated in a low energy state of crystallographic register, but such associations will be continually breaking up and reforming. The polymer molecules will be free to diffuse or flow as required in the liquid state. There is no difference in principle here from the behavior of simple liquids composed of small molecules. However, at a lower temperature, there can be a difference. Small molecules cannot have any memory of their previous association, but the polymer chain segments can. If the density of associations is great enough and the rate of breakup is slow enough, the chains will not achieve complete mobility, because, when any particular segment becomes free, it is held in position by the association of neighboring segments. The material will have changed from a liquid to a solid.

In going through such a phase change, the structure has formed what Hearle [41] calls a *dynamic crystalline gel*. Close to the melting point the entropy will be high, and the material will be soft because the segments in register will be in a state of dynamic equilibrium, but at lower temperatures it will be a firmly bonded semi-crystalline solid. Hearle associates this idea with a thermodynamic explanation of the phenomena of multiple melting found in nylon and polyester fibers by Bell et al. (1968, [42–44]) which was reviewed by Hearle and Greer [45]. The dynamic crystalline gel is identified with Bell's Form I, which is formed on rapid quenching from the melt. Annealing leads to Form II, which Hearle supposes to be a micellar structure. Like all structural models, the dynamic crystalline gel is not easily represented in diagrams, but a possible view is the more continuous structure shown in Figure 14.10.

The state of uncertainty over the complexity between continuous and discontinuous structures is well illustrated by the two different pictorial representations shown in Figure 14.11 from the same book on high-speed fiber spinning. Both are from a series intended to show the differences in the structure of polyester fibers spun at different speeds. Even allowing for the fact that these are schematic representations, Heuvel and Huisman [46] appear to favor a continuous structure whereas Shimizu [47] inclines to the existence of well-defined crystalline blocks.

14.2.4 Other Complexities

Two other ways in which the common working model may be lacking in complexity must be mentioned.

One feature of the model, even in the more complex form of Figure 14.8 is that the crystallites are represented as perfect and that the chains emerge from the ends and not the sides of the crystallite. In simpler polymers, such as polyethylene, the work of Reneker and Mazur [48] has shown the importance of crystal defects. In particular, a combination

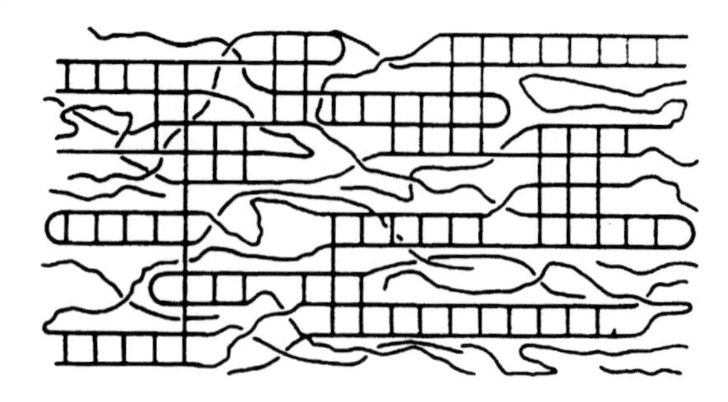

Figure 14.10 An alternative view of fine structure shown by Hearle and Greer [45].

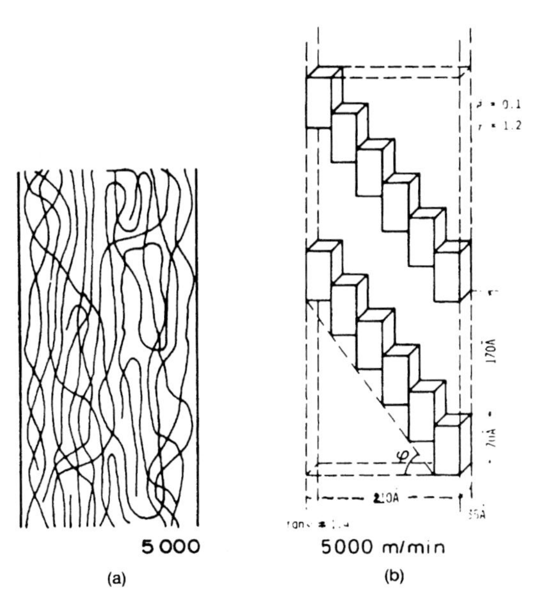

Figure 14.11 Two views of the structure of PET filaments spun at 5000 m/min from the same book on *High-Speed Fiber Spinning*. (a) By Heuvel and Huisman [46]. (b) By Shimizu et al [47].

of twisting and bending of the chain causes an additional unit, which may be mobile, to be accommodated within the structure. I have taken the view that such defects could not occur with the long repeats in nylon and polyester fibers, since the displacement of one $-CH_2-$ group would cause a misfit all along the molecule and the presence of an additional full repeat would cause a major structural disturbance. However, there is one type of defect that could occur. This consists of a chain moving sideways out of the crystal, as shown in Figure 14.12(a). The structural disturbance would not be appreciably more than in polyethylene defects. Defects of this sort could not occur in large crystals, because they would accumulate into an excessive disturbance with increasing distance from the center of the crystal; but this would not be a problem with crystallites which are only a few chains wide, as in typical melt-spun fibers. The occurrence of this type of defect could fit in with the interpretation of X-ray diffraction data as being influenced by both crystal size and perfection, but would also influence the pattern of interconnections within the structure. Figures 14.12(b) and (c) contrast schematically the emergence of tie-molecules from the ends of crystallites with molecules emerging from the sides of crystals. Figures 14.12(c) and (d) show forms which diverge more from the common working model. There is great potential complexity when the mind-set implicit in the conventional view is overcome.

The common working model also shows perfect crystal orientation, which is not compatible with the X-ray diffraction evidence. There are various ways in which misorientation could be accommodated in real fibers. The most likely is a division into zones, which have perfect local orientation but different orientation directions in different zones. This view is supported by some electron diffraction studies by Bebbington [49] on nylon fibers, which show variations in orientation over a fiber cross section. It is unlikely

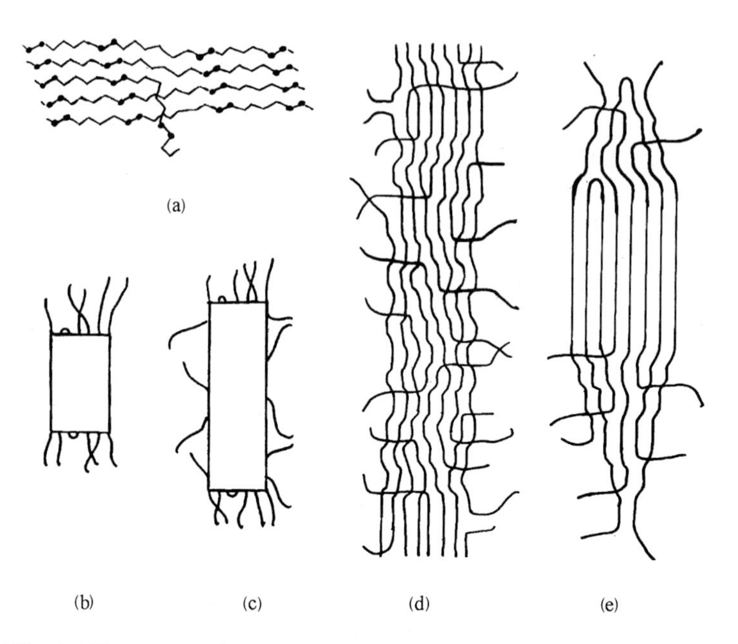

Figure 14.12 A hitherto unconsidered defect, which might occur in polyester and nylon fibers. [a] A defect allowing a polymer molecule to emerge from the side of a micelle. [b] The conventional view of the emergence of tie-molecules from a crystallite. [c] Contrasting view with tie-molecules emerging from the sides of the crystallite. [d] A possible fibrillar form with these defects. [e] Another possible form.

that there are substantial differences of orientation of crystallites within a fibril, but another possibility is a wandering of the pseudofibrils among each other.

As a result of the diverse attempts to represent the structure of a polyester or a nylon fiber, the many different details of packing that have been suggested, the unreality and limitations of the pictures, and what might be expected when a tangled mass of polymer molecules collapses into a partially crystalline form, there can be no doubt that the fine structure is complex, and that there may also be complexities at a larger scale. Attempts, such as that by Hearle [41], which led to Figure 14.10 above, to treat the formation of such structures by classical equilibrium thermodynamics are clearly of limited value when the structures form far from equilibrium. This leads to modern developments of the science of complexity.

14.3 The Science of Complexity

14.3.1 Far-from-Equilibrium

Nothing in the universe is in equilibrium, defined as a state of minimum free energy. Any large object could reduce its potential energy by collapsing onto the ground or increase its entropy by disintegrating into rubble – but this requires an earthquake or a bomb to cause movement towards a lower energy state. On a microscopic scale, small crystals can grow into larger ones. However, most of the systems studied in classical physics show deep wells of low free energy, separated by high barriers from other potential wells. At most, as in the work of Eyring, the transitions from one minimum to another over an energy barrier is analyzed.

In stark contrast to this, the collection of polymer molecules in a polyester or nylon fiber exists in a landscape full of minima. It is a matter of chance which semi-stable structural state with a local minimum of free energy is the one into which the system collapses as it cools or is stretched. It is unlikely that chance will lead to a particularly "tidy" form of structure, and changes to other semi-stable states will be easy.

Furthermore, the structures are not static. At room temperature, polyester and nylon lie between the two parts of the glass-to-rubber transition. Considerable molecular motion is possible in noncrystalline regions due to the freedom of rotation at the bonds between the half-dozen atoms in the aliphatic parts of the molecule. Above about 100 °C, larger motions occur due to the lack of permanent interactive bonding between benzene rings in polyester or peptide links in nylon. The melting-point of polyester and nylon 66, defined as an operational change from solid to liquid, is usually given as about 260 °C, but from around 180 °C the phenomena of heat-setting must involve other freedoms in the crystalline/non-crystalline assembly, though these are not well understood. They are probably associated with some mobility in crystallites, but, in polyester, could be due to chemical transesterification altering the pattern of tie-molecules.

These are obviously problems which would benefit from the application of the ideas emerging of the science of complexity. However, there is a major difficulty. Complexity tends to defeat itself. The ideas are presented in relation to simpler systems. We see their relevance to the formation of fiber structure, but we do not know how to apply them in any realistic detail. Qualitative comments can be made, but the quantitative analysis, which would yield scientific understanding and technological progress, remains to be achieved.

Perhaps appropriately for the final sections of this book, all that can be done is to outline some of the developments which may lead to future progress. There are a number of advances which appear to have relevance.

14.3.2 Characterizing Structure: Computer Graphics

Three features make melt-spun fibers particularly complicated systems. The first is that a partially ordered material is difficult to describe, because there are so many possible combinations of order and disorder, which are different and behave differently. The second is that polymers interact over short distances between neighboring molecules and over long distances along molecules. The third is that the structures form and change in very short times.

A prerequisite for any detailed advances is the development of better ways of describing structure. In the past, I have advocated the use of six structural features: degree of order; degree of localisation of order; aspect ratio of localised units; degree of orientation; size of localised units; molecular extent. Others have suggested other measures, such as amorphous orientation. The mechanical modelling mentioned above brings in a larger set of parameters, which are listed in Table 14.1. However, although these approaches may be useful, a limited number of parameters cannot characterise the complex detail of the real structures. They all impose an artificial ordering in the descriptions.

The advance of computer power provides a way forward. A computer graphics model could define both the gross morphology of an irregular crystalline/non-crystalline assembly and keep track of the paths of individual molecules over meaningful distances. There have been major advances in computerised molecular modelling, particularly for the pharmaceuticals industry. A similar attack could be made on fiber structures. However, although, in contrast to most of the other approaches discussed below, there is no great difficulty of principle, a strong R&D input would be needed.

A glimpse of what might be achieved can be seen in the illustrations used by Hearle [50] in a book celebrating the achievements of polyester. These pictures, which are shown in Figure 14.13, were obtained in about one day by using commercial software. The models of the polymer chain and of the crystal lattice merely make easier what can be done with physical models. However, the representation of the amorphous structure is beyond the capability of being built; the interlacing of the molecules would be too complex to assemble, at least without a CAD plan to follow. Even such a quick approach offers useful insights. It provides an impression of the contrast between localised close packing and larger spaces between molecules. If the mind adds in molecular motions, it becomes possible to envisage the movement of dye molecules or the deformations due to entropic rubber elasticity. However, the view has appreciable limitations. The structure may still be far from equilibrium packing and the size of the repeat, about a quarter of the picture in Figure 14.13(f), is small. The research challenge is first to enhance the reality of the separate regions and then to link more and less ordered regions together in a representation of the fine structure of a melt-spun or other fiber.

14.3.3 Polymer Dynamics

Some advances in polymer science can be applied to the problems. There is increased understanding of the melt. It is clearly recognized that the other molecules form a perfect theta-solvent for any particular molecule being considered. Consequently, the molecules in a quiescent melt should take up the theoretically predicted unperturbed conformations. Experimentally, this has been confirmed by neutron scattering studies. So we now have a sound basis for following through the overall distribution of molecules as they proceed through the fiber formation process. The precise pattern will be affected by crystallization, but the coarse path will not be. This is important in predicting molecular extent, which has a major influence on fiber strength. However, whether it is derived from continuous polymerisation or from the melting of solid chips, the melt may not have reached an equilibrium state and the conformations will certainly be changed by the flow processes of melt spinning. The departure from the unperturbed conformations will be greater if polymerisation reactions are still taking place.

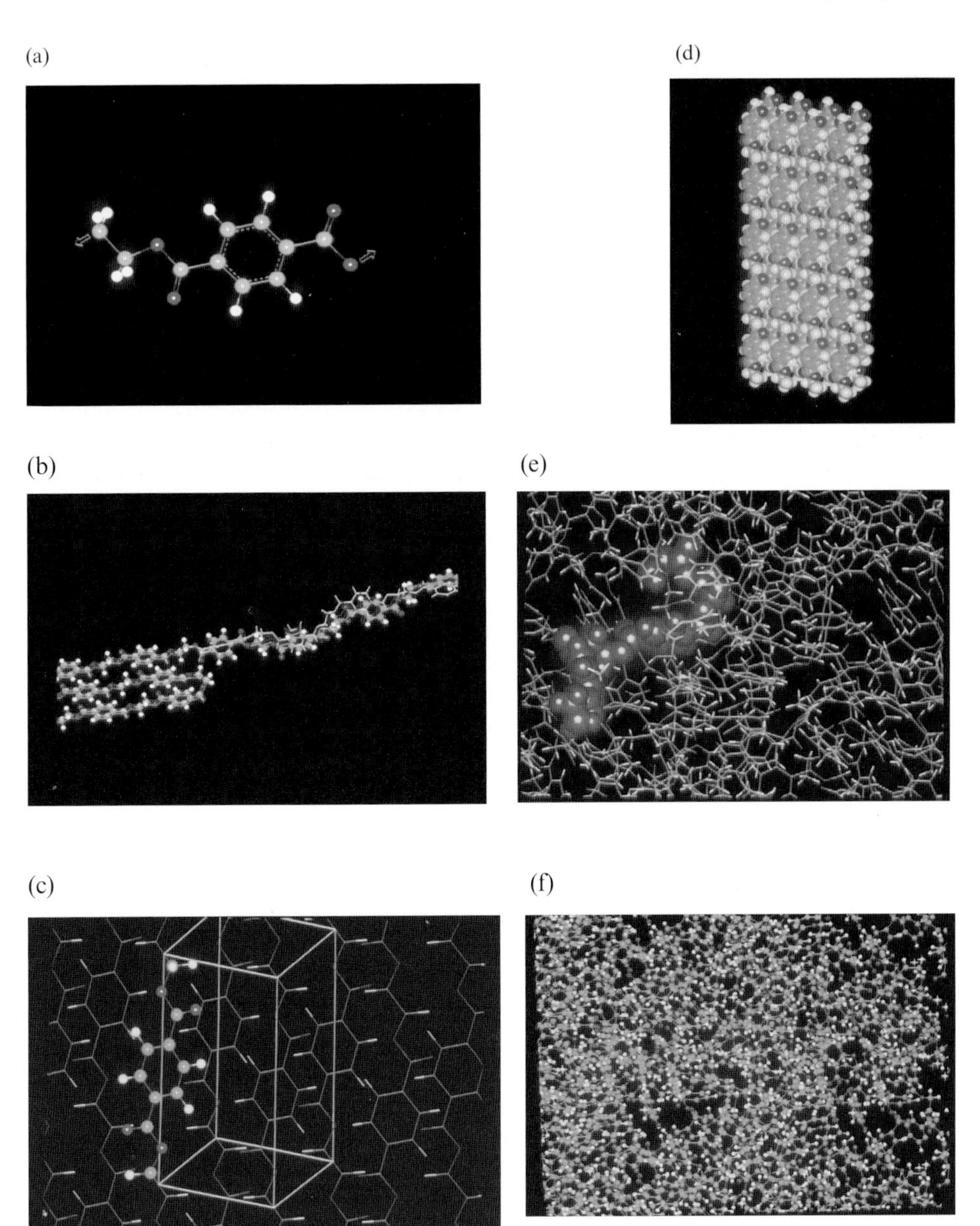

Figure 14.13 Computer graphics applied to the problem, from *Polyester: 50 Years of Achievement*, Brunnschweiler and Hearle [50]. [a] The repeat unit of the PET molecule. [b] A molecule going from crystalline packing to an amorphous chain. [c] The crystal lattice and the path of one molecule. [d] Solid packing in the crystal. [e] The amorphous form and the path of one molecule. [f] Solid packing in an amorphous region. The illustrations were produced by Dr Andrew Tiller of BIOSYM Technologies Inc., using their molecular dynamics and other graphical polymer modelling programs.

What happens in the molten threadline is determined by the balance between affine elongation of the molecular assembly, relaxation back towards an equilibrium state of lower orientation, and flow of molecules past one another. In principle, theories of polymer dynamics provide a means of attack on these problems. However, a closer examination shows that the theories, despite their considerable mathematical complexity, contain simplifying idealizations which limit their useful application. For example, the widely acclaimed idea of reptation, described by de Gennes [51] and Doi and Edwards [52], which is certainly useful in explaining some aspects of polymer behavior, has the limitation that the polymer motion is artificially constrained to wriggling within a tube. In reality we need to consider both the faster motion along the tube and the slower interchange of the tube walls. The theory then becomes more arbitrary or much more complicated. Older approaches such as Bueche's [53] theory of one molecule dragging along others or Lodge's [54] transiently crosslinked network may still have relevance, as do the approaches of Bird et al. [55], which avoid explicit use of the tube.

Yet another approach is to eschew any attempt to characterize the systems by equations expressed in terms of abstract concepts, but instead to try the numerical approach of computational molecular dynamics. The collection of atoms is modeled as an assembly governed by Newton's laws of motion. If we know the forces (potential functions) between the constituent atoms, then the evolution of the system from a given starting point can be explicitly computed. Such methods have given excellent predictions for rather uninteresting materials like argon gas and useful results for some systems of small molecules. However, for fibers, and other polymer materials, there are serious difficulties. First, there are many potential functions to take into account along and between chains, and simplifications may prevent correct or useful results being obtained. Second, a realistic model of the fine structure, which includes a number of ordered regions mixed with disorder, would contain at least 100,000 atoms and its evolution should be followed for at least a second (see also Section 4.4.3). This is beyond the capability of current computing. A brute force approach looks impossible as a way of generating a complete model, but this does not mean that nothing useful can be done. The way ahead with this approach will involve the clever formulation of specific situations, which will lead to information applicable to more general models. A hierarchy of computer models at different scales is needed. Pictorial representation can be highly instructive. For example, the reality of thermal motion in nylon crystals has been demonstrated by Wendoloski et al. [56].

If high talent in the theory and practice of polymer dynamics can be combined with insight into the problems of fiber formation, then major advances may be made, but the way does not look easy. However, we should also look at what has been happening in the wider world of the science of complex systems.

14.3.4 Fractals

The first development is concerned with geometry. Mandelbrot [57] has shown that many systems are better described by a fractal geometry than by a Euclidean geometry. A mathematical example is a Koch curve, which is self-similar over a range of magnifications. If self-similarity is accepted on a statistical basis, many natural phenomena like the coastline of Britain or the pinnacles of Bryce Canyon have a similar geometry. One consequence of such a geometry is that the enclosed area is not proportional to the square of the linear dimension, nor the volume to the cube. Instead fractal dimensions are involved.

The form taken by a random polymer molecule is clearly a fractal. This was implicit in de Genne's approach to reptation in the melt: the constraining tube approximates to a random walk on a larger scale than the chain itself. The mass of the chain is proportional to the number of repeat units but the volume occupied is proportional to the 3/2 power. In other applications, Meakin [58–59] has shown that both clustering of particles and the breakage of materials commonly follow fractal geometries. He suggests that it might be possible to obtain a better understanding of complex growth and aggregation processes from simple models, which capture the most essential elements, instead of trying to include "as much as possible of the known physics in models that are as difficult as the real systems."

In general terms, it seems likely that the fine structure of fibers will show fractal features. The pictures of fine structure in Figures 14.2 to 14.5 and 14.7 to 14.11 show that some authors, perhaps unconsciously, draw tidy Euclidean forms but others draw forms that look more fractal. We are then led into the dilemma that seems to permeate all these discussions. While the ideas seem relevant, they are not directly applicable in the form in which they have so far been promulgated. In the frequently demonstrated computer-generated patterns, it is easy to recognise fractal geometry, and this is also true of the common examples mentioned above. But it is not obvious how to apply the ideas to the assembly of linear polymer molecules in a semi-crystalline fiber. One is left with the uncomfortable thought that neither the Euclidean nor the fractal idealization is correct, but that the structures have forms which lie between regular and self-similar geometries.

14.3.5 Kinetics of Nonhomogeneous Processes

Fractals are a static descriptive geometry of form, sometimes including the sequential steps leading to the fractal form, but we would also like to understand the dynamics which leads to these forms. The basic ideas concerning nonlinearity considered in the next section may offer a clue, but there is a more physical approach, that has yielded useful predictions in other fields and is clearly related to fractals. The ideas are not easy to bring out in a brief statement, because they involve the coming together of many disciplines. Perhaps, it is best to quote from the first and last paragraphs of Freeman's [60] preface to *Kinetics of nonhomogeneous processes: a practical introduction for chemists, biologists, physicists and materials scientists*:

> Half of the book describes processes induced by radiation. The other half describes thermal processes in structured systems that model parts of nature... Some of the thermal systems are initially highly structured. In others, structure grows from randomness. Membranes, micelles, vesicles and biological cells are highly structured

> and have properties in common. Reactions that involve them are nonhomogeneous. The formation of these structures from random distributions of monomeric molecules, and the production of chemical waves in certain homogeneous chemical systems, bring us closer to wondering about the processes of life.
>
> The concept of fractals was conceived in mathematics; it now permeates parts of physics and will rapidly spread through areas of chemistry and biology. Keep an eye out for growth at tips!
>
> Nonhomogeneous processes occur throughout nature. We focus attention on special properties that nonhomogeneity can create. The most magnificent special property, which remains far beyond understanding, is life itself.

These last words obviously have a direct relevance to the formation of natural fibers through living operations, but also stimulate us to think of how we can achieve the right conditions to produce desirable structures by the control of manufacturing systems. The closest approach to such problems comes in the chapter by Merrill and Tirrel on dynamics of polymer melts, but, as well as emphasizing limitations of the reptation model, the applications are considerably simpler than melt-spinning and drawing. However, another quotation from Freeman expresses our aims well:

> The physical arrangement (structure) of a system has an enormous effect on its behavior. That truth is self-evident for an object as small as a single molecule and for a very large object such as an animal. For objects of intermediate size, such as agglomerations of a few tens to a few millions of molecules, the effects of arrangement can also be very large and are only now receiving great attention.

14.3.6 Nonlinear Irreversible Thermodynamics

Until recently, science has progressed with amazing success by concentrating on those aspects of material behavior that can be approximated by simple (usually linear) mathematical forms. But now, partly influenced by the power of modern computers to handle more complexity, there is increasing awareness of the characteristics of more complex, nonlinear systems, and new ordering features are being explored.

First to emerge were the developments in thermodynamics, described by Glannsdorf and Prigogine [61] and Nicolis and Prigogine [62] in mathematical terms and by Prigogine and Stengers [63] in more popular terms. Classical thermodynamics (really "thermostatics"), which was developed over a century ago, predicts equilibrium conditions at the minima of free energy, and works well for systems characterized by a few well-defined states. There are considerable problems with fibers because these exist in a phase space with a vast number of local shallow minima of free energy. The next development, Onsager's reversible thermodynamics, dealt with changes close to equilibrium where linear relations apply: this is still a long way from the complex reality. Later Prigogine tackled the problem of nonlinear irreversible thermodynamics, and attempted to define entropy in a way which would allow a description of the evolution of systems through states far from equilibrium. However, the detailed application of Prigogine's ideas has so far proved to be limited in any field.

One can feel certain that the formation of melt-spun fibers, and perhaps other fibers, is an event happening far from equilibrium, dominated by the flow of energy and the movement and organization of material. This is Prigogine's world, as correctly pointed out by Lindenmeyer [64]. However, there are problems with Lindenmeyer's contribution. First, as pointed out by Bullough [65], detailed examination has not been able to provide the connected theoretical structure by which Lindenmeyer's imaginative ideas could be made into an applicable theory. There are no means as yet of implementing these ideas in practice. Second, there is a more basic problem. We are unable to describe the systems, the assemblies of thermally active molecules in the forming fibers, in a way that allows the specific application of general theoretical relations.

We are left with one important qualitative result of Prigogine's work. He has shown that systems far from equilibrium, which dissipate heat, evolve as organized forms, called dissipative structures. Frequently quoted as a macroscopic example are the Benard convection cells in a liquid heated at the bottom. In our thinking about the formation of fiber structures, we must recognize that it can evolve in such a regime, and that the geometrical patterns may not be predicted from the classical descriptions of crystal nucleation and growth. As Lindenmeyer pointed out, the dissipative structures will have a characteristic scale, but he was not able to show how to calculate this or how it would be influenced by control parameters during fiber formation. Nevertheless, the energy flows will be a determining factor, and so Lindenmeyer's suggestion that fluctuating conditions would produce new structures has practical merit.

14.3.7 Chaos Theory

Other approaches to nonlinearity start from mathematics and not physics. Since 1967, analytical methods have been found for solving integrable nonlinear partial differential equations. The direct numerical approach to the study of chaos has become more popularly noticed, for example in the books by Gleick [66] and Waldrop [67]. There is also related work on catastrophe theory, described by Thompson [68]. Chaos is easily demonstrated on a personal computer with a graphics output. In brief, rather simple nonlinear relations, such as $x \rightarrow rx(1-x)$, yield fixed points on a single curve over a certain range of the parameter r, but then, as r gets larger, the patterns increasingly bifurcate, switch between different solutions, and become chaotic. Surprisingly, the chaos can then show domains with more order.

We must recognize that in many natural and industrial processes, the nonlinearity is at a level that will lead to chaos. One of the origins of the development was in Lorenz's attempt to model weather patterns. It seems possible that the circumstances involved in the formation of melt-spun fibers will be another manifestation of the effects of nonlinearity leading to chaos. However, although the demonstration of the features of chaos is easy with simple equations, the application to complex real systems in terms of many variables is much more difficult. For example, it is known that turbulent flow is governed by the comparatively simple Navier-Stokes equations, which would lead to chaos with a characteristic pattern of eddies, but computers have not been large and fast enough to provide numerical solutions. For the formation of melt-spun fibers, we cannot even write down the equations. Once again, we are left with a feeling of relevance, but no way forward for quantitative application. But there are qualitative lessons to keep in mind.

First, we must recognize that complexity of structure is not only due to complex causes, but can be the result of relatively simple forms of nonlinearity. Second, we must recognize that there can be organization within chaos. Third, and most important, is the observation that very small change in initial conditions can lead to enormous differences in the final outcome, as the solutions are deflected on to different paths. The often quoted example is that the flutter of a butterfly's wings in Peking today can transform a storm system in New York next month. Similarly, in fiber manufacture, we can expect that small changes in conditions might lead to changes in structure, which could be exploited commercially. This should encourage a greater search for ways to control fiber structure, even if there is little guidance on where to explore.

14.4 The Possible Role of Quantum Theory

14.4.1 Quantum Effects

Implicit in the above discussion, and indeed in virtually all approaches to the subject, is the assumption that, difficult though the problems of fiber structure may be, they lie almost entirely in the realm of classical physics. Quantum physics would determine the potential functions between atoms and so would be relevant to *ab initio* calculations of geometry and bonding, but the structure at any larger level would be dependent, to a good approximation, on classical applications of the potential functions. This is the basis for dynamic molecular modeling based on Newton's Laws.

This view was questioned by Hearle [33]. The doubts were raised by a quotation from a wide-ranging book by Penrose [69], which finally addresses the problem of consciousness as a quantum effect. Penrose writes:

> Recall, first of all, that the descriptions of quantum theory appear to apply sensibly (usefully?) [not] only at the so-called *quantum* level – of molecules, atoms or subatomic particles – but also at larger dimensions, so long as energy differences between alternative possibilities remain very small.

The final qualification describes just the situation that exists in a polymer assembly as it organises itself into the solid fiber.

Later in the book, Penrose refers to the quasiperiodic tilings, which he discovered and which *almost have fivefold symmetry*. Hearle [70] showed that a passage in Penrose, which describes how certain alloys can crystallize in this form, can be rewritten in a way which could describe the formation of the fine structure of semi-crystalline polymers by changing a few words. *Tiling patterns* changes to *polymer fine structures*; *atoms* to *polymer segments*; *atomic* to *molecular*; *quasi-crystal* to *fine structure*.

The essential features are due to the fact that the structure formation is non-local. It is necessary to examine the state of the structure far from the point of assembly, in order to avoid error in fitting the pieces together. Penrose acknowledges that there is controversy surrounding quasi-crystal structure growth. He speculates that the quasi-crystalline substances are highly organised. More tentatively, he suggests that their assembly cannot be achieved by the local adding of atoms (polymer segments) according to classical crystal growth, but that the non-locality involves quantum-mechanical effects.

Penrose regards the state of lowest energy as difficult to find. It cannot be reached by adding atoms [polymer segments] one at a time in the hope that this will solve the minimisation. The global problem must be solved by cooperation between many atoms [segments] acting all at once. This must be achieved quantum mechanically. Many different arrangements are "tried" simultaneously in linear superposition. Finally, there is *selection of an appropriate (though probably not the best) solution to the minimising problem......which would presumably occur when the conditions are just right.*

In classical crystal growth, atoms, molecules or polymer segments add on individually at growth fronts on crystal faces. Each crystal is a well-defined entity. In contrast to this view, Penrose says that there must be *an evolving quantum linear superposition of many different alternative arrangements*. The alternative arrangements *coexist in complex linear superposition*. Some of the alternatives will grow to larger sizes until the quantum superposition reduces and eventually collapses to *a reasonable-sized quasi-crystal* [region of defined fine structure].

If Penrose's views are correct, and if they also apply to polymer crystallization during fiber formation, it will be necessary to face up to the uncertainties concerning *Quantum Magic and Quantum Mystery*, to quote the title of a chapter in Penrose [69]. What causes the wave functions of the quantum world, with many superposed states, to collapse to the single states of the classical world – or, on the many worlds hypothesis, do they all exist but we are only aware of one? Where, and what, is the classical/quantum boundary? Some of these points are touched on by Hearle [70], but they are more extensively discussed in many books, such as *The Mystery of the Quantum World* by Squires [7].

Penrose argues for an interaction with gravity, and suggests that there will be collapse of the wave function when a one-graviton level, at which the masses involved cause appreciable space-time curvature, is reached. His first crude estimate is 10^{-5} gram, but, for a cloud chamber experiment, he refines this to 10^{-7} gram. In a later book, Penrose [72] correlates size with reduction time for the collapse from quantum to classical. The correlation involves the energy differences between states, and gives a *kind of "half-life" for the superposed state*. He then calculates that the superposed states of a speck of water of radius 0.1 μm would last for hours, but a 10 μm drop would stabilise in less than 1 microsecond. Clearly these are sizes and times that are relevant to the formation of fine structure in fibers.

In many macroscopic systems, such as a volume of gas, the consequences of a quantum approach will not make an appreciable difference to the practical results. Almost all the possible arrangements, which will have very small energy differences and which will be in a state of dynamic equilibrium, will have virtually the same properties. Only the extreme possibilities, such as all the atoms being in one half of the volume, will be lost in quantum superposition. This situation will also apply to amorphous polymers. The different arrangements will tend to have the same properties. But, in semicrystalline polymers, such as nylon and polyester fibers, the many possible structures with a density midway between crystal and amorphous, may behave very differently. There will be subtle differences in mechanical properties, probably greater differences in the poorly understood behavior in heat setting, and differences of major technological importance in dye diffusion. For example, a fine structure with large crystallites will behave differently to one with small crystallites, but their free energies may not be very different. And there are many subtler variations in structure with similar free energies. Consequently, the evolution of structure may well be different when viewed as a quantum process than when treated as a classical crystallization process.

14.4.2 Melt-Spun Fiber Formation

If the validity – in general terms, even if not in every detail – of Penrose's approach is accepted, it is possible to make comments on what might happen in the formation of fiber fine structure. However one can only start by presenting the discussion in terms of the current working model, although in reality the proper concepts might be quite different. The multiplicity of close energy levels is not in question; and it is also certain that the determination of energy levels is non-local. In contrast to the dropping of a small molecule into a crystal lattice from a neighboring molten state, which has little effect at a distance greater than a few atoms, the incorporation of a polymer segment tightens the associated length of tie-molecule in the non crystalline matrix, puts stresses on the crystalline entity at the other end of the tie and thus changes its energy level, and so on for a considerable distance.

The same point can be expressed in another way. How large a volume of polymer must be considered in order to get realistic estimates of minimum energy levels? One would guess that this would have to contain at least 100 crystallites, so that if, in modelling, the total structure was built up by repetition of a single cell, as is done for the amorphous structure in Figure 14.13(e) and (f), any crystallite would be about 5 crystallites away from its repetition, which then has no separate freedom for independent structural adjustment. Changes within such a cell would certainly have an effect over the whole cell, so that Penrose's arguments would apply. Even this size, with a linear dimension of about 50 nm, may be too small for sufficiently accurate calculations of energy levels.

There thus appear to be convincing arguments that the formation of the semicrystalline polymer fine structure should be treated in quantum terms. This implies a superposition of quantum states, until one state has achieved a dominant position and reduction of the state vector occurs. There is also the possibility that the quantum situation is reactivated, when a rise in temperature in subsequent processing alters the distribution of energy levels.

What would be the consequences of treating the problem in quantum terms? This is a difficult question. What I have referred to as the Lindenmever syndrome comes in, the relevance is reasonable but explicit application is difficult. There is no way, at present, of giving a mathematically or computationally adequate description of the state of a piece of fiber fine structure containing millions of atoms – not to mention of solving the wave equation for such a system. Some clever conceptual and theoretical approaches are needed. However, one can hazard a few suggestions.

The coexistence of many states opens up the variety of forms of fine structure to be considered. Almost all the states will be in transition between the clearly defined structures. which, as described earlier, we try, not very successfully, to describe. The transitions between states will be easier. They will not be blocked by high energy barriers, because, to use an older terminology, quantum tunnelling can occur. Movement of defects and other structural changes, which would be forbidden in a classical system, could thus occur.

The freedom given in this way removes the problem in statistical thermodynamics of how, once some crystallization has taken place, the structure can change, except in minor ways to new forms. In classical terms, the energy barriers in going from small crystals to large ones would be too great. The problem does not arise with atoms or small molecules, and perhaps not with polymers like polyethylene with small repeat units, because changes can occur through the thermally activated movement of small localized defects over small energy barriers. With longer repeats, as in PET and nylon, change would involve a major disturbance with passage through a very unfavorable energy state. Two particular consequences of invoking quantum freedom can be given.

In a simple picture, crystal nucleation should occur randomly and lead to growth of randomly distributed crystallites with sizes determined by crystal growth rates. However, with those restrictions, the degree of crystallinity is limited by the problems of packing in a random way. In order to get to 50% crystallinity, we are led into proposing structures in which crystallites are regularly stacked in fibrillar arrays in a form approximating to a super-lattice. In classical terms there would seem to be great difficulty in proposing a route from the initial random arrangement to the final more ordered one, but this would be a natural consequence of a quantum mechanism.

Annealing poses similar problems in classical terms. How can an assembly of small polymer crystals transform itself into an assembly of larger crystals without complete melting and reorganization? Yet we know that this sort of change does take place.

14.4.3 Clusters and Quantized Energy Levels

The above discussion brings in the difficult philosophical questions of the interpretation of quantum theory, which have been mentioned above. However another approach, which has been described by Berry [73] is applicable to semi-crystalline melt-spun polymer fiber structures. This approach involves no more than Planck's original postulate of quantised energy levels.

Berry describes the behavior of small clusters of atoms, which show a transition from solid to liquid over a range of temperatures, instead of at a sharp melting-point. This is confirmed by computations of theoretical models and by experiments on collections of between about 5 and 100 argon atoms. In the classical argument, the melting point is a sharp transition, which occurs when the free energy $(U - TS)$ is equal in the crystal and the melt. The simple theory is modified for small crystals, whose melting point is reduced due to surface energy, and a similar effect occurs in imperfect crystals. It is at the very smallest sizes that Berry's arguments apply. At the lowest temperatures, the internal energy U is still dominant and the only available potential well is for the crystalline cluster. At the highest temperatures, the entropy S dominates and the available potential well is for a liquid cluster. However, over an intermediate range of temperature, there are accessible minima in both potential wells and the clusters can flip between the crystalline and liquid forms.

In principle, the above seems to be applicable as a classical argument. However, in Berry's analysis, quantum effects are brought in, because there are different densities of quantised energy levels. The crystal is a deep trough with few energy levels; the liquid is shallow with many energy levels. This increases the chances of the cluster being in the liquid state even at a higher energy level. The results of Berry's computations, which are confirmed by experiments with a probe atom present, are that over a substantial range of temperature the cluster flips between one state and the other. For some time it is a crystal, but then it changes to a liquid for a period, before jumping back to the crystalline form, and so on. Over several degrees Kelvin, the two forms coexist, interchanging about 10^{12} times per second.

Berry's research depends, of course, on being able to isolate small clusters of atoms, but it may have relevance to the fine structure of a rapidly quenched polyester or nylon fiber. A small crystallite could be viewed as a cluster of ordered polymer segments trapped between zones of amorphous entanglement in the matrix, as illustrated schematically in Figure 14.14. A simplistic view would assume that, below the melting point, the chains would be packed in a stable crystalline lattice, as in Figure 14.14(a), and that, above the melting point, the clusters would disperse completely in the mobility of the melt. However, because of the special features of long polymer chains, there could be intermediate states. While still trapped in a cluster, the polymer segments could develop a degree of liquid-like motion in a less ordered arrangement, as suggested in Figure 14.14(b).

The quantization of energy levels would be similar to that described by Berry: very few in the crystal form and very many in the liquid form. Consequently, it is possible that there will be a range of temperature below the observed melting point over which the polymer clusters are flipping between crystalline and liquid states. In this range, the material as a whole will be solid because it will be held together by the clusters that are crystalline, some of which will always be present though their identity will be changing. Other clusters, also changing in identity, will be liquid. This picture has affinities with Hearle's [41] explanation, which was discussed above, of the annealing of Form I fibers, which were viewed as a dynamic crystalline gel, to Form II, which was regarded as a micellar structure.

In contrast to Berry's isolated clusters of atoms, the polymer clusters are not sharply segregated from the rest of the material. During the periods in the liquid-like state, segments could move so as to give a limited interchange between the cluster and the matrix. If stresses are present, this would allow relaxation to occur. Over long times, the relative movement of molecules, by reptation or in other ways, would allow flow and self-diffusion of the total system. This would be the liquid-like behavior known as creep.

The above mechanisms may influence the way in which structure develops during the thermo-mechanical processing of polyester, nylon and other fibers. In particular, this may

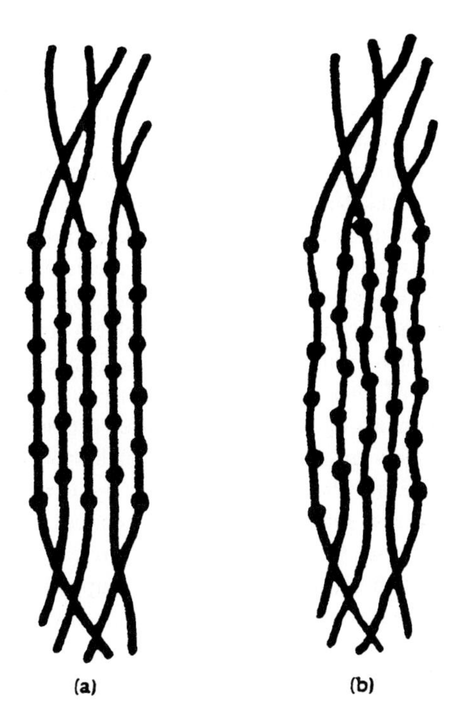

Figure 14.14 A cluster of molecular polymer segments trapped by entanglements. [a] In crystalline register. [b] In a disordered liquid-like form.

be what happens during heat-setting of polyester and nylon 66 between about 180 °C and 240 °C (when fibers start to stick together), which is of great commercial importance but has never been satisfactorily explained scientifically,

14.5 Conclusion: Scientific and Engineering Opportunity

In the early years of the 20th century, many confused ideas were under discussion over the structure of natural fibers, such as cotton and wool, and of the nature of the solids formed when some liquids, such as vinyl chloride, were heated. The mind-set of organic chemistry was on molecules with a precisely defined constitution. Many chemists thought that the complexity of these natural and artificial materials was due to their being colloidal associations of small molecules.

Then the polymer hypothesis overcame these problems, and a valid science emerged. The chemistry of cellulose and, at least in basic terms, of keratin was established; addition reactions led to vinyl polymers; and Carothers initiated condensation polymerisation. The uncertainty of chemical constitution disappeared from homopolymers and became comprehensible in proteins and synthetic copolymers. Commercially, this led to the chemical design of fiber materials. Du Pont made the first polyamides. Whinfield and Dixon recognised the value of a benzene ring in PET. Later, in the second generation of high performance synthetic fibers, the aromatic polyamides (aramids) were created, the potential of fully aromatic copolyesters was recognised (Vectran), and, most recently, the double ring of naphthalene has been introduced in PEN fibers and the triple ring in PBO and similar materials.

In this way, the chemical design of polymers has dominated innovation in the production of polymeric fibers. Each synthetic fiber type is designated by a different chemistry. The one exception is high modulus polyethylene, where gel-spinning has led to a new structure for a fiber made from a simple polymer. This is rightly regarded as in a different category to simple melt-spun polyethylene monofilaments. Interestingly, HMPE has one of the least complex of fiber structures.

During this era, the development of fine structure, which provides a range of variants of each basic fiber type, was largely empirical. Control of melt-spinning, and the other fiber production methods, was mainly dictated by engineering convenience and economic advantage. Only a few features, such as orientation related to extent of draw, were designed in an engineering sense. Other structural features, such as the larger size of crystallites in fibers spun at high speeds, were found to happen. The detail of the paths and packing of molecules through a partially ordered assembly was lost in the obscurity of the speculations of scientists described in the first part of this chapter.

One aspect of fiber structures attracted strong scientific investigation and was well worked out: the crystal lattices. Unfortunately, these usually have little influence on fiber properties. The crystallites provide rigid blocks, whose mechanical properties have only a secondary influence on overall fiber mechanics. What needs to be better understood are the properties of the disordered polymer material and the way it links the whole structure together. Interestingly, these are the aspects of the generally well-documented structure of wool fibers about which least is known, but which has a large effect on properties.

The second part of the chapter introduces the ideas of the science of complexity, but their detailed relevance and application is still in a state of obscurity resembling that of the chemistry at the beginning of the 20th century. The useful, if confused, ideas need to be clarified and applied to polymer assemblies. The fine structure of fibers could then, in the 21st century, be put on as sound a scientific basis as polymer chemistry in the 20th century. Inevitably there will be more complexity, but computers provide a means for handling the mass of data needed to describe a structure adequately.

If the science can be formulated, what are the prospects for commercial exploitation? For this, it will be necessary to understand not only the fine structures but also the formation mechanisms, and then to develop engineering techniques to manipulate the formation processes.

Current melt-spinning is a relatively simple process. In its simplest form, the emerging molten threadline is cooled and stretched monotonically. Some changes of temperature and stress may be applied in subsequent processing. However, the indications from the various aspects of the science of complexity are that the fine structure will be influenced by fluctuating and localised changes of temperature and stress through their actions on energy and mass movements. Highly creative research is needed to put this on a sound and detailed scientific basis, which enables formation sequences to be specified.

The engineering prerequisite is available for application. Advances in computer control of machines would enable complex sequences, once they have been designed, to be followed. If science and engineering can come together in a creative synergy, the 21st century may see a new generation of polymeric fibers, which result from control of the complexities of the formatiion of fiber fine structure.

14.6 References

1. Dunbar, J.J., Prevorsek, D.C., Tam, T.Y-T., Weedon, G.C., and Winckelhofer, R.C. European Patent 0 205 960 (1990), Chem. Abstr. 106:215460t
2. Meyer, K.H. and Mark, H. *Der Aufbau der Hochpolymeren Naturstoffe* (1930) Leipzig
3. Seifriz, W. Amer. Nat. (1929) 63 p. 410
4. Meyer, K.H. *Kolloid-Z.* (1930) 53, p. 8–19
5. Nageli, C. *Micellartheorie*, Original papers reprinted as *Ostwalds Classiker*, No. 227 (1928) A. Frey (Ed.) Leipzig
6. Nodder, C.R. *Trans. Faraday Soc.* (1933) 29 p. 317–324
7. Staudinger, H. *Die Hochmolekularen Organischen Verbindungen* (1932) Springer-Verlag, Berlin
8. Hermes, M.E. *Enough for one lifetime. Wallace Carothers, inventor of nylon* (1996) American Chemical Society
9. Gerngross, O., Herrmann, K. and Abitz, W. *Biochem. Z.* (1930) 228 p. 409–425
10. Hearle, J.W.S. In Fibre Structure. Hearle, J.W.S. and Peters, R.H. (Eds.) (1963) The Textile Institute, Manchester, pp. 209–234
11. Herrmann, K. and Gerngross, O. *Kautschuk* 8 p. 181–185
12. Kratky, O. and Mark, H. *Z. Physik. Chem.*(1937) B36 p. 129–139
13. Frey-Wyssling, A. *Submikroskopische Morphologie des Protoplasmas und Seiner Derivate* (1938) Gebrüder Borntraeger, Berlin. Revised and translated as *Submicroscopic Morphhology of Protoplasm*, 1st ed. (1948), 2nd ed. (1953) Elsevier, Amsterdam
14. Kratky, O. In *Die Physik der Hochpolymeren*, Vol. 3, Stuart, H.A. (Ed.) (1956) Springer-Verlag, Berlin
15. Mark, H. *J. Phys. Chem.* (1940) 44 p. 764-785
16. Meyer, K.H. and van der Wyk, A.J.A. *Z. Elektrochem.* (1941) 47 p. 353–360
17. Hermans, P.H. *Kolloid* Z. (1941) 97 p. 231–237
18. Hess, K. and Kiessig, H. *Z. Physik. Chem.* (1944) A193 p. 196–217
19. Götze, K. *Chemiefasern nach dem Viskoseverfahren* (1951) Springer Verlag, Berlin
20. Hess, K., Mahl, H. and Gütter, E. Kolloid-Z (1957) 155 p. 1–19
21. Hearle, J.W.S. *J. Polymer Sci.* (1958) 28B p. 432–435
22. Bunn, C.W. In *Fibres from Synthetic Polymers.* Hill, R. (Ed.) (1953) Elsevier, Amsterdam, pp. 240–286
23. Mark, H. In *Cellulose and Cellulose Derivatives*, Part I, 2nd edition (1954) Ott, E., Spurlin, H.M., and Grafflin, M.W. (Eds.) Interscience Publishers, New York, pp. 217–230
24. Sherer, P.C. In *Matthews' Textile Fibers*, 6th ed. (1954) H.R. Mauersberg (Ed.) Wiley, New York, pp. 53–101
25. Alexander, P., and Hudson, R.F. *Wool: its Chemistry and Physics* (1954) Chapman and Hall, London
26. Stuart, H.A. *Die Physik der Hochpolymeren*, Vol. 3 (1956) Springer-Verlag, Berlin
27. Jane, F.W. *The Structure of Wood* (1956) Black, London
28. Hearle, J.W.S. *Skinner's Silk and Rayon Record* (1956) 28, 354–359
29. Boulton, J. *Textile Res. J.* (1958) 28 p. 1022-1030
30. Battista, O.A. *Fundamentals of High Polymers* (1958) Reinhold, New York
31. Hohenstein, W.P. and Ullman, R. In *Unit Processes in Organic Synthesis*, 5th ed. Groggins, P.H. (Ed.) (1958) McGraw-Hill, New York, pp. 856–942

32. Cumberbirch, R.J.E. and Harland, W.G. *J. Textile Inst.* (1959) 50, T311–T334
33. Hearle, J.W.S. *J. Appl. Polymer Sci.: Appl. Polymer Symp.* (1991) 47 p. 1–31
34. Keller, A. In *Fibre Structure*. Hearle, J.W.S. and Peters, R.H. (Eds.) (1963) The Textile Institute, Manchester, pp. 332–390
35. Morton, W.E., and Hearle, J.W.S. *Physical Properties of Textile Fibres*, 2nd ed. (1975) 3rd ed. (1993) The Textile Institute, Manchester
36. Schultz, J.M. In *Polyesters in Solid State Behavior of Linear Polyesters and Polyamides* (1990) Schultz, J.M. and Fakirov, S. (Eds.) Prentice-Hall, Englewood Cliffs, NJ, pp. 75–130
37. Hearle, J.W.S. and Greer, R. *J. Textile Inst.* (1970) 61 p. 240–244
38. Prevorsek, D.C., Harget, P.J., Sharma, R.K. and Reimschussel, A.C. J Macromol. Sci. Phys. (1973) B-8 p. 127–156
39. Murthy, N.S., Reimschussel, A.C. and Kramer, V. J. Appl. Polymer Sci. (1990) 40 p. 249–262
40. Hearle, J.W.S., Prakash, R. and Wilding, M.A. *Polymer* (1987) 28 p. 441–448
41. Hearle, J.W.S. *J. Appl. Polymer Sci.: Appl. Polymer Symp.* (1978) 31 p. 137–161
42. Bell, J.P., Slade, P.E. and Dumbleton, J.H. *J.Polymer Sci. A-2* (1968) 6, p. 1773–1781
43. Bell, J.P. and Dumbleton, J.H. *J.Polymer Sci. A-2* (1969) 7, p. 1033–1057
44. Bell, J.P. and Muryama, T. *J.Polymer Sci. A-2* (1969) 7, p. 1059–1073
45. Hearle, J.W.S. and Greer, *R.Textile Progress* (1970) 2, Number 4, The Textile Institute, Manchester
46. Heuvel, H.M. and Huisman, R. In *High-Speed Fiber Spinning*. Ziabicki, A. and Kawai, H. (Eds.) (1985) Wiley, New York, pp. 295–331
47. Shimizu, J., Okui, N. and Kikutani, T. In *High-Speed Fiber Spinning* (1985) Ziabicki, A. and Kawai, H. (Eds.) Wiley, New York, pp. 429–483
48. Reneker, D.H. and Mazur, J. *Polymer* (1983) 24 p. 1387–1400
49. Bebbington, E.M.O. PhD thesis (1989) University of Manchester
50. Hearle, J.W.S. In *Polyester: 50 Years of Achievement*. Brunnschweiler, D. and Hearle, J.W.S. (Eds.) The Textile Institute, Manchester, pp. 6–9
51. de Gennes, P-G. *Scaling Concepts in Polymer Physics* (1979) Cornell Univ. Press, Ithaca, NY.
52. Doi, M. and Edwards, S.F. *The Theory of Polymer Dynamics* (1986) Oxford Univ. Press, London.
53. Bueche, F. *Physical Properties of Polymers* (1962) Wiley, New York
54. Lodge, A.S. *Rheol. Acta* (1968) 7 p. 379–392
55. Bird, R.B., Hassager, O., Armstrong, R.C. and Curtis, C.F. *Dynamics of Polymeric Liquids* (1977) Wiley, New York
56. Wendolowski, J.J., Gardner, K.H., Hirschinger, J., Miura, M. and English, A.D. *Science* (1996) 247 p. 431–436
57. Mandelbrot, B.B. *The Fractal Geometry of Nature*. (1977) Freeman, New York.
58. Meakin, P. In *Phase Transitions* 12 (1988) Academic, New York
59. Meakin, P. In *Encyclopedia of Polymer Science and Engineering*, Supplementary Volume, 2nd edition (1989) Wiley, New York, 323–342
60. Freeman, G.R. (Ed.) *Kinetics of Nonhomogeneous Processes* (1986) Wiley, New York
61. Glansdorff, P. and Prigogine, I. *Thermodynamic Theory of Structure, Stability and Fluctuations* (1971) Wiley, New York
62. Nicolis, G. and Prigogine, I . *Self-Organisation in Non-Equilibrium Systems* (1977) Wiley, New York
63. Prigogine, I. and Stengers, I. *Order Out of Chaos* (1984) Bantam Books, New York
64. Lindenmeyer, P.H. *Textile Res. J.* (1980) 50 p. 395–406
65. Bullough, R.K. *private communication* (1990)
66. Gleick, J. *Chaos* (1987) Viking, New York.
67. Waldrop, M.M. *Complexity* (1992) Simon and Schuster, New York.
68. Thompson, J.M.T. *Instabilities and Catastrophes in Science and Engineering* (1982) Wiley, New York
69. Penrose, R. *The Emperor's New Mind* (1989) Oxford Univ. Press, London.
70. Hearle, J.W.S. *Polymer Engin. and Sci.* (1994) 34 p. 260–265
71. Squires, E. *The Mystery of the Quantum World* (1994) Institute of Physics Publishing, Bristol
72. Penrose, R. *Shadows of the Mind* (1994) Oxford University Press, London.
73. Berry, R.S. Scientific American (1990) 263 (2) (Aug.), p. 68–74

Index

David R. Salem is Director of Research at TRI/Princeton, which he joined after receiving his Ph.D. from the University of Manchester (UMIST), U.K. During a two-year leave of absence from TRI/Princeton, he conducted research at the central laboratories of Rhône-Poulenc in Lyon, France. Dr. Salem is best known for his numerous publications on structure formation and microstructure characterization of oriented polymers, and he has also contributed to the literature on fluid transport in porous materials. He has received various honors including the Award for Distinguished Achievement in Fiber Science (1996), and he has served on the Governing Council of the Fiber Society. He is a Member of the Institute of Physics (U.K.) and a member of the American Chemical Society, and he serves on various national and international advisory boards.